AN INTRODUCTION TO
ELECTRICAL ENGINEERING

WITH LAB ACTIVITIES

AN INTRODUCTION TO
ELECTRICAL ENGINEERING
WITH LAB ACTIVITIES

Gary H. Bernstein

JENNY STANFORD
PUBLISHING

Published by

Jenny Stanford Publishing Pte. Ltd.
101 Thomson Road
#06-01, United Square
Singapore 307591

Email: editorial@jennystanford.com
Web: www.jennystanford.com

British Library Cataloguing-in-Publication Data

A catalogue record for this book is available from the British Library.

An Introduction to Electrical Engineering with Lab Activities

Copyright © 2026 Jenny Stanford Publishing Pte. Ltd.

ISBN 978-981-5129-30-4 (Hardcover)
ISBN 978-1-003-71345-6 (eBook)

DOI: 10.1201/9781003713456

Dedication

To my wife and three sons, all of whom continue to inspire me.

About the Author

Gary H. Bernstein received a BSEE from the University of Connecticut, Storrs, with honors, in 1979 and an MSEE from Purdue University, W. Lafayette, Indiana, in 1981. During the summers of 1979 and 1980, he was a graduate assistant at Los Alamos National Laboratory and, in the summer of 1983, interned at the Motorola Semiconductor Research and Development Laboratory, Phoenix, Arizona. He received his PhD in electrical engineering from Arizona State University, Tempe, in 1987, after which he spent a year there as a postdoctoral fellow. He joined the Department of Electrical Engineering at the University of Notre Dame in 1988 as an assistant professor and was the founding director of the Notre Dame Nanoelectronics Facility from 1989 to 1998. He received an NSF White House Presidential Faculty Fellowship in 1992. Promoted to the rank of professor in 1998, he served as associate chairman from 1999 to 2006. Prof. Bernstein was named the Frank M. Freimann Professor of Electrical Engineering in 2010 and served as the associate director of the Notre Dame Center for Nano Science and Technology from 2013 to 2017. He was named a Fellow of the IEEE in 2006. He received the Innovation Excellence Award from the Indiana Economic Development Center and Forbes Summit Group, Indianapolis, in 2014 and the 1st Source Commercialization Award in 2016. Prof. Bernstein was named to the National Academy of Inventors in 2020. He is also a cofounder of Indiana Integrated Circuits, LLC.

Contents

Preface for the Instructor

About the Course and the Book

I hope you will find this textbook and the accompanying lab activities to be a refreshing take on introductory electrical engineering – something that really captures your students' interest and adds value to their learning experience. At Notre Dame, we noticed that many of our electrical engineering students felt they were not getting a clear sense of what the field was really about early on. This course was created to change that.

The philosophy underlying this new lab course is not only to introduce them to some of the more relevant topics in EE but also to predispose (or "pre-expose") them to more complicated concepts generally reserved for later semesters – to start the induction process earlier. Sometimes it takes a long time for concepts to sink in, and the math does not always help all that much at the beginning (e.g., Maxwell's equations, frequency domain, Fourier transforms, etc.). In typical courses, there often is not enough time to both present a breadth of material with all of the mathematical foundations and let it sink in well enough for the students to become comfortable in their understanding. This course is an opportunity to spend some time playing with the concepts – to build a foundation of intuition on which to build rigorous courses later. The most mathematical part of this course is phasors, which cannot be taught in any depth without some exposure to complex numbers; other than that, the math is fairly minimal.

The book recognizes that students bring to their early college education a variety of skills, knowledge, and confidence and is designed to make students at all levels feel comfortable. At no point is the intent to talk down to students, but rather to bring along those students who may be weak or forgetful in certain areas. For example, where a key concept in trigonometry is needed, it is presented as a review. My sense is that it is better to have everyone on the "same page" so that everyone starts at the same place when new material is presented.

Some basic circuit concepts are covered in this text, but it is not a book on circuits, *per se*. The circuits-related material presented here serves to support material in subsequent chapters and is not intended to be a substitute for a "real" circuits course. There is a lab for each chapter. Most students can complete most labs in two sessions, and often in one. Labs can be trimmed at the discretion of the instructor in order to fit into an allotted time, generally with the most basic material covered first, and the last steps adding more flavor. If your curriculum teaches a separate circuits course either before this course or in parallel, it will be possible to teach perhaps eight chapters with their labs, providing a wider breadth of material. Beyond the first five chapters, additional chapters may be chosen to suit the taste or needs of the instructor.

Overall, the goals of the course are to give students the following experiences:

- Hands-on skills with lab equipment
- Exposure to (some) breadth of EE
- Appreciation of the importance of EE in society
- Understanding of time and frequency domains

- Practical experience in electrical systems
- EE-related aspects of the physical world around them
- Ability to relate technically to some of the pervasive technologies that they use every day

Given that the Institute of Electrical and Electronics Engineering (IEEE) with its over 400,000 members has a total of 39 technical societies and seven technical councils, it clearly is not possible to cover every area in EE or that students might have an interest in. The topics covered in this course break down roughly evenly between circuits/systems and devices. Since physics and electrical engineering are so closely related, no attempt is made to hide the physics behind engineering technologies. Most of the underlying concepts are found universally in electrical engineering, or engineering in general, for that matter. This textbook is *not* intended to be a collection of descriptions of EE careers, but rather a collection of concepts that will support later courses and that are difficult enough that an additional, first exposure is warranted. For the sake of this course, it is often enough that students be simply exposed, with explanations, to new terminology, new devices and new concepts, with the bulk of the lab activities providing a kinesthetic sense for those concepts.

If this text is used in a course without the lab, the lab activities may still be used as teaching material. Many of the lab sections show results of various measurements that illustrate basic concepts, and may be used for deeper discussions. Several of the lab activities, designated with "*," use custom-built hardware and can, in some cases, be either replicated or substituted with similar equipment. The topics covered in the 10 chapters and associated labs are:

1. Chapter 1: Components, manufacturing and soldering (solder AM radio kit)
2. Chapter 2: Use of lab bench equipment
3. Chapter 3: Power transmission, house wiring, and electrical safety*
4. Chapter 4: Time and frequency domains
5. Chapter 5: Complex arithmetic, phasors and resonance
6. Chapter 6: Radio transmission, Maxwell's equations, amplitude modulation (using the AM radio from Lab 1)
7. Chapter 7: Semiconductor concepts, diodes, transistors, op-amps
8. Chapter 8: Energy conservation and alternative energy – light bulbs and solar cells*
9. Chapter 9: Digital music – ADC and DAC, Nyquist theory*
10. Chapter 10: Batteries and power supplies*

About the Lab Sessions

The instructions provided for the lab steps are fairly detailed. This reflects the reality that labs are to be completed in less than six hours, and as few as three, so confusion would waste precious time. Since not all students have perfect recall, some repetition on the equipment use is appropriate. Also, every lab introduces new features of the lab equipment, and instructions are provided in the activities part rather than being relegated to a manual or appendix. For  example, being stuck on which button, menu, and submenu to find a certain function could take, say, 15 minutes to correct. Multiply that by only a couple of times and you have a prescription for frustration. My decision was to err on the side of caution and provide rather detailed instructions. It is unlikely that you would use much of any of the same equipment, so adjustments must be made to suit every situation.

Lab activities contain several types of information, including introductory material to point out relevance; instructions on how to set up the activity; instructions on how to use the equipment; and questions pertaining to that lab step. Each step begins with text in **boldface** that summarizes what that step is about. The black text alerts the student to actual steps that must be performed. The terms in **boldface** indicate the words of a function that can be found on the test equipment. The green text alerts the student that the material is explanatory in nature or is a question. These questions, asked in many of the steps, are designed to foster thinking and engagement, and the answers are required in the lab reports. Students may choose to discuss these questions and answer them in their notebooks during lab, or may choose to answer them later. The amount of weight given to these answers for their grade is up to you, the instructor.

At Notre Dame, I set up this course's lab sessions to minimize the possibility that students will break into passive and dominant modes in which one student does all the activity and the other just watches. I am aggressive at combating this phenomenon by keeping the groups to a size of only two. Most importantly, students are required to state in their notebooks who is "pilot" and "co-pilot" at any given time. The pilot operates the equipment, turning the knobs, adjusting the circuits, etc., while the co-pilot follows and reads the instructions and takes notes. The two lab partners are required to switch off every 30 minutes, stating so in their notebooks, with the obvious change in handwriting. This way, the co-pilot must pay attention to the details of the work at all times in order to be able to take over from the pilot, and each student gains roughly the same experience operating the equipment. Simply increasing the group size to three breaks this mode of interaction, and it is a near certainty that one of the three will not learn. In my opinion, it is better that a student work alone and take longer than to work in a group of three. At Notre Dame, students are always free to schedule additional time in the lab in case any time-related issues arise.

Furthermore, students do not choose their own lab partners. In fact, the two students are randomly paired for each new chapter's lab. I feel this has many advantages, including teaching them to work better in teams from the beginning, as every chapter's lab creates a new relationship. Also, students tend not to fall into poor work habits, e.g., where only one student learns to use the equipment or one student just answers questions, and more responsibility for preparation is taken by each student each week. In spite of minor difficulties for the students, I feel it increases overall learning. A secondary benefit is that students get to know more of their classmates during this early semester than they would if they had a dedicated lab partner. It is not our role as instructors to manage social interactions, but I feel it is good for them to know more of their classmates, thus potentially reducing feelings of isolation and increasing camaraderie among the class cohort. Another benefit is that if there are any conflicts between students, then, like the proverbial weather in South Bend, students just need to wait a bit and it will change. In 10 years, there have been no student interactions that were bad enough to be brought to my attention, although likely some have been better than others.

About the Lab Activities and the Equipment

In order to complete all 10 lab activities, the course requires all of the basic, but modern, lab bench equipment commonly found in undergraduate electronics labs, plus some non-standard items, as well as custom systems developed for some of the lab activities, denoted by the asterisks above. This text is based on the equipment on Notre Dame's lab benches. As mentioned above, the lab instructions teach the use of the equipment, which is problematic for facilities at different institutions using different equipment. Modern equipment is highly software-based; many instruments operate basically the same way with similar menus and hopefully can be easily translated for your institution. This textbook can be read to great effect without doing the lab activities. If the student is interested enough to read through the lab activities, they will be rewarded by many examples of experimental results that add further understanding to the main topics of the chapter.

The following are the basic tools needed for this course.

1. 2-channel DC power supply with floating grounds.

2. 2-channel function generator capable of amplitude and frequency modulation, and frequency sweep.

3. Digital multimeter with temperature probe.

4. 4-channel digital oscilloscope with at least 25 MHz bandwidth, 16-bit digital probes, frequency-spectrum capability and internal math functions. One of the major themes of the course is frequency domain, so an excellent frequency-spectrum mode is preferred.

5. Hall current probe with sensitivity to a few mA. This item makes current measurements fast and simple, without overly complicating the circuits. However, shunt resistors could be added instead, and are added where necessary.

6. Professional-quality, 31-band single-channel audio graphic equalizer. Inexpensive home-audio equalizers are discouraged, as they add considerable distortion to the waveform. Only professional-quality ones are relatively distortionless. One dual-channel instrument can be shared by two adjacent lab benches.

7. Inexpensive lux meter, UV light meter, and AC power meter.

8. LCR/ESR meter that can report the magnitude and phase of impedance. The least expensive versions cannot do that.

9. Various breadboards, audio connectors, cables, probes, solar cells, DC motors, etc.

10. Curve tracer. One or two may be sufficient for an entire classroom, and older ones (such as the Tektronix 571 used at Notre Dame) are fine for this course.

Custom trainers developed at Notre Dame for this course include:

1. Lab 1: Each student is given an Elenco AM-780K AM radio kit.

2. Lab 3: Low-voltage 3-phase power supply and set of four transformers having appropriate sockets and markings.

3. Lab 3: A custom "load board" for facilitating wye and delta power configurations was developed to make things easier but can also be set up by students on a breadboard. Lab 3 is especially "wire-intensive" and the instructor should do everything possible to help the students easily troubleshoot their setups. This includes color coding all of the phase lines where possible.

4. Lab 8: Hardware for doing shielded light measurements on the benchtop.

5. Lab 9: Analog-to-digital / digital-to-analog trainer board adjustable from 1 to 8 bits of resolution and sample rates up to hundreds of kHz. The board has audio and BNC ports for use with a signal generator or music source.

6. Lab 10: Microprocessor-controlled relay for creating short-duration short circuits to measure internal resistance of batteries. This was what was on-hand at Notre Dame, and simpler timers could be easily fabricated.

What Is Not in This Textbook

This book is not about careers in electrical engineering. It is, though, about the globally critical role that EEs play, and thus motivates careers in EE. It is not intended to address the breadth of EE, since that would make the book and the course much longer. Chapters 1, 2, 4, and 5 are core to all of EE. Chapter 3 plays an important role in introducing many of the lab skills, and the concepts of phase, frequency, and complex impedance. It can be considered to be one of the core chapters; if power transmission and/or residential wiring is not one of your preferred topics, you may go back to Chapter 3 to introduce complex impedance. This book is limited in scope; a longer version would ideally cover the following topics in rough descending order of importance and/or interest to the students, *in my opinion*:

- Image processing
- Control systems
- Digital electronics
- Embedded systems
- Antennas
- Microwave circuits and waveguides
- Robotics
- Quantum computing

In conclusion, I feel that this course material will give your students a unique perspective on the importance of electrical engineering and a good start toward the more advanced material that they will see in their future courses. I genuinely hope that you enjoy teaching, and students enjoy learning, from the material presented in this book.

Gary H. Bernstein
Notre Dame, IN
August 2025

.

Preface for the Student

Welcome to the world of the electron and the photon! Regardless of your major, be that electrical, mechanical, computer, or any other engineering or science field, you will find the material presented in this book to be relevant to what you are learning now in other engineering or physics courses as well as upcoming electrical engineering courses, and provides an introduction into the wonders of your technical environment. This material is designed to introduce you to technologies that you interact with daily, such as wall outlets, power lines, light bulbs, solar cells, and batteries. It is also intended to introduce you to difficult concepts that you will be taught in later courses. Advanced material that will challenge you in upcoming courses, such as frequency domain, frequency spectra, electrical filters, phasors, complex impedances, Maxwell's equations, semiconductor fundamentals, radio waves, and properties of light, are often difficult to grasp. This course serves to break the ice, giving you a chance to gain a sense of them, as well as to play with them in the lab before you are subjected to the mathematical baptism by fire. I believe that it is one thing to read about something, and another thing to experience it in the lab, doing measurements and seeing firsthand how the system responds to changes. The lab activities are designed to reinforce and help you understand the most difficult aspects of the chapters.

This course is not about careers in electrical engineering – employment opportunities will be obvious after learning about the topics. The book is much more about understanding difficult concepts in science and engineering. You will find that most of the chapters take the following arc: background on why what you are about to learn is important, relevant to your life, and relevant to society; introduction to several different concepts that may seem unconnected at first but must all be laid out for later collection and use; a bringing together of the disparate information to allow an important and perhaps arcane concept to become clear; examples of how those concepts are important for courses that you are likely to take within the next year or so.

Some of the practical things that you will be able to do after this course include:

- Recognize electronic components and know what is inside of them, and how they behave in a circuit.
- Build prototype circuits on breadboards.
- Understand and operate a wide variety of electronic bench equipment, including power supplies, meters, function generators, oscilloscopes, and more. By the end of the semester, you will feel at home in the electronics lab.
- Look up at power lines and know not only what you are looking at but also how various power electrical components are used in the vast national power grid.
- Look around your home and understand how power is delivered inside the home from the distribution panel to power outlets and understand the dangers presented by household electrical voltages.
- Understand more deeply your own electric power bill and how energy efficiency can benefit you.

- Understand the related concepts of frequency and phase, as well as the basic underlying mathematics.

- Understand how your radio is able to provide news and music via electromagnetic waves.

- Understand the role of materials in electronics and how semiconductors are different from other materials.

- Understand how light is produced in various types of lightbulbs, how your eyes react to various wavelengths of light, and why fundamentally some sources of illumination are more energy efficient than others. You will also be able to interpret more than what is provided by superficial advertising of lighting products.

- Listen to the music played on your cell phones and visualize the entire path that it has taken from the studio through the process of converting to digital files and back again to music.

- Look at batteries and know about the various types, sizes, and internal construction, and to make determinations of why particular batteries are appropriate to their applications.

- Understand, design, and build, if you choose, a power supply that you could use on your own benchtop.

If you are daunted by heavy mathematical treatments, you will be glad to know that this book is mostly about concepts and ideas, and not so much about the math. Chapter 5 is about a mathematical technique called "phasors," so obviously it must be somewhat mathematical, but is still heavy on concepts. Wherever possible, I anticipate where you might not have a complete grasp of material that you have already been exposed to (i.e., were supposed to have learned, but ummm, maybe you did not quite). I know that happens to most, if not all, students at some point pertaining to at least some materials, and maybe even a lot, so reviews are liberally spread throughout the chapters. No skipping them... it does not hurt to review, even if you think you know it perfectly already.

Here is some personal advice:

To my knowledge, most of the lab activities are unique to this course and give you a chance to explore engineering principles in ways that are, in some cases, not available in even more advanced courses. Learning comes from reading (the chapters), listening (to lectures), and doing (homework and the labs). Each component is important, and none is as effective alone. I understand that students do not read as much now as my generation did, but most lab steps are there to illustrate a principle; without the basic knowledge of the chapter, you will likely not get much out of the lab.

Students learn in their own ways; some read everything carefully, and others prefer to do the readings only to answer homework questions or to study for exams, if at all. I would like for you to consider how important it is to take the time, i.e., slow down, relax, and concentrate on the readings, before you go to the lab. This advice holds for all of your academics as well. Reading is generally a lost art, but that should not be the case in *your* education. If you do not like to do slow and thoughtful readings, try it and see how much your understanding and grades improve. If you have read this far, then you must already be a reader. Spread the word to your friends.

The most common question that I get during exam-review sessions is, "How should I study for the exam?" I tell my students that they should know what they did in the labs and why, and to know the concepts covered in the chapters. This stuff is not easy, and being an electrical engineering student takes a lot of effort. Been there, done that. With careful attention to the material presented here, you will be successful in this course, and likely in future courses.

I do have another suggestion for how to prepare for exams: Teach. Teach yourself. Teach other students. Teach your friends and parents. The best way to learn is that as you read, pretend that you are teaching everything you learn as you learn it. Take notes. Explain things to yourself as if you were teaching a future you who may not remember the material. I do that whenever I read new material. It is much more effective to study as if you have to teach it than it is to study for an exam. Teaching takes honesty because as soon as you start to teach, you learn very quickly what you do not know. If you just try to guess what will get you through an exam, you can easily fool yourself into believing that you know more than you do. It is not very likely that you will guess correctly, and it is a bad strategy.

If you have read this far, I am very optimistic about your success! I sincerely hope that you enjoy this book, the lab activities, and the course in general. It will be a success if at the end you feel you have made a big step toward your career goals!

Gary H. Bernstein
Notre Dame, IN
August 2025

Acknowledgments

This textbook has been 10 years in the making. Over that time, dozens of individuals, mostly students, have contributed in some way, so it is not possible to thank each of them by name. Many undergraduates helped me test out the laboratory experiments, craft assignments, and create figures. Many graduate teaching assistants have also helped either directly or through suggestions.

A few people, however, have provided considerable assistance during the preparation of this book and its associated laboratory. I would like to thank Mr. Clint Manning, who did an amazing job creating the laboratory training tools and managed the laboratory classes and teaching assistants. I would also like to thank Dr. Alfred Kriman, who was my counselor and coach in this adventure. Many topics benefited from his enormous breadth of knowledge; his command of physics and writing, along with his willingness to share, has made all the difference. I also thank Prof. Patrick Fay, who has been there for me at key moments to offer suggestions about course material and labs, and to help me through the topics with which I was unfamiliar and that I found confusing. He is a very busy man, and his patience with me throughout all of this is deeply appreciated. I also thank Dr. Chris Martino for giving me help and advice improving the treatment of power systems. In spite of all the help I received, I am sure that errors remain, and for those I take full responsibility.

Finally, I thank all of the Notre Dame undergraduates who have taken this course and have been understanding and encouraging of this effort, regardless of its flaws. It is for them that I do this, and without them it could not, and would not, have happened.

Chapter 1

Electronic Components and Assembling an AM Radio Kit

1.1 Introduction and Overview

In this chapter, you will:

1. Become familiar with various common electrical components
2. Understand the concept of electric and magnetic fields
3. Learn about modern methods of electronic manufacturing
4. Gain practical experience in soldering

Throughout this book, when new technical terms are introduced, they will be in **boldface text**. Sometimes 'single quotes' will be used for simple emphasis. Fluent speakers of a language also use slang. The technical slang used within a profession is called jargon, and 'single quotes' will be used when jargon is introduced (among other uses.) There is some overlap between technical terms and jargon, but both are terms of art, and it is good to become familiar with them since, in electrical engineering as in every profession, insiders recognize each other by their use of specific technical language. Therefore, you should learn the language of EEs.

1.1.1 Electronic Components

Sometime over a century ago, the world was electrified. It was one of the few really major turning points in human history. Life changed profoundly and accelerated. You stand on a threshold of your education; you are about to enter a community of practitioners who are akin to magicians and wizards. In this week's activity, you have before you all the tools to make a radio – to reenact history and to take your first steps toward the technological future. You are entering into communion with technical revolutionaries. As you proceed, keep in mind what you can see all around you: the revolution continues.

This chapter is about the electronic components that you will assemble into your AM radio kit, shown in Fig. 1.1. (AM stands for "amplitude modulation," which is the subject of Chapter 6). There is no attempt in this chapter to relate those devices to the circuits in which they will be used – that is left for later chapters. The assembly of your radio should, however, be more than just an exercise in checking off boxes. As you progress through your EE education, you will encounter many kinds of

An Introduction to Electrical Engineering with Lab Activities
Gary H. Bernstein
Copyright © 2026 Jenny Stanford Publishing Pte. Ltd.
ISBN 978-981-5129-30-4 (Hardcover), 978-1-003-71345-6 (eBook)
www.jennystanford.com
DOI: 10.1201/9781003713456-1

electronic devices and will need to understand how they behave. Here, you will meet the most common players in the vast league of electronic components used in electronic systems. When done, you should be able to identify each component used in the kit, and what it does.

The radio you build is a simple example of one of the biggest trends in electronics over the past half-century, which is ever-increasing levels of miniaturization. Miniaturization is a two-pronged issue: the size of the actual electronic devices and the size of the system in which they are placed. We discuss how one shrinks electronic devices and how they are connected to each other. We discuss the transition from wires to printed circuit boards and soldering to integrated circuits, and we show how at each step the system becomes more compact.

Soldering is a way to connect electrical components by using small amounts of melted metal, called **solder**. (Solder as both a noun and verb is pronounced SAH-dur in the U.S.). Your Model AM-780K AM Radio Kit from Elenco contains roughly 25 separate components, most of which you will manually solder to the provided circuit board. Solder and soldering are discussed in detail in Section 1.3. In Lab 6, we will discuss the underlying concepts behind the operation of the radio, but in this chapter, we are concerned only with the physical devices that go into the kit, and the skills needed to assemble it.

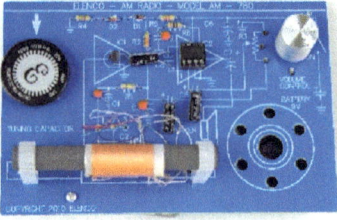

FIGURE 1.1 Your AM radio kit before and after construction. Be sure to look over the manual before you come to lab.

1.2 Current, Voltage and Wires

You may or may not be familiar with currents and voltages in electrical circuits. **Current** is the movement of electrical charges. Charge is generally represented by the variable Q, and current is represented by the variable name I or i, depending on whether it is constant or changing, respectively. In this text, we will use upper case letters for the general case, and lower case when it is specifically necessary to invoke changing quantities.

To understand current, you must harken back to your introductory chemistry course. Charges can have either positive or negative sign. The unit of **electric charge** is **coulombs**, with symbol "C." We represent the elementary, i.e., smallest division, of charge as q, where $q = 1.602 \times 10^{-19}$ C. There are, therefore, 6.24×10^{18} individual charges in 1 C (a very large number, indeed). An arbitrary amount of charge is represented by the variable Q.

Water and its properties make an excellent analogy for many aspects of the flow in electrical systems. Such comparisons will be used throughout the text. Here, the flow of charge is much like the flow of water in a pipe. Current is the rate at which charge moves past a point within, say, a metal wire, and has units of coulombs per second, C/s, known as **amperes**, whose symbol is "A" ('amps' for short):

$$i = \frac{dQ}{dt} \tag{1.1}$$

One amp is one coulomb flowing past a plane (for example, the cross-section at some point along a wire) in 1 s. If the current through an electric fan is 2 A, then in 1 s, 2 C, or 1.25×10^{19} charges (electrons in this case), pass a plane along the wire. That amount of current is not in any way remarkable, so it is clear that single electrons carry very little charge.

Protons have positive charge and electrons have negative charge. Atomic charge in the form of positive ions rarely moves in solid materials, such as metals, so in the vast majority of practical cases, it is electrons that carry current. Notable exceptions in which positive charges carry current are current in semiconductors, as discussed in Chapter 7, and current through wet media such as, for example, salt water or batteries, as discussed in Chapter 10.

In practice, most of the current in electric circuits is carried by metals and therefore is due to the movement of electrons. Current has an arbitrary sign associated with it – current in one direction may be labeled, like a vector, as positive in one direction and negative in the opposite direction. By convention agreed upon by everyone (since Benjamin Franklin suggested it), electrons moving in one direction (in a wire and elsewhere) corresponds to positive electrical current in the opposite direction. Were positive charges to move, that direction would correspond to positive current.

But what makes charged particles such as electrons or ions move in the first place? Things move due to forces. Charges move due to the force of an **electrical potential**, or **voltage**, whose variable name is V or v, and whose unit is **volts**, V. You are likely familiar with the notion of potential because mass, e.g., rocks, drop from higher gravitational potential to lower gravitational potential. A rock falling off of a cliff in the presence of gravitational force is a reasonable analogy to electrical current and voltage. Note that the rock falls over a *difference* in elevation (and thus of gravitational potential). Similarly, in electric circuits, current (charge) flows in response to the force that arises from a **potential difference** or a **voltage difference**. The notion that electric fields cause those potential differences, and the force on the charges, is discussed in more detail) in Section 1.4.2.

Let's extend this analogy. A rock falling through a potential difference gains kinetic energy from the loss of gravitational potential energy, U_g. We can say that the change in potential energy, ΔU_g, due to falling across a difference in height, h, is $\Delta U_g = mgh$, where m is the mass of the rock and g is the acceleration of gravity (recall that on Earth, $g = 9.81$ m/s^2). Similarly, some charge, Q, moving through an electrical potential difference gains energy

$$U = Qv. \tag{1.2}$$

To extend the analogy, we relate the mass, m, to the charge, Q, and product of the height and acceleration of gravity, gh, to the electrical potential, v.

Applying Eqs. 1.1 and 1.2, we see that the power, or energy flow per unit time, is

$$P = \frac{dU}{dt} = \frac{d(QV)}{dt} = \left(\frac{dQ}{dt}\right)v = iv. \tag{1.3}$$

Equation 1.3 is extremely fundamental and important in electrical systems. It tells us that if a current of i amperes flows through any electrical device across which there is a potential difference of v volts, it transfers a power $P = iv$ watts. That power can be used to either do some kind of work, e.g., raise an elevator, or just cause the device to get hot. As an example, the effective voltage from a U.S. wall outlet is 120 V; a 1200 W microwave oven draws 10 A from that wall outlet. As we will see in Chapters 3, 4 and 5, the energy can even be stored inside the device. We will see that relationship extensively and in many forms throughout the book.

1.2.1 Conductivity and Resistivity of Materials

The *amount* of current that flows in response to a given potential difference depends on the particular material in which it flows. **Conductivity** is the (intensive) property of materials that characterizes how easily electrons move when subjected to a voltage difference. For example, copper conducts current well and tungsten does not. Conductivity is represented by the Greek letter sigma, σ.

An electrical or electronic **circuit**, for example, a doorbell or cell phone, is a collection of devices connected to each other by wires. It is called a 'circuit' because current must *circulate*, or make a complete *loop*, whenever current flows. What makes circuits possible is that there are materials, e.g., metals, that have much higher conductivity than the surrounding air. Hence the current is channeled through the electrical circuit rather than flowing around it through the air. The electrons must flow *inside* the circuit, and never *outside* the circuit, which is one reason that electronics should never be allowed to get wet. The confinement of current to the circuit is loosely akin to cars driving on a highway rather than through an impassable forest alongside it.

The conducting paths that connect electrical components together are generally called **wires**. Wires are simply long, narrow pieces of metal. They are often characterized, or **modeled**, as the simplest elements of a circuit, but you cannot just take them for granted. In practical circuits, the wires may exhibit properties that cannot be ignored, as in the residential wiring discussed in Chapter 3, or in their behavior in advanced systems, such as integrated circuits, radar, or cell phones. Additionally, wires must not touch each other because current would likely flow in unwanted ways. Wires that inadvertently touch would be the equivalent to a leak in a water pipe in which the water could flow anywhere, and not just through the piping system. Since electrical components act as if to slow the electrons down in the circuit, i.e., use up some of the energy or voltage, removing some components would leave that work to fewer components, and they might experience too much voltage drop, and could heat up excessively according to Eq. 1.3 – too much energy would be converted to heat, which would be dangerous to the circuit, and some components could fail!

Electrical and electronic systems might use large currents (let's say 10 A to 100 A) or very small currents (more like 10 *micro*amps, μA), and often both in different sections of the same system. Obviously, each of these cases presents different requirements for the dimensions and material of the wires.

Conductivity describes how well a material conducts – in other words, how easily charge moves through it. The inverse of conductivity is called **resistivity**; it is represented by ρ (rho), and

> By "large" and "small" currents, I mean what is commonly experienced sitting at a lab bench or found in a common home appliance. There are special cases in which you might interact with nanoamps, but that is difficult and requires specialized equipment. Likewise, some systems, such as power transmission lines, operate at hundreds or thousands of amps. Common prefixes are:
>
> nano, n = 10^{-9}
> micro, μ = 10^{-6}
> milli, m = 10^{-3}
> kilo, k = 10^{3}
> mega, M, = 10^{6}
> giga, G = 10^{9}
>
> *Note*: The Greek letter mu, μ, is the SI abbreviation for micro, meaning 10^{-6}. It used to be acceptable in electrical engineering to use a "u" instead of a "μ" because in the era of typewriters, μ's had to be written in by hand. It is still common to use 'u,' but with modern computers, there is little excuse not to use the correct Greek letter μ.

$$\rho = 1/\sigma. \tag{1.4}$$

Resistivity tells us how much the material *prevents* charges from flowing, i.e., how much it *resists* the flow of current. Conductivity and resistivity are two sides of the same coin; they are equivalent descriptions of the same property of how easily charge moves through material. The most common device used in electronics is the **resistor**, which is basically a piece of material having a certain resistivity and therefore **resistance**, whose unit is the **ohm**.

The metal with the lowest resistivity (and the highest conductivity) is silver. Common materials used in electronics are

> *Important*: **Gauge** is a measure of something, here wires, for which a number stands in for some size scale. Examples of the use of 'gauge' include **wire gauge** as well as for train tracks, pocket-watch movements and shotgun-shells. In all cases the object size decreases with increasing gauge number. Examples of American wire gauge, AWG, are 12-gauge wire, 2.305 mm diameter, and 18-gauge, 1.024 mm diameter.

(in units of ohm-cm, discussed below) copper ($\rho = 1.68 \times 10^{-6}$), aluminum ($\rho = 2.82 \times 10^{-6}$), gold ($\rho = 2.44 \times 10^{-6}$) and silver ($\rho = 1.59 \times 10^{-6}$). Nickel/Chrome alloy ($\rho = \sim 10^{-4}$) is used in wires that are intended to get hot, as in a toaster or stove-top element. Simple wires are either **solid core** (**single-stranded**), **stranded** (many very thin wires running next to each other), or **braided** (thin wires woven together). These have varying levels of gauge, i.e., thickness, as well as flexibility, reliability, and overall conductivity.

An important consideration is the choice of insulator to surround wires. Clearly, wires must be insulated from each other and the rest of the world. In no case do you want the wires to accidentally touch each other, or for you to become the wire! Insulators are materials that are the opposite of conductors in the sense that they do not conduct electricity. Air is an insulator, but air does not prevent bare metals from touching each other. The rubber insulation on old wires (picture a vintage radio from WWII) can be brittle and crack, exposing bare wires. In house wiring and many other instances, this is a real hazard if not attended to.

> *Practical*: You may have to use a **wire stripper** to remove the plastic insulation around wires without damaging the metal. There are many types of wire strippers out there. In a pinch, you could use a jack-knife, but that is not recommended for reasons of quality and safety. Basic wire strippers have multiple openings in the cutting surface to match the wire gauge (https://learn.sparkfun.com/tutorials/working-with-wire/how-to-strip-a-wire). Self-adjusting wire strippers grab the wire, tighten around the insulation, and pull to a specified length – the Cadillac of wire strippers! **Magnet wire** is insulated by a thin lacquer coating so that it can be wound very tightly into a coil without touching (for an inductor, to be presented later). Such insulation is removed with something like sandpaper.

The usual choice of insulator is colored plastic, as shown in Fig. 1.2. Different colors are used to distinguish wires from each other. This is especially useful in troubleshooting.

The big brother to the basic wire is the **cable**. The word *cable* refers to more than just your TV provider, but there is a connection between the usages of the term. Electrical 'cable' is something more than a wire. In power distribution (Chapter 3), cable refers to a very thick wire that carries lots of power. In electronics, cable refers to an assembly of conductors that carry high-speed or low-noise signals (e.g., for cable TV). The simple wire discussed above does not work well at high frequencies or at low signal levels. A simple wire can be affected by its electrical environment, so weak signals can be swamped by environmental **interferences**. When very weak or high-frequency (such as radar) signals are required, the center core, which is a single wire, can be surrounded by a metal jacket, like a surrounding tube of metal. This **shield** creates a protected space inside of which the signals can be transmitted free of interference or losses. The **co-axial**, or 'co-ax', cables used in the lab are very common.

FIGURE 1.2 Wires come in many forms, from thin films of metal on printed circuit boards to cables that carry large currents or high frequency signals.

> *Of interest*: An area of extensive research is **superconducting** wires, which have literally zero resistance at low temperatures. In Chapter 3 we will discuss how this might someday transform the national power infrastructure. Also, **graphene**, a special form of carbon that is extensively researched, has very low resistivity and may one day be used in practical circuits.

If ordinary, plain wires were used instead, your lab experiments would be degraded by all of the electrical interference coming from the lights and other sources in the room!

1.2.2 Short Circuits and Open Circuits

As long as we are talking about wires, let's talk about **short circuits** and **open circuits**. These concepts are such a frequent part of conversation that they are usually shortened to just 'shorts' and 'opens.' Look at Fig. 1.3. Imagine that you connect a piece of wire between two random places in that circuit. The conductance of a wire is very large – let's call it infinite. Any current that *can* flow *does* flow, but only through the wire, entirely bypassing everything else between its two connection points. This is a short circuit. We call that 'shorting out' the circuit. Clearly, the circuit will no longer function correctly. Often, a short circuit results in catastrophic current flow somewhere else in the circuit, damaging components or even causing a hazard.

An open circuit is the opposite. If current is supposed to flow in a circuit, but a connection point is broken, for example, by a bad solder connection (Section 1.3) in Fig. 1.3, then the current flow at that point will be zero because electrons cannot cross the air gap. As an analogy to water flow, an open circuit is like a pipe that has been cut and capped on both ends of the break – the water can't pass the cap, and pressure just builds up. An open circuit is less likely to damage the circuit since current flow is generally reduced, rather than increased, somewhere else.

Of interest: In 2019, the historic Notre Dame cathedral in Paris suffered a disastrous fire attributed to a short circuit.

FIGURE 1.3 Printed circuit board populated with many surface-mount devices, which are soldered directly to the traces on the board and do not go through holes. The small surface-mount devices, SMDs, are a variety of resistors, capacitors, inductors, and other simple devices. The large square parts are integrated circuit packages with semiconductor devices embedded inside. The small bright spots are solder joints that let current flow from the board to the devices.

1.2.3 Printed Circuit (or Printed Wiring) Boards

When an electronic system uses many components, there may not be enough space to include all of the separate wires. Also, those wires would have to be extremely small, and all that wire and assembly would be costly and possibly unreliable. A modern cell phone is more powerful than the supercomputers of the 1990s and could not be built with ordinary wires. Even if it were possible to build a cell phone using ordinary wires for every connection, it would be more like the size of a refrigerator. Put that in your pocket!

Complex electronics are assembled on **printed circuit boards (PCB)** (sometimes called **printed wiring boards** (**PWB**)), which incorporate the wires as very thin, flat, copper **traces** on the board. Figure 1.3 shows several integrated circuit chips wired together by the traces on a PCB. Some traces are too narrow to make out in the figure. You will build your AM radio on a simple PCB.

A PCB is a flat structure consisting of a nonconducting **substrate** having conducting traces embedded on or within insulating layers. Older PCBs used holes to mount and connect to the electronic devices on the surface. These are appropriately called **through-hole** PCBs. Newer ones, like that of Fig. 1.3, support devices attached directly onto the surface to the traces, called **surface-mount** devices. The substrate can be stiff or flexible. **Flexible PCBs** (one of which is shown at the top and center of Fig. 1.3) are used in consumer electronics where they need to bend around physical structures, as on the inside of a laptop computer. PCBs can be either single-sided, double-sided, or multi-layer. Your AM radio is a single-sided, through-hole PCB with bare copper exposed on only one surface, to which you will melt your solder and affix your electronic components, as described in the next section.

Most circuits are sufficiently complex that traces must cross over each other without touching. **Double-sided** PCBs make this possible by having traces on both faces of the board, thus using the board as an insulator between traces. A further extension of this concept is the **multilayer PCB**. The most complicated type, it comprises stacks of alternating insulating layers and traces, having vertical tunnels, called **vias**, to conduct the signals among the layers. Multilayer PCBs can be made with as many as 100 layers.

1.3 Solder and Soldering

The electronic components in a circuit must make low-resistance contact to each other through the PCB, and must also be reliably fixed to the PCB in the presence of vibration, shock, temperature changes, etc. The universal solution to the challenge of connecting metal to metal is **soldering**. Solder is a low-melting-point metal that bonds to both the component and to the PCB traces. It is not necessary that the component itself touch the PCB trace since the solder creates a low-resistance path, as if it were an extension of the device itself. Solder must melt at a reasonably low temperature, say around 200°C (but not so low that it cannot withstand temperatures that it might encounter under use, such as under the hood of a car). It must also coat the parts uniformly, and not crack or de-adhere from the parts under use. Figure 1.4 shows the process of hand-soldering a component to a PCB (in this case, your AM radio). The figure shows solder being applied with a **soldering iron**. This has a tip that is hot enough to melt the solder (but not so hot as to damage any of the other components).

Most solders are a combination of the metal solder material with organic compounds (at the core of the solder wire) called **flux**. The word "flux" comes from the Latin *fluxus* for "flow" (and is related to the word "fluid"). Most metals grow a thin outer layer of oxide that forms from the air. Solder, when heated in air, also forms oxides. Molten solder does not stick to the oxides that form on its surface, the copper traces and the component, so the oxide prevents the solder from adhering to the component and PCB. The organic flux in the solder mixture chemically removes the oxides, allowing the solder to flow. If you use the same piece of molten solder on the iron's tip for too long, the flux is used up and the solder won't flow well onto the parts, so you should clean it off and melt new solder on it, a process referred to as **tinning**.

At the top is a hand-made prototype processor circuit from about 1975, when it was not efficient to make a PCB for a prototype circuit, even as complicated as this. The picture below shows all the wiring used to interconnect the integrated circuits. Computers have made PCBs accessible. Circuit courtesy of N. Dotson.

FIGURE 1.4 A soldering iron and solder in the process of soldering a component to a PCB. Note from the inset that the tip is not melting the solder directly. The damp sponge can be seen at the upper left.

Manual soldering is a skill and takes practice. This AM radio may be your first soldering experience. Building the radio is not just an excuse to do some soldering; you need the radio for later labs, so you should carefully read about soldering techniques in the radio instruction booklet and in Appendix 1 in this chapter. Also, you should start slowly and work patiently as you gain experience. The main thing to keep in mind is that the tip should be shiny, and you should be careful to heat both the component and the PCB trace so that the solder melts, i.e., **wets**, onto both of them and not just the tip.

> *Of some importance*: Historically, common solders included lead, usually alloyed with tin. Sn/Pb in mass ratio of around 60/40 is common. In 2006, the European Union passed the Restriction of Hazardous Substances (RoHS) directive that phased out the use of lead in solders, and the U.S. is slowly following suit. This change has caused industry to seek other alloys to perform the same task, albeit at a higher melting point and resulting in other problems. The most popular of these is Sn-Ag-Cu (tin-silver-copper), or 'SAC.'

1.3.1 Safety

A note about safety: You should always wear safety glasses when soldering, as well as in just about every other lab activity where things are very hot or energy might be released suddenly. In our case, this is not just a *pro forma* exercise. Firstly, cover the wires with your hand when you cut them off of the components after soldering, lest they fly halfway across the room into someone *else's* eye. Secondly, solder is hot when molten, and presents a potential eye hazard. In fact, once I was soldering when I had the bright idea to use the tip of the soldering iron to move a piece of wire. The wire sprung free of the tip, and a quarter-sized blob of molten solder instantly flattened itself against my eyeglass lens – dead center over my pupil. It happened faster than I could have blinked. I shudder to think of the effect on my cornea had I not been wearing eyeglasses. Let this be a lesson to you about your ability to plan not to have an accident – because *you can't!*

1.4 The Big Three: Resistors, Capacitors, and Inductors

Now we turn our attention to the devices that will be soldered into the radio. We call the devices discussed in this section the "Big Three" because you will be mostly concerned about these circuit components in your first circuits course. In case you are unfamiliar with them, we will introduce (or review, as the case may be) them here. All three are **passive two-terminal** devices. *Passive* means that the device does not require an external power source to do its function. (Devices that are not passive, and do require external power, are said to be **active**.) A *two-terminal* device has two wires connected to it.

Circuits are represented by drawings called **circuit diagrams, schematic diagrams**, or simply 'schematics.' In these, single lines represent simple, ideal wires. (Recall that an ideal wire has no resistance and has no effect on the electrical properties of the circuit, other than to connect devices together.) In the following sections, we will introduce physical and electrical properties of the Big Three, starting with resistors.

1.4.1 Resistors

In a previous section, we introduced the notion of the resistivity of a material. All materials have some resistivity. If we use a piece of material having reasonably high resistivity to act as a circuit element whose purpose is to limit current flow, then we call it a **resistor**. Resistors have an older and newer graphical notation used in schematics, both of which are in common usage. The old notation uses a zig-zag line, and the new notation uses a rectangle, both shown in Fig. 1.5.

A resistor is characterized by its **resistance**, *R*, measured in units of **ohms**. Ohms is abbreviated by the Greek letter Ω (capital omega). Resistors obey **Ohm's Law**, which is one of the central equations of electrical engineering:

$$v = iR,\qquad\qquad(1.5)$$

where *i* is the current *through* the resistor, and *R* is its resistance. *v* is the voltage between the ends of the resistor, referred to as the voltage *across* the resistor, or simply **voltage drop**. Ohm's law tells us that the voltage drop is proportional to the current through that resistor. As a water analogy, it takes more water pressure (*v*) to push the same rate of water flow (*i*) through a narrower hose (higher R).

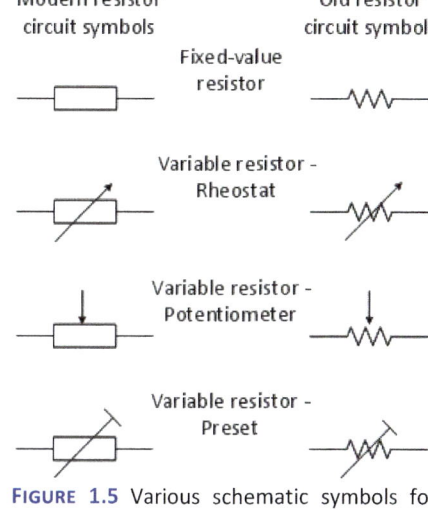

FIGURE 1.5 Various schematic symbols for fixed and variable resistors.

Just as materials can be described by their resistivity, or equivalently by their conductivity, similarly resistors can be described by either their resistance or **conductance**, *G*. (There is *not* a corresponding device called a "conductor.") Conductance is the inverse of resistance: *G* = 1/*R*. The unit of conductance is **siemens**, abbreviated as S; S = 1/Ω, or Ω^{-1}.

It would be great if metals had zero resistivity and wires had no resistance whatsoever. Rather, the resistivity of metals used in wires oftentimes cannot be ignored. As we will see in Chapter 3, the power loss due to wires costs the U.S. billions of dollars per year. The mathematical prediction of currents and voltages in a circuit is called **circuit analysis**. In performing circuit analysis, we frequently ignore the resistance of wires, i.e., assume it is zero ohms. Sometimes we cannot, so we must be able to calculate the resistance of wires.

The resistance of a long wire is (Pouillet's Law):

$$R = \frac{\rho L}{A},\qquad\qquad(1.6)$$

where *i* is the resistivity of the metal, *L* is the length of the wire, and *A* is its cross-sectional area (assumed constant along the wire in our case). If it helps, you may think of a resistive wire as a water hose packed with gravel: the longer it is (larger *L*), the narrower it is (smaller *A*), or the more gravel it has (higher *ρ*), the less water flows through it for a given water pressure. Here, water pressure drop is analogous to the voltage drop across the resistor.

Consider a home electrical appliance like a vacuum cleaner with an electrical cord. Current flowing through a resistance causes heating, so it is normal for appliance wires to get warm, but in the extreme, bad things could happen. If the wire were too resistive (*ρ* is high, *A* is small, or *L* is large), the wire could restrict the current, which could cause the vacuum cleaner to be underpowered. More seriously, if the resistance were too high for the amount of current flowing in it, the cord could overheat, causing a fire. In any case, any unnecessarily high-resistance path will, at the very least, waste energy.

Resistivity has a very broad range of values, allowing materials to be selected as good conductors or good insulators. For example, the resistivity of fused quartz (a type of glass) at

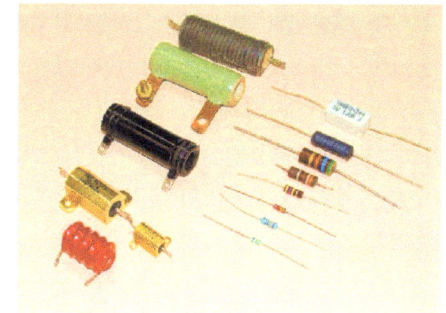

FIGURE 1.6 Resistors of various power ratings. Largser resistors can dissipate more heat without failing.

7.50×10^{17} Ω-cm is fully 23 orders of magnitude greater than that of copper at 1.68×10^{-6} Ω-cm! Figure 1.6 shows a variety of resistors made of different resistive materials. On the left is a variety of power resistors that are built to dissipate considerable amounts of heat. The ones on the right can dissipate heat in decreasing amounts moving down the column.

Many two-terminal devices, including most resistors, are called **axial**, i.e., the two leads come out at the ends along the axis. All of the resistors on the right side of Fig. 1.6 are axial. Resistors like those in the figure are made of many materials, usually a composite of carbon in the form of graphite and insulating fillers – the more carbon, the lower the resistance. **Carbon-composite** resistors (middle four resistors on the right in the figure) are formed, attached to leads, baked, coated, and painted with stripes that encode the value of the resistance.

Another common type of resistor is **wire-wound** (top two resistors on the right in the figure). Wire-wound resistors employ a length of resistive wire (think $R = \rho L/A$ that sets a precise resistance value and can withstand high temperatures. Carbon-composite and wire-wound axial resistors often look alike, but differ in their **tolerances**, i.e., the accuracy of the coded value.

Yet another family of resistors is that of **carbon-film** or **metal-film** resistors, which are the lowest two resistors on the right in Fig. 1.6. Three carbon-film resistors with the outer coating scraped off are shown in Fig. 1.7. These resistors start out as a carbon film on an insulating ceramic cylinder. Then a spiral pattern is cut into the film to create a thin ribbon. Finally, caps with axial leads are applied to the ends of the film, which after coating with the outer skin gives it a dog-bone-like appearance, as illustrated in Fig. 1.8. The resistor is much like a wire-wound resistor since its resistance depends on a thin length of resistive material that can be tailored by the choice of thickness and ribbon width (yep, $R = \rho L/A$). The tolerances of carbon-film

For clarity: The word spelled "lead" is used in this chapter as a noun as the metal connector to a component (pronounced 'leed'), and also as the metal lead, Pb (pronounced 'led').

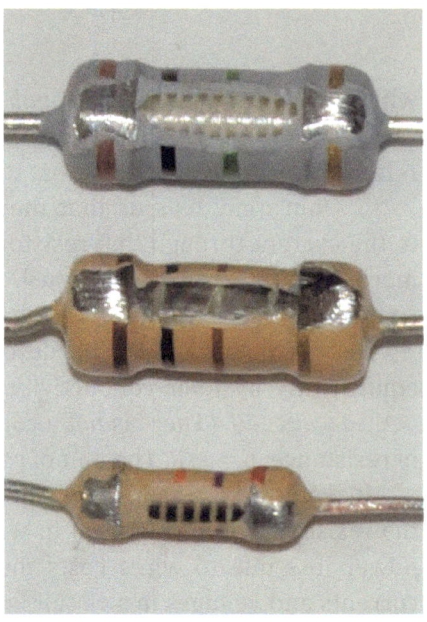

FIGURE 1.7 Carbon-film resistors with patterned film and caps exposed. Cutting the film converts into a stripe that rotates around the body of the resistor. Various stripe widths are visible, and different thicknesses and composition increase the range of resistance values. The top resistance value is 2 MΩ and the middle resistance is 100 Ω. The top film has been patterned to have a longer path and narrower ribbon between end caps. Both have tolerance of ±5%.

and metal-film resistors are intermediate between those of wire-wound and carbon-composite resistors.

The resistance value of small resistors is often indicated by small, colored **bands**. Most of the bands represent a digit from 0 to 9, as shown in Fig. 1.8. The simplest resistors are labeled by three bands and have ±20% tolerance. Reading from the band that is closest to an end of the resistor, the first two bands are interpreted as a two-digit number that is multiplied by 10 raised to the power represented by the third band. For example, yellow, violet, orange represents "473," which is understood as 47×10^3 Ω, or 47 kΩ (kΩ is pronounced "kil-ohm", although it might seem more natural to say "kil-a-ohm").

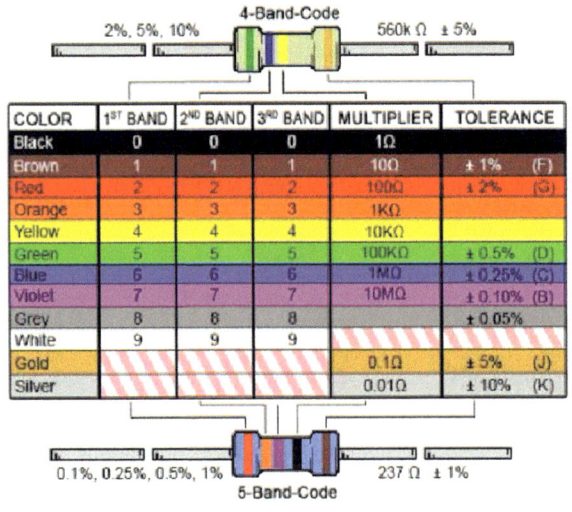

COLOR	1ST BAND	2ND BAND	3RD BAND	MULTIPLIER	TOLERANCE	
Black	0	0	0	1Ω		
Brown	1	1	1	10Ω	± 1%	(F)
Red	2	2	2	100Ω	± 2%	(G)
Orange	3	3	3	1KΩ		
Yellow	4	4	4	10KΩ		
Green	5	5	5	100KΩ	± 0.5%	(D)
Blue	6	6	6	1MΩ	± 0.25%	(C)
Violet	7	7	7	10MΩ	± 0.10%	(B)
Grey	8	8	8		± 0.05%	
White	9	9	9			
Gold				0.1Ω	± 5%	(J)
Silver				0.01Ω	± 10%	(K)

FIGURE 1.8 Universal electronic color code.

The tolerance color band tells you the maximum possible error in the value indicated by the code. Three bands separated by equal gaps (i.e., no tolerance band) implies a tolerance of 20%, which puts the actual value in our example 47 kΩ somewhere between 37.6 kΩ and 56.4 kΩ. If there is no tolerance band, read from the end closest to the wire lead toward the center.

A fourth band is used for somewhat tighter tolerances, most commonly 10% (silver) and 5% (gold).

> *This might be you*: About 8% of males and 0.5% of females are color-blind. I know from personal experience that being color-blind can make reading color codes challenging. Some tricks to make it easier include using a very bright white light, looking through a good magnifying glass, and keeping an ohmmeter handy for backup! The tolerance colors of gold and silver are easily distinguished.

A wider space usually separates the tolerance band from the resistance values. (See page 2 of the radio kit for another explanation of the color codes used in this lab.) Other resistor codes can use up to six bands for even tighter tolerances but are less common and will not be discussed here.

1.4.1.1 Power Dissipation in Resistors

Every resistor heats up to some extent while passing current. This can be understood by the sliding rock analogy. When a rock slides downhill there is friction with the ground. That friction heats the rock and the ground. As a consequence, the rock does not go as fast as it would if it dropped off of a cliff with no contact to the ground. This transfer of energy from gravity to heat happens at a certain rate of energy per time. The rate of transfer of energy per unit time is power, *P = iv*. (Eq. 1.3) Using Ohm's law, the power dissipated by a resistor is therefore given by

$$P = iv = \frac{v^2}{R} = i^2R. \qquad (1.7)$$

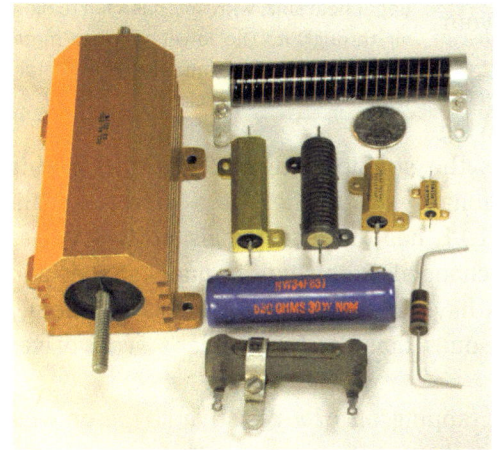

FIGURE 1.9 Several power resistors with a quarter for scale. The gold-colored ones are wire-wound embedded in ceramic material. These are designed to be bolted to a heat sink that draws heat away. The long tubular ones are air cooled rather than heat-sinked. The one in the lower right is a carbon composite resistor capable of dissipating a few watts, but is not generally referred to as a 'power resistor.'

> *Practical info:* So, just how much *is* one watt, after all? Here are three examples. One watt of music from a stereo is loud enough for a small party with a fair amount of background noise. One watt of laser power would do significant damage to your skin and obliterate your retina. (Laser pointers operate at a maximum of 5 mW). Finally, 1 W of heat from a ¼ W resistor would be enough to cause burns on your skin if you were unlucky enough to touch it as it fails.

If *i* is in amps, *v* is in volts, and *R* is in ohms, then *P* is in watts. If *P* is too large, then resistors will overheat and smoke, which is the universal sign of component failure. Every experienced electrical engineer knows that electrical components stop working if you let the smoke out (☺). (This is the oldest joke in all of electrical engineering!)

But seriously, heat is released from surfaces, so larger resistors having more surface area can transfer more heat power to their surroundings. Resistors have **power ratings** in terms of the number of watts that they can dissipate. The actual size of the resistor is an indication of its power rating. In fact, people refer to resistors by their power ratings, as in "a quarter-watt resistor." The resistors in your radio kit are all ¼ watt resistors. The carbon-composite resistors shown in Fig. 1.6 have power ratings ranging from ¼ W up to a couple of watts. Figure 1.9 shows several **power resistors.**

Depending on their size, power resistors can dissipate up to hundreds of watts and get very warm in the process. The smallest one shown can dissipate 25 W, and the largest one can dissipate 250 W.

Figure 1.10 shows two **heat sinks**, (or one word as **heatsinks**), which are (usually) metal structures that carry heat away from a device to protect it from failing. The figure shows two heat sinks used on the PCBs to transfer heat from high-power devices to the surrounding air. If lots of heat is to be dissipated, cooling fans, as shown, are used as well.

1.4.1.2 Variable Resistors

One important class of resistor is the **variable resistor**, whose resistance is adjustable by the user. Figure 1.11 (left) is a rendering of how a simple variable resistor is constructed. In the olden days (before digital circuits were ubiquitous), variable resistors were used for things like volume controls on radios, as it is for your AM radio kit, or motor speed, as in an old sewing machine. Symbols for the variable resistor are shown in the third row of Fig. 1.5. Those symbols correspond to the conceptual drawing

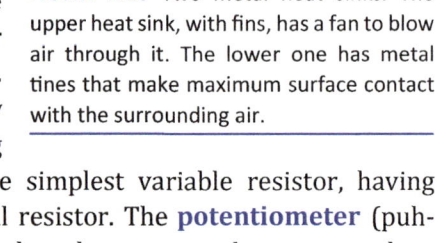

FIGURE 1.10 Two metal heat sinks. The upper heat sink, with fins, has a fan to blow air through it. The lower one has metal tines that make maximum surface contact with the surrounding air.

on the left side of Fig. 1.11. The **rheostat** (REE-oh-stat) is the simplest variable resistor, having only two terminals (the wiper and one end), as in a conventional resistor. The **potentiometer** (puh-ten-shee-AH-mit-ur) or simply 'pot,' shown in Figs. 1.5 and 1.11, has three terminals, so two values of resistance change simultaneously as the wiper position is changed. Pots come in many shapes and sizes, as shown in the right side of Fig. 1.11. Potentiometers have knobs ranging from large enough to turn with your whole hand (the big one), to really small ones that need a jeweler's screwdriver (used to make electrical adjustments on PCBs).

As you now know, $R = \rho L/A$; in this case, changing L by tapping off of a resistive stripe or wire whose length is L_{max} at different positions allows the selection of a resistance value. The **wiper** in the interior of the pot, connected to the lead labeled "W" in Fig. 1.11, is in physical contact with a stripe of carbon or a part of a wound piece of resistive wire. As the wiper moves along the stripe, the effective value of L between the wiper and both ends changes. If the wiper is all the way to the left in Fig. 1.11, then $R_{AW} = 0$ Ω, and R_{WB} is the maximum value, R_{AB}, corresponding to L_{max}. Functionally, a potentiometer separates the stripe at the wiper into one distinct resistor on each side. For example, a carbon-film potentiometer uses a stripe of carbon composite running from A to B (e.g., 100 kΩ), and the wiper slides on the surface to access the stripe, as shown on the left side of Fig. 1.11, to split that 100 kΩ into two pieces whose total length is L, and therefore whose total resistance is $R = 100$ kΩ.

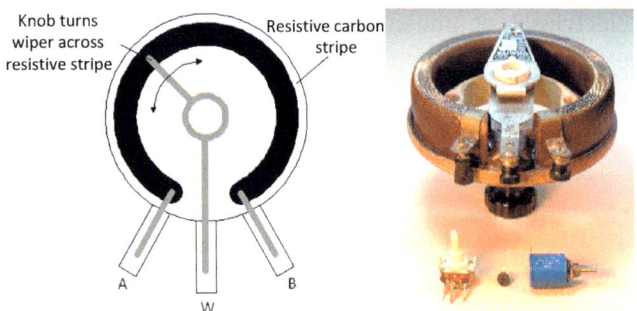

FIGURE 1.11 Left side: Drawing of the inside of a carbon composite potentiometer showing the carbon stripe and contacts to the wiper and two ends. Right side: Photo of four different potentiometers including an example of high-power, wire-wound (top), carbon stripe like the drawing on the left (bottom left), PCB-sized (bottom middle), and ten-turn for accuracy (bottom right).

Figure 1.12 shows a simple circuit that illustrates the operation of a potentiometer. The left and right leads at points A and B have a voltage applied across them. If the wiper is all the way to the right, then $R_{WB} = 0\ \Omega$, and R_{AW} is R_{AB}. By changing the relative values of R_{AW} and R_{WB}, the user may select, or 'tap off,' at the wiper any voltage from 0 to v_{AB}. Such a function is referred to as a **voltage divider**, discussed in Chapter 2.

The largest potentiometer in Fig. 1.11 is wire-wound and is shaped just like the left-hand drawing of the inside of the pot. The pot at the lower left of the photo is very much like the one used in the AM radio kit. It too shares the same arrangement with a wiper and carbon stripe. The

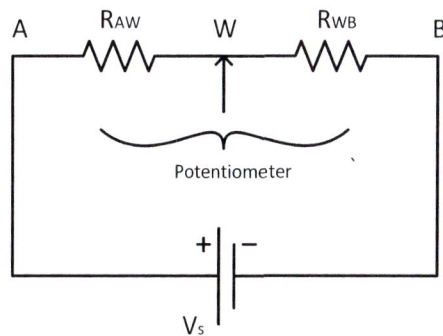

FIGURE 1.12 Schematic drawing of a simple voltage divider made with the potentiometer of Fig. 1.11.

small black one is for mounting on PCBs and is adjusted with a small screwdriver. The one on the right is a 'ten-turn pot' that employs a threaded screw to move the wiper over the full range of travel by turning the knob ten times, thus enabling precise control of the selected resistance, and hence wiper voltage.

Note that if all you want is a variable resistor, or rheostat, then you can just use two terminals, A and W or W and B; now you can select a resistance value anywhere from 0 to R_{AB}. Can you imagine how the rheostat or the pot might be used as a volume control? By changing the resistance, the properties of an audio amplifier could be changed to make the volume louder or softer, as on your AM radio. Also, the pot in your kit incorporates an **on-off switch** at one end of the travel so that your battery is not being drained when the radio is not in use. Not all pots have this feature.

1.4.2 Electric and Magnetic Fields in Capacitors and Inductors

1.4.2.1 Electric Fields

These next sections about electric and magnetic fields may seem like a deviation from a discussion about devices, but they are essential to understanding upcoming capacitors and inductors. Electric and magnetic fields are central not just to those devices, but to nearly all areas of electrical engineering, so it is appropriate to spend some time introducing, or perhaps reviewing, these concepts.

Fields at their most general are functions of position in space and time. The ones we are interested in here are vector-valued functions. In other words, defining an **electric field E** is simply saying that at every point **r** in space there is a vector $E(r)$ called "the electric field at **r**." (The bold font used for E and r and other single letters implies that they are vectors – not to be confused with our use of bold font to introduce new terminology.) The space we're usually interested in is three-dimensional space, or 'three-space' for short. It is often useful to use 'two-space' for visualization or illustration, as in Fig. 1.13. At every point in space, there is a force on a charge due to the field, discussed below. That force pushes the charge in a particular direction, i.e., the direction with a certain magnitude, so it is a vector. To visualize

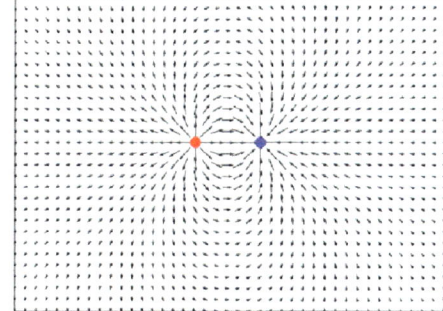

FIGURE 1.13 Force on test charge and electric field produced by two stationary point charges. The field is defined everywhere, but to make the illustration humanly readable, arrows show the magnitude and direction of the field only at a few regularly spaced points. github. com/tomduck/electrostatics.

the field, arrows are drawn at regularly spaced points. There are many ways to draw the fields, but often the length and direction of each arrow indicate the magnitude and direction of the vector field *at that point,* as shown in Fig. 1.13.

One kind of field is the **force field**. The value of the force field at a point is simply the force exerted at that point. An important example of a force field is the **Coulomb force**, also known as the **electrostatic force**. The Coulomb force exerted by a single charge Q_s at the origin, acting on a small **test charge** q at the point r, is given by Coulomb's law as

$$F(r) = \frac{rqQ_s}{4\pi\varepsilon_r\varepsilon_0 |r|^3}. \tag{1.8}$$

Recall from your physics courses that r is a vector having magnitude $|r|$. So, dividing the original vector r by the scalar quantity $|r|$ you normalize it to the value of '1' in whatever units it has. Now it is a **unit vector** that retains its original direction. Thus, the Coulomb force has a direction $r/|r|$ (the directional information resides in the r term) and magnitude:

$$F(r) = \frac{rqQ_s}{4\pi\varepsilon_r\varepsilon_0 |r|^3} = \frac{qQ_s}{4\pi\varepsilon_r\varepsilon_0 |r|^2}. \tag{1.9}$$

You may say that the force is caused by an electric field at every point in space, so the electric field is also a vector quantity:

$$F(r) = q \cdot E(r) \tag{1.10}$$

so,

$$E(r) = \frac{F(r)}{q} \tag{1.11}$$

In words, you would say that the electric field is the "force per unit charge" at a point. If the charge of the test charge is doubled, the force doubles. Since the field is the force per charge, it is a constant that does not depend on the specific charge that is placed in the field, whereas the force field does depend on the value of that charge. When viewing the vector field as in Fig. 1.13, you may think of it as either a force field or electric field. They both have the same properties.

In Eqs. 1.8 and 1.9, ε_0 is the **permittivity of free space** and ε_r is a dimensionless number called the **relative permittivity**. ε_0 characterizes how the electric field and Coulomb force are related in free space, i.e., a vacuum, and ε_r characterizes how the electric field and Coulomb force are related inside of a real material that does not conduct current, often called a **dielectric**. ε_r has the value 1 in free space, since there is no material there (thus leaving the total permittivity as ε_0), and ranges to about 15 for most common materials. These properties of the dielectric arise because the atoms of the dielectric material are composed of electrons and protons (nucleus). These atoms with their charge carriers come with their own internal fields; the external field causes the electrons to shift positions within the atoms, a phenomenon called **polarization**. The polarization changes the total electric field due to the charges. Hence, we need ε_r to account for that.

We say that the Coulomb force around a point charge is an "inverse-square force" or that it "falls off as $1/r^2$," where r is $|r|$. A "test charge" is an imagined charge; it is *not* shown in the image. Imagine that if you magically place a very small (negligibly small – approaching zero in the limit) positive charge at a point, it would feel a force in the direction of the arrow with a magnitude of force that is proportional to how long the arrow is drawn. But doesn't the addition of another charge change the field distribution? Well, normally it would, but we assume that the test charge is so small that it does not cause any change to the force field produced by the other,

Important: There are a lot of details in this section about fields. It will all be important in later courses. The main point for now is that the fields are where the energy is stored. This will be important as we discuss capacitors and inductors.

actual charges. Figure 1.13 shows the force field (here due to one negative charge and one positive charge) that would act on a very small positive charge if you were to put one in to "test" the field. Since the electric field is the force in the limit of zero test charge, the same figure shows the electric field. E is more general or convenient than F in the sense that one can talk about E without knowing the value of q in advance, and then compute the force F by simple multiplication if we care about that.

One extremely important fact is that the *electric field stores energy*. Let's see how we can prove that to ourselves: If you move some charges and change the electric field, the change will *eventually* act on distant charges and may perform some work on them. The effect is to transfer some energy from the mechanical motion of the nearby charges to distant ones. But information about the changes in the electric field can't travel faster than the speed of light. After you've stopped moving the nearby charges, and before the news reaches the distant charges, you are done doing all the work that you are going to do. However, the distant charges haven't yet gained that energy! So, until then, where is the work energy that you put in? Answer: *In the field*.

Another thing about electric fields and the energy they store is that you have to do work to create them. You don't just get fields for nothing. Based on Eq. 1.8, when the charges have the same sign (e.g., two electrons) the force is positive, which is repulsive in this sign convention. If we create an electric field by bringing two electrons close together from infinity, we do work putting energy into the new, larger, electric field around them. They were perfectly happy not having any force on each other, being at infinity and all, but we had to cram them close together due to their mutual repulsion, which overall creates a larger field in most of the space around the two electrons. The energy we used to push them together is contained in the new electric field.

From here on, we use U (as do others) to represent energy in order to distinguish it from the magnitude of electric field, E. (Unfortunately, there simply aren't enough ways to write the letter E to distinguish all its many technical meanings.) We can know how much energy is stored in an electric field based on how concentrated it is:

$$\text{Energy density in an electric field} = U_E = \frac{\varepsilon_0 \varepsilon_r E^2}{2} \text{ J/m}^3, \tag{1.12}$$

where $E = |E|$ is the magnitude of the electric field. That energy density is defined at every point in space, so you can find all of the energy in a field by integrating, or adding up, U_E over all of the space where the field is, i.e., perform a volume integral.

There is another very important thing you need to know about electric fields. Remember that electrostatic potential is analogous to gravitational potential. Well, there was a ski run at Killington Ski Area in Vermont that was six miles long from the summit to the base (at the time, the longest in the East). (Just stay with me on this one.) As you can imagine, it was not very steep being that long to the bottom. In fact, I found it to be rather tedious since I could hardly pick up any speed at all. However, those double-black-diamond runs that go straight down the mountain along the lift line can be downright dangerous because they are so steep. Analogously, the "steepness" of the change in voltage with distance is the electric field. Mathematically that is written as:

$$E = -\frac{dv}{dx} \text{ (volts/meter)}, \tag{1.13}$$

where the derivative is the "steepness" of the voltage change. (In future courses, you will call this steepness in three-space the **gradient**.) Put into words, this equation says that "The electric field is how quickly the voltage changes along a path."

Why the minus sign? The minus sign is a convention. With this definition, the electric field points in the direction that a positive charge is forced, which for all charges is in the direction of maximum decreasing potential. For the capacitor (discussed below), a positive test charge q placed between the plates, one with positive charge and one with negative charge, would move toward the negative plate, which by analogy is "downhill" for it.

By the fundamental theorem of calculus (i.e., by integrating both sides of Eq. 1.13), we get

$$v = -\int E dx \text{ (volts)}, \tag{1.14}$$

which could be read as "the voltage difference between two points within an electric field can be found by summing, or integrating, or adding up, all of the field along a path between those two points." Put another way, moving a charge against the force of an electric field moves the charge to a higher electrical potential. Or, conversely, if you let a charge move due to the force of an electric field, it will move to lower electrical potential.

You may have noticed so far that this has been reading like a physics course, and not an electrical engineering course. That is because fundamentally, electrical engineering is based on physics and math, and first we must form the foundation. Along those lines, we must ask, "How does electrical potential translate into energy?" The energy gained or lost by a charge Q due to the electric field and electrical potential, or voltage, difference v is given by

$$F = QE; \quad \text{energy } U = \int F dx = Q \int E dx = Qv \text{ (joules)}, \tag{1.15}$$

In words this says, "Pushing a charge Q against the field for a voltage difference v costs energy Qv in joules." Or, conversely, if you allow the charge to move due to the force of the field, the charge gains energy Qv from the field in the form of kinetic energy as the charge accelerates. Do you recall that current flowing in a resistor causes it to get hot? That's because the electrons inside the resistor are accelerating for small bits of the resistor voltage drop before colliding with the atoms, giving up to heat that little bit of energy gained from the field at each collision. So, those electrons don't ever get going very fast, but they do give up Qv to heat the resistor as they fall through the full voltage across the resistor. Think of it as bouncing down a flight of stairs, one step at a time, rather than leaping down the staircase in one jump.

1.4.2.2 Magnetic Fields

We saw that electric charges generate electric fields that are tied to forces on other electric charges. You know that magnets, such as refrigerator magnets, act on other magnets. Thus, similarly, there is a **magnetic field**, a vector B, that describes that force. (The magnitude of B is B.) However, you may not already know that magnets *also* exert forces on *electric charges*. Magnetic forces on charges depend on the magnitude of the electric charge, just as electric forces do, but they are more complicated. For one thing, magnetic forces act only on charges that are *moving* – the magnitude of the force is proportional to the speed of the charge. For another, the direction of the force is perpendicular to both the field and the velocity. These relationships are encapsulated in the **magnetic force**:

$$F = qv \times B \tag{1.16}$$

where *v here* is the electron's velocity and "×" is the **cross product**. The cross product of two vectors is a new vector that is perpendicular to both of the factor vectors. You will see cross product again in Chapter 6 when you learn about wave propagation and radio transmission. It is certainly possible, and common, to have both an electric field and magnetic field present at one place at the same time. The electric field accelerates the charge by the Coulomb force, and the magnetic field causes it to turn in the direction of the cross product, i.e., perpendicularly to both the velocity and magnetic field. The total of the electric-field force plus the magnetic force is called the **Lorentz force**:

$$F = q(E + v \times B), \tag{1.17}$$

which is the sum of the electric and magnetic forces.

B is a field defined throughout space, just like *E*. Again, this will be very important: The magnetic field *stores energy just as an electric field does*. The energy density of the magnetic field is given by a formula similar to the previous one for electric field:

$$\text{Energy density in a magnetic field} = U_B = \frac{B^2}{2\mu_r\mu_0}, \tag{1.18}$$

where μ_0 is the **permeability of free space** and μ_r is a dimensionless number called the **relative permeability**. Whereas the relative permittivity ε_r describes the effects of charges in a dielectric material on the electric field, μ_r arises from electrons in a material that act like atomic-scale magnets.

If the only way to get a magnetic field were to use a permanent magnet, then technology would be stuck before the year 1819 when Oersted noticed that current in a wire turned a compass needle. Today, we well understand that we can generate magnet fields by simply flowing current in a wire. The magnetic field generated by a current in a wire goes around the wire as shown in Fig. 1.14, described by **Ampere's Law**. Ampere's Law is discussed in more detail in Chapter 3, but here we can say that the amount of magnetic field around a wire increases with the current in the wire, and decreases with distance away from the wire. The significance of magnetic fields occupying the space around a

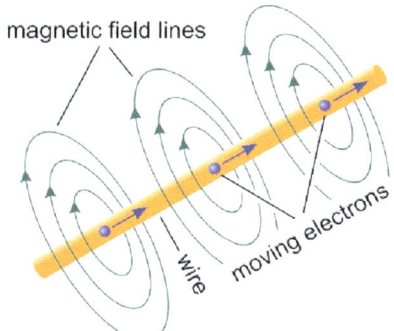

FIGURE 1.14 Current flow in a wire and the resulting magnetic field lines around it. This phenomenon is described by **Ampere's Law**. faithfulscience.com/classical-physics/electric-and-magnetic-fields.html.

current cannot be overstated. It makes possible nearly every electromagnetic device, including motors, inductors, and **transformers**. Transformers are essential to electric power generation all over the world, and they are also explored in depth in Chapter 3.

1.4.3 Capacitors

Now that you have been introduced to fields, we may call on the electric field to explain the operation of **capacitors**, the next member of our Big Three. A capacitor, whose electrical symbol is shown in Fig. 1.15, stores electric charge on metallic **plates**, or flat electrodes, that are separated by an electrical insulator. The plates have equal and opposite charges, $+Q$ on one plate and $-Q$ on the other plate. The separated positive and negative charges give rise to an electric field that points from the plate with positive charge on it and across the gap to the plate with negative charge on it. With the plates

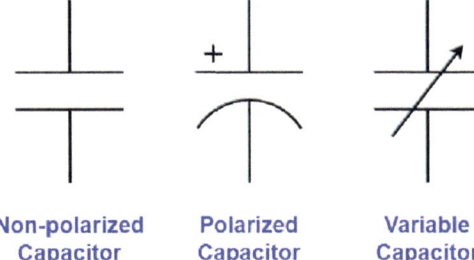

| Non-polarized Capacitor | Polarized Capacitor | Variable Capacitor |

FIGURE 1.15 Common capacitor symbols. https://www.derf.com/how-tantalum-capacitors-work/.

very close together, the electric field is confined almost entirely to the space between them. A word about Q: Of course all materials are made of atoms, which normally have in them equal amounts of positive and negative charge. When there is negative charge on a capacitor plate, we think of that as "more electrons" than usual. When there is positive charge on a plate, we think of that as "fewer electrons" than usual, thus leaving net positive charge from the nucleus behind. Again, electrons move around circuits; positive charges generally do not (except special cases like semiconductors and fluids, Chapters 7 and 10, respectively).

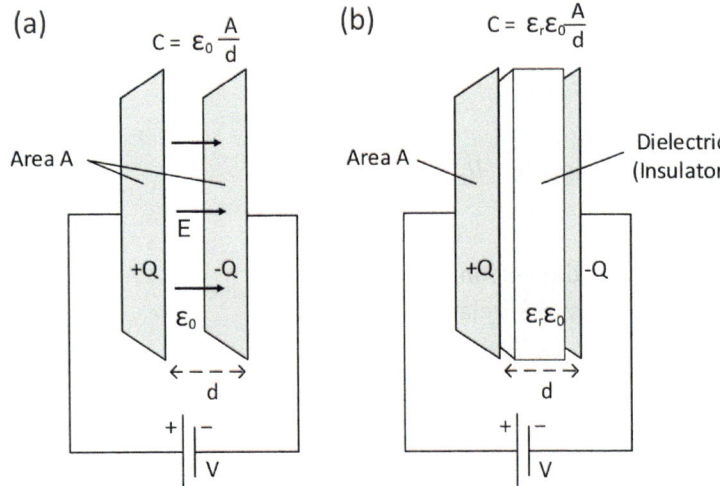

FIGURE 1.16 Diagram of two simple parallel-plate capacitors. (a) Air dielectric with permittivity of 1 and (b) insulating material with higher relative permittivity.

Equations 1.8 to 1.15 tell us that the force is due to charges, and therefore the electric field between them, are proportional to the amount of charge present. Let's call the voltage specifically across the capacitor v_C. (The subscript in v_C is a temporary crutch to remind you that the voltage we are concerned with here is only that across the capacitor itself, although there may be other voltages in a circuit.) If we assume that the plates of the capacitor, as shown in Fig. 1.16, are fixed in place (they don't move), then the voltage between the two plates, being the integral of the electric field between them, is also proportional to the charge on the plates, $+Q$ and $-Q$. So opposite charges on the plates give rise to a voltage that is proportional to the amount of charge on the capacitor. The constant of proportionality between charge and voltage is the **capacitance** of the system, having symbol "C":

$$Q = Cv_C \tag{1.19}$$

The capacitance depends on the physical size, shape, and materials of the capacitor. When v is in volts and Q is coulombs, C is in units of **farads**. Electrical engineers tend to use batteries or power supplies (Chapter 2) to apply voltage to circuits, which results in charges appearing on the capacitor.

Whether you think of a voltage causing the charge on a capacitor, or the charge causing the voltage across the capacitor, Eq. 1.19 is always true. These are simply alternate ways to think of the charge–voltage relationship.

Figure 1.16 shows two examples of capacitors, one (a) with an air or vacuum insulator, and one (b) with a solid insulator. Insulators are often referred to as dielectrics, a word used to emphasize the behavior of an insulator in the presence of an electric field. Figure 1.16(a) also shows the field inside the space between the plates. The field begins on the positive charge and ends on the negative charge. The entire space inside the gap is filled with electric field; the energy associated with this field is discussed below.

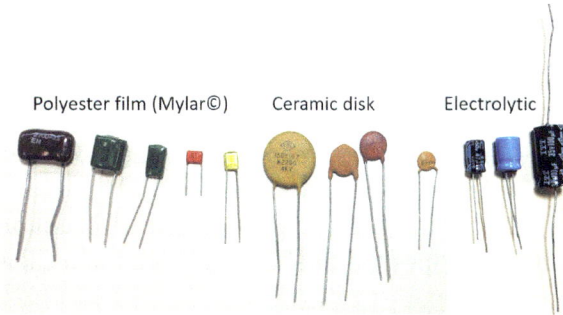

FIGURE 1.17 Various types of small capacitors. Note that unlike the idealized capacitor in the previous figure, the plates in real capacitors are not usually flat, but rather are either layered or rolled up into a can or other package to pack more area into a smaller volume.

Figure 1.17 shows a variety of different types of capacitors. These capacitors vary in the shape of the plates, but more importantly in the material used for their dielectric. The AM radio kit uses two of those shown in the figure, namely **ceramic disk** and **electrolytic** capacitors.

Although the unit of capacitance is the **farad**, most practical capacitors are small, and exhibit only small fractions of a farad, like micro- (10^{-6}), nano- (10^{-9}), or pico- (10^{-12}) farad (**µF**, **nF** and **pF**, respectively). The capacitance of a simple parallel-plate capacitor is

$$C = \frac{\varepsilon_r \varepsilon_0 A}{d},$$

(1.20)

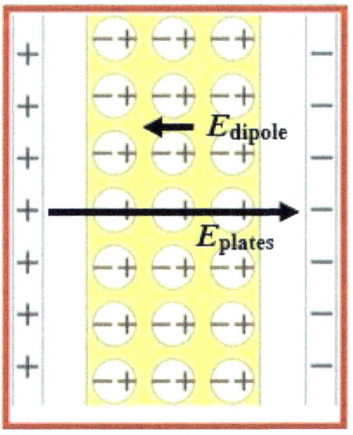

FIGURE 1.18 Parallel plate capacitor with a dielectric material inside the gap. The atoms of the dielectric are polarized due to the charges of the plate. Their electric fields oppose that of the external charges.

where A is the area of a capacitor plate, and d is the distance between the plates. As stated above, the factor ε_0 is the permittivity of free space, and has a value $\varepsilon_0 = 8.854 \times 10^{-12}$ F/m. ε_r is the relative permittivity or **dielectric constant**. Like the term *dielectric*, the terms *permittivity* and *dielectric constant* refer to the fact that charges at the atomic scale inside the material change their positions due to an electric field. You can think of this as the negatively charged electrons being shifted away from the positively charged nuclei, creating a tiny **dipole moment** of electric field at every atom. Figure 1.18 shows how the atoms inside the dielectric material separate slightly to form a dipole moment, each with its own internal electric field, E_{dipole}. Those tiny electric fields all oppose the electric field that arises from the charges on the plates, E_{plates}, and thus overall decrease the total electric field inside the capacitor for a given amount of plate charge. Keeping this in mind, Eqs. 1.19 and 1.20 tell us that in the presence of a dielectric material inside the capacitor, a higher value for capacitance is the same as saying that more charge, Q, is stored on the plates for the same voltage. Now we can see why the word "capacitance" is derived from "capacity".

As discussed, electric fields store energy. The energy stored in the field inside the dielectric of the capacitor is the energy deposited by the circuit into the capacitor as it 'charges up,' i.e., accepts excess charge from the circuit. The energy *density per unit volume* of the electric field, U_E, in a dielectric is $\varepsilon_r \varepsilon_0 E^2/2$ (Eq. 1.12). Notice that U_E depends not only on E but also on ε_r. That tells us that energy is stored both in the electric field between the plates and in the 'stretching,' or polarization, of the charges within all of the atoms. If allowed to, the electrons would move back to their neutral positions, so energy is needed to keep them in their polarized state. The field that causes the polarization is also part of the total electric field, so we can still say that energy is stored in the fields within a capacitor. Whenever the term ε_r appears in an equation, it tells us that the dielectric material is playing a role in storing energy.

Since the electric field is (nearly) uniform within the gap, the *total energy* volume integral that describes the total energy stored in a capacitor, U_C, simply becomes volume energy density × gap volume:

$$U_C = \frac{\varepsilon_r \varepsilon_0 E^2}{2} \times Ad \tag{1.21}$$

For the typical capacitor design, the electric field between the plates is constant throughout the dielectric. For a capacitor with a voltage v_C across it, the total electric field is, therefore,

$$v_C = -E \cdot d \text{ and } E = -\frac{v_C}{d}, \tag{1.22}$$

so, from the circuit point of view, the total energy can also be expressed as

$$U_C = \frac{C v_C^2}{2} \left(\text{or } U_C = \frac{Q^2}{2C} \text{ since } Q = C v_C \right). \tag{1.23}$$

If you like, you may think of capacitors as a kind of battery in which energy is stored in the field, and you may get that energy back to do some work. That is the basis for next-gen regenerative braking in some electric cars: the brakes charge a capacitor and the capacitor later briefly runs the motor.

If the plates are far apart, then the total electric field between the plates (taking into account both the charges on the plates and the fields in the dielectric material) is small (recall the long ski slope). If the plates are close together, the same voltage gives rise to a large total electric field, since, by Eqs. 1.13 and 1.22, the field is integrated over a shorter distance to the same voltage.

> *Of interest*: There are devices called **supercapacitors** or **ultracapacitors** with values in the many-farad range. Tens of farads operating at several volts is common, with one company offering 165 F at 48 V! Supercapacitors have lower energy density than rechargeable batteries, but are quickly charged and discharged, and can be cycled many more times. You can use the equation $U = \frac{1}{2}CV^2$ to calculate the maximum energy stored in a capacitor, in joules, where V is the maximum rated voltage of the capacitor. Currently, batteries store large amounts of energy for electric vehicles, but supercapacitors can drain very quickly and provide temporary energy storage for accelerating. They might someday totally replace batteries.

Although we haven't had any need to use this, it is important to point out that electric fields can be described by two different, intimately related quantities that are often interchanged, namely the vector quantity D, having units of C/m^2, and the vector quantity for electric field, E. D goes by the names of **electric flux density, displacement, free charge**, or **surface density**. Whew. Why all the names? Because it is a bit hard to pin down exactly how to think of D, and the quantity implies several different things. Let's stick to a simple parallel-plate capacitor for this discussion, as it is easy to visualize.

First, D at every point is, in general, related to the electric field E by

$$D = \varepsilon_r \varepsilon_0 E. \tag{1.24}$$

Although **E** is always called the electric field, often the generic use of the word 'field' encompasses both **D** and **E**. The word "displacement' implies the movement or shift of position of something. Shift of what in this case? It is the separation of positive from negative charge. Displacing them relative to each other results in an electric field, but that field, as discussed, depends on what material is between those charges. The term 'electric flux' came from the original, historical, view that something "flowed" from the positive free charges on one plate to the negative free charges on the other, and that flow has a density to it. As charges shift apart (are displaced), a field arises. It is often useful to think of **D** as the flux density, as we can integrate it across a surface, and that tells us the amount of free charge or surface density on the plates that caused it. For the same amount of free charge, or **D**, as ε_r increases, **E** decreases. This **E** is the total field, say, inside the capacitor, so it seems that the dielectric is "using up" the electric flux to do its polarization thing, and leaves some lesser amount of total electric field inside the gap.

1.4.3.1 Practical Capacitors

All two-terminal devices have some relationship between the current through them and the voltage across them. This current–voltage relationship is called the **current–voltage characteristic**, or **i–v characteristic** (pronounced 'eye-vee'). The i–v characteristic for a resistor is the simple relationship $v = iR$, Ohm's law.

For a capacitor, things get somewhat more complicated. To derive the characteristic for a capacitor, we start with Eq. 1.19, $Q = Cv_C$, and invoke the definition of current as the movement of charge past a point, Eq. 1.1, $i = dQ/dt$. Recall that in this equation, small 'd' is the derivative (not dielectric thickness) and t is time. Substituting Eq. 1.19 into Eq. 1.1, we get the current–voltage relationship:

$$i_c = C\frac{dv_c}{dt} \tag{1.25}$$

This is the equation that relates the voltage across to the current through a capacitor at every point in time. This equation means that as current flows through a capacitor, its voltage changes continuously, or alternatively that current flows through the capacitor if and only if the voltage across it changes. Of course, the voltage cannot change without the charge on the plates changing.

Let's make use of our water analogy now using water flowing from a tap into a bucket. The water flow is analogous to the charge (Q) flow or current $\left(i_c = \frac{dQ}{dt}\right)$, and the height of the water in the bucket is the voltage $\left(v_c = \frac{Q}{C}\right)$. The area of the bucket is analogous to the capacitance $\left(C = \frac{\varepsilon_r\varepsilon_0 A}{d}\right)$ – a larger area means it takes more water, or charge, to reach a given height, or voltage. (We can take this analogy a bit further: The height of the bucket's brim is analogous to its voltage rating. If it fills too high, it will overflow, which corresponds to destroying the capacitor by rupturing the dielectric layer.)

The beginning electrical engineer might reasonably look at the structure of a capacitor (Fig. 1.16) and ask, "Why is this not an open circuit? *How can current flow through the empty space inside a capacitor?*" You may think of a capacitor *plate* as a bucket of charge that is filled by the electric current flowing into one side of the capacitor from the rest of the circuit. The positive charge on that side of the capacitor exerts an electric force on the charges on the other side, repulsive for positive charges and attractive for negative charges. This pulls negative charges, i.e., electrons, onto the other plate, i.e., into the 'bucket.' You could also think of this as forcing positive charge off of the other plate. Either way you look at it – whether negative electrons move in or positive charges move out – it makes the charge on the second plate more negative. Making $+Q$ more positive makes $-Q$ more negative *by the same amount*. Hence, the same current flows into the capacitor through one lead and out of the capacitor through the other lead, *even as no actual charge moves through the space between the plates*. Because the current in the leads is the result of the separation of charge, the current through the capacitor is called **displacement current**.

Generally speaking, we want electronics that are smaller, lighter, and cheaper. The variables we have to play with are the ones in Eq. 1.20: ε_r, A, and d. For a given capacitance, we can decrease the overall plate area, and therefore size, by increasing ε_r. The lowest relative dielectric constant is that of vacuum (Fig. 1.16(a)), whose ε_r is 1. In **air capacitors**, there is nothing between the plates but air, so air is the dielectric, where ε_r is close to 1. Other dielectric materials include quartz ($\varepsilon_r = 3.9$), Mylar (a polyester with $\varepsilon_r = 3.1$), paper ($\varepsilon_r \sim 1.4$), mica ($\varepsilon_r \sim 3$ to 6). Ceramic capacitors use a range of higher-ε_r materials such as barium titanate with ε_r in the thousands.

Continuing with our capacitance-shrink exercise, we are interested in both linear dimension and volume. Often, very large capacitances are needed, so a strategy for controlling their physical size is important. Equation 1.20 suggests that large C might demand very large parallel plates (Fig. 1.16) whose extent might outstrip the size of the whole electronic system. One way to keep A large and the overall dimensions small is to pack the plate area into a compact stack, like a multi-decker sandwich. This is called a **multilayer capacitor**. Another way to pack the area compactly is to use long plates rolled up into the form of relatively small cylinders called **rolled capacitors**.

An important way to shrink the area of a capacitor while maintaining a high capacitance is to decrease the thickness, d, of the dielectric. One way to make d small is to use the oxide on a metal plate as dielectric, because most metal oxides are insulators. Note that a very thin oxide, i.e., one with small d, allows two plates to be *extremely* close together. Merely placing one plate up against an oxidized plate would surely leave extra space between the metal plate and the oxide, and this space would contribute to the value of d, thus decreasing the capacitance. Figure 1.17 shows the form factor of several common types of capacitor. The very common **electrolytic capacitor**, shown in Figs. 1.17 and 1.19, uses a 'trick' that solves this problem. An electrolytic capacitor, often referred to simply as an 'electrolytic,' starts with two plates separated by an **electrolyte**. An electrolyte is a material that conducts electricity through ions rather than electrons. The electrolyte used here is liquid, gel or soft solid – the point being that it deforms to completely fill the space between the plates. Next, current is passed between the plates and through the electrolyte, causing the negative plate to grow a very thin layer of oxide on it. The oxide is an insulator, and the conducting electrolyte is now an extension of the other plate. Hence, the only dielectric contributing to d is the ultrathin oxide. But wait! There's more! During the formation of the oxide, manufacturers use a heat process to make the surface of the metal plates rough, rather than smooth. This roughness increases the overall surface area, A, while maintaining the extremely small separation, d. You may now understand why electrolytics have very high values of capacitance in relatively small packages.

FIGURE 1.19 Electrolytic capacitors on a printed circuit board. Electrolytics come in metal cans wrapped in an outer label. In this picture, there are about 15 of them, mounted vertically with black or yellow plastic wrappers. On some of them you can see that the metal is scored (deeply scratched) at the top to ensure that if they fail, they pop open without exploding.

FIGURE 1.20 Top: large radio tuning capacitor bank. Bottom: small radio tuning capacitor, from your AM radio kit. The capacitor case has been opened to reveal the plates, magnified in the inset at upper right.

Applying a polarity opposite that used to create the oxide will erode it, eventually developing holes that short-circuit the capacitor. This current can heat the electrolyte causing it to boil, thus increasing the internal pressure to the extent that the cylinder could explode. For this reason, the outer metal can is scored (partially cut through) in order to weaken it so that it fails at less pressure and less dangerously. Therefore, one must be mindful of the correct polarity when installing these devices. A minus sign labels the side or the end of the capacitor with the negative lead. In addition to soldering, this is one reason to use safety glasses while working in the lab.

All devices can be destroyed by applying too much voltage. Thus, capacitors have a voltage rating in addition to a capacitance value. When the voltage rating is exceeded, the dielectric **breaks down** and real charge current flows through it, destroying it (i.e., you let the smoke out – don't forget the smoke). (Advance warning: in Chapter 7, Semiconductors, we will discuss a different 'breakdown' process in reverse-biased diodes, but that breakdown is not destructive.)

Capacitance values are indicated differently across the various types of capacitors. For example, electrolytic values are printed directly on them. Disk capacitors have a simple code: the first two of the three digits are a number that is multiplied by ten raised to the third digit, as in the case of the resistors, and the letter indicates the percent error tolerance. The whole value is in picofarads, or pF. (The unit of pF is often pronounced 'puff' by EE insiders. Nanofarad, nF, is *not* pronounced "nuff.") See page 2 of the kit instructions for examples of capacitor value codes.

Given everything said above about capacitors, there is one more capacitor in your kit that is almost quaint in its simplicity. The tuning of this radio is performed by adjusting the capacitance of a **variable capacitor** (Fig. 1.20) that is made from plates that slide past each other without touching, of course, since that would short out the capacitor plates. This is an example of an air capacitor. As discussed above, a simple capacitor model is that of two parallel metal plates, and this variable capacitor is nearly exactly that, except for the ability to move the plates relative to each other, thus varying the plate area A in Eq. 1.20, $C = \varepsilon_r \varepsilon_0 A/d$. When the plates overlap completely, the capacitance is at a maximum, and when they are separated, the capacitance is at a minimum. The semi-transparent case on some radio-tuning capacitors, like the one in your AM radio kit, allows you to see the plates move past each other.

1.4.4 Inductors

The last of the Big-Three circuit elements is the **inductor**, whose circuit symbol is shown in Fig. 1.21 and for which some examples are shown in Fig. 1.22. A basic inductor is simply an insulated wire that is wound into a coil. Just as resistors are characterized by resistance (R) and capacitors by capacitance (C), inductors are characterized by their **inductance**, L. The unit of inductance is the **henry**, abbreviated "H". Generally, physically larger inductors have larger inductance. Inductors tend to be big and heavy, and are often the largest components in a circuit, such as an electrical filter

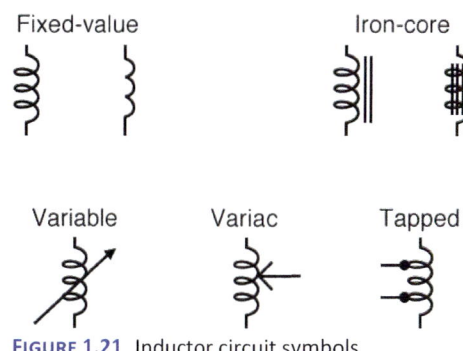

FIGURE 1.21 Inductor circuit symbols.

FIGURE 1.22 Various small inductors. Most shown are toroidal and a few are cylindrical. All have magnetic cores.

like that found inside of stereo speakers, used separate the high and low frequencies of the music. Chapter 5 discusses speaker filters in some detail.

Just like capacitors, inductors are energy-storage devices. Capacitors store energy in electric fields, while inductors store energy in magnetic fields, as illustrated in Fig. 1.23. The wire coils are in the shape of a straight wire circle which, according to Ampere's Law, (as shown in Fig. 1.14) generate a magnetic field as current flows through the wire. Very analogously to electric flux and electric field, magnetic field is expressed as either magnetic flux density, B, having units Wb/m^2 (Wb is the magnetic flux unit of weber) or magnetic field, H, having units of A/m, where

$$B = \mu_0 \mu_r H. \tag{1.26}$$

Are you curious to know what the equivalent of polarization is in a magnetic material? Imagine that every single atom is a tiny magnet. We say that each atom has a tiny **magnetic dipole moment**, represented by an arrow, or vector, in Fig. 1.23, that can point in any direction. Taking those directions and vector quantities into account, the dipole moments can either add to or cancel each other, and we say that the density of the magnetic dipole moments is the **magnetization**. The total magnetization in a volume is also called the **magnetic moment**, which, however, is now the vector sum of all the atomic magnetic dipole moments. There is an enormous amount of variation in how atomic magnetic dipole moments can behave in materials, so we will assume the simplest of the magnetic materials, namely

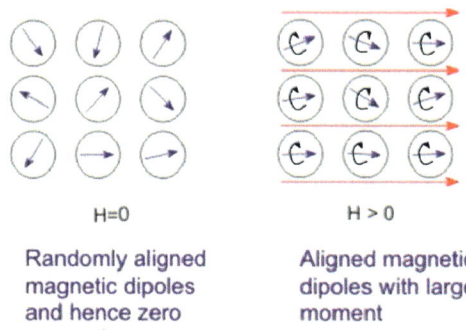

H = 0

Randomly aligned magnetic dipoles and hence zero moment

H > 0

Aligned magnetic dipoles with large moment

FIGURE 1.23 (Left) Magnetic dipole moments in an unmagnetized material with no external magnetic field. (Right) Magnetic dipole moments aligned due to the presence of an external dipole moment.

ferromagnetic materials, like iron or nickel. When a piece of iron is *not* acting like a common **permanent magnet**, or in the vernacular a "magnet," it is **unmagnetized**. In that case, all of those atomic magnetic dipole moments are randomly oriented and cancel each other out. As an external magnetic field is applied, say to an iron nail, then moments (the arrows) rotate, and the field due to the atomic moments start to align opposite to the external applied field, just as in the dielectric where the field of the polarizations is opposed to the external electric field.

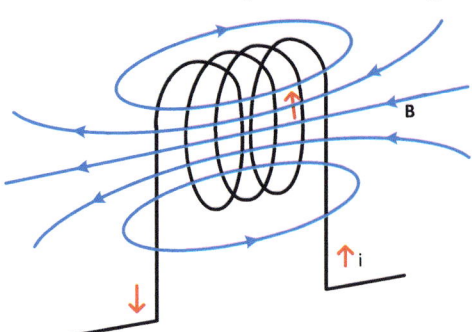

FIGURE 1.24 Magnetic field lines in an Inductor.

This results in the same sort of relationship in which more energy is stored in a magnetic material for the same amount of total internal field, H, due to the rotation of the magnetic dipole moments, which store extra energy. The value of μ_r is 1 for vacuum or air, and for some magnetic materials can be very high, often in the *thousands*. The turns of an inductor are wound around some space. That space can be filled with air, or with a magnetic material. Just like relative permittivity of a dielectric, a magnetic core has a relative permeability, μ_r, that increases the inductance L and thus increases the stored energy.

As in dielectrics, magnetic material uses up some of the magnetic flux due to an external source, such as current in a wire. Energy is stored in the magnetic material in holding the magnetic dipoles in position, whereas they would like to shift back to their original, disordered state. The energy density stored in a magnetic field is

$$U_B = \frac{1}{2}\mu_0\mu_r\boldsymbol{H}^2.$$

(1.27)

Inductors are constructed as coils to maximize their energy storage. When the wire is curved into a single loop, the magnetic field inside the loop becomes concentrated and larger. When many such loops are placed together to form a cylindrical coil, also called a **solenoid**, the magnetic field is multiplied approximately by the number of wire turns and is concentrated within the cylinder. If you knew U_B (Eq. 1.27) everywhere in and around the coil of an inductor (that is not as simple as in the case of the parallel-plate capacitor), and integrated it over all of space, you would get the total energy in the magnetic field. The thing to note is that the energy density of a magnetic field ($B^2/2\mu_r\mu_0$) goes as B^2, so concentrating the field in this way increases the total stored energy. The total energy stored in an inductor is expressed as

$$U_L = \frac{Li_L^2}{2},$$

(1.28)

where i_L is specifically the current through the wire of the inductor, or just "through the inductor."

Each loop of the wire in an inductor coil is called a **turn**. The entire assembly is called the **winding**. Most inductors have many turns. All the turns, being in series, have the same current flowing through them. As current increases, the magnetic field increases, as does the energy stored in the magnetic field.

Note that Fig. 1.23 suggests that the magnetic field of an inductor tends not to be uniform in space, which makes it harder to find the total stored energy from the energy density. Also, it makes it difficult, in general, to calculate values of L from inductor geometries. Capacitors used in circuits almost always closely resemble parallel plates, but inductors come in any number of shapes, sizes, and materials, and it is difficult to calculate the inductance from simple formulas. When winding a coil, each turn takes up some space, so all turns are not equal in contributing to the total field inside the inductor, which adds to the difficulty of doing simple calculations of inductance. That all said, the inductance of an inductor that is very long and narrow, with each turn making the same contribution to the field is

$$L = \frac{\mu_0\mu_r N^2 A}{l} \text{ (H)},$$

(1.29)

where N is the total number of turns, and A is the cross-sectional area of the core, or the open space inside the coil, and l (lower-case 'el') is the length of the very long coil. μ_r tells us the ability of the core material to contribute to the overall energy stored inside of the coil, increasing the inductance.

> *Important:* Recall that power is the rate of energy flow over time. Summing up the power over a time interval yields the energy that is transferred during that interval. Put succinctly: $U = \int Pdt$.

As mentioned above, the power that heats a resistor is the current times the resistor voltage: $i \cdot v$. For the inductor (and capacitor), the power provided by the circuit is still $i \cdot v$, but instead of being turned into heat, it is stored in the magnetic (electric) field. That stored energy can be returned to the circuit and causes no heating of the inductor. This analysis assumes an **ideal inductor**, i.e., it assumes that the wire resistance is zero, so no energy is lost to heating the inductor wire. The relationships between current, voltage and power in inductors and capacitors are explored in Chapters 3 (power grid) and 5 (phasors).

The relationship between the current through, and the voltage across an inductor is

$$v_L = L \frac{di_L}{dt}. \tag{1.30}$$

(Again, the subscripts in v_L and i_L are just temporary crutches to remind you that the voltage and current we are concerned with here are those of the inductor.) When current i_L is positive and *increasing*, $v_L = L di_L/dt$ implies that $i_L \cdot v_L$ is positive, so energy is transferred from the circuit into the magnetic field; when current is positive but *decreasing*, $i_L \cdot v_L$ is negative (the same sign as from a battery or discharging capacitor), transferring energy from the decreasing magnetic field back into the circuit. *Note that even when a large constant current flows ($di_L/dt = 0$ A), creating a high, constant magnetic field in the inductor, there is still no voltage drop across the inductor, i.e., $v_L = 0$ V.* In that case, the energy stored in the magnetic field is large and constant, and no power is being transferred either to or from the inductor's magnetic field. Now we can see why the terms "inductor" and "inductance" are used to describe magnetic devices like coils: it is because a voltage is induced across the inductor, whose magnitude depends on its inductance.

1.4.4.1 EMF versus Voltage

Now for something rather subtle. Equation 1.30 is written here, and elsewhere, in terms of the "voltage drop across the inductor, v_L." When an inductor is experiencing a change in current, it is actually acting like an electric generator or a battery, and stores energy inside of it in its magnetic field; it is really acting like a source of electrical potential, whose value is the voltage v_L. Equation 1.30 is true, and most often written in this form, but you might see in your continued learning about circuits that v_L is sometimes called the **electromotive force**, **EMF** of the inductor or the **back EMF** of the inductor. As far as we are concerned, we can call it voltage.

But what, then, do we strictly mean by "voltage?" The word "voltage" simply implies any difference in potential between two points, regardless of what causes that difference in potential. In the case of a resistor, current flows, and the voltage drop across the two ends appears. That voltage drop is not an EMF, since it only dissipates energy – none is generated or stored in the resistor. So, an EMF has a potential difference, in volts, but is a source of energy or stores energy. You will see in Chapter 3 that a coil of wire, i.e., an inductor, is the basic element of electrical generators, and does in fact convert the stored magnetic energy into power generation, so the potential difference across it is an EMF, but frequently written as a voltage only, as in Eq. 1.30.

1.4.5 Ferrite Rod Antenna

In general, antennas work best when they are on the same order of size as the wavelength, which is several meters long, and which just does not do for a portable AM radio like the one you will build in the lab, and myriads like it. To solve that problem, a **ferrite rod antenna**, or **loopstick antenna**, is used in portable AM radios. The ferrite rod antenna like that shown in Fig. 1.25 is kind of unique in the antenna world. Unlike most antennas, which are made

FIGURE 1.25 Ferrite rod antenna used in your AM radio kit.

of long metal strips or wires, this device is an inductor wound on a paper tube that is placed over a core made of high-permeability (i.e., large μ_r), low-conductivity material called a **ferrite**. We see it mostly in AM radios, and not in very many other places. Your ferrite rod antenna is shown on page 1 of your AM radio-kit instructions. (Be careful handling and soldering the ferrite rod antenna because the wires are thin and fragile.)

Space does not allow much to be said here about antenna theory, but to gain some insights, you can refer to the fundamental wave equation

$$\lambda f = c, \tag{1.31}$$

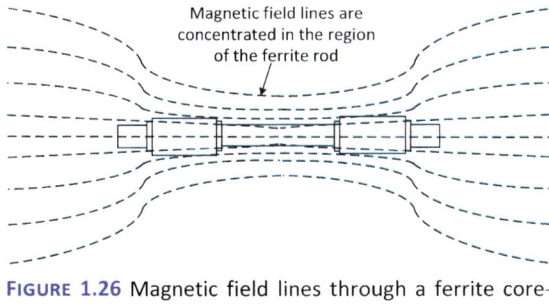

Magnetic field lines are concentrated in the region of the ferrite rod

FIGURE 1.26 Magnetic field lines through a ferrite core-based antenna are concentrated axially.

where c is the speed of light, λ is the wavelength of a radio (or any other) wave, and f is its frequency. The **frequency** of anything that oscillates, cycles, or goes back and forth, is the number of times it makes the trip per second, and is described in units of "cycles per second", or the more modern unit of hertz, Hz. You may have learned in physics class that radio waves, and all other electromagnetic radiation, consist of crossed electric and magnetic fields. (Don't worry, this will be discussed in more detail in Chapter 6.) Metal-wire antennas 'couple to' (i.e., are affected by) the electric field component of the wave, but, uniquely, the ferrite rod antenna is designed to couple to the magnetic field component. (Figure 1.26 illustrates the coupling.) The ferrite core fills the otherwise empty space inside the coil and increases the sensitivity of the antenna by, having high μ_r, concentrating the magnetic field at the antenna.

The inductance value of the ferrite rod antenna cannot be easily changed, but a variable capacitor in combination with the antenna inductance creates an electrical filter (Chapter 4) that selects the frequency of the desired radio station from all of those that are passing by. Broadcast AM radio in the U.S. covers frequencies from 525 kHz to 1705 kHz (1 kHz = 1000 Hz). (You might wonder why there are four wires coming from the rod antenna – that is to select different amounts of the coil to better tune over the full AM radio frequency range.) The operation of AM radios is the subject of Chapter 6.

1.4.6 The Speaker

Many of you are somewhat familiar with the ubiquitous 'speaker' component of your radio. Shown in Fig. 1.27, the speaker is a kind of electric motor, very much related to the kind that makes a Tesla Model S Plaid go from zero to 60 in 2.1 s. However, instead of rotating, the speaker 'motor' just goes back and forth (reciprocates), pushing and pulling the air at the same frequency as the original sound waves, reproducing the original sound. The speaker is one example of a **transducer**,

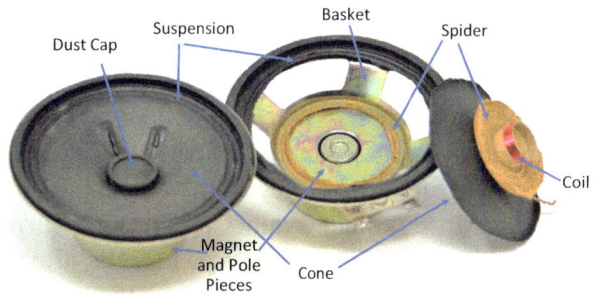

FIGURE 1.27 View of a loudspeaker and its interior components.

which is a device that converts energy between different forms, which in this case is from electrical to mechanical. Other common transducers are microphones (which are essentially speakers run in reverse), electric motors, antennas, light sensors, and light bulbs.

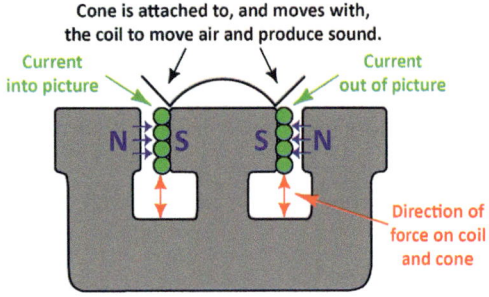

FIGURE 1.28 Cross-section of a loudspeaker showing the directions of current, magnetic field and resulting forces at a moment in time. The magnetic part of the Lorentz force due to electrons moving in the AC current causes the coil to move up and down, thus moving the air.

Figure 1.28 shows the direction of the forces on a **speaker coil**. Let's assume that the current is **alternating current**, **AC**, meaning that it goes back and forth, or alternates between positive and negative. (This is contrasted with **direct current**, **DC**, which is current that flows at either a constant value, or may change, but flow only in one direction.) A speaker has a winding (like an inductor) wrapped around a permanent magnet, but on a moveable sleeve so that it can physically slide back and forth over a reasonably long distance. The moveable sleeve is attached to the large speaker 'cone' that physically pushes volumes of air at the frequency of the alternating current, toward your ears, making sound at that frequency. If the waveform is more complex like, say, music, then the speaker will attempt to recreate that sound.

Speakers are driven by the magnetic part of the Lorentz force, $F = q(v \times B)$. This magnetic force is responsible for the speaker's action. The cross product means that the electrons in the current in the speaker coil are pushed in the direction of the arrows in Fig. 1.28 as they move in the wire within the magnetic field shown. The tiny forces from each of those many electrons that move in the wire add up to a significant perpendicular force, pushing the wire and causing the speaker coil and sleeve that is attached to the cone to move, pushing on the air; those air-pressure waves are the sound.

The radio kit has a small speaker that can generate sounds from about 200 Hz to 15 kHz, but that is merely an estimate. You can experiment with that in the lab if you want. You will be able to do it yourself after the Chapter 2 lab, in which you will learn how to use the waveform generator. You can drive the speaker directly from the generator, or through an amplifier if necessary, and see what frequencies you can hear from the speaker. (This presumes you are of typical college age, since many older folks cannot hear sounds as high as 15 kHz.)

1.5 Semiconductor Devices

Semiconductors are a special class of materials that will be introduced in Chapter 7. Semiconductors are covered in depth in several upper-level courses. The world of semiconductor materials and devices is vast, and is one of the most economically important areas of electronics. Just about any circuit you are likely to encounter in any modern product uses semiconductors.

A semiconductor is often described as having conductivity somewhere between a metal and an insulator, but, although correct, barely tells you anything useful. The real utility in a semiconductor is how it can be made to conduct current in only one direction, and its conductivity can depend on the voltages that you apply to it, like a switch.

The radio kit has two **diodes** and two integrated circuits, all made of semiconductors. Diodes are about the most basic semiconductor device; they allow current to pass in only one direction. In the diode symbol of Fig. 1.29, current flows easily in the direction of the arrow, i.e., into the **anode** (triangle) and out of the **cathode** (the stripe). (See box that discusses these terms.) Figure 1.30 shows an assortment of diodes from high-power (left) to low-power.

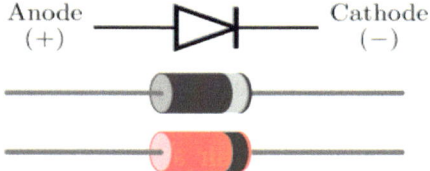

FIGURE 1.29 Schematic and drawing of a semiconductor diode.

Transistors are *three*-terminal semiconductor devices that can allow or stop current flow, like a light switch, by electrical signals connected to them. Transistors, in turn, can switch other transistors on and off. It is possible to manufacture many transistors into a single piece of semiconductor to make an **integrated circuit**, or **IC**. Before you know what hit you, you've got a 100-billion-transistor microprocessor IC in a supercomputer discovering new drugs, analyzing protein folding mechanisms, solving global weather and climate models, or working as an artificial intelligence interface to answer your questions. Transistors are the paradigmatic example of **active devices**, which are usually made of semiconductors. Semiconductors and ICs will be discussed in more detail in Chapter 7, where you will learn more about diodes and transistors, and also in Chapter 8 where you learn about light-emitting diodes and solar cells.

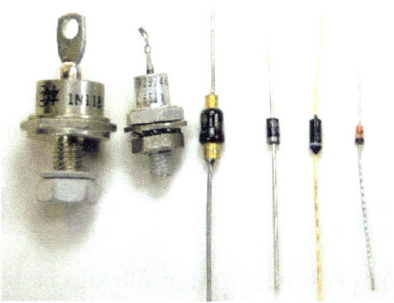

FIGURE 1.30 Photograph of various diodes. The bolts on the leftmost two are used to connect them to heat sinks in order to dissipate a lot of power. The third from the right is a common diode package. The rightmost diode is a small glass package.

Important: The terms *anode* and *cathode* deserve some explanation. The **anode** is the terminal of a polarized device through which conventional current (positive charge) enters (electrons exit). The **cathode** is the terminal through which current exits (electrons enter). 'Polarized' here means that it matters inside the device whether negative or positive charges are flowing, so voltage polarity matters. Diodes and batteries are polarized devices. Resistors and wires, for example, are not.

An interesting aspect of ICs is that the individual devices on them are not all that high quality. Transistors, resistors, diodes, inductors, and capacitors on ICs are not as functional as **discrete** (i.e., separate and not integrated) ones, but the nature of engineering is to make compromises. The fact that IC devices are oh-so-very-tiny more than compensates for their inferiority as individual devices, making the IC the most important electronic system in today's technology-based world.

One of the two simple ICs in the kit, the **audio amplifier**, powers the speaker so that you can hear the radio without earbuds or headphones. The other one does all the heavy lifting for receiving the AM signals. ICs are usually very complex, and they provide the computing capability for all of today's digital devices. In the vast majority of cases, the small flat structure that you can hold in your hand is actually the IC **package** that protects a small silicon **chip** and brings metal leads out to the world where electrical contact can be made to it. ICs are often referred to as 'chips' because they are relatively small pieces cut from a large thin disk of semiconductor called a **wafer**. The chip is safely nestled inside the package, as shown in Fig. 1.31.

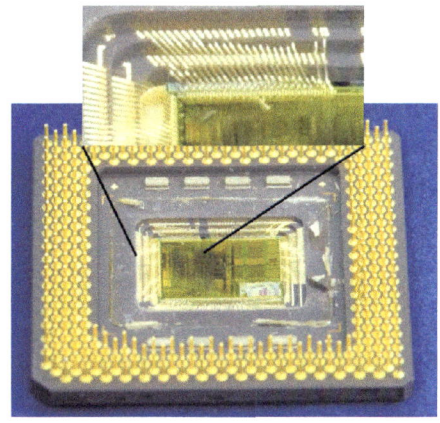

FIGURE 1.31 Photograph of an older packaged microprocessor integrated circuit (IC) that has been de-lidded. This particular chip has perhaps 100 million transistors on it, and is connected through the package to the pins by hundreds of gold **wire bonds** that are about 25 microns in diameter, about ¼ the width of a human hair. The wire bonds are easier to see in the expanded inset at the top.

ICs are the marvels of the 20th-century electronics revolution, and they continue to evolve. An IC chip contains many transistors, from tens of devices to hundreds of billions in the most advanced ones. The chips in the radio kit contain a few tens of transistors and other components. The 484/7642 AM radio chip contains all of the components that process the received radio signal into one that you can listen to. The LM-386 chip is a miniature amplifier that takes the low-voltage output from the radio and boosts it to a higher power level that can 'drive' the speaker. These concepts are explored in Chapters 6 and 7.

1.6 Manufacturing of Electronic Systems

It is useful to know how real electronics, such as TVs, computers, cell phones, and so much more, are manufactured. These next sections serve as a short introduction to some of the core technologies.

1.6.1 Surface Mount Technology (SMT)

Your simple AM radio kit is based on PCBs with old **through-hole** technology. The components are attached to the PCB by wires, or leads (pronounced "lEEds"), that are inserted through holes and soldered to the traces. Although through-hole is still popular for simple projects, the most common manufacturing technology for attaching devices and IC packages is called **surface mount technology, SMT**. In contrast to **leaded** through-hole components, surface-mount resistors, capacitors and inductors are very small, **leadless** rectangles, about the size of a grain of rice or even smaller, that have a tiny, metalized area for soldering at each end. See Figs. 1.32 and 1.33. The smallest **surface-mount devices**, **SMDs**, are less than ¼ mm on a side – smaller than the period at the end of this sentence.

FIGURE 1.32 Several surface mount devices. A broad array of devices are available in the SMT form factor. Shown is a selection of resistors, capacitors, LEDs and integrated circuits. The package-size designation "0805" means that it is approximately 8 hundredths of an inch by 5 hundredths. The reticle shown is in mm. The smallest available SMT package size is the 01005 package, which measures a mere 0.4 mm x 0.2 mm (or 0.016 x 0.008 inches). For comparison, the finest resolution of human vision is generally considered to be about 0.1 mm.

FIGURE 1.33 PCB with SMDs. This PCB needs no through holes because everything is surface mounted, including the ICs. All major device types, including ICs, resistors, diodes, capacitors, inductors, diodes, and more, are available as SMDs.

SMDs are leadless and are mounted on the surface of boards. To mount SMDs, **solder paste**, which is an amalgam of solder and flux, is printed onto the surface of the PCB. As discussed below, devices are placed on the board, and the entire board is heated to melt all of the solder in one step. Working with SMDs can be done by hand but requires special tools and more skill than that needed to build your AM radio kit.

1.6.2 Automated Manufacturing of Electronic Systems

You might have noticed in Fig. 1.33 that modern electronic systems have a lot of components, numbering perhaps in the many hundreds per board, and that they can be very tiny. If not for sophisticated mass-production methods, the consumer electronics that most people take for granted would be prohibitively expensive. The good news is that individual components are mass-produced and sold by the millions. They are delivered to electronic systems manufacturers relatively inexpensively on reels (Fig. 1.34), like old-fashioned tape recorder reels. These parts can

FIGURE 1.34 Modern high-speed pick-and-place tool with multiple reels of parts.

be as cheap as a tenth of a penny each. Then, automated assembly makes the construction of the system affordable.

High-volume assembly is done with robotic **pick-and-place** systems that locate and fetch parts from the reels and place them on the PCB in fractions of a second apiece, with several pick-up tools working simultaneously. Prior to the pick-and-place, the solder paste is applied to the board so that the small parts will stay in place. After the board is 'populated' with all the parts, the solder is **reflowed**, or melted, to form the solder bond.

Alternatively, if parts are inserted with wire leads, like your AM radio kit, **wave soldering** can be used in which the board is placed in contact with a moving wave of molten solder (Fig. 1.35), and those metal parts that are exposed attract just the right amount of solder to make all of the connections at once. Wave soldering can also be used for SMT, but that has been largely replaced by the reflow method.

FIGURE 1.35 Wave soldering, in which the boards come in contact with a wave of molten solder (circled).

1.6.3　Economics of Electronics Manufacturing

Just a short word about the economics of producing consumer electronics products. As alluded to above, the cost of the components is just one of the many factors, and not by any means the greatest, that go into the final price of a product. These factors include research and development (R&D), intellectual property (IP) licensed from other companies, component costs, assembly and manufacturing, testing, packaging, software development, risk (liability and product returns), packaging (i.e., the cardboard box, etc.), shipping, distribution, retail sales, and the profits that must go back to the company whose product it ultimately is after the overhead it takes to run the company, including personnel, space rental, legal fees, cost of borrowing money, etc. Engineers should never take their eyes off the economic realities of what they do, since often those take precedence over the technical issues that they face every day. Many, if not most, decisions ultimately come down to the economics of a product, rather than the elegance of the engineering approach. Except for some limited spaces, such as the military, if it is too expensive, it just won't be a commercial success.

1.7　Lab 1 Activities

Before the lab session, read Appendix 1 about soldering and soldering techniques. Download and look over the kit instructions before your lab session. In the lab, follow the kit instructions and build the AM radio. You will not have a lab partner for this lab, but you will be sharing a bench.

As you are inserting parts into the printed circuit board, take a moment to think about each one, and what its electrical characteristics are. Why does each part look physically as it does? What are the markings, and what do they mean? Check to ensure it is the correct part BEFORE soldering it on the circuit board. Discuss with your bench partner or TA if you are unsure.

FIGURE 1.36 Bench vise shown holding PCB as a "third hand." It is positioned in front of an air filter.

As you insert the parts that have long leads, bend them tightly to the solder-side of the board so that they are held in place. If you can, and want to, you can put a few in at a time, bend them tightly, solder all

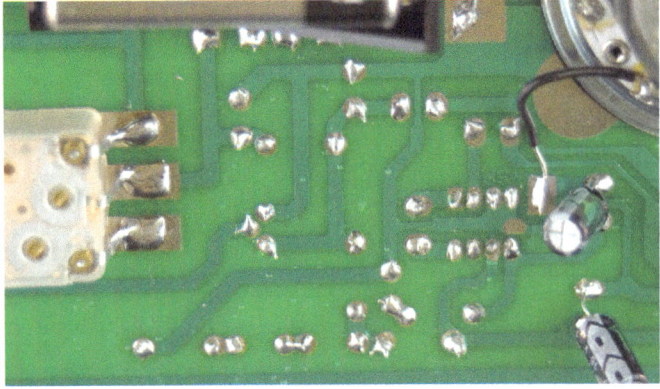

FIGURE 1.37 Example of completed radio with good soldering overall.

of them, and cut off all the leads in one step. If that is too difficult, you can choose to just do one component at a time. Also, it might be helpful to use a small vise (Fig. 1.36) that holds the board for you while you do the soldering. Figure 1.36 also shows an air filter that captures the fumes. Soldering should be done in a well-ventilated environment, or the air should be filtered as shown. Also, be prepared to help your bench mate by holding various parts when needed and ask for help when you need it. Part of the finished radio PCB is shown in Fig. 1.37. The solder joints shown are of acceptable quality. If you look carefully, you will see that those solder joints have all wet well (Appendix 1), and don't use excessive amounts of solder. From the angle of the picture, there seem to be no 'cold' solder joints.

Have the TA test and sign off on your completed radio kit. They will be grading based on completeness, quality of workmanship, and functionality. There is no lab report due for this lab, but you should note any questions and ask the TA or lab instructor.

1.7.1 Appendix 1: More about Soldering and Good Soldering Technique

Why Solder?

In the manufacture of electronic systems, it is necessary to interconnect all the various devices, such as those in your AM radio kit. As you have learned, the PCB is a good way to provide small, low-resistance wires very cheaply and reliably. The PCB electrically connects the components and provides a firm substrate to support them. It would be great if we could just glue the components onto the PCB, but unfortunately, most glues are poor conductors. That is what soldering is for.

Soldering Irons

Solder is melted by applying heat from a **soldering iron**. Soldering irons come in several forms, from powerful soldering 'guns' to inexpensive, low-power soldering 'pencils,' like that shown in Fig. 1.38. A soldering gun would be used to deliver, say, 140 W of heat flow from a broad metal tip to large metal workpieces. One feature of the guns is that they heat up quickly when a trigger is pulled, and the tip cools when the trigger is released. A soldering pencil with a fine metal tip, typically 20 W to 40 W, would be preferred for soldering smaller wires and components. Soldering irons generally take longer to heat up, and they are left plugged in and hot continuously. As such, they are docked in a base while hot. Soldering pencils are most commonly referred to simply as soldering *irons*; soldering *guns* are usually referred to as such. More advanced soldering irons incorporate temperature controls to better conform to the solder type and size of wires to be soldered. Soldering

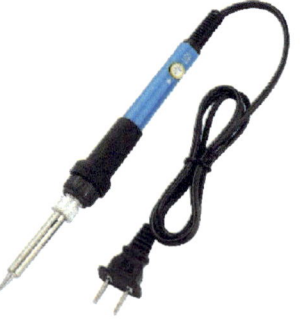

FIGURE 1.38 A typical soldering iron or pencil for small electronics projects. They often cost well less than $10.

irons with special, finer-pointed tips are used to solder SMT devices to the surfaces of PCBs. Soldering irons cost anywhere from a few to several hundreds of dollars, depending on quality and features.

Solder Flux

Oxides are the enemy of soldering. The oxides that form on solder are hard insulators that cover the surface and prevent the solder from sticking to the workpiece. When an oxide coating forms on the surface of molten solder, it also acts like a straitjacket on the surface, which constrains it from flowing and prevents metal from atomically coating the wire and PCB trace. This coating process is in general referred to as **wetting**. Fortunately, the oxide is very thin and can be dissolved with a common substance, called **flux**, mixed in with the solder. The word "flux" is derived from the Latin *fluxus* meaning "flow," which is entirely appropriate in this case.

Flux is a **reducing agent**, meaning that the metal atoms in the oxide gain electrons and the oxygen is released, so the oxide is dissolved. Flux takes the form of a brownish liquid or paste. One common variety of solder is **flux-core** or **resin-core** solder in which the flux fills the core of a hollow solder wire and then flows as the solder is melted. Flux composition is clearly noted on the solder label. If the solder does not incorporate flux, you should apply flux to the wire and board, usually with a small brush, prior to applying heat. Most types of common flux can be removed with isopropyl alcohol, and some can be removed with water.

Properties and Types of Solder

Solder comes in various physical forms. In bulk, it comes in blocks, but for manual soldering it comes in wires wound on a spool. The width of the solder wire should be appropriate to the size of the soldering iron tip and workpiece. Thick solder can need a lot of heat to melt, and it is difficult to control how much solder is applied to the component and board. Solder thicknesses vary from very thin wire, such as 27 gauge (0.38 mm) to 9 gauge (3.05 mm).

Although most pure metals melt at high temperatures, the compositions of solders are chosen so they melt at intermediate temperatures. A good soldering material has a melting point that is high enough that it can survive hot environments, but low enough that it can flow at temperatures that will not damage the board or components. It also must adhere well to metal surfaces, and be strong enough to survive physical stress and shock.

Solder is used in more than electronics, including plumbing, jewelry, and other metal-working. If the melting point is less than about 450°C, it is called a **soft solder**. Higher than that it is called a **hard solder**, and the process is called **brazing**, rather than soldering. Here we discuss only soft solders and soldering.

Many types of solder are used in various applications and conditions. A particular solder **alloy** composition is chosen for its melting point, brittleness, reactivity with other metals, cost, toxicity, and more. An alloy is a solution of metals that has properties that are different from those of its components. Solders are often alloys with metal compositions at or near a **eutectic point**. At the eutectic point the alloy has a sharp freezing point, meaning that it abruptly goes from molten to solid as it cools. Otherwise, you would have to hold it very still during cooling. The eutectic composition often has a lower melting point than any of its components. For example, lead, Pb, melts at 328°C and tin, Sn, melts at 232°C; a common Sn-Pb solder alloy of 60/40 (60%/40% by mass) melts at 188°C. That composition is close to the Sn-Pb eutectic composition of 63/37 by mass, which melts at 183°C. Both compositions were commonly used in electronic soldering.

Lead-Free Solders

Lead has been such a common component of solder that any solder that doesn't have it is called 'lead-free.' As you may know, lead is toxic even in low doses, causing damage to neurological systems (especially in children), and much more severe symptoms in large doses. For that reason, there is a worldwide effort to eliminate lead from all electronics manufacturing processes, including solder.

Fortunately, or unfortunately, leaded (here we mean Pb) solders are great – they flow well, are inexpensive, are ductile, and have useful melting points. It has not been easy finding lead-free solders that work as well, so the industry has had to adapt. Many common lead-free solders are based on the tin-silver-copper alloy system, collectively referred to as 'SAC' for Sn-Ag-Cu. (Aren't engineers clever? If nothing else, they don't like to waste time using several words when they can instead use TLAs, i.e., three-letter acronyms.) In general, lead-free solders have higher melting points and don't flow as well, making them harder to use, but they can be more resistant to failures than lead-based solders. The eutectic combination of SAC is 96.5% Sn (m.p. 231.9°C), 3.0% Ag (m.p. 961.8°C), and 0.5% Cu (m.p. 1084.6°C). The eutectic melting point of this composition of SAC is 217°C.

Soldering Nuts and Bolts

Just like painting or playing a musical instrument, manual soldering is a skill in which practice improves technique. Your first attempts will likely be somewhat sloppy, but you will improve with practice. Fortunately, soldering is not like playing the violin – just a little experience goes a long way. That said, here you will learn certain rules that will provide good, dependable results:

1. **Safety**. First and foremost, understand that you are working with very hot surfaces and molten metal. These provide not inconsequential safety risks. MOST IMPORTANTLY: WEAR EYE PROTECTION! You should never solder without wearing safety glasses. Molten solder can, and does, splatter, and can certainly cause injury if it lands on your eyeball. The workspace should be uncluttered. The soldering iron should be in a comfortable position in your dominant hand. Use an air filter to capture the fumes. Breathing the flux fumes can cause asthma and should be avoided. Also, lead solder is still around: if you do use it, be sure to wash your hands thoroughly when done, and do not touch your face or mouth with your hands until you do.

2. **Tinning**. Tinning is the process of putting a layer of fresh solder on the hot tip. Without this layer, the tip will oxidize and corrode. Keep a damp sponge wetted with clean water near the soldering iron. After a soldering step, simply place the soldering iron aside, or in the safety holder. Just before starting the next solder joint, wipe the tip clean on the damp sponge. It will be shiny, but will have only a thin coating of solder. Immediately tin the tip by melting some solder on the tip to create a thicker layer, which will help protect the tip and help transfer heat to the workpiece. You are now ready to continue soldering.

3. **Heat transfer**. Press the tip against both the device lead and the metal trace on the PCB in order to transfer heat to both parts, as shown in Fig. 1.39. If the tip is tinned properly, heat will flow adequately to both.

4. **Flowing the Solder.** Touch the solder to the parts to be soldered and not directly to the soldering iron. The parts should be hot enough to make the solder melt and flow across the junction between the device lead and the **solder pad** on the PCB. The power, i.e., wattage, of the soldering iron, must be high enough to heat the parts to melt the solder directly. The temperature setting is typically much higher than the melting point of the solder because you have to heat a lot of material besides the solder. A typical temperature is 600 °F (315°C). If the iron is underpowered, or the solder is too thick or has a high melting point, the solder won't melt easily. Make sure that the tip is sufficiently tinned to allow good thermal contact.

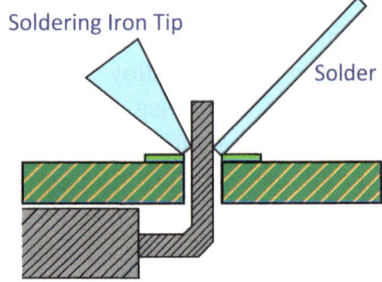

FIGURE 1.39 Proper tip and solder placement. Notice that the component, not the tip, melts the solder.

5. **Wetting angle.** The solder left over at the solder joint is called the **fillet** (pronounced "FILL-it"). The solder wets the part at a **wetting angle**, defined by the angle formed at the edge between the fillet and the solder pad. The fillet will ideally have a very small wetting angle, about 30° or so, as shown in Fig. 1.40. If any oxide is present, the wetting angle will be closer to 90° or more. Such an oxide layer is bad – it can cause an open circuit in the short term, or poor reliability in the long term. Unreliable connections due to poor metal-to-metal contact are called **cold solder joints**. If this happens, you should reflow it using new solder or new flux. If that does not improve it or leads to other problems, you should rework the joint, as discussed below.

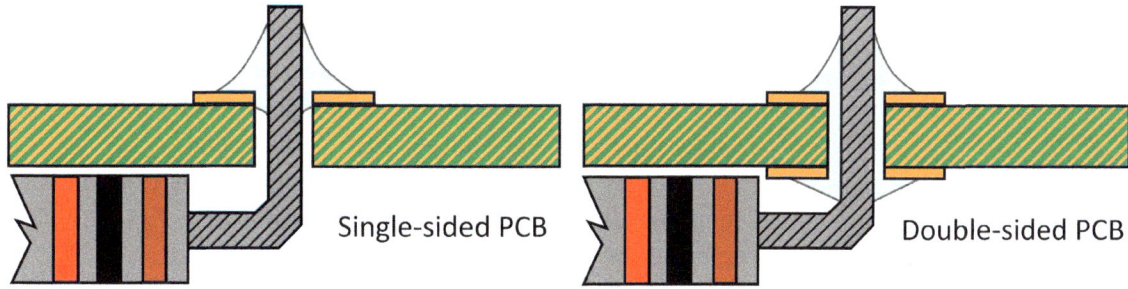

Single-sided PCB   Double-sided PCB

FIGURE 1.40 Side view of ideal solder joints.

6. **Moving while freezing.** It is inadvisable to let the lead move during the process of freezing (cooling until solid). This can result in a rough, or **matte**, surface, but is not in itself a defect.

7. **Solder defects.** Examples of unsuitable solder joints include insufficient solder at the solder pad; insufficient solder over the entire joint; cold solder joint due to the presence of oxides; bridging between two or more solder joints (an obvious non-starter); too much solder, which can lead to bridging or mask a cold solder joint. See Fig. 1.41 for some examples.

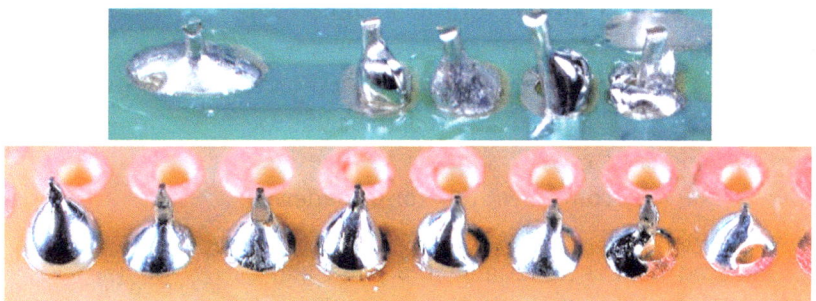

FIGURE 1.41 Examples of good and bad solder joints. Top left: good solder joint with good wetting and just about the right amount of solder. Top, third from right: matte finish due to wire moving. Top, second from right: insufficient pad wetting. Top right: cold solder joint. Bottom left: too much solder. Bottom, second from left: good solder joint. Bottom, all others: insufficient solder or inadequate solder and pad wetting.

8. **Solder rework.** If soldering is unsuccessful, you may have to **rework** the joint. Sometimes, just reflowing the solder can help achieve wetting. Or you may add some new solder or flux before reflowing, but then you might end up with too much solder, which might bridge to adjacent joints. To remove solder, you can touch a clean tip to the solder and let it wick onto the tip, thus drawing off some solder.

FIGURE 1.42 Solder wick made of braided copper. The clean braided end is put in contact with the molten solder, which wicks into it and is removed from the board. That part of the braid is cut off and discarded.

It is fairly common to just give up on a joint altogether and start over, but to do so, you need to remove most of the solder. One way to do this, *not necessarily advised*, is to melt the joint and immediately give the board a good hard rap on the tabletop. If done right, most of the solder will fly off, leaving a clean surface to start on. However, this is the lazy person's method of reworking. A better way is to use a roll of finely braided copper called **solder wick**, shown in Fig. 1.42. This material is ideal due to its excellent wettability (copper) and large surface area (braids). The solder wicks into the fabric of the copper braid, leaving a clean surface behind. This may have to be repeated for complete solder removal. The end of the solder wick having the excess solder is cut off using side cutters and discarded.

Another inexpensive (just a few dollars) way to remove solder is to use a manual solder vacuum tool, as shown in Fig. 1.43. This tool is called a **solder pump**, **desoldering pump** or even **solder sucker** (really). It is so much fun to use that you might be tempted to rework a joint just to get to use it, but it does take practice. Press down on its spring-loaded piston. Melt the joint with the soldering iron while holding the tool with your other hand. The tip of the solder sucker is held nearly up against the molten solder. In one quick, skilled motion, move the soldering iron out of the way and push the tip of the solder sucker against the molten joint. Pressing its trigger releases the piston, sucking the molten solder up inside the tube. It is emptied after many uses. This method takes a bit of coordination. Caution should be taken not to touch the tip of the solder sucker to the tip of the iron. It can take some heat but will melt on direct contact.

FIGURE 1.43 Solder pump or 'solder sucker.' The plunger is pressed down, the tip is placed next to the molten solder to be removed, and the trigger is pressed to release the plunger and suck the solder off of the board.

The professional way to rework a joint is to use a **rework station**. Since some devices have many leads that must all be desoldered together, variations of rework stations involve heat lamps, hot air, or ovens. One very convenient tool for desoldering single leads is called a **desoldering iron**, shown in Fig. 1.44. This tool provides a heated hollow tip to melt the joint, and a vacuum pump to simultaneously pull the solder away through the tip. One version of this is the 'desoldering gun,' which is in the shape of a gun, of course, and provides vacuum while heating the solder.

FIGURE 1.44 Desoldering iron. This functions like the solder sucker in that molten solder is drawn up through the tip by vacuum. Here, the heat and vacuum are applied through the same tip. The vacuum is created by an internal pump, and the tip is stainless steel.

Problems

1. The point of this question is to show that Coulomb forces are very strong.

   (a) How much force is there between two electrons that are effectively 10 nanometer, nm, apart? (1 nm is 10^{-9} m). (This assumes that electrons are a point in space, which they are not.) Your answer should be in N (newtons).

(b) If we assume they are in free space (not in a material), recalling $F = ma$ (force equals mass × acceleration), what is the acceleration due to the Coulomb force? The rest mass of an electron is 9.11×10^{-31} kg. Do you think this is a large value for acceleration? Does it demonstrate that Coulomb forces are very strong?

(c) If you turn on a battery-operated radio that uses a current of 50 mA, how many Coulombs and electrons will have passed through the battery after 30 minutes? (Notice that it is 'pass through' and not 'given off by' because for every electron that leaves the battery, one returns to it instantly. That is what KCL tells us.)

(d) Let's just say that the electrons in the previous question did NOT return to the battery, but instead are somehow deposited onto a point 1 meter away. Now you have positive and negative charge Q separated by 1 m. (Treating the battery like a point as well.) Without giving away the previous answer, let's say that the charge is *only* 1 C (as if this were possible, which it definitely is NOT). Note that 1 pound of force is 4.45 N. How many pounds of force is there between that negatively charged point and the positively charged battery ?

(e) How many times is that force compared to the weight of the largest oil tanker ever built, the Seawise Giant, which weighed 180 million pounds?

2. A printed circuit board has a rectangular copper trace (i.e., metal line) that is 5 cm long, 0.2 mm wide, and 0.035 mm thick.

(a) What is the resistance of the copper trace?

(b) How much current would have to flow in it to drop 10 mV on that wire?

3. The device shown in the following figure, called a shunt or sense resistor, is used to measure very large currents by measuring a voltage drop across a very small resistance. Knowing the resistance and voltage drop provides the current by $V = IR$. The shunt resistor is made of a copper strip connecting two blocks of copper with bolts in them, as shown in the figure.

(a) The resistivity of copper is 1.68×10^{-6} Ω-cm. The copper strip is 1 cm wide and 3 cm long, and 0.5 mm thick. Ignore the resistance of the large blocks at the ends. What is the resistance of the copper strip? (*Hint*: Always pay attention to units to be sure that the units that you use all cancel correctly to yield the desired units of your answer.)

(b) When I = 100 A (a very large current) passes through the shunt resistor, what is the voltage drop from one end to the other?

(c) What is the current when the voltage drop is 0.03 V?

4. House wiring in the U.S. is most often rated to carry a maximum current of 15 A from the where it enters the home at the breaker box to the wall outlets. (Chapter 3 discusses house wiring in detail.) Suppose the length of wire is a total of 100 feet to an electric heater from the breaker box, and then 100 feet back to the breaker box for a total wire length of 200 feet. The heater draws 12 A without considering the additional house wiring. You are concerned about the extra resistance of the house wiring both to and from the heater and the associated power lost in the wires. House voltage is nominally 120 V.

   (a) What is the resistance of the heater while it is operating, ignoring any effects of the house wiring?

   (b) What is the power drawn by the heater without any wire losses?

   (c) House wiring is typically copper having a diameter of 1.63 mm (14 gauge). What is the resistance of each length of 100 feet? The wire has a circular cross section.

   (d) What is the total resistance of the 200 ft. of wire?

   (e) Using $V = IR$, what is the current to the heater including the wire?

   (f) Now, what is the power output of the heater including the effects of the house wiring?

   (g) What is the percentage decrease of the heat power out from the heater due to the wires? (Define percentage change as 100% × (OLD – NEW)/OLD)

5. Determine the nominal value of all of the following resistors based on their color bands. Also determine the range of possible values of the actual resistance as specified by the tolerance band. Note that these are somewhat 'typical' resistor values. Your answer should look like this (in kΩ or MΩ): (nominal value) kΩ, ± (percentage value) %, [lower bound, upper bound] kΩ.

   (a) Brown-Black-Red-Gold

   (b) Red-Violet-Red-Gold

   (c) Orange-Orange-Yellow-Silver

   (d) Brown-Black-Green-Silver

   (e) Red-Red-Orange-None

6. What features of a power resistor allow it to dissipate more power than a carbon resistor?

7. It is crucial to not overheat electrical components since elevated temperature can degrade or destroy them. You can imagine heat flowing from a hot electrical device to a heat sink, and then flowing into the air and away from the device as if the heat were electrical charges, and the lowest temperature, i.e., the air, were the lowest voltage in the circuit. So, heat flow and charge flow are analogous to each other. As you have learned, the resistance of a piece of material is $R = \rho L/A$. The heat-flow analogy to electrical resistivity, ρ, and total resistance, $R = \rho L/A$ uses thermal resistivity, θ, and thermal resistance, $R_T = \theta L/A$. The thermal resistivity of metals is generally very low. Since both equations pertain to a flow that depends on a medium in which to flow, they have the same form and same dependence on length and area.

   (a) Based on the above information, look at Figure 1.10 and explain why the two heat sinks are designed with many thin fins or spikes rather than just being a solid block of metal, or not being there at all, for that matter.

   (b) Would a heat sink made of plastic or cardboard be very effective? Why or why not?

(c) A thermally conductive white paste, called "thermal paste," is often applied between the device and the heat sink. Why do you think it is used, and how does it do its job?

8. Consider the potentiometer shown in the following figure:

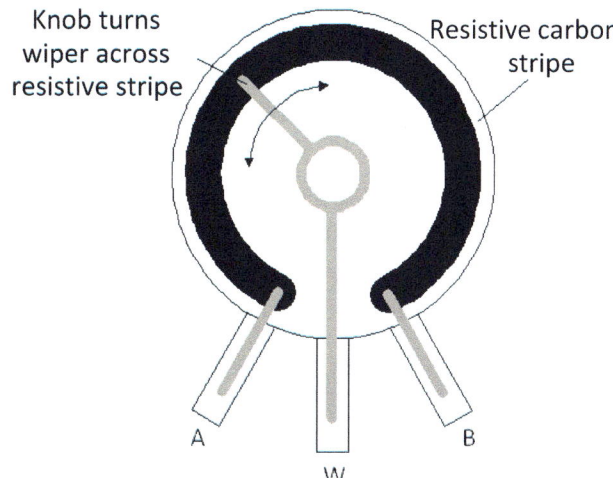

The stripe is made of carbon having a resistivity of 10 Ω-cm, and has a rectangular cross-section. It is 50 μm thick, 0.5 mm wide and 20 mm long between A and B. Using your best estimate of fractional lengths along the stripe, what is the resistance between the leads from:

(a) A to B?

(b) A to W?

(c) W to B?

9. Sue and Mary need a 2200 μF capacitor. They are sorting through a bin of aluminum electrolytic capacitors, as in Figs. 1.15 and 1.16, all of which are the same type and voltage rating, whose values range from 47 μF to 3300 μF. Mary starts to use a capacitance meter working from the smallest to the largest to find the value that she needs. Sue says, "Wait, Mary, to save time, start looking at the largest ones first."

(a) Is that good advice?

(b) Why might Sue give such a suggestion? Use an equation and explanation to back up your answer.

10. The plates of a capacitor that has an area A = 10 cm × 10 cm are d = 0.01 mm apart. There is air between the plates. A 12 V battery is placed across the plates and charges them up.

This might be helpful: The electric field is $E = -dV/dx$. When E is constant you can write that $E = -\Delta V/\Delta x$ and $V = -\int E(x)dx = -E\Delta x$. (The minus sign simply indicates which way the electric field is pointing – from positive to negative charge.) The field is constant between the plates of a capacitor.

(a) Draw a cross-sectional diagram of the system including the battery, the capacitor, the charge on the plates and the electric field with the correct direction.

(b) What is the electric field inside the space between the capacitors? Your answer should be in V/cm.

(c) What is the capacitance of the capacitor? (Hint: What is the relative permittivity of air?)

(d) How much charge is on each plate? Your answer should be in coulombs.

(e) How much energy is stored in the electric field of the capacitor? Your answer should be in joules.

(f) Capacitors always have a nonconducting dielectric material between the plates – that was air in part (a). Now you will place a dielectric of tantalum pentoxide between the plates. Do a web search to find the commonly given dielectric constant of tantalum pentoxide used in "tantalum" capacitors. What is the capacitance now?

(g) For the same 12 V battery, what is the charge on each plate with the dielectric? Is it larger or smaller than what you calculated in part (d)?

11. Imagine it is 1903 and you are an electronics hobbyist building your own crystal radio set. You need a capacitor of fixed value, so you use some large sheets of tin foil and writing paper.

For a full-size piece of writing paper, 8.5×11 in^2, what capacitance might you expect to make? What assumptions do you have to make to fill in missing factors?

12. Based on the current equation for a capacitor, if a constant voltage is applied to the capacitor for a long time, what would be the current across the capacitor? Does this resemble an open circuit or a short circuit? Explain your answers.

13. Assume that an inductor has a certain current flowing through it. How does the energy stored in the inductor change for a larger current? How about for a smaller current? Explain your answers.

14. You have a home-made, air-core wire-wound inductor and measure its inductance using an LCR meter (to be used extensively in Chapter 5 lab). You remove the ferrite rod for the AM radio kit and insert it into the center of the inductor. What do you expect to happen to the measured inductance, and why?

15. For an inductor, $V_L = L dI/dt$, where V is the voltage across the inductor, and I is the current through it. Let $L = 100$ mH.

(a) Is there a voltage across the inductor when current is increasing linearly, say, at 1 A/s?

(b) What is V during this time?

(c) The current levels off and remains at 5 amps. What is the voltage across the inductor now?

(d) Here is a hard question: A wire is disconnected from the inductor while a large current is flowing through it. What will the voltage be (ideally) across the inductor, and what will *actually* happen where the wire is pulled away from the contact?

16. A rough estimate for a wire-wound inductance of an air-core inductor is $L = d^2 \times n^2/(18d + 40l)$, where L is inductance in *microhenrys*; d is coil diameter *in inches*; N is number of winding turns; l is its length *in inches*.

(a) Approximately what is the inductance of an inductor that is 3 inches long, 0.5 inches diameter with 800 turns of 32 AWG copper wire, which is 0.21 mm diameter?

(b) What is the approximate resistance of the wire? Do you think that resistance could be important in a circuit?

17. (a) How does a speaker recreate sounds? Your answer should be about the motion of the cone and the effect on air. (b) Describe briefly how electrically a speaker works. Include discussion of the magnetic field and the current in the windings.

18. Suppose that you drop a 3 kg weight a distance of 2 meters, and in so doing you drive a perfectly efficient electrical generator. That generator charges a capacitor having a capacitance of 200 microfarads.

    (a) How much energy is stored in the electric field of the capacitor? (*Hint*: In a previous physics course, you probably learned that $U = mgh$ for the potential energy of a mass m at a height h.)

    (b) How much charge will be stored on its plates?

19. Consider an ideal diode with some non-zero voltage across it. If the current through the diode is a few mA, then in which direction is it flowing? Use the anode and the cathode to describe the direction of the current.

20. Sometimes in ICs, a semiconductor is used as the material for wires. Would this wire generally have a greater or lower resistance than a metal wire with the same dimensions? Explain your answer.

21. Here are some notes that Fred entered into his lab notebook while soldering:

    (a) Always wear safety glasses when soldering.

    (b) Swipe the iron tip across the damp sponge to clean it.

    (c) Always make sure the soldering iron tip is properly tinned and therefore shiny.

    (d) Use the tip to melt the solder directly so that the soldering can be finished quickly without damaging the PCB.

    (e) Always cover the component leads with a hand when cutting them off the components after soldering.

    Did Fred make any errors while soldering? If so, which ones, and what did he do wrong?

Chapter 2

Basic Electronic Test Equipment: What's on *Your* Bench?

2.1 Introduction and Overview

This chapter's laboratory activity uses series and parallel circuits to give you practice operating a variety of electronic test and measurement equipment that you will use throughout your EE education. You've probably seen series and parallel circuits in another course or two. You may even remember the formulas for computing the equivalent resistances. Nevertheless, we're going to review series and parallel circuits, because it's not just about the formula – it's also about developing a good physical intuition about their behavior.

In this chapter, you will learn about:

1. Kirchhoff's voltage and current laws
2. Series and parallel circuit elements
3. Voltage and current dividers
4. Breadboards
5. Power supplies
6. Output resistance
7. Electrical grounds
8. Digital multimeters
9. Oscilloscopes

This is not a text on circuit analysis. We present only those materials that are essential to understanding the larger concepts presented here. It is assumed that you have been, or will be, exposed to a full treatment of circuits elsewhere.

2.2 Kirchhoff's Current and Voltage Laws

2.2.1 Kirchhoff's Current Law

Figure 2.1 shows a simple electrical circuit in which a voltage source, such as a battery, drives current through several resistors. We consider a circuit to comprise a collection of **lumped** elements, where

An Introduction to Electrical Engineering with Lab Activities
Gary H. Bernstein
Copyright © 2026 Jenny Stanford Publishing Pte. Ltd.
ISBN 978-981-5129-30-4 (Hardcover), 978-1-003-71345-6 (eBook)
www.jennystanford.com
DOI: 10.1201/9781003713456-2

the term 'lumped' means kind of what you think it might – that all of the properties of the element are confined to one place in space. All of the devices discussed in Chapter 1 were lumped elements. If the properties of an element are spread across a wide region of space, such as the resistance and continuous voltage drop along the entire length of a wire, that element is called **distributed**, which is relevant to courses in electromagnetism, and won't be considered here.

Wherever two or more wires join to

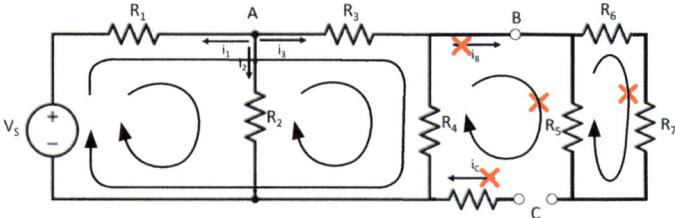

FIGURE 2.1 A voltage source, such as a battery, is on the left, labeled "v_s". One node is labeled "A", and the three current components are shown as i_1, i_2 and i_3. Neither current labeled i_B or i_C can flow, since there is an open circuit at C, and current must flow in loops. In other words, there is no return path for the current i_B. Since no current flows past nodes B and C, no current flows in the loop containing R_5, R_6 and R_7.

connect two or more lumped elements, we call that point a **node**. Nodes are connected by branches, where a **branch** is any path that connects two nodes. The details of the circuit are not important at this time, except for the following. The word "circuit" refers in general to some path that comes back to its start. It is for a very important reason that electrical circuits are called that; it is due to the fact that charge may not pile up at a node – it must pass through that node. Even for a capacitor, the current flows *through* the capacitor in the form of displacement current, and does not build up on the node, even though it accumulates inside the device on the plates. In fact, if charge *were* to build up at a node, then it *would be* a capacitor connecting to some other node, and we should include that capacitance in the circuit diagram. However, ideal nodes are assumed to have zero capacitance, so no charge accumulates at nodes.

As shown in the circuit diagram, current must therefore flow in **loops**, where a loop is a closed path. Current must have a return path. This implies that current cannot simply flow into a node that is connected to open air or space. Figure 2.1 shows a continuous wire at B and a broken wire at C. You might at first think that current flows into and around or through the resistors R_5, R_6 and R_7, but it cannot. Any current flowing out to node D can return only through point C, which is broken. Therefore, current cannot flow in a loop, and both i_B and i_C are zero; the right part of this circuit is completely irrelevant to its operation.

The circuit properties discussed above are encapsulated in **Kirchhoff's Current Law (KCL)**, which states that all the charge that flows into a node must also flow out of the node, so all the current that flows in must flow out. Charges don't build up, as discussed above. In circuit theory we may arbitrarily choose which direction we consider to be positive current, for example, current out of a node may be positive (electrons flowing into the node), and current into the node is negative (electrons flow out of the node). Or, heck, we could choose the opposite. We just need to be consistent, and not change our minds halfway through an analysis of a circuit. In this book, we will use the first convention – positive current flows out of a node. So, we may summarize all of this by simply stating KCL as

$$\sum_n i_n = 0 \tag{2.1}$$

where i_n is the current in the n-th branch attached to the node. In Fig. 2.1, at node A, $i_1 + i_2 + i_3 = 0$, which implies that at least one of the currents must be out of the node, which, again, we will call positive, and at least one other must be into the node, which we call negative, in order for them to sum to zero.

One important conclusion from KCL is shown in Fig. 2.2; if current from one branch, in this case i_0, flows into a node, node A here, and splits into more branches, then the resulting split currents, into, say, i_1 and i_2, must add up to the current i_0 flowing into that node. Where those circuit elements rejoin, at node B, the currents add back up to the original current, i_0. That is called a **current divider**.

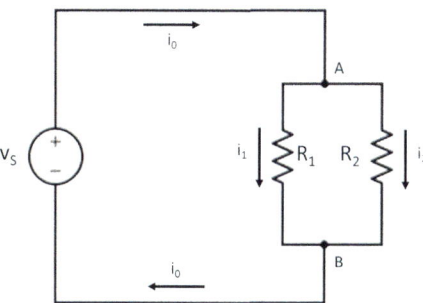

FIGURE 2.2 Current i_0 splits at node A into branch currents according to KCL. The sum of the branch currents must equal the incoming current. At node B the currents rejoin to sum to i_0.

2.2.2 Kirchhoff's Voltage Law

In Section 1.2, we discussed how voltage, or electrical potential, is analogous to gravitational potential, and that a rock at various elevations is like an electron at various voltages. Let's continue to extend this analogy even further (because it's a really good one!). Consider a tall building with 50 floors. Keeping track of your gravitational potential, you take the elevator to the 50th floor. Then, you take the elevator down to the 10th floor and walk the rest of the way down the stairs to the ground floor. You are now back at the same elevation, and potential, as when you started. In other words, the potential you gained going up the elevator was given up by lowering yourself to the 10th floor (using the elevator) and then lowering yourself again by ten more floors to the ground (by taking the stairs). You made a loop that brought you back to the same elevation. Let's add a couple of variations while we are at it. You might have chosen to go back up to the 15th floor from the 10th floor before heading back down again. Or, you may have chosen to visit the basement, which is below ground, during this trip. When you get back to the ground level, you are back to the elevation at which you started; the net gain in elevation is zero.

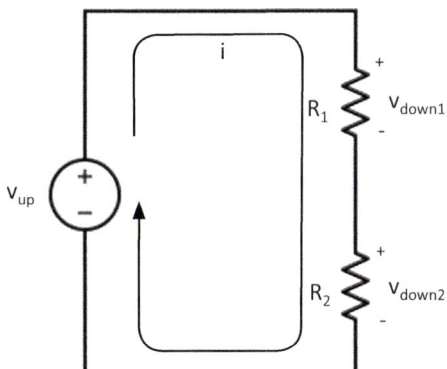

FIGURE 2.3 A voltage source and current in two series resistors. The two resistor voltage drops must sum to the supply voltage according to Kirchhoff's voltage law, KVL. v_{down1} and v_{down2} can be found using the voltage divider relationship.

We can relate all of these changes in potential to the simple electrical circuit shown in Fig. 2.3. The trip up in the elevator corresponds to the battery or voltage source having a voltage v_{up} that raises the potential of the charges so that current may flow. That is the highest electrical potential of this circuit, equivalent to being on the 50th floor. Then, there is a voltage drop of $v_{down1} = iR_1$, which corresponds to the elevator ride to the 10th floor, and then another voltage drop of $v_{down2} = iR_2$, which corresponds to moving down the stairs to ground.

Just like the building trip up and down, in making the circuit loop, the sum of all the up voltages must equal the sum of all the down voltages, which in this case is $v_{up} = v_{down1} + v_{down2}$. In fact, we can generalize this example to **Kirchhoff's Voltage Law (KVL)**, which states that the sum of all voltages of the circuit elements going around a loop in an electrical circuit must be zero; you must go up and down the potentials in such a way that you get back to where you started. KVL applies to a circuit of lumped elements, and is formally written as,

$$\sum_n v_n = 0. \qquad (2.2)$$

Notice that whether you took the elevator or the stairs, a voltage drop is a voltage drop; they must still add up to zero around the loop. This is analogous to saying that the voltage drops can be across voltage and current sources, resistors, capacitors, inductors, semiconductor devices, and even open circuits. All those devices must combine their electrical potentials to get back to the starting point after going around the loop. Note: what we call 'wires' are assumed to be ideal; wires have no resistance and drop no voltage. As discussed in Chapter 1, real wires do have some resistance, but that should be low in order to function well as a wire. If anywhere in this book you are supposed to know that a wire has some nonideal resistance, it will be stated explicitly.

In Fig. 2.3, the up voltage is the battery, and the down voltages are across the resistors. Since the resistor voltage drops add up to the battery voltage, how are those values apportioned to each resistor? Since there is only one current, $v_{up} = iR_1 + iR_2$; $v_{up}/(R_1 + R_2) = i$; so, $v_{R1} = iR_1 = v_{up} \times R_1/(R_1 + R_2)$ and $v_{R2} = v_{up} \times R_2/(R_1 + R_2)$. In general, when resistors are in series, *the voltage drop across one of them is the same fraction of the total voltage drop as that resistance is to the total of all the resistances.* This voltage divider relationship is expounded on in Section 2.3.3.

2.3   Series and Parallel Devices and Circuit Elements

2.3.1   Series and Parallel Circuit Elements

An essential part of electrical engineering is calculating voltages and currents in a circuit. Modern circuit simulators eliminate the drudgery of actually doing so for large circuits, just as pocket calculators have reduced the need for doing long-hand arithmetic. Without such simulators, advanced circuit analysis would be impossible. (A free, yet reasonably powerful, circuit simulator is that of LTspice, provided by Analog Devices at https://www.analog.com/en/design-center/design-tools-and-calculators/ltspice-simulator.html. It is recommended that you get that tool and try it out.)

However, simulation codes just simulate – they don't understand. An engineer needs to understand in order to *design*. So how does one gain understanding? An important skill is that of recognizing circuit patterns, such as the series and parallel combinations presented in this section. A simulator only needs to know the behavior of each individual branch element in order to run the simulation. You, the engineer, must have a more-global understanding of how things work, and their purpose, by recognizing *how elements work together*. You will be able to assess how a circuit is likely to behave without doing detailed calculations or relying on computer simulations. One goal of this chapter is to help you internalize series and parallel circuit structures, so that you won't need to scrupulously analyze the circuit properties to know how that part of the circuit behaves. You will need to have these basic circuit skills to understand how electronic bench test tools presented in this chapter are used, how they function with a circuit, and what their limitations are.

Let's consider just resistors for this discussion, but be aware that it applies as well to capacitors and inductors, and all other two-terminal elements, although with modified rules. (The behavior of capacitors and inductors in filter circuits is the topic of Chapters 4

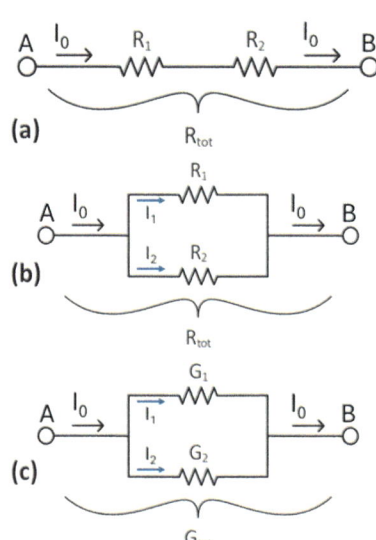

FIGURE 2.4 (a) Resistors placed in series, and (b) in parallel. (c) Conductances in parallel. A resistor can be represented either as a resistance or conductance, so (c) is the exact same thing as (b) but with the resistors represented as conductances.

and 5.) When two elements, such as resistors, are attached end-to-end, connected at only a single node as in Fig. 2.4(a), they are said to be connected **in series**. Another way to say this is that if current can flow through one element into only a second element, then those two elements are in series.

You can think of two resistors in series as being one resistor with a larger equivalent length L that appears in $R = \rho L/A$ (Eq. 1.6), so having two resistors in the series combination increases the total resistance. It is therefore reasonable that the total, or equivalent, resistance is the sum of all of the series resistances. The series combination of two resistors having resistances R_1 and R_2 behaves like a single resistor with the equivalent resistance:

$$R_{eq} = R_1 + R_2 \qquad (2.3)$$

Figure 2.4(b) shows two resistors **in parallel**. Two elements are in parallel if their ends are connected at the same two nodes. According to KCL, current entering multiple branches splits apart and is shared by them, as discussed above. As more parallel current paths are added to the circuit, you can think of this as increasing the area, A, of a resistor, so the total resistance decreases. Two resistors in parallel have an equivalent resistance:

$$R_{eq} = \frac{1}{\dfrac{1}{R_1} + \dfrac{1}{R_2}} = \frac{R_1 R_2}{R_1 + R_2} \qquad (2.4)$$

Both series and parallel combinations of resistors obey the same characteristic equation as a single resistor, $v = iR$, but with $R = R_{eq}$. What do we mean by 'v' in this case? We mean the voltage dropped across the entire group of resistors. Example: For a 10 Ω resistor in parallel with a 20 Ω resistor (Fig. 2.4(b)), $R_{eq} = (10 \times 20)/(10 + 20)$ Ω = 6.67 Ω. If the current i is 2 A, then the voltage dropped between nodes A and B would be $v_{AB} = 2\ A \times 6.67\ \Omega = 13.34\ V$.

2.3.2 Conductance

It is often useful to think about a circuit element in terms of its ability to *conduct* current rather than *resist* it. It can be more convenient algebraically to think of the resistor as a conductor with **conductance** $G = 1/R$ that causes current $i = Gv$ to flow in response to an applied voltage v. Conductors are the identical devices as resistors, but, instead, treated in terms of how well they pass current rather than restrict it. Suppose you want to rank the height of a group of trees. It is normal to think of the height of a tree as its 'tallness,' and you could rank them in order of "least tall to most tall." However, you might just as well think of a tree as having a certain "shortness" rather than "tallness." You could then rank them in order of how short they are, rather than how tall they are. In the end, they are still the same trees. It is the same thing with resistors and conductors, but one of them (probably resistance since that is more commonly used in circuit analysis) may seem more intuitive than the other. So, you may think of the same piece of carbon as a conductor having a conductance G if that helps you do the algebra more easily.

Since G is the inverse of R, its units are the inverse of ohms, and are called **siemens**, S: 1 S = 1 Ω^{-1}. Example: A 100 Ω resistor has a conductance of 0.01 S. Again, conductance is a natural extension of the concept of resistance. If you put several current paths in parallel, then the overall conductance is increased (Fig. 2.4(c)), or equivalently, the total resistance is decreased. Makes sense. Likewise, if you put several resistors in series, the overall conductance is decreased, i.e., resistance is increased. You will see series and parallel resistances and conductances in many of your courses, so get used to thinking in terms of both.

On the basis of the preceding arguments, the following rules for the equivalent values of series and parallel resistances and conductances should make intuitive sense:

Series resistors:
$$R_{eq} = \Sigma R_n \qquad (2.5)$$

Two Resistors in series:
$$R_{eq} = R_1 + R_2 \qquad (2.6)$$

Parallel resistors:
$$1/R_{eq} = \Sigma 1/R_n = 1/R_1 + 1/R_2 + \ldots \qquad (2.7)$$

Two Resistors in parallel:
$$R_{eq} = \frac{1}{\dfrac{1}{R_1} + \dfrac{1}{R_2}} = \frac{R_1 R_2}{R_1 + R_2} \qquad (2.8)$$

Parallel conductances:
$$G_{eq} = \Sigma G_n \qquad (2.9)$$

Two conductances in parallel:
$$G_{eq} = G_1 + G_2 \qquad (2.10)$$

Series conductances:
$$1/G_{eq} = \Sigma 1/G_n = 1/G_1 + 1/G_2 + \ldots \qquad (2.11)$$

Two conductances in series:
$$G_{eq} = \frac{1}{\dfrac{1}{G_1} + \dfrac{1}{G_2}} = \frac{G_1 G_2}{G_1 + G_2} \qquad (2.12)$$

Example: For the 10 Ω and 20 Ω parallel resistors, we can add the conductances, as in Eqs. 2.9 and 2.10: G_{eq} = 1/10 S + 1/20 S = 0.15 S. Then, if we want to get R_{eq}, we do the same thing as the calculation of the left side of Eq. 2.8: R_{eq} = $1/G_{eq}$ = 1/(0.15 S) = 6.67 Ω.

You might notice that *the equivalent parallel resistance is always less than that of either of the two resistances that are in parallel.* Why? Because adding a parallel conductance path can only lower the resistance, not increase it. For example, let's put a 100 Ω resistor in parallel with a 5 Ω resistor. I may not be able to instantly calculate the equivalent resistance in my head, but I do know instantly that the parallel combination is a little less than 5 Ω (calculate it) because: (1) the result is less than either of them, and (2) the 100 Ω resistance is a lot bigger than the 5 Ω resistance, so only a little bit of the total current will flow through the 100 Ω resistor compared to that through the 5 Ω resistor, and the equivalent resistance is hardly changed from that of the 5 Ω resistor alone. An easy and useful fact that you should simply know is that placing n identical resistances in parallel yields $1/n$ times the resistance of any of them, and in particular, *placing two identical resistances in parallel yields half the resistance of either.* (You can easily prove that to yourself.)

Of use later: Inductances combine the same way that resistors do. You can imagine that placing two identical coils end to end would behave like one coil that is twice as long, and has twice the inductance, so this makes intuitive sense. For two inductors in series, $L_{eq} = L_1 + L_2$. For two inductors in parallel, $L_{eq} = L_1 L_2 / (L_1 + L_2)$.

Adding two capacitors in parallel is like increasing the total area of a single capacitor, so the total capacitance is higher. So, capacitances combine in the same way that conductances do. For two capacitors in parallel, $C_{eq} = C_1 + C_2$. For two capacitors in series, $C_{eq} = C_1 C_2/(C_1 + C_2)$.

Note that the right side of Eq. 2.4 is often referred to as the **"product-over-sum"** (P/S) representation of two parallel resistances. Although that form applies only to two resistances, in this text, most of the time you will be considering only two resistances in parallel. It is less efficient to calculate two reciprocal numbers, add them together, and then calculate that reciprocal to get your answer; it is faster and easier to just use P/S, which can often just be done in your head. It is my experience that some students seem to have an aversion to using P/S, but I have never been able to figure out why. I recommend that you use P/S whenever you are working with just two parallel resistances.

2.3.3 Voltage Dividers

The next critical concept is that of the **voltage divider**, which is an application of series resistances, and was briefly introduced above. Consider Fig. 2.5, which shows a group of ten resistors in series. Because there is only one current path, KCL tells us that the same current, i, flows through every resistor. Therefore, each resistor will 'drop' a voltage of $v_n = iR_n$, and the voltage drop for each resistor is proportional to its resistance. Furthermore,

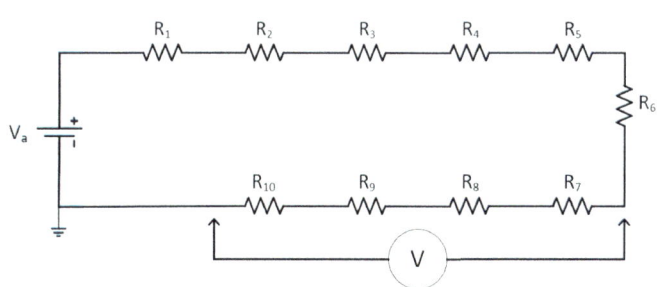

FIGURE 2.5 Voltage divider with ten resistors.

according to KVL, the total voltage across all the resistors must be that of the applied voltage, V_a ('a' is for 'applied').

If we put several resistors in series, with an applied voltage V_a across the entire group, each point along the current path has a different voltage. The input voltage is divided among series resistances according to the voltage divider equation, as follows. The voltage drop, V_n, across each resistor, R_n, depends on the applied voltage and the total, equivalent resistance as

$$v_n = v_a \frac{R_n}{R_{eq}}. \tag{2.13}$$

For example, if there are 10 resistors as shown in Fig. 2.5 of equal resistance, then each resistor will drop 1/10 of v_a. This can be calculated using Eq. 2.13 with $v_a = 5$ V and $R_n = 100$ Ω, v_m (m for measured) is 5 V × (100 Ω/1000 Ω) = 0.5 V. Figure 2.5 shows a voltage measurement (using a voltmeter discussed in detail below) of the voltage drop across four resistors. In that case the meter would read 5 V × (4 × 100 Ω/1000 Ω) = 2 V. As another example, let's say that the nth resistor has a value of $R_n = n$ Ω (e.g., $R_8 = 8$ Ω). Then the voltmeter of Fig. 2.5 would read $v_m = v_a \left(\sum_{n=7}^{10} R_n / \sum_{n=1}^{10} R_n \right) = 5 \times 34/55 = 3.09$ V.

At the voltage source, at left in the picture, a voltmeter would read 5 V with respect to (w.r.t.) the negative terminal of the battery, and after the next resistor it would read 4.5 V, etc. as we move the terminal of the voltmeter toward the right. Each resistor drops some fraction of the total voltage, and does so in proportion to its resistance.

A couple of more important examples will be useful. Consider Fig. 2.5 where all of the resistances are 100 Ω except for, say, R_5. If R_5 is very large, say 1 megohm, MΩ, (1 MΩ = 10^6 Ω) then *almost all* of the applied voltage will drop across R_5 ($v_{1\ M\Omega} = v_a \times (10^6\ \Omega\ /1.0009 \times 10^6\ \Omega)$).

Strategy: A useful habit to get into when analyzing a circuit is to ask, "What if this parameter were very large, like infinity?" In the first example, I used 1 MΩ since it was much larger than 100 Ω, and the calculation was easy. Then, you can repeat the process using a very small number, like zero, if you want. I chose 1 Ω, since zero would have been slightly more difficult to mentally process, and it still makes the point. Extrapolating to the limits often gives a basic understanding of circuit behavior that reveals most of what you need/want to know.

On the other hand, if R_5 were very small, say $1\ \Omega$, then *almost none* of the source voltage would drop across R_5 ($v_{1\,\Omega} = v_a \times (1\ \Omega\ /901\ \Omega)$). These are important examples for quickly estimating the effect of large and small resistances in voltage dividers.

In summary, for a voltage divider, the voltages dropped by each of the resistors in series add up to the applied voltage. Also, *each resistor has a voltage drop that is the total voltage drop times that resistor's fraction of the total resistance*. Therefore, as combinations of series resistances, the larger resistances drop more voltage than the smaller resistances. From now on, when you see a voltage divider, try to intuit its behavior as I have described in the 'Strategy' text box above. You will often need it going forward here and in your other EE classes.

> *Of use later:* If you use a DC voltage source just as in Fig. 2.2, where the resistors are replaced by various values of inductance, the voltage division follows the same rules as for resistors. However, for capacitance, the small capacitances will drop more of the voltage, as if small capacitances were larger resistances. Of course, no current will flow at DC through a capacitor, but you can still measure a DC voltage across each of the capacitors because of the charge on the plates. The voltage drop across one capacitor in series is $v_n = v_{\text{total}} \times (C_{\text{total}}\ /C_n)$. (Note that based on the equivalent series capacitance given above, C_{total} is less than any of the individual capacitances.)

2.3.4 Current Dividers

Just as voltage is divided across series connected resistors (due to Kirchhoff's voltage law), current is divided through parallel-connected resistors, as shown in Figs. 2.4(b) and 2.4(c). Here is a good spot to be thinking of resistors as conductors, whose conductance is $G = 1/R$. A large resistance is a small conductance, and a small resistance is a large conductance. When resistors are in parallel, the current flows most easily through the one that has the lowest resistance, or highest conductance, and progressively less current flows through those that have lower conductance. The equivalent, or total in this case, conductance is the sum of all the parallel conductances (Eq. 2.9), since more paths make it overall easier for current to flow for a given (parallel) voltage drop. Note here that G_{eq} is greater than that of any of the parallel resistors, so as in the parallel resistance discussion, the overall resistance is less than that of any of the resistors alone.

So, it seems reasonable that when current i_0 enters a network of n parallel resistors it is split amongst them like the voltage divider according to the conductances as

$$i_n = i_0 \frac{G_n}{G_{\text{eq}}}. \tag{2.14}$$

In words, *each resistor passes a current that is the total current entering times that resistor's fraction of the total conductance*. That said, the current divider equation can be written in a form with only resistances, whose derivation is left to you. For the special, but very common, case of *two resistors in parallel*, it is often easier to use the form

$$i_1 = i_0 \frac{R_2}{R_1 + R_2} \quad \text{and} \quad i_2 = i_0 \frac{R_1}{R_1 + R_2}. \tag{2.15}$$

In words, a resistor gets its share of the total current according to the ratio of the *opposite* resistor to the total of the two resistances. You will find this is the easiest form of the current divider to use in practice. Notice that $i_1 + i_2$ in Eq. 2.15 equals i_0, as it should.

Another extremely important property of parallel resistances as a result of KVL can be seen in Fig. 2.4(b). Let's use Eq. 2.15 to find the voltage drops across R_1 and R_2, which we call v_{R1} and v_{R2}. $v_{R1} = i_1 R_1 = i_0 \dfrac{R_1 R_2}{R_1 + R_2}$. Likewise, $v_{R2} = i_2 R_2 = i_0 \dfrac{R_1 R_2}{R_1 + R_2}$, so $v_{R1} = v_{R2}$. That parallel voltages are equal must be the case since no matter how we go from node A to node B, i.e., whichever path we take (through R_1 or R_2), we must start and end at the same potential. Hence, we know that NO MATTER WHAT, the voltage across ANY parallel branches MUST be EXACTLY the same. This fact is used endlessly in circuit analysis. It is one of the first go-to facts used in any circuit analysis that you will make.

2.4 Potentials along Electric Field Lines

You have seen electric fields in Chapter 1 and will have a deeper dive in your electromagnetics course. This section may provide further insight into electric fields, and why a voltage divider breaks the circuit into discrete voltages between the two endpoints.

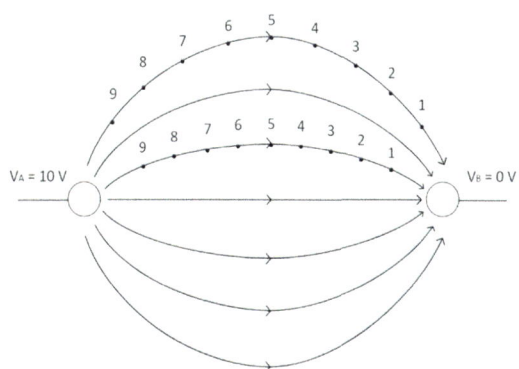

Suppose you set a voltage v_A at a point A in space and a voltage v_B at another point in space, B. Furthermore, let $v_B = 0$ V, or ground. Figure 2.6 shows two electrodes at the two potentials, v_A and v_B, and the field lines between them. Following any path connecting A to B will eventually take you through every voltage between v_A and v_B, analogously to going up and down the building through different elevations. No matter which path you take (different stairways, for example), you will start at the top of the building (v_A) and end up at the bottom ($v_B = 0$ V). Also, you will have gone through every elevation on the way down, regardless of the way you did it. *You cannot skip an elevation.*

FIGURE 2.6 Two electrodes, A and B, having potentials at 10 V and 0 V, respectively. Any path taken from A to B must pass every possible voltage between 10 V and 0 V.

If you could use a meter to measure the potential at every point in space relative to that at point B (voltmeters are discussed later in this chapter, and do *not* work this way), you would measure a voltage v_A at A (top of the building) and a voltage $v_B = 0$ V at B (bottom of the building). In the absence of any more charge found along the way, the voltage must vary continuously along any path from A to B. As you move along any path from A to B, your hypothetical voltmeter will read every voltage until you get to B, where it will read 0. This would be true no matter which path you chose to get from A to B, as shown in the figure. This is the basis for Kirchhoff's voltage law, which says that the voltage across parallel paths in a circuit is the same, just as the elevation difference is the same no matter which stairway you take from the top of the building to the bottom.

Electric field lines at a point in space point along the direction of fastest descent in voltage. (Analogously, if you let a ball roll down a hill, it will start out along the path of the largest change, or gradient, in potential, akin to the electric field.) In particular, the voltage decreases monotonically along a field line. The field lines within a circuit run along wires. If the path is through a series combination of resistors, you will likewise have to go through every single potential from one end to the other in sequence. Why do you see only the discrete voltages provided by the voltage divider equation? Because you cannot get inside the resistors! If you could, you would essentially have a potentiometer, and then you could read every voltage from V_B to V_A!

2.5 Breadboards

In this and future lab courses, you will need to build **prototype** circuits (i.e., prototype them) quickly. It would be exceedingly inconvenient to solder hundreds of wires, or make a PCB, every time you want

FIGURE 2.7 Components neatly placed on a solderless breadboard.

to test a new circuit idea. **Solderless breadboards** (Fig. 2.7) make prototyping easier. A breadboard (Figs. 2.7, 2.8 and 2.9) is like a peg board that easily connects wires and devices in an orderly fashion without solder. Wires and components are inserted into **tie points**. Tie points are equivalent to nodes in a circuit diagram in that they are the location that elements connect together. Figure 2.8 shows an unpopulated breadboard where all the tie points can be seen. Breadboards come in various sizes, from around an inch to a foot, with varying numbers of tie points and layout details, but the tie point spacing is standardized so that certain integrated circuit packages and other devices can precisely straddle columns of tie points, as seen in Fig. 2.7.

You can see in Fig. 2.8 that the short sections at the interior of the breadboard have five holes, and in Fig. 2.9 you can see the metal connector strips under the plastic surface. Embedded under each set of five holes is a single bar of metal with five clips (Fig. 2.10). One group of five clips forms a circuit node, i.e., a point of intersection of wires. When two or more wires or device leads are inserted, they are connected electrically as if by a solder joint, except that you can easily pull them apart. If you need more than five tie points connected together, simply add a short wire 'jumper' to an adjacent set of five tie points, and you will have a new total of 8 tie points at the node (why 8?).

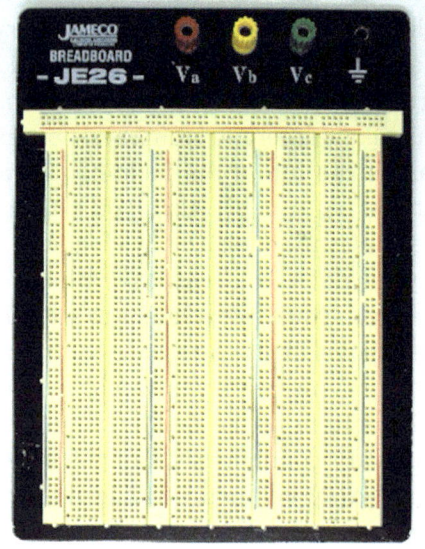

FIGURE 2.8 Standard solderless breadboard.

FIGURE 2.9 Breadboard made with clear plastic showing metal strips inside. On the right is a magnified view of the upper left corner. Each horizontal row of five tie points is one metal strip, and so creates one electrical node. Each long vertical column between the red and blue lines is one large tie point. These are typically used for positive and negative power supply voltages and grounds.

You will also notice that there are long rows of 5-hole sections that have red and blue lines alongside in both of the figures. All of those rows of five are connected together, so they make one large node. This long row of tie points is useful for providing a voltage, say from a battery or power supply, that can be easily accessed from any part of the breadboard. That way, you can conveniently provide a particular voltage source or ground to any interior point of your circuit with short, neatly placed wires. The maximum and minimum voltages available on a circuit are typically called **voltage rails** (or just **rails**), and on the breadboard they actually look like rails.

When breadboarding a circuit, be neat and organized. Be careful not to let bare wires inadvertently touch and, thereby, create short circuits. Make sure that the wires and

Something practical: What is better, a breadboard or a PCB? Breadboards are great for temporary circuits, but besides being more expensive, large and heavy, they are delicate – it is easy for wires to inadvertently pop out. Also, the electrical properties are not as good as well-designed PCBs, so circuits will not work as well on breadboards at higher frequencies, say above a few MHz.

device leads are firmly inserted in the holes so that you don't have any open circuits. The circuit in this chapter's lab is very basic, so you should not have any problems, but it will give you some experience with breadboarding. Notice in Fig. 2.7 how neat and organized that layout is. You won't be trimming wires to fit so precisely in this course. However, when you work on a more complicated project over a long time, you will want to keep things neat and organized in order to avoid mistakes or to help you locate any mistakes you might make.

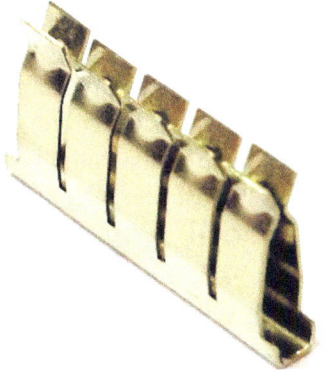

FIGURE 2.10 A single breadboard node comprising five tie points.

2.6 Ideal and Real Power Supplies

The term "power supply," or just "supply," usually refers to an electronic source of one or more **DC** (**direct-current**, i.e., constant) voltages that power a circuit. Some schematic symbols for a power supply are shown in Fig. 2.11 (top). For example, a supply might provide **fixed** DC voltages of 15 V and 5 V, or **variable** (meaning that you can change them at will) DC voltages up to 30 V.

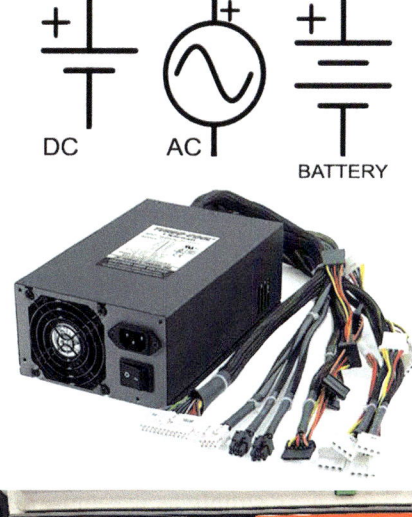

Figure 2.11 (middle) shows a typical desktop computer power supply. The various connectors provide power to the computer components, such as the motherboard and solid-state drives, with voltages of typically ±3.3 V, ±5 V and ±12 V. The total power consumption of desktop computers is about 400 W to 1000 W, so the power supplies are rather large, and have fans to help keep them cool.

Your cell phone is obviously more useful if you can carry it around rather than keep it plugged into the wall all the time, so it uses a battery. Since it would be inconvenient, expensive, inefficient, and environmentally unfriendly to just use batteries in the lab, you will almost always use a benchtop power supply. (In Chapter 10 you will see that batteries don't make very good power supplies.) As an example, the Keithley 2231A-30-3 triple power supply (Fig. 2.11, bottom) can provide three independent, adjustable DC voltages. It is rated at up to 30 V and up to 3 A per channel on Channels 1 and 2. Channel 3 has a maximum voltage setting of 5 V. The front panel allows you to set the voltage output and read out either

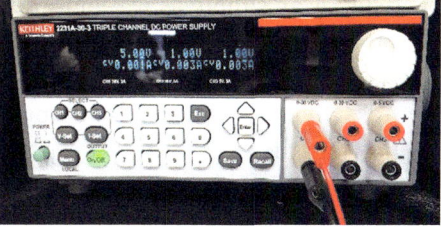

FIGURE 2.11 (Top) Schematic symbols for DC and AC power supplies and battery. (Middle) Typical desktop computer power supply. (Bottom) the Keithley 2231A-30-3, which is the DC power supply you will use in lab.

the voltage or current from each supply. Notice also that the outputs are color-coded red and black to indicate the positive and negative output terminals, respectively. That color code is universal in electronic systems.

2.6.1 Norton and Thevenin Equivalent Circuits

A power supply as presented in this section has a very specific purpose – to power an electrical circuit at a constant voltage. If the output voltage of a power supply in a radio were to change when you turned up the volume, it might cause the station to change. That would be annoying. (True story: I once had a car that accelerated if I honked the horn when the cruise control was on.)

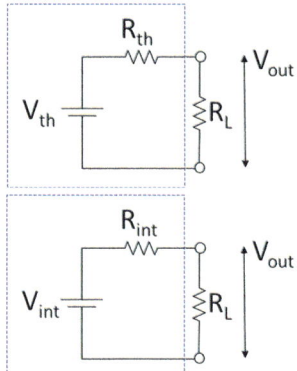

FIGURE 2.12 (Top) Thevenin equivalent of a circuit with voltage sources, current sources, and resistances. (Bottom) Same circuit but relabeled to reflect terminology of a real power supply.

You can think of any part of a circuit in some way as a power supply in that it interfaces to other circuits and must provide a voltage and a current, that is to say drives them in some way. Every subcircuit provides a voltage and current to something else further along in the signal path.

Knowing the behavior of a subcircuit is important. One method of reducing a complicated circuit to a simpler one for the sake of analysis is to form something called a **Norton or Thevenin equivalent circuit**. Since this is not a circuits book, we will not get into the weeds of how that analysis is performed, but it is instructive to think in such terms. Suffice it to say that through a series of transformational steps, many complicated circuits (for completeness: only linear, time-invariant circuits, which is what is covered in most introductory courses) can be represented by a simple voltage source, the **Thevenin voltage** v_{th}, in series with a resistance called the **Thevenin resistance** R_{th}. (We will ignore Norton equivalent circuits in this discussion.) Figure 2.12 (top) shows the form of the extremely simple Thevenin equivalent circuit as it drives a load represented by R_L. The **load** is whatever the circuit is providing voltage and current to at the output terminals.

Any pair of terminals at which you are interested in the voltage and current may be referred to as a **port**. You can have the **input port** where the input to a circuit is, and the **output port**, where the output is, as in where the load is placed on the power supply. The voltage at the output port and across the load is v_{out}. In fact, *you can think of the entire power supply as a Thevenin circuit comprising v_{th} and R_{th}*. Using that as the **model** of the power supply, you will be able to predict its behavior.

(Heads up: this section is all about voltage dividers, so you need to be very comfortable with that concept.) Now you are ready to understand the implications of the Thevenin equivalent circuit. It really *is* as simple as it seems. Referring to Fig. 2.12, how much current flows when R_L is removed? Another way to say that is that the output is **open circuited**, i.e., the output terminals are not connected by anything other than air. (Draw that for yourself on a piece of scrap paper.) In other words, what is v_{out} when the output is open? In that case, $i_{out} = 0$ A. No current flows. Since no current is flowing, then the voltage drop across R_{th} is zero, and v_{out} is simply v_{th}, which is the same as if R_{th} weren't there to drop any voltage. This is not intuitive, so give yourself a few minutes to think about it. Another way to think of this is that if R_L is infinite, then as a voltage divider with two resistors, namely R_{th} and $R_L = \infty$, v_{out} gets all of v_{th}. See? It's actually pretty easy.

Now, if $R_L = 0$ Ω, i.e., the output is **short circuited**, i.e., the output terminals are connected by a zero-resistance path, say, by a thick wire, then what would happen? The load current, i_L, would be v_{th}/R_{th}. Depending on the value of R_{th}, i_L might be a small current or a large current. Regardless, i_L would not be infinite; in this case the current is limited by R_{th} and is the largest it can possibly be.

Now let's add a finite load resistance R_L. In this new case, the current will flow through both R_{th} and R_L having a value of $i_L = v_{th}/(R_{th} + R_L)$. Also, v_{out} is that of a voltage divider, determined by $v_{out} = v_{th} \dfrac{R_L}{R_{th} + R_L}$. It's a very simple circuit that you will have no problems with, but here is an example anyway. If we choose R_L to be R_{th}, then using simple voltage divider analysis, $v_{out} = v_{th}/2$. What current then flows through R_L? $i_L = v_{th}/2R_{th}$.

Now, recall in the above examples that when R_L was very large (infinite for an open circuit), then the output voltage was v_{th}; when the load was the same as R_{th}, then the output voltage dropped to ½ of v_{th}. So you see, when R_{th} and R_L were of the same order, the output voltage changed as R_L changed, and the circuit could not provide a constant voltage that was independent of R_L.

Beyond its utility in circuit analysis, it is very helpful to *think* of real systems in the same terms – that inside of a power supply or other complicated circuit, comprising even many thousands of components, are just these two components v_{th} and R_{th}. Even though real circuits have many different kinds of components and, therefore, do not necessarily yield to strict Thevenin analysis, their behavior is often very much like the Thevenin equivalent circuit. Keep in mind that you cannot actually access the Thevenin voltage source and resistor inside the power supply — they are only models of the circuit, and you can only infer their effective values from outside of the box.

For these real circuits, we may refer to the internal components as the **internal voltage source,** v_{int}, and an **internal resistance**, R_{int}, as shown in Fig. 2.12 (bottom). This is not a different concept – just different terminology. The two circuits in Fig. 2.12 are identical except for what we choose to call the internal voltage source and series resistance. You may think of the output load as being attached to the red and black wires of Fig. 2.11 (bottom). For a real power supply or voltage source, the internal voltage is often called the **open-circuit voltage**, and the internal resistance is often referred to as the **output resistance**. These terms are more or less interchangeable, except that any device (e.g., an inductor or capacitor) can have internal resistance, but not necessarily output resistance, as in a power supply.

If the circuit of Fig. 2.12 (bottom) is a power supply, it might not be a very good one, depending on the value of R_{int}. As you saw above, R_{int} limits both the output current and output voltage. An **ideal power supply** has no internal resistance ($R_{int} = 0\ \Omega$), and can provide any amount of current, no matter how large. As a consequence, the ideal supply would provide exactly the rated voltage (say 5 V) regardless of the load resistance. Also, there is no upper limit to the amount of current that an ideal voltage source or power supply could provide. It is assumed that v_{int} in Fig. 2.12 is itself an ideal voltage source. Saying that a power supply is ideal is equivalent to saying that it can provide whatever current you ask of it. A **real power supply** is not like that, of course. There are limitations due to its internal components that limit how much current and power it can provide to a load. Keeping a voltage divider in mind, if R_{int} were very, very small, then the power supply could provide a lot of current to a load before v_{out} dropped significantly below v_{int}. That understanding is essential to this chapter.

Let's say you put a short circuit of 10 mΩ resistance across an *ideal* 5 V power supply. Then you might naively expect to get 500 A from the supply (assuming the wire doesn't melt or the insulation doesn't burst into flames, which it would). That would be a power output of $P = iv = 2500$ watts (about two large microwave ovens). Such power supplies exist, but they are not very common. A typical **benchtop** or 'bench' supply might provide, say, 100 W, but even that would be on the large side. So, what would happen in our hypothetical case if the power supply were real and not ideal? The output voltage of the supply might drop substantially because R_{int} is of some realistic value, and the power supply cannot provide that much current. Another possibility is that internal protection circuitry would disconnect the output to protect itself from damage. Either way, no 500 amps at 5 V for you today.

If you have a *real* power supply at 5 V with, say, a 1 Ω output resistance, and you short it out with the same 10 mΩ wire, what current will pass through that wire? First, let's ignore the 0.01 Ω and approximate the total resistance as just the internal 1 Ω. Then we calculate that the output current, i_{out}, would be 5 A, which is pretty respectable for a small bench supply.

A power supply that can provide precisely the selected voltage up to some maximum load current is said to be **regulated**. Regulated power supplies monitor v_{out} and make internal adjustments to keep v_{out} constant. If you continuously decrease R_L at the output of a non-regulated power supply, its output voltage would drop, or 'sag', as i_{out} increases. For a regulated power supply, v_{out} stays constant while i_{out} increases as R_L decreases. At some point, the supply will simply let you know that it has exceeded the maximum allowed value. Nearly all modern benchtop power supplies are regulated. Generally, you can also set the maximum current, up to some maximum value, called the **compliance limit**, that it will output in order to protect your load from being overdriven.

2.7 Ground

Now we discuss the extremely important concept of ground and ground potential. Suppose you are standing on the first floor of a building, and you drop a bowling ball on your toe. It will hurt the same as if you did the same on the second floor. The ball falls out of your hand and onto your toe a couple of feet below. All that matters is how far it falls. Recall that the ball gains kinetic energy mgh where m is its mass, g is the acceleration of gravity, and h is how far it falls, i.e., from your hand to your foot. It does not matter which floor you are on, but rather just the difference in height, h, that the ball falls.

It is important to know that all voltages are likewise differences in potential between two points. It is meaningless to say that a *point* is at a particular potential, either electrical or mechanical, without knowing "with respect to the potential where?" When we say that $v = iR$ for a resistor, we mean that the voltage 'drop', or difference, *from one end of the resistor to the other* is the current times the resistance. Here, v, energy per unit charge, is analogous to gh, which is energy per unit mass.

A voltmeter is a tool that measures potential *difference between two leads*. If we just state a voltage, such as the voltage at node B, 'v_B,' without asking with respect to *where*, then it is implicitly with respect to the node that we define to be at zero volts, which is referred to as **ground**. This is the standard term in the U.S.; in most English-speaking countries it is called **earth**.

The concept of ground as a reference voltage is very important. By "ground" we sometimes do mean actual dirt-like ground, or earth. The physical ground that you walk on is electrically the **earth ground** for all of the power systems in buildings as well as the power grid that delivers power to us. The metal frame of a building is attached to earth ground via one or more metal stakes driven several feet into the dirt near the building. The round socket of a modern (U.S.) wall outlet connects through the walls to earth ground. (Much more on this in Chapter 3.) That one little hole is an electrical portal to every piece of structural metal used to build every building on your campus (and more, of course), plus almost every piece of exposed metal on any piece of electrical test equipment that is plugged into a wall socket.

Improving comprehension: The Mariana Trench, in the Pacific Ocean north of Australia and SSE of Japan, is the deepest place in the ocean at about 36,070 feet below sea level. Although it wouldn't be very useful, it might make some kind of sense to refer to that as zero elevation, the same way that electrical ground is typically set to the lowest potential in a circuit. In this case, all elevations on Earth would have a positive sign. Now sea level would be at an elevation of +36,070 feet. It sounds high, but it would make no difference. In this case, the elevation of the University of Notre Dame near South Bend, Indiana, in the United States would be 36,805 feet.

When we use sea level as the zero of elevation, we commonly refer to things that are above the surface of the ocean as 'above sea level' having a positive sign, and things below the water as 'below sea level.' If we were to calculate potential differences, we would assign a negative value to elevations below sea level. (Death Valley, California, is at an elevation of 282 feet below sea level, or negative 282 feet.) You can see that it is possible for elevations to have both positive and negative elevations when sea level is chosen as the zero, or reference, of elevation. Likewise, we can choose a point in a circuit that we call 'ground' that is assigned the voltage value of zero, and there can be voltages both higher and lower than ground potential in the circuit. Where we choose to define zero volts is arbitrary, but it does affect how easy it is to think about and solve for circuit behavior.

Let's consider circuits like those in portable consumer electronics. They also have a designated node that serves as the reference for zero volts, but it is not connected to the ground outside. In this case, the circuit ground is referred to as **common** or **reference**. Not every piece of electrical equipment that is plugged into an electrical outlet is connected to earth ground. Those circuits will have a common node that is not connected to earth ground as well.

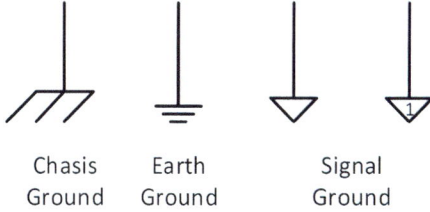

Chasis Ground   Earth Ground   Signal Ground

FIGURE 2.13 Various ground symbols used in schematic diagrams.

Could you just call the voltage at some arbitrary point in a circuit 'zero volts'? Yes, and it would be correctly called "ground". If you define a point somewhere on a hill to be at zero gravitational potential, then the potentials would be positive above and negative below. However, the relative differences in elevation on the hill (analogously voltage) wouldn't change due to your arbitrary choice. You could think of sea level as "ground" for elevations quoted on maps, as discussed in the sidebar about the Mariana Trench. There is a reason that ski areas are compared by their "vertical drop" – it is because skiers care only about how far they can ski without having to get back on the lift, and not the

height above sea level. Likewise, if you designate any arbitrary point in a circuit to be at ground, then higher voltages are positive and lower ones are negative. That said, there are guidelines for picking the most convenient ground point in a circuit, which is discussed in a traditional circuits course. Here, we will discuss the implications of a grounded node only from a practical point of view, and not how it affects our choice of circuit-solving strategy.

Figure 2.13 shows a few symbols for grounds used for different purposes in electrical **circuit diagrams**. A circuit diagram is also called a **schematic**, but EEs tend to reserve that term for more formal electrical diagrams printed by equipment manufacturers or for more complicated circuits. A variety of symbols for ground, such as those in Fig. 2.13, tend to be needed for such complex circuits to reduce ambiguity during troubleshooting. **Chassis ground** (pronounced CHASS-EE) is the voltage of the metal frame inside many electronics cases. It is also often connected to earth ground through the round prong, as discussed above.

Consider a meter stick floating in the air. It is not touching anything, and since it has no relationship to sea level (because it is just floating around up there) it can be used only to measure the length of things floating around in the air with it, such as a hummingbird that flies by and hovers next to it (hey, go with me on this). Now, put one end of the meter stick on the ground at sea level. It is now able to measure the elevation of things relative to ground. Although it is still the length of the hummingbird, the head of the hummingbird standing on the ground is at its elevation. As another example, if an insect were perched on top of a plant that is 75 cm tall and planted in the ground at sea level, then the insect would be at 75 cm above sea level.

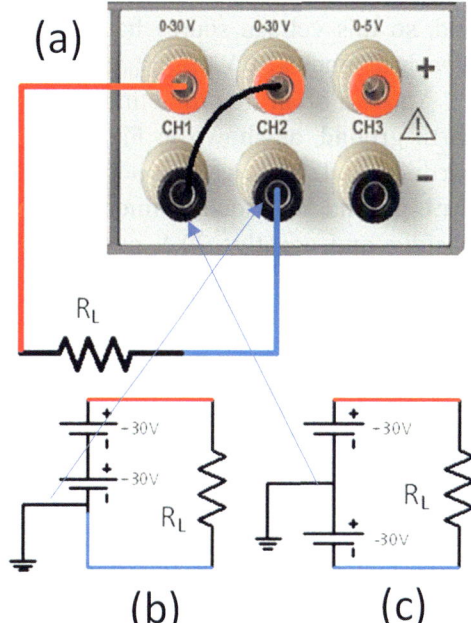

FIGURE 2.14 Three ways to get 60 V from two 30 V channels of a power supply. Connecting them in series can create either (a) a floating single 60 V source, (b) a grounded +60 V or −60 V source, or (c) +30 V and −30 V sources, depending on whether and how you connect to ground in your circuit. In all cases, the red line is at the highest voltage in the circuit and the blue line is at the lowest voltage. This fact is independent of where we choose to call ground; the current through the resistor would be the same in all circuits. The arrows show where to connect to the physical supply to ground the circuit in each of the schematics. In this circuit, the ground symbol indicates that the supply connects to earth ground somewhere. (The Occupational Safety and Health Association, OSHA, considers voltages > 50 V to be unsafe and must be shielded. You should not work in the lab with voltages that large without supervision. Take it from me that even 35 V can hurt – a lot).

Now to the electrical analog of the meter stick. If you hold a simple 1.5 V battery in your hand, then it is **floating** (and here, floating is a technical term), meaning that it is not connected to any reference or ground voltage. Consider an AA-size (double-A) battery floating in the air, unconnected to any circuit. The voltage difference between the two ends would be 1.5 V, as that is built into the battery chemistry (see Chapter 10). That is analogous to the length of the meter stick. However, the battery itself has no electrical relationship to anything around it. Now, touch one end of that battery to, say, the exposed metal of an automobile. Generally, any exposed metal parts on a vehicle are grounded, i.e., are attached to the common points of all the circuitry inside. Now the positive terminal of the battery is at 1.5 V above the vehicle's ground (reference voltage), just as the top end of the meter stick was one meter above sea level.

All of the above matters a lot in the lab because whenever you apply or measure a voltage, you must be aware of what that voltage refers to, or what it is referenced to. Figure 2.14(a) shows the outputs of a power supply with two channels connected in series, the positive terminal of one source to the negative terminal of the second source. There is a load resistor, R_L, driven by the full voltage of the power supply. Because there is no ground connected, the output voltage floats, again meaning that it isn't referenced to any other potential outside the circuit. Voltage sources in series add, so this voltage source has a total voltage of $v_{tot} = v_{CH1} + v_{CH2}$ between the blue and red lines. If each channel is 30 V then v_{tot} is 60 V from blue to red.

You are often free to define your own ground for your circuit. If you 'tie' the negative terminal to earth ground, as shown in Fig. 2.14(b), then it fixes the other terminal as positive with respect to earth ground. Or, to get a **bipolar power supply**, i.e., one that has both positive and negative voltages w.r.t. ground, you could connect ground between the supplies, as shown in Fig. 2.14(c), and get both positive and negative 30 V outputs. Often, bench-top power supplies provide a convenient terminal for earth ground to the earth-grounded wire of the power cord, but not always.

2.8 Digital Multimeters

The **digital multimeter** (DMM) is one of the most versatile tools in an EE's toolbox. It is a multifunctional tool that can measure voltage, current, and resistance, and sometimes more. The DMM is the equivalent of a stethoscope for a medical doctor in that it is the one instrument that you want to have close at hand. (In fact, you might keep a DMM draped over your shoulders to impress your friends.) The DMM is a multimeter that specifically displays the measurement as digits rather than as the position of a needle indicator. Although the old analog meters are cooler to look at (being retro and all), they are not as functional as DMMs, nor do they have high resolution.

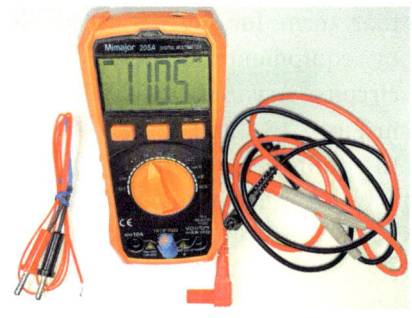

FIGURE 2.15 Handheld DMM with probes. Besides the standard voltage, current and resistance, this one can measure capacitance, frequency, and temperature.

FIGURE 2.16 Fluke 45 digital multimeter (DMM). Fluke 45 digital multimeter (DMM)

DMMs are often handheld units (see Fig. 2.15), but they come in a benchtop style as well, such as the Fluke Model 45 (Fig. 2.16). Handheld DMMs can perform measurements and tests that include the requisite AC and DC current and voltage and resistance, as well as additional measurements such as capacitance, inductance, frequency and temperature, diode test, and transistor test. (Note: DMMs do not measure power – that function is reserved for somewhat specialized

wattmeters.) They can also hold maximum and minimum values and log data to a computer; maybe some will shine your shoes as well. Not every meter has every function, and there is a huge variety out there for well under $100 that could last your entire professional career. Benchtop DMMs usually have more digits of resolution than do typical handheld DMMs. Basic handheld DMMs can sometimes be purchased for less than $5, or with an impressive array of features for less than $20.

2.8.1 Voltmeter and Ammeter Functions

Remember that voltage is the difference in electrical potential between two points. A **voltmeter** measures this voltage drop between two points of a circuit. If the two points are on opposite ends of a component, it is said to measure 'across' the component. An **ammeter** reads current *through* a component or branch of a circuit. Both are shown in Fig. 2.17. Sometimes current is referred to as **amperage** the same way that a potential difference is referred to as voltage.

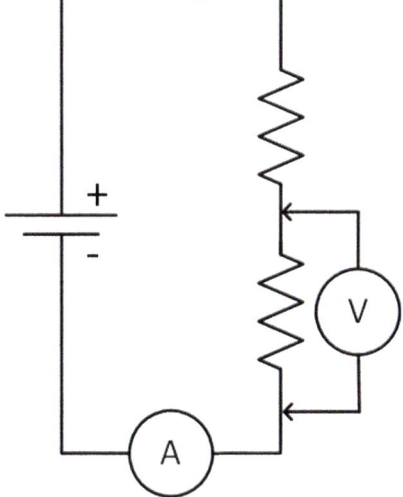

FIGURE 2.17 Voltage measurements are made in parallel and current measurements are made in series. The voltmeter acts as a high-resistance shunt resistor to the circuit.

Both current and voltage can be designated as either alternating current (AC) or direct current (DC). "AC" means that the current or voltage changes direction repeatedly, and "DC" means that the sign of the voltage or current does not change with time. (The terms 'alternating voltage' and 'direct voltage' do not exist. Those, paradoxically, are called AC voltage and DC voltage. Huh.) In Fig. 2.16 you can see the symbols on the buttons on the left side of the panel for AC, "~", and DC, a solid dash over a broken dash. The Fluke Model 45 shown as an example has **automatic ranging**, or **autoranging**, for voltage, current and resistance measurements. This means that you don't have to tell it what range of values it is measuring, e.g., tens or thousands of ohms – it picks the digits to display by itself and indicates the multiplier. Nowadays, even inexpensive handheld DMMs do autoranging. In fact, some of the cheapest meters are autoranging because they operate with fewer buttons, which may cost more than the functionality of the extremely inexpensive integrated circuits inside! (What a world we live in!)

Here is important information about using voltmeters and ammeters. When you attach a meter to a circuit *it becomes a part of the circuit, so the circuit behavior can change when you do measurements.* We can approximate, or **model**, the effect of meters on a circuit as if they were resistors. You are interested in the behavior of the circuit without the added resistance, so the meter must have as little effect as possible on the circuit.

As shown in Fig. 2.17, voltmeters are connected in parallel with the element or part of the circuit for which you wish to know the voltage drop. What do you think would happen if the voltmeter behaved like, or 'looked like', a *low* resistance? This resistor would behave as a **shunt**, i.e., an additional current path that diverts current from the circuit through the meter, thus changing its behavior. Wouldn't it be "ideal" if voltmeters looked like infinite resistance so that no current were shunted? Well, now that you mention it, a hypothetical **ideal voltmeter** looks like an open circuit (infinite resistance). An ideal voltmeter doesn't **load down** the circuit, meaning it doesn't draw so much current out of it that it changes the voltage at the terminals.

The **input resistance** of any electrical circuit or device is the resistance 'looking into' it, meaning how much equivalent resistance it has at its input port. A real voltmeter, such as the

Fluke 45, has a relatively high **input resistance** (here, 10 MΩ) in voltmeter mode, which may or may not be high enough to ignore, depending on your circuit and the accuracy you need. Many voltmeters have input resistance of 1 Mohm or even less, and some specialized voltmeters can have input resistances well into the tens of gigaohms (GΩ, 10^9 ohm).

Conversely, an ammeter has to be connected *in series* with the circuit so that it can measure the current flowing through it. The voltage divider rule tells us that a real ammeter should drop little voltage while doing its job of measuring the current through itself. If it looked like a high resistance, it would impede the current that it was trying to measure, and it would change all of the series voltage drops as in a voltage divider. Therefore, a hypothetical **ideal ammeter** looks like a short circuit (ideally zero resistance).

Note that to read current with an ammeter, you have to break the circuit and insert the ammeter at the break as if it were a wire in the circuit. This can often be a nuisance, but that's how it goes. In some of the laboratory activities in this book, however, it is assumed that you have access to a specialized piece of equipment called a **current probe**. The current probe simply wraps around a wire and uses magnetic fields generated by the current flowing in the wire (Ampere's law) to sense the current going through it, then displaying it on an oscilloscope. There are basically two types of current probes, namely those based on the Hall effect, which can measure DC currents down to a few mA, and those based on sensing AC magnetic fields, which are not as sensitive or expensive, but are indispensable for working on power systems that pass high AC currents.

Unfortunately, the Hall probes are quite expensive and not very common. In the usual case when a current probe is not available, you will insert an ammeter into the circuit. Since inexpensive DMMs working as ammeters can cost less than twenty bucks, you have to ask yourself if that convenience is worth two grand. For a student, no. For a busy professional, perhaps yes. One more thing: current probes are not as sensitive as fairly inexpensive DMMs, so there is that to consider.

Another alternative for measuring current is by inserting a low-value resistor – low enough that it doesn't much affect the circuit performance – and measuring the voltage across it. Using $v = iR$, you can calculate the current. This is not always possible, but it can be handy. That is referred to as a **shunt resistor**. Several of the labs in this book rely on the use of shunt resistors to know the current.

2.8.2 Input Plugs and Jacks

Note that on the Fluke 45 DMM, there are separate inputs (upper left corner of Fig. 2.16) that are used to measure voltage and resistance, current up to 100 mA, and current up to 10 A. All tests are done with respect to the **com** lead. "Com" is short for "common," which, again, is another way of saying "ground" for the DMM. In the Fluke 45 DMM, the COM connection is *not earth ground*. That is to say, the meter is floating, as mentioned in Section 2.7. Since the meter is floating, you may touch any of the leads to any point in the circuit, regardless of where the circuit ground has been defined, even if the circuit ground is earth ground. Later you will see that this is NOT true of the oscilloscope because its ground is earth.

Unfortunately, the technical term that historically describes a socket is **female**, and the term that describes a plug is **male**. Connectors are still frequently described as the 'male' or 'female' end of the connector. I try to avoid the use of these truly inappropriate and anachronistic terms in this book, and there is a relatively recent effort to do away with them. Accepted terms include "plug" and "socket," "pin" and "receptacle," or "header" and "housing." See https://www.clynemedia.com/PAMA/InclusiveLanguageInitiative/PAMA_InclusiveLanguageInitiative.html.

The connectors in Fig. 2.16 accept **banana plugs** (bold font because this is its technical name!) that are inserted into the sockets. A **plug** is an electrical connector that has one or more prongs; a **jack** or **socket** accepts the plug, i.e., is a receptacle. Sometimes the word 'jack' is used to mean 'plug.'

These terms are often interchanged. Figure 2.18 shows a coax cable with banana plugs (yes, they look kinda like tiny bananas – cute?) that plug into banana jacks. The banana jacks in most bench equipment are spaced a standard distance apart to fit pairs of plugs, as shown in Fig. 2.18, in most bench equipment. On the dual banana plug connector shown, the small plastic tab visible on the left-hand plug indicates the prong that is connected to the outer shield of the cable (see discussion of coax cable below), which is often connected to earth ground at the instrument.

FIGURE 2.18 Banana plugs like those that connect to the Fluke 45 and Keithley 2231A-30-3 power supply. The tab seen on the left indicates the prong that is electrically connected to the outer shield of the coax cable, which is usually grounded.

2.8.3 Fuses

As long as we are talking about the current measurements, now is a good time to introduce the humble **fuse**. A **fuse** is a very small resistance placed in series with a circuit that you want to protect from destructively high current. A few small fuses are shown in Fig. 2.19. Because the resistance is small, it drops very little voltage and has negligible effect on the circuit being protected even while relatively large current flows. The fuse is designed to be the weakest link in the chain: it breaks first when large currents flow, "falling on its sword," so to speak. Fuses are rated in amperes. When the rating is exceeded, the fuse **fuses** (that is, melts) and 'fails open,' i.e., becomes an open circuit, preventing the imminent destruction of the circuit.

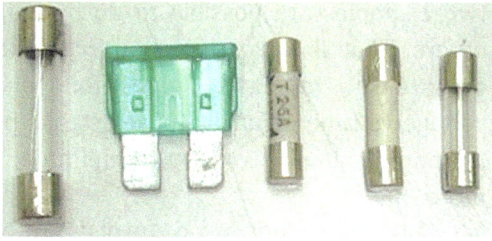

FIGURE 2.19 Examples of some small fuses. A fuse protects an electronic device or circuit from excessive current. The fuse will 'blow,' i.e., melt and create an open circuit, if too much current passes through a circuit. The fuse must then be replaced.

Fuses come in at least two varieties, including **fast-blow** (or **fast-acting**) and **slow-blow**. Fast-blow fuses open nearly instantly. These are the most common, but in some situations the circuit can tolerate some brief **overcurrent**. Slow-blow fuses are designed to tolerate some excess current for a longer time to allow the problem to resolve before failing open. Fuses are used, for example, in the inputs to the DMM and oscilloscope, the power cords from a wall outlet, inputs to the power supply that drives audio amplifiers, at the outputs their speakers, various circuits in motor vehicles (there are several fuses under the hood or dashboard of most cars and trucks), etc.

2.8.4 Resistance Measurements

The **ohmmeter** function of a DMM is used to measure the resistance between its two inputs. It does so by applying its own voltage, measuring the current, and deducing the resistance from their ratio according to Ohm's law. If you want to measure the *resistance* of part of a circuit, *the first thing to do is turn off the power to the circuit.* Otherwise, current from an energized circuit would flow through the meter, possibly causing damage.

When measuring the resistance of a resistor that is in a circuit, don't forget that the rest of the circuit is in parallel with it, which would likely cause errors in the measurement. Therefore, you must go to the trouble of disconnecting one end of the resistor so that current cannot flow out of the meter to the rest of the circuit.

Some ohmmeters offer a **continuity** test feature that is used to simply test if a connection is good or bad. In this setting, if the resistance is below a low value set by the meter, it beeps. This allows you to find short circuits without having to turn your head to read the display. Most DMMs include that feature. (It is quite handy.)

2.9 Waveform Generator

Imagine that you take hold of the output knob of the power supply and manually ramp the voltage up from 0 V to 10 V (the **amplitude**) and then back down to 0 V, and you repeat this cycle every 10 s. You would be generating a voltage varying periodically in time. It has a **period** of 10 s and **frequency** of 1/(10 s) or 0.1 Hz, where **Hz** is short for **hertz**, which is the unit of **cycles per second**. The relationship between frequency and period is

$$f = 1/T \tag{2.16}$$

Now change the period to 1 s so the frequency is 1 Hz... or do it at 1 kHz... or 100 kHz. Well, the first two examples are possible to do by hand, but not the others. That is what the **waveform generator (WG)**, or **function generator**, does for you. It creates a voltage-vs-time signal with selectable properties such as shape (called **waveform**), amplitude, frequency, and more.

Time-dependent waveforms are referred to as 'AC', which stands for "alternating current," as mentioned previously. There is some ambiguity in the term "alternating." "AC" can refer to any time-varying

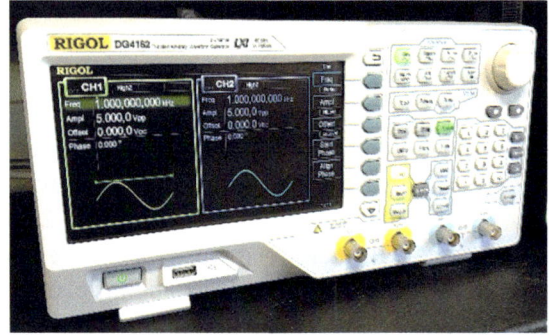

FIGURE 2.20 The Rigol DG4162 waveform generator (WG) that you will use in lab.

waveform, but often means simply sinusoidal. Let us define a voltage as $v(t) = A \sin(t)$, ignoring units; you know that $A \sin(t)$ is positive half the time and negative the other half, and is therefore clearly alternating. However, if we add a **DC offset voltage** of 3 V so that $v(t) = 3 + A \sin(t)$ V, then if $A = 1$, then $v(t)$ is always positive. Is it ever called 'alternating' even if the direction does not change sign? Yes, the term 'AC' *does* include time-varying waveforms that *do not change sign*, but you should be careful on a case-by-case basis to be clear about the context.

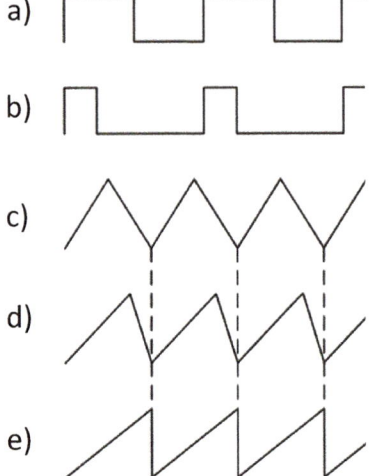

FIGURE 2.21 Examples of waveforms. (a) Symmetric square wave (i.e., duty cycle of 50%); (b) asymmetric square wave; (c) triangle wave; (d) sawtooth wave; and (e) ramp.

Figure 2.20 shows a Rigol model DG4162 WG. Most waveform generators offer at least three waveform options: **sine wave**, **square wave**, and **ramp**. The names describe the general shape of the voltage as a function of time. The sine wave function outputs a sinusoidal voltage $v_{out} = A \sin(2\pi f t)$, where A is the **amplitude** or **peak** voltage, v_p, and f is the frequency in Hz. (Since the argument for harmonic functions (sines and cosines) is an angle expressed in radians, you must multiply the frequency f, in cycles per second, by 2π radians/cycle to convert it to radians per second.) The total voltage from the greatest to the least is called the **peak-to-peak voltage, v_{pp}**. For sine waves, $v_{pp} = 2v_p$.

Another way of describing a voltage waveform is its **root-mean-squared** voltage, v_{RMS}, which is a kind of average. RMS voltage and current are discussed more in Chapter 3 in the context of average power carried by a waveform.

Square waves are shown in Figs. 2.21(a) and 2.21(b). The fraction of time that a square wave is high, expressed as a percentage, is called the **duty cycle**. For the special case when the square wave has equal time 'high' and 'low,' (Fig. 2.21(a)) it has a duty cycle of 50%. When those times are different (Fig. 2.21(b)), the duty cycle is some other value greater than 0% and less than 100%.

The ramp function on the panel accesses triangular waves. When the rise and fall times are equal (Fig. 2.21(c)), it is called a **triangle wave**. **Sawtooth waves** are triangular with unequal rise and fall times, as shown in Fig. 2.21(d). Furthermore, when either the rise or fall is instantaneous the waveform is called a **ramp** (Fig. 2.21(e)).

The frequency and amplitude of any waveform can be adjusted over a wide range. The Rigol waveform generator can produce waveforms with frequencies as high as 160 MHz. The DG4162 can perform other functions, such as **amplitude modulation**, which you will use in Chapter 6. It also provides the capability to define your own output waveforms, so it is also referred to as an **arbitrary waveform generator**.

2.10   Analog and Digital Oscilloscopes

Finally, we are at the most important single piece of equipment in the test arsenal – the **oscilloscope**. The oscilloscope has been around in its basic form since about 1930. It allows imaging of time-varying waveforms, such as those created by the function generator. It is often critically important to be able to visualize the time-dependent shape of a signal in order to diagnose circuit problems.

An oscilloscope is basically a graphing device – it creates a plot of voltage versus time for a given voltage input. Vertical and horizontal grid lines allow voltage to be read vertically and time horizontally. (The grid lines are technically called the **graticule**.) Large grid lines set about one centimeter apart, seen in Fig. 2.22, are called **divisions**; the **scale** might be set to, say, 2 V/div in the vertical and 1 ms/div in the horizontal direction. These larger divisions are usually subdivided by 5 **tick marks** for better precision.

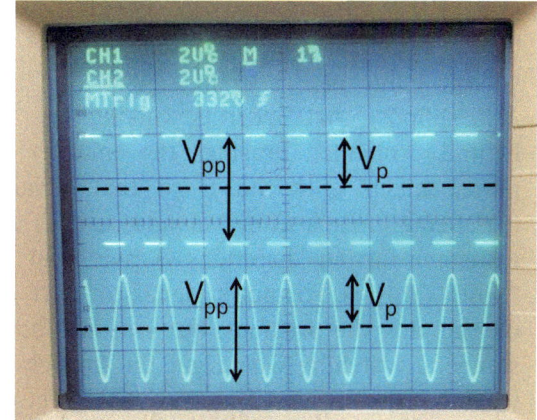

FIGURE 2.22 Sine and square waves displayed on two channels on an analog oscilloscope. The vertical scale on each channel is set to 2 V/div, so the waveforms have peak-to-peak voltage of about 5 V and peak voltages of about 2.5 V.

Although a DMM can provide a small bit of information about your voltage waveform, the 'scope' allows you to observe the shapes and details of the waveforms. It is important for you to understand the basic operation of an oscilloscope. In this section, you will learn how older-style analog scopes work, and then we will move to the newer digital scopes.

Although it is a less-sophisticated, older scope technology, we will start with how **analog oscilloscopes** work. Analog devices process signals directly without storing them as digital bits, so that the waveform on an analog oscilloscope is continuous. Figure 2.22 shows a sine wave (bottom on the screen) and a square wave (top) on an analog scope. The **trace** of the waveform on the screen is formed by an electron beam that strikes the inside surface. The inside of the screen has a **phosphor** coating that glows where the electron beam lands. The **persistence** of the phosphor is very short, meaning that it glows for only a very short time as the beam moves along, so it basically disappears immediately after it starts to glow. It is worth pointing out that the window can hold information for only a limited, and often very short, slice of time. If the trace were instead made (more slowly) on a piece of paper, it could go on for hundreds of feet, and a long record of the waveform could be analyzed. Although that technology used to be common in the form of **strip-chart recorders**, it isn't practical for an oscilloscope, so the beam must trace over itself inside of the relatively small screen. Therein lies much of the technology of how an oscilloscope operates and is discussed below.

Here is an analogy of how an analog oscilloscope works. Suppose that Yolanda wants to trace the shape of a triangle wave on dry-erase board as it is being generated. The image of the waveform is called the trace. She starts at the left edge of a board and walks along it while drawing with a marker. Her lab partner, Valerie, shouts out instantaneous voltage values of the waveform while Yolanda is walking, tracing out the shape as her arm moves up and down to record the position of the instantaneous voltage value of the triangle wave. Also suppose that Pedro is walking behind Yolanda, erasing the marker as she moves along. In effect, the viewer sees pretty much only the trace that Yolanda just finished, with no marker behind her, since it is being erased. After Yolanda reaches the end of the board, she leaps back to the beginning and waits for further directions. Travis waits until the triangular waveform exactly repeats and yells "Now!" This is Yolanda's signal to start drawing the waveform again at the same speed, exactly overlapping the previous, but erased, trace. If the team did this repeatedly and at superhero speed, someone watching would get the impression that the board had on it a fixed trace of the triangle wave because it was constantly being refreshed at a speed that is faster than the persistence of the human eye.

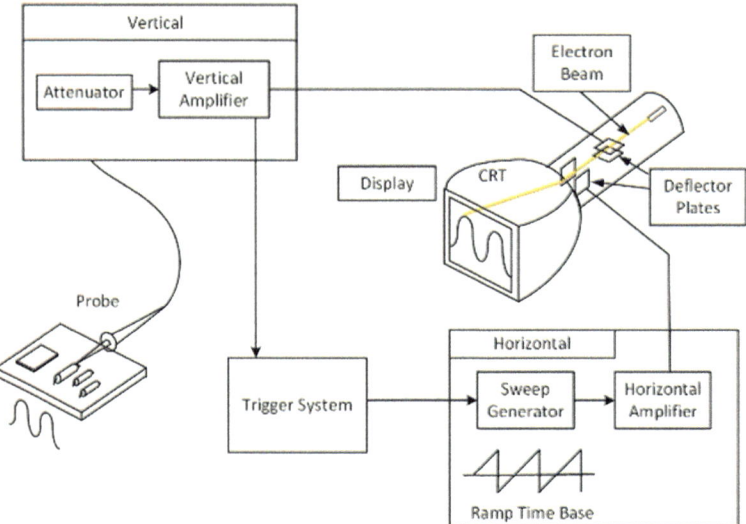

FIGURE 2.23 Trigger and sweep system in an analog oscilloscope.

That pretty much describes how an analog scope works. In this analogy, the whiteboard is the glowing screen (**cathode ray tube**, or **CRT**) (Fig. 2.23). Yolanda's marker is the narrow beam of electrons that performs the y-axis trace, and Valerie is the **vertical amplifier**, or voltmeter, that tells Yolanda how high to direct the beam that Valerie is reading from the **probe**, which is attached to the circuit. Furthermore, Yolanda's feet walking along the whiteboard are the ramp signal coming from the **horizontal amplifier**. The speed at which the beam moves across the screen, analogous to the speed that Yolanda walks along the board, is set by the **sweep generator**, or **time base**.

By erasing the whiteboard as it is being drawn, Pedro's erasing the board represents the short persistence of the screen. Travis yells "Now!" to function as a circuit called the **trigger**, which causes the trace to begin moving from left to right each time at exactly the same point in the waveform. Without Travis doing the triggering, the scope trace would look like a tangled jumble of overlapping lines and would not be of much use. Displaying a signal without a trigger is called **free running**.

Analog scopes have a wide range of controls to adjust the vertical voltage range displayed, the total amount of time that it takes for the beam to sweep across the screen, the value of the voltage and slope that it triggers at, called the **trigger level** and **rising** or **falling edge** (exactly when Travis yells "Now!"), the type of input (AC or DC), time delays, multiple inputs for viewing several signals at a time, image brightness, screen grid brightness, and much, much more.

Figure 2.22 showed two waveforms from two inputs, also called two **channels**, where the height of each waveform can be independently selected, and the trace can be placed anywhere on the screen that you like using **vertical and horizontal offset** controls. In CRT-based scopes, there is typically only one beam, and therefore one time base. Dual signals input to two scope channels can be displayed in one of two modes. In one, the **Alternate** mode (or sometimes **Dual** mode), the scope displays one channel at a time during each horizontal trace and switches between the two channels every other trace. In **Chop** mode the vertical amplifier switches (chops) between inputs every few milliseconds to display both traces seemingly at the same time.

One more word on analog oscilloscopes. The most sophisticated of all analog scopes were **storage** oscilloscopes. They could not actually save the measurement data, but they could keep the trace glowing on the screen for up to a few minutes, allowing the engineer to inspect it after the waveform had stopped running. These were among the most expensive scopes of their day, but they were quite crude in comparison to modern digital scopes, discussed later.

2.10.1 Oscilloscope Probes

You should think of an oscilloscope as a fancy voltmeter. As such, it is used in the circuit the same way as a voltmeter, i.e., connected to the two points of the circuit across which we wish to know the voltage drop. Therefore, we must be concerned about its input resistance, just as we were concerned about the input resistance of a voltmeter. Recall that adding an oscilloscope to the circuit could change the circuit properties, and thus give us incorrect readings if the scope's input resistance is on the same order as, or lower than, the resistance of the circuit.

The input resistance of standard oscilloscopes is 1 MΩ. (Certain high-frequency scopes use a 50 Ω input. Note that many digital scopes can switch between 1 MΩ and 50 Ω.) The 1 MΩ input is sufficiently high to present little change to most circuits under test. The probe (Fig. 2.24) uses a **BNC** plug (Fig. 2.25) to attach to the scope's input, which is a **BNC** socket. A BNC connector is always attached to a **coaxial cable**, or 'coax,' which carries the signal on an inner conductor that is surrounded by an outer, grounded **shield** conductor. The purpose of the shield is to prevent **electrical interference** (for example, from the overhead lights or a nearby fan motor) from going through the cable and interfering with the measurement.

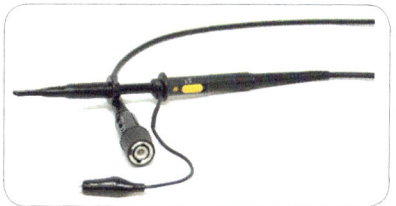
FIGURE 2.24 Oscilloscope probe used to measure waveforms in circuits. Modern probes mostly use BNC connectors to attach to the oscilloscope. The narrow end has a small, retractable clip that attaches to component leads. The attached wire is an alligator clip that goes to ground in the oscilloscope.

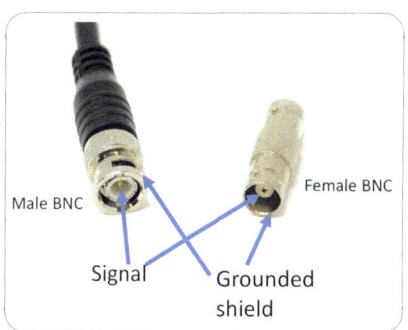
FIGURE 2.25 BNC connectors are used to connect probes and signals to the oscilloscope and the function generator.

FIGURE 2.26 Coaxial cable with a BNC connector on one end and miniclips (top) or alligator clips (bottom) on the other. Miniclips are also called EZ-clips.

Any type of probe that has no appreciable resistance is called a **1×** probe. In this case, the load on the circuit is the 1 MΩ of the oscilloscope. If this is not a sufficiently high load, i.e., there is not enough resistance to prevent excessive current from flowing out of the circuit, then the user can select a **10×** probe. A 10× probe adds an additional 9 MΩ, causing the probe plus scope to 'look like' a total of 10 MΩ to the circuit. In this case, the voltage read by the oscilloscope is decreased by 10 times due to the voltage divider effect (1 MΩ of the scope in series with 9 MΩ of the probe). Voltage measurements on some older scopes must be mentally multiplied by 10 to obtain the correctly measured values. In many newer scopes, there is a "10×" setting that automatically compensates for the extra 9 MΩ and displays the correct values. Nowadays, if the probes are designed for it, some scopes automatically detect the 10× probe and present the correct values on the screen.

One need not use a conventional scope probe like that shown in Fig. 2.24 to make a measurement. The 1 MΩ input is presented by the scope and not the probe itself. Even a simple wire that has a BNC connector on the end could be used, but it might also pick up some interference from the environment and suffer from other problems. A coaxial cable with either **miniclips** or **alligator clips** on the ends (Fig. 2.26) is often used as a probe, but it might have relatively high inductance, which could distort the waveform. The cleanest signal with the least distortion is obtained using a conventional scope probe that is shielded right up to the tip, especially at high frequencies.

It is very important to note that scope probes incorporate both signal and earth ground leads. The signal lead goes through the center of the coax cable to the center pin on the scope BNC socket. The ground lead surrounds the center lead on the coax cable and goes to the outer ring on the scope BNC socket. The outer ring makes direct contact to the earth-grounded metal on the scope, and hence, by connecting through ground wires in the wall, to every other piece of earth-grounded electronic test equipment. *You may touch the signal lead to any part of the circuit to measure its voltage, but the earth-grounded lead must only go to earth ground in the circuit if it has one.* If there is not earth ground in the circuit, then your probe's earth ground **defines** earth ground in the circuit. By 'define' we mean that a floating circuit becomes referenced to earth ground by touching the earth ground lead to any point of that circuit, just as we discussed placing one end of the meter stick on the ground. That point on the circuit is now at earth ground, and the circuit is no longer floating.

Let's try to prevent some problems before they start. **If you were, wrongly, to use a second scope channel, or a grounded waveform generator, or some other source of earth ground, and connect two earth grounds to separate points on the circuit, you would be shorting those two points through the ground leads.** This could cause incorrect measurements that may be difficult to recognize since the error is in the probe placement and not the circuit itself. *So, be careful not to place earth ground at more than one point of a circuit. This might happen whenever you use an oscilloscope, so you need to understand it.* You will get lots of practice doing this correctly in the Chapter 3 lab.

2.10.2   Digital Oscilloscopes

Fast forward to, like, 1980. Around that time, **digital oscilloscopes** were introduced. Digital oscilloscopes are basically computers whose job is to input the measured values, and process and display them on the screen very much as if it were an analog oscilloscope. A circuit called an **analog-to-digital converter** (**ADC**) is used at the inputs to convert the analog signal to digital numbers comprising 1s and 0s (bits) that can be understood by the oscilloscope. (Analog-to-digital conversion is discussed in Chapter 9.)

Digital oscilloscopes are all storage oscilloscopes (because like computers, they operate on stored bits), and so are also called **digital storage oscilloscopes** (**DSO**). Digital scopes do everything that analog scopes do, and a whole lot more. The input voltages can be displayed as a graph of voltage versus time on a screen on the unit or on a computer. The trace is now just the display of the digital storage of voltage data for a set time interval. Nowadays, almost all new scopes are digital, but you might still run across analog scopes in a lab setting.

Figure 2.27 shows the LeCroy HDO4104 DSO used for all of the lab activities in this book. A DSO does not process the analog voltages that appear at the inputs, but rather works only on the digital representations of the voltages after they have been converted to digital 1s and 0s by the ADC. Every voltage measured is called a **sample**. The sampled data are stored in memory, and therefore are not temporary, as in an analog scope. Once the data are

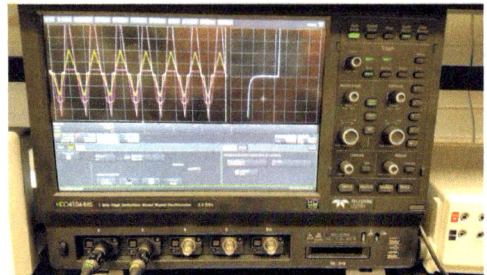

FIGURE 2.27 Digital storage scopes convert analog signals to digitally stored values.

stored, they can be displayed in any desired format, and even more importantly, can be processed. For example, you can display the data directly, or you could display, for example, the logarithm of the data. Also, it is much easier on a digital scope to acquire and analyze fast events that happen only a single time, and then display the waveform on the screen and manipulate the data. You can perform other math functions as well, such as adding or subtracting two channels after the data has been acquired, or you can define any math function you need. Furthermore, you can export the data to a computer for further analysis and reporting.

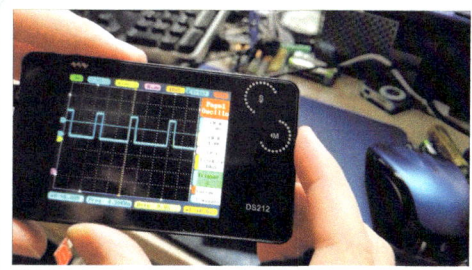

You can download free oscilloscope apps for your phone, but to my knowledge they don't accept electrical inputs – only audio signals from the microphone. You should download one and play around with singing, humming, and whistling different frequencies, as well as using a musical instrument. There is a lot to learn from that simple experience. At some point, check out pocket-sized oscilloscopes. One of these makes a great complement to a handheld DMM. They are often less than $50 including probe.

The capabilities of any scope, analog or digital, are partially expressed in terms of the **bandwidth** of the vertical (voltage) input circuitry. Bandwidth pertains to a range of frequencies, which for an oscilloscope includes DC, or 0 Hz. For example, if you want to view a radio signal that has a frequency of 100 MHz, then the electronics within the oscilloscope must be able to react to at least that frequency, or the signal cannot be acquired, so the bandwidth must extend beyond 100 MHz.

Analog traces are smooth with no steps, whereas the traces on digital scopes are sequences of steps, like staircases. Therefore, the metrics that we use to describe the DSO are different from those of the analog scope. You can imagine that the voltage samples must be fairly close in time for the DSO to display a correct waveform. So, we must know a DSO's **sample memory depth** and the **sample rate**.

The HDO4104 has a bandwidth of 1 GHz (gigahertz, or billion oscillations per second), a memory depth of 12.5 million samples per channel, and a sample rate of 2.5 GS/s (giga or billion samples per second), which is many samples per cycle at the lower frequencies, and more than two samples per cycle at the very highest frequency of 1 GHz. As we will see in Chapter 9, this is slightly greater than the minimum theoretically required sample rate in order to reproduce the signal with minimal risk of losing signal information.

As noted earlier, the point of the digitization process is that the voltage at each moment of the waveform is stored as a number that can be processed by the computer. Thus, functions become possible that are not available on analog scopes. One of the most important of these is the **fast Fourier transform (FFT)**. This is an algorithm that takes samples of a waveform over a window of time and computes the frequency content of the sampled waveform segment – i.e., it converts the waveform segment into the sum of many sines and cosines of different frequencies and computes 'how much' of each frequency is present. The simplest example would be a sine wave at a particular frequency – only that one frequency is present – but other signals will reveal a richer collection of frequencies. The discussion of functions in terms of frequency components is said to be in the **frequency domain**. The FFT allows you to observe its frequency components directly. Dedicated instruments that do this are called **frequency spectrum analyzers**, and they are generally more capable than just using the FFT on a scope. Frequency domain is the subject of Chapter 4, and is extremely important not only to electrical engineering, but to science and engineering in general.

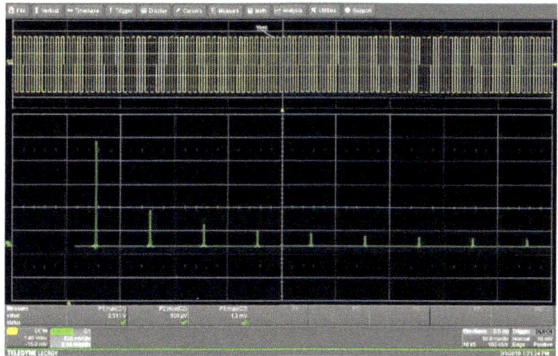

FIGURE 2.28 Frequency spectrum of a square wave having 50% duty cycle.

Let's take just a moment now for a brief introduction to the frequency domain. You know from your past trig classes that a sinusoidal wave is characterized by its amplitude, frequency, and phase, as in

$$v(t) = A\sin(\omega t + \phi), \tag{2.17}$$

where A is the **amplitude**, ω is the angular frequency, and ϕ is its phase. The **frequency spectrum** is a plot of how much of each frequency is present. For a sine wave, there is only one frequency present, namely its frequency that you selected from the waveform generator. Let's say that's 1 kHz (1000 cycles every second). The frequency spectrum would consist of a single peak at 1 kHz and no other peaks anywhere else. The larger the amplitude of the sine wave, the higher the peak. More complicated waveforms have more frequency components, i.e., more peaks in the spectrum (Fig. 2.28). In this lab, you will practice using basic frequency spectrum analyzer functions in preparation for later labs where you will do much more with it.

2.11   Equipment Tutorials

This section is not just a collection of information about how to operate various pieces of electronics test equipment. Embedded throughout this section are important discussions about how they operate, which are based on simple electrical principles as discussed earlier in this chapter. It is highly recommended that you pay close attention to this section even though you may not be using the very same pieces of equipment in your course.

2.11.1   Keithley DC Power Supply

This section is not just a collection of information about how to operate various pieces of electronics test equipment. Embedded throughout this section are important discussions about how they operate, which are based on simple electrical principles as discussed earlier in this chapter. It is highly recommended that you pay close attention to this section even though you may not be using the very same pieces of equipment in your course.

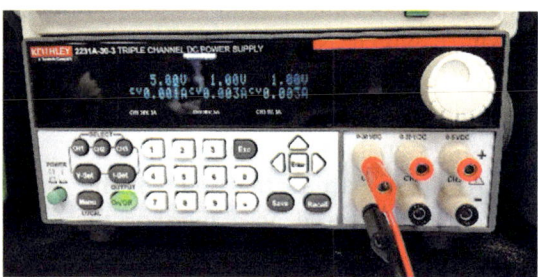

FIGURE 2.29 The Keithley 2231A-30-3 DC power supply you will use in lab.

Figure 2.29 shows the Keithley 2231A-30-3 Triple Channel DC power supply. It is a regulated supply, of course. The top row of digits shows the set and output voltage on each of the channels, which are labeled CH1, CH2 and CH3. The set voltage is displayed *only* while you are adjusting it and then reverts to zero.

The lower row of digits shows the maximum allowed and actual output currents. When the **Output** light is off, the outputs are all zero. When **Output** is pushed, the outputs will go to their set values. Note: You may not remember nor know the output voltages until you press that button, and by then you might damage your circuit. Pressing the **V-Set** shows you the set voltage for each of the three channels on the top row. Pressing the **I-Set** shows on the lower row the maximum current that can be produced by the outputs. That maximum current selected is called the **current compliance limit**. The Keithley power supply can provide up to 3 A on each channel.

Here is how you set up the supply for a selected voltage and compliance current. Press **V-Set** to choose the output voltages. Selecting CH1 to CH3 selects the voltage on that channel – the voltage value to be changed flashes. The desired voltage is entered using the numeric keypad, or it can be set using the cursor buttons around the **Enter** button. Choosing a numeric value and pressing **Enter** locks in that value. CH1 and CH2 can be set from 0 V to 30 V. CH3 has a maximum of 5 V.

Selecting **I-Set** and a channel causes the lower number to flash. Now you can set the current compliance limit. The output will not exceed an output voltage that causes the maximum set current to be exceeded. Because it is a regulated power supply, you can be sure that the set voltage is maintained for any load resistance that requires less than the compliance limit.

Be sure to choose a current compliance that is reasonable for what you expect you will need for your application. This prevents excessive current from possibly causing damage in the event of an accidental slip of the probe, miswiring, or miscalculation. A good place to start (and this is only an example) is by knowing the power rating of your resistors and choosing a compliance that will not exceed that rating. Only resistors in series with the power supply will get the full current, and it might be tedious to calculate every resistor's maximum rated current. So, you will have to use your judgment setting the compliance limit, but certainly the 3 A available is much larger than anything you will need throughout this course. A typical compliance limit used in the lab activities would be around 100 mA. Start on the lower end, and if you need more current to run your experiments you can increase it. If you leave it at the maximum setting, you are taking your chances of making any mistakes, and that is not advisable.

2.11.2 | Waveform Generator

Figure 2.20 shows the Rigol DG4162 waveform generator (WG). As discussed above, a WG produces voltage waveforms, most commonly as sine, square, and triangle waves. The WG has all the same source properties discussed in Section 2.6.1, but the output is AC instead of DC. Figure 2.30 is the simple equivalent circuit, or model, of what you can think of as inside the WG box, as discussed in Section 2.6.1. As already presented, you can reduce the guts to an ideal WG having a voltage v_{int} in series with a resistor R_{int}. A more complicated form of resistance is its AC form, called **impedance**, represented by Z rather than R. Impedance is the subject of later chapters. You will often see 'output resistance' written as 'output impedance.' For now, if you see the word 'impedance,' just think 'resistance' and you will be fine.

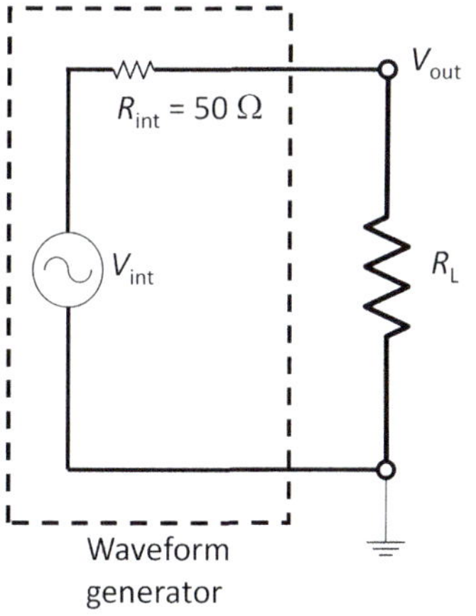

FIGURE 2.30 Equivalent circuit model for a waveform generator having internal, or output, resistance of 50 Ω.

Basically, a real voltage source always has some internal resistance (again, which we often call 'output resistance') in series with its ideal, internal voltage source. The output resistance limits the maximum amount of current that can be delivered to the load, and also causes v_{out} to change with the load resistance due to the voltage divider effect. The following discussion can all be reduced to saying simply that the WG is not a regulated voltage source. You must understand that the actual output voltage is not monitored to assure that $v_{out} = v_{set}$ – you can set the value of the internal voltage source, which may be the *desired* amplitude of the WG, but it is not regulated; if you try to drive too much current by using a *large* load (*small* resistance), the actual voltage at the output will decrease, or 'sag,' or be 'loaded down' due to the output resistance of the WG. Typically, you select a voltage that you want to appear at the output. Let's assume that v_{int} is the value displayed on the screen of the WG, which we call here v_{set}. So, for now, v_{int} provides a voltage at the selected voltage v_{set}. The actual voltage measured across a load is called v_{out}. Due to all the reasons discussed in Section 2.6.1, v_{out} is generally not going to be exactly the desired voltage, v_{set}. Rather, v_{set} is only the voltage that you set it to – what you *want* to get from it; v_{out} is less than v_{set}, and often considerably so.

Why is that? Most WGs have an output resistance of 50 Ω, as shown in Fig. 2.30. (The choice of 50 Ω is a standard for communications test equipment.) In order for v_{out} across the load to be close to your desired v_{int} (i.e., not sag), according to the voltage divider relationship, the load resistance must be much (at least 10×) larger than the R_{int} = 50 Ω. For example, if R_L = 500 Ω, and v_{int} is set to 10 V, then $v_{out} = v_{int} \times \dfrac{R_L}{R_{int} + R_L} = 10 \times \dfrac{500}{550} = 9.091\text{V}$. This may or may not be acceptable for your purposes.

The Rigol DG4162 WG has two modes that you should be aware of lest you make a mistake: "**HighZ**" and "**50 Ω**." (Again, the "**Z**" refers to an impedance, which for you at this time is resistance.) Under the **Utility** menu, you can choose **CH1Set**, then **Resi**, and then **HighZ** or **50 Ω**. Consider the three terms that represent voltages: v_{set} – the voltage that you see on the display that you want to have at the output; v_{int}, the actual voltage of the internal voltage source; and v_{out} – the actual voltage measured at the output with the load attached. In **HighZ** mode, the display shows you $v_{set} = v_{int}$. In this setting, the WG assumes that you will have such a large R_L that the v_{out} will be very close to v_{int}. However, if R_L is not very large, as said above, v_{out} will be less than $v_{int} = v_{set}$ because of the voltage divider. So, in **HighZ** mode, you better have a relatively large load resistance if you want to measure the voltage that you chose with v_{set}.

Now, what is the **50 Ω** mode? You need to know that many devices in the field of electronic communications are effectively 50 Ω loads. In that case, the load driven by the WG is 50 Ω, or R_L = 50 Ω; the voltage divider equation tells you that $v_{out} = v_{int}/2$. (This is the same as the example used in Section 2.6.1) In the **50 Ω** mode, the WG *assumes* that you are using R_L = 50 Ω to save you from having to (shudder) divide by 2; v_{set} *shows* $v_{int}/2$ on the display in order to precompensate for using a load resistance of 50 Ω, i.e., so that the actual v_{out} will be the same as the voltage shown on the display. If you have the WG set to the **50 Ω** mode by mistake, your output voltages will be larger by nearly 2× than what you intended. The moral of this tale: If something seems wrong, first make sure you have the WG set to **HighZ** mode.

What if you need to drive a larger load (less resistance) than, say, 500 Ω, or you need more current than the WG can provide? In some of our lab activities you will use a Rigol PA1011 power amplifier between the WG and the original load. Figure 2.31 shows how adding the power amplifier largely solves the output resistance problem. The output of the WG goes to the input of the PA1011 and original load goes at the output of the PA1011. The PA1011 takes as its input the output

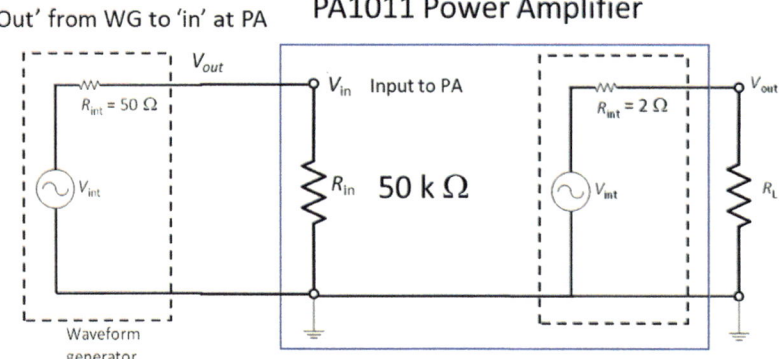

PA1011 Power Amplifier

FIGURE 2.31 Circuit diagram showing models of the WG and power amplifier. The WG "sees" an input resistance of 50 kΩ at the input to the PA, so the output of the WG is barely loaded down. Because the PA has a low output resistance of 2 Ω, it can drive a much lower load resistance without sagging.

from the WG and then outputs to the load the same voltage as its input, but with the ability to source more current. (In this case of $v_{out} = v_{in}$, we say that its **gain** is 1, where gain pertains to the voltage change when passing from input to output of a device or circuit.) The PA1011 has a relatively high **input resistance** (50 kΩ) (the effective load presented to the WG when connected to the PA1011). Therefore, it does not load down the WG because 50 kΩ is much larger than the WG's R_{int} = 50 Ω (again, think voltage divider). In turn, the amplifier's output resistance is only 2 Ω and therefore can source more current into a larger load (lower resistance) before being loaded down or sagging. The PA1011 isn't regulated, but it is much better than using the WG to directly drive a large load.

Heed these words: A common mistake in troubleshooting when something goes wrong is that you may forget to turn on the output. On the Rigol DG4162 WG, the **Output1** button must be lit. This happens more frequently than you might think; expect to make that mistake several times throughout the course.

The **Sync** BNC output on the front panel of the WG allows easier scope triggering. Sometimes you have a noisy input to the scope or a small voltage that is difficult to trigger on. Usually, the signal of interest is periodic with the same frequency as the output of the WG. The **Sync** output provides a clean square wave whose period corresponds to the output waveform. This external sync can be used as input to the scope's **Ext** trigger input. Choosing to trigger off of the **Ext** input of the scope can solve many triggering problems.

Parameters such as frequency, amplitude, voltage offset and more can be set by pressing digits or by using either the left and right buttons under the knob in the upper right corner to select a decimal place on the screen's value, and then rotating the knob. For example, to set the frequency to 2 kHz you can press the arrow buttons to select the 100 Hz position and rotate the knob. Or you can press '2' and then choose the kHz unit. The first method is good for manually sweeping a frequency, and the second is faster for just entering a frequency.

2.11.3 Digital Multimeter

A digital multimeter, or DMM, can perform a variety of measurements. The basic ones are resistance, AC and DC current and voltage; the most common additional functions are measurements of frequency, capacitance, diode and transistor properties (discussed in Chapter 7), and frequency. Here are a few things to know about DMMs in general, and the one you will use in particular, that will help you get the most out of them.

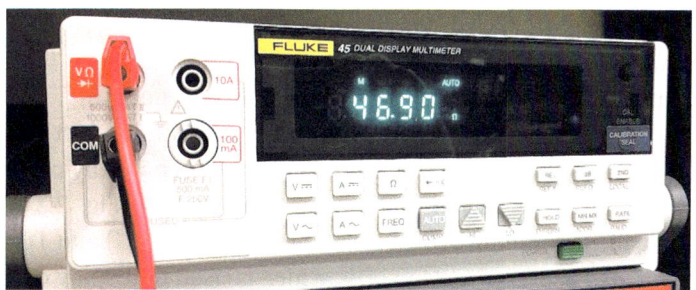

FIGURE 2.32 Fluke 45 digital multimeter.

The Fluke 45 Dual Display Multimeter has four banana-plug inputs, visible on the left in Fig. 2.32. The black **COM** socket is *not* connected to earth ground, but rather is floating. That implies you can put it anywhere in the circuit without worrying about shorting to earth ground. **COM** is used with all of the three inputs. The upper left **VΩ** socket is used in measuring volts and ohms and running a diode test. (Diodes are semiconductor devices, and are discussed in Chapter 7.) The other two sockets are for measuring current in two different ranges, namely 100 mA maximum and 10 A maximum. Voltage and current can be measured in either DC or AC mode, denoted by the buttons showing a straight line over a dashed line or the ∼ wave symbol, respectively. Resistance is only DC and is selected by the Ω button. **Auto** automatically selects the **range** to display. The range refers to the placement of the decimal point within the displayed digits, and the units (e.g., Ω or kΩ). When used as a voltmeter, the input resistance of the Fluke 45 is 10 MΩ. See the next section for a discussion of input resistance.

If you need to measure very small resistances, less than 10 Ω or so, you may wish to subtract out the wire and contact resistances to improve accuracy. By shorting the leads without the **device under test (DUT)** (must I actually define what that is?), you will likely measure some small resistance. Pressing the **REL** button **re-zeros** the meter, i.e., resets the display to 0 Ω, so that only the DUT resistance is measured without the contribution of the leads.

One button has symbols of both a diode and sound waves. The sound waves indicate that when a low resistance is measured, it beeps. This is useful for testing for **continuity**, i.e., a good electrical connection, without looking up from your work. In this mode, it also tests diodes for functionality and for turn-on voltage. (See Chapter 7 if you want to understand this part.) By swapping the test leads across a diode in one direction and then the other, you can determine (a) whether the diode is working correctly, (b) which ends are anode and cathode, and (c) how many volts it takes to make them conduct. When the positive, red, lead is attached to the anode, it will indicate a number, typically between about 0.7 V and 2 V, which is the turn-on voltage. When the black lead connects to the anode it simply reads **OL**, or "overload," indicating a resistance that is higher than can be measured.

2.11.4 Oscilloscope, Probes, and Input Resistance

The LeCroy oscilloscope runs beneath the hood under the Windows environment. It looks like an oscilloscope, but it is actually a Windows computer with the associated oscilloscope hardware built in. The front panel, Fig. 2.33, contains both a touch-screen display and control buttons along the right-hand side. Either of these interfaces can be used to input parameters to configure how the input is analyzed. Some of the buttons are straightforward. For example, the **Power** button turns the scope on, but in order to turn the scope off, it is preferable to go

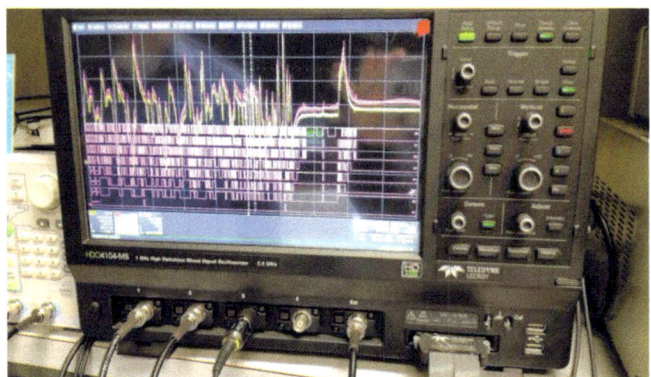

FIGURE 2.33 LeCroy HDO4104-MS.

to the **File** menu from the touch screen and select **Power Off** to ensure that the software closes properly. The touch screen can be turned on or off by pressing the **Touch Screen** button.

There are various types of scope probes (Fig. 2.34) that may be connected to a channel input. The most common are 1× (pronounced "one – ex") and 10× ("ten – ex") passive probes. In this course you will also use a **current probe**, which is used to display current rather than voltage on the oscilloscope.

The oscilloscope presents to the circuit under test, i.e., has an input resistance of, or "looks like," 1 MΩ. (See text box.) This means that putting the scope into the circuit is like inserting an extra, unwanted 1 MΩ resistor in parallel with your circuit. If you are analyzing the voltage between two points of, or looking somewhere into, a circuit that has an equivalent (or Thevenin) resistance of, say, 100 Ω, then this would not present a problem. However, if the circuit has a resistance between the two test points of, say, 300 kΩ, then the new 1 MΩ component (meaning the scope) would cause appreciable current to flow out of the circuit and into the scope, changing the voltage across those two terminals, and therefore the measurement done by the scope would not be very accurate. A 1× probe,

> *Thinking like an EE:* What does "sees" or "looks like" the 1 MΩ of the oscilloscope mean? This phrasing implies a lot about how circuits can be understood. Recall that circuits are described in terms of current and voltage. What does it mean to 'see' a resistance? It means that it is enough to know that the circuit can be represented, or modeled, by the current/voltage behavior of such a resistor. In short, we mean that if you imagine that the entire oscilloscope with all of its complexity behaves as if it were a single 1 MΩ resistor, then we don't have to worry about any of that. In a circuits course, this would be called the equivalent resistance of the circuit.

which is either a conventional 1× scope probe or a bare wire or a coax cable (shown in Fig. 2.34(a)), will add essentially no extra resistance to the input resistance of the scope, and the scope will look like a total input resistance of 1 MΩ. One variety of dedicated oscilloscope probes like that shown in Fig. 2.34(b) are 1× probes. They are better than a simple wire or cable in that their capacitance can be adjusted to match the measurement conditions across a wide range of frequencies, and they are better at shielding external electrical interference.

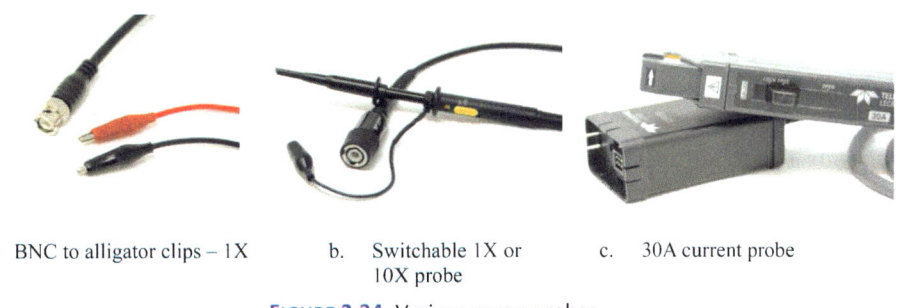

a. BNC to alligator clips – 1X   b. Switchable 1X or   c. 30A current probe
                                       10X probe

FIGURE 2.34 Various scope probes.

In order to get higher resistance looking into the scope, we can substitute a different probe, namely a 10× probe (also Fig. 2.34(b)), which presents a total (probe resistance plus scope resistance) of 10 MΩ to the circuit. (You often cannot tell a 1× probe from a 10× probe without reading the fine print on the probe itself. Some, like that shown in Fig. 2.34(b), are switchable between the two.) The 10× probe-plus-scope resistance is 10 MΩ, and hence will tap off ten times less current from the circuit, but will **attenuate** (i.e., decrease) the signal to the scope by a factor of 10, meaning that the voltage signal at the scope input will be 10× smaller. (It also improves the frequency characteristics, but that is not the point here.) Therefore, the signal into the probe is divided by ten at the scope, so without any corrections, the scope display will show the signal to be ten times smaller than the actual signal. Using a 10× probe with basic scopes often requires that the user do the multiplication by 10 and not forget that the higher-attenuation probe was used. Many modern scopes, including the HDO4104 and its matching 10× probes, adjust the display automatically to show the correct signal voltage without user intervention.

All scope probes have a BNC plug (Fig. 2.35) that matches with the BNC sockets on the scope. To connect, push it on and twist it clockwise. You should feel a slight click when it locks on. The HDO4104 is a four-channel oscilloscope, so four probes on four channels can be used simultaneously. There is a tiny adjustment screw at the BNC connector on many scope probes that is used to adjust the capacitance to match the scope input characteristics. It is used by inputting a square wave and adjusting the screw with a small screwdriver to make the square wave look flat on the top of the waveform. Most scopes provide a convenient square wave for this purpose, which in this case is a tang marked **CAL** found at the lower right corner of the front panel of the scope.

FIGURE 2.35 BNC connectors. Shown on the foreground side of these connectors are the plug or pin side found on the scope probe and the socket or receptacle side found on the scope input. See Figs. 2.33 and 2.34 for their use on oscilloscopes. The left side here shows an adapter from an RCA connector, common in audio applications, to the BNC plug; on the right is an adaptor from BNC plug to another BNC plug.

The signal end of the probe hooks to your circuit. Both the simple 1× cable and the probe have a black alligator clip, which is your common or ground point. All of the metal parts on the outside of the scope (except the **CAL** tang) connect to the scope **chassis**, which is earth ground. The chassis of an electrical system is the metal frame inside the outer plastic case. The chassis supports all of the hardware and is typically its local common ground. The black clip on the probe is attached to the shield (outer ring metal) of the BNC connector to the chassis and then to the ground wire of the power cord and finally to earth ground through the wall outlet to the outside of the building. Attaching the black lead to any part of your circuit makes a short circuit directly to earth ground, so be sure to know where in your circuit you intend ground to be. For example, if you use a scope probe to do a simple measurement of a sine wave in a circuit whose input comes from the WG (whose BNC shields also go to earth ground), you must attach the black probe lead to the grounded lead from the WG, or alternatively, use only one ground connection for the entire circuit.

Most oscilloscope probes provide a choice of connections to your circuit – a sharp point and a **miniclip**. The point is used to reach tiny test points, and the miniclip is used to attach to small wires, freeing your hands for other work. The sharp tip is permanently connected to the signal line of the coax probe cable. The miniclip is an attachment that slides over and covers the sharp tip. You can see the miniclip at the left end of the probe in Fig. 2.34(b). The miniclip hook is exposed by pulling back on the hood covering the miniclip. The miniclip attachment just slides on and off the end of the probe over the tip, but you need to be careful to put it on straight, or you could break the tip off the probe.

The **active current probe** (Fig. 2.34(c)) measures current and converts the current into a voltage that the scope recognizes. The current in the wire causes a surrounding magnetic field according to Ampere's Law. That magnetic field induces a voltage by the **Hall effect** in a special device called a **Hall probe**. (You may learn about the Hall effect in a semiconductors course.) The Hall probe easily measures both AC and DC currents without breaking the circuit or adding a shunt resistor. The current probe is different from the other probes. Be especially careful when using this probe as it costs upwards of $3000. To attach it, slide the sensor jaws open, hook them around the wire whose current you are measuring, and then slide the probe closed. Push it firmly to lock. Be careful not to force anything, and to treat it delicately. The LeCroy oscilloscope will automatically recognize it and display the current directly. Note that the sign of current measured will correspond to the direction of the arrow (positive) on the probe clamp.

2.11.4.1 Using the Front Panel

Once you have a probe or two (or three or four) hooked up, you will need to set up the channels. On the right side of the panel (Fig. 2.36) are the channel selection buttons. Select a channel 1–4 to turn on or off the trace for that channel. It is a **toggle** switch: the first press turns the trace on for that channel and the second press turns it off. The button is lit when the channel is on.

The **Vertical Scale** knob changes the voltage scale for the *active* channel. Each channel has its own vertical scale setting. The vertical scale is expressed in volts per division (V/div) (except when using the current probe). The screen of an oscilloscope is divided into vertical and horizontal divisions, and you can see a grid pattern on the screen with major and minor gridlines. For example, if your setting is at 1 V/div, this means that from one major gridline to the next is 1 volt. Each of the five minor gridlines in between would represent 0.2 volts. You have several choices of displaying the major and minor axes as lines or **tick marks**, which are small marks that replace the lines and make a less-cluttered display.

Adjusting the vertical scale changes the scale (V/div), and therefore the vertical size, of a wave on the screen. Changing the vertical scale

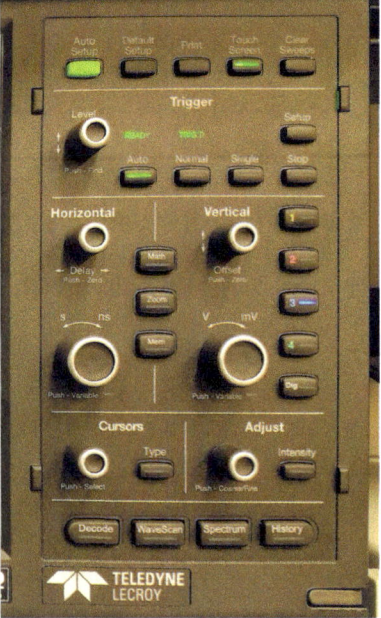

FIGURE 2.36 Front panel of the LeCroy HDO4104-MS.

will have no effect on the shape of the waveform other than to change its height. Sections of the wave can be enlarged by selecting the **Zoom** button. The **Vertical Offset** knob can be used to adjust the position of the waveform on the screen. This is useful when attempting to view multiple waveforms at one time without overlap. When changing the vertical offset, you must keep track of where 0 V (ground) has moved to. That is indicated by a small yellow line at the left of the screen labeled by its corresponding channel number.

The **Horizontal** section controls the **time base** of the scope, i.e., the duration of the waveform displayed on the screen. If this were an analog scope, the horizontal scale would be how quickly the electron beam scanned across the screen, from left to right. In a digital scope, it is an actual plot of voltage versus time, since the data is stored internally before being displayed. The horizontal axis is labeled, from slower to faster, in units of s/division, ms/div, μs/div, etc. Each individual channel can have separate vertical scales, **but all of the channels have the same horizontal scale**, simply because all four channels are acquiring data at the same time. The **Horizontal Scale** knob changes the time base for all four channels. Turning this knob will stretch or compress the waveform horizontally, but it will not change its overall shape.

Triggering initiates the start of the trace across the screen (or data collection in a digital oscilloscope). The user may select the voltage and the slope at which that happens. Any one of the four channels can be selected to initiate the trigger event. If a trigger point is hard to identify for noisy signals, you may instead choose the external sync input as the trigger event. The **Horizontal Offset** knob controls the location of the trigger point on the display screen. Turning this knob shifts the waveforms horizontally. You may want to use this to align a feature of a signal with a major gridline to help in measuring. Or you might care more about what happens before or after the trigger event. So, if you move the trigger point to one side of the screen or the other, you can see more pre- or post-trigger parts of your waveform. In order to set the voltage at which the waveform will trigger, you need to adjust the **Trigger Level** knob. When the signal having the correct slope crosses this voltage, the trigger will be initiated. A constant trigger point will make the waveform appear as a stable, clean trace on the display.

For more accurate measurements, it helps to use horizontal and vertical cursors when determining values on the display. A cursor picks a position and provides a readout of the value. This is much more accurate than simply eyeballing. Two horizontal and vertical cursors can be selected to measure differences, indicated by the Greek letter capital delta, Δ, in voltage or time. To use cursors, press the **Cursor** button repeatedly. This will cycle through the options (one or two horizontal, vertical, and both). Turn the **Cursor** knob to move the cursor to the desired feature. Once one cursor is in the proper place, press the knob to lock it and adjust the position of another cursor.

2.11.4.2 Using the Touch Screen

All of the functionality of the front panel shown in Fig. 2.37, and much more, is available on the touch screen. (Screen touches can be done with either a finger or using the stylus attached to the front panel.) For example, the channel parameters can be accessed through the touch screen. To do this, after a channel is activated through the front panel or from the **Vertical** dropdown menu, press that channel's box on the bottom of the screen. To turn the display off, uncheck the trace on the menu. Offset can be adjusted by pressing the arrows next to **Offset**. It can be zeroed back to the center of the grid by pressing **Zero**. The waveform can also be offset by touching the screen and dragging the waveform. The vertical scale can be adjusted by pressing the arrow buttons under the **Vertical Scale** options. To zoom on a portion of the waveform, trace out a zoom box over the portion of the wave that you are interested in examining further.

In this course, you will need to apply math functions to waveforms. To do this, select the **Math** menu from the menu bar at the top of the screen or from the button on the dialog tab that shows up on the bottom of the screen when you select a parameter. This pulls up a menu of possible functions. For example, to display a frequency spectrum (which you will learn about in Chapter 4), you will find the **Fast Fourier Transform** option on the menu, which turns on the spectrum analyzer. These can be displayed on a separate grid by selecting the **Auto Grid** option in the **Display** menu at the top of the screen. This can also be turned on by selecting the **Analyze** menu at top of the screen. This will display the waveform in the frequency domain. This function will be used extensively in later labs.

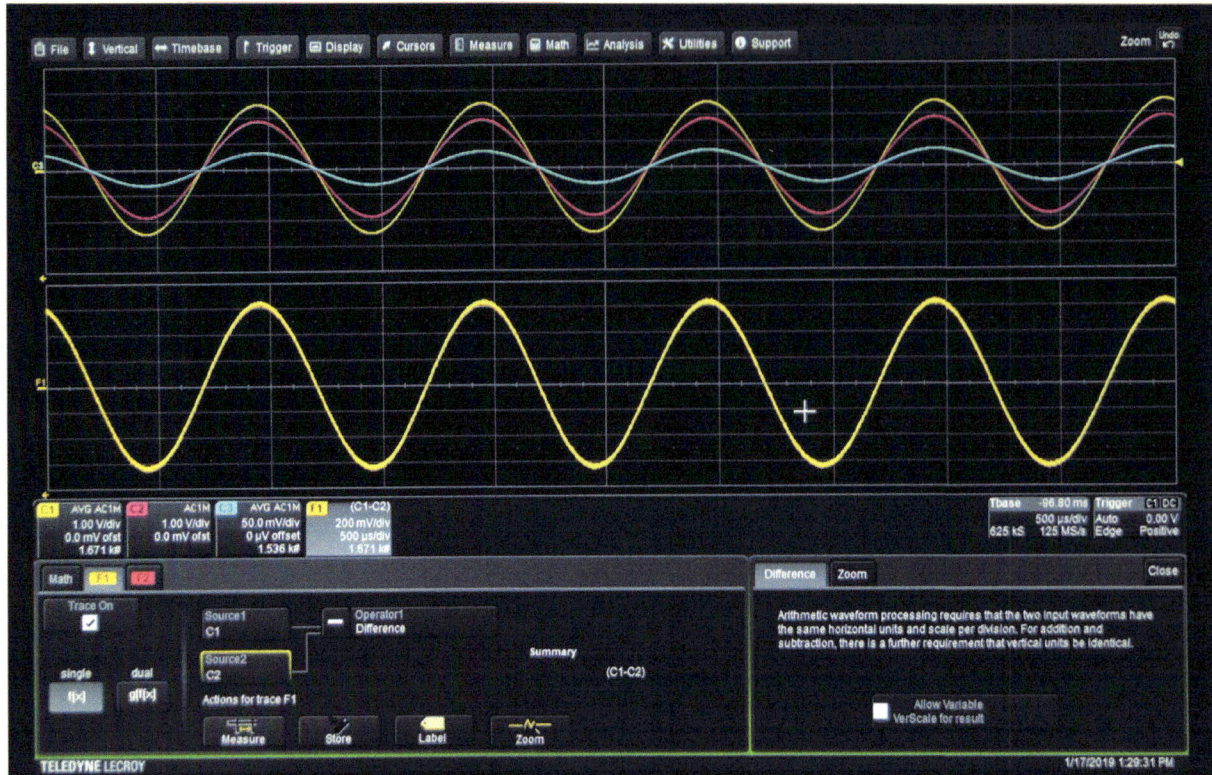

FIGURE 2.37 Touch screen display of the LeCroy HDO4104-MS.

As an alternative to using cursors to perform simple measurements, and also to perform other, more complex measurements, you can select the **Measurement** option on the bottom of the screen. This pulls up a menu with many possible characteristics that can be determined from the waveform. This pulls up a table above the **Channel Descriptor** box. This can display multiple measurements that you can select from the **Measurement** menu on the bottom. This is the quickest and easiest way to obtain good measurements.

Your screenshots can be saved from the **File** menu at the top of the screen. You can create a new lab notebook by selecting **New Lab Notebook**. Once this is created, you can add screen shots by selecting **Lab Notebook** in the **File** menu. The files can either be printed or saved to a flash drive. To save files to a flash drive, insert it into the USB port on the front of the oscilloscope. Instructions for saving screen shots are in this chapter's Lab Activity section. *When using the lab notebook with a flash drive, check frequently during the lab session that you are saving the files, and that they are safe before you leave the lab. Don't find yourself in the position that you didn't actually save any of your precious data.*

2.12 Lab 2 Activities

Green text indicates that the material is explanatory. **Black text indicates the actual steps to be performed.** In this and every lab to follow, you will alternate 'pilots' every half hour or so. The 'pilot' is the lab partner who handles the main tasks, while the 'co-pilot' mostly reads instructions and takes notes but always participates in decision making. It is, unfortunately, natural for one person to fly the plane most, if not all, of the time, so we are all going to make sure that doesn't happen. Therefore, you are required to switch off the pilot and co-pilot every 30 min or so, and *indicate in your* *notebook your roles and the time that you switch.* It is enough to write something like, "Switched to Gary as the pilot, 2:34 pm" or, "Gary built circuit while Clint took measurements," or even "2:34 pm, Pilot: Gary." You will hand in a joint lab notebook for a grade, so plan on working together that week on the lab and the lab reports. Also, most steps have an associated question that is intended to make you think more deeply about what you are doing. **You must answer all of those questions in your lab report**. If you have time, you may answer them in the lab notebook as you work, or just jot down some thoughts and answer them more thoroughly later. You are encouraged, though, to at least take a few minutes to discuss them with your lab partner.

In all of the Lab Activity sections, words on instruments, either on buttons, in software menus, or printed on the instrument are designated by **boldface** text. Context will allow you to distinguish between the use of boldface text for instrument details or in the introduction of new terminology, as used in the rest of the text. All calculations, answers to questions, and analysis must be recorded in the notebook as part of your lab report.

1. **Measuring and removing the resistance of the leads.** Use banana to alligator clip leads, one red and one black. Refer to Section 2.11.3 to review instructions on operating the DMM. Turn on the Fluke DMM by pushing in the power button on the lower right corner. It defaults to the **Auto** mode, meaning that it automatically selects the range of measurement values. Insert the black banana plug into the **COM** jack and the red banana plug into the **VΩ** jack. Of course, the color of the lead does not affect the measurement but does allow you to keep their polarities in mind. Press **Ω** to access the ohmmeter function. Clip the leads together, measure their total resistance, and record the value. This is a good place to start using the ohmmeter. Looking ahead to the resistor values in **Fig. 2.38**, do you think you need to take this resistance value into account in subsequent measurements? Explain.

   While the leads are still connected to each other, press the **REL** button to zero out the meter with leads for subsequent measurements. Note that after you leave the ohmmeter mode and return for subsequent resistance measurements, you will have to repeat the **REL** function if you need it.

2. **Learning about the breadboard.** Find the breadboard in the "Group #" box. Use the ohmmeter and a couple of wires inserted into the breadboard to test your understanding of the breadboard connections, making note of which connections make an open circuit and which make a short circuit. Make observations in your notebook.

3. **Skin resistance, safety, and the effect of holding probes with fingers.** Be careful not to spend too much time on the following step. It is not central to today's lab. You will be using the DMM as an ohmmeter to measure your resistance values for the circuit shown in the figure. You will, of course, be tempted to hold the probe leads in your hands and press them against the resistor leads. If you hold a resistor in your fingers during the measurements, then the resistance of your body will be in parallel with the **device under test**, or **DUT** (i.e., the resistor) and could affect the accuracy of the measurement. Draw in the notebook the circuit diagram for the DUT in parallel with your body. Label the resistances and include the ohmmeter as well. In general, it is a good idea to be mindful of what you touch with your fingers around electrical circuits. In fact, your skin resistance is much greater than that of the big, conductive blob of salt water that is your body, so it is your skin that does the most to protect you from electric shock. **Safety note**: the most dangerous electrical shock is hand-to-hand since the current takes a path through the heart, which is the lowest resistance because it is the shortest path. In this case, ohmmeters produce very low voltages and current, so there is no danger.

   So, let's see what the resistance of your body (mostly skin) is. Hold the two leads between two fingers on each hand and measure the resistance between your hands. (Note that MΩ is megohms and kΩ is kohms.) Try squeezing different ways and with different pressures to get a range. The resistance presented by your skin depends on many conditions, such as degree of perspiration or stress level, or time spent recently in the shower. Compare your skin resistance with your lab partner's. How do the measured resistances compare to the values of the resistors that you are measuring, as shown in **Fig. 2.38**? If you did a measurement with your skin resistance in parallel, how much (lots or a little) would it affect your measurements of these resistors? How large would you say a resistor would have to be before your skin resistance would significantly affect the accuracy of the measurement? (A ballpark guess here is fine.)

Select one of the resistors shown in the circuit of Fig. 2.38 as a DUT. (Do not build the circuit yet.) Draw a schematic diagram of your body resistance in parallel with the resistor under test along with the meter and calculate about how much your body resistance affects the measurement of the resistor's value.

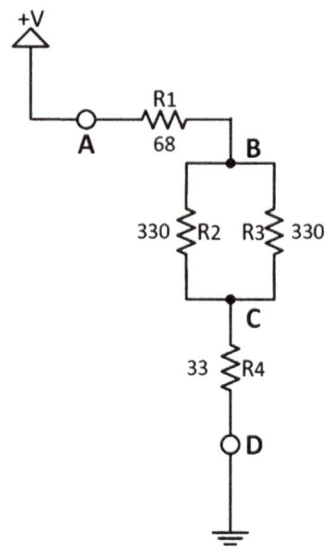

FIGURE 2.38 Voltage divider circuit with parallel and series resistors.

4. **Comparing measured and marked (bands) value of resistors.** Refer to the circuit schematic diagram ('schematic') in Fig. 2.38 Find the resistors using the color codes along with a meter to check. After zeroing the leads using the **Rel** mode, measure the values of all four resistors. Record the actual and color-coded (bands) values in your notebook. In general, the clips make a better electrical contact through the surface oxides, so that is the advisable way to make the measurements. Assume that the meter is calibrated and accurate so that the measured value is the true resistance. Calculate the tolerance of the resistors as $|R_{meas} - R_{marked}|/R_{marked}$ (i.e., the absolute value of the difference divided by the intended value). Use the color bands to determine the tolerance of your resistors. Are the actual resistances within the tolerances marked on the resistors?

5. **Building your circuit.** Build the circuit shown in the schematic Fig. 2.38 on your breadboard (as shown in Fig. 2.39). Do not attach the power supply yet.

6. **Learning the ohmmeter.** Calculate the total expected resistance from A to D. Use the DMM ohmmeter to measure the total resistance from A to D by placing the leads of the ohmmeter at points A and D of Fig. 2.38. Let's now define *percent difference* (PD) between two measurements as the difference between two values divided by their average (i.e., 100%×(A–B)/[(A+B)/2]. What is the PD between the measured and predicted resistance calculated from the bands? Is it better than that indicated by the resistor tolerances (assuming they are all the same)? If it is (and it may not be), that might be because random errors tend to cancel.

FIGURE 2.39 (Step 5) Sample layout of Fig. 2.38 circuit. Your layout does not have to look exactly like this. Here, the horizontal red wires go out to ground and power strips. You can omit those if you like.

7. **Learning the power supply – current compliance limit.** Refer to **Section 2.11.1** for brief instructions on the operation of the power supply. Set channel 1 (CH1) of the DC power supply to 5 V (press 5.00 **Enter**), but do not connect it to your circuit or turn on the output yet. Draw the circuit and include

the power supply, indicating the current out of the 5 V supply. Based on the total resistance in Step 6, use the equation $v = iR$ to calculate the approximate current that you expect through your circuit from the supply. Set the overcurrent protection (also called "current compliance limit") to about 20% greater than what you calculate the current to be. Record that current value in your lab notebook. The current compliance can be checked any time by pressing I-Set.

8. **Checking the power supply voltage reading.** Plug another set of banana plug/alligator clips into the CH1 power supply jacks. Do not yet connect the clips to your circuit. Set the DMM to DC voltmeter (**V** button with lines and dashes). Connect your voltmeter to the leads. Turn on the output from CH1 of the power supply. Compare the voltage reading on the DMM with the voltage reading from the power supply itself. How much do the readings differ? What PD do you measure? *Note*: there is no obviously correct answer to the following questions. Just give it a little thought. Which one do you trust more? For critical applications, bench equipment is often sent out to repair facilities to be recalibrated. Should your voltmeter be recalibrated? Why or why not?

    Read and record the current shown on the Keithley 2231A-30-3 power supply. Do you expect to read any appreciable current given that the voltmeter is the only load on the supply (other than air, if you want to think of it that way), and it has an input resistance of 10 MΩ? Explain. Turn off CH1.

9. **Powering your circuit.** Leaving all connections in place, add two more leads to connect the power supply to the circuit. Turn on CH1. Is there an appreciable difference in the voltage set point and actual output voltage? If there is, you have set the current compliance limit too low. If so, increase it and recheck.

10. **Reading current from the power supply.** Record the current reading from the power supply. Calculate the PD of the actual current compared to the current you calculated in Step 7. If it isn't a small percentage, check for errors. Discuss if necessary. Turn off the power supply output.

11. **Using a resistor to determine current.** Use the DMM voltmeter to measure the voltage drop across R4. Recall from this chapter that the **COM** lead is floating, i.e., not connected to earth ground. (Things will work just fine without an earth ground until you start using the oscilloscope, where the probes are always referenced to earth ground. Then you must be more careful.) Put the positive, red, **VΩ** lead at C and the black, **COM**, lead at D. Reversing these will change the sign of the measured voltage, but will otherwise not be a problem. Try it.

    Turn on the power supply output. Determine the current from the measured voltage and the measured value for R4. Record qualitatively how well your current measurement agrees with the current reading from the power supply.

12. **Using the DMM DC ammeter.** Turn off the output of the power supply. Note that the DMM jacks for measuring current, i.e., as an ammeter, are different from those used to measure voltage. Plug the red lead into the jack marked **100 mA**. Decide where you would break the circuit and then insert the DMM leads to measure current from the supply. If you want, use extra wires and breadboard test points to help make the contacts, or just connect alligator clips to each other. On the DMM, press **A** with the solid line over the dashed line to measure DC current. Turn on the power supply output and measure that current with the ammeter. When done, turn off CH1. Now you have three different measurements of the power supply current. Calculate the PD between the ammeter reading and that which you measured with R4 in the previous step. Take out the ammeter and reconnect your circuit.

13. **Analyzing a voltage divider.** Using the voltage divider equation and actual resistance values, calculate the expected voltages from A to B, v_{AB}, from B to C, v_{BC}, and from C to D, v_{CD}. Of course,

you should see that $v_{AB} + v_{BC} + v_{CD} = v_{supply}$. Remove your lead from the **100 mA** jack, and switch from the ammeter setting to voltmeter. Using the DMM as a voltmeter while in the ammeter mode can result in a blown fuse (and who needs that headache?) Draw the schematic diagram of that and explain why it could blow the fuse. Measure v_{AB}, v_{BC}, and v_{CD}. Do you see that the voltage divider taps off the voltage according to each resistor's share of the total voltage? Does the sum of the three measured voltages equal the measured supply voltage?

14. **Setting up the oscilloscope.** Turn on the waveform generator (WG) and oscilloscope. You may choose to use the stylus that is inserted on the lower right edge of the front panel for pressing the screen. Before using the LeCroy HDO4104 oscilloscope, you should know how to save your screenshots for your lab reports:

    - Insert thumb drive into USB slot
    - Select **File** tab in the top left of the screen
    - Select **Print Setup**
    - In the bottom left corner select **File**
    - Set **Colors** to **Print** in order to conserve toner when printing
    - Set file type, name, and location where it will be saved. To save to the flash drive, set the directory to E:\
    - Set **Hardcopy Area** to **DSO Window**
    - Change the file name. Each print will increment that name by an appended digit. You can go through this process and select new file names for each screenshot, so you don't get confused later trying to figure out what each one was.
    - Close the print setup menu, and set up the scope screen to how you want it to look like in your report
    - Save the screenshot by pressing **Print** on the top row of buttons on the right panel.

15. **Using a keyboard to enter data.** Rather than using the touchscreen, you may use the keyboard to enter data directly into the oscilloscope. To do this, select the oscilloscope button, #2, on the white junction box on the lab bench. Or, you may plug a USB keyboard into a USB port on the scope.

16. **Displaying a sine wave on the oscilloscope.** Find a BNC-to-BNC cable in the "BNC Cables" box. Plug the output of Channel 1 (CH1) of the WG into CH1 of the scope, as shown in Fig. 2.40. Recall from the readings that BNC cables plug in and then turn clockwise to lock. Reverse the order to remove.

    Choose the **Sine** output (top row of buttons). Using the buttons to the right of the display, select 5 V peak-to-peak, or 5 V_{pp}, and a frequency of 1 kHz. Turn on the output of CH1 of the WG by pressing **Output1**, which lights up when active. (This is the most common reason for things not working – forgetting to turn on the output of the WG. Everyone does it at some point.)

    On the oscilloscope, you may have to adjust the horizontal time base and the vertical scale to see a full sine wave on the scope screen, as shown in Fig. 2.40. Vertical scale is the larger of the **Vertical** knobs. Pressing it switches between large and small increments of vertical scale. You might see the sine wave rapidly moving horizontally on the screen rather

FIGURE 2.40 Rigol function generator with sine output going to CH1 of the HDO4104 Teledyne Lecroy oscilloscope.

than showing up as a fixed waveform. This means that the scope is not triggered properly. For beginners, the easiest thing to do is press the green **Auto Setup** button at the top right of the front panel and confirm the selection. (See https://www.youtube.com/watch?v=hT4daNnsGJM for a video on triggering.) This will stabilize the waveform and select reasonable values of horizontal time base and vertical scale. Play around with changing the time base (left large knob) and vertical scale (right large knob), to see what they do to the appearance of the sine wave. This sine wave is an example of an *AC signal* as opposed to the *DC* power supply voltage. One more adjustment: press the **C1** box at the lower left of the screen and **Zero** the **Offset.** That way the sine wave is centered around the middle grid line.

Now, at the WG, select **Ampl** and change the voltage amplitude, (making note of the peak amplitude, v_p, or the peak-to-peak amplitude, v_{pp}) and **Freq** to change frequency, f; view the waveform on the scope. Sweep f (i.e., change it continuously) on the WG by choosing the variable digit with the left/right cursor buttons (arrows) and observing f on the display as you turn the knob, which now changes f with the resolution of that digit. Do the same for v_{pp}.

Remember that frequency is $1/T$ where T is the time for one period of the waveform, which here is a sine wave. For one selected amplitude and frequency, measure v_{pp} and f by using the V/div and sec/div values (see Fig. 2.41 – one division is one large rectangle) that you estimate from the screen. What is the PD between the WG setting and your measurement? What might give you errors doing the measurement this way?

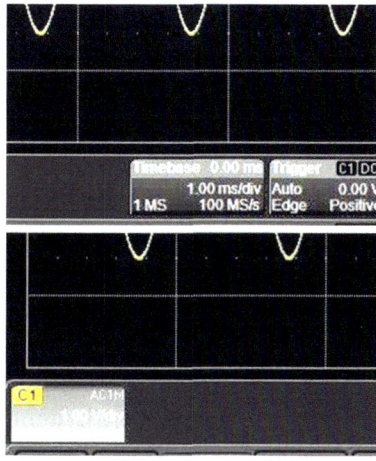

FIGURE 2.41 (Step 16) (Top) Lower right and (bottom) left parts of scope screen showing the time base and vertical scale settings, respectively.

17. **Using measurement functions on the scope.** The scope can automatically measure the amplitude and frequency of a waveform, among many other things, without having to manually measure points on the screen as you did in the previous step. Press **Measure** at the top of the touchscreen, then **Measure Setup, P1** and scroll through with the touchscreen to select **Frequency.** Make sure the source you are measuring is C1. You can figure out how to change that. You can choose more parameters, **P2**, etc. up to **P8**. Set **P2** to measure **Amplitude**. The measurements are shown automatically in the **Measure Value Status** bar, as seen in Fig. 2.42. What is the PD between the amplitude and frequency measured on the scope and selected by the WG? You may toggle between **Auto** trigger and **Stop** trigger to read the measure- ments boxes.

18. **Learning to use cursors.** Cursors are vertical and horizontal lines that mark a specific point of time or voltage allowing you to directly read the period, frequency, amplitude and, later, phase. Turn on the cursors by pressing **Type** next to the **Cursors** knob on the lower right part of the front panel. Pressing once lights up the button, and two vertical white lines appear. Now, below the **Time base** and **Trigger** boxes in the lower right corner of the screen (Fig. 2.42)

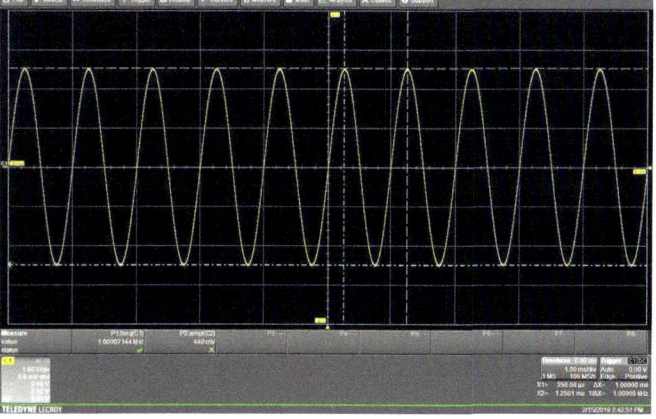

FIGURE 2.42 (Steps 17 and 18) Screen trace showing horizontal and vertical cursors. The horizontal cursor information is at the lower right, and vertical cursor information is at the lower left. Sine wave shown is 5 V_{pp} at 1 kHz. (P2 measures the amplitude of C2, which is not shown.)

are shown **X1=** and **X2=** information that correspond to the locations of the two cursors. Pressing the **Cursors** knob selects the left or right cursors, or both, in sequence. Rotating the knob moves the selected cursor(s). The time difference between the cursors is indicated along with inverse time corresponding to the related frequency.

Pressing the **Type** button again brings up horizontal cursors used to measure voltage differences. After pressing **Type** the second time you must press the cursor's **Select** knob once to switch to the horizontal lines. Now, the information from the horizontal cursors is displayed at the lower left inside the channel information boxes. Note that the two cursors will be overlaid, so you have to turn the **Select** knob to separate them. For one selected amplitude, measure v_{pp} by using the cursors. Of course, if you already know the voltage, you are prone to set the cursors to match it. If you are working with a lab partner, one of you should set a voltage and cover the screen while the other of you does an actual measurement. How close did you get? What is the PD between the WG setting and your measurement?

The frequency based on the cursors (reciprocal of the time difference between vertical lines) is shown at the lower right-hand corner. Measure one selected frequency using the cursors. What is the PD between the WG setting and your measurement? Again, one of you should hide the setting while the other does the measurement. How close did you get?

19. **Using both channels of the WG.** Turn off all the cursors to clean up the screen. Attach both CH1 of the WG to CH1 of the scope and CH2 of the WG to CH2 of the scope. Select CH2 of the WG and generate two identical sine waves of 5 V_{pp} and 1 kHz (don't forget to turn on both outputs) and select both channels on the scope to observe both. Make them appear the same size by adjusting the vertical scale. For the selected channel, use the smaller vertical knob, labeled **Offset**, to move the position, or voltage offset, of the sine waves. Or press the **Offset** knob to zero the offset. Can you make the sine waves overlap perfectly? If not, press **Align Phase** on the WG. (When they are aligned, one of the traces may disappear under the other one.)

20. **Measuring phase.** Select **Start Phase** on the WG for either of the channels and change the phase across a wide range. See Fig. 2.43. *Phase is the time difference between two waveforms of the same frequency expressed as a fraction of a period in degrees or radians. One period is 360° or 2π radians. This will be discussed in depth in Chapters 3 and 5.* Select a phase of 90°. *Does it look like sine and cosine functions? It should.* Explain using a simple trig identity.

21. *Using measurement function for phase.* Select **Measure**, **P3**, and then **Phase**. Make sure that you have several periods of the waveform displayed on the screen. If not, use the **Horizontal** knob to slow

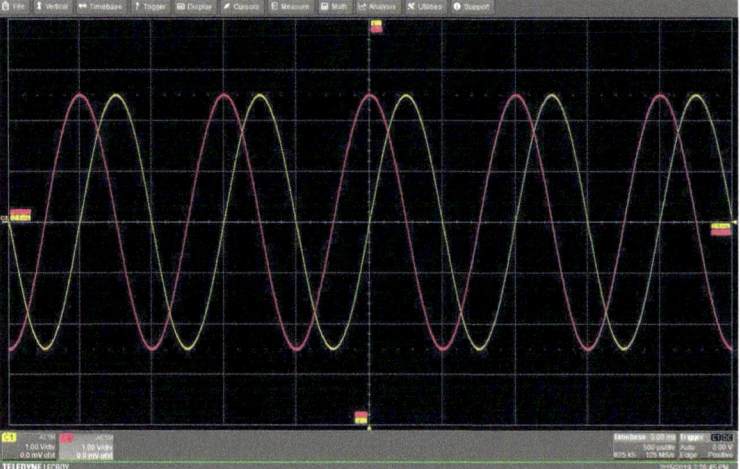

FIGURE 2.43 (Step 20) Two sine waves out of phase by 90°, or $\pi/4$ radians, just as sine and cosine functions differ by $\pi/4$ radians.

the time base, i.e., shrink the waveforms and include more on the screen. See Fig. 2.44. Change the phase across a range to see what it looks like, and mentally relate it to the measured values. How well does the phase measurement on the scope agree with that of the WG? Does this help you get a better understanding of what "phase" means?

22. **Learning about other waveforms.** Try changing waveforms (sine, square and ramp) and their parameters in both channels while displaying the two waveforms. Play with the settings to get to know the WG and scope better. Change their positions, scale, cursors, and automatic displays as you wish. Take some screenshots for your lab notebook. If one of the waveforms does not look stable, that means it is not triggering properly. Choose the **Setup**

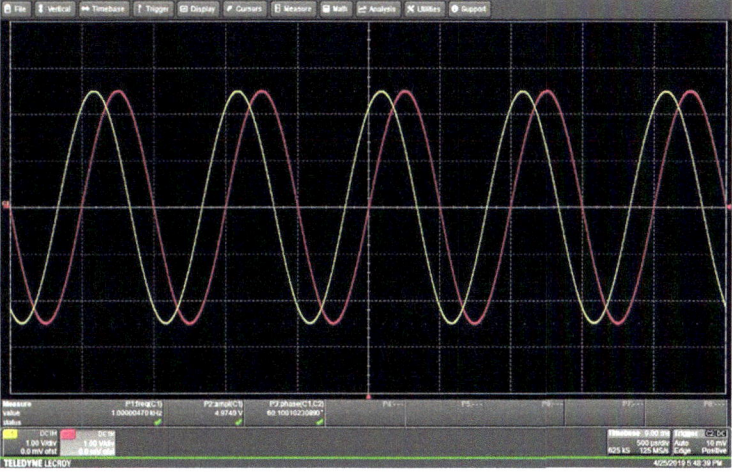

FIGURE 2.44 (Step 20) Two sine waves out of phase by less than 90° (60° is shown).

button of the **Trigger** panel on the scope and select the relevant **Source**, i.e., the channel that is informing the trigger. Rotate the trigger knob to move the trigger point vertically onto the trace until the trace becomes stable. Triggering can be a tricky thing, since it can be affected by various selected parameters, noise, averaging, and other factors. Did you have any problems and then get a stable trace? Briefly write in your notebook what you did. You don't need all of the details.

23. **Making AC measurements.** Now you will use the WG to drive your circuit. Turn off the WG outputs. Remove the DC power supply from the circuit and shut it off. Connect the BNC-to-mini-clips cable (Fig. 2.45) from CH1 of the WG to your circuit in place of the DC supply. *Do not attempt to insert the clip end into the breadboard. Miniclips are delicate and expensive, so be careful not to bend them for any reason. They should be used ONLY to clip onto thin wires, as shown in the figure.*

   Apply a sine wave of 6 V_{pp} and 1 kHz to the circuit in place of the DC supply. The red lead should go to point A while the black connects to D. The mini-clips can attach either directly to one of the resistors, or to an additional wire sticking out of the breadboard. Cables can often pull components out of the breadboard, so the additional wire is often preferred.

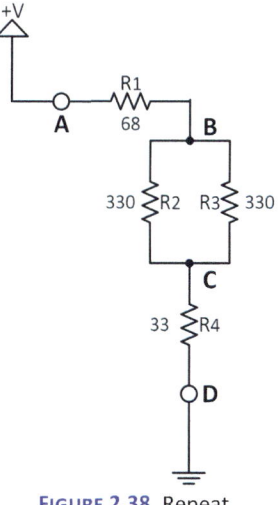

FIGURE 2.38 Repeat

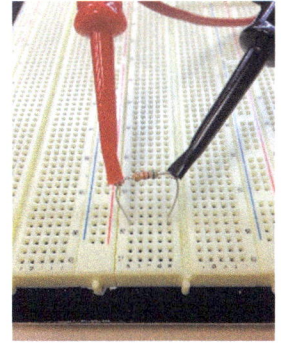

FIGURE 2.45 (Step 23) Mini-clips or EZ-clips. Be careful not to damage their delicate ends. They are costly.

Important conceptual information: Point D is now set to earth ground through the outer shield of the cable, which leads to the outer shield of the BNC connection on the WG, which in turn is earth ground. Reminder: A ground connection to a circuit is a single node whose voltage is called 'zero' volts. It is one electrical point, ideally having no resistance along the whole piece of ground metal. If you connect two separate points of the circuit to earth ground, that is the same as shorting those circuit points between the two ground leads. All outer BNC connections and nearly all exposed metal on any of the instruments are at earth ground. (The one exception is the metal point marked **Cal** on the oscilloscope, which outputs a square wave to help match the capacitance of oscilloscope probes to that of the oscilloscope. The metal port next to the **Cal** port that is marked with the symbol for earth ground is the same as all of the outer shields on the BNC connectors.) If you are confused by these important details, now is a good time to ask your TA to explain.

24. **Connecting oscilloscope probes to the circuit.** Connect the BNC end of three oscilloscope probes to channels CH1, CH2 and CH3. The black clips on each probe are earth ground, so connect them to ground (D) of Fig. 2.38. Do not forget that the black alligator clip on each probe goes to the outer shield of the BNC connector and are therefore all the same earth ground. Although electrically it is quite correct to use only one black clip, since that effectively connects the ground on all of the probes to that point, it is not advisable. The danger is that a loose ground connection dangling near your circuit might come in contact with an energized circuit element, and possibly short it out, causing damage. (Also, the black clips are removable, but please don't remove them, or they may get lost.) Therefore, good practice is to first connect all the ground leads to some earth-grounded location in your circuit, and then connect the signal leads of the probes. Now, pull on the probe hood and attach the probe clips to points A (CH1), B (CH2) and C (CH3). You might choose to attach the probes to extra pieces of wire on the breadboard in order to make your circuit less cluttered and to avoid pulling out the resistors with the probes, which can exert considerable force on the leads.

25. **Making multichannel measurements and taking into account output resistance of the WG.** Turn on all three channels and set them all to the same scale (say, 1 V/div). Use your favorite method of measuring the values of v_p for all three channels. Make careful note of the CH1 reading, which is the direct output of the generator. You will notice that CH1 is noticeably less than the 6 V_{pp} (3 V_p) that the WG indicates as its output voltage. That display is not an actual meter reading, but just the setting of the output, as discussed in Section 2.11.2 regarding output resistance. In this case, the 265 Ω load of the full circuit is low enough to modestly **load down** the WG, so that the output voltage 'sags' to less than 6 V_{pp}. Obviously, the WG does NOT behave as if it is a regulated supply, which would otherwise guarantee the set voltage. What is the actual source voltage, and by how many volts did the WG get loaded down? Use the reading of the actual output voltage from the WG for your calculations in the next steps.

    The WG acts like an ideal voltage source in series with its own output resistance of 50 Ω. Using a voltage divider equation that includes the internal resistance of the WG in series with your circuit, you can calculate the expected voltage at A to be [3 V_p x R_{AD}/(50 Ω + R_{AD})], where R_{AD} is the measured total resistance of your circuit. What is the PD between your calculation and the WG output voltage as measured by the scope?

26. **Using the voltage divider equation for AC measurements.** Arrange the three traces from the three channels with the same reference (0 volts) and suitable scales to observe all three waveforms, as in Fig. 2.46. Ignoring the internal resistance of the WG, the voltage readings at B, and C will satisfy the voltage divider equation using the voltage measured at A on CH1 (Step 25) as the input to the circuit. Using your voltage divider knowledge, are the measured voltages what you expect? Show your calculations.

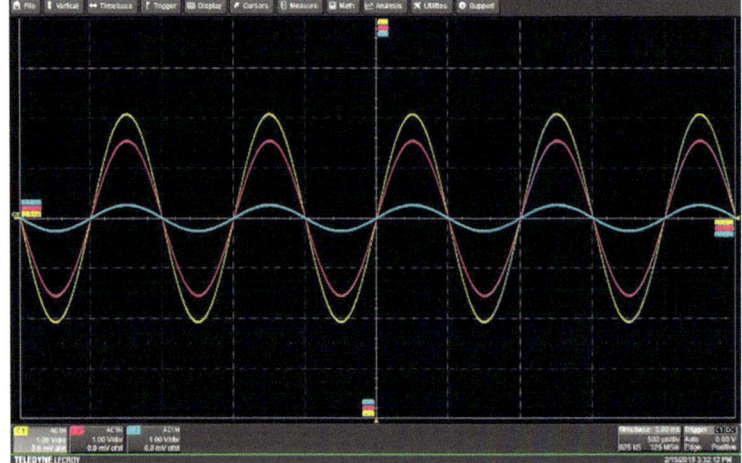

FIGURE 2.46 (Step 26) Scope traces showing voltages at A, B and C on CH1, CH2 and CH3, respectively.

Given that the probes are 10× type and therefore "look like" 10 MΩ resistors to your circuit, is that large enough to be neglected in your circuit measurements?

27. **Learning to use a math function.** Now you are going to use the scope to measure the AC current from the WG by using $v = iR$. In order to do this, you will use one of the resistors and its voltage drop to calculate the current as you did in Step 11, but now for an AC waveform rather than the DC used there. Again, the resistor is acting as a shunt. You could do this easily with R4, but that wouldn't teach you anything new. Instead you will learn to use a math function.

    Note that the voltage drop across R1 is $V_{AD} - V_{BD}$. That will tell you the current into the circuit based on $v_{AB} = iR1$. However, if you tried to use a scope probe to measure directly across R1 while the ground connection at D is connected to earth ground of the WG, you would short out the circuit from B to D. Here is how to get around that: The scope can display the difference between two voltages using two channels, two probes, and a function that subtracts one channel from the other. Place a probe on CH1 from A to D (earth ground) and a probe on CH2 from B to D (again, earth ground) and press **Math** at the top. Select **Math Setup**, **F1**, in the trace window, and **Difference** in the prompt window. Now select the sources that you want to subtract. That function will display along with your other selected channels. You can now use the cursor to measure its peak voltage, which you can convert to current. Note that you may select any trace by touching the information box, which then allows you to change the vertical scale. Show your calculation of the current.

    If the signal is 'noisy' you can make it cleaner to provide a higher-resolution measurement. On the screen, touch the **C1** box, and in the **Pre-Processing** menu set **Averaging** to 3 sweeps. More averaging is often unnecessary and can add long delays. Alternatively, in the **Noise Filter (Eres)** box, you can pick some number of bits to average, e.g., 2. The meaning of this will not become clear until Chapter 9, but you can try that out as well. What voltage do you measure and what current do you calculate from it?

28. **Learning to use the current probe.** Find the current probe in the flat, padded box labeled "CP030." (Not to be confused with C-3PO of Star Wars.) Carefully remove the probe. **Be careful since it is very expensive ($3000).** Now, very carefully align and then insert the current probe into CH4. (You can choose any channel, but CH4 keeps it out of the way of the other inputs. For the sake of consistency, we will use CH4 for current measurements in future labs, but in no way is that necessary.) *Never force anything mechanical that does not operate smoothly.* It is connected simply by pressing it into the jack. Now, pull back on the button near the jaws to open them, and wrap them around the positive lead from the generator. Push the button forward to lock the jaws in place, as shown in Fig. 2.47. Turn off the

FIGURE 2.47 (Step 28) Current probe clipped onto wire. The direction of the arrow shows the direction of positive current.

unwanted traces by pressing the colored buttons along the righthand side (or by unchecking them on the screen), and turn on CH4. It will automatically detect the current probe and display current, and you can now select its sensitivity, in A/div, and time base as you did with the voltage probes. Use whatever method you want to measure the current. Now it is very relevant to calculate the PD between this measurement and the one you did in Step 27. Do you trust the current probe for accurate, low-current measurements? Do you trust your resistor-voltage current calculation more?

29. **Learning about the Rigol power amplifier.** Now, to provide an output voltage, you will use an amplifier that has a high input resistance and low output resistance. Because of these

properties, as explained in Section 2.11.2, the input voltage to the amplifier is very close to the set voltage, and the output voltage from the amplifier, which now drives the circuit, does not sag as much.

Locate the Rigol PA1011 power amplifier (Fig. 2.48) It should already be plugged in; it has no power switch and will be on when energized. Do not unplug it, or it might change its programming. If it is not already plugged in, plug it into the power strip.

FIGURE 2.48 (Step 29) The Rigol PA1011 power amplifier. Bandwidth is 1 MHz. A Rigol waveform generator is needed to control for gains of 1× and 10× and other functions. The white BNC connector is the input and the yellow BNC connector is the output. It has no power switch – it comes on when plugged into a wall outlet.

In this course we will not use any of its capabilities other than as an amplifier having a gain of 1× ("unity gain"). Such an amplifier is called a **buffer amplifier**. What good is an amplifier whose gain is 1×? Why not just use the original signal? Because the output of the amplifier can produce more current than the input source (the WG in this case) can, so the output will not sag as much. The PA1011 can also be set to a gain of 10× using the settings on the WG. (It has a bandwidth of 1 MHz.)

Be aware that the PA1011 should retain its settings even after unplugging the power. Depending on its previous use, you may have to reset it. Also, sometimes it does not reboot properly, and you must either reset it or, if it does not reboot, use a different unit. See your TA if there is a problem. If you must restart the amplifier, plug in the USB cable to the back of the PA1011 and into the front panel of the WG. Press **Utility** and using the third arrow from the top, next to the screen, select **PA Setup**. Now select **Gain x1**. **Switch** must be set to **ON**. Pressing **Utility** gets you out. If after all of this you cannot get a working amplifier, skip this step and make a note in your notebook.

Assuming all works as it should, notice that it has an input and output BNC connection. Run a BNC cable from the output of the WG to the input of the amplifier, and a cable from the output of the amplifier to the circuit. Lift one lead of the 68 Ω resistor to stop the current flow. What is the actual output voltage from the PA? Is the gain really 1×? Note that voltage. Now record the PA output voltage when you put the resistor back in, and the PA drives current through the circuit. Has the voltage sagged as much as it did when the WG was used to directly drive the circuit (Step 25)? Note that the specifications for the input resistance of the PA is 50 kΩ and the output resistance is <2 Ω. Qualitatively describe why it behaves as it does in terms of input and output resistances and voltage dividers.

30. **Introduction to the frequency spectrum analyzer.** In this section you will be introduced to the scope's ability to display the frequency components of a waveform. For today, you may think of the frequency spectrum as something like a Taylor series expansion of the waveform, except that instead of adding a bunch of powers of x (i.e., the Taylor series), you are adding sine waves of many different frequencies. The **frequency spectrum** is the display of the amounts of each frequency component. The frequency spectrum will be important in Chapters 4, 5, 6, 7, and 9. This will be only a very short introduction. Set the power amplifier aside, but do not unplug it from the electrical power outlet. Disconnect all leads from all channels except for a BNC cable from CH1 of the WG to CH1 of the scope. Output a sine wave at 5 V_{pp} and 1 kHz. Select **Measure** again and turn all measurement displays off, i.e., clear all selections to clean up the display.

Press the **Spectrum** button on the lower right corner of the panel. You need to set up the range of frequencies being displayed. To do this, first select **Mode** to be **Normal** and **Persistence** off (unchecked). Then select on the trace window the **Start Stop** button; start at 0 kHz and stop at 10 kHz. (Making changes to the display often changes the start and stop values, so you will probably have to go back and reset them several times throughout this exercise.)

Select the **Output**, under the **Scale** section, to be **V$_{RMS}$**. Go to **Peaks/Markers** and turn off the table for now. If you need to, you can move the trace to the left or right by (a) pressing the **Horizontal** knob and turning it slowly, or (b) touching the spectrum on the screen and dragging the spectrum. Select CH1 and change the time base so that there are so many periods on the screen that they cease to be distinct from each other, as shown in Fig. 2.49, or even tighter than what is shown.

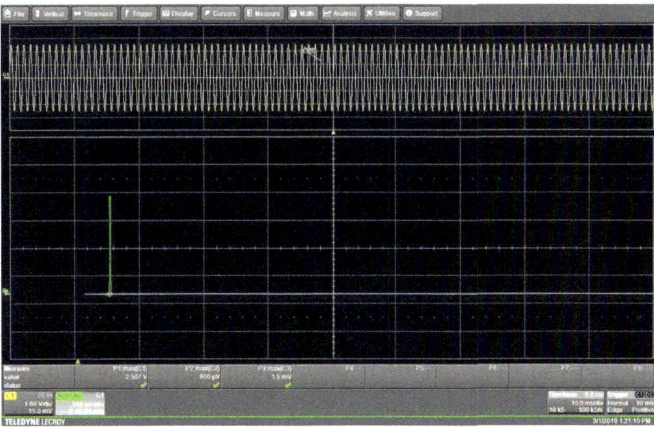

FIGURE 2.49 (Step 30) Frequency spectrum of a sine wave. The top of the image is the trace of the sine wave with a very slow time base, so the waveform is not visible. This makes for a sharp peak in the frequency domain.

The spectrum will show you one large peak and possibly several smaller ones. The large peak represents the one sine wave frequency needed to make up the original sine wave, i.e., the frequency component at the frequency of the sine wave. Other peaks really should not be there, and are a measure of the lack of perfection of the sine wave created by the WG.

The peak should be rather sharp as shown in the figure. Now change the time base to increase and decrease the number of periods seen on the screen and observe that the peak narrows and widens, respectively. This is an important feature of the FFT algorithm: The algorithm operates on the data that is shown in the trace window, not on what is stored in the memory of the oscilloscope. A good frequency spectrum requires many periods to be visible on the screen from which it uses the data to perform the calculation. The fewer the number of periods displayed, the wider the frequency peaks appear. Also, the full vertical waveform must be visible in the window and cannot be partially off the top or bottom.

Increase the frequency of your sine wave by 0.1 kHz at a time and watch the largest peak move across the screen on the spectrum analyzer. Describe the results in your notebook.

31. **Observing the frequency spectrum of a square wave.** Continue looking at the spectrum but change from sine to square wave from the WG (Fig. 2.50). Change the frequency span to start at 0 and stop at 50 kHz. Turn on **Peaks** in the **Table** menu in **Peaks/Markers.** Again, the frequency spectrum of the square wave shows the frequency components that added together make up the square wave. You don't need to understand that now, but you will surely understand it after Chapter 4. Increase the frequency of your square wave by 0.1 kHz at a time and watch the peaks move across the screen on the spectrum

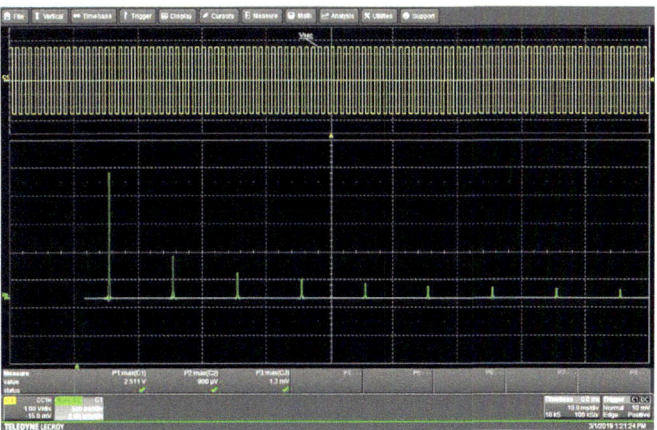

FIGURE 2.50 (Step 31) Frequency spectrum of a square wave having 50% duty cycle.

analyzer. Briefly, describe what happens to the spectrum as you change the square wave frequency.

32. **Learning about potentiometers.** The following activities demonstrate the operation of the potentiometer. You might need to review it from Chapter 1. Attach the DMM ohmmeter to the outer two leads of the pot as shown in Fig. 2.51. That is the maximum resistance due to the full length of the stripe of resistive material. What value do you read? Call it R_{max}.

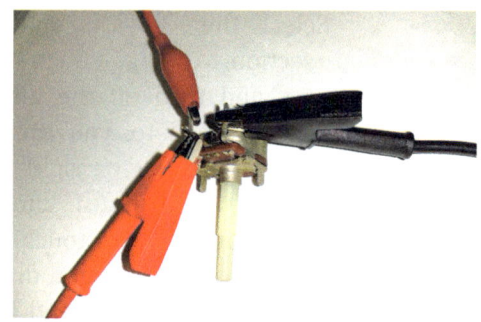

33. **The wiper.** Attach the ohmmeter to one outer lead and the middle lead, which is the wiper. Rotate the knob. What resistance range do you read? Explain. Repeat the measurement with the middle lead and the other outer lead. Compare the range and also the direction of moving the wiper for the observed change in resistance. Explain that behavior.

FIGURE 2.51 (Step 32) Potentiometer with outer leads connected to the ohmmeter or power supply and the middle lead connected to the DMM voltmeter.

34. **The potentiometer as a voltage divider.** Turn off the output of the DC power supply. Set the output voltage to 10 V. Set the current compliance to 100 mA if it isn't already that value. Connect the outer two leads to the DC power supply to the outer two leads of the pot so that you will have a voltage across the full resistance of the stripe inside the pot. (How much current will you draw from the PS?) Do not turn on the output yet.

Set the DMM to measure DC volts. Place the positive lead of the voltmeter to the middle lead, which is the wiper. Connect the negative lead of the voltmeter to the potentiometer at the negative terminal of the power supply. Draw a schematic of this circuit. Note: you do not need an earth ground in this configuration. Turn on the output of the PS. Rotate the knob clockwise and counterclockwise. You should be able to select any voltage from about 0 V to 10 V. Does it work? Explain what is happening in terms of a voltage divider and your schematic drawing.

Congratulations! You have finished your first electronics lab of your new career! Hopefully you have gained a basic understanding of what each of the bench tools does, and how to operate the basic functions. You will hone your skills considerably throughout the rest of the semester.

Problems

1. Show that the following equations are equivalent:
   (a) For resistors in series: $R_{tot} = \Sigma R_n$; For conductors in series, $1/G_{tot} = \Sigma 1/G_n$
   (b) For resistors in parallel, $1/R_{tot} = \Sigma 1/R_n$; for conductors in parallel: $G_{tot} = \Sigma G_n$

2. Calculate the equivalent resistance shown in the figure.

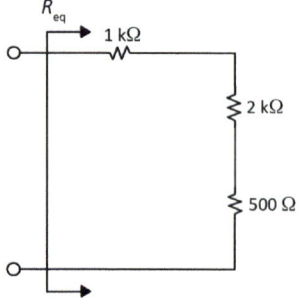

3. Calculate the equivalent resistance shown in the figure.

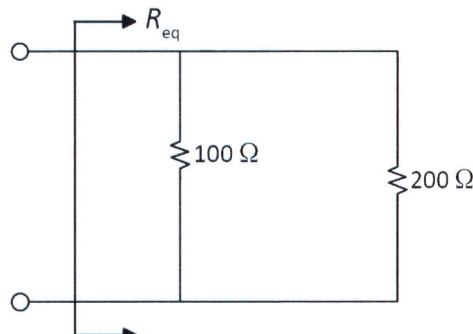

(a) Using the reciprocal of sums.

(b) Using the product over sum.

4. Calculate the equivalent resistance shown in the figure.

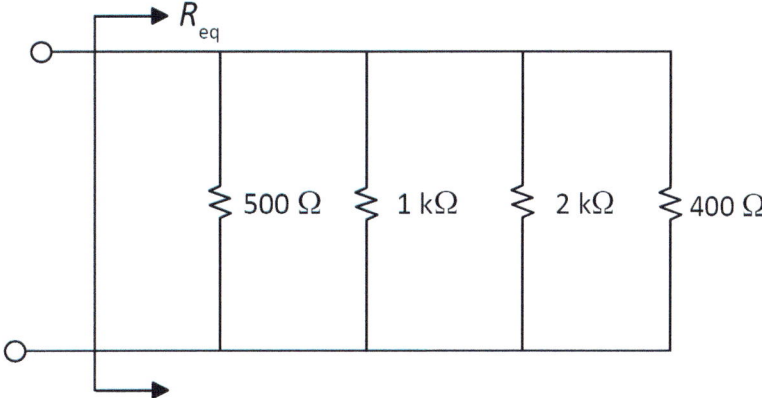

5. (a) What single resistance could you use to replace R2, R3 and R4 in the circuit below that would give the same voltage at node A with respect to ground? (b) What is the voltage at node A with respect to ground?

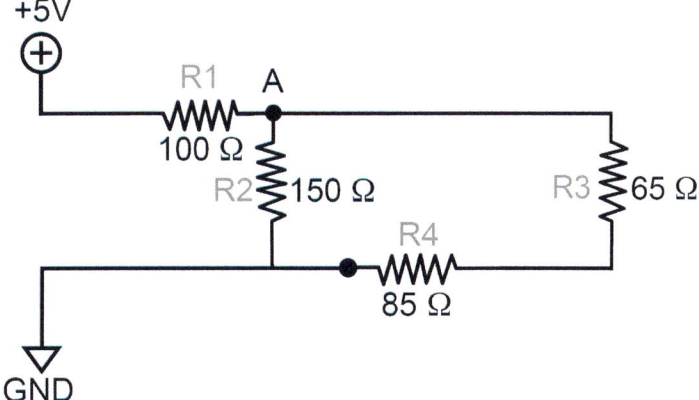

6. For the following combinations of resistors, what is the equivalent resistance, R_{eq}? Your answers are in ohms.

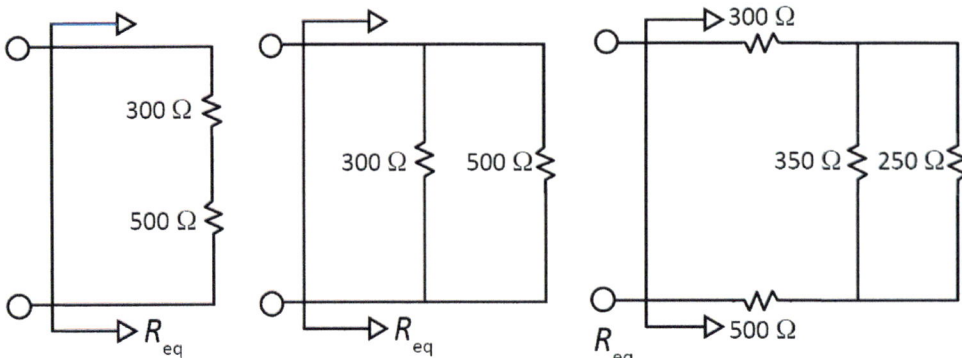

7. (a) What is the resistance R_{AB} between nodes A and B?

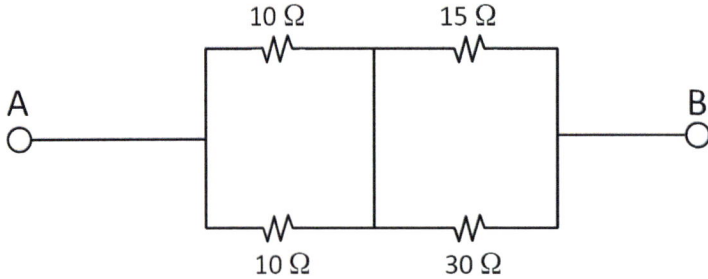

(b) And for this circuit?

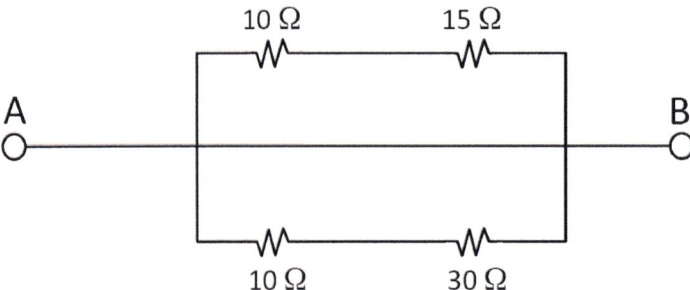

8. You are measuring the resistance of a resistor having a known, accurate resistance of 10.000 kΩ. The ohmmeter is calibrated to NIST traceable standards. (This means simply that the measurement is as accurate as you are going to reasonably get: https://www.beckman.com/support/faq/industry-standards/nist-traceable-standards).

   (a) You hold the leads against the probes with your bare fingers. Draw a circuit diagram of this situation including the resistance of your body.

   (b) Estimate the value of resistance that you measure. *State your assumptions.*

   (c) You repeat the measurement using miniclips and leads from the meter to the resistor and zero out the meter to remove the lead resistance. Answer the same question as part b), again *stating your assumptions.*

9. Suppose that a student measures a 100 kΩ resistor while it is connected in a circuit. The ohmmeter reads 50 kΩ. Assuming that neither the ohmmeter nor the resistor is broken, what mistake did the student make?

10. (a) What voltage V_{AB} is measured across nodes A and B in the circuit?

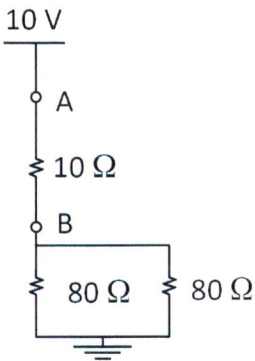

(b) Use the voltage V_{AB} to determine the current that flows from node A to node B.

11. Consider the following circuit with values $V = 15$ V and $R = 120$ Ω. The voltmeter is inserted between points A and B.

(a) What voltage will an ideal voltmeter read across A and B? (*Hint*: Insert the equivalent resistance of the voltmeter, which is very large. Treat it like a voltage divider.)

(b) What current will an ideal ammeter inserted between A and B read? (*Hint*, insert its equivalent resistance.)

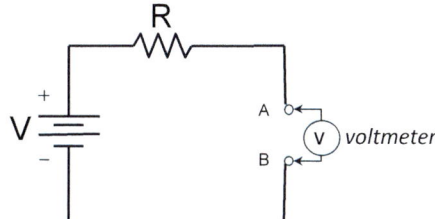

12. Find all four voltages V_a through V_d. Show all equations used to get your answers. Your answer must be in volts. Check by showing that they all add up to 20 V.

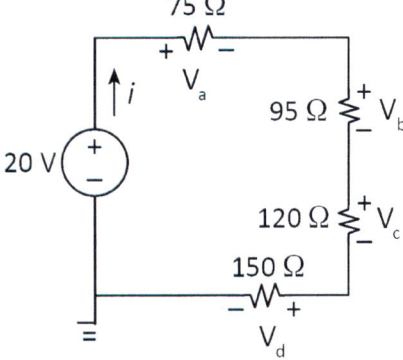

13. For the given circuit:

(a) What is R_{1-2}, the equivalent resistance of R1 and R2 combined?

(b) What is R_{3-4}, the equivalent resistance of R3 and R4 combined?

(c) What current flows through the power supply?

(d) What voltage, V_B, is at node B referenced to ground?

(e) What is V_D, the at node D referenced to ground?

(f) What is V_{AB}, the voltage across AB?

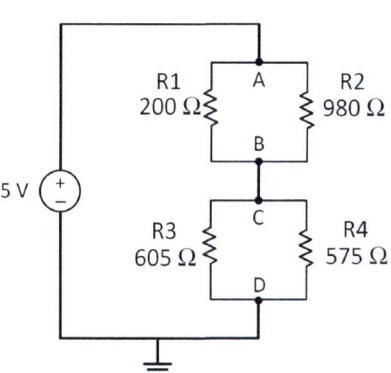

(g) What is P_{tot}, the total power provided by the power supply?

(h) If the wire that connects node B with node C is removed from the circuit, what is VB?

(i) If the wire that connects node B with node C is removed from the circuit, what is the voltage at node C?

(j) When you built your circuit on a breadboard in the lab, you accidentally connected a wire between nodes A and D. If the internal resistance of the voltage source is 50 ohm, what current would flow through the 200 ohm resistor?

14. Review the way that voltage is divided along the stripe in a potentiometer. The figure shows a battery having a voltage of 6 V between A and B at the ends of the resistive stripe. The wiper, W, makes contact at points along the stripe. The voltage at the wiper is V_W. Remember that V_W is the voltage on the wiper with respect to ground, which is zero volts.

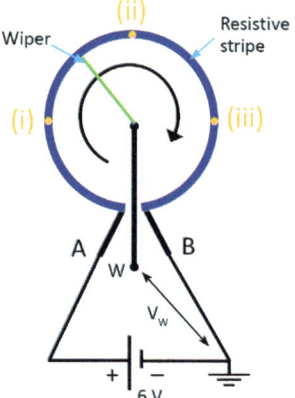

(a) Draw a diagram of this circuit including the voltage source and resistors on each side of the wiper position. Show V_W on the circuit diagram.

(b) What is V_W when the wiper is at points (i), (ii), and (iii)? Your answer should be a close estimate based on the positions of the points along the stripe.

15. In the figure, calculate I_S two ways.

(a) Use the equivalent resistance to find I_S.

(b) Find I_1 and I_2 and add them together to get I_S.

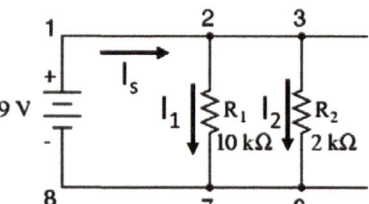

16. Consider the following circuit. $R_1 = 200\ \Omega$ and $R_2 = 300\ \Omega$. $V_s = 10$ V.

(a) What are G_1 and G_2?

(b) What are R_{eq} and G_{eq}?

(c) Find I_{tot} and use the current divider equation for conductances to calculate I_1 and I_2.

(d) Use the current divider equation for resistances to calculate I_1 and I_2.

17. (a) For the following circuit, redraw it replacing the resistance values with the conductance values. Show your work in calculating the conductances. Calculate the total conductance of the four conductances.

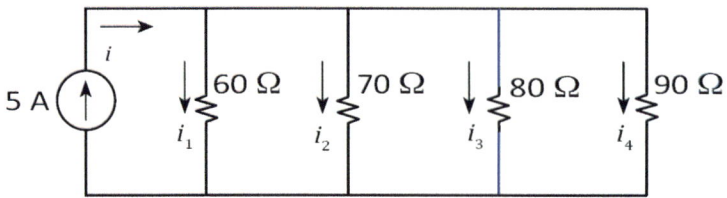

(b) What are the four currents i_1 to i_4? Use the current divider formula to solve this. Check by showing that they add up to the current from the current source. Show your work.

18. Use the current divider equation to calculate the current through the following resistors. (Yes, you could do this more than one way. You may wish to check your answer another way.)

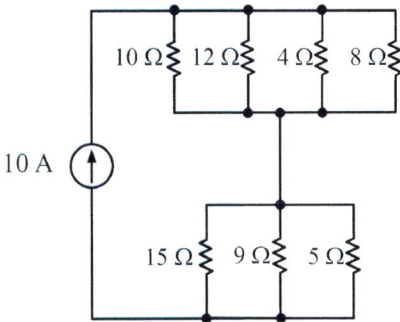

(a) Current through the 10 ohm resistor.

(b) Current through the 9 ohm resistor.

19. For the following circuit, answer the questions.

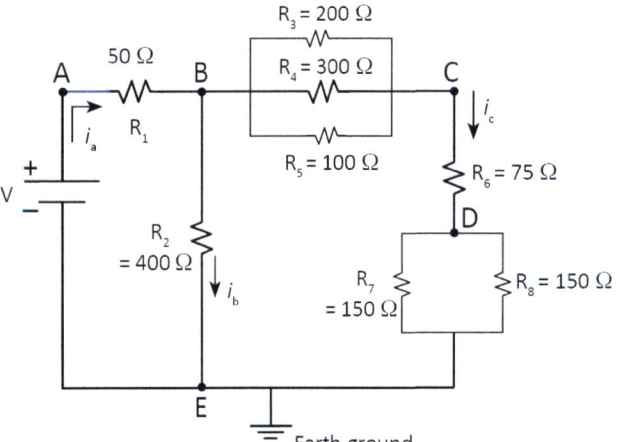

(a) What is the equivalent resistance between nodes B and C, R_{BC}.

(b) What is the equivalent resistance between nodes B and E, R_{BE}? Note: you must take into account all branches between B and E.

(c) What is I_a?

(d) What is V_{BE}? Use the voltage divider equation.

(e) What is V_{DE}? Again, use a voltage divider.

20. Think of wall electrical outlets as being powered by a single power supply. Suppose you plug in several electrical appliances into the wall outlets in a room. Would you guess that these devices are in series or parallel with each other? Why?

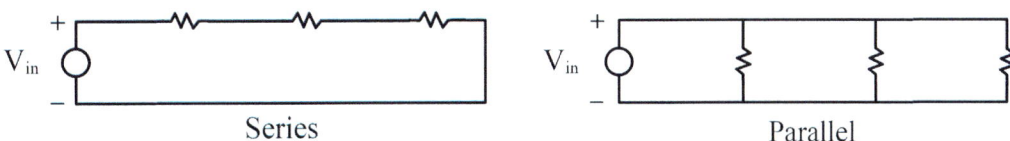

Series   Parallel

21. A very simple, but practical, real project. This may require some thought on your part.

My computer tends to get hot. To keep it cooler, I rigged up a fan with a wall-plug DC power pack, but it blew too much air. Sometimes I want more air, and sometimes less, so I put a variable resistor, or rheostat (two leads of a potentiometer), in series with the fan to cut down the current, and hence fan speed, to some fairly small fraction of the fan speed without the rheostat.

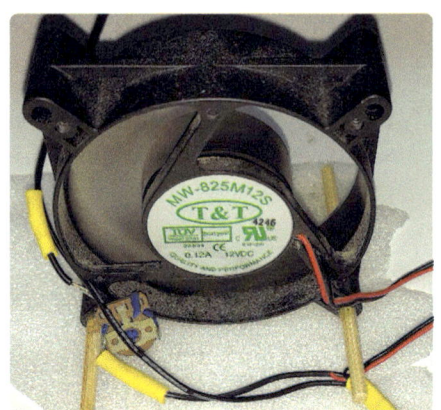

(a) As shown on the label, the fan runs on 12 VDC and draws 0.12 A. What is the DC resistance, DCR, of the fan?

(b) Draw a schematic diagram of the circuit. Include all component values as you know them so far. *Hint*: it is a simple circuit diagram of only three components.

(c) I did not have much of a choice of pots (potentiometers) laying around my house. I had a 10 Ω (0 to 10 Ω) pot, but that wasn't very useful. Why?

(c) I also had a 1 MΩ pot (0 to 1 Mohm) like the one shown, but that also wouldn't have been very useful. Why in this case?

(d) What value of pot resistance would you recommend? Why?

22. (a) An inductor has a DC current flowing through it. The wire is thin and there are a lot of windings (several hundred). If the DC resistance of the inductor (called DCR) is 30 ohms, and 150 mA flows through it, after a long time (dI/dt = 0), what DC voltage will there be across it?

(b) A toaster is rated at 800 W. You know the wall outlet voltage is 120 V. You may treat that as a DC voltage, even though it is AC (that is called its "RMS" voltage). What current is flowing through the toaster?

(c) What is the resistance of the toaster when it is hot?

23. The following question illustrates the fundamental principles of how computers operate, except that we are using simple switches instead of more-complicated transistors that behave much like switches. We can think of the supply voltage as equivalent to logic value "1", and think of 0 V, or ground, as logic value "0".

Now, consider the following circuit that contains 2 switches. The picture on the left shows two open switches and the picture on the right shows two closed switches. (Note that neither of these two cases is an actual logic configuration – it only illustrates switch positions.) When a switch is closed (pushed to the right in the picture so that it touches the two circles, making electrical contact to each) the switch behaves like a simple wire, and when it is open (pushed to the left away from the circles) the switch behaves like a missing wire, or open circuit. All of the wires are connected simply as shown – no tricks.

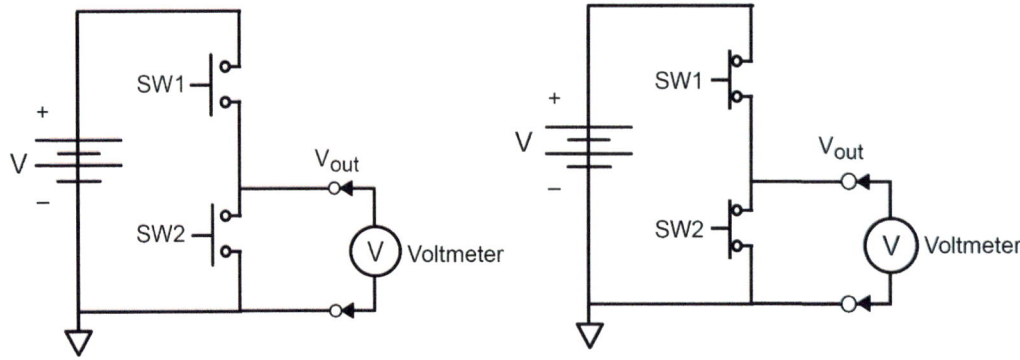

(a) When switch 1 is closed and switch 2 is open, what will the voltmeter read? What is the logical equivalent to the value read by the meter? Explain.

(b) When switch 1 is open and switch 2 is closed, what will the voltmeter read? What is the logical equivalent to the value read by the meter? Explain.

24. (a) Sketch on the figure below how you might breadboard the circuit in Fig 2.38. (There is not a single correct answer.)

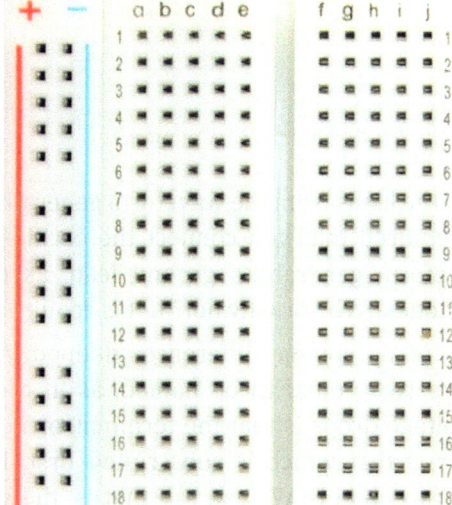

(b) In the circuit in Fig 2.38, if the source voltage is set to 5 V, what is the voltage across R2?

25. A student tries to measure the current through a resistor using an ammeter so they use the circuit shown below. The ammeter reads zero. What did they do wrong?

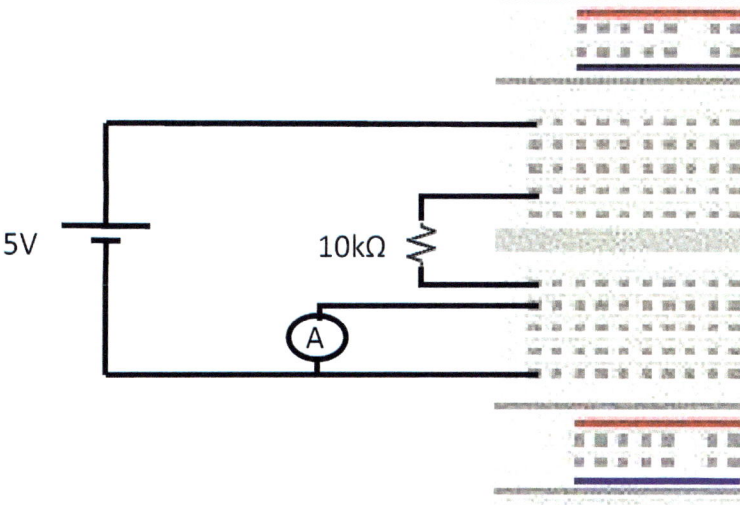

26. A student connects the voltmeter to the circuit shown in the figure, but the meter reads 0 V. What is wrong?

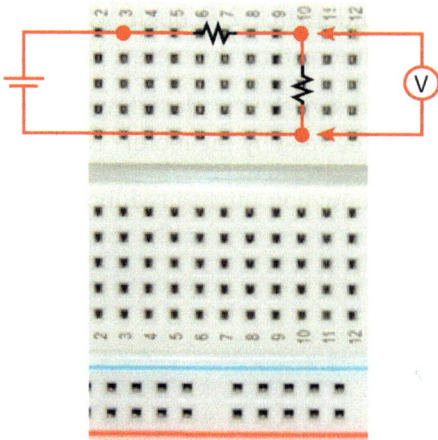

27. Suppose you connect both the leads from the waveform generator and the leads of an oscilloscope probe to a circuit. Is it required that you connect the oscilloscope's ground lead to the same circuit node as the waveform generator's ground lead, as opposed to connecting the two ground leads to different nodes? Why or why not? If it is required, what happens if you don't?

28. The following circuit has many voltage differences defined. As discussed in the text, voltage differences always require two points to be named, as in the voltage from A to B, V_{AB}. The convention for naming voltage drops is that the terminal, or node, that is defined to be positive with respect to the other is that which is subscripted first. For example, V_{AB} is the positive voltage drop from node A to node B. We can say that is 'the voltage from A to B'. V_{AB} has the opposite sign as 'the voltage from B to A', V_{BA}: $V_{AB} = - V_{BA}$. In the case of sources, the positive terminal is always listed first.

When a grounded node is identified, that node is taken as zero volts, and any node having only one subscript is understood to be the voltage difference between that node and ground, or zero. For example, if we write V_A we mean 'the voltage at node A' but it is assumed it is the voltage at A with respect to ground, which is the difference between the voltage value at A and a voltage value defined to be zero. $V_A = V_A - 0$. The zero is suppressed. We can see that the difference in voltage between any two nodes is actually the differences between each of the nodes and ground: $V_{A0} - V_{B0} = V_A - 0 - V_B + 0 = V_A - V_B = V_{AB}$. Now you will practice your understanding of this concept using the following circuit (which is not intended to be solved in any way).

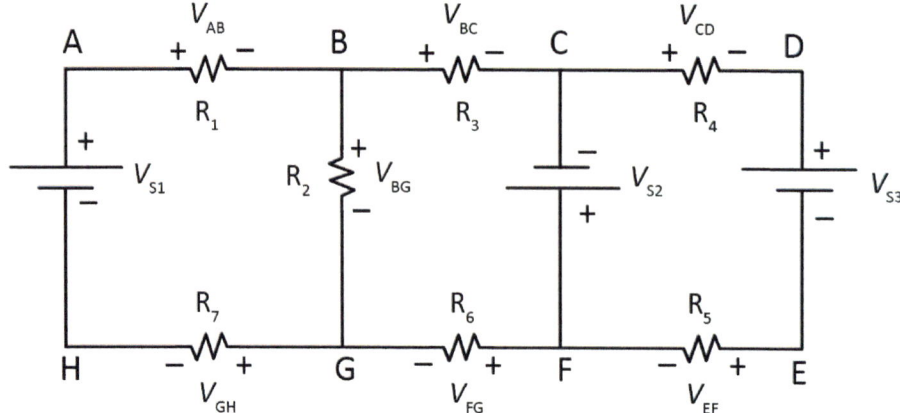

True or false (T/F):

(a) With no ground identified, you may not use any single-letter subscripts to describe any voltages.

(b) The voltage from B to C is V_{BC}.

(c) The voltage from C to B is V_{CB}.

(d) $V_{BC} = -V_{CB}$.

(e) The magnitudes of the voltages V_{BC} and V_{CB} are the same.

(f) $V_{AC} = V_{AB} + V_{CB}$

(g) $V_{DE} = -V_{S3}$

Now insert a ground at F. This can be either a chassis ground or earth ground – the principles are the same.

(h) $V_{EF} = V_F$

(i) $V_{FG} = V_G$

(j) $V_A = V_{AB} + V_{BC} + V_C$

(k) $V_A = V_{S1} - V_{GH} + V_G$

(l) $V_C = V_{S2}$

(m) What is the voltage at node B called?

(n) Write at least two ways of expressing V_D.

(o) Can you add another ground besides the one you already have at F? For example, can you also put a ground at node H? Explain.

(p) Would the circuit work correctly if you put a wire between nodes F and H so that both are now ground?

(q) Looking very carefully, with the wire between nodes F and H, what is the resistance between node G and node F/H (since they are now the same node due to the short circuit)?

29. Consider the situation of two 1.2 V batteries used in series to create a power supply, as shown. Draw the circuit diagrams showing how you would configure them with the ground to create the following three voltage supplies made of NiMH or NiCd batteries, which have $V_{battery} = 1.2$ V each: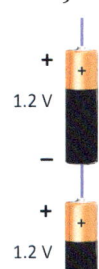

(a) A single +2.4 V source

(b) A single –2.4 V source

(c) A source having available both +1.2 V and –1.2 V

30. Look again at the circuit in Problem 2.18. Now you will do measurements of voltages and currents using both the DMM and the oscilloscope.

(a) You want to use a multimeter, like the Fluke 45, to measure the DC voltage between B and C, or V_{BC}. Can you do that, when earth ground is at node E? Explain your answer.

(b) You are using an oscilloscope whose probes are all referenced to earth ground.

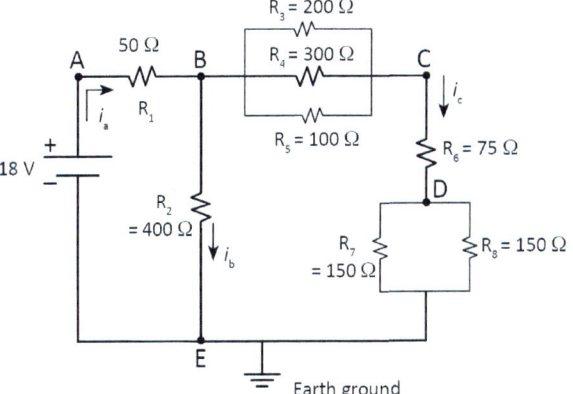

You place one probe at C to measure V_{CE} and at the same time you want to measure V_{BC} by placing another probe at B, with its ground at C to measure V_{BC}. Can you do that? Explain your answer.

(c) You want to use the oscilloscope to measure the voltage from B to C, V_{BC}. Looking forward to the lab instructions, explain how you would do that. (*Hint*: it requires a math function.)

31. (a) What is the input resistance of an ideal voltmeter? In other words, what should an ideal voltmeter '*look like*' to a circuit when measuring a voltage?

(b) What should an ideal ammeter look like to a circuit when measuring current? Why?

32. (a) What is the actual input resistance of the Fluke digital voltmeter? This is representative of most digital multimeters.

(b) Is an oscilloscope a voltmeter? Discuss.

(c) What is the input resistance of the Teledyne oscilloscope plus probe using a 1× probe? This is representative of most oscilloscopes.

(d) What is the input resistance of the Teledyne oscilloscope plus probe using a 10× probe?

33. Suppose you use an ohmmeter to measure a correctly labelled 100 kΩ resistor while it is connected in a circuit. The ohmmeter, however, reads 50 kΩ. Assuming that neither the ohmmeter nor the resistor is defective, what mistake did you make?

34. Ammeters ideally have negligible resistance while voltmeters ideally have infinite resistance. Answer the following questions for this circuit.

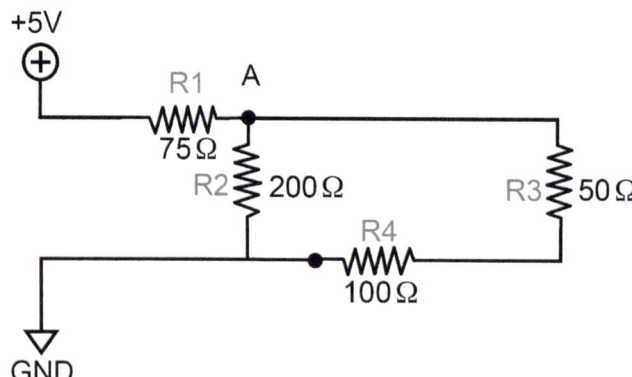

(a) How can a voltmeter be used to measure the current through a known resistance?

(b) What single resistance R_{eq} could you use to replace R_2, R_3 and R_4 in the circuit that would give the same voltage at node A with respect to ground?

(c) What is the current from the 5 V source to ground ?

(d) What is the voltage at node A with respect to ground?

(e) Now add a real ammeter in between R1 and node A and redraw the circuit with the ammeter represented as a resistor RA.

(f) The ammeter has a resistance of 10 Ω. What is the total resistance between the 5 V source and ground?

(g) What is the current from the 5 V source with the real ammeter inserted?

(h) Repeat part (c) for the ideal ammeter.

(i) What percentage change in the current did you cause by adding this real ammeter opposed to an ideal ammeter? Change = $(I_{ideal} - I_{real})/I_{ideal} \times 100\%$.

35. The Keithley 2231a-30-3 DC power supply is representative of other professional power supplies. It has three outputs, A, B and C, all of which can be used independently. Channels A and B are variable from 0 to 30 V. Channel C is variable up to 5 V. The output current can be as high as 3 A on all channels simultaneously. (Remember, there is no earth ground on any of the outputs – they are all floating.)

(a) What is the maximum power that can be supplied by each of the channels?

(b) Can you drive a 4 ohm resistor with 2.5 A from Channel A? What would be the voltage output from the source? Explain your answer.

(c) Refer to the large power potentiometer (or rheostat in this case) shown on the top right of Fig. 1.11. If you set Channel A to 30 V, what is the lowest resistance you can drive current through that variable resistor before the supply protects itself by going into overload? (*Hint*: Use your answer from part a) above.)

(d) Using the Keithley power supply, could you wire up the power supply to provide 65 VDC if you needed it? (That exceeds the OSHA limit of "safe" DC voltages of 50 V. 50 V would be quite painful, and still possibly dangerous. Do NOT do this.) Draw a circuit diagram of wiring the power supply to get that voltage.

36. Consider the following circuit. Note the position of earth ground in the circuit. The meter shown in the diagram is an oscilloscope. Answer the following questions.

(a) The input resistance of the oscilloscope is $R_0 = 1$ MΩ, as if you used the oscilloscope with a 1× probe. For $R_1 = R_2 = 100$ Ω, what is V_{BC} (the voltage from B to C)?

(b) Now let $R_1 = R_2 = 300$ kΩ. Now, what is V_{BC}?

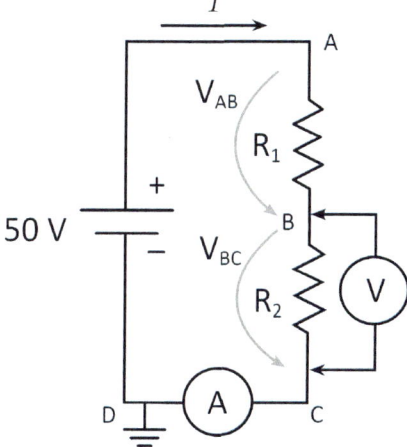

(c) Did you get a different value for V_{BC} with the larger resistor values in part (b)? Explain your answer to part (b).

(d) What is the current through the ammeter for $R_1 = R_2 = 300$ kΩ) *without* the oscilloscope in the circuit?

(e) Now, what is the current *with* the oscilloscope in the circuit?

(f) Explain why your answers for parts (d) and (e) are different with the oscilloscope in the circuit.

(g) Repeat part (b) for $R_0 = 10$ MΩ, as if you used the 10× scope probes. What is V_{BC} now?

(h) What is the current through the ammeter as shown for this case ($R_0 = 10$ MΩ)?

(i) If you are using the Fluke DMM instead of the oscilloscope to measure the voltage, can you use it to directly measure V_{AB}? Yes or no? Why or why not? (*Hint*: This question is about grounding.)

(j) If you are using the oscilloscope, can you use it to directly measure V_{AB}? Yes or no? Why or why not?

(k) If ground were placed at node C instead of D, would the power supply's negative terminal still be effectively at ground potential? Yes or no? Why or why not?

(l) Can you measure the current by breaking the wire at A instead of at C and D, as shown? Why or why not?

37. The waveform generator in the lab is used to power the following circuit with a sinusoidal voltage set to 5 V_p. Your answers will also be in V_p or peak amperes, A_p.

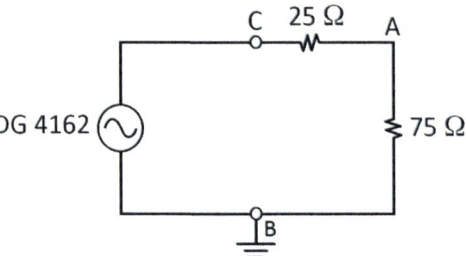

(a) Redraw the circuit to include the 50 Ω internal resistance of the WG. What is the voltage that you will read at nodes C and A, i.e., V_C and V_A?

(b) By mistake you apply the earth grounds at both nodes B and C while at the same time using the current probe at node C to measure the *current out of the WG*. What current I_{WG} from the WG will the current probe read?

(c) You open node C and measure the voltage into the WG. We will call that $V_{open\ circuit}$, or V_{OC}. What voltage V_{OC} will you read?

(d) If we call the current from the source in part (b) $I_{short\ circuit}$ or I_{SC}, what is the ratio of V_{OC}/I_{SC}? Include the units.

(e) Based on these results, can you make a general statement that relates V_{OC}, I_{SC} and R_{int}?

38. This is a continuation from the previous question. Now you will insert the power amplifier (PA) between the WG and your load comprising the 25 and 75 Ω resistors: Connect the output of the WG and its internal resistance, without the load, to the input of the PA1011 power amplifier having gain = 1 (its output voltage is the same as the input voltage, for an open circuit at the output). Remember, its input resistance is 50 kΩ and its output resistance is 2 Ω. Now drive the two resistors, 25 Ω and 75 Ω, from the output of the PA.

(a) Draw a diagram of the entire circuit including all internal resistances and source voltages and their values, and the two load resistors, much like the circuit in Fig. 2.31. Label the output of the WG as node C, as in the previous problem. Label the output of the PA node PA, and the voltage there with respect to ground is V_{PA} (here, PA stands for power amplifier). Node A is still across the 75 Ω resistor, as it was in the previous problem.

(b) What is the voltage at node C, V_C, at the output of the WG? (Hint: This should be a simple voltage divider, based on your circuit diagram.) Compare to the voltage you calculated in the previous problem.

(c) What are V_{PA}, the voltage output of the PA, and V_A, the voltage across the 75 Ω resistor? Note: The input voltage of the PA is the voltage V_C that you calculated for node C, since that is what the PA sees at its input. That voltage is the internal source voltage of the PA. You still have the two resistors, 25 Ω and 75 Ω, between node PA to ground as the load that the PA is driving. V_A is still across the 75 Ω resistor.

(d) Compare and explain the differences between the answers to part (c) and those of part (a) of the previous question.

39. An "ideal" voltage source can supply any amount of current (for example a billion amps) demanded of it by the circuit that it is part of. "Real" voltage sources, however, are limited in how much current that they can actually produce (for example one amp, 10 amps or maybe hundreds of amps). Common batteries such as those that you might find in a watch, an MP3 player, a laptop computer, or a car, can have very different capacities to drive current. These days it seems that just about everything has a battery in it, so it is useful and important to understand some basic ideas regarding them.

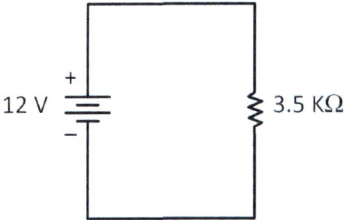

Battery and radio shematic

Consider two different batteries that drive a radio that has a resistance of 3.5 kΩ, as shown in the circuit diagram. In one case, the battery is 8 AAA cells (or individual batteries) in series, each 1.5 V, making a voltage source totaling 12 V weighing 90 grams, as shown below. In the second case, the battery is a 12 V car battery weighing 40 pounds, like that shown in Fig. 10.9.

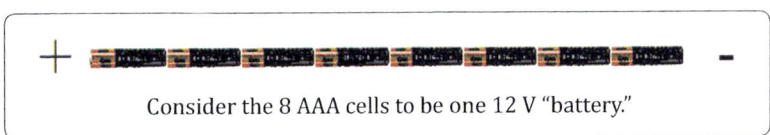

Consider the 8 AAA cells to be one 12 V "battery."

(a) Ignoring any internal resistance of the batteries, calculate the current in the 3.5 kΩ resistor for each voltage source.

(b) Suppose the 3.5 kΩ radio is changed to 0.1 Ω representing a gasoline car engine starter. This is an electric motor that rotates the engine until it runs on its own. What current do you calculate for each battery now?

(c) When you compare the size of each battery, is your answer for part (b) reasonable? What does your intuition tell you about this result?

(d) A real voltage source, like a battery, can be modeled as an ideal voltage source in series with a resistor, as shown in the figure. Consider the revised circuit that includes the "internal resistance," Ri, of the battery. The internal resistance is inside the battery, so you cannot take it out of the circuit. If you short circuit the battery leads, then the current will be limited due to the internal resistance in series with the ideal voltage source.

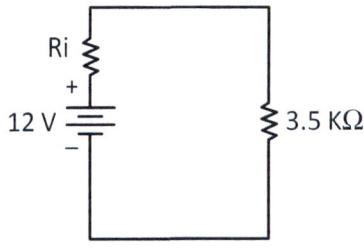

For the car battery assume an internal resistance of 0.015 Ω and for the AAA batteries, assume an internal resistance of 8 Ω. What is the short-circuit current for each of the batteries now? (*Hint*: to calculate the short circuit current, replace the 3.5 kΩ resistor with a wire of zero resistance.)

(e) In order to start running, internal combustion engines need an electric motor (the 'starter') to turn the engine so that the gas combustion process can take over. Suppose you use the car battery having internal resistance as provided above to start a car having a 0.1 Ω starter motor. What voltage will appear across the starter? This is a voltage divider question. Draw the circuit and calculate the voltage drops.

(f) How much current will flow through the starter?

(g) Repeat parts (e) and (f) for the 8 series-connected AAA batteries, including the circuit diagram. Why do you suppose we don't use 8 AAA batteries to start our gas-powered cars? Explain.

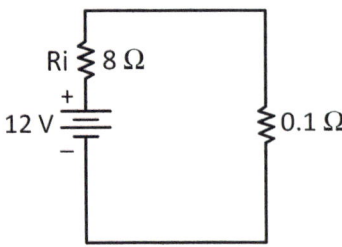

40. Is an oscilloscope a voltmeter? Discuss.

41. What is the input resistance of the Teledyne oscilloscope (and most others) including a 1× probe?

42. What is the input resistance of the Teledyne oscilloscope including a 10× probe?

43. For whatever reason, you've decided to use your oscilloscope as a voltmeter to measure a DC voltage and sure enough you see a straight line across your oscilloscope's screen. Other than that line's location on the screen, what two pieces of information do you need to know the voltage measured?

44. Name some advantages of using the current probe rather than an ammeter.

45. Does it matter what the input resistance of a current probe is?

46. Suppose you use both the DMM and the oscilloscope in the same circuit. Do you have to connect the COM port of the digital voltmeter to the same ground as the oscilloscope probe? Why or why not?

47. What is meant by the trigger level of an oscilloscope?

48. You generate the waveform shown below.

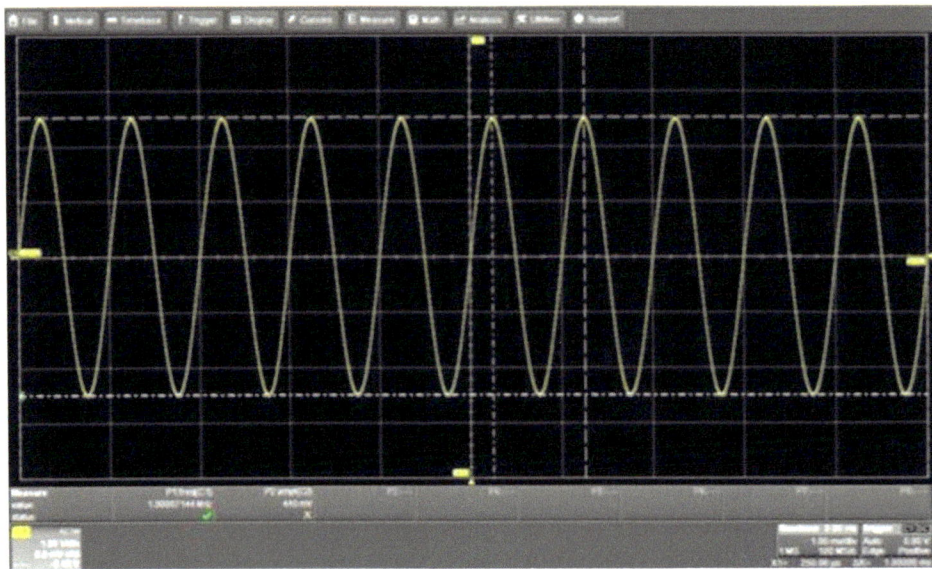

(a) If the scale is set to 5 V/div, what is the peak-peak voltage of this waveform? (*Hint*: ignore all of the cursors and use the grid lines.)

(b) What is the peak voltage?

(c) If the time base is set to 1 ms/div, what is the frequency of this waveform?

Chapter 3

Power Transmission and the Power Grid

3.1 Introduction and Overview

In this chapter you will learn about:

1. The national power grid and power generation
2. EMF, inductors, and transformers
3. Three-phase power
4. How power is delivered to large buildings and industry
5. How power is delivered to our homes
6. How houses are wired
7. How power is delivered efficiently (power factor)
8. How power is delivered to large commercial customers.

3.1.1 Power Generation and Transmission

On August 14, 2003, a blackout began that knocked out much of the northeast of the United States and Ontario ... no traffic lights or refrigerators ... 56 million people affected – the largest power outage in U.S. history. And the cause of all this mayhem? A power line touched a tree branch in Walton Hills, Ohio ... and a software glitch caused 265 power plants to go offline within a few minutes. In the United States, we take for granted power 24/7 with no interruptions. But when a power failure does happen, it is disruptive and dangerous. In fact, many emerging countries have little or no power away from large cities, and it is intermittent in many large cities.

A basic knowledge of power generation and transmission is an essential part of a complete electrical engineering education. Moreover, the electric power industry is undergoing an unprecedented transformation, so modern power engineers will be in high demand. Those future careers will require state-of-the-art knowledge. In this chapter you will learn about power systems all the way from the hydroelectric generator to the switch that turns on a lamp.

An Introduction to Electrical Engineering with Lab Activities
Gary H. Bernstein
Copyright © 2026 Jenny Stanford Publishing Pte. Ltd.
ISBN 978-981-5129-30-4 (Hardcover), 978-1-003-71345-6 (eBook)
www.jennystanford.com

DOI: 10.1201/9781003713456-3

3.1.2 The Field of Power Engineering

The discipline of electrical engineering is broad. There are 38 technical societies comprising the Institute of Electrical and Electronics Engineering (ieee.org) in fields ranging from aerospace systems to information theory to magnetics to vehicular technology. The Power and Energy Society members are concerned with the generation and delivery of power from power plants over a complex network of wires to their final destination in homes and industry. Those members notwithstanding, many experienced electrical engineers do not fully appreciate the miles of giant transmission lines and massive substations, or the significance of transformers scattered around their own neighborhoods and up on utility poles. By the end of this chapter and lab, you will see with different eyes, and gain an appreciation for all of the electrical infrastructure that surrounds you.

3.1.3 The U.S. Power Grid

The term **power grid** refers to the collection of hardware and software that makes possible the generation and delivery of electrical power. Power grids tend to connect large areas, ranging up to the size of countries and even continents. The voltage oscillations are sine waves (as in Chapter 2) whose **line frequency**, also known as **utility frequency**, is 60 Hz in North America and in a few countries elsewhere. It is 50 Hz in the rest of the world. Power transmission around the world is carried on three wires whose voltage oscillates at either 50 Hz or 60 Hz, but with each wire voltage differing in **phase** from the other two wires. Worldwide, power is distributed by three wires with sine waves on each wire that are separated in time by 1/3 of a cycle. This is called **three-phase power**, which will be discussed at length in this chapter.

The contiguous 48 states of the United States are served by three separate, huge power grids breathing, electrically speaking, 60 times per second (i.e., 60 Hz) over vast networks of power lines, each having three phase lines synchronized over hundreds of thousands of square miles. This giant array of wires has many thousands of sources contributing power to the grids, and many millions of loads taking power off of the grids.

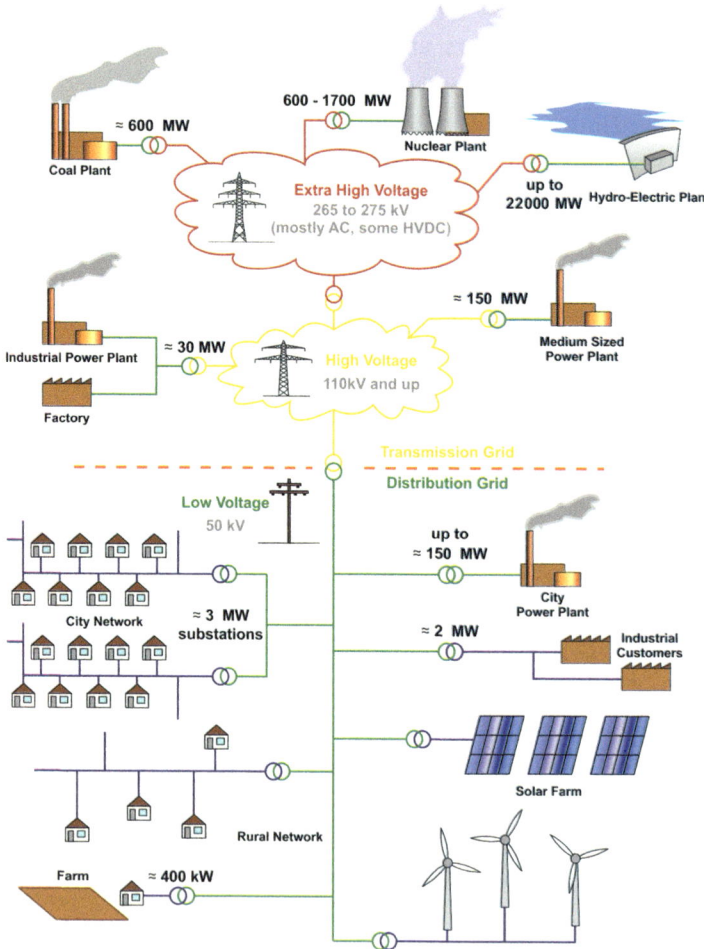

FIGURE 3.1 Power grid overview. *Source*: https://commons.wikimedia.org/wiki/File:Electricity_Grid_Schematic_English.svg, published under Creative Commons Attribution CC BY 3.0.

The main components of a power grid (Fig. 3.1) are: (1) various types of power generators, (2) transmission lines for moving power over long distances, (3) devices (transformers, introduced below) to step voltages down from very high transmission voltages to much lower customer line voltages, and (4) consumers of the power, i.e., the customers. We shall discuss each of these components.

Important: The term 'load' is extremely general. A **load** is anything that is driven at the output of any circuit. It is completely up to you to define what circuit you are talking about, what the output is, and what it means to 'drive' the load.

3.1.4 Electrical Energy Usage in the United States

The United States uses about 4 trillion kWh (kilowatt-hours) of electrical energy per year, i.e., 4×10^{15} Wh (watt-hours) (www.eia.gov). What is a "kilowatt-hour?" For that matter, what is a "watt-hour?" It is common among nontechnical folks to conflate the notions of energy and power, so you must be clear on these concepts. Power is the transmission of energy *per unit time*. Energy is expressed in joules (MKS units), and power is expressed in watts. The integral of power over time yields the quantity of energy used over that time. (In MKS units, 1 W × 1 s = 1 J.) Often, the integrand (power) is constant, so the integral (energy) is simply power × time.

Energy is reported to customers by the power industry in multiples of kWh, which we pay for on our electric bills. The cost of 1 kWh of electricity (or more precisely, *electrical energy*) is roughly between 10 and 15 cents depending on location and time of day. It is estimated that about 6% of all power supplied to the grid is lost as heat in the transmission lines. Based on these numbers, you can calculate that the annual cost of losses in the national grid's transmission lines is on the order of several tens of billions of dollars. This might explain why many companies are developing superconducting (i.e., zero resistance) transmission lines. (See: https://www.eenewseurope.com/en/cutting-the-complexity-of-superconducting-power-lines/.)

Of interest: Now that you are sensitized to the difference between watts and watt-hours, you might start to notice this common confusion in non-technical articles about renewable energy.

Now would be an excellent time to review the basic laws of electromagnetism. You have probably seen them in a high school or college physics course. If you unclear about what they mean, they will be explained as they are used in this text.

Gauss's Law for electric fields: $\oint \vec{E} \cdot d\vec{A} = \dfrac{Q_{enclosed}}{\varepsilon_0}$

Gauss's Law for magnetism: $\oint \vec{B} \cdot d\vec{A} = 0$

Coulomb's Law: $F = \dfrac{1}{4\pi\varepsilon_0} \dfrac{q_1 q_2}{r^2}$

Ampere's Law: $\oint \vec{B} \cdot d\vec{\ell} = \mu_0 I$

Biot-Savart Law: $\mathbf{B(r)} = \dfrac{\mu_0}{4\pi} \displaystyle\int_C \dfrac{I d\vec{\ell} \times \mathbf{r'}}{|\mathbf{r'}|^3}$

Faraday's Law of Induction: $\oint \vec{E} \cdot d\vec{\ell} = -\dfrac{d\Phi_B}{dt}$

Lenz's Law:

Lorentz Force: $\vec{F} = q\vec{E} + q\vec{v} x \vec{B}$
  Electric   Magnetic
  force   force

3.2 Magnetism in Power Devices

3.2.1 Inductors

The last circuit element of the Big Three discussed in Chapter 1 was the inductor. Recall that an inductor is basically a coil of wire, with or without a magnetic material as its core. The inductor forms the heart of many important devices used in power systems. In an inductor, the magnetic field resists changing currents according to

$$v_L = L \frac{di_L}{dt}, \tag{1.30}$$

where v_L is the potential difference across the ends of the inductor, and i_L is the current though it. As discussed in Section 1.4.4.1, v_L is actually an EMF, but is frequently simply referred to as a voltage.

Equation 1.30 tells us that the faster the current changes (higher di_L/dt), the greater the voltage across the two ends of the inductor. The inductance, L, determines how much voltage is induced for a given change in current. It is worth noting that the term "inductance" is actually shorthand for **self-inductance**, which implies that its v_L is due to its *own* magnetic field. When we get to the transformer, which is two coupled inductors, the magnetic field in a coil will not be due only to that generated within itself.

The basic equations underlying the behavior of inductors are Ampere's Law and **Faraday's Law**. Ampere's Law (see Section 1.4.2.2) tells us that any current causes a related magnetic field to form instantaneously – if there is current then there is magnetic field, with no time lag whatsoever. Faraday's Law is a more fundamental way of representing Eq. 1.30 in terms of the fields. It says that a changing magnetic field (whether due to a changing current or a moving magnet) causes an EMF. Faraday's Law is written as

$$\text{EMF} = -\frac{d\phi_B}{dt}, \qquad (3.1)$$

The EMF is the voltage (integral of electric field) around a single loop surrounding the total enclosed flux, Φ_B, as shown in Fig. 3.2 and discussed in the sidebar below it. Voltage is the integral of electric field, which in this case forms a loop going in the same direction, like a snake eating its tail. That integral is written as

$$v = -\int E dx \ (volts). \qquad (1.14)$$

Integrating around a full loop, since the integral ends at the point that it begins, you might think that the voltage around the loop should integrate to zero by Kirchhoff's voltage law. KVL is vastly useful in performing circuit analysis, and it is valid even if there are magnetic fields inside circuit elements such as inductors. How is KVL not required when summing the voltages around the loop of wire shown in Fig. 3.2? Basically, KVL pertains to lumped EMFs (a source of voltage as discussed in Section 1.4.4.1) combined with lumped dissipative circuit elements in which the EMFs raise and/or lower potentials, and the net potential is dropped at the dissipative elements. To put it more simply, the answer comes down to the nature of an EMF versus a voltage drop. The lumped circuit is a combination of the two, whereas the electric field around the loop is purely an EMF. The circuit has 'ups' and 'downs' in potential depending on if one passes an EMF or a dissipative voltage drop, whereas the loop has only 'ups' in potential while traversing around the loop.

In order to illustrate these ideas more fully, let's transform Fig. 3.2 into a collection of lumped sources of EMF. Imagine you take twenty 1.5 V batteries and connect them end-to-end in a loop, with the polarities all in the same direction. The total voltage drop would be 30 V adding voltages from a starting point, around and back to itself. There would be an electric field internal to the batteries that would go around continuously (like that snake eating its tail). If the batteries had almost no internal resistance, then the current in the loop would be very large. If the batteries

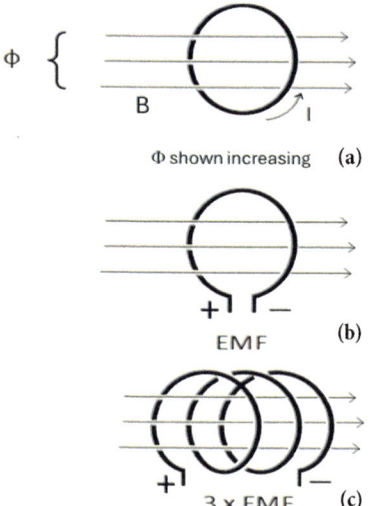

FIGURE 3.2 Electromotive force, EMF. (a) Magnetic flux shown increasing through the interior of a closed wire loop. Current flows due to the electric field from the EMF in the direction that opposes the increasing flux (Lenz's law). (b) Here, the wire has a break in it so no current flows and the EMF appears across the ends of the wire. The sign of that EMF is such that it would drive current externally to satisfy Lenz's law. (c) Three turns in series with the same increasing flux, and 3 times the EMF appears across the turns. This increase in total EMF due to increasing number of turns underpins the operation of transformers.

had very high internal resistance, then the current would be near zero. Either way, the electric field internal to all those batteries would have to integrate to a value of EMF$_{turn}$ = 30 V, or to put it really simply, $\pi D\boldsymbol{E}$ = 30 V, where $\boldsymbol{E}$ is the electric field, D is the diameter, and πD is the distance around the loop.

Now let's consider a single closed loop, or turn, of wire in space (Fig. 3.2(a)) and create our $d\Phi_B/dt$ by pushing a bar magnet toward the loop. As we do so, the resulting EMF around the loop will cause a flow of current equal to v/R in that wire, where v is the EMF and R is the wire resistance. The current will flow in such a way that its magnetic field, due to Ampere's Law, always opposes the change of the flux due to the bar magnet. That relationship is called **Lenz's Law**. Can you imagine what would happen if Lenz's Law did not exist? The resulting current would add more magnetic flux, which would

> *Important:* In this treatment we use the terms magnetic 'field' (B or H) and magnetic 'flux' (Φ_B). These do not mean the same thing, but they are related and often used interchangeably. The field strength is a measure of the forces between two magnets. Refer to Fig. 3.2. When we draw lines of magnetic field, we denote field strength by the separation of the lines – closer lines are stronger fields. Now if we collect all those lines, like drinking straws, that cross through a closed surface (say the area of a loop in Fig. 3.2) we are talking about the number of straws as the total flux crossing that surface or passing through the loop (units of webers, Wb). So, flux refers to how much total magnetic field crosses a device. We could have said that the magnetic flux is the surface integral of the magnetic flux density ('straws per unit area,' in units of tesla), but that kind of talk is reserved for junior-level courses in electromagnetics.

add more current, etc., until that single current loop *destroys the entire universe*. We couldn't have that now, could we? Seriously, if the current could increase the magnetic flux, then more energy would end up in the total magnetic flux at no cost, and this would violate conservation of energy, which simply *never* happens. The magnetic field due to the current opposes the direction of the increase of magnetic field in order to maintain the energy in the magnetic flux in the coil at that source of the current.

Now further imagine that instead of using a closed loop of wire, we make a tiny break in it (create an open circuit, Fig. 3.2(b)). Pushing the bar magnet toward the coil, now no current will flow, but a voltage of EMF$_{turn}$ = $-d\Phi_B/dt$ will appear instantaneously *across the ends of the loop* as the magnetic field changes. That tiny break effectively adds a resistance of infinite ohms in series with the induced EMF$_{turn}$. If current did flow in some load between the two ends, it would flow due to the polarity of the EMF shown in the figure. Now, further imagine that the one broken turn continues into a coil of N connected turns (Fig. 3.2(c) N = 3), all threaded by the same magnetic field, but still having a break at the ends. Each one of the N turns will develop a voltage of EMF across it, so N turns in series will sum to a total voltage of

$$\text{EMF}_{coil} = N \times \text{EMF}_{turn} \tag{3.2}$$

across the two ends of the entire coil. Each turn within the coil behaves as if it were a battery having a voltage of EMF$_{turn}$, and N of those in series sums to a voltage of EMF$_{coil}$.

In a coil used in a circuit, the changing magnetic field usually comes from current flow rather than from a moving permanent magnet or some other changing magnetic field. (The notable exceptions are transformers and generators, discussed below.) **Ampere's Law** states that an electric current, I, in a wire is surrounded by a magnetic field. (For completeness, the fuller form of Ampere's law is $\oint_c \boldsymbol{B}.d\ell = \mu_0 \iint_s \boldsymbol{J} \cdot d\boldsymbol{S} = \mu_0 I_{enc}$, but you probably won't understand it until an electromagnetics course.) A changing current through the coil causes a changing magnetic field and a total EMF$_{coil}$.

We can now repeat the previous discussion of inductors in terms of changing current rather than changing magnetic field alone. Current through the wires of an inductor generates a proportional magnetic field (Ampere's Law) with a changing total enclosed magnetic flux; as current changes, magnetic flux changes instantaneously along with it; *changing* magnetic flux gives rise to a voltage that is proportional to its rate of change (Faraday's Law); the relationship between the original change in current and the ultimate voltage collapses to a total EMF of $v_L = L\frac{di_L}{dt}$ (Eq 1.30).

The formula $v_L = L\frac{di_L}{dt}t$ is just a retelling of Faraday's Law, including the inductance, which depends on the physical properties of the coil of wires. Equation 1.30 tells us that a voltage (really an EMF) develops across the inductor *only* when the current, and hence field, are *changing*. A *constant* current

($di/dt = 0$) will *not* cause a voltage drop, no matter how much that current is. For a given di/dt, the more turns that the inductor has (i.e., larger number of turns, N, hence larger L), the stronger the magnetic field within it, and the higher the EMF. Unlike capacitors, there are no terms that increase the component value, in this case L, by shrinking. Given a certain core material with a fixed permeability, the only way to increase L is to add more turns. Therefore, inductors tend to be big and heavy, and are often the most massive components in a circuit, such as an electrical filter like that found inside speaker cabinets (to send the highs to the tweeter and the lows to the woofer, Fig. 3.3).

FIGURE 3.3 Passive crossover network for a high-end stereo speaker. The inductors and capacitors send the low frequencies to the big driver (woofer) and the high frequencies to the small driver (tweeter).

3.2.2 Generators

You are probably at least somewhat aware of electric power plants that use fossil fuels (coal, natural gas, or oil), nuclear reactors, and hydroelectric turbines. Each of these is simply a different source of energy used to spin an electric generator. A **generator** is basically an electric motor running in reverse: it converts mechanical energy into electrical energy. The energy source causes the magnetic field of a **rotor** to turn past multiple stationary coils, also called **windings**, collectively called the **stator**, as shown in Fig. 3.4. A simple generator having only three coils is shown in Fig. 3.5, left. The rotor is very much like a permanent bar magnet that has N and S poles. The rotor is turning on an axis so that the poles sweep out a circle. As a result, at every point on the circle the magnetic field changes sinusoidally with time. The stator is a set of coils placed on that circle that experience that changing magnetic field.

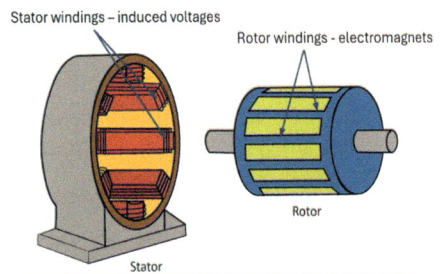

FIGURE 3.4 Rotor and stator parts of an electrical generator. (Top): Drawing of rotor and stator. (Bottom): Generator with rotor suspended overhead. The rotor has many magnetic poles, and the stator produces 3 phases of power using multiple smaller windings connected in series. *Source*: army. mil/article/189754/crew_installs_249_ton_ hydropower_rotor. Public domain.

According to Faraday's Law, a voltage is induced on a coil when a magnetic field through that coil changes. The rotating magnetic field of the rotor is constantly changing at the stator coils, and so induces a voltage on each stator coil. For three-phase power generation, the stator could comprise exactly three independent coils. These would be evenly spaced around the circle and swept by a magnetic rotor having one North, N, and one South, S magnetic **pole**, as shown on the left in Fig. 3.5. This produces three separate sine waves, one from each stator coil, at the appropriate phases for distribution on the power grid, as shown on the middle of Fig. 3.5. Note that each stator coil operates independently of all the others, each producing a sine wave. As a magnetic N pole sweeps past a coil it produces voltage in one direction, and then the magnetic S pole produces voltage in the other direction. A complete sinusoidal voltage cycle takes a full change from N to S and back to N of the magnetic poles. Typical generators use multiple coils for each of the three phases, as shown at the right of Fig. 3.5.

In practice, bar magnets are not used to make up a rotor. Instead, rotors use wire-wound electromagnetic coils, or **windings**, as shown in Fig. 3.4. The two magnetic poles on the rotor, called a **pole-pair**, are simply the ends of the rotor windings with constant current flowing through them. Using a single pole-pair for a *huge* power generator would require it to make 60 revolutions in a *second* (3600 revolutions per minute, rpm), creating *massive* physical forces that could tear it apart. Instead, very large generators use multiple, equally spaced pole-pairs on a rotor that sweep past multiple sets of three stator coils. This configuration of multiple rotor and stator windings is depicted at the top of Fig. 3.4. Figure 3.5, right, shows a generator that has

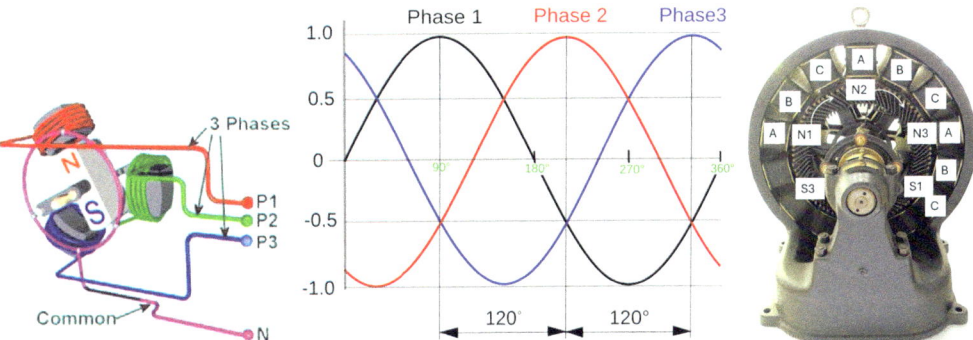

FIGURE 3.5 (Left) Three-phase power generator and (middle) the resulting sine waves that come from each set of coils. Notice that the waves are identical but displaced from each other by a third of a period (120°). (right) A 3-phase generator having 5 sets of phase windings on the stator. Every fourth winding is part of one of the phases. All of the N poles on the rotor are at the same relative location on one of the phases, phase A shown here. The corresponding S poles are on the opposite side of the axle. Sources: Left—https://electronics.stackexchange.com/questions/279857/simple-3-phase-generator; middle—https://commons.wikimedia.org/wiki/File:3_phase_AC_waveform.svg, public domain; right — https://commons.wikimedia.org/wiki/File:Three-phase_alternating_current_generator,_ASEA,_1890s,_view_1,_used_in_Hellsjo_power_station,_TM2907_-_Tekniska_museet_-_Stockholm,_Sweden_-_DSC01478.JPG

15 total windings arranged in three sets of five windings, each set labelled as A, B or C, comprising one of the three phases in 3-phase power. Each winding in a set of three provides a part of the total voltage for one of the three phases. Hence, there are three times more windings on the stator than there are pole-pair windings on the rotor. The full stator still produces three output voltage phases, but each voltage phase is the sum of the many smaller voltages from the set of many separate sub-windings wired in series. All of the same polarity (N or S) poles of the rotors fly past each of the sub-windings of the phases at exactly the same time at the same relative location. This is shown in Fig. 3.5, right, as N1, N2 etc. Therefore, all the sub-windings are synchronized for each phase and add the same amount of voltage to the total phase voltage.

The rotational speed of the rotor, f_{rot} in RPM, is related to the number of pole-pairs, p, and electrical frequency, f in Hz, by

$$f_{\text{rot}} = \frac{60 f}{p}. \tag{3.3}$$

FIGURE 3.6 Hoover dam on the Colorado River between Arizona and Nevada. *Source:* https://commons.wikimedia.org/wiki/File:Hoover_Dam_09_2017_6048.jpg

As an example of how using multiple poles and stator windings results in a slower overall rotation speed, consider a generator having 60 pole-pairs and 180 stator windings generating three phase voltages at 60 Hz. In this case, the rotor will spin at a much more manageable rotational speed of 60 × 60 Hz/60 pole-pairs = 60 RPM = 1 rev/s compared to 60 rev/s or 3600 RPM for a single pole-pair.

The output voltage is 'phase-locked' to the rest of the grid so that the generator produces electricity at precisely the right frequency and phase to put power onto the grid in synchrony with the power lines that serve a huge area. Line frequency is kept very close to 60 Hz, deviating by about ± 0.1 Hz.

When you hear "steam-driven," you might think of a 19th century train locomotive. But most power generation today really is steam-driven. In steam-driven generators, either fossil fuels are burned, or nuclear reactions create heat to boil water into steam that drives a **turbine**. The shaft of the turbine is directly attached to, and drives, the rotor of the generator. In hydroelectric power plants, the dam (Fig. 3.6) holds back large volumes of water whose surface is at a higher elevation than the bottom

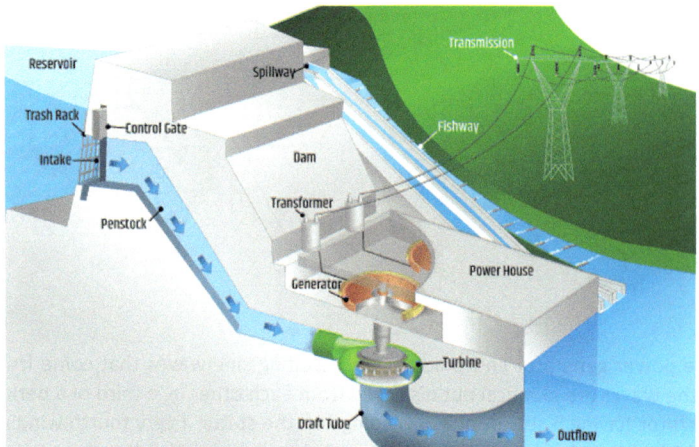

FIGURE 3.7 Dam structure and operation. *Source:* https://www.energy. gov/eere/water/types-hydropower-plants.

of the dam, so water is forced through turbines connected to generators (Figs. 3.7 and 3.8). The source of energy is "renewable" in the sense that the sun causes precipitation to fill the dam with water, and we use that stored energy to create electricity. The top of Fig. 3.8 shows a cutaway view of a turbine/generator, and the bottom shows the generators inside the bottom of Hoover Dam. (If you are heading out that way sometime, consider taking a tour of the facility.)

Alternative energy sources, most commonly wind and solar, can also be used to generate electricity that is supplied to the grid and delivered to customers. Wind turbines incorporate electric generators that are directly driven by wind, but as with many other alternative-energy sources, the power is 'conditioned' to match frequency, amplitude, and phase to the grid before it goes out. Solar farms can generate electricity from thousands of solar panels (the subject of Chapter 8) and can also use sunlight reflected from thousands of mirrors to store thermal energy in a molten material (such as sodium nitrate). The molten material is used to boil water to run a generator, or the heat may be used to perform high-temperature manufacturing processes, such as melting steel or generating hydrogen fuel.

How is electric power put onto the grid from a generator in synchrony with the rest of the grid? Imagine two people turning two jump-ropes as one of those impressive double-Dutch rope-jumping teams. The second jump-rope must be turned in synchrony with the first one. The phase and frequency of the second rope with respect to the first rope must be correct. Similarly, a power generator spins up to high voltage at the right frequency, amplitude, and phase while off the grid. When it is running at precisely the desired values, a set of large switches is closed to connect the generator and the grid so that power flows smoothly from the generator. Very slight changes to its timing (i.e., phase) relative to the grid cause either more or less power to flow from the generator. Once this happens, the generator's frequency becomes 'locked' to the grid. In this way, power plants can be brought on- and off-line over the course of the day to satisfy varying power demands, such as air conditioners on a hot day.

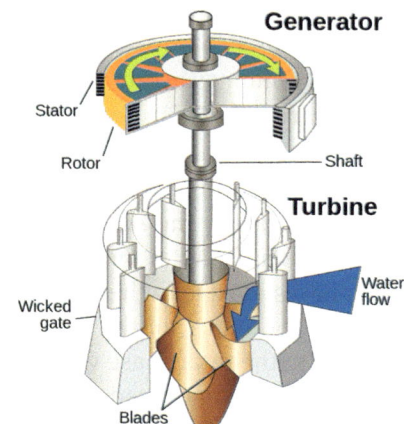

FIGURE 3.8 (Top) Drawing of a hydroelectric water turbine. *Source*: https://en.wikipedia. org/wiki/Water_turbine#/media/File: Water_turbine_(en_2).svg. (Bottom) Hoover Dam generators. Photo: C. Manning.

3.2.3 Transformers

Our discussion of power transmission systems continues with an essential electrical device called a **transformer**, which is not to be confused with the mechanical creature from another planet. *Our* **transformer** is an extension of the inductor, and it is an essential component of the power grid, as well as most electronics that plug into a wall outlet. The transformer comprises two inductors that share a common magnetic field. The two inductors are almost always wound on a common core made of magnetic material, frequently a **ferrite** (which is a ceramic magnetic material) (see Fig. 3.9). As an AC voltage and current are input on one side, another AC voltage is induced on the other side. The induced voltage can be higher, lower, or the same, depending on the transformer. This is critical to the advantages of AC transmission systems because it is necessary to transform voltages from low to high and high to low. 'High' is often hundreds of thousands of volts on power lines and 'low' can be the 120 V used in homes.

The two windings are most often not electrically connected, but rather just share their magnetic fields within the magnetic core, as mentioned above. The fraction of total magnetic flux that is shared between the two coils is the **coupling coefficient**, *k*. In order to visualize what is meant by *k*, imagine magnetic flux lines in Fig. 3.9 passing only through the core (for *k* = 1) or partially in the space outside, but close to, the core bypassing the coils (for *k* < 1). Here, we assume that the core is extremely good, and *k* = 1. This means that 100% of the flux passes through both coils. It is the very high permeability of the core that confines the fields to the core, just as in the AM radio antenna in Chapter 1.

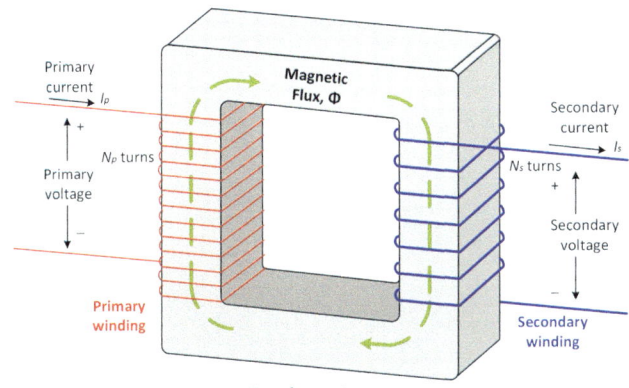

FIGURE 3.9 Transformer with primary and secondary windings. In this configuration, the windings are separated for easy viewing, whereas in most transformers, the windings overlap each other and share the core material within both windings. This is a step-down transformer because $N_S < N_P$.

In an ordinary coil, current creates a magnetic field, and that magnetic field induces a voltage in itself. A *transformer* uses the field from one winding to induce a voltage in the *other* winding. The winding normally used as input is called the **primary winding**, or just **primary**. The winding normally used as the output, is called the **secondary winding**, or just **secondary**.

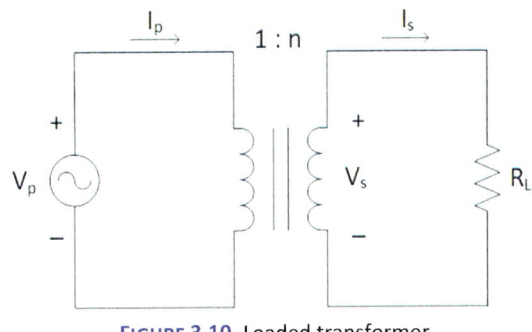

FIGURE 3.10 Loaded transformer.

The effect of current in one coil on the voltage produced in the other coil is characterized by a constant called the **mutual inductance**, *M*, whose name correctly implies that the effect is the same in either direction. If the secondary is open (i.e., no load, i.e., infinite resistance), no current will flow in the secondary, so no magnetic flux can pass back to the primary. Now 'hang' a load resistor on the secondary, as shown in Fig. 3.10. As current flows through the resistor on the secondary side it produces a magnetic field felt by the primary, yielding a voltage equation

$$v_p = L_p \frac{di_p}{dt} + M \frac{di_s}{dt} \qquad (3.4)$$

where subscript "*p*" denotes the primary side and "*s*" denotes the secondary side. First consider R_L to be very large, like an open circuit. Then no current flows in it, so $i_s = 0$ and no magnetic field is generated by the secondary, so no extra magnetic flux affects the primary. Thus, an unloaded transformer looks like an inductor having the inductance of the primary, and we recover only the ordinary inductor equation $v = Ldi/dt$ from Eq. 1.30.

Because both windings share the same core and we assume that the flux threads only through the core (*k* = 1), the windings share exactly the same flux. Since Faraday's law applies the same way to each winding, the voltage induced in each winding is proportional to the number of its turns, as given by Eq. 3.2. We see that for N_p primary turns and N_s secondary turns, where EMF$_{turn,1}$ and EMF$_{turn,2}$ are the EMFs due to a single turn in coils 1 and 2, respectively,

$$\mathrm{EMF}_{\mathrm{coil},1} = v_p = N_p \times \mathrm{EMF}_{\mathrm{turn},1}, \tag{3.5}$$

$$\mathrm{EMF}_{\mathrm{coil},2} = v_s = N_s \times \mathrm{EMF}_{\mathrm{turn},2}. \tag{3.6}$$

However, because the flux is the same in both coils,

$$\mathrm{EMF}_{\mathrm{turn},1} = \mathrm{EMF}_{\mathrm{turn},2}, \tag{3.7}$$

so

$$\frac{v_p}{N_p} = \frac{v_s}{N_s} \tag{3.8}$$

$$\frac{v_p}{v_s} = \frac{N_p}{N_s} = n \tag{3.9}$$

$$v_p = v_s \times \frac{N_p}{N_s} = nV_s \tag{3.10}$$

and

$$v_s = \frac{v_p}{n}. \tag{3.11}$$

N_p/N_s is called the **turns ratio**, which we denote by n. If $N_s > N_p$, then the secondary voltage is greater than the primary voltage, and it is called a **step-up transformer**. If $N_p > N_s$, then the secondary voltage is lower than the primary voltage, and it is called a **step-down transformer**. For example, if N_p = 1000 turns and N_s = 250 turns, and the input voltage is 20 V, then the secondary open-circuit voltage will be 5 V, where voltage is expressed in V_p, V_{pp}, or V_{RMS}. Of course, the signal must be sinusoidal at a reasonable frequency, say at least 10 Hz or so, to create enough of a time-varying flux for the transformer to operate. DC simply won't do it, and any other waveform creates distortion, so these equations are not strictly correct for any waveform other than a sinusoid that is at a frequency that is neither too low nor too high. At very high frequencies, the inductor would not pass any current due to its inductance, which is the subject of Chapter 4.

If $N_s = N_p$, then the transformer is called an **isolation transformer**. An isolation transformer has **unity gain** in voltage. It might seem that unity gain has no use, but the word "isolation" is a clue. As with all other transformers, the secondary is electrically separated from the primary, and their coupling is only through the magnetic field – electrons cannot get from one coil to the other. It is often useful to electrically separate the primary and secondary circuits, for example to define two separate grounds. For the isolation transformer, since the output voltage is the same as the input voltage the purpose is only isolation. One other aspect of an isolation transformer is that any DC voltage present on the primary that is in addition to the AC voltage will not show up on the secondary since $di_p/dt = 0$, so the DC voltage is effectively removed. Total electrical isolation occurs when the primary and secondary are connected to their own grounds that float relative to each other. They can also be connected so as to share a common ground. (Safety considerations of ground are discussed later in this chapter.)

What about currents? An *ideal* transformer does not waste any energy in transforming voltage from primary to secondary – it doesn't get warm no matter how much power flows through it from primary to secondary. Practical power transformers are about 95% to 99% efficient. Wasted energy in the form of heat is caused by '*i-R* losses' and by currents flowing in the core material. The *i–R* losses (pronounced "eye-are" for current and resistance) are due to current in the resistive windings; currents flowing in the magnetic core are called **eddy currents** and are due to the magnetic field lines that cut through the core as per Faraday's Law.

For an ideal transformer, power out from the load = power into the primary because there is no place else for the power to go. It is somewhat of an approximation, but in most applications of transformers, the effect of the inductances on the circuit behavior is much less than the effect of the load resistance. Given power conservation, it turns out that the currents and voltages are then simply related by:

$$P_{out} = P_{in}. \tag{3.12}$$

Therefore,

$$v_s i_s = v_p i_p, \tag{3.13}$$

and

$$\frac{v_p}{v_s} = \frac{i_s}{i_p} = n. \tag{3.14}$$

However, transformers can be more complicated than just two intertwined coils. A single transformer can provide more than one ratio of voltage transformation, *n*, by the addition of wire **taps** along the secondary coil. Here, the word 'tap' is used in the sense of drawing off something from an interior place, as in tapping a tree for maple syrup. An important case is the **center-tapped transformer**, which has the normal two ends of the secondary coil plus an additional wire, the **center tap**, in the middle. The center-tap wire is attached to the middle of the secondary coil wire, as shown schematically in Fig. 3.11. Half of the total voltage appears to either side of the center tap. This is exactly equivalent to a transformer

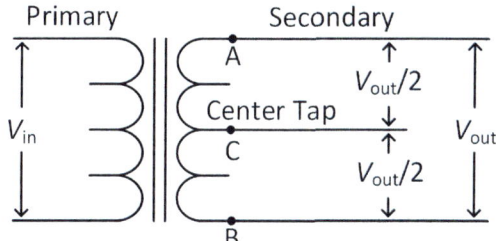

FIGURE 3.11 Center-tapped transformer. The output voltage, V_{out}, is $V_{out} = V_{in} \times (N_S/N_P)$, as usual, where N_S is the total number of windings on the secondary. The voltages between the ends and the center tap are half of V_{out} because the number of secondary windings between the ends and the center tap is half that of the total number of windings.

with two secondary coils connected in series. As such, those two coils can be used independently or together, and a ground can be defined at either end or the middle. Some transformers have many secondaries for a wide range of accessible voltages. It is important to understand that the two (or more) secondary coils are all electrically isolated from one another, and multiple grounds could be defined for several different circuits, or they can be connected together to share a ground.

Because the flux passes in the same direction through both AC and CB coil sections in Fig. 3.11, v_{AC} and v_{CB} have the same sign. When subscripts are used to define a voltage, it is conventional to place them in order of positive to negative. For example, v_{AC} is the voltage at A with respect to (w.r.t.) the voltage at C, or $v_{AC} = v_A - v_C$, where v_A and v_C are defined w.r.t. a ground point. Based on this argument, $v_{AC} = -v_{CA}$. If this confuses you, you should probably review the discussion of how voltages are always defined as the differences of two potentials in Section 1.2.

Furthermore, since the center tap has equal number of windings on each side, v_{out} is split equally, so $v_{AC} = v_{CB} = -v_{BC}$. Thus, the voltages as seen from the center tap (point C), are equal in magnitude and opposite in sign. Put another way, if you monitor the voltages v_{AC} and v_{BC} with respect to C, you will see two sine waves of the same amplitude that are 180° out of phase. On an oscilloscope, that looks like one sine wave is upside down compared to the other one. All of this will play a part in the discussion of how voltages are distributed throughout U.S. homes.

Again, transformers are often the largest or heaviest component in an electronic system. If you ever have the chance to use a quality vintage stereo amplifier, the first thing you notice is how heavy it is. Some from the 1970s weigh up to 80 pounds, due largely to the transformer that can support hundreds of watts of output power. In this chapter, you will learn that several stages of massive transformers are used to convert voltages in transmission systems. Voltages on the transmission lines can be as large as about 1 MV and are stepped down using transformers to hundreds of kV, and then stepped again to about 10 kV and finally down to hundreds of volts for residential use. Without transformers, power transmission would have been limited to Edison's original DC power transmission scheme and would have been vastly less efficient.

3.3 Power Transmission Systems

3.3.1 Transmission Lines

Figure 3.12 shows common transmission lines that carry power across the landscape from power plants to customers at distant locations. Figure 3.13 represents the entire transmission grid from the **power plant** (or **power station**) (A), through step-up transformers to the transmission lines (B), through step-down transformers at receiving stations to subtransmission lines (C) to distribution substations, stepped down again (D) and then sent on the distribution grid to neighborhoods, and then stepped down by distribution transformers to the customers (E). Voltages at (E) and (F) (where you plug stuff into your wall outlets) are discussed in Section 3.4.

FIGURE 3.12 Common transmission lines. *Source*: https://commons.wikimedia.org/wiki/File:Three_Phase_Electric_Power_Transmission.jpg. Published under Creative Commons Attribution CC BY-SA

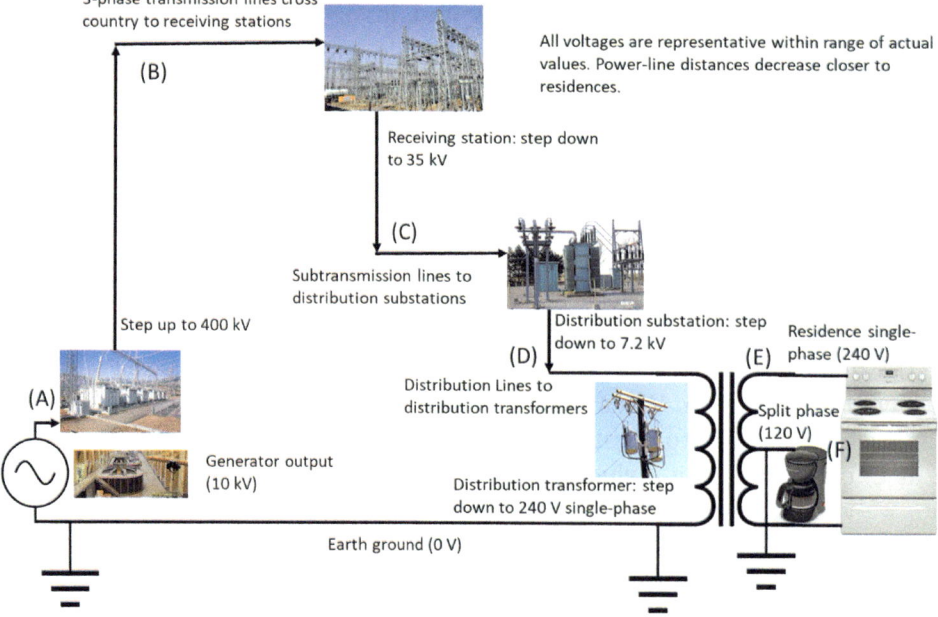

FIGURE 3.13 Power grid with representative voltages at each stage.

Power plants must transfer huge amounts of power to their customers with minimum losses along the way. Large power plants typically produce around a couple of gigawatts (GW) of power. As of this writing, the largest in the world is the Three Gorges Dam in China, which can produce more than 20 GW. The power available for use in a typical residential home is about 25 kW with average use of about 2 kW. So, a 2 GW power plant can service up to about one million average U.S. homes, but fewer at peak times.

<aside>*Nomenclature:* 'High voltage' is sometimes referred to as 'high tension,' especially in some European countries.</aside>

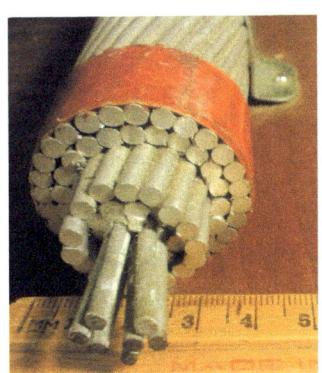

FIGURE 3.14 Transmission line up close. *Source*: https://commons.wikimedia.org/wiki/File:Three_Phase_Electric_Power_Transmission.jpg. Published under Creative Commons Attribution CC BY-SA 3.0.

As seen in Fig. 3.13, the voltage created at the generators is passed through a step-up transformer and boosted to hundreds of kV on the transmission lines (Fig. 3.14), which carry current at these *high voltages* in order to deliver electrical power most efficiently. For a given amount of power delivered, the *i–R* power lost in the transmission lines can be minimized by pushing the current at the highest possible voltage.

This strategy can be understood by comparing the power absorbed by a resistive load at the customer ($P_{load} = iV$) and the power lost as heat on the transmission lines ($P_{transmission\ loss} = i^2 R_{lines}$ with the same current *i*). For a fixed power delivered by the transmission line (P_{load}) to all of its customers, increasing *v* decreases *i*. Hence, transmission loss ($i^2 R_{lines}$) is decreased quadratically. In summary, lower current at higher voltages translates to lower *i–R* losses on the transmission lines, leaving more power for the customers, more profits for the power company, less energy wasted to the environment, and fewer resources needed to satisfy the needs of the population.

Equation 1.13 can be expressed as $E = \Delta v / \Delta x$, which says that the electric field, *E*, between two objects having some voltage difference, Δv, increases as the two objects get closer together, i.e., Δx decreases. In high-voltage systems, wires are not insulated like the wiring discussed in Chapter 1. Instead, the bare-metal cables are electrically isolated from the surrounding structures by high-voltage **insulators** made of glass or ceramic (Fig. 3.15).

The purpose of the insulators is to keep the high-voltage lines from touching any grounded objects, such as the metal tower, and also to keep a large enough distance between them so that the electric field does not rise above that which will cause **arcing** (also called **arc flash** or **flashover**) due to **breakdown** of the air dielectric. This arcing occurs when electrons are stripped off of molecules in the air, resulting in a large

FIGURE 3.15 Insulators on a powerline transmission tower.

<aside>*Of interest:* The breakdown field for air depends on humidity and can be as low as 10 kV/cm. Breakdown occurs when the high electric field rips electrons off of air atoms, leaving behind ions. So when you get a spark from your finger to a doorknob after walking on your carpet in the winter, you are actually charged up to thousands of volts. In fact, the current is also rather high (amps), but fortunately the discharge time is exceedingly short (nanoseconds), so very little energy is deposited on the tip of your finger. Still, it can hurt.</aside>

current, much like a lightning strike. The stacked disk shapes on the longer high-voltage insulators serve two functions: they increase the distance along the surface on which charge might flow (recall Eq. 1.6, $R = \rho l / A$), and they are cup-shaped like small umbrellas that keep inner parts dry, decreasing the likelihood of having a complete conduction path formed by rainwater.

Even at the high voltages intended to minimize current flowing in transmission lines, the current can still be upwards of a thousand amps. When power lines carry too much current on hot days, the resulting extra heat can cause them to weaken and sag, which can cause flashover to trees or other grounded objects. This can put an extra burden on the local grid, causing systems to shut down. *Because high-voltage wires are not insulated – if you ever see one on the ground, STAY AWAY!! Keep at least 35 feet away, warn others, don't touch anything nearby that could be energized by the power line, and call 911.*

Small problems can grow into big problems in a phenomenon called **cascading failure**. When one substation shuts down, it may bring other substations with it as they try to take up the slack and exceed their own capacities. Also, power plants must keep a balance between the power that they are producing and the loads that draw the power. If this balance is not maintained, the generating equipment can be damaged, and the plant must shut down to protect itself. In the worst case, other power plants may go offline. Ultimately, this can lead to large parts of the country suffering power outages like the one that introduced this chapter. It is important to note that a generating plant cannot easily come back online without the assistance of the rest of the grid. Therefore, bringing the grid back from a widespread outage can take a lot of time.

Nomenclature: A recurring theme in electrical engineering is the power wasted as heat due to current flowing through stuff. The resistance of that stuff gives rise to energy loss of the electrons as they collide with atoms in the material. This energy loss is called by many terms, including '*I–R* losses,' '*i²R* losses,' 'resistive heating,' 'resistive losses,' '**Joule heating**,' and others. In fact, the whole reason for using very high voltages on power transmission lines is to minimize this wasted energy due to heat.

3.3.1.1 Three Phases

Now we come to the "phase" in "three-phase." There is an expression that "all good things come in threes," and power lines are no exception. Look again at the large transmission-line towers in Fig. 3.12 – transmission towers support power lines in sets of three. Each of the three lines corresponds to one source of power at one of the three phases.

Even if you feel you already understand, it is worthwhile to think more deeply about what we mean by "phase." In general, **phase** refers to the fraction of a cycle we are referring to, where a cycle is 360° (2π radians), as you were taught in your trig classes. Any repetition of an activity can be thought of in angular measure if you think of one cycle as 360°. You would be at your engineering nerdiest if on April 1 you told your friends that we were 90° through the year. Any phase value requires some arbitrary point to be taken as 0°, and you had obviously considered January 1 to be 0° in your calendar. If you were in competition with an accounting major for the 'nerdiest student,' she would point out that October 1 is 0° in the U.S. government's fiscal year, so New Year's Day comes at 90° and not 0° as you asserted; then, April 1 would be 180° (through the government's fiscal year).

The lesson is this: In electrical signals, a phase *difference* is the time difference between two points of a waveform *as a fraction of one period*, expressed in degrees or radians. The conversion from a time difference to a phase difference, ϕ, depends on the frequency, f, or period, T, as

$$\phi = 360 \cdot f \cdot \Delta t \,(\text{degrees}) = 2\pi \cdot f \cdot \Delta t \,(\text{radians})$$
$$= 360 \cdot \Delta t / T \,(\text{degrees}) = 2\pi \cdot \Delta t / T \,(\text{radians}). \tag{3.15}$$

For example, a 1° phase difference corresponds to a time difference of 1/360th of a period. In the calendar example above, the period, T, is 52 weeks, so one week, Δt, is a phase difference of about 6.9°. As another example, a cosine function having frequency f = 60 Hz changes in phase by 21.6°, or 0.377 rad, in Δt = 1 ms. However, if the frequency is different, say 50 Hz, then that same 1 ms is a different phase – 18° or 0.314 rad. The upshot of all of this is that a phase difference is a function of the shape of the waveform – a fraction of a cycle. For two different frequencies, the same time difference represents two different phases, or conversely, the same phase difference is two different time differences.

Recall that a voltage is always a difference between two arbitrary values or a value relative to a point of zero reference (ground). So too, "phase" expresses some difference between two arbitrary phases or an implicit time difference relative to some presumed point of zero phase. Put another way, the point at which you start to measure Δt is the point of zero phase. Where you start measuring Δt is up to you. The "zero phase" reference does not have a formally defined term like "ground." So, to understand the use of the term "phase," *you must be clear what point in time is taken as 0° of phase whenever you discuss phase values for any cyclical event.*

So far, we have been discussing phase in the context of a single signal, where phase refers to distance (in time, expressed as an angle) along the waveform. Often one compares a phase difference between two different waveforms, where one waveform is shifted relative to the other. However, *the concept of relative phase is meaningless unless the two waveforms have the same frequency.*

Historical: **War of the Currents**: In the late 1880s, a battle broke out between Thomas Edison's DC-based and George Westinghouse's AC-based electrical grids. The 'war' became quite caustic, with Edison participating in the killing of animals to prove how deadly AC was. It was the development of transformers by Westinghouse that won the day for AC. After some 150 years, high-voltage DC systems are coming back due to the advent of modern, efficient DC-to-DC converters that can replace the use of transformers to increase and decrease line voltages. Advantages of DC over AC are the lack of power factor considerations, no synching of frequencies or phase between power grids, less loss, no skin-depth effect (not covered in this text), and more. DC is the future (but it will take time).

In electrical power transmission, each transmission line is a sine wave with frequency 60 Hz (T = 16.67 ms) and amplitude equal to the maximum voltage of the transmission line (e.g., hundreds of kV). All three power lines carry identical (sinusoidal) waves at the same frequency, but shifted in time, and therefore phase, relative to each other. You may arbitrarily denote the time when the sinusoidal waveform of one of the transmission lines crosses 0 V as it rises as "0°." One of the other two lines is offset by 120°, and the other by another 120°.

Figure 3.16 shows oscillographs of three, 120 V sine waves separated by 120°, which is characteristic of three-phase power found in a small industrial facility, such as a laboratory or small manufacturing site. Three-phase power is the norm in industrial settings, and uncommon in residential settings. Transmission of three-phase power occurs at much higher voltages than that shown in Fig. 3.13, but the waveforms are identical. If you were to attach oscilloscope leads to 765 kV power transmission lines and live to tell about it, that is the shape of what you would see. (You would also see the three phases if you attached the scope to lower voltages, like 240 V in a commercial building. As we will discuss later, 240 V, and even 120 V, can be quite dangerous.)

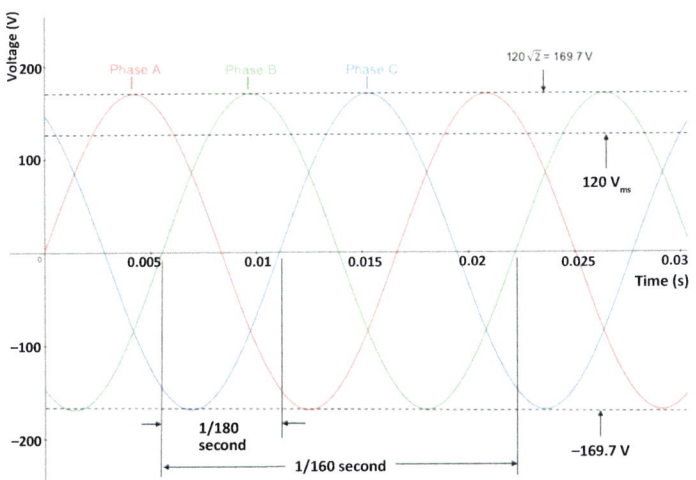

FIGURE 3.16 Three sinusoids at 60 Hz and 120 V_{RMS} separated by 120°.

An aside on voltage units: When we say that a sine wave is "120 V" it is important to specify what exactly that means. In almost all power transmission discussions, a "voltage" is implicitly the "**root-mean-squared**," or "**RMS**," value. Let's describe a cosine function as $v(t) = v_p\cos(\omega t + \phi)$, where v_p is the amplitude, also called the **peak value**. For *a sinusoid of any phase,*

$$v_{RMS} = v_p \big/ \sqrt{2},$$

(3.16)

as seen in Fig. 3.15 (The sinusoid has a peak voltage that is about 1.41 times the RMS value.)

More generally, any function has an RMS value that can be calculated from

$$f_{RMS} = \sqrt{\frac{1}{T_2 - T_1} \int_{T_1}^{T_2} [f(t)]^2 \, dt}.$$ (3.17)

The equation says, "Square the function, find its mean, and then take the square root." That is the opposite of 'root-mean-squared.'

A waveform is a time-varying voltage, current, or power, and can be described in different ways. Periodic waveforms are described by their time average (typically indicated by a subscript avg: v_{avg}), peak voltage, v_p, peak-to-peak, v_{pp}, or RMS values.

The time average voltage of a sine wave is 0 V since it is positive as much as it is negative. But that doesn't mean it doesn't provide power to a load: after all, toasters do get hot! You may think of the RMS voltage of a sine wave as the equivalent DC (or battery) voltage that you would need to make a toaster heating element just as hot as the sinusoid used to heat it. For that reason, the RMS value of a voltage or current is called its **effective value**.

One often indicates the RMS value by saying "volts RMS," but the units are still just plain volts. Here we follow general practice and indicate the RMS average of a voltage variable by "V_{RMS}." A 120 V_{RMS} wall outlet measured on an oscilloscope has a peak voltage of about 170 V. A DMM indicates the RMS value of the AC voltage and would show "120 V". In most discussions about power, voltages may be assumed to be RMS values.

Using an oscilloscope at a home electrical outlet, you would see a sinusoid like any one of the sine waves in Fig. 3.16, but you wouldn't be able to say anything about its phase relationship to any of the other of the three phases the came to your neighborhood. In this chapter's lab, you will have a low-voltage version of three-phase power to experiment with, and will be able to interact with, and get to know more about, these phase relationships, as well as three-phase-industrial and single-phase-residential power.

3.3.2 Power Transformers and Substations

Power systems rely on transformers to transform the transmission voltages from low-to-high (step-up) and high-to-low (step-down). Power transformers vary from merely *really* big (Fig. 3.17) to absolutely *humongous* (Fig. 3.18) depending on where in the power distribution sequence they occur.

FIGURE 3.17 Pole-mounted power transformers. There are three transformers on this pole, one to step down each of the three phases and send that single phase out to nearby homes.

A substation (Fig. 3.19) is an installation of transformers and switches that step-down voltages for power distribution and protect the grid from potential problems. Potentially serious events that may lead to blackouts include excess power consumption (as in heavy use of air conditioners in heat waves), trees contacting power lines, lightning strikes, vehicular accidents, and even small animals contacting high-voltage equipment.

Figure 3.17 shows a "pole-mounted" **distribution transformer** used to step down the distribution voltage of one "leg" (or phase) from several thousand volts, typically 7.2 kV, to a couple or few **hundred volts**, typically 480 V

FIGURE 3.18 Humongous substation power transformer. *Source:* https://commons. wikimedia.org/wiki/File:High-voltage_ transformer_750_kV_%D0%A2%D1%80%D 0%B0%D0%BD%D1%81%D1%84%D0%BE% D1%80%D0%BC%D0%B0%D1%82%D0%BE %D1%80_750_%D0%BA%D0%92.jpg

for commercial and 240 V for residential customers.

Perhaps you have seen boxes in your neighborhood like the one shown in Fig. 3.20 and wondered what they are. They are "pad-mounted" distribution transformers. The one shown in Fig. 3.20 serves the same function as a pole-mounted transformer. It steps down one of the three phase lines, often called a **leg conductor** or just "leg," and can serve several houses. Figure 3.21 shows the specifications and wiring of a pad-mounted transformer, indicating that it steps down the 7200 V distribution voltage to 240 V and 120 V, and can supply 50 kVA. Apparent power, indicated by the unit kVA, will be explained later.

FIGURE 3.19 Distribution substation. *Source*: ohsa.gov/SLTC/etools/electric_power/ illustrated_glossary/substation.html. Public domain.

FIGURE 3.20 Step down distribution transformer (and Duke).

FIGURE 3.21 Transformer specifications.

3.4 The Customer

Now we come to you, the customer. Large installations or buildings require three-phase power, which is supplied at 480 V (on each leg), 600 V, or higher. There are many standard voltages in use for distribution of power to homes and manufacturing plants.

Three-phase power is used in buildings and factories to run heating, ventilation, and air conditioning (HVAC) units, motors for water pumps, manufacturing equipment, and other large loads. Each individual leg of the three phases can power a subsystem. Each leg is referred to as **single-phase** since the loads see only that one phase. Distribution transformers that supply a residential building typically step down to single-phase 240 V or 120 V. This is also called "split-phase" for reasons discussed later.

3.4.1 Safety

A broad view of the power distribution grid from the generator out to your house has been presented. Now we delve into residential wiring in greater detail. The dangers of line voltages found in homes come under two categories: fire hazards and the risk of electrocution. Even 120 V can be deadly under the right conditions. Many people have experienced an electrical shock in their homes or hotels and the like, but were not seriously injured because dry, healthy skin has sufficiently high resistance to protect them from excessive current ($i = v/R$). Humans are mostly big bags of electrolyte, i.e., conductive fluid. If there

> *Safety:* The Occupational Safety and Health Administration, OSHA, considers voltages higher than 50 V to be dangerous.

is a cut or the skin is wet, current has a low-resistance pathway through the heart. It takes only about 10 mA through the heart to cause cardiac arrest in about half a second. Some 250 people die each year in residential electrocutions.

This section provides some caveats that come under the category of "a little knowledge is a dangerous thing." A label actually found on some Halloween costumes says, "Caution: Cape does not enable wearer to fly." Likewise, "Caution: This chapter does not enable students to work on house wiring." Do not assume that just because you read this chapter carefully and do well on the exam that you are prepared to work on line-voltage electrical systems. Do not ignore the dangers of residential electrical wiring. Besides the dangers of working around line voltage, there may be required permits, effects on homeowners' insurance, risk of fire hazards, and violation of electrical codes. There is a good reason that electricians are highly trained and must pass numerous certification exams.

Formal training includes some basic rules that should be followed when working around any voltages that could be potentially dangerous: Do not stand in water when using electricity. Do not use electrical systems with wet hands. Avoid using both hands at the same time when working on electrical systems. Remove metal jewelry from your hands and wrists. Wear insulating gloves and shoes, as well as eye protection. Test for voltages rather than assuming that circuits are turned off. Again, knowing these things does not constitute training to do electrical work.

3.4.2 House Wiring

When you go home this semester for Thanksgiving break, your Uncle Fred will say, "You're studying electrical engineering? Tell me about wall outlets." Well, you don't want to look bad in front of Uncle Fred, but it wouldn't necessarily be your fault. As a new student of electrical engineering, you may not realize that the subject of house wiring is overlooked in many college curricula. Many engineers pick this stuff up along the way – and some never do. One purpose of this textbook is to fill in such gaps, especially where such information can serve as an introduction to, and support the understanding of, some important concepts in electrical engineering.

3.4.2.1 Distribution Panel, Breakers and GFCIs

FIGURE 3.22 (Left): Typical breaker box. (Right): expanded view of upper left breakers.

Breaker boxes (Fig. 3.22) directly protect against fire hazards (and indirectly against electrocution). Exceeding the current rating of a wire can make it so hot that it could start a fire inside the wall. Power enters the house at a **distribution panel**, or "breaker box." From this central point, power is distributed throughout the house on **branch circuits** like branches extending out from the trunk of a tree. The breaker box of Fig. 3.22 contains many **circuit breakers** (Fig. 3.23), each of which protects a branch circuit.

Imagine you are late preparing for a Christmas party. You are trying to run your 1250 W microwave oven, 1500 W space heater, 1500 W hair dryer, 1000 W toaster oven, and 1000 W vacuum cleaner, from one circuit. That would be about 52 amps. Circuit breakers prevent fires that might arise from excessive current (like 52 amps in 15 amp wiring) and subsequent heating of the wires in the walls and ceilings.

FIGURE 3.23 Single circuit breaker.

When the current is too high, the breaker "trips" to an open circuit so the current is stopped entirely. When a breaker trips, the switch position flips from the inside toward the outside. A glance at the panel shows which breaker is affected. (Find the tripped breaker in Fig. 3.22.) Once the problem is corrected (such as by unplugging the microwave oven, space heater, hair dryer, and toaster oven, but leaving the vacuum cleaner), everything returns to normal, and your Christmas party is a huge success! (And you didn't have to invite the fire department!) Seriously, if a circuit breaker trips, you should investigate why it did so as it could represent a hazardous condition.

Historical: In the old days, fuses, rather than circuit breakers, were used to protect house wiring. A fuse such as in the figure above is a piece of wire (inside a glass bulb for viewing) that is designed to melt when a certain current is exceeded. The screw-in fuses were about the size of a coin, and when burned out due to an overloaded circuit, needed to be replaced. The stereotypical "penny in the fuse box" was used when a fuse was not available, but sometimes at the cost of human lives, since a penny could pass far more than the rated current of the house wiring, and could lead to fires.

The capacity of the breaker must match the wire thickness, or gauge, so that the maximum current does not exceed the wire's rating, i.e., the wire does not overheat. Circuits in homes are typically rated at 15 amps for wall outlets that run small appliances such as light fixtures, toasters, etc. Twenty, thirty, and even fifty amps are used for larger appliances like ovens and electric dryers, and the wire thickness increases (gauge number decreases) accordingly.

As alluded to above, circuit breakers also provide some protection against electrocution, but not in *all* cases (how they do so is discussed in Section 3.4.2.2). The best protection against electrocution is provided by a **ground fault circuit interrupter** (GFCI or just GFI). GFCIs are usually located at an individual electrical outlet (or "receptacle), as shown in Fig. 3.24. GFCI receptacles are the ones with "test" and "reset" buttons. A GFCI detects small amounts of current (about 5 mA) flowing incorrectly outside of the intended circuit, which could indicate an ongoing electrocution. The GFCI is an additional breaker that trips within a couple of cycles of the current, limiting the potential victim's exposure to a fault. A GFCI can also replace an ordinary circuit breaker in the breaker box in order to protect its entire branch circuit. Within homes, GFCIs are required in all locations that have a water source, such as bathrooms and kitchens, as well as outdoors.

FIGURE 3.24 A GFCI wall receptacle stops current when it detects a small amount of current flowing incorrectly, thus protecting people from possible electrical shock. The "TEST" button causes a current to flow that should trigger the GFCI. After the test, it is reconnected by pressing the "RESET" button. Now that you know what it is, you should follow recommendations and test them every few months.

Residences (or areas of larger buildings) are nearly always powered by one of the three phase legs. As you know, these are AC voltages. Look carefully at Fig. 3.25. The single-phase power comes to the residence from a distribution transformer on three **feeder lines** A, B, and **neutral**. These are thick cables that can carry a lot of current from the outside distribution transformer. The neutral line comes from the center tap of the distribution transformer; the voltage between line A and line B is 240 V; line A to neutral is 120 V; and neutral to line B is 120 V. The neutral line is connected to earth ground and splits the 240 V supply into two 120 V legs. This configuration that uses the neutral line and its connection to ground to break the 240 V into two 120 V legs is called **split-phase** power.

Breaker boxes generally have a set of large breakers at the top, called the **main breakers**, often 100 amps or 200 amps, that prevents the house from drawing too much total current. A "100 amp breaker box" has a 100 amp breaker on **feeder** line A and another 100 amp breaker on feeder line B. When the main breakers are in the on position, beware, since the distribution box is energized. Do not work inside an open breaker box without proper training, as lethal voltages are present.

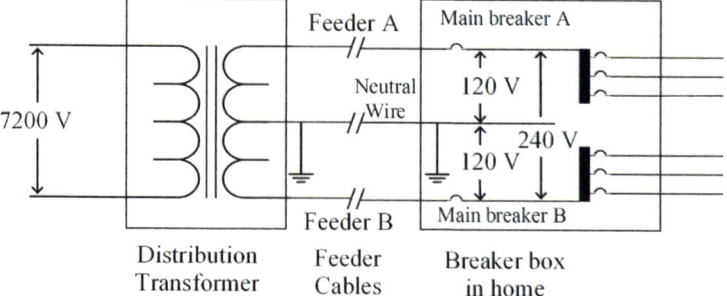

FIGURE 3.25 Center-tapped distribution transformer outside a home connected to home breaker box. Two busbars and several breakers are shown on the right. Hot wires leave the breaker box. Each branch circuit has a returning neutral wire (not shown) that attaches in the breaker box to the neutral feeder line.

Figure 3.26 shows a fully populated breaker box, and Fig. 3.27 shows a partially populated breaker box. The two thick cables coming into the breaker box shown at the top of Fig. 3.26 are the feeder lines A and B, and feed **busbars** A and B (shown at (a) and (b) of Fig. 3.27). The terms "busbar" and "bus" in this sense refer to heavy pieces of metal to which other wires are connected for power distribution. Busbars are thick metal plates in the breaker box that

Nomenclature: The word 'bus' is used in many contexts in electrical engineering. Like a bus that carries people, it holds connections (the seats?) for many inputs and outputs (the people?).

carry current to the breakers. Each busbar has flat tangs, (d) and (e), that extend perpendicularly from the busbar; breakers snap onto the tangs, making electrical contact. A breaker has been flipped over to show the contact points on the bottom (c). Two breakers, (f) and (g), are shown at an angle just as they are about to be snapped into place onto tangs. Breakers (h) and (i) are flat because they have already been snapped in.

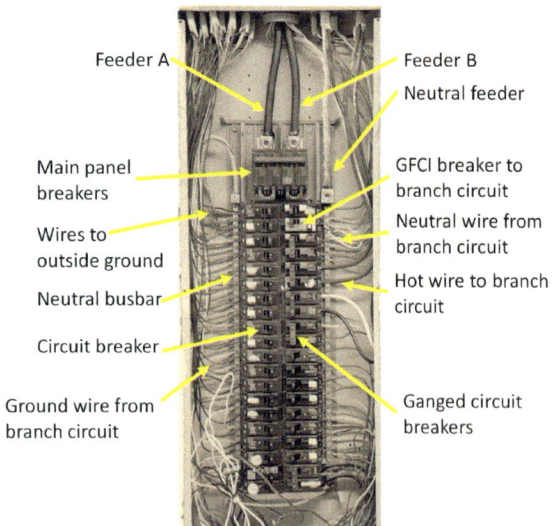

FIGURE 3.26 Hot, neutral, and ground wires in Bernstein's home breaker box. Neutral wires and ground wires are connected together at the neutral busbar.

Busbars carry the full current from the feeder lines. Each 120 V circuit breaker carries some of the total current from either busbar A or B. The branch circuit breakers tap current off the busbars to energize, often referred to as 'hot,' wires that feed various wall outlets, ceiling fixtures, and appliances throughout the home. Hot wires can be seen leaving the breakers in Fig. 3.26. It is worth mentioning that a hot wire can be called a 'live' wire, but a hot wire is only 'live' when it is energized. You can have a 'hot' wire that isn't 'live' if the power is disconnected.

Remember that all of the current is AC at 60 Hz, meaning that its direction changes 60 times per second. It can be helpful to *imagine* it as DC, since Kirchoff's laws must be obeyed at every instant in time. Using your intuition about DC circuits will be helpful, and you may transition to thinking in terms of AC when the discussion warrants it. That said,

Kirchhoff's current law tells us that no current will flow unless the circuit is closed. In this case, current flows "out of" (using a DC mindset) the breaker box on the hot wires (Fig. 3.26), through the loads, and returns to the breaker box on neutral wires. All of the neutral wires return to the breaker box and connect to a common metal bar called a **neutral busbar**. In the case shown, two neutral busbars can be seen in Fig. 3.26, one each along the left and right sides of the breakers. The neutral busbars connect to the neutral feeder line, which returns to the center tap of the distribution transformer.

Residential 120 V branch circuits typically use either 15 A (most commonly) or 20 A breakers (with the 20 A circuit requiring thicker wire). This implies that the most power that could be delivered by 120 V into a single 15 A branch circuit is $P = iv = 15$ A × 120 V = 1800 W. So how do we provide power for very large appliances (such as dryers, stoves, and ovens) that might use an excess of 10 kW? Recall that power to a home is provided in the form of a single-phase 240 V sine wave configured as two 120 V circuits that are 180° out of phase (the split-phase system). We can drive the larger loads by connecting the appliance across the full 240 V that is between feeders A and B. This provides twice as much power for the same current.

A 240 V circuit is protected by two breakers placed side by side in a "ganged pair," i.e., with their switches

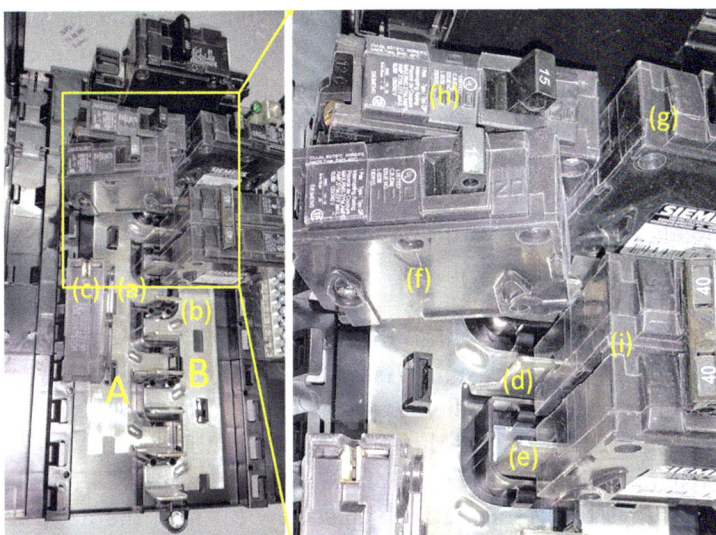

FIGURE 3.27 Partially filled breaker box (left) that shows how breakers attach to busbars. The right side is a closeup of the breakers. (a) and (b) indicate busbars A and B. (c) is a loose breaker flipped upside-down to expose the contact point just above the letter (c). Busbar tangs from A and B are indicated by (d) and (e), respectively. Breakers (f) and (g) are about to be snapped into place. Breakers (h) and (i) are flat, i.e., already snapped into place. Two 240 V double-pole breakers, rated at 40 A, are indicated by (i).

joined by a single bar (again, as seen in Figs. 3.26 and 3.27); the two breakers flip on and off together. These are called **double-pole** breakers and are the norm for large appliances, like ovens or air conditioners that run off 240 V. If there is a problem, or **fault**, in one of 120 V legs, tripping its breaker, the connecting bar trips the other breaker as well. These ganged breakers are each rated at currents higher than 20 A, ranging up to about 50 A for typical large home appliances (again with appropriate wire thickness). The 50 A current flows out of Bus A through breaker A, through the appliance, and back through breaker B and Bus B. No current flows on a neutral wire, and none is needed. Now the breaker box is sourcing up to 50 A at 240 V for a typical maximum power of 12 kW, which is much more than 120 V branch circuits alone normally provide.

This raises another question: "Why can't we just use the 120 V circuit, but with more current?" Well, you could, but just as in the power transmission lines, running loads at twice the voltage requires half the current, and therefore can use thinner wires, which can really save money in a whole building.

An additional consideration of split-phase power has to do with **balancing the loads**. A trained electrician populates the breaker box from top to bottom in such a way as to make the currents drawn from the A and B busbars as close to equal as possible, specifically within 10%. From a simple circuits aspect, balanced loads imply that all of those branch-current resistances across each of the A and B legs have the same equivalent resistance. All those lights, appliances, etc. are just resistances, after all, and they are all in parallel. Plugging in another doohickey into an outlet is exactly the same as adding another resistor in parallel on that leg.

So why is it good to balance the loads at all? First, when a large appliance (such as a dryer) is connected across the 240 V legs, current flows through the appliance without using the neutral conductor, and the load is automatically balanced. This is not necessarily the case for split-phase where current could split at the neutral conductor. When the two busbars at 120 V each are used to power their own sets of loads (e.g., think a high-power space heater on Bus A and a low-power lamp on Bus B, but many more appliances), unbalanced loads would cause the two busbars to source different amounts of current; in this case, one of the main breakers might trip before the full current capacity of the breaker box was utilized. Second, Kirchhoff's current law will tell you that if the loads are perfectly balanced, current flowing in one 120 V leg conductors will be exactly the same as that in the other leg conductor. Therefore, current will flow back to the transformer on the other leg conductor, and no current will flow on the neutral conductor (refer to Fig. 3.25). This is why a good electrician will estimate the expected loads on each bus as the branch circuits are put into a new home.

Let's imagine that Bus A is driving a 20 W lamp and a TV set, for a total of 170 W. (You may figure out what that equivalent resistance is.) Also, suppose that Bus B is driving a 1200 W microwave oven. (Again, what is that resistance?) Clearly, this represents unbalanced loads. When the loads from each of the 120 V legs are not equal, KCL says that some current must flow through the neutral conductor. It is expected that some current will normally

> *Nomenclature of wiring and colors:* House wiring is color-coded in standard ways so that all technicians know the purpose of every wire. The wires that carry the voltage are most often referred to as **hot** or **live** wires. More rarely they are referred to as **energized**. Here are some observations to help you remember the wire colors. Hot wires are typically black (suggesting death) or red (suggesting danger), although blue, yellow, and orange are also used. "Ground" is either green (alliteratively or suggesting bucolic earth scenes) or is a bare wire, and neutral is usually white (or can be silver). (White is a "neutral" color.) The ground wire can be uninsulated because exposed metal is normally grounded anyway.

flow on the neutral conductor back to the transformer, but it should not be excessive. Although the feeder lines are quite thick, they still have some, albeit small, resistance. The neutral conductor, therefore, develops a voltage and wastes energy when current flows through it. Bear in mind that each phase leg is 120 V, which is across the loads and the neutral conductor back to the transformer. Any voltage dropped on the neutral conductor is not available to drive the loads, and the voltage at the load will be reduced. Balancing the loads, therefore, helps to keep the voltages on the two busbars from sagging (i.e., dropping much below 120 V) due to the voltage drop on the neutral conductor. You might have noticed in your homes that the lights might dim when a high-power device, like a pump, starts up, which is due to these kinds of effects.

As an illustration of the role of the neutral wire in split-phase, consider this true story. My home developed a fault on the neutral conductor from the house. It was an intermittent problem, so the power company was hard to convince, and I lived with it for several months before it got frequent enough to repair. Without the neutral wire, balancing the loads relative to each leg mattered a lot more than just the case of unbalanced loads just discussed. Without the neutral, the series combination of the loads on both legs formed a simple voltage divider between Leg A and Leg B; those loads now split the 240 V across the series combination of the two legs, as the ground at the neutral wire was removed. The resistance of the earth ground at the house back to the ground at the transformer, which is in parallel with the underground neutral wire, was not nearly low enough to replace the neutral wire. When a high-power microwave oven (1200 W) was turned on, it acted as another resistor in parallel in its leg, and, by the voltage divider relationship, the voltage in that whole leg dropped considerably. At the same time, the voltage in the other leg increased considerably in order to add up to the 240 V provided by the transformer. Instead of the normal 120 V, my wall outlets were providing between 100 V and 140 V depending on what was running at the time. So, some things in the house got very bright and motors ran very quickly (risking a failure) while others got dim or ran slowly. This was a bit concerning. I came up with the following temporary fix for my microwave oven problem using a second microwave oven: When I was cooking food using one phase leg, I ran another microwave oven on the other phase leg, thus keeping the loads more-or-less balanced, and the voltages stayed fairly constant. So, when I needed to, I ran both microwave ovens to keep the voltages within a normal range. Although inconvenient, this worked until the power company finally admitted there was a problem and dug up the ground a couple hundred feet from my home. The problem was solved after only six months of complaining.

3.4.2.2 Neutral and Ground are not the Same Things

Notice in Fig. 3.25 that the neutral wire connects to ground in the breaker box. So, is the neutral wire actually ground? You may think of neutral as ground for the split-phase system, but there are some subtle yet important differences between neutral and ground, which will be discussed later. There are several different locations within the split-phase system that are all referred to as "neutral." The thick conductor or wire that goes under the ground or in the overhead lines to a residence is the neutral feeder line. Inside the breaker box there are neutral busbars. Coming from those and extending throughout the house are the neutral wires in every branch circuit that returns current from every electrical outlet and fixture (again, thinking DC is easier). Finally, every fixture or appliance that is run from the receptacle has a neutral wire on which return current flows toward the breaker box.

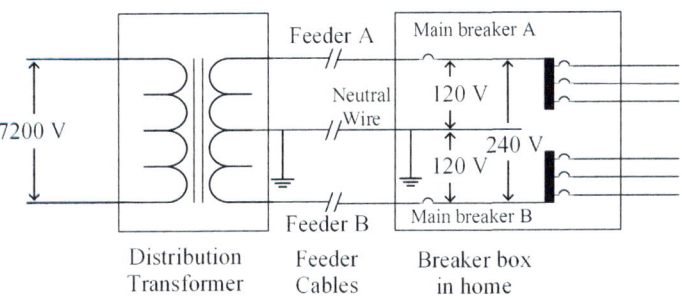

FIGURE 3.25 Repeated for convenience.

A quick summary: The neutral wire originates as the center tap of the distribution transformer. The center-tap neutral is the reference wire for the 120 V lines at the midpoint of the incoming 240 V, as shown in Fig. 3.25. (Like ground, you may think of the neutral anywhere in the system as being at 0 V, but that is not strictly the case.) Thus, there are two different sources of 120 V w.r.t. the neutral wire at your wall outlets; they are 180° out of phase relative to each other (as noted in Section 3.2.3). When you plug something into a wall outlet, you don't need to know which of the A or B busbars you get. Both A and B electrical outlets work the same way as far as any appliance is concerned.

Current is supplied to household loads using two wires, since current must flow out of, and back to, the breaker box. Kirchhoff demands this. For 120 V, the two wires are line A *or* line B, *and* neutral. For 240 V they are lines A *and* B, *without* neutral. A third wire, when present, is the ground wire; *it carries current only when there is a fault.*

Let's focus only on 120 V branch circuits, which feed most wall outlets and light fixtures. Figure 3.28 indicates the hot, neutral, and ground **slots**, or **terminals**, for a U.S. wall receptacle or "outlet". The neutral wire connected to the neutral slot returns current (but of course, it's AC, so it flows in both directions) from the load back to the distribution panel. There it is connected to earth ground, as indicated in Figs. 3.25 and 3.26. Recall from Chapter 2 that the earth ground wire is connected to a metal stake driven into the planet. The neutral wires in the home are physically connected to the ground wire only at the breaker box. *Everywhere else in the home, neutral is a wire separate from ground.* Figure 3.29 shows a 15/20 A grounded plug that mates with U.S. receptacles. The metal parts are called **prongs**, and the flat ones are called **blades**.

A normally small resistive voltage drop (up to a few volts) develops along the length of the neutral wire due to the return current. Hence, the whole neutral wire is considered safe, as all of its length is nearly at ground potential. Remember, the current flows all the way from the appliance, e.g., a coffee maker, through its power cord, then through the wires in the wall, then through the feeder

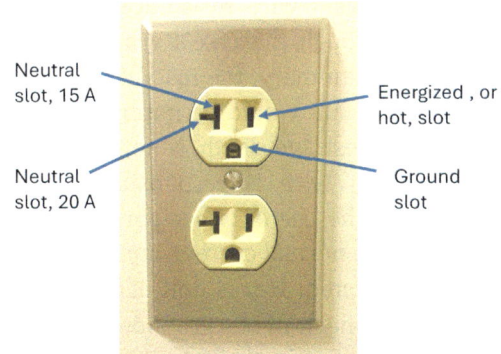

FIGURE 3.28 Wall electrical receptacle with two outlets. Each outlet has a tall vertical slot for neutral and a shorter vertical slot for hot. The half-round hole is earth ground. This receptacle is typical of those found in commercial buildings because it has an additional horizontal slot to accommodate a 20 A plug with a horizontal neutral blade.

lines back to the transformer. If the current is high enough, the resulting voltage drop on the wires could cause a noticeable effect on the operation of appliances. In the case of the neutral wire, the voltage drop is usually low enough to not be dangerous, or even noticeable. This phenomenon is the same as the notion of output resistance of a power supply discussed in Section 2.6.1, where the ideal power supply is the transformer and output resistance is that of the wires that feed the house and the branch circuits. When too much current flows, the load voltage drops.

As mentioned at the start of this section, you might find the use of two wires that both go from an appliance to ground, namely the neutral wire and the ground wire, to be strange and ask, "Why then is neutral not simply ground, and why is a dedicated ground wire used in addition to the neutral wire?" That is a reasonable question. The ground wire provides a current path to ground that is *independent* of the neutral return path, which *also* goes to ground. Using two paths in this way decreases the possibility of electric shock. How so? First, you should understand that it is always safe to touch ground. After all, you are walking on it all the time. Much of the metal in structures and appliances is connected to ground. As long as you are at ground and don't contact any other voltages, no current will flow through you. Many appliances (think microwave oven) have a metal outer case that is grounded through a ground wire. It is safe to touch the metal parts of the oven *because of the ground wire*.

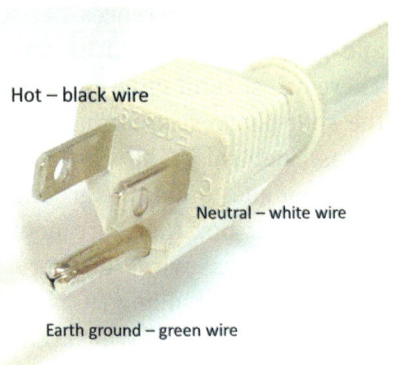

FIGURE 3.29 Standard three-prong wall plug.

Let's remove the grounding from the microwave oven's outer metal case, which is a potentially dangerous situation. That is relatively easy to do – just snap off the ground prong, Fig. 3.29, from the plug (*not* advised) and plug it in. It will operate as normal, but now the case is electrically floating.

Without being grounded, if, through some defect or fault, an internal 120 V (*i.e.*, hot) wire were to somehow contact the outer metal case, then the case would also be (electrically) hot. No current would flow from the case because there is an infinite resistance between the case and ground. This condition would not have any effect on the operation of the appliance, and it would go unnoticed – for a while anyway; it would be an accident waiting to happen. When, not if, someone touched the metal case, they would become an alternate path to ground and receive a dangerous shock. The person becomes the load for 120 V! However, with the case properly grounded, this situation could not arise. The instant that the hot wire touched the grounded case, the load to the power bus becomes a short circuit (just the small wire resistance). A large current from the errant hot wire would instantly short to ground through the case's ground wire and instantly trip the breaker, thus protecting the potential victim. In summary, the instant that the fault occurred inside the microwave oven, the breaker would trip. The result would be nothing worse than an annoying tripped circuit breaker, instead of possible cardiac arrest and death. Note, then, that *the ground wire carries current only during a fault condition*; the neutral wire is responsible for carrying the normal appliance current when it is working properly. The use of two separate wires, neutral and ground, serves the two different functions – one to carry the normal current, and one to carry current due to a fault.

This system works only if the ground prong, as shown in Fig. 3.29, is properly plugged into the ground connection of a grounded wall outlet. Houses built before 1974 were not required to have three-prong receptacles. It was common for people (clearly not those who have taken this course) to break off the ground prong in order to access older two-prong outlets. Alternatively, they used 3-to-2-prong adapters, or "cheater plugs" (Fig. 3.30), which can be plugged into a two-prong outlet; there is no ground prong, but rather the ground connection is a tab or wire sticking out. The tab allows for a connection to ground through a wall-plate screw, but it was rarely used. Many older homes still have the old, ungrounded outlets, or have only two wires running to three-terminal outlets whose ground hole is not attached to a ground wire. This is the equivalent of a cheater plug.

At all modern wall outlets in the U.S. (Fig. 3.28), one slot is shorter and is the hot side, and is connected to either phase leg A or leg B; the other slot is longer and is the neutral terminal; and the small half-round hole is ground. To distinguish hot from neutral, many modern, *ungrounded* appliances with plastic, nonconducting cases have two-blade plugs on them with one wide and one narrow to ensure that the plug is inserted in only one possible way. This kind of plug is called **polarized** and is shown on the cheater plug in Fig. 3.30. The polarized plug forces the appliance to "know" which wire is hot and which is neutral.

Polarized plugs make appliances safer when turned off by connecting the hot wire, rather than the neutral wire, to the on/off switch. That way, when the switch is off, the live voltage is confined to a small region up to the switch, and the rest of the appliance is not energized. If the blades were reversed in the socket, the whole circuit could be live (a 50/50 chance without a polarized plug) and more likely to expose the user to an electric shock. This is especially true for a common screw-in Edison-type light-bulb socket, called the Edison base (Fig. 3.31). It was designed in the late 1800s, and as such, is a horrible design that would never be considered today. A light bulb passes current from the small **center contact** at the bottom through the bulb and back through the outer, threaded **screw shell**. The screw shell is large and easily contacted by a finger. For safety's sake, a polarized plug should be used to ensure that the screw shell is connected to neutral while the hot wire is connected through the switch to the smaller center contact. Wired incorrectly, inadvertent contact with the shell will cause a shock, even with the switch off. With the hot wire going directly to the shell, it is always energized. Because the shell is so prominent, often sticking out above the edge of the bulb socket, contact is very possible, and the danger of a shock is real, especially when changing a lightbulb, or reaching into the light shade to find the switch. To preserve its safety benefits, a polarized plug should never be modified (but by now you don't need to be told that).

Now that you understand how current flows on the hot and neutral wires, and how a ground wire protects from electric shock in some cases, you are ready to understand how a GFCI is different from a circuit breaker, and how it protects specifically against electrocution, but not overheating. A GFCI compares the current leaving the hot wire with the current returning on the neutral. With no other current path, KCL tells us that those currents should be exactly the same. If those are different, even by a few mA, then current must be flowing outside of the hot and neutral wires due to some kind of fault condition. This could indicate a dangerous ongoing shock to a person. When the current difference is about 5 mA for a very short time, about 20 ms, the breaker in the GFCI trips, thus protecting the potential victim. It takes only 10 mA to stop a heart, so the GFCI must be working properly, and therefore should be tested regularly.

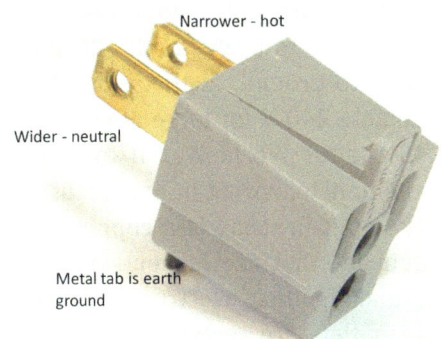

FIGURE 3.30 'Cheater plug' for connecting to older two-prong outlets. Note that the wider blade is neutral, and narrower blade is 'hot,' or 'live.'

FIGURE 3.31 Common light-bulb socket. Current passes from the center contact through the light bulb and back through the large threaded outer surface inside the insulator, called the screw shell. It is easy to accidentally touch the screw shell. A polarized plug should be used so that when the light is turned off, the screw shell is connected to neutral rather than hot to help prevent an electrical shock.

Historical: James Watt in the late 1700s developed the steam engine. He defined the unit of horsepower, hp, to be 550 ft-lb/s (which is now equivalent to 735.5 watt). In fact, a horse can develop as much as 15 hp over a short time. Presumably, Watt's intent was to make a fair comparison between a steam engine and a horse working at a normal pace. He certainly didn't want to define a unit that would harm a horse.

One more note about safety: It isn't all that easy to contact 240 V in the home because you would have to bridge both 120 V legs. Instead, it is more likely that you would get a shock *only* from 120 V to ground. In that way, the U.S. system uses 120 V when 240 V is not needed, but still has available 240 V for large appliances. Most of the rest of the world standardized single-phase 220 V to 240 V after World War II to save money on the cost of the extra copper needed for a split-phase system. Also, wiring in Europe uses a higher gauge (thinner wire) than is typically used in the United States (again using less copper). Now can you see why they can get away with using thinner wire? Countries that use higher single-phase voltage have compensated for the more-serious safety issue by making their electrical outlets much harder to contact with fingers and such. European plugs and outlets are recessed and shuttered, making it nearly impossible for accidental contact with the electrical terminals. In contrast, it is relatively easy to insert metal objects into unshuttered U.S. outlets, and to contact the metal blades when a plug is hanging loosely in the socket. The one redeeming feature of the U.S. system is that 120 V is, in fact, less dangerous than 240 V, but that safety advantage is somewhat squandered by the neolithic design of the plugs and outlets. The only thing that prevents more accidents is the 'common knowledge' that electrical outlets are dangerous. You, however, now know many more of the details.

3.4.3 Commercial Customers: Wye and Delta Circuits

Commercial customers generally have higher power requirements than residential ones. The split-phase system used by residential customers is based on three wires but comes from only a single phase of the three-phase distribution system. In contrast, consumers such as factories make use of all three phases at the same time.

Let's imagine that we are running a three-phase electric motor of 50 horsepower: that's 37 kW – *whoa Nelly*! Each of the phase legs is synchronized with the others to provide magnetic force to internal coils that are physically spaced 120° apart inside the motor, somewhat like the generator in Fig. 3.5, but with windings on both the rotor and the stator. The combination of the three stator coil phases creates a rotating magnetic field that couples to the rotor,

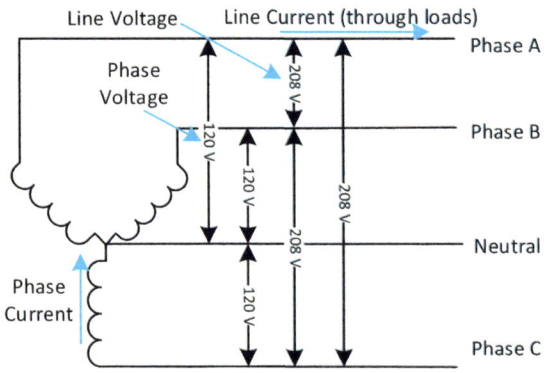

FIGURE 3.32 Wye configuration of three phases, A, B, and C. In the figure, all three sources are the secondaries of individual transformers being driven 120° apart. The wye configuration uses a neutral wire. The voltages between phases are √3 times the voltage of a single phase.

pushing it with just the right timing to propel the motor around. Details of such machines are left to a course in power systems.

The high voltages of transmission lines aren't of direct practical use in any applications, so step-down transformers are used at various points of the transmission grid. The three phases anywhere in the grid come from three separate transformer secondaries. These sources can be wired together in two basic configurations, namely **delta** or **wye**. Figure 3.32 shows the **wye**, or **star**, configuration, so called because the circuit schematic of the sources on all three phases make a "Y" or "star" shape. The three sources are connected by three **lines** to three distinct loads. The loads may also be in either the delta or wye configuration, independent of the configuration of the sources. As shown in Fig. 3.32, the voltages across the sources, i.e., the voltage of the individual sources, are called **phase voltages**. Voltages between two lines are called **line voltages**. Currents through each voltage source are **phase currents**; currents that flow from the sources to the loads through the lines are **line currents**.

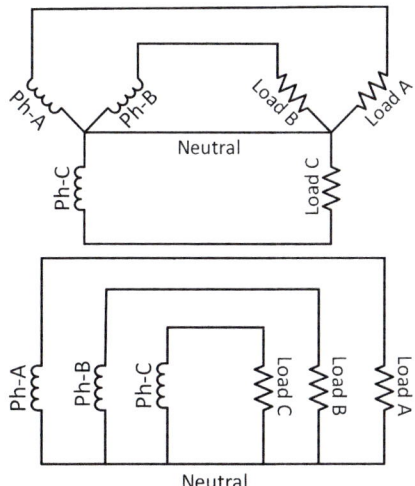

FIGURE 3.33 (Top): Wye source and wye load configuration, called wye-wye. (Bottom): Note that this is really just each source phase driving its own load, and all three in parallel are connected at the neutral wire.

The top of Fig. 3.33 shows the sources and the loads both connected in the wye configuration, a circuit topology called **wye-wye**. The wye-wye configuration isn't as complicated as it looks – each phase simply drives its own load, as shown in the bottom of Fig. 3.33. The voltage drop across each load is the same as the phase voltage that drives it. For wye-wye, the phase and line currents are the same because the same current flows through both the source and the line, as shown in Fig. 3.32.

When three two-terminal devices are joined in the shape of a triangle, the circuit is in the **delta configuration**, after the Greek letter delta, Δ. Figure 3.34 (top) shows the three-phase **delta-delta** connection, so called because the sources and loads are both laid out in deltas. Again, the topology is drawn differently in Fig 3.34 (bottom) to bring out important relationships. By Kirchhoff's Voltage Law, each of the delta source voltages must at every moment be exactly equal to the sum of the other two source voltages. Remember that each of the three phases is out of phase by 120° w.r.t. the other phases, so this and other voltage and current relationships are somewhat complicated.

In wye-wye, the phase and line currents are the same, but the phase and line voltages are different. In delta-delta, it is the phase and line *voltages* that are the same, but the phase and line *currents* are different.

By now you are wondering, "Considering the complexity, why is three-phase power preferred over single-phase?" The answer lies partly in how three sine waves interact when connected to loads in the delta and wye configurations. Here we use wye-wye to illustrate, and we will leave delta-delta to the laboratory activities. It is important to understand that **adding any number of sine waves of the same frequency, regardless of their phases or amplitudes, results in a new, perfect sine wave of the same frequency but having some new amplitude and/or phase**. The three, three-phase voltages and currents are all of the same frequency, namely 60 Hz, as shown in Figures 3.5 and 3.16. Therefore, whenever two or three voltages or currents add together, they combine to form a sine wave having the same frequency, but some new phase and amplitude that must be determined. In short, everything going on is a sine wave at 60 Hz. The mathematical method for determining the resulting voltages and currents is called **steady-state sinusoidal analysis**, which is introduced in Chapter 5. For now, we will rely only on some of the basic concepts.

Refer to the top of Fig. 3.33. Note that the central node at the sources and the loads each have four wires. Per Kirchhoff's current law, at any moment in time all the current flowing into a node must equal the

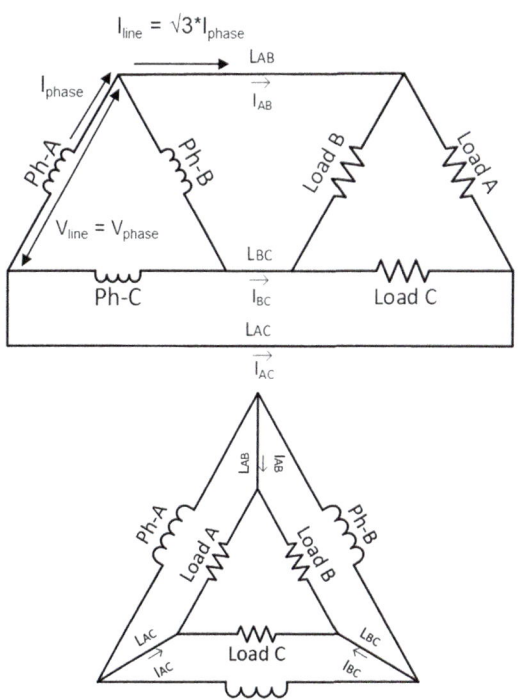

FIGURE 3.34 Delta-delta circuit. Top: conventional topology. Bottom: alternative topology showing better the relationship between each phase and the loads. Notice that there is no neutral wire, and all current returns to the sources through the loads. Also notice that each phase voltage is in parallel with the sum of the other two phases.

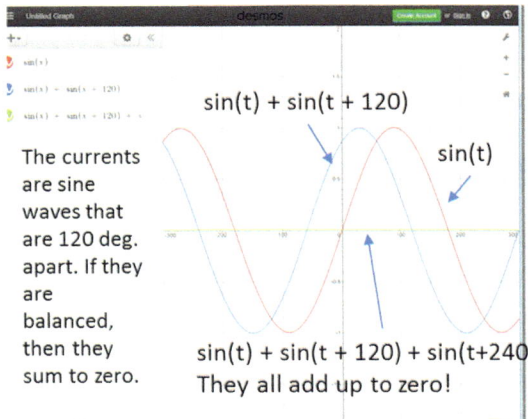

The currents are sine waves that are 120 deg. apart. If they are balanced, then they sum to zero.

$\sin(t) + \sin(t + 120)$

$\sin(t)$

$\sin(t) + \sin(t + 120) + \sin(t + 240)$
They all add up to zero!

FIGURE 3.35 This plot shows that when we add three sinusoids together, each of which has the same amplitude but are out of phase by 120°, then they add up to exactly zero. This is why all the current in the balanced wye-wye configurations flows only in the three loads, and none of it flows back to the transformers on the neutral wire.

current flowing out. If the three-line currents at the loads entering that node add up to zero at all times, there will never be any current flowing on the neutral line back to the voltage sources.

Just as electricians try to balance the A and B loads in a split-phase system so that no current flows on the neutral line back to the transformer, they attempt to equalize the three loads in three-phase systems. If the three loads are identical, i.e., **balanced**, then the (sinusoidal) current flowing in any phase A, B, or C will be exactly equal and opposite to the sum of the currents flowing in the other two at any moment. *Stated another way*, all the line currents add up to zero (Fig. 3.35) at every moment. *Stated another way*, each wire acts as the return wire for the other two phases at every moment. *Stated another way*, no current flows on the neutral wire. Those huge, three-phase transmission lines use only three thicker line cables and one thinner, neutral cable because the loads are assumed to be nearly balanced.

So far, we have assumed that the loads are perfectly balanced, so that the line current in any one leg of the circuit is precisely balanced by the line currents in the remaining two legs, and no current flows on the local neutral line. What if they are not perfectly balanced? What if two legs serve quiet residential neighborhoods, and at the third there is a huge block party with really loud rock bands and lots of bright lights? What happens then? Well, that is what the neutral line is for. Any unbalanced load current will flow back to the source by way of the neutral line, which is generally used where balancing is not assured. Current on the neutral line wastes some energy and uses up some of the voltage that would normally go to the loads, just as in the split-phase neutral cable.

In some industrial applications, the loads are well controlled, and a neutral line is not required. This is the case for three-phase motors where the loads are the three windings inside the motor, and those are set by the manufacturing process. In both wye and delta systems, if the loads are unbalanced and there is no neutral, excess current flows in the sources and loads, and voltages can fluctuate, just as it did in my home, as discussed above (but in a complicated way due to the three phases).

3.5 Complex Loads and Power Factor

3.5.1 Introduction to Energy Storage, Reactance and Impedance

This and following sections in this chapter introduce the concepts of where power flows in capacitive and inductive circuits and systems, and the mathematics of that is the topic of Chapter 5 where you will be introduced to *phasors*. It will be helpful for you to review the materials presented below when learning the material in Chapters 4 and 5.

As part of the overall development of power flow, this section introduces a new concept called impedance. Impedance serves the same role as resistance in a circuit, but with an additional, critical, difference, which is that capacitors and inductors store energy in their fields. It takes time to store that energy, which comes from the voltage and current of the devices. As such, there is some time delay between the passing of current and the storing of energy, which affects the voltage. The bottom line is that the current and voltage do not track each other perfectly –

their waveforms are separated in time. Since resistors dissipate energy but do not store any of it, their current and voltage waveforms track each other perfectly.

Let's consider the resistor. Using a sinusoidal source, if you plot the current through, and the voltage across, a *resistor*,

$$v(t) = i(t) \cdot R, \tag{3.18}$$

on two separate channels of an oscilloscope, you will see that the current and voltage sinusoidal waveforms will overlap perfectly in time (if you scale them vertically). This is called being **in phase**, meaning there is no time or phase difference between current and the voltage.

For a moment, consider the word "resistor." Well, it's based on the word "resist." What is resisting what? In this context, the resistor resists the flow of current due to a voltage. We might have used the similar word "impede." We could have said that the **impedor** impedes current due to a voltage, but we didn't. Instead, we assign the word **impedance** to the properties of a circuit element that limits current (impedes it), AND *whose current and voltage are not in phase*. There really is a word "impedor," which is a device that has impedance, but it is rather obscure

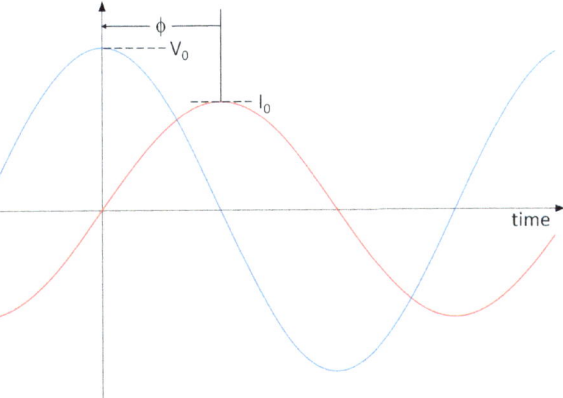

FIGURE 3.36 Sinusoidal waveforms for an inductor. There is a 90° phase difference between voltage and current. Voltage is said to lead current because the voltage peak happens before the current peak. For a capacitor, switch the positions of the current and voltage waveforms. Current leads voltage for a capacitor.

and rarely used. An impedor has impedance, but we will just say "an impedance" as everyone does. To summarize, the current and voltage of an impedance differ in time, as shown in Fig. 3.36. Such elements must store energy, and include capacitors and inductors, as well as combinations of them.

Of use to you: Here, Iceman is NOT the Marvel superhero. In fact, prior to the advent of refrigerators, people cooled food in iceboxes, which used ice delivered by an iceman. When this mnemonic was coined, everyone would have known about the iceman. A less-common mnemonic is CIVIL where 'for a capacitor, current leads voltage' (CIV), and 'voltage leads current for an inductor' (VIL). This may be better than ELI the ICEman because electrical engineers are especially CIVIL folks.

If you plot the current through an ideal *inductor* and the voltage across it, you will see two sine waves that have a phase difference of 90° between them, as shown in Fig. 3.36. The *voltage* **leads** the *current* in an inductor. The leading waveform shows up to the left on the scope screen or graph because it happens first. Conversely, a waveform that **lags** happens later, i.e., appears to the right on the screen. (It would be a good idea to get comfortable with the terms leading and lagging right now, as it will be used a lot going forward.)

A similar sort of thing holds for a capacitor, but the *current* leads the *voltage*. The leading and lagging relationships hold because $i_C = C \cdot dv_C/dt$ and $v_L = L \cdot di_L/dt$. For the capacitor, if $v_C = V_0 \sin(\omega t)$ then by differentiation $i_C = \omega C V_0 \cos(\omega t)$, which can be expressed as $\omega C V_0 \sin(\omega t + \pi/2)$, or $\omega C V_0 \sin(\omega t + 90°)$, demonstrating that current leads the voltage. A similar calculation explains the inductor behavior. (Note that I find degrees easier to visualize than radians, so I often use degrees to represent phases when I have the plot in mind, rather than the mathematics.)

Any circuit element whose current and voltage are 90° out of phase is called **purely reactive**, or just **reactive**. Ideal inductors and ideal capacitors are assumed to have no internal resistance and are purely reactive. There is a well-known mnemonic that has helped generations of electrical engineers: "ELI the ICEman" stands for "voltage (*E* for EMF) of an inductor (*L*) leads current (*I*), and current (*I*) of a capacitor (*C*) leads voltage (*E*)." This mnemonic will be used extensively going forward.

In general, impedance is a complex number represented by the symbol **Z**, which has units of ohms. **Z** is a combination of two terms: a real term represented by the symbol *R* and an imaginary term, represented by the symbols *jX*. The *R* term contributes no energy storage in the impedance while the *jX* term indicates that energy is being stored. We write the total impedance as

$$\mathbf{Z} = R + jX, \tag{3.19}$$

R is the **real part** of **Z**, *j* is the complex unit $i = \sqrt{-1}$, and *jX* is the **imaginary part** of **Z**, where *X* is called the **reactance**. Electrical engineers use "*j*" instead of "*i*" because "*i*" is already taken. Note to the aspiring cognoscenti: "*jX*" is the accepted order of the symbols – not "*Xj*."

Reactance, X, in itself is real, i.e., not imaginary; *jX* is imaginary. In the general case, devices include both resistance and reactance, so *R* and *X* are both nonzero. For a resistor, *X* = 0 Ω, and for an ideal capacitor or inductor, *R* = 0 Ω. You might be tempted to think of "*R*" as an actual resistance in a circuit, which can sometimes be the case, but in general "*R*" stands for "real" and is a mathematical term that represents the part of the voltage and current that is in-phase and comprises the effects of other circuit elements, not just resistors. This is a complicated concept that you cannot yet easily understand but will become clear during Chapter 5. Also, in Chapter 5 you will see that **Z** is much like a vector having real and imaginary components *R* and *jX*. Much more on this later.

3.5.2 Power Transmission

3.5.2.1 Passive Sign Convention

The phase difference between current and voltage is important in power transmission (as it is for every circuit). Power is the rate of the flow of energy. The **instantaneous power** going into a device (which is the load in our case) is given by:

$$P(t) = i(t) \cdot v(t). \tag{3.20}$$

where *i*(*t*) is the current through that device and *v*(*t*) is the voltage directly across that device. Again, energy is stored as a magnetic field in an inductor and as an electric field in a capacitor.

At this point, we need to make some decisions about how we mean "positive" and "negative" w.r.t. current, voltage, and power. Resistors can only absorb power (or energy) and convert that power to heat. That absorbed power is taken in by the resistor and must be released to the environment, or the resistor will overheat and be destroyed. Ideal capacitors and inductors both absorb power from the circuit and release that power back to the circuit at various points in the AC cycle; they do not dissipate any power as heat. In order to either absorb or release power, the direction of currents or voltages change. For a capacitor, the voltage polarity stays the same, but the current changes direction to release power. For an inductor, the current direction stays the same while the voltage polarity changes. What about voltage sources? Don't they *always* provide power? Nope. Think of a battery charger. Current flowing one way charges the battery; in which case the battery absorbs power. Current flows the other way when the battery is providing power. The terminal polarities do not change between these two cases. This is also true of a power supply in a circuit, even though it doesn't need to charge, like a battery – it can still absorb power.

It is critical in circuit analysis to keep all the signs correct; one wrong sign can really mess things up. The thing about signs is that there are only two of them – positive and negative, but everyone must agree on which of them is which. If half the world uses equations for which up is positive and the other half considers up to be negative, then they had better learn to talk to each other, or engineering systems will be very difficult. The same is true of circuits, and how we define positive current and positive voltage. If you personally consider up to be positive, and stick with that in all of your calculations, you will be perfectly fine. However, you might get tied up when you work with other people or read a book for which up is negative. Therefore, there are sign conventions that are well-established that you should also use.

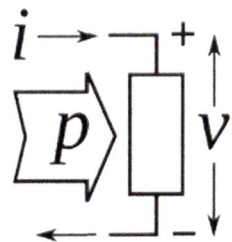

FIGURE 3.37 Passive sign convention. This diagram shows that when positive current flows into the positive terminal of an element, it absorbs power.

The widely accepted convention for the sign of power flow is called the **passive sign convention**, PSC. In circuit analysis, the direction of the current arrow is the definition of positive current, just as the placement of the '+' sign defines the positive terminal of a voltage drop. Referring to Fig. 3.37, the PSC tells us, firstly, that *positive current into a resistor, or any passive device, flows into its positive terminal.* If you see positive current flowing into a resistor, you know that terminal is positive, and you should label it as such. What about a power supply or battery? Those terminal polarities are fixed; you can have current either flowing into or out of the positive terminal.

Now to power. Power is defined as positive when energy flows into the circuit element. Since power is defined as $P(t) = i(t) \cdot v(t)$ (Eq. 3.20), when positive current flows into the positive terminal, the PSC says that *power goes into* the circuit element, and its *sign is positive*. Since positive current always flows into the positive terminal of a resistor, then both $i(t)$ and $v(t)$ are positive, and $P(t) = i(t) \cdot v(t)$ is positive. When current flows out of the positive terminal (without changing the direction of the arrow, which defines the positive current), as, for example, from a battery, then that current is negative, $P(t)$ is also negative, and power is, therefore, supplied by the battery.

How can we think of power being supplied by an element versus power consumed by an element? Well first, recall that current is universally defined as positive in the direction of the motion of positive charge, even though current is actually carried by negative electrons in a metal wire. In a battery, chemical energy is used to raise the potential energy of the charges against the internal electric field, creating the voltage on the battery. Those positive charges are repelled from the positive electrode, driving current through the circuit, providing power. When those positive charges enter, say, a resistor, they lose their energy by falling through the resistor's internal electric field. In so doing, the charges collide with atoms, giving up their energy as heat, and the resistor consumes power.

3.5.2.2 Phase between Current and Voltage due to Impedance

Summarizing the previous section: Remember that capacitors and inductors can charge, taking in energy, and then discharge, providing energy. For AC current, energy may at any particular time move into or out of a circuit element during the course of an AC cycle. During those times when the current and voltage have the same sign, i.e., when $P(t)$ is positive, energy and power flow into a circuit element. When $i(t)$ and $v(t)$ have opposite signs, i.e., $P(t)$ is negative, and energy power flow from the element back to the circuit. At those moments, the circuit element is acting like a power source, i.e., it is a source of EMF, as discussed in Section 1.4.4.1.

For the special case of a resistor, current and voltage are always in phase, i.e., always the same sign, so electrical energy always enters, or is absorbed by the resistor. That energy has to go somewhere;

it is often converted to heat, as in a resistor, but other work could be done, like raising an elevator or creating a lot of sound energy at a Taylor Swift concert. For an ideal capacitor or inductor, the current and voltage are out of phase by 90° (refer to Fig. 3.36), so $P(t)$ is positive during half the cycle and negative during the other half. During the positive half cycle, all of the energy that flows in is temporarily stored in the electric field (capacitor) or magnetic field (inductor); during the negative half cycle, that energy is returned to the circuit.

These ideas can be applied not only to simple electrical devices and circuits, but much more broadly. A factory is a complicated electrical load, so it is useful to introduce the concept of a black box, as shown in Fig. 3.38. A **black box** is something whose detailed contents are not known but can be included in a circuit model in terms of its external behavior. In Fig. 3.38, the black box could be anything, from a resistor to an entire factory. All you know is the voltage waveform across, $v(t)$, and current waveform into, $i(t)$, the black box. That information is enough to tell you its impedance. If you think of a factory as a black box that behaves like a complex load with real and imaginary parts, then as it uses power to do its thing, it also stores some energy on one half cycle (due to its reactance) and returns that energy on the other half cycle to the power lines, as if the load (here, the factory) were momentarily a small power company. It isn't necessary to know what is inside that entire factory that is storing and releasing energy – and the power company doesn't care. They only care that the factory is an impedance, having both real and imaginary parts.

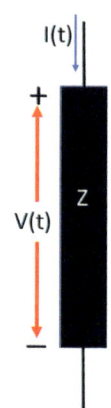

FIGURE 3.38 A "black box" represented by an impedance, Z. From the outside we can see $i(t)$ and $v(t)$, and deduce everything we need to know about Z, but we don't know what is actually inside of it.

If the power grid had no resistance, then temporarily storing energy in the factory would not be so much of an issue. However, there is always resistance. It takes current to get the energy to the complex load where it is stored by the reactance before it is returned to the grid. These currents flow both ways, causing i^2R losses on the power lines and requiring larger transformers, etc. to carry that power, all of which are more costly to the power companies. Industrial customers would be happy if the power companies charged only for the energy they use, which is due to the real part of their load. However, the power companies charge extra to industrial customers whose loads look in part like capacitors or inductors.

Important: In complicated circuits inside of a black box, the values for R and X in $\mathbf{Z} = R + jX$ can contain combinations of resistances, inductances, and capacitances, because the currents flowing around the circuit can be affected by current flow through all of the components. So, R is not simply the resistances in the circuit, although it can be just the resistance in very simple circuits.

Thus far, we have considered purely resistive (real) and purely reactive (imaginary) loads. Let's build on these concepts. In order to consider capacitors and inductors on a similar basis with resistors, we need something more like resistance to describe the reactive devices in order to put values into $\mathbf{Z} = R + jX$. The reactance of an ideal inductor is

$$X_L = \omega L,\tag{3.21}$$

which is positive. Notice that X_L depends on (angular) frequency, ω – the higher the frequency of the AC signal, the higher the reactance. Since reactance is part of the impedance, and impedance impedes current for a given voltage, it follows that higher frequency currents do not flow as easily through inductors.

The reactance of an ideal capacitor is

$$X_C = -1/\omega C, \tag{3.22}$$

which is negative. The negative sign is about the leading and lagging nature of the current w.r.t. the voltage, and it will become important later in the discussion. The magnitude of X_C is what determines how much current flows. In this case, as ω increases, the magnitude of X_C decreases, so current flows more easily through capacitors at higher frequencies. These basic facts are central to the behavior of electronics, and specifically underpin the concepts of filters presented in Chapter 4 and phasors in Chapter 5.

Note that the reactance, X, is not by itself imaginary, but is multiplied by j as jX and becomes the imaginary part of $\mathbf{Z} = R + jX$. Any *positive* value X, say that of a black box, can be thought of as, or **modeled**, as reactance due to an inductor having $L = X/\omega$. This positive reactance causes the voltage to lead current, as with ELI. So, referring to the black box, any impedance for which voltage leads current is called **inductive**. Similarly, any *negative* reactance causes current to lead voltage, as in ICE, and the impedance is called **capacitive**. Because reactance changes with frequency, the reactive nature of the impedance can change between inductive and capacitive as a function of frequency when energy is stored in both magnetic and electric fields.

Refer to Fig. 3.39 showing the current and voltage out of phase by some value ϕ. The phase angle between $v(t)$ and $i(t)$ is in general given by

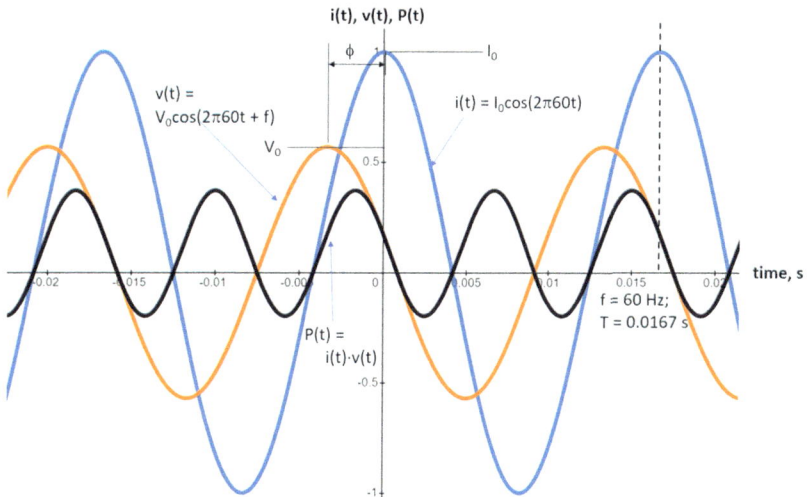

FIGURE 3.39 Sinusoidal current and voltage waveforms due to a complex impedance. The current and voltage are separated in time with a phase difference φ. Here, voltage leads current, so the impedance is inductive. The plot is shown for a frequency of 60 Hz. Note that the product of current and voltage is the instantaneous power, which can change sign at various times.

$$\phi = \arctan\left(\frac{X}{R}\right) \tag{3.23}$$

The *magnitude* of the impedance, $|\mathbf{Z}|$, is

$$|\mathbf{Z}| = \sqrt{R^2 + X^2} \tag{3.24}$$

$|\mathbf{Z}|$ includes the effects of both the real and imaginary parts, and it tells us by *how much $\mathbf{Z}$ blocks the current flow*. The larger $|\mathbf{Z}|$ is, the less current flows for a given voltage amplitude. $|\mathbf{Z}|$ acts like a resistance in ohms law:

$$|v| = |i|\,|\mathbf{Z}|. \tag{3.25}$$

$|v|$ and $|i|$ are the amplitudes of the voltage and current sine-wave functions, respectively. We define a few new variables: $|v| = V_0$; $|i| = I_0$; $I_{0,RMS} = I_0/\sqrt{2}$; $V_{0,RMS} = V_0/\sqrt{2}$. From Eq. 3.25 we see that

$$V_0 = I_0 |Z|, \tag{3.26}$$

and by dividing both sides by $\sqrt{2}$,

$$V_{0,RMS} = I_{0,RMS} \cdot |Z|. \tag{3.27}$$

In power systems, it is assumed that voltage and current values represent their RMS equivalents. For example, in a discussion of residential voltages you will most-often see a number like "120 V". It is understood to be the RMS voltage. However, if you are dealing with a random AC circuit and see a voltage given as "8.3 V," you need to find out what is meant by that value, whether it represents the peak or RMS value.

Equations 3.24 and 3.25 tell us that if there is a large resistance *or* a large reactance, then less current will flow for a given voltage. The impedance, $Z = R + jX$, tells us how much the circuit element stops current from flowing, Eq. 3.25, *and* the phase shift between the voltage and current (Eq. 3.23). As a reminder, *if X is nonzero, there is a phase shift between the current into the impedance and the voltage across the impedance.*

ϕ is exactly –90° or 90° *only* for a **pure** capacitance or inductance, i.e., $R = 0\ \Omega$. (Plug $R = 0\ \Omega$ into Eq. 3.23 to see what you get for a pure inductance and a pure capacitance.) When the impedance of any circuit element, circuit, system, or black box is complex, having nonzero values of both R and X, then the phase between the current and voltage, ϕ, is not ±90°, but instead must be somewhere between those values, as shown in Fig. 3.39.

As you know from your trig classes, trig functions are periodic in integer multiples of 2π; this means that you can add 2π to any angle ϕ and the result will be the same. However, we use phase angles ϕ that are limited to ±180° since any value of $|\phi|$ that is greater than 180° can be replaced by a value of –(360° – ϕ). In other words, when you look at two peaks of the sinusoids, the phase angle is taken as that time difference whose phase is less than 180°. For example, current leading voltage by 240° is the same as current lagging voltage by 120°, which is how that relationship would be described. You don't need to go through all of this rigor. All you have to do is look at both sides of a peak or zero crossing of one of the sinusoids and pick the smaller phase. If on one side the phase is greater than 180° then on the other side it must be less than 180°, and that angle is ϕ and determines what leads or lags what.

By the way, don't sweat the words "imaginary" or "complex." Imaginary numbers are not imaginary. Mathematicians at first just needed a word to describe a different kind of number than the usual "real" numbers. Also, complex numbers are not so "complex." In this context, *the use of complex numbers is just a way of incorporating a phase difference between the voltage and the current.* Complex numbers as applied to impedance and phase, and how they behave over a range of frequencies, are discussed in Chapters 4 and 5.

3.6   Complex Power Flow

You have been introduced to the concept of complex impedance for which voltage and current are out of phase. This whole discussion could be based on the mathematical method of phasors, which is the subject of Chapter 5. As you will see there, phasors eliminate the need for using trigonometric identities, which, except in the simplest of cases, get very cumbersome. However, you as beginners should have a foundation in what underlies all of the "complex-ness" (my word) of complex impedances, and the properties of their power consumption. To this end, the following sections provide the underlying properties of the voltage, current and power waveforms using trigonometry upon which the more-advanced phasor treatments are based.

Let's begin with a short review of impedances, and then continue to the nitty-gritty of power in complex loads. The load could be a device element, a circuit, a giant heater or motor, or even an entire manufacturing plant, and can be treated as a black box. As always, $Z = R + jX$, whose magnitude is $|Z| = \sqrt{R^2 + X^2}$, and whose phase angle between current and voltage is ϕ = Arctan(X/R). To obtain the instantaneous power delivered to, or returned by, the load, we multiply the instantaneous current by the instantaneous voltage: $P(t) = i(t) \cdot v(t)$ (Eq. 3.20). If $i(t)$ and $v(t)$ are not perfectly in phase, they will change signs at different moments (as in Fig. 3.39). As a consequence, their product $P(t)$ will change sign over a period, meaning that *power will alternately flow into and out of the load*. In this and the next few sections, we will break down the components of power flow and provide a foundation for the power that flows to the load and is taken up by the load or stored and returned from the load. We will start with the general form of the power flow and then investigate each component.

A generalized sinusoid has the form $A\cos(\omega t + \phi)$, where A and ϕ are constants. The prefactor A is the amplitude of the sinusoid (not the RMS value of the sinusoid), and ϕ is the horizontal shift of the sinusoid relative to $A\cos(\omega t)$. Again, ϕ is an angle that is some fraction of the total period; ϕ depends on frequency as in Eq. 3.15. In the following, we will arbitrarily choose current to have a zero-phase shift:

$$i(t) = I_0 \cos(\omega t). \tag{3.28}$$

Its phase is taken as zero, which is the same as choosing to start the clock when the current is a maximum at $t = 0$.

Again arbitrarily, voltage is taken as

$$v(t) = V_0 \cos(\omega t + \phi). \tag{3.29}$$

Its phase is ϕ relative to the current, meaning that it is shifted horizontally on the time axis. The shift is to the left for $\phi > 0$, which is leading, or inductive, and to the right for $\phi < 0$, which is lagging, or capacitive.

With these choices of current and voltage waveforms, power is a product of the two cosine functions. The following trig identity is central to our discussion of power systems:

$$\cos(\alpha) \cdot \cos(\beta) = \frac{1}{2}[\cos(\alpha - \beta) + \cos(\alpha + \beta)] \tag{3.30}$$

Let's call $\alpha = \omega t + \phi$ and $\beta = \omega t$, which represents the time variations of the voltage and current, respectively, and the phase difference between them. Now,

$$\cos(\omega t + \phi) \cdot \cos(\omega t) = \frac{1}{2}[\cos(\phi) + \cos(2\omega t + \phi)]. \tag{3.31}$$

This trig identity tells us that the product of two cosine (or any two sinusoidal) functions of the *same* frequency is always equal to a sinusoid of *twice* the frequency, $2\omega t$, plus a *constant*, which here is $\left(\frac{1}{2}\cos(\phi)\right)$.

The instantaneous *total* power flow into a generalized load $\mathbf{Z} = R + jX$ that causes a phase shift ϕ can be written as

$$
\begin{aligned}
P(t)=P_C(t) &= I_0\cos(\omega t)\cdot V_0\cos(\omega t+\phi) \\
&= \frac{1}{2}I_0V_0[\cos(\phi)+\cos(2\omega t+\phi)] \text{ (applying trig identity)} \\
&= I_{0,\mathrm{RMS}}V_{0,\mathrm{RMS}}[\cos(\phi)+\cos(2\omega t+\phi)] \text{ (convert to RMS)} \\
&= S[\cos(\phi)+\cos(2\omega t+\phi)] \\
&= P+S\cos(2\omega t+\phi)
\end{aligned}
\tag{3.32}
$$

where we define

$$
P = S\cos(\phi).
\tag{3.33}
$$

The instantaneous total power, $P_C(t)$, is formally called the **complex power**, for reasons discussed later. Complex power is the *total* power that flows to and from the load, *comprising that used by the load and that stored and returned*. We have defined S as the *amplitude* of the sinusoidal component of the complex power, expressed in RMS volts and amps:

$$
S = I_{0,\mathrm{RMS}}\cdot V_{0,\mathrm{RMS}} = I_{0,\mathrm{RMS}^2}|Z| = \frac{V_{0,\mathrm{RMS}^2}}{|Z|}.
\tag{3.34}
$$

Let's see what happens when we expand the $S\cos(2\omega t+\phi)$ term using another trig identity,

$$
\cos(\alpha+\beta) = [\cos(\alpha)\cos(\beta)-\sin(\alpha)\sin(\beta)].
\tag{3.35}
$$

This leads to

$$
\begin{aligned}
P_c(t) &= P+S\cos(2\omega t+\phi) \\
&= P+S[\cos(2\omega t)\cos(\phi)-\sin(2\omega t)\sin(\phi)]
\end{aligned}
\tag{3.36}
$$

$$
\begin{aligned}
&= P+P\cos(2\omega t)-Q\sin(2\omega t) \\
&= P(1+\cos(2\omega t))-Q\sin(2\omega t)
\end{aligned}
\tag{3.37}
$$

where we define

$$
Q = S\sin(\phi),
\tag{3.38}
$$

P, S and Q are constants, and not functions of time. P is called the **real power**, **true power**, or **active power**. S is called the **apparent power** and Q is called the **reactive power**. In this text, of the three common choices, we choose to use the term "active power" for *P* because "real" is used way too often, and "true" is just, well, a little too poetic. Also, "active" shows the contrast to the word "reactive," and helps distinguish their two distinct properties.

Consider the term $P(1 + \cos(2\omega t))$ in Eq. 3.34. That is a cosine function that is shifted up by exactly its amplitude, and which is always positive. Since that is always positive, $P(1 + \cos(2\omega t))$ is the waveform, i.e., time dependence, of the power absorbed by the load. Taking the time average of the absorbed power, we see that since the time average of $\cos(2\omega t)$ is zero, the active power, *P*, is the average power delivered to the load.

Now consider Eq. 3.36. If *P* were zero, then $P_C(t) = S\cos(2\omega t + \phi)$, which is a sinusoid centered around $P_C(t) = 0$. On average, no power would be absorbed by the load – it would all be returned. That said, $S\cos(2\omega t + \phi)$ is still a transfer of power from the circuit or the power lines to the load, regardless of the value of *P* and what the load decides to do with it – keep it or return it. That power must be transported to the load so that it can be stored and then returned. So, *S* is the magnitude of the total power waveform transported to the load, and *P* shifts it up or down to determine how much of that power is in the positive range (absorbed by the load), and how much is left in the negative range (returned by the load.) For this reason, *S* is called the *apparent* power, which implies that it *seems* to be all the power that the load absorbs, but it really *isn't* because some of it is returned, depending on the value of *P*.

Since the complex power $P_C(t)$ is the sum of the absorbed power, $P(1 + \cos(2\omega t))$, and returned power, we conclude that $Q\sin(2\omega t)$ in Eq. 3.37 must be that returned power. You know by now that returned power is due to the reactive part of the impedance. Since *Q* is the amplitude of the power waveform that is transported to the load and then returned, it is called the "reactive power."

We would be tempted to just use watts units for all of the various powers P, S, and Q, but it isn't that simple. *P* is the active power, and power is expressed in units of watts. Yep, just watts – the same watts you always thought you understood. If you want to get a job *done*, then *P* is what it's all about.

To avoid confusion with power that is used by a load, *S* is expressed in units of **volt-amperes**, or **VA**. Yes, apparent power *is* power, but to distinguish it from *P* and *Q*, it has voltage units of a different name. Finally, the reactive power, *Q*, is called "**volts-amps reactive**," or **VAR**, pronounced "VAHR."

We have used the word "absorbed" to describe power that is taken in by the load and is not returned, for whatever reason. The real part of the impedance could be a simple resistance, in which case the energy is absorbed as heat. Heat is a great thing if your intention is to heat something, but often it is just wasted energy, in which case it isn't actually "used." Alternatively, the real part of the impedance could be used by an electric motor that performs work – for example, to lift a weight by a crane. That energy would be converted from electrical to gravitational energy just as it would be converted to heat in a resistor. As far as the power source is concerned, no energy is returned, so that part of the load impedance is real. In short, complex power flow comprises some absorbed power, the details of which we don't really care about, plus some stored and returned power.

Now, let's investigate each term in the second to last line of Eq. 3.32. The double-frequency term is $S\cos(2\omega t + \phi)$. Because a sinusoid is positive and negative for equal times, the net transfer of energy is zero; it represents only energy that is stored and returned. The constant term, $P = S\cos(\phi)$, is the time average of the power that is absorbed by the load. *P* is also the fraction $\cos(\phi)$ of the apparent power, *S*. The constant $\cos(\phi)$ pretty much tells the whole story of the effect of the phase, so it has a name – the **power factor**, **PF**.

$$\text{Power Factor (PF)} = \cos(\phi). \tag{3.39}$$

We have seen that the active power is *P*, which is due to the real part of the impedance. To complete the complex power, $S\cos(2\omega t + \phi)$ must carry the power that is due to reactance. Again, that power is stored in the electric and magnetic fields and is returned to the circuit, or in the case of the power grid, the power lines. The stored-and-returned power is called the "reactive power" for that reason. Here is a simple memory aid: $\underline{a}$ctive power is $\underline{a}$bsorbed, and $\underline{r}$eactive power is $\underline{r}$eturned.

All power transfer is alternating, or AC, at the frequency 2ω, or twice the frequency of the voltage and current. P is not a DC voltage, but rather the time average of the AC complex power. Referring to Eq. 3.37, the AC waveform that transfers active power is $P(1 + \cos(2\omega t))$, which is a shifted sinusoid that is always positive; the AC waveform that transports reactive power is $-Q\sin(2\omega t)$, which is not shifted, and is equally negative as it is positive.

We can combine Eqs. 3.33 and 3.38 to get the relationship between S, P and Q,

$$P^2 + Q^2 = [S\cos(\phi)]^2 + [S\sin(\phi)]^2 = S^2\left[\cos^2(\phi) + \sin^2(\phi)\right] = S^2,$$

or,

$$S^2 = P^2 + Q^2. \tag{3.40}$$

The impedance of the load, $\mathbf{Z} = R + jX$, might represent, or model, a transformer or motor, or a nonideal inductor where $X = \omega L$, as in the preceding section. For now, you can think of the real part of $\mathbf{Z}$, i.e., R, as a DC resistance, such as the resistance of the wire that forms the turns of a non-ideal inductor. This additional resistance shifts the phase ϕ back from $\pi/2$, where it would be for an ideal inductor, toward zero, where it would be for a purely resistive load.

It is the real part of $\mathbf{Z} = R + jX$ that absorbs power, while the imaginary part contributes

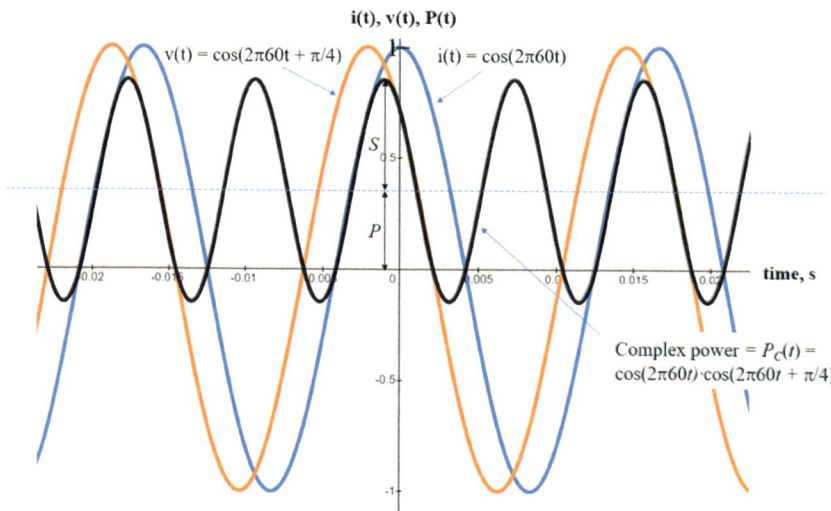

FIGURE 3.40 Voltage and current waveforms and the instantaneous, complex power for an impedance of $\mathbf{Z} = 0.707 + j0.707\ \Omega$. $i(t)$ and $v(t)$ are out of phase by 45°. The product of current and voltage is the complex power, which here is both positive and negative at various times signifying power flow into the load and out of the load, respectively. The amplitude of the complex power is the apparent power, S. The vertical offset of the complex power is the active power, P, which is the time average of the complex power.

nothing to power absorption. As current flows through, say, a nonideal inductor, it might get warm as the wire resistance absorbs some power, but the magnetic field of the inductance just stores and returns energy.

Figure 3.40 shows two cosine waveforms that represent current, proportional to $\cos(\omega t)$, and voltage, proportional to $\cos(\omega t + \pi/4)$. (Don't forget that the units in Fig. 3.40 are not the same for power, current and voltage.) In this example, the phase difference between the current and voltage is arbitrarily chosen to be $\pi/4$, or 45°, which is the case for $R = X$. Also, the amplitudes of the current and voltage are arbitrarily chosen as 1 A and 1 V, respectively. What values of R and X will cause a 45° phase shift and current and voltage amplitudes of 1 A and 1 V, respectively?

Because $I_0 = 1$ A, $V_0 = 1$ V, and $V_0 = I_0 \cdot |\mathbf{Z}| = 1$ V$_p$, then

$$|\mathbf{Z}| = \sqrt{R^2 + X^2} = 1 \ \Omega.$$

Since $X = R$, we must then have

$$R = 0.707 \ \Omega, \text{ and } X = 0.707 \ \Omega, \text{ i.e., } \mathbf{Z} = 0.707 + \text{j}0.707 \ \Omega$$

so that $|\mathbf{Z}| = 1 \ \Omega$.
 Since $I_0 = 1$ A,

$$I_{0,\text{RMS}} = 1/\sqrt{2} \ A_{\text{RMS}} = 0.707 \ A_{\text{RMS}}.$$

Since $V_0 = 1$ V,

$$V_{0,\text{RMS}} = 0.707 \ V_{\text{RMS}}.$$

In this example,

$$S = 0.707 \ A_{\text{RMS}} \times 0.707 \ V_{\text{RMS}} = 0.5 \text{ VA and } P = 0.5 \times \cos(45°) = 0.354 \text{ W}.$$

Look carefully at Fig. 3.40. The black curve,

$$P_C(t) = i(t) \cdot v(t) = \cos(\omega t) \cdot \cos(\omega t + \pi/4) = P + S \cos(2\omega t + \phi)$$

is the total power into this particular load at time t. Note that $P_C(t)$ is sinusoidal, having twice the frequency of the current and voltage. The amplitude of the sinusoid is the apparent power, S. The sinusoid is shifted up by $P = S \cos(\phi)$. The plot shows that the amplitude $S = I_{0,\text{RMS}} \cdot V_{0,\text{RMS}}$ is, in fact, 0.5 VA and that the shift upward is $P = 0.354$ W.

Equation 3.32 says that the phase of $S \cos(2\omega t + \phi)$ should be the same as that of the impedance, $\mathbf{Z}$. Looking carefully, you will see that indeed, the complex power curve is shifted by 45° from $t = 0$ after shifting down by P watts. Note, though, that 45° for the power curve does not represent the same time delay as that between the current and voltage curves; in this case, ϕ is inherited from $\mathbf{Z}$, but it is now the phase shift of the double-frequency $S \cos(2\omega t + \phi)$ term.

Because energy = power × time, the integral of the power curve, i.e., the area under the power curve, is energy. For much of the time $P_C(t) > 0$, and for some of the time $P_C(t) < 0$; these signs correspond to the energy going into the load and back out of it, respectively. In general, the energy that flows back from the load is less than the total energy that flows in, as some of it is absorbed, as seen in Fig. 3.40.

Let's look at this mathematically. Notice that $P_C(t)$ is a constant plus a sinusoidal term having frequency 2ω. The time average of the sinusoidal part of the power flow is zero. This term does not do any work because just as much power flows out as in every half cycle. The upward shift is due to the constant term, $P = S \cos(\phi)$, the active power, which is always positive, so active power flows only one way – in. That is the power that never comes back, so active power is the part of the apparent power that does all of the work on the load. In Fig. 3.40, the sinusoid is shifted up by $P = S \cos(\phi) = 0.5 \cos(45°) = 0.354$ W.

The unit of VA is a variation of watts, and is used to specify the apparent power, which is the magnitude of the waveform of the total power delivered to the load. That power sloshes around at the load, some of which is used up by the load as heat or work (active power) and some of which is returned back (reactive power) to the power lines, or any circuit for that matter. The negative part of the black curve in Fig. 3.40 is the power that returns to the power company.

If the load is "purely resistive," then $R \neq 0\ \Omega$ and $X = 0\ \Omega$. If, rather, $R = 0\ \Omega$ and $X \neq 0\ \Omega$, then the load is "purely reactive", i.e., purely inductive ($X > 0\ \Omega$) or purely capacitive ($X < 0\ \Omega$). However, loads in the real world are generally partly real and partly reactive. You'll learn more about this in Chapter 5. For now, just keep in mind that *when X is not zero, and therefore* $\mathbf{Z}$ *is complex, there is a phase difference between the current and voltage of the load, and at least some power is stored and returned from the load.*

3.6.1   Absorbed Power

In this section we investigate only the power that is absorbed by a resistive load and is dissipated as heat or work. Now $\mathbf{Z}$ is real, so $\phi = \text{Arctan}(0/R) = 0$. Plugging $\phi = 0$ into the trig identity Eq. 3.30 gives the instantaneous *active contribution* to $P_C(t)$, which we call $P_A(t)$:

$$\begin{aligned}
P(t) = P_A(t) &= I_0\cos(\omega t)\cdot V_0\cos(\omega t) \\
&= \frac{1}{2}I_0 V_0\big[\cos(0)+\cos(2\omega t)\big] \\
&= \frac{|I_0||V_0|}{2}[1+\cos(2\omega t)) \\
&= I_{0,\text{RMS}}V_{0,\text{RMS}}\big[1+\cos(2\omega t)\big] \\
&= S[1+\cos(2\omega t)] \\
&= P + S\cos(2\omega t)
\end{aligned} \qquad (3.41)$$

Since "active power" refers specifically to the constant value P, we can call the time-varying function $P_A(t)$ the *absorbed* power.

As stated above, apparent power, S, is the magnitude of the total AC power that flows into the load that is either absorbed or stored and then returned. In the previous section, we arbitrarily chose a value of $\mathbf{Z}$ to make that example easy to understand: $\mathbf{Z} = 0.707 + j0.707$, which gave unity current and voltage amplitudes. To analyze the power that is absorbed in this load, which is due only to the real part, we use $\mathbf{Z} = R = 0.707\ \Omega$, which is real, and for which the current and voltage are in-phase: $\phi = 0$. For a current of $I_0 = 1$ A and $|\mathbf{Z}| = 0.707\ \Omega$, we get $V_0 = 0.707$ V.

$$\begin{aligned}
P(t) = P_A(t) &= I_0\cos(\omega t)\cdot V_0\cos(\omega t) \\
&= \frac{1}{2}I_0 V_0\big[\cos(0)+\cos(2\omega t)\big] \\
&= \frac{(1\,\text{A})(0.707\,\text{V})}{2}[1+\cos(2\omega t)] \\
&= 0.354[1+\cos(2\omega t)] \\
&= S[1+\cos(2\omega t)]
\end{aligned} \qquad (3.42)$$

Comparing Eq. 3.42 to Eq. 3.32 using Eq. 3.33 with $\phi = 0$, we see that for a real load, $P = S$; the active power is the same magnitude as the apparent power. Equation 3.42 becomes

$$P_A(t) = S + S\cos(2\omega t)$$
$$= P + P\cos(2\omega t)$$
$$= P[1 + \cos(2\omega t)], \qquad (3.43)$$

which we saw previously as the absorbed power in Eq. 3.37.

Figure 3.41 shows the relationship between current and voltage for $\mathbf{Z} = R = 0.707\ \Omega$ and the resulting power flow. The figure shows three sinusoidal waveforms. The blue curve is the current of unity amplitude and the orange one is the voltage, which is the current scaled by the resistance, $R = 0.707\ \Omega$. The black curve shows the resulting power waveform, which has twice the frequency, $2\omega t$, and is shifted up by the '1' term such that now the power is always positive. The upward shift is required by the fact that the square of a function, such as $\cos(\omega t)$, is always positive. This demonstrates that all of the power that flows into a real load is absorbed by the load. As far as the power lines are concerned, that power is gone, never to return.

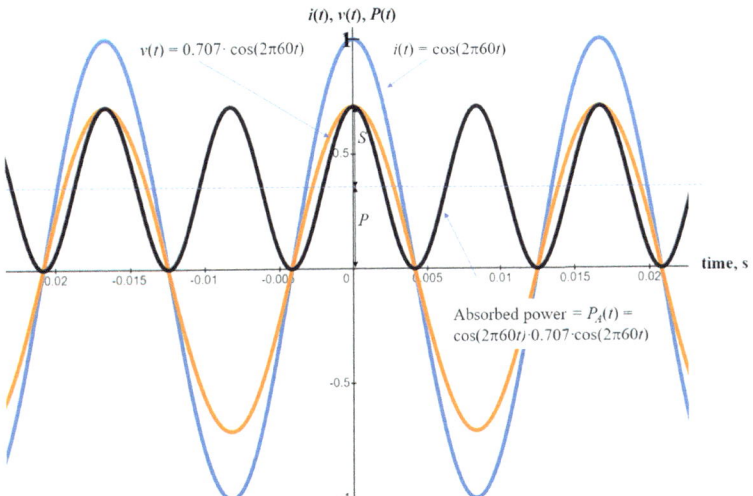

FIGURE 3.41 Voltage and current waveforms and the absorbed power, $P_A(t)$, for an impedance of $\vec{\mathbf{Z}} = 0.707\ \Omega$. For a real load, $i(t)$ and $v(t)$ are in phase, and their product is the power absorbed by the real load. For a real load, the power factor, PF, is 1, and power flows into the load, i.e., is positive, at all times. Notice that for a real load, $S = P$.

As in the case of a complex load, the sinusoidal power waveform having amplitude S is shifted up by the constant, P. For this example, in which $\mathbf{Z} = R = 0.707\ \Omega$, the resulting waveform in Fig. 3.41 has an amplitude $S = 0.354$ VA and is shifted up by the active power, which is $P = 0.354$ W. That is good, because this example demonstrates that the absorbed power due to the real part of the load impedance is the same with or without any reactance from the complex load. It also demonstrates that $P = S$, so the power waveform is always positive.

3.6.2 Returned Power

In this section, we investigate power that is temporarily stored in the electric and magnetic fields of the load and then returned back to the source. Now we discuss purely reactive loads ($\mathbf{Z} = 0 + jX$), like an ideal capacitor or inductor. Recall ELI the ICEman, who tells us that voltage leads the current in an inductor and current leads the voltage in a capacitor. For the sake of consistency, let's discuss an inductor as the load. We choose the current to have a phase of zero, so ϕ for the voltage due to an ideal inductor is 90°, or $\pi/2$ radians, as shown in Fig. 3.36. Here we expand on the ideal case of Fig. 3.36, $X > 0$ and $R = 0\ \Omega$. In the ideal case, $\phi = \arctan(X/0) = \arctan(+\infty) = 90°$ or $\pi/2$.

To analyze the power that is returned in this reactive load, we use $\mathbf{Z} = jX = j0.707\ \Omega$, which is imaginary, and for which the current and voltage are 90° out of phase: $\phi = 90°$ or $\pi/2$. For a current of $I_0 = 1$ A and $|\mathbf{Z}| = 0.707\ \Omega$, we get $V_0 = 0.707$ V. Plugging those into the trig identity Eq. 3.30 gives the instantaneous *reactive contribution* to $P_C(t)$, which we call $P_R(t)$.

Since "reactive power" refers specifically to the constant value S, we can call the time-varying function $P_R(t)$ the *returned* power.

$$P(t) = P_R(t) = I_0 V_0 \cos(\omega t) \cdot \cos\left(\omega t + \frac{\pi}{2}\right)$$

$$= \frac{1}{2} I_0 V_0 \left[\cos\left(\frac{\pi}{2}\right) + \cos\left(2\omega t + \frac{\pi}{2}\right)\right]$$

$$= \frac{1}{2}(1A \times 0.707V)\left[\cos\left(\frac{\pi}{2}\right) + \cos\left(2\omega t + \frac{\pi}{2}\right)\right] \tag{3.44}$$

$$= 0.354\left[0 + \cos\left(2\omega t + \frac{\pi}{2}\right)\right] \text{ (converting to RMS)}$$

$$= S\cos\left(2\omega t + \frac{\pi}{2}\right)$$

Comparing Eq. 3.44 to Eq. 3.32, we see that for a reactive load, $P = 0$, i.e., the active power is 0 W, indicating that none of the power is absorbed by the load. Figure 3.42 shows the relationship between current and voltage for $\mathbf{Z} = jX = 0.707\ \Omega$ and the resulting power waveform. The figure shows three sinusoidal waveforms. The blue curve is the current of unity amplitude and the green curve is the voltage shifted and scaled by the impedance, $jX = 0.707\ \Omega$. The black curve shows the resulting power, which has twice the frequency, $2\omega t$. For the purely reactive case, the power waveform is not shifted away from an average power of 0 W. As can be seen, the area under the power curve is the same above and below the zero-power axis. This demonstrates that all of the energy that enters the load is returned to the "circuit," which in this chapter is the power lines. None of that energy or power is absorbed by the load. As far as the power lines are concerned, it gets all of the power back. As said earlier, the power company would lose money due to all of the expense of generating and delivering power that was not used; therefore, they charge industrial customers for the reactive power.

Notice that S is the same for the reactive case as the real case; the amplitude of the sinusoidal power waveform is the same, but it isn't shifted away from the average value of 0 VAR. For this example in which $\mathbf{Z} = jX = j0.707\ \Omega$, the resulting waveform in Fig. 3.42 has an amplitude $S = 0.354$ VA, for which $Q = S\cos(90°) = 0.354$ VAR. Keeping in mind the case of the real load, this example demonstrates that the part of the power waveform below its average value due to the imaginary part of the load impedance is the same with or without any real part of the complex load. It also demonstrates that $P = 0$ so the power waveform is equally positive and negative. In the next section, we will look more closely at the combination of the absorbed and returned powers, and how they add up to the complex power by including the phase.

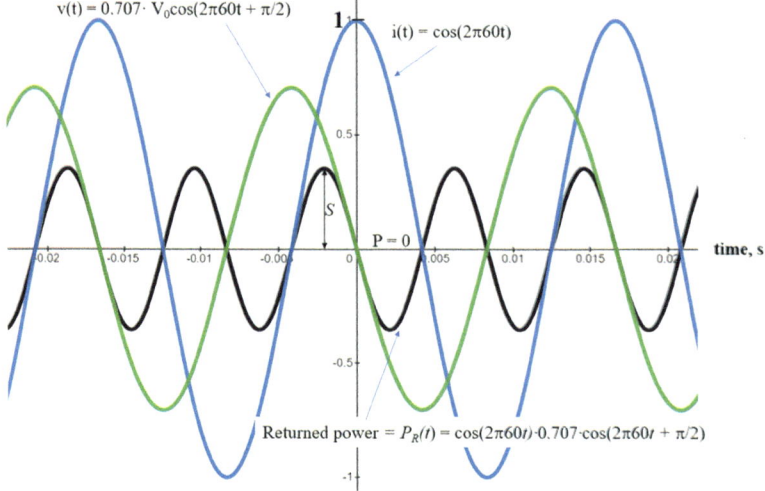

FIGURE 3.42 Voltage and current waveforms and the returned power, $P_R(t)$, for an impedance of $\mathbf{Z} = j0.707\ \Omega$. $i(t)$ and $v(t)$ are 90° out of phase, and their product is the power transported to and returned from the load. For a purely reactive load, the power factor, PF, is 0, and power flows into and out of the load equally every half cycle.

3.6.3 Absorbed and Returned Components of Complex Power

In this section we put all three of the results obtained above for **Z** = 0.707 + j0.707 Ω to show graphically that $P_C(t) = P_A(t) + P_R(t)$. Let's summarize what has been done so far: Figure 3.40 shows complex power for **Z** = 0.707 + j0.707 Ω by directly calculating $P_C(t) = I_0\cos(\omega t)\cdot V_0 \cos(\omega t + \phi)$. Then, using the real and imaginary parts of the **Z**, we plotted the absorbed component of the complex power in Fig. 3.41 and the returned component in Fig. 3.42.

This begs the question, "Do the two power components

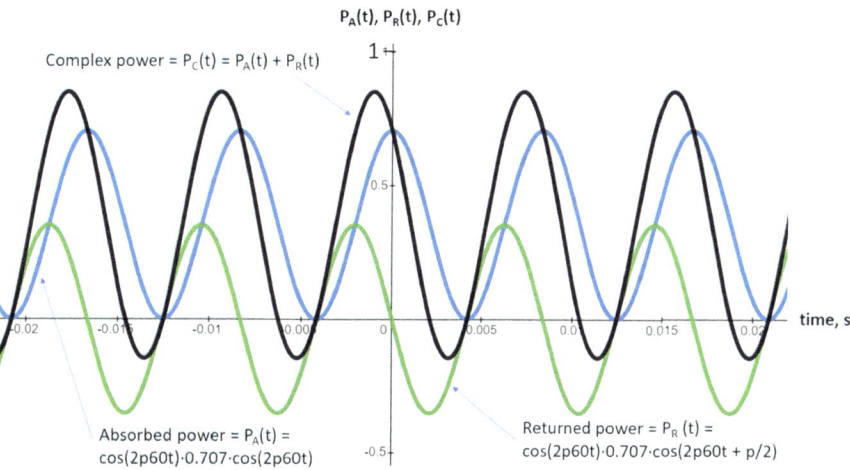

FIGURE 3.43 Absorbed power, returned power, and complex power due to a complex impedance. Here, the complex power is calculated by summing the active and reactive power using $P = 0.5\cos(\pi/4)$ and $Q = 0.5\sin(\pi/4)$. Comparing the complex power, black, with that in Fig. 3.40, which was calculated using Eq. 3.32, we see that they are the same. This demonstrates graphically that the apparent power is composed of the active and reactive power components.

really add up to give the complex power as found by multiplying current and voltage waveforms?" Figure 3.43 adds the two power components to form the total, complex power. Comparing the black curve of Fig. 3.40 with that of Fig. 3.43 shows that indeed, the complex power comprises the absorbed and returned components.

We conclude this section with a perspective on these basic concepts. The words "true" and "real" used to describe true, real or active power does not imply that the other powers, reactive and apparent, are somehow 'false power,' or are not actually power. Reactive power is really there – it's just not really *absorbed*. Power is the flow of energy; complex power flows from the power company and ends at the load where it moves on to become other forms of energy. Complex power sloshes back and forth within the load; some of it is absorbed, and some of it goes back to the source, being a general circuit, or more specifically, a transformer, or the power lines. Remember that although the units are expressed differently, i.e., *S* in VA, *P* in watts, and *Q* in VAR, these are all power, but under different guises to highlight their different effects on the power delivery from the power company to the load.

3.6.4 Power Triangle

Power engineers use the **power triangle**, shown in Fig. 3.44, to capture the relationships of the complex impedance and complex power that we went to such great trouble to explain in the previous few sections using trigonometry. The power triangle is a simple graphical way to encapsulate all of the relationships among apparent power, *S*, active power, *P*, and reactive power, *Q*, where *S* is the hypotenuse, *P* is the horizontal component, and *Q* is the vertical component. Remember that *S*, *P* and *Q* are the magnitudes of the time-varying power components of the total, absorbed, and returned powers.

Before delving more into the power triangle, it would be useful to motivate how it comes about. Notice that in Eq. 3.34, $S = I_{0,RMS}^2 \cdot |\mathbf{Z}|$, can be written in a form that retains the complex form of the impedance as

$$\vec{S} = I_{0,RMS}^2 \, \mathbf{Z} = I_{0,RMS}^2 (R + jX) \qquad (3.45)$$

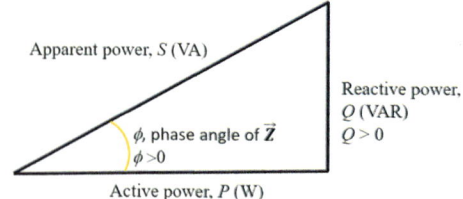

Apparent power, S (VA)

Reactive power, Q (VAR) $Q > 0$

ϕ, phase angle of $\vec{\mathbf{Z}}$ $\phi > 0$

Active power, P (W)

Power triangle for inductive impedance

Now, $\vec{S}$ is the complex form of the apparent power. Equation 3.45 implies that $\vec{S}$ has the same magnitude and phase properties as $\mathbf{Z}$, since it is just scaled by the constant value $I_{0,RMS}^2$; the active and reactive components of the complex power are related in exactly the same ways as the real and reactive components of the impedance. As shown in the power triangle, the active power is the projection of the apparent power, S, along the horizontal axis:

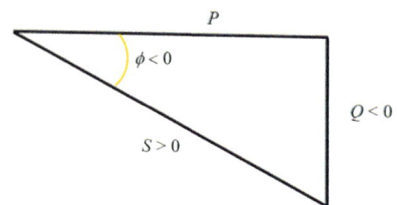

P

$\phi < 0$

$Q < 0$

$S > 0$

Power triangle for capacitive impedance

FIGURE 3.44 (Top) Power triangle showing the relationship between the apparent power, reactive power, and active power. The phase and relative magnitudes derive from those of $\mathbf{Z}$. (Bottom) Power triangle for capacitive reactance. ϕ and Q are negative values.

$$P = S \cos(\phi) \qquad (3.46)$$

and reactive power is the projection of the apparent power in the vertical direction,

$$Q = S \sin(\phi) \qquad (3.47)$$

These are the same relationships that were already derived for the complex power. Since the power triangle is a right triangle, we see that $S^2 = P^2 + Q^2$ as derived back in Eq. 3.40, and

$$|S| = \sqrt{P^2 + Q^2}. \qquad (3.48)$$

The angle between S and P is ϕ, the phase angle of $\mathbf{Z}$. Again, $\cos(\phi)$ is the power factor. When ϕ is positive, as shown in Fig. 3.44, *the power factor is said to be lagging*; when ϕ is negative, *the power factor is said to be leading*. But, what

> *Clarity:* 'Eddy' as used here is not someone's name. It is a rotation of water that leads to the formation of a whirlpool.

is leading or lagging what? The convention for the power triangle is the phase of the current w.r.t. the phase of the voltage. In our examples, we had assumed that $i(t) = I_0\cos(\omega t)$ and $v(t) = V_0 \cos(\omega t + \phi)$, so the voltage waveform is shifted to the left, which leads the current. In the parlance of the power triangle, the current lags the voltage, so this power factor is *lagging*, which is drawn as a positive value of both ϕ and Q, i.e., the reactive power points upwards. For what type of impedance does current lag voltage? ELI, so a lagging power factor applies to inductive impedance.

It may have seemed until now that we were never going to discuss capacitive reactance, but really everything is the same except that $X_C = -1/\omega C$ is a negative reactance. On the power triangle, that means Q is negative and ϕ is negative, as shown in the bottom of Fig. 3.44. Hence, the power triangle is flipped so that Q points down, i.e., is negative for the negative ϕ. Everything else is the same, including the fact that S and P stay positive, regardless of the sign of ϕ.

It is useful to notice that P, Q and S have lengths that are proportional to R, X and $|\mathbf{Z}|$, respectively, and ϕ is the phase angle associated with $\mathbf{Z}$. So, the power triangle is almost nothing more than a statement about $\mathbf{Z} = R + jX$. However, *underlying all this seeming simplicity is the more-complicated time dependence demonstrated by Fig. 3.43 and everything discussed in the previous sections!!*

We are finally able to understand why the transformer in Figs. 3.20 and 3.21 is rated in VA and not in W, V, A or VAR. As stated, apparent power is total power that flows through the transformer, some of which goes to power the loads in the house, and some of which returns to the power lines. The transformer neither knows nor cares (the height of anthropomorphizing) what happens to all the power that flows through it to the load. In fact, even if all the power were returned power, it would still be power flowing through the transformer, in spite of the fact that none of it is absorbed by the load. All that transported power takes a toll on the transformer in the form of heating its wires and its magnetic core. The magnetic core is heated by **eddy currents**. Eddy currents are real currents that are due to the magnetic fields and flow in circles inside the core. These eddy currents cause joule heating, as usual. Therefore, the transformer is rated in terms of how much power it can move through it, in units of VA. So, apparent power is 'real' in the ontological sense (for you philosophy minors), and too much of it can destroy a transformer.

3.6.5 Power Factor

Now, we look more closely at the power factor, which is an important quantity for power companies and large industrial facilities. The power factor is the cosine of the phase of the load impedance, $\mathbf{Z}$, i.e., $\cos(\phi)$, or simply P/S. It is the fraction of all the power transported toward the load that gets absorbed by the load and not returned. Since $-\pi/2 < \phi < \pi/2$, the PF is always positive and must be between 0 and 1. Again, when ϕ is negative, Q points down, and S is beneath P on the power triangle. However, P and S are always positive.

> *Power factor = 1*: If the load is purely real, then the current and voltage are perfectly in phase, so all of the power delivered to the load is used by the load, i.e., none of it is returned to the lines. In this case, $\phi = 0$, $\cos(\phi) = 1$, and the power triangle reduces to a line segment on the horizontal axis for which $S = P$. All of the power is active power. This is often not the case in actual industrial plants.
>
> *Power factor = 0*: If a load is purely reactive, then ϕ is $\pm 90°$, $\cos(\phi) = 0$, $P = 0$, and the power triangle reduces to a vertical line segment, pointing up (inductive) or down (capacitive). The apparent power is all reactive, and $S = Q$.
>
>   *Because $\cos(\pi/2) = 0$, the PF = 0*. As mentioned many times, no power is used by the load. Now the apparent power is all reactive power, and all of the power sent to the customer is returned toward the power company every cycle as if the load were a tiny power company itself for half of a cycle. That power is not completely useable by other customers due to the i^2R losses on the lines. This condition would never actually occur in industry because there would be no power used by the load to do anything meaningful, so why bother?
>
> *Power factor between 0 and 1*: As mentioned above, a load is complex if it is either capacitive or inductive, and has a real component. In that case, ϕ is between –90 and +90°, and $\cos(\phi)$ is between 0 and 1. For this case, the power triangle looks like that of Fig. 3.44. This is commonly the case and is not efficient for anyone – the power company or the industrial customer. Hence, corrective action should be taken to bring the PF close to 1.

3.6.6 Power Factor Correction

Now comes the point of this whole story: Industrial sites have power factors that are typically between about 0.7 and 0.9. The large extra current that flows back to the power company causes i^2R losses in the power lines and transformers, so the power company does not recover all of that energy, and they would lose money. Also, the power company's equipment must be made larger to be able to tolerate larger apparent power. Put another way, the customer is a black box to the power company, and that black box should appear to the power company as a real load. Due to these extra expenses, power companies charge extra fees to industrial customers that have power factors that fall below a certain threshold. Therefore, the customers are motivated to correct for low power factors.

Since ELI the ICEman says that $i(t)$ leads $v(t)$ in a capacitor, and $v(t)$ leads $i(t)$ in an inductor, it may seem reasonable to you that combining capacitors and inductors in some way can cancel these phase differences, moving the power factor closer to 1. Inductances and capacitances naturally cancel each other because current leads voltage in the

> *For understanding:* So, do you and I pay for active power, in watts, or apparent power, in volt-amperes, when we pay our electric bills? The answer is generally watts. The electric company would prefer to charge for VA, but they are not allowed to do so. If the power factor drops below a certain value, then they start charging for VA.

capacitor (ICE) and current lags voltage in the inductor (ELI), so putting them in series and parallel combinations allows the phases to cancel under appropriate values of C and L. Put another way, adding an equal reactance of the opposite sign changes the impedance, canceling out the reactive term: $\mathbf{Z} = R + jX_L - jX_C = R$.

You might think that a large industrial heater, like a blast furnace in a steel mill, would be mostly real and have a power factor close to 1. However, such large loads often have considerable inductance and have lower power factors. A big motor with lots of windings can look like a large inductor (with the resistance of the actual wire windings included). Modern three-phase motors are designed with power factor correction built in.

How about big industrial capacitive loads? Interestingly, there aren't very many examples of that. This is a good thing because it is cheaper to make big capacitors to cancel big inductive loads than it is to make big inductors to cancel big capacitive loads. Large factories switch capacitor banks in and out of circuits to cancel the inductance of their giant machines in order to save on energy costs by bringing the power factor back toward 1.

What happens to the power when a capacitor bank is used to bring the power factor back to 1? When power is returned by an inductive load and it is in a circuit with an appropriately sized capacitor, the energy flows to the capacitor on the return cycle, rather than flow back on the power lines. This is a short trip for the current, so i^2R losses are minimized. The capacitor stores this energy and then returns it to the inductive load in the next cycle, so the power company is taken out of the circuit as far as power factor is concerned. Power flows from the power lines to replenish the power that is absorbed by the load, including the power factor correction capacitors, but no reactive power flows back to the power lines. This concept of energy transferring from an inductive to capacitive element is central to many circuit concepts that you will see in Chapter 5 and later courses in circuits, electrical machinery and power systems.

3.7 The "Smart Grid"

The reason that widespread power failures can happen isn't that much of a mystery. In 2003, back when the massive failure discussed in the introduction occurred, a human would have had to see the problem happening, make a decision, and head it off before it grew out of control. The speed at which substations and power stations shut down to protect themselves from overloading, and maybe ruining much of their equipment, was simply too fast for the operators to react as power company followed power company in going offline like dominoes. Also, there were few sensors to notify operators of voltages and currents along the grid relative to the vast distances involved.

Even today, there are relatively few sensors along the United States' 400,000 miles of high-voltage transmission lines and many thousands of substations. Also, keep in mind that new transmission lines must be built to carry wind and solar power from their sources to customers in big cities across America, and new transmission lines are being added continuously. According to a *CNBC report*, transmission lines are being expanded at a rate of only about 1% per year, while the number of wind and solar projects that depend on those transmission lines grow at a much faster rate (see Chapters 7 and 8).

The new "smart grid" currently in development will incorporate sensors and digital controls to make automatic decisions about how power is routed, and how problems are handled. In order to accomplish this, new ways of using the power lines as communications links, new sensors, and new actuators that control high voltages and high currents, as well as new ways to use the grid efficiently, will all need to be developed. You should understand that it is not trivial to build an electronic device that can handle hundreds of thousands of volts and hundreds of amps. This is a major area of research and development today.

The next-gen smart grid will more-efficiently send power to where high demand is anticipated and will allow sporadic power sources such as solar and wind to be used to the greatest effect. Also, better control will help make the grid more robust in the face of natural disasters or in the (hopefully unlikely) event of a terrorist attack. With a new, smart grid also comes the danger of an internet attack, so security is even more important.

The smart grid may even allow communication from the grid to the consumer, such as a factory or even home appliances, in order for power stations to balance loads and increase their efficiency. Understand that power stations cannot start up instantaneously, so if more generators are needed, some amount of time is required to make them operational and putting power onto the grid. Already, smart thermostats are used by power companies to make individual homes warmer or colder to lessen the demands on their power plants. Big data from more customers will help energy providers maximize their efficiency and reduce their environmental impact. This is, indeed, the beginning of a new age in the power industry.

3.8 Lab 3 Activities

Note: Before coming to lab, you will need to generate a short, simple Desmos or Matlab script to add three sine waves that are 120° apart, but whose amplitudes vary. You should set this up so that you are ready to run it when you get to those steps. It will take more time to set it up and test it than to actually run it, and you don't want to waste that time during the lab session.

Welcome to the power industry! As the new owner of your very own nuclear power plant, you'll need to become familiar with the main aspects of power generation. However, in order to perform activities with three-phase power, naturally you must have a three-phase power source. While normally this is provided to the building at line voltages, that would be very inconvenient, and not a little bit dangerous. Therefore, for the purposes of this lab, the wide box (bottom center of Fig. 3.45 on top of the three transformers) will serve as a working model of a three-phase power plant like your nuclear reactor and generator, or some other part of the three-phase power distribution system.

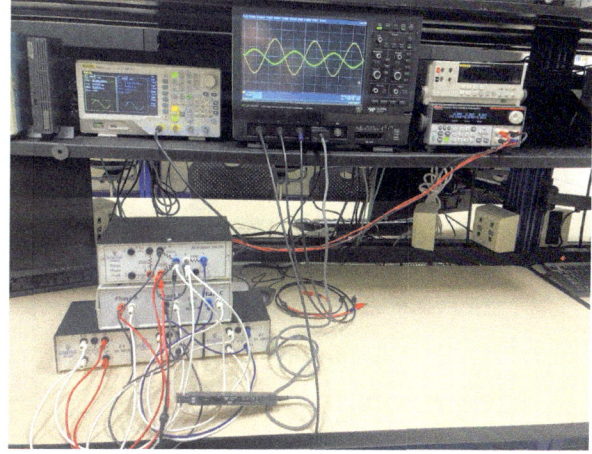

FIGURE 3.45 Lab 3 model power grid. The bottom three boxes are transformers, and the wide box on top of them is a low-voltage 3-phase generator.

This laboratory is intended to be a miniature power grid on your lab bench. We provide the low-voltage source of three-phase power that is your 'power plant,' wires for transmission lines, and transformers that are your distribution stations. The voltages will be stepped down as from the transmission grid to the distribution grid (skipping the subtransmission station), and then stepped down again from the distribution grid to the customers. As you do the lab, try to keep in mind what part of a real power grid you are working with in your model system. Have fun!

There are a lot of wires used in today's lab, so you must be organized. Place the three transformers side-by-side as shown in Fig. 3.45. We will call them T1, T2 and T3, from left to right. Place the load box on top of the transformers and then the three-phase generator (3PG) on top of the load box. This way, all the Ph-A, Ph-B and Ph-C phase voltages and T1, T2 and T3 are vertically lined up and the outputs from the transformers are next to the loads. That will help keep things organized.

1. *Setting up the bipolar power supply to drive the 3PG.* Starting out, you will need to power the 3PG and adjust the amplitudes and relative phases. Set up the DC power supply as follows. See Fig. 3.46. Set CH1 and CH2 outputs to 16 V. Now connect a short black banana-to-banana jumper between the positive terminal of CH2 to the negative of CH1. This is just like back-to-back batteries with ground placed in the center, creating a single **bipolar power supply** to run the three-phase generator, **3PG**. Without this, you could not have both positive and negative voltages and currents at the output of the generator. Set the current compliance limit on both CH1 and CH2 to 300 mA: Select CH1, press **I-set**, and then '.3'. Repeat for CH2.

Refer to Fig. 3.46 and attach a longer black lead from the center common point to the black ground connection at the back of the **3PG**. Attach the +16 V from the positive terminal of CH1 with a red cable to the red +16 V connector on the 3PG. Now, Attach the –16 V from the negative terminal of CH2 to the blue –16 V connector of the three-phase generator. *Have the TA check your wiring before you turn your power supply outputs on.*

Do not adjust the voltage or current settings again for the entire lab; the outputs should always be set to 16 V. *Use the OUTPUT On/Off button to turn power on and off to the generator whenever you make changes to the wiring.*

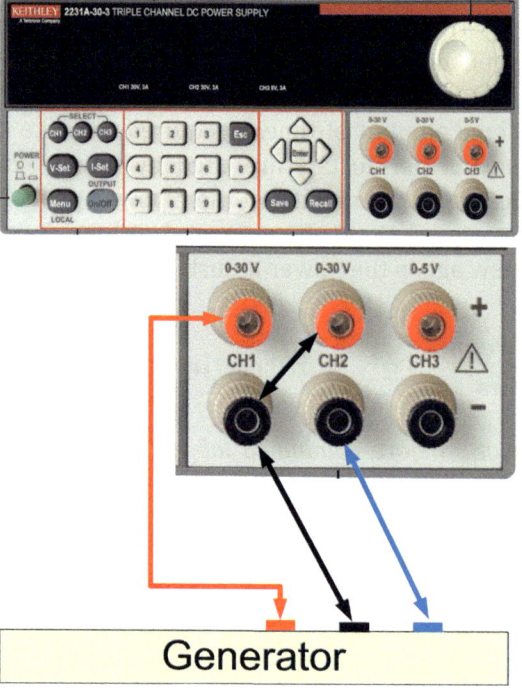

FIGURE 3.46 (Step 1) (Top): DC power supply (Bottom): Power wiring diagram to the 3PG.

2. **Operating the 3PG.** The 3PG will run at the frequency provided by the waveform generator, WG. Set the WG to a 60 Hz sine wave with an output voltage of 5 V_{pp}, and connect it to the **IN** BNC terminal of the 3PG with a BNC cable. This lab uses 1× scope probes made especially for it. The probes are BNC to banana plug and are color coded (with tape) to match phases on the **3PG**.

3. **DC resistance of transformer.** The next piece of your power grid is the transformer (Fig. 3.47), the workhorse behind power transmission. On your bench are step-down transformers with a turns ratio of 2:1 (here N_p/N_s) from primary (H4, H2) to secondary (X6, X10), so that the secondary voltage is half the primary voltage. In actual power transformers, the voltage ratio is much larger than 2:1. Those in today's lab model the function of the step-down transformers from, say, a transmission voltage of 400 kV to a distribution voltage of 35 kV. (The Risk Management and Safety Office prefers that you not use those actual voltages on your benchtop, so you will have to use your imagination.)

Perform resistance measurements using the Fluke DMM (do *not* use an LCR meter): Connect the leads together and press **REL** to remove the lead resistance as you did in Lab 2. Measure and record the resistances of the primary and secondary between (a) H2 and H4; (b) X6 and X10; (c) X8 and X10; (d) X8 and X6. In each case explain what the resistance is due to. Draw a diagram of the transformer and the windings to explain your readings. Explain their relative values (i.e., which is largest, next largest, etc.). Hint: look at the drawings on the front panel to see how the windings are related to the sockets.

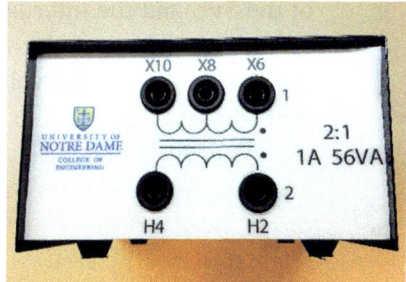

FIGURE 3.47 (Step 3) The transformer used in Lab 3.

4. **Isolation of primary and secondary.** Measure the resistance from (e) H2 to X6. What is it? Can you explain that value of resistance?

   You have set the current compliance from the DC power supply to 300 mA per channel. In an abundance of caution, for the rest of today's lab, whenever you change the wiring, turn off the outputs from the power supply to the 3PG in order to protect it in the event of a mistake in the wiring. When you are ready to power it up, re-activate the outputs to 16 V. The power supply should read 16 V each time you activate the outputs. If you have a major problem, the voltage will read less in order to limit the current, which protects the 3PG.

   Also, it is a good idea to not unplug inductors or transformers while powered and current is flowing. A large di/dt can develop a large voltage, and arcing can occur. (Although this will not happen in today's lab, it is a good practice.)

5. **Step down voltage comparison.** The most important characteristic when deciding which transformer to use is its turns ratio; thus, it's a good idea to confirm the turns ratio of our transformers. Starting now, and for the rest of the lab, use color-coded wires to aid the eye in troubleshooting. Match the colors to the 3PG that are the same as the standard color codes for three phases. This will help keep things organized as the wiring gets more complicated.

   Attach Ph-A from the 3PG to the primary (H2 and H4) of the Ph-A transformer. *From now on going forward, for consistency, use H4 for neutral (the white wire from the 3PG) and H2 for "hot" inputs. Use X10 as the negative terminal and X6 as the positive terminal of your secondary. This will help keep all the relative phases correct, which is critical to today's lab.*

   Use CH2 of the scope to measure the amplitude of the voltage (press **Maximum**) across the secondary (X6 and X10) of Ph-A. Given that the turns ratio is the number of turns on the primary divided by that of the secondary, and that this ratio is directly related to the ratio of respective voltages, what is the turns ratio of this transformer? Write down $V_{secondary}$ for all the transformers. It will become important later that they be close in output voltage for the same input voltage.

6. **Confirming resistances and output resistance of sources.** Before you continue with the load box, use your DMM with the **REL** function (to remove the lead resistance) and confirm that the resistance values on the load box shown in Fig. 3.48 are all within 1% of 250 Ω. If they are not, make a note of it in your notebook and notify the lab manager.

   Now you will check to see if the 250 Ω resistor loads down, i.e., decreases, the output voltage of the transformer. Use a 250 Ω resistor on the "load box" (see Fig. 3.48) to serve as the load on the secondary at sockets labeled X6 and X10. Using CH1, measure the output voltage X6-X10 before and after adding Load-A (points 5 to 3). Is the output voltage with the load the same as the output voltage without the load? Redraw Fig. 3.48 including both the output resistance of the 3PG and the internal resistance of the secondary of the transformer secondary. What is the 'voltage source' that drives the load in this case? Thinking back to Lab 2's discussion about output resistance, in general terms only, why might the voltage from X6-X10 drop a little bit when the load is attached?

7. **Center-tap voltage.** One feature found on the secondary of these transformers, just like real distribution transformers, is the center tap. Connect the 250 Ω load across X8-X10. What is the new voltage across the load? Explain the result in terms of the turns ratio of H2-H4 to X6-X10. Repeat the measurement for X6-X8. How do your measurements confirm that it is, indeed, "center tapped?"

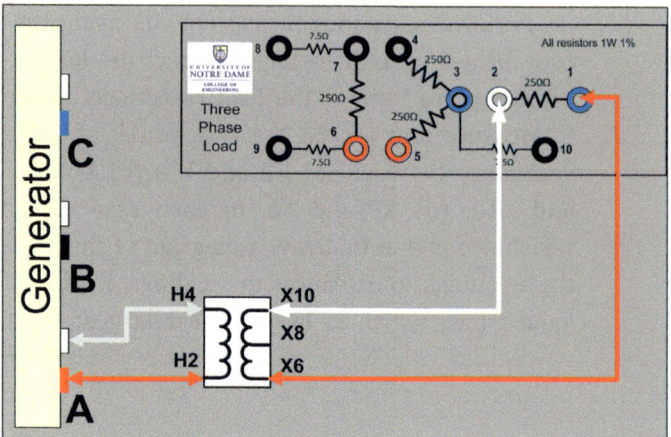

FIGURE 3.48 (Steps 6 and 7) (top) "Load box," and (bottom) drawing of the circuit of Step 7 using the load box. Four 250 W resistors allow both wye and delta configurations.

8. **Step up voltage.** Run the transformer in reverse, now as a *step-up* transformer by doing the following. Remove the load resistor. Attach Ph-A to X6-X10. Record the voltage at H2-H4. Explain the result in terms of the turns ratio of H2-H4 to X6-X10.

9. **Stepping down from transmission-line to residential voltages.** Later you will build models of wye-wye and delta-delta circuits. For now, you will start at the level of house wiring and demonstrate some concepts of split-phase voltage. You will use two of your transformers to model power flow from a receiving station to the home.

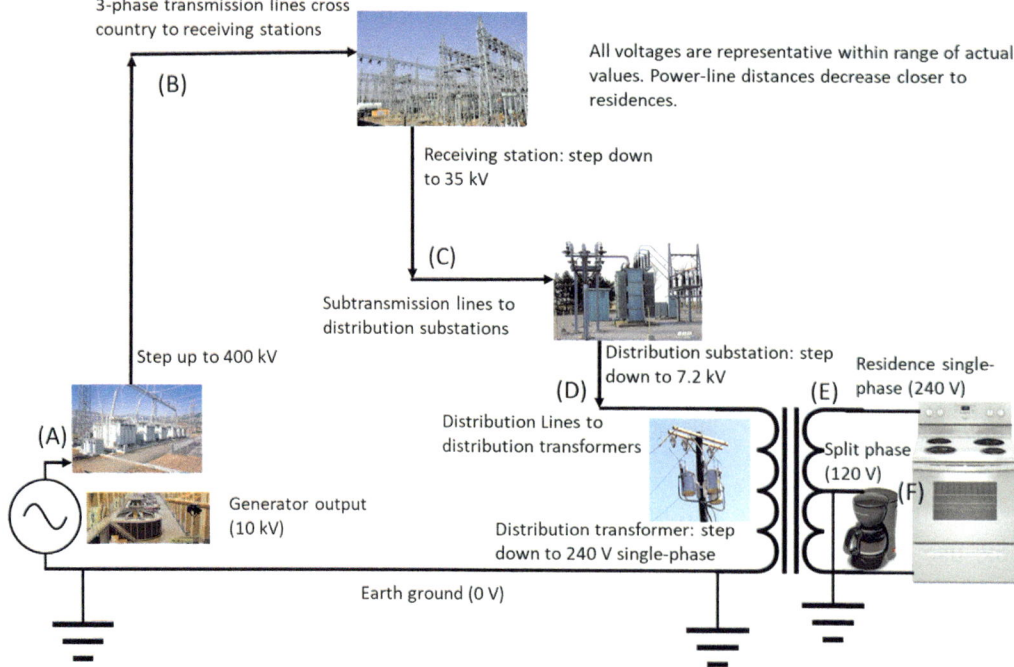

FIGURE 3.13 (Step 9) Power grid with representative voltages at each stage.

Referring to Fig. 3.13, imagine that the 3PG is the three-phase subtransmission lines at 35 kV (C); one of your transformers (T1) is one phase of the distribution substation with 35 kV at the primary, and 7200 V (D) at the secondary. The secondary of T1 feeds the primary of T2, which represents a distribution transformer that steps down to 240 V (E) and provides power to several homes ((E) and (F)). In short, you will replace the 250 Ω resistor in Step 7 with the primary of T2. The resistors that come off the secondary of T2 will model home appliances. You will use only Ph-A, since homes are only single-phase and there is no need to include the other phases, which would normally power a different part of a neighborhood.

An important note about grounds and grounding in this lab activity. Transformers couple primary voltages to secondary voltages through magnetic fields. Electrons do not flow between coils. Therefore, even if the primary is connected to ground, the secondary is floating – it has no ground defined until you put one in. When you build your circuits, you are free to ground any one of the secondary wires. That will affect what you see on the oscilloscope. The waveforms won't be different, but they can be shifted by 180° or equivalently, "flipped vertically." So, how you view your waveforms will depend on where you put your grounds.

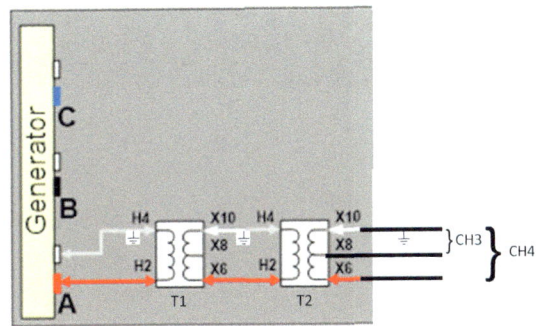

FIGURE 3.49 (Step 9) Wiring diagram showing how to measure the modeled 35 kV transmission line voltage, the 7200 V subtransmission voltage, and both the 120 V leg and the 240 V leg of the split phase system. Note the grounds at the top of all of the windings, H4 and X10. This keeps the AC phases the same.

Each step going forward tells you where to place your grounds. Every coil, be it a primary or secondary coil, must be grounded at one end or the other (or the center tap), and following the directions will ensure that you get the same results as everyone else, and that your results are more understandable. Of course, you must be sure that the oscilloscope does not have the channel "inverted" or that will be as if you flipped the ground and signal line at the input. Look for that if things are upside-down.

Refer to Fig. 3.49. You already have the output of the 3PG Ph-A connected to H2-H4 of T1. Connect the secondary of T1 to the primary of T2 in the following way: X6 of T1 to H2 of T2, and X10 of T1 to H4 of T2. The three outputs of T2, i.e., X6-X8, X8-X10, and X6-X10 now represent the 240 V/ 120 V split-phase voltages inside the breaker box of your home, as shown in Fig. 3.11.

Display on CH1 the output of the 3PG at Ph-A (the modeled 35 kV legs) with the earth ground scope lead connected to white; display on CH2 the output of T1 X6-X10 (the modeled 7200 V output of the 35 kV to 7200 V distribution transformer) with ground to X10; on CH3 put the output of T2 X6-X10 (modeled 240 V legs of your breaker box), with ground at X10. Finally, using the fourth BNC-to-banana cable, display on CH4 the voltage X8-X10

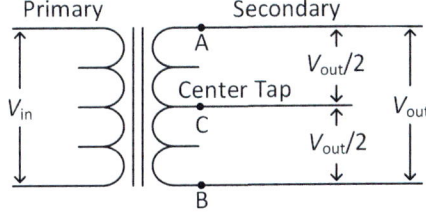

FIGURE 3.11 Repeated (Step 9).

(one of the two modeled 120 V legs). Make sure the grounds are connected at X10. Power up the 3PG. Set all four of the displays to 2 V/div. Now, the voltages on CH1 > CH2 > CH3 > CH4, and they should all be in phase. Your waveform should look like Fig. 3.50. Record all the voltages and record the trace. Explain the relative voltages in terms of step up/step down and center-tapped transformers. Explain what each value represents relative to your model power grid.

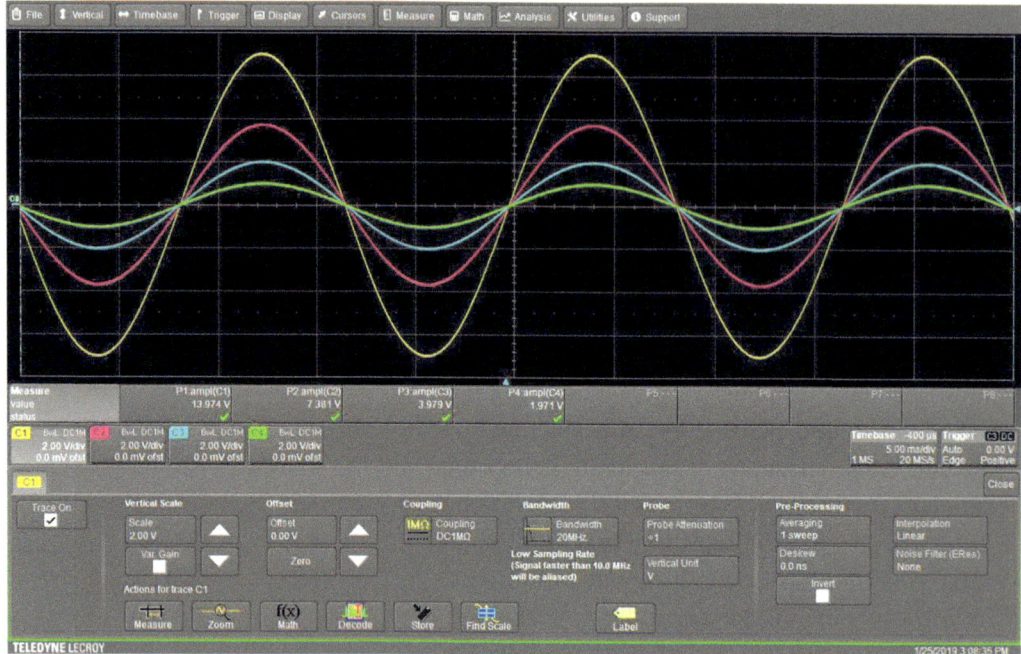

FIGURE 3.50 (Step 9) Voltages stepping down to the home. CH1: output of 3PG (35 kV subtransmission grid); CH2: output of distribution substation, T1 (7200 V at neighborhood); CH3: output of distribution transformer, T2 (240 V at home); and CH4: one leg of center tap (120 V leg in home).

10. **Observing the two 120 V legs and the combined 240 V leg.** Refer to Fig. 3.51. Move the ground from X10 of T2 to X8 of T2. Now you see the two 120 V lines 180° out of phase. This shows the phase of each 120 V leg relative to ground. Add a function that shows CH3 minus CH4. Record your results. This demonstrates how the two 120 V legs combine to provide a larger voltage when needed for large appliances.

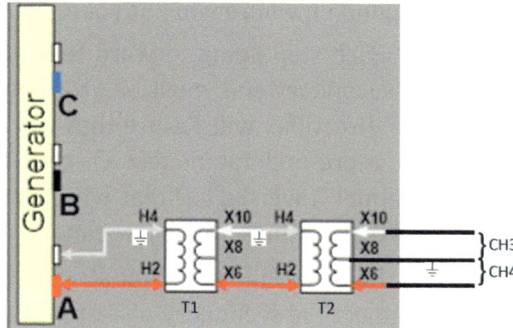

FIGURE 3.51 (Step 10) Wiring diagram for Step 10. Note the ground at X8 of T2.

11. **Higher voltage requires less current for the same power.** In this step, you will see why delivering power at higher voltage can be more economical in terms of the current needed, i.e., why transmission lines are high voltage, and why U.S. homes have available both 120 V_{RMS} and 240 V_{RMS}. Our model here is a built-in space heater inside of a home. The goal is to deliver the same total power but with less current by doing so at a higher voltage. Since $P = iv = i^2R$, using a higher voltage requires less current and therefore allows the use of thinner wires. This is the same concept as that of the efficiency of ultra-high-voltage transmission lines, which waste less energy for the same wire thickness (i^2R is less).

Imagine that two 250 Ω resistors are separate heater elements in an electric heater in your home. Each resistor is one element, and they may be connected either in series or parallel, but must use the same power, i.e., do the same job heating the room.

This is the 240 V_{RMS} output version of the secondary driving your two heater elements: Turn off the DC supplies. Connect T2 X6-X10 (the modeled 240 V_{RMS}) to two 250 Ω resistors in series (you can use ports 4 and 5 on the load box). What is the total resistance of the heater? Insert the DMM ammeter (**300 mA** and **COM** ports) in series with the load and set to AC current. Remember that it reads in mA$_{RMS}$. Connect the scope CH1 across the series resistors. What voltage and current do you measure? (You might want to change P1 to read **RMS** and not **Maximum** to make things easier. This will make it easier to compare to the ammeter reading; also, power is almost always expressed in W_{RMS}.) If the ammeter reads a small offset from zero when open circuited, subtract that off. Draw the schematic diagram of the circuit including the meters. In addition to measuring current, convert the voltage to A_{RMS} by dividing voltage by 500 ohms. As in Lab 2, the measurement should be close. Since the load is a resistance, i.e., real, i and v are in phase ($\phi = 0$), the power factor is 1, and the power consumed by the 'heater' is just $P = iv$. What power are you using to heat your 'home?' Your answer will be in W_{RMS}.

Now, you will use the modeled 120 V to achieve the same power, i.e., heat, from the 'heater.' Connect the resistors in parallel and hook them up between X8 to X10 of T2. Draw this schematic. This is the version of power that uses one leg of your home breaker box to run the two heater elements so that they are now producing the same heat as the 240 V_{RMS} case above. Now what is the resistance of the heater? What voltage and total current do you now read? How much power are you now heating with? Is the power the same with both voltages? Do you see that using 120 V_{RMS} instead of 240 V_{RMS} requires more current at a lower voltage as long as you set up your load to consume the same power? Again, the lesson is that using higher voltage allows thinner and less-expensive wiring to be used to deliver the same power. (However, you must use two breakers in your breaker box to use 240 V.

12. **Matching the 3PG amplitudes.** We now begin studying 3-phase power. Disconnect everything you built in the previous steps. Plug the banana plugs of the three color-coded banana-plug-to-BNC cables into the three phase-voltage outputs at the front of the 3PG. The black banana plugs are the cables' outer-shield earth ground.

Important: The **black banana plugs must go into the white sockets** because all the white sockets are tied to earth ground through the WG's BNC cable. Plug the BNC ends into their respective channels on the scope. For organization, put Phase A to CH1, Phase B to CH2, Phase C to CH3. Now, all the earth grounds are connected together for the WG, three BNC cables, scope inputs, and neutral sockets on the 3PG.

Turn on the outputs from the DC supply and the WG and view CH1, CH2 and CH3 on the scope. *Be sure to turn on the CH1 output of the WG, or the 3PG will not work.* You can now set the amplitudes of all the channels from the 3PG using the **Amplitude** potentiometers and the screwdriver. You will use a plastic "alignment tool" to make the adjustments through the small holes. **Be very careful not to damage the delicate potentiometer screws, so do not turn anything that doesn't move easily.**

Use the measurement capability to help you make accurate readings of voltages. Amplitude first: Select **Measure** along the top menu, then **Measure Setup**. The list is shorter if you select **Vertical**. For **P1**, select **Maximum** under **Measure** and **C1** under **Source**. You will see the measurement under the trace window. Once the measure bar shows up, you can just press the boxes. Repeat for **P2** for CH2, and **P3**, for CH3. Using the potentiometers and the alignment tool, set the three phase voltages as close to 7 V_p as possible. See Fig. 3.52 for the correct waveforms.

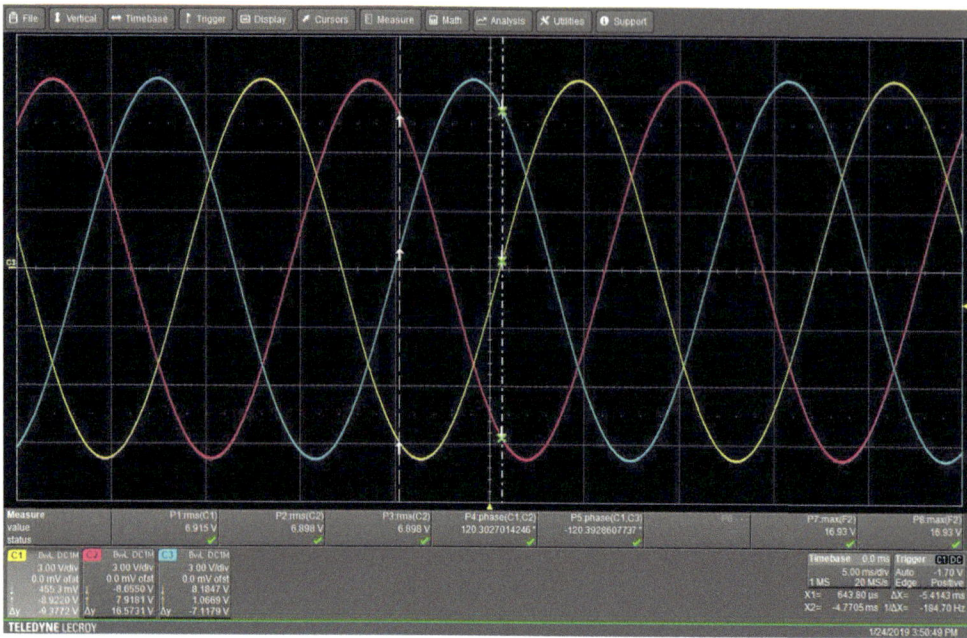

FIGURE 3.52 (Steps 12 and 13) Correct waveforms of three voltage phases. Your voltages should be close to $V_p = 7.0$ V.

13. **Aligning the 3PG phases.** Now, the phase between each leg isn't yet aligned to the required 120°. You will configure the phases by adjusting potentiometers for phases B and C until the phases between Ph-A to Ph-B and Ph-A to Ph-C are as close as possible to ±120°. If you are having trouble, check to make sure that **Invert** is not checked on any of the channels. Use a measurement function to measure the phase differences between A and B as well as A and C as follows. Using the menus, select **P4**, **Horizontal** (to make the screen less cluttered) and slide down to select **Phase**. Now you can select the two channels and relative directions to compare. Select the phase voltages A to B, i.e., **C1** and **C2**. Set **P5** for **C3** to **C1** (if you get the order backwards, it will read negative phase). Carefully turn the potentiometer for Phase B until the phases are 120° apart. Now adjust phase C until 120° is shown and the waveforms look like Fig. 3.52.

A note about 3-phase current measurements: Currents flowing on the neutral line are sensitive to the phase and amplitude settings from the generator. You should try to have nearly 120° and identical waveforms to get good current measurements. Keep that in mind as you perform these steps. You may have to recheck your settings before doing sensitive measurements.

14. **Checking voltages at 3-phase transformer outputs.** Now you will start to build your benchtop model of the United States' 1000-gigawatt power grid as shown in Fig. 3.13. You will take measurements from all the phases at the outputs of the transformers.

Refer to Fig. 3.53 when making the following connections. Connect the output of Ph-A from the 3PG to H2-H4 (with the white/neutral connection to H4). Do the same for all three 3PG phases and transformers T1, T2 and T3. Use color-coded wires whenever possible. Use the BNC-to-banana

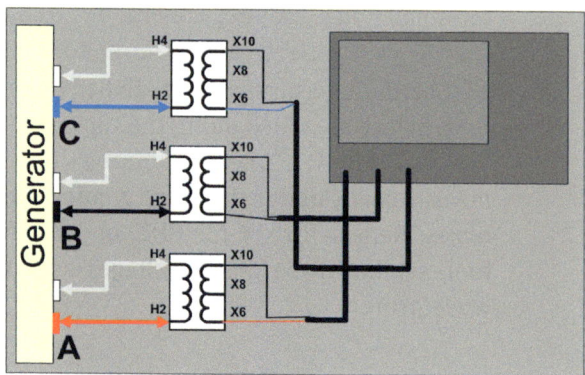

FIGURE 3.53 (Step 14) Connections for each Phase to the oscilloscope. Connect all the X10s to grounds (black) on the scope probes.

cables to connect all three secondaries to the scopes at their respective ports X6-X10. Note: All three secondaries are floating (think about why and discuss with your lab partner), so they must each have their own ground connection to the scope. To keep the relative phases intact, *use X10 for the ground connection* to the scope on each transformer (black banana plug). Be consistent in using scope CH1 for viewing Ph-A, red lead with no tape, CH2 for Ph-B, red lead with black tape, CH3 for Ph-C, red lead with blue tape (CH4 used for current measurements) so that you easily know what each channel is showing.

You should see all three phase voltages with about the same amplitudes, but this time with voltages from the secondaries. Measure and record the amplitudes of all three channels. If the measured voltages are off by more than about 0.1 V, go back and carefully adjust the amplitudes at the outputs of the 3PG so that you get the same amplitudes at the outputs of all of the transformers. Check the phases while you are at it. (These transformers add no phase shift at the low frequency of 60 Hz.)

15. **Building the wye-wye circuit.** Turn off the outputs of the DC power supply while you build the next circuits. Build the wye-wye circuit (Fig. 3.54), using the load box resistors that are configured like a wye (ports 1 to 5).

Referring to Figs. 3.54 and 3.55, use the shortest black banana-to-banana wire as the jumper at the center of the wye (ports 2 to 3). Connect all the neutrals together at the X10s of the transformers with short white wires; use only one white neutral wire to connect to the center of the wye at the loads (the black jumper), just like a real transmission line.

Now imagine that each of the transformers on the wye-source subcircuit is a substation transformer that steps down from 35 kV to 7200 V ((C) to (D) in Fig. 3.13). Each line heads out to a neighborhood (called 'NA,' 'NB,' and 'NC') where it will be connected to distribution transformers from 7200 V to 240 V ((D) to (E)). In your model system, the 250 Ω load resistors represent one of these 7200 V to 240 V distribution transformers that might serve a few homes. The purpose of the neutral wire at the transformers is to provide a return path to the 35 kV substation. Ideally, for a wye-wye balanced load, no current flows on that neutral wire. When the loads are unbalanced, some current will flow in the neutral, which will be tested in the next few steps.

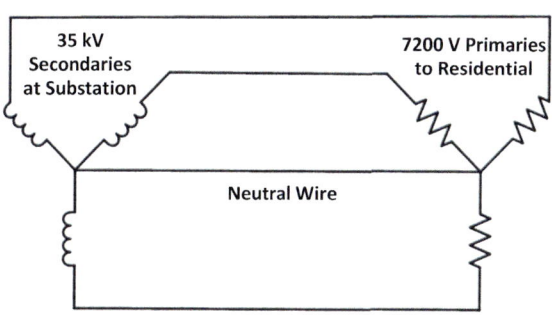

FIGURE 3.54 (Step 15) Wye-wye circuit diagram. Your trans-former secondaries are on the leftand the resistors used in the model are on the right.

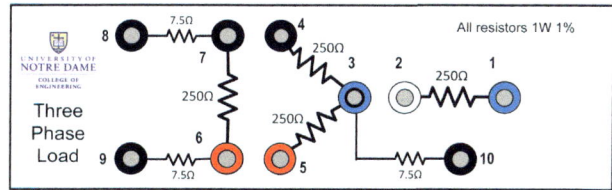

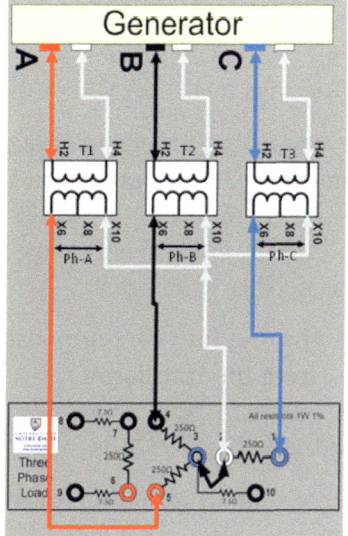

FIGURE 3.55 (Step 15) Wiring diagram for Fig. 3.46 and load box. You do not use the 7.5 W resistors.

16. **Phase voltages are load voltages in wye-wye.** Connect three channels of the scope to their respective phases at the load: CH1 to Ph-A, etc. In this step, "Ph-A" refers to the phase voltage created at the output of X6-X10 of T1, etc. Note that you need only one probe ground to go to the neutral wire at the center of the wye since they are all the same at the scope, but hook them together anyway just to give them a safe place to be. Turn on the DC supplies and measure the voltages and relative phases across each load. The voltages across each load resistor (i.e., a distribution transformer) are the same as their corresponding Ph-A, Ph-B and Ph-C phase voltages because they are in parallel, as explained in Fig. 3.33, and that you saw in Step 6. Record the voltages.

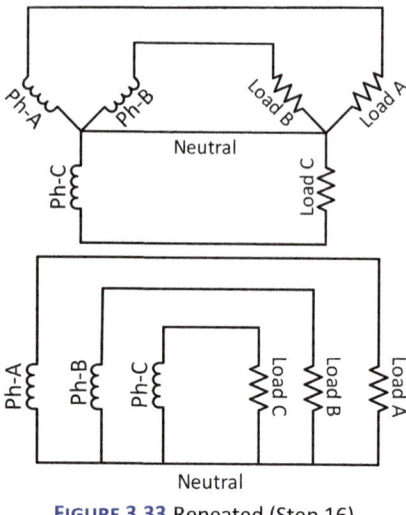

FIGURE 3.33 Repeated (Step 16).

17. **Measuring current through each load.** So far you have a balanced load of 250 Ω per leg, which should result in the same line current in each leg. Calculate the expected line current into each leg. Use the DMM ammeter to measure each load current separately. Note that the ammeter shows RMS current. What are the currents in each leg? Comment on the differences between the currents in the legs.

18. **Checking the neutral current with balanced loads.** Measure the current flowing in the neutral wire for the balanced loads. It should be much lower than any of the individual currents, for balanced loads?

19. **Plotting current on neutral line with balanced loads.** Use your Desmos or Matlab script that you have prepared to plot all three phase currents using the amplitudes that you measured in Step 17 as well as the correct phases. Each current will be of the form $I_{0,RMS}\sin(360 \cdot 60t + \phi)$, where ϕ is the phase (in degrees) you are measuring between the three phases, which you should still have up on your display. The phase of Ph-A is $\phi = 0$. Be sure that you use ϕ in degrees and set the angular units to degrees in the calculation. Make a plot like that of Fig. 3.35 using the actual currents and phases that you have measured; the three currents should sum to nearly zero. This calculation demonstrates that the resulting current in the neutral wire is nearly zero when loads are balanced. Does the calculated amplitude of the neutral current agree somewhat with your measured value? Save the plot to include in your lab report.

20. **Investigating unbalanced loads in wye-wye.** Now you will unbalance the load by putting the remaining 250 Ω resistor on the load box in parallel with Load-A. What is the calculated total resistance for the two resistors in parallel? Measure the load voltages as in Step 15. Have the load voltages changed? There might be small changes in the load voltages due to voltage drops in the transformer windings.

21. **Comparing line currents.** Using the DMM, measure the line current into each leg of the loads, i.e., from the transformer. Is it the same in each leg as it was before? Which one is different? Explain the change in line current by understanding that each source drives its own individual load, as in Fig. 3.33. How has the current in the new load changed?

22. **Current on neutral line for unbalanced loads in wye-wye.** Measure the current on the neutral line. Is it as low as it was when the loads were balanced? Compare it to the current in Load-A after unbalancing Load-A.

23. **Plotting current on neutral line with unbalanced loads.** Repeat Step 17 with these new measured line currents to see if the measured and calculated values of the neutral-line current agree. Confirm that the *excess* line current in Ph-A now appears on the neutral. The excess current will be the difference between the balanced load current and the unbalanced load current. Include the plot in your report.

24. **Losing one phase of three-phase wye-wye circuit.** Remove the extra 250 Ω resistor to rebalance the loads. Unfortunately, it is 119° F today, everyone is running their air conditioners on high, and a cooling fan on a distribution transformer seized up, so one of your transformers just failed, leaving you with only two working Phases to feed the neighborhoods. To simulate this, remove the "hot" lead (X6) at the Ph-A transformer. Leave everything else in place. What line currents do you now read in Ph-A, Ph-B, Ph-C and neutral? Explain what you see in terms of what happens to the current into the different neighborhoods (NA, NB and NC) fed by the three phases. You should see that the currents are still the same in the remaining loads (demonstrating how a wye-wye system can fail softly), but now the neutral returns the extra current that cannot be returned in the missing load of Ph-A.

25. **Plotting current on neutral line with only two sources.** Repeat Step 17 with the three currents, including the zero current on Ph-A. The sum of the three currents is the current flowing in the neutral line. Comment on what you have found.

26. **Adding an inductor to Load-A.** Now you will experiment with reactive power. As mentioned in the lab readings, your personal power company doesn't deal solely with real loads. In fact, as a power provider, a good portion of your business comes from the industrial sector, which is composed primarily of inductive, and hence, reactive loads. Thus, an analysis of reactive loads and power factor is necessary before you can

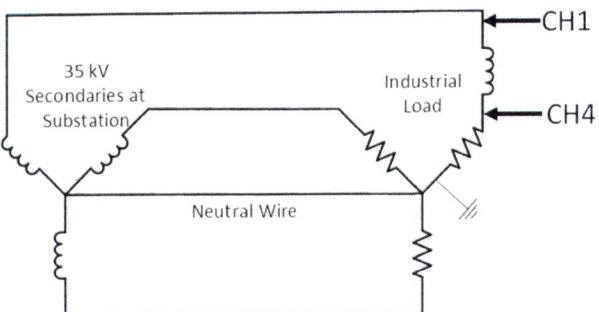

FIGURE 3.56 (Step 26) Three phases including one inductive load.

profitably and efficiently sell power over the power grid. In the following steps you will add an inductor in series with the 250 Ω resistive load to model an inductive industrial load, such as a large electric motor.

Turn off the DC power to the 3PG. In this step you will add an inductor in series with the 250 Ω resistive load to move the power factor away from 1. Find the 50 mH inductor – it is black and about the size of a pinky finger, and marked with '503.' Measure the DC resistance of the inductor wires using the DMM. Make a note of it and keep that for later. Use alligator-clip leads to insert the inductor into the circuit as shown in Fig. 3.56.

27. **Observing phase shift between voltage and current in inductive load.** The total load is now the series combination of the resistor and inductor, so its impedance has both real and imaginary components. Copy the schematic of the source and inductive load from Fig. 3.56 into your notebook. To calculate the components of the power triangle for Load-A, including power factor, you will need to measure the phase between the current and voltages at the load as follows: CH1 across Ph-A is the *RMS* voltage across the entire Load-A; CH4 across just the resistor provides the *RMS* current through the load $v = iR$. Note: *do NOT connect any grounds to the load board at 2,3* because the ground wire will provide an alternative current path around the neutral line, and later you will want to know the current in the neutral line.

Big change here: because the available inductor has only a moderate inductance value (large ones are expensive), *change the WG frequency to 600 Hz.* Remember that $X_L = \omega L$, so the inductive reactance increases with frequency. Therefore, 600 Hz will cause a larger phase shift between the current and voltage and make for a larger change in power factor, $\cos(\phi)$. Unfortunately, you will now need to reset the amplitudes and phases from the 3PG. Hopefully, you have the functions still set up and this will go quickly.

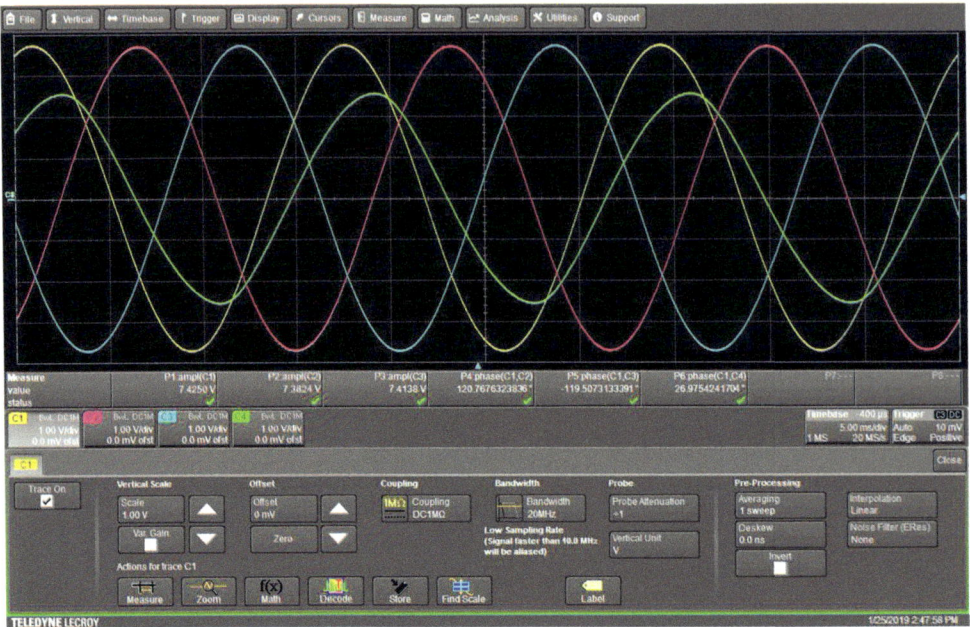

FIGURE 3.57 (Step 27) Three phases including voltage across the loads with inductance in series with Load A.

Done? Good. Now, use measurement functions to read the Phase-A voltage (CH1), the Load-A current (derived from CH4), and the angle between the Phase-A voltage and the Load-A current. See Fig. 3.57 as an example. What are your measurements of the Phase-A voltage and the Load-A current? The phase between Phase-A voltage and the Load-A current is the angle ϕ that determines the power factor, PF = $\cos(\phi)$. What is the phase angle ϕ? Is the current leading or lagging? Does that agree with ELI? What is the PF for the load?

Draw the power triangle including S, P and Q. What are the apparent power, S, active power, P, and reactive power, Q? *Make sure you are using RMS values of i and v!* Recall that $S = I_{0,RMS} \cdot V_{0,RMS}$ with no phase angle required. Once you have that, you can use the triangle to calculate P and Q from $\cos(\phi)$ and $\sin(\phi)$. Draw the triangle with all values on it.

Now for a comparison to test if the triangle really represents the correct power absorbed by the load, P, do the following: Calculate $i_{RMS}^2 \cdot R_{Load}$. What should you use as the load resistance? Since the inductor is part of the load, add the inductor wire resistance to the 250 Ω resistance to give you an R_{tot}. Does the active power calculated from $i^2 R_{tot}$ come close to that derived from the apparent power and the phase angle? Good job! You have done your first power triangle measurement.

28. **Plotting current on neutral line with inductive load.** Insert the DMM ammeter in the neutral line path. (This is why you couldn't have the ground there, i.e., it would have made an alternate current path.) Measure the current on the neutral line. You should see that there is an increase in current on the neutral line due to the reactive load. Now you will visualize all the relevant

waveforms. Repeat Step 17 to see that the change in the phase and magnitudes of the currents through the loads leads to an appreciable current on the neutral line. This will be a little trickier since you have to shift the ϕ in Load-A current by your reading – get the sign correct! Leave the other two phases at 120° and 240° since they are referenced to the Ph-A voltage, which has not changed.

29. **Correcting the power factor of inductive load.** This current going back to your power company results in wasted energy on the neutral line and power returning to the transmission lines. As a major supplier of the nation's electrical power, you have a fiduciary responsibility to your shareholders, and you want to be a good steward of the Earth's resources. Therefore, you will want to minimize this waste of power (and money) due to a power factor that is less than 1 by adding a capacitive load in series with the inductive one. First, turn off the outputs of the power supply. Now, choose a capacitor around 1.5 μf and put it in series with the inductor at the end closest to the transformer to compensate for the inductive reactance by adding capacitive reactance, $X_C = -1/\omega C$.

   Turn the supply outputs back on and measure the current and phase of the load again. What has happened to the phase between the current and voltage of Load-A, which now consists of the capacitor, inductor and resistor? Is the power factor improved? You may not have achieved perfect compensation due to our limited choice of capacitance values, but it should be improved. Also, note that the voltage across Load-A has not returned to the full phase voltage. This is due to the series resistance of the inductor.

30. **Building the delta-delta circuit – grounding.** Now that you have a fundamental understanding of transformers and wye-wye circuits, it's time to test out the delta-delta power configuration.

   Looking at Fig. 3.58, imagine that each of the phase voltages on the left delta source circuit is the secondary of a transformer in an auto manufacturing plant that drives a 500 horsepower (373 kW) three-phase motor. Notice that because there is no neutral wire, all the current must return on the three phases. That will be investigated further.

   Turn off the outputs of the power supply. Disconnect all wires from your wye-wye steps except for those running from the 3PG to the primaries of T1, T2 and T3.

   Reset the frequency to 60 Hz at 5 V_{pp} and reset the 3PG to make the outputs at X6-X10 of T1, T2 and T3 as close as possible to the same amplitude and correct phases.

   You must be organized; using the same color wires and tie points as everyone else will help you and the TAs spot any problems. Go slowly and carefully. Build the circuit of Fig. 3.58 using the delta group of load resistors (labeled 3 to 7, Fig. 3.59) according to the following instructions:

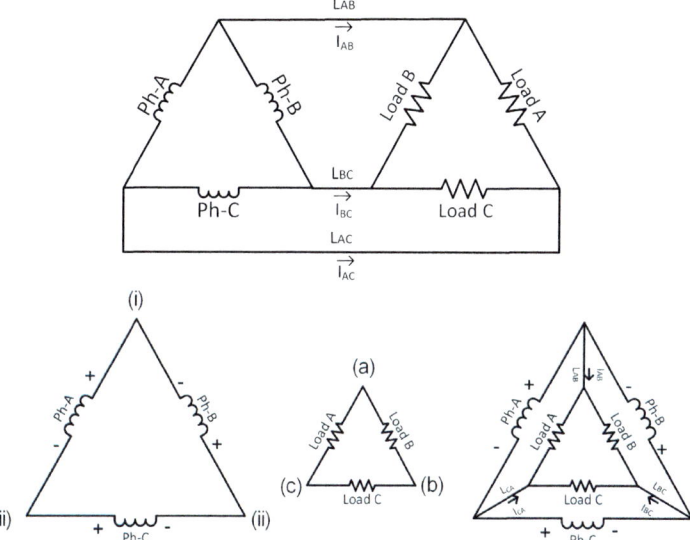

FIGURE 3.58 (Step 30) Delta-delta circuit. Top: conventional topology. Bottom: alternative topology showing better the relationship between each phase and the loads. Notice that there is no neutral wire, and all current returns to the sources through the loads. Also notice that each phase voltage is in parallel with the sum of the other two phases.

First, build just the outer triangle connecting the outputs of the transformers as shown in Fig. 3.58 (bottom left). Use a red wire from X6 of Ph-A to X10 of Ph-B. Using black and then blue wires repeat two more times to make the series connection of T1, T2 and T3 as shown in the figure.

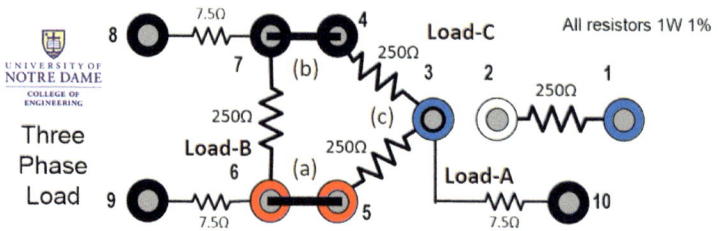

FIGURE 3.59 (Step 30) Use loads 3–4, 3–5 and 6–7 for the delta-delta circuit. Put jumpers as shown.

Next, build just the inner triangle by itself using the load resistors configured in the delta shape on the load box. Refer to Figs. 3.58 (middle) and 3.59. To do this, just use two short, black jumper wires to connect the three loads into a delta, as shown. (Ignore the 7.5 W resistors. Those are for measuring currents on L_{AB}, L_{BC} and L_{CA} without using a current probe, which we have.)

Finally, complete the circuit of Fig. 3.58 (bottom right) by inserting lines L_{AB}, L_{BC} and L_{CA} connecting the two triangles together as follows (refer to Fig. 3.59):

L_{AB}: Red wire from (i) (X6 of T1) to (a) (connections 5,6 on load box);

L_{BC}: Black wire from (ii) (X6 of T2) to (b) (connections 4,7 on load box);

L_{CA}: Blue wire from (iii) (X6 of T3) to (c) (connection 3 on load box).

Loads A, B and C are indicated in Fig. 3.59. This wiring scheme ensures that each phase is correct going around the loads and will simplify the following analysis.

A note about grounds: The phase voltages, Ph-A, Ph-B and Ph-C, shown in Fig. 3.58 are the secondaries of the transformers T1, T2 and T3, respectively. Also, the loads and their phase voltages are in parallel, so their voltages are the same. For example, Ph-A is in parallel with Load-A, so they are the same voltage at each instant in time.

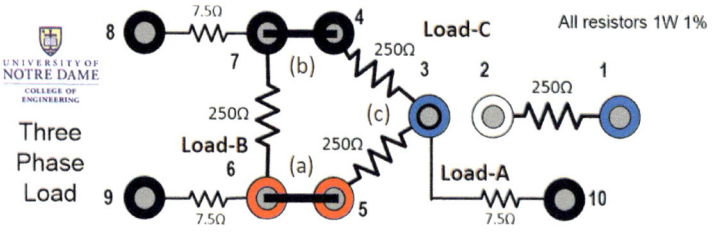

FIGURE 3.59 (Step 31) Repeated.

Everything connected to the secondaries is floating independently of the primaries, which are grounded. Put another way, there are no earth grounds in the delta-delta circuit until we connect one with a scope probe. Also, even when you want to insert an earth ground connection, there is no single point shared by all voltages, so no single point can be ground for all the voltages at the same time. If you were to use the scope as before to look at the three voltages at the same time, then each of the points (i), (ii) and (iii) would be grounded, and therefore every voltage would be shorted out. You *can* look at two voltages that share a common point, and designate that common point as ground, but the relative phases will be reversed. There is no *direct* way to observe all three voltages at once using the oscilloscope. Be sure that you understand this concept about grounding. In the next step you will use some scope tricks to observe all three voltages at once even with only one ground point in the circuit.

Turn the power supply back on. Using only CH1 of the scope, measure the voltages on each of the load resistors, but only *separately – not at the same time*. That way, you earth-ground only one load at a time. You should find that the voltages across the loads are nearly the same.

Careful inspection of Fig. 3.58, bottom right, will tell you that each load resistor is in parallel with a phase voltage, so the load voltages should be the same as the phase voltages. If necessary, readjust the outputs of the 3PG to provide equal amplitudes at the loads.

31. **Measuring three-phase delta sources.** It is not possible to directly view all three load voltages at the same time because you would need three separate places for ground, and that is not allowed, as discussed above. The following exercise shows you how to observe all the phase and load voltages with only one ground.

    You can easily view *two* of the voltage legs at the same time. Refer to Fig. 3.59 (repeated). Put CH1 across Load-A where ground is at point (a); put CH2 across Load-B, keeping the two earth grounds at the same jumper at (a). Since the polarity of CH1 is reversed relative to the source, you must invert it to see the correct phase. This is easy to do: Select the CH1 menu and turn on the **Invert** check box at the lower right corner. That is the equivalent of switching the two CH1 leads, which you cannot actually do because you need to keep the grounds together. Now you can see that there is a 120° phase difference between the Load-A and Load-B voltages.

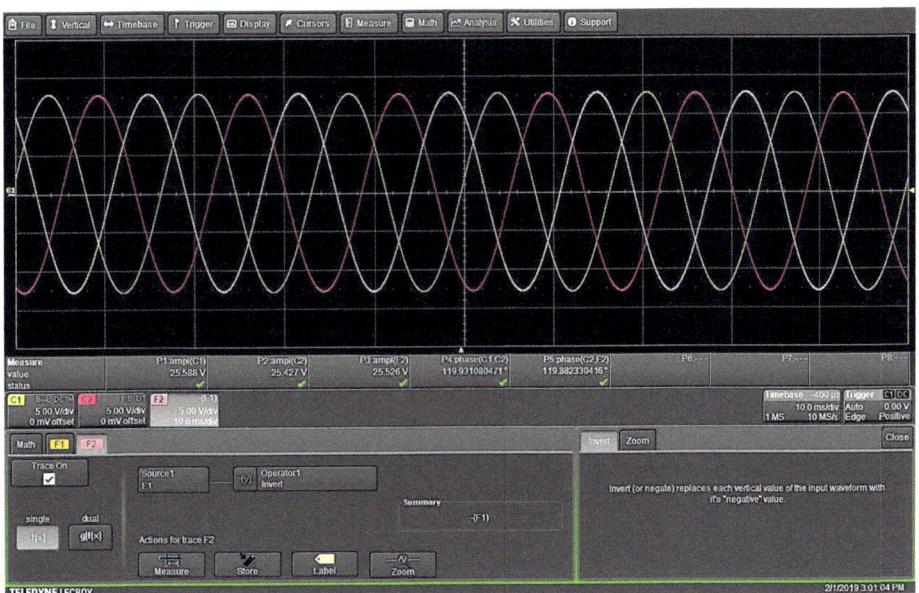

FIGURE 3.60 (Step 31) Three load voltages in the delta configuration. The third trace is a math function showing the differential measurement of Load C.

To plot the Load-C voltage, notice that it is the difference of the other two voltages so you can use math functions, as follows: Press **Math** and **F1 Setup**, and set function F1 to be the inverse of CH1. Next, set F2 = F1 − C2 = C1 − C2, which is the voltage across Load-C. Select to display CH1 (which is inverted), CH2 and F2. To put them into the same window on the display press **Display**, **Display Setup**, and **Single**. Note that to change the time base for all three traces, you must select either CH1 or CH2. Finally, you should see the three phase voltages as in Fig. 3.60, which shows that the three voltages are simply the same three phases that you have already observed. No surprise there, but you just learned a lot of scope techniques.

32. **Measuring the line currents.** You will use the current probe to measure the line currents i_{AB}, i_{BC} and i_{CA} in Fig. 3.58 right (repeated here) and their relative phases. Disconnect the probes on CH2 and CH3 as you no longer need them.

Here is how you will measure the phases of the line currents I_{AB}, I_{BC} and I_{CA}: Use CH1 voltage as a phase reference by putting a probe across points (i) to (iii) – ground. Be sure to trigger off of CH1. You will leave CH1 up on the scope during the current measurements.

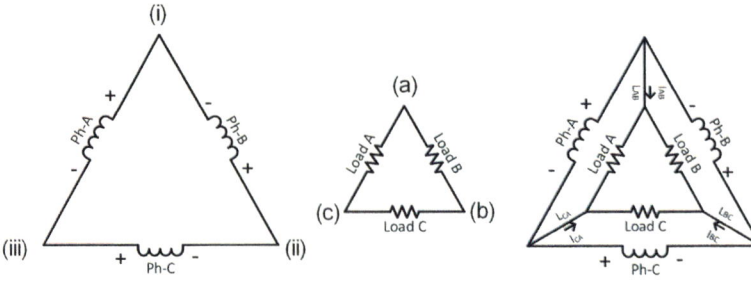

FIGURE 3.58 (Step 32) Repeated.

It serves only as a reference for phase since you can only observe one line current at a time. (If we had wanted to drop another $6k on each lab bench, we could have bought two more current probes to measure all the line currents at once, but we didn't. Also, you may have noticed the 7.5 Ω resistors on the load board. The purpose of those is to serve as shunt resistors to measure the currents in the lines using the methods of Step 29, but you will not use those as long as you have the use of one current probe.)

Use CH4 for the current probe. Set the averaging to 3 to reduce some of the noise. Set a measurement function to read the current in CH4 and another to read the phase from CH1 to CH4. Set **Coupling** of CH4 to **DC** (or else you cannot zero the probe, coming up next.) Select CH4 tab and then the tab labeled **C4CP030** to open the menus for degaussing the probe and zeroing its offset. Follow the instructions. *The DC offset tends to drift, so rezero it every time you make a current measurement.*

Use the current probe three separate times for the currents from the transformers into the loads: i_{AB}, i_{BC} and i_{CA}. To make sure that phases are consistent, point the arrow on the current probe toward the loads and lock the jaws by *gently* pressing forward on the latch. For each measurement, you should get a waveform like that of Fig. 3.61. Make a table for your data. For each line current, note the amplitude and its phase relative to the voltage source.

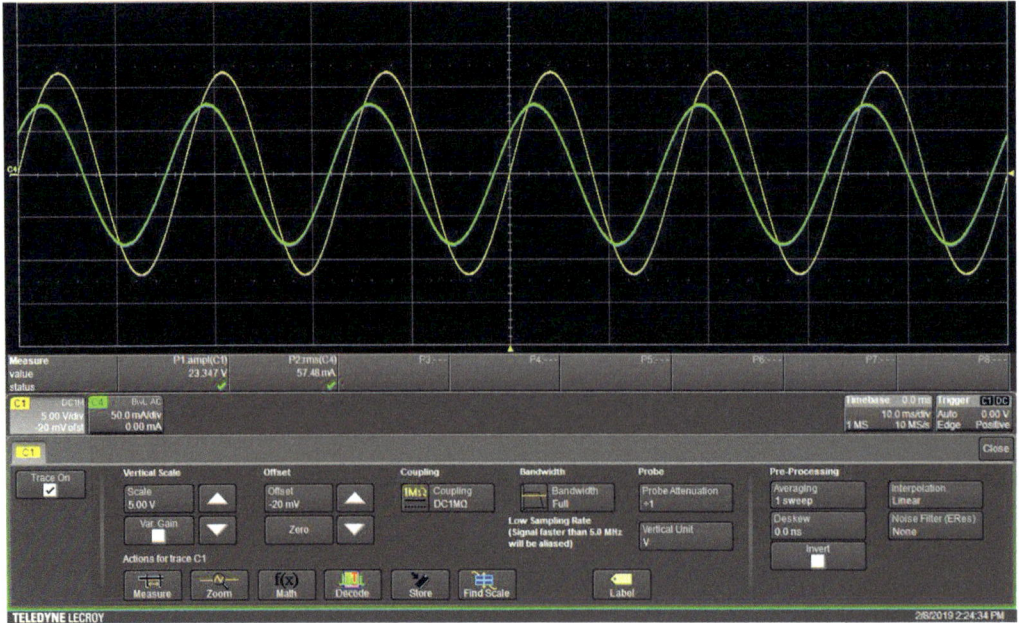

FIGURE 3.61 (Step 32) Waveforms of I_{AB} and measurement of I_{CA} using the current probe.

Repeat Step 17 for each of the currents to see how close the sum of the three currents comes to zero. Again, make a plot like that of Fig. 3.35; it should show that the three line-currents sum to nearly zero using the phases that you measured.

How did you do? Include a printout showing all the waveforms in your lab report. If done correctly, the three currents should sum to zero for all times. Thinking of the inner triangle of loads as a single (super)node, Kirchhoff's current law demands that the three currents must sum to zero at all times. What is important here is the load currents flow equally through each load while at every time flowing back to their sources through a combination of the other loads.

[Extra info that you can skip if you like: The current in a load, e.g., i_A, which you are not measuring, is equal to its voltage drop, which is its associated phase voltage, divided by its resistance, or $i_L = v_{Ph}/R_L = v_{Ph}/250 \; \Omega$, and it is in phase with its phase voltage. The current in line L_{AB}, which is i_{AB} that you measured, is the sum of the Load-A and Load-B currents. Because the load currents add with their phases, as you have been doing in your Desmos or Matlab calculations, I_{AB} should lag the Ph-A voltage, as indicated in Fig. 3.61. What phase lag do you measure? In fact, this lag should be 30°, but you probably measured more like 2°. This discrepancy is a limitation of the current probe, whose current reading leads the actual current by about 7° at 60 Hz. Oh well.]

33. **Measuring the phase currents in balanced delta-delta loads.** Recall that in the wye-wye circuit the phase currents (through the transformers) and line currents (to the loads) were the same because they were in series. Now from Fig. 3.59, right, you can see that in delta-delta the phase currents split at the loads and other sources, so the phase and line currents are not the same. Measure and record the phase currents of each transformer and their phases as you did for the line currents in Step 32. The phase currents are on the single wire that carries all the current into the transformer at X6 or X10. For balanced loads, the amplitudes of the phase currents will be identical. By now you know what the sum of the currents is, so you don't need to repeat Step 17.

34. **Unbalanced loads in delta-delta.** Unbalance the loads by adding the fourth 250 Ω resistor in parallel with Load-A. Repeat the measurement, table, and plot as you did in Step 32. Also, compare the new line currents i_{AB}, i_{BC} and i_{CA} to the previous balanced ones. By now you should be getting good at this and it should go quickly. Now you can see how the individual currents in the lines change as the loads change. Include a printout showing all the waveforms in your lab report.

35. **Observing phase currents.** Measure the phase (i.e., transformer) currents as you did in Step 33. Do you expect the phase currents to now be identical? How do those compare to the case of the balanced load in Step 33?

36. **Losing one transformer in delta-delta and testing phase voltages.** Look yet again at Fig. 3.59, left side. Notice that Ph-A is in parallel with the series combination of Ph-B and Ph-C. That means that at every moment the sum of Ph-B and Ph-C must be the same (or opposite if we insist on preserving the signs) as Ph-A. So, *if Ph-A were removed, there would be no change to the voltage between (i) and (iii)*. This is called "**open delta.**" As a consequence, you could expect that none of the loads would experience a change in their driving voltages or currents. However, now the current would be carried by just two of the three phase voltage sources, and they would be working harder. You will test this out in this step.

Voltage measurements for loss of T1: Now you will see what happens when one of the transformers 'fails' due to a hot day. Disconnect the 4th load so all three are balanced again. Remove Ph-A from the circuit by unplugging the wire(s) from Ph-A at X6, but leaving everything else connected as before. The circuit now looks like Fig. 3.62. Measure the voltages across all the loads. Does your three-phase motor lose power or not? Ideally, all of the load voltages will still be the same and the motor should still run. Is that what happens? If there are significant differences, go back and check the balancing of the 3PG at the transformer outputs. You can expect some extra voltage drop because the remaining transformer secondaries source more current to make up for the loss of Ph-A, and some extra voltage is lost to the internal resistance, i.e., the windings.

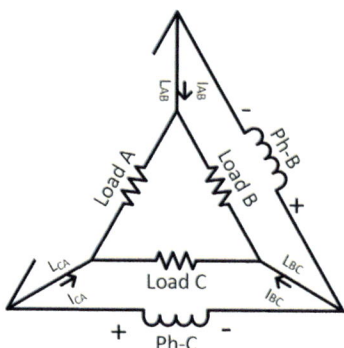

FIGURE 3.62 (Step 36) Fig. 3.53 repeated but with Ph-A removed.

37. **Measuring line currents with open delta.** Measure the line currents i_{AB}, i_{BC} and i_{CA} with Ph-A still disconnected. How do those compare to the case with all three transformers intact found in Step 32? They should be the same since the load voltages are still the same. System reliability is one advantage of the delta-delta circuit.

38. **Measuring phase currents with open delta.** With Ph-A still open, again measure the phase (i.e., transformer) currents as in Step 33. Of course, there is zero current for Ph-A. What happened to the current in the other two transformers? Does that make qualitative sense? Enough's enough! You are done!

Congratulations! You have finished investigating the entire power grid from the high-voltage transmission lines all the way to your wall outlets, and you even heated your home. You have also delved into wye-wye and delta-delta circuits. This experience should have given you a good introduction to how electrical power is delivered from a power plant to your wall outlets and how wye-wye and delta-delta circuits fundamentally behave. With students like you, the power grid will be in good hands!

Problems

1. Use Fig. 3.1 and the text of Chapter 3 to answer this question. Provide representative voltages for the following sectors of the national power grid from generation to the home.

   (a) Transmission grid: Step up from .... to ....

   (b) Distribution grid: Step down to ....

   (c) Substations to neighborhoods: Step down to ....

   (d) Neighborhood transformers to homes: Step down to ....

2. High-voltage transmission minimizes wasted energy by lowering I^2R losses. What "R" are we referring to?

3. "Joule Heating" happens whenever current flows through anything that has resistance, including any wires. Compare two identical transmission lines. One supplies power at a voltage V_1 and the

other at voltage $V_2 = 3V_1$. Let's say that the loads for the two transmission lines are transformers 1 and 2 that step that voltage down to whatever is needed. For the same power delivered to each of the transformers, how does the power lost in the transmission wires change? Call the resistance of the wires R_{wire}, and call the power lost in the wires P_{wire}.

4. Some of the lab activities ask you to imagine that you have the power grid on your benchtop. Describe what each part of your model power grid represents and its corresponding voltage. Refer to Fig. 3.13 (repeated in Step 9). Look through Step 9 of the lab. You are given a model that corresponds to the simple case of the transmission lines to the subtransmission grid. Now, you fill in the table for Step 9.

Example: Transmission lines to subtransmission grid			
Grid hardware	Grid voltage (RMS)	Your model's hardware	Model's voltage
Transmission lines	say, 750 kV	3PG	7 Vp or 5 VRMS
Output of a receiving station	35 kV	One transformer, secondary, end-to-end	3.5 Vp or 2.5 VRMS

Step 9:			
Grid hardware	Grid voltage (RMS)	Your model's hardware	Model's voltage

5. You can calculate the EMF from one end of a coil to the other using Faraday's Law in MKS units by using magnetic flux density in units of tesla and total magnetic flux in webers. One tesla is one weber/m^2. (FYI, the field at the surface of a powerful permanent magnet is around 0.5 tesla.)

Consider a single circular loop of wire, like that in Fig. 3.2, middle, with diameter of 5 cm. It is opened over a very small gap. It has a total resistance of 25 Ω (obviously not made of copper). A magnet is brought toward the loop and creates a uniform magnetic flux density, B, across the area of the loop that changes linearly from 0 to 1.75 tesla in 0.75 seconds.

(a) What is the magnitude of the EMF developed across the gap in the loop (Fig. 3.2 middle)? Show your work, including a drawing of the loop, the B-field through it, and the total flux Φ. Apply Faraday's Law (ignore the sign from Lenz's law here).

(b) Now suppose the gap is removed by touching the ends of the wire to each other (Fig. 3.2 top), so it is now a continuous loop of wire. What is the magnitude of the current that will flow in the loop during the 0.75 s change in field strength?

(c) If you open up the gap again, and connect 150 loops (or turns) in series (Fig. 3.2, bottom) to make a coil, and move the magnet through it so that the flux density changes the same across all the turns, what will the voltage be across the ends of the coil (ignoring signs)?

(d) Now short out the ends of the coil by connecting the ends together with an ideal wire (i.e., having no resistance). What is the current through the coil (ignoring signs)?

(e) With the coil shorted, what is the power, P (in watts), consumed by the resistance of this 150-loop coil during the 0.75 s?

(f) What is the energy, U (in joules), consumed by the wire resistance over the full 0.75 s?

6. This question is about Lenz's law and is intended to demonstrate why inductors give the same voltage drop for a changing current regardless of which way they are wound. Consider two coils that are identical in every way other than that the left-hand coil (LHC) is wound clockwise looking in from the left, and the right-hand coil (RHC) is wound CCW looking from the left.

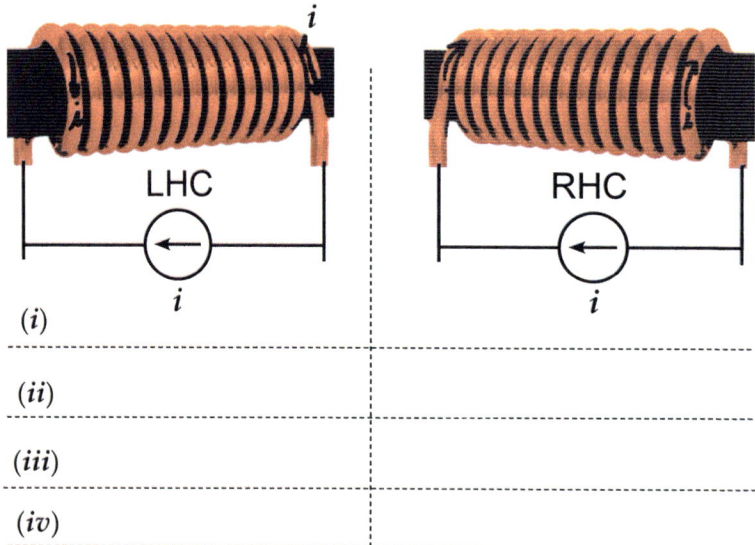

(a) Review Ampere's Law as presented in Figs. 1.14 and 1.24. They show the right-hand rule for the direction that flux wraps around a current: When you point the thumb of your right hand in the direction of the positive current, your fingers curl in the direction of the positive magnetic flux. For both the LHC and RHC, draw on the pictures the flux, **B**, given the direction of current shown in the picture of the coil. Show the flux, with arrows, both inside and outside the coils.

(b) The EMF induced by Lenz's law is that source of voltage that would push current externally in such a way as to oppose the change in flux or current. In the figure, imagine that the current source is a resistor. Ask which way the current would flow to oppose the change in current. The voltage 'drop' that appears across an inductor due to a change in current is that EMF. Assume that the current changes sinusoidally, as shown. Regions of the current

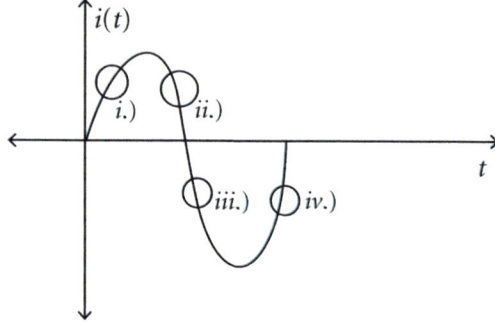

are labeled (i) current is positive and increasing, (ii) current is positive and decreasing, (iii) negative and increasing in magnitude and (iv) negative and decreasing in magnitude. For each of these regions, which side of the LHC and RHC will be positive and negative during the increase? On the table below the pictures, fill in "+" or "−" in the spaces indicated. There are 16 spaces in total.

7. How fast, in revolutions per minute RPM, must a three-phase-generator rotor turn if it has 9 sets of three-phase stator windings (27 total windings) and the output is 50 Hz?

8. For the following center-tapped transformer:

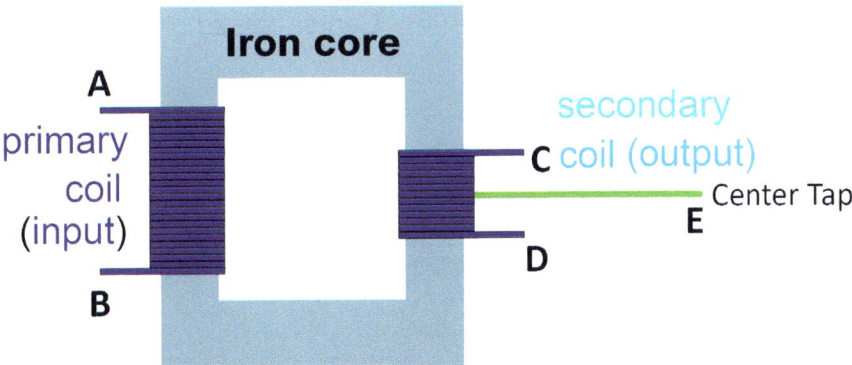

(a) The primary coil has 300 turns between A and B and the secondary coil has a total of 75 turns between C and D. Apply 240 V_{RMS} from A to B. (All voltages here are in RMS.) What is the voltage between C and D, V_{CD}?

(b) What are V_{CE} and V_{ED}?

(c) Keeping in mind that when writing "V_{AB}", the A-side is the positive terminal and the B-side is the negative terminal. What is the relationship between V_{CE} and V_{ED}? Are they in-phase or out-of-phase, and what are their relative magnitudes?

(d) What is the magnitude of $V_{CE} + V_{ED}$? The answer is in V_{RMS}. What is the relationship of this answer to V_{CD}?

(e) If you provide 10 V_{RMS} between C and D, what will be the voltage between A and B? *Hint*: You may apply a voltage to either the primary or the secondary, but generally transformers are designed for the expected primary and secondary voltages and currents.

(f) Now let's consider the center tap at point E. For 10 V between C and E, what is the voltage between A and B?

(g) For 10 V between D and E, what is the voltage between A and B?

(h) For 240 V_{rms} between A and B, and a load of 1 kohm between C and D, find both the current out of C and the current into A.

(i) What is the voltage between C and D if you apply a *DC voltage* of 240 V between A and B (after you wait for it to settle)? Explain your answer.

9. What are the following voltages V_1 to V_7? Express all answers in RMS values.

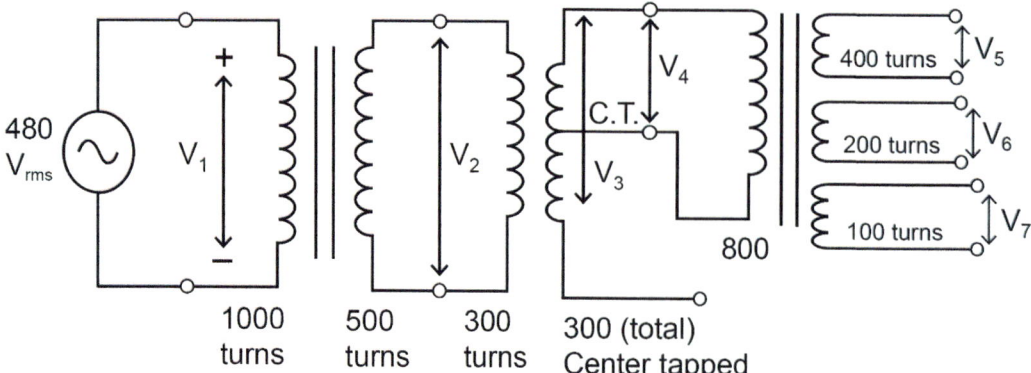

10. Look at Step 3 of the lab activities.

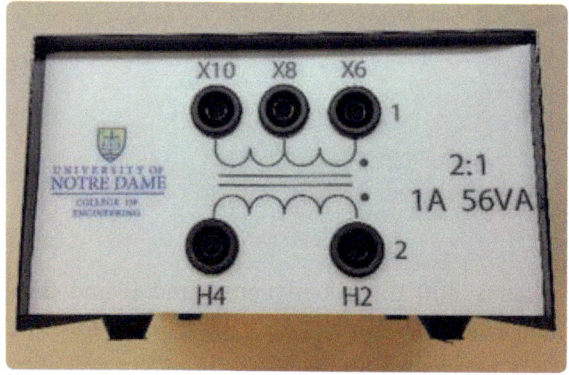

(a) If you connect the primary voltage of 14 V_{pp} between H2 and H4, what voltage do you expect between X6 and X10?

(b) What about between X6 and X8 or between X8 and X10?

(c) If you want to step up the voltage by 4 times, how would you connect the primary and secondary?

11. (a) At the instant that the voltage on one line of three-phases is at the peak voltage, V_p, what voltages (in terms of V_p) are on the other lines? *Hint*: they are 120° apart, as you expect for three-phase.

(b) Provide a plot that shows the relationship from part (a).

12. The equation for the phase (in degrees) between two signals is

$\theta = 360 \times f \times \Delta t$

f = Frequency in hertz

Δt = Time between peaks of the two signals

The period, $T = 1/f$.

(a) Given this, what period and time delay between 3-phase voltages are there in North America?

(b) In Europe?

13. For the following waveform:

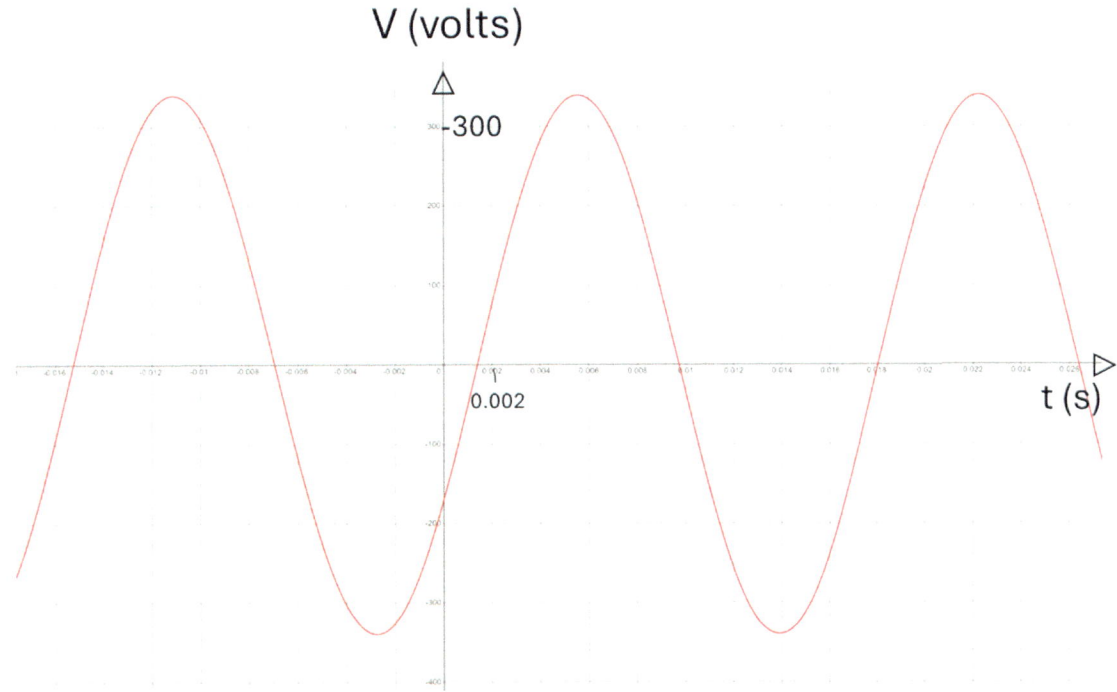

V (volts)

-300

0.002

t (s)

(a) What are V_p, V_{pp}, V_{RMS}, V_{ave}, T, f, and ω? Draw on the plot above and take your answers off of the plot.

(b) What is the equation for the waveform? Your answer should be a cosine function, expressed in degrees (not radians). You need the angle, ϕ. Remember leading and lagging: Find the angle that is less than 180°. The waveform that comes first in time (on the left) 'leads.' The other waveform 'lags.' Is this waveform leading or lagging a cosine function having zero phase? Express the frequency part in cycles/sec, degrees/sec and time and include the phase in deg.

14. Consider a square wave that is at the peak voltage, V_p, for a time t_h and zero for a time t_l, where $t_h + t_l = T$, the period. The duty cycle of this square wave, D, is the fraction of the time that its value is high expressed as a number between 0 and 1; $D = t_h/T$. What are T, t_h and t_l, and D for the following waveform?

(a)

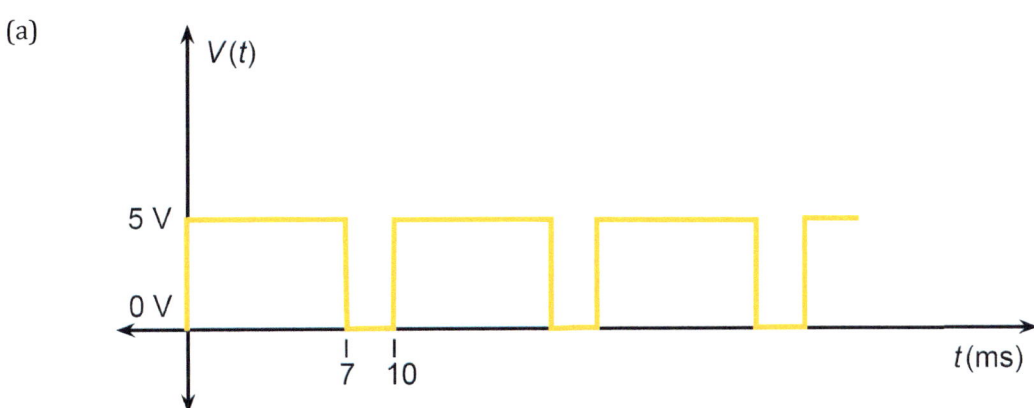

V(t)

5 V

0 V

7 10

t(ms)

(b) Starting from the definition of the RMS value of a function, using the symbols T, t_h and t_l, and D, show that the RMS value of the square wave in part (a) can be expressed as $V_{RMS} = V_p\sqrt{D}$.

(c) What is the RMS value of $v(t)$ for the waveform in part (a)?

(d) Use the integral expression for the RMS value of a voltage to calculate the RMS value of $v(t)$ below. Note: It is *not correct to use the equation given in part (b)*. You must calculate the integrals.

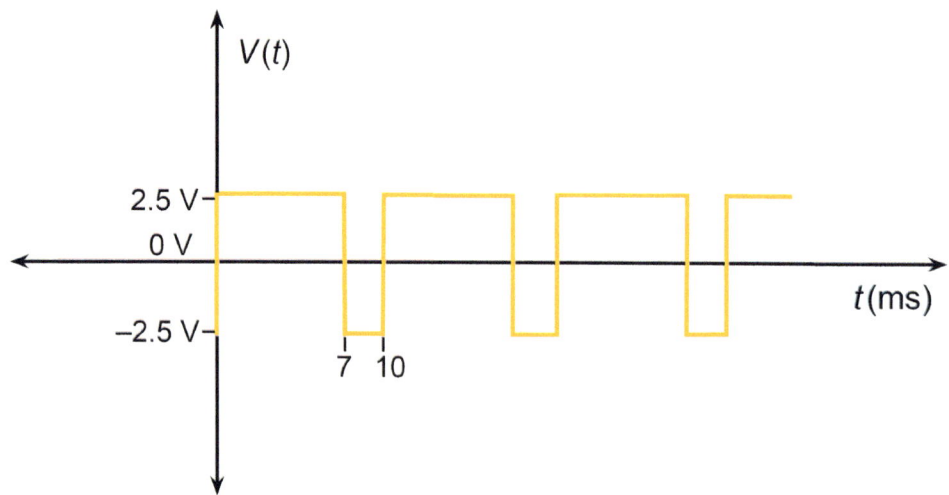

15. When we speak of the voltage on busbar A, what is that with respect to?

16. Suppose at $t = t_0$ the voltage on busbar A is 63.5 V, what is the voltage on busbar B at that moment?

17. As you know, 240 V appliances (such as a stove) run off busbars A and B with no neutral back to the center tap. There is a breaker on each busbar, so two breakers are always used for 240 V appliances. If the stove wiring is rated at 50 A, what size should the breakers be? 25 A or 50 A? Why?

18. When we say that no current flows on the neutral wire with balanced loads, which neutral wire do we mean – in the walls or under the ground (or up in the air) going back to the transformer?

19. How do polarized plugs increase safety in small appliances?

20. All quantities are in RMS. Suppose busbar A is running a coffee maker (900 W), a TV (150 W), a computer system (200 W) and several rooms of lights (350 W), while busbar B is running a microwave oven (1400 W), a space heater (1500 W), several rooms of lights (350 W), and a dishwasher (800 W).

    (a) How much current flows on the neutral wire?

    (b) Suppose the neutral wire has a total resistance of 0.1 ohm going back to the transformer from the breaker box. What voltage is dropped on the neutral wire, and how does that change the voltage to your home?

21. (a) What role do breakers play in residential wiring?

    (b) Suppose someone intentionally replaces a 15 amp breaker with a 30 amp breaker in a service panel. Is this a good idea, bad idea, or really bad idea? Why?

    (c) What would be a very easy way to tell what size breaker a branch circuit is designed for?

22. What is the role of the neutral wire in a home? What is the role of the ground wire? Where are they connected? Why are they not the same wire everywhere?

23. The diagram shows split-phase residential wiring. One goal of arranging the loads on the branch circuits is to have the minimum current flowing on the neutral wire going to the transformer. On the diagram, draw in the following 5 appliances **in such a way that the current on the neutral wire to the transformer is minimized.** Place the resistors and connecting wires on the diagram as illustrated in the figure by the example resistor labeled 'x'. Resistor x plays no role in your answer, other than to illustrate how to place your appliances.

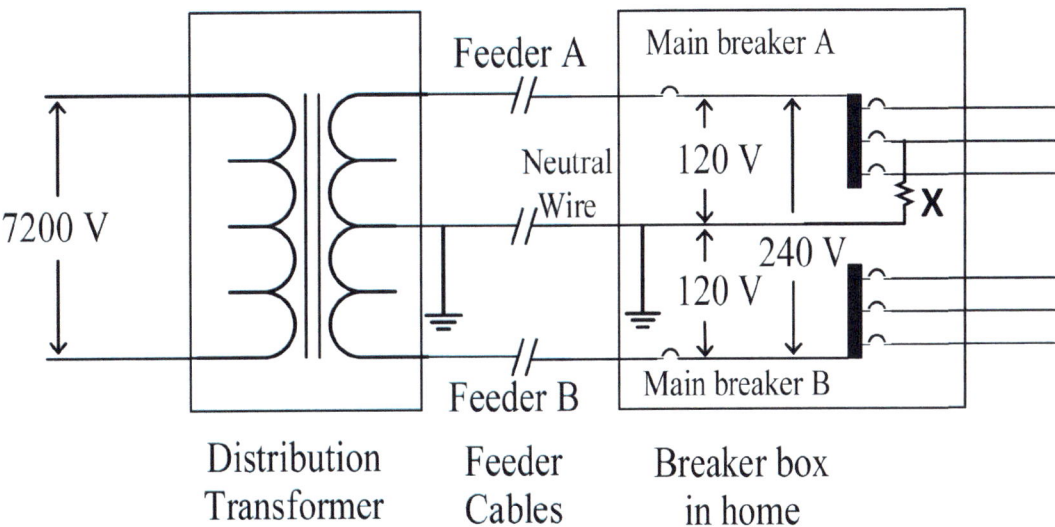

Draw all the appliance resistor symbols onto the diagram between line voltage and neutral:

120 V, 800 W coffee maker: CM

120 V, 1200 W microwave oven: MO

120 V, 1000 W toaster oven: TO

120 V, 1000 W hot air fryer: HAF

240 V, 8000 W electric dryer: ED

24. (a) Step 11 asks you to draw two schematic diagrams for powering a space heater with two different resistor configurations at two different voltages and two different currents, but the same power. Draw those schematics here, including the ammeters.

    (b) Suppose that the 250 Ω resistors that you use in Step 11 are actually powered by 120 V_{rms} and 240 V_{rms} as discussed in the previous question (of course you will have lower voltages on your bench.) What current would flow from the sources in both cases in Step 11? What power would you produce by the heaters in both cases of Step 11?

25. Suppose that the neutral wire in your home is 125 feet long from the breaker box to the furthest part of your home. It is made of 14 gauge copper wire and is on a 15 A breaker. You manage to touch the neutral wire at the furthest point with 15 A flowing through a toaster oven and a space heater and then through it.

    (a) Draw a schematic diagram of the situation including both the oven and the heater. Assume that they have polarized plugs, so also show the on/off switch on the schematic diagrams.

    (b) When the current is 15 A (just before the breaker would trip), what voltage would you touch on the neutral wire, and how dangerous is it? *Hint*: you will have to go back to Chapter 1 and review wire gauge and resistances.

26. Refer to Fig. 3.25. Supposing someone thinks he is being clever and plugs one end of a long wire from the hot slot of one wall outlet and attaches the other end to the hot slot of another wall outlet somewhere else in the home, far from the first one. This is a risky thing to do because one of two things will happen:

    (a) Suppose that nothing at all happens when he pushes the wire into the hot slot of the second outlet. Explain how this could happen.

    (b) Now suppose instead something else happens: there is a big spark, the rubber is melted near the end of the wire, and the breakers trip in the breaker box. Explain how this could happen.

27. A typical breaker in a home trips at 15 A.

    (a) Assuming nothing else is taking current out of that breaker, what is the most power in watts that a wall outlet can provide to something that you plug into it?

    (b) One horsepower, HP, is 745.7 W. What is the maximum HP anything plugged into a common wall outlet can provide (as, for example, a power washer or an air compressor) when in use?

    (c) A typical electric range, with four burners on the surface and one element in the oven, uses about 8000 W, and typically runs off of 240 V. What must be the rating of the breaker in order to allow the oven to run with all the elements turned on at once? Assume that high-amperage breakers come in increments of 10 Amps, e.g., 30, 40, 50... amps.

    (d) An electric power washer is rated at 2500 psi (pounds per square inch) at a flow rate of 1.55 GPM (gallons per minute). All of that high-pressure water takes power, which can be calculated from P (power in watts) = Q (flow rate in cubic meters/second) × pressure, p (in pascal, where pascal is an SI unit of pressure).

    Nothing is perfectly efficient, so the power washer will draw even more power from an outlet. What is the power drawn from a 15 A wall outlet (in W) if this machine can deliver the pressure and flow rate advertised and is perfectly efficient? *Hint*: You must do some research to perform unit conversions.

    (e) How many watts is there to spare before the breaker trips? Do you believe the manufacturer's claims?

28. The following is a simplified split-phase diagram. R_1 represents the total resistance of all the appliances running in parallel off of Bus A and R_2 is the same for Bus B. All of the branch circuits on each of the buses are in parallel.

(a) If you have a 700 W coffee maker, a 1200 W microwave oven, and a 1000 W toaster all going at once being driven from multiple branch circuits that are all tied to Bus A, what is the resistance, R_1, seen by Bus A to neutral? Assume that the wiring is all ideal, i.e., it has zero resistance.

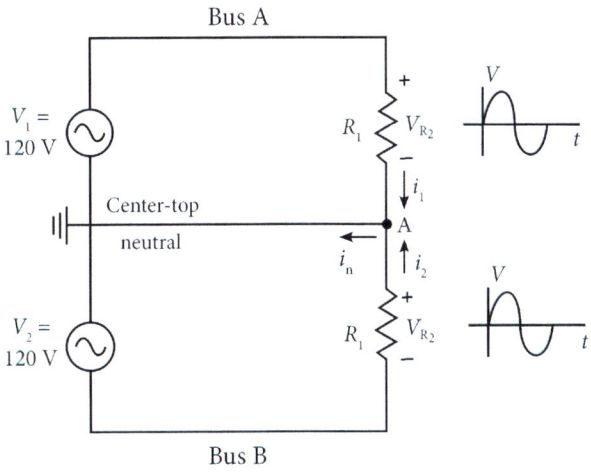

(b) What is the RMS value of current i_1?

(c) Now assume that there is a 1500 W space heater, a 1300 W vacuum cleaner, and a 1200 W toaster oven being run off of Bus B. What is R_2? You can solve it the same way as (a).

(d) What is the RMS value of current i_2?

(e) Do not forget that the two power sources are the top and bottom of the secondary of a center-tapped transformer. At the instant that $V_1 = V_{R1}$ reaches its positive peak value of +170 V (with respect to ground), what is the voltage and polarity of $V_2 = V_{R2}$? *Hint*: The answer is embedded in the circuit diagram above. Look carefully.

(f) Does all of the current in Bus A flow through Bus B? Why or why not? If not, explain (do not calculate yet) what is the difference in those currents. *Hint*: Use Kirchhoff's current law at node A.

(g) Now calculate: What is the RMS value of i_n?

(h) What would i_n be if $R_1 = R_2$? That is called "balanced loads."

29. Use Desmos or equivalent to show that the line currents in the wye-wye connection, Fig. 3.33, add to zero when the loads are balanced, so that there is no current on the neutral line. Include your plot and explain what you did and why you got the result that you did.

30. Following from the previous question, increase the loads on Ph-A by 3 times (so that the current is one third) and repeat the plot to show the current that flows on the neutral line is no longer zero. Explain what you did and why you got the result that you did.

31. Look at Fig. 3.33, the wye-wye configuration. Suppose that the loads are neighborhoods. (a) What happens to neighborhoods B and C if the power transformer that is Ph-A fails? (b) What happens to the current on the neutral line? (c) Provide a plot that supports your answer to part (b) for balanced loads.

32. In the delta configuration, the series combination of two of the voltage sources are in parallel with the third. This is true for any combination of the three sources, no matter how you look at them. Since ideal voltage sources can never be at any voltage other than what they are made to provide (even if they provide an unlimited amount of current), that means that the sum of any two of the delta sources must, by Kirchhoff's voltage law, exactly equal the voltage of the third source, or else an infinite amount of current would flow between the two sources and the single source. Instead, no current flows because the two in series always sum exactly equal and opposite to the third.

(a) Draw the three-phase delta circuit diagram and show on that what is meant by the statements in the introduction to this problem.

(b) Provide a plot that proves that three phase power provides this condition.

33. For each of these impedances, let $R = 180 \ \Omega$, $L = 0.3$ H, $C = 15 \ \mu$F.

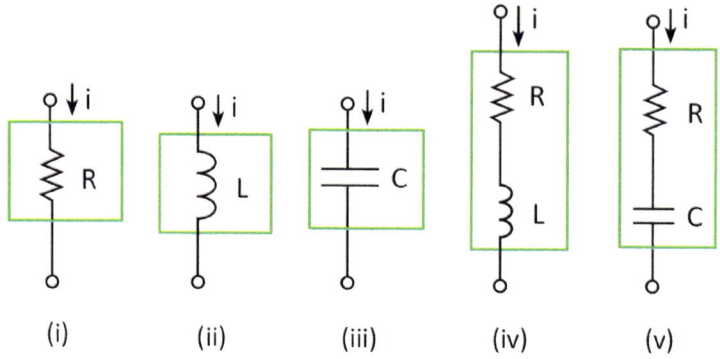

(i) (ii) (iii) (iv) (v)

Fill in the tables. Angle ϕ must be in degrees. $\mathbf{Z}$ is in the form $\mathbf{Z} = R + jX$ and $|\mathbf{Z}|$ is the magnitude of $\mathbf{Z}$ in ohms. There is room to write in the calculations.

(a) $f = 60$ Hz

	$\mathbf{Z}$	$\|\mathbf{Z}\|$	ϕ
i			
ii			
iii			
iv			
v			

(b) $f = 1$ kHz

	$\mathbf{Z}$	$\|\mathbf{Z}\|$	ϕ
i			
ii			
iii			
iv			
v			

34. The entire circuit to the right of the voltage source has a complex impedance $Z = R + jX = 130 + j\omega L$. Assume $f = 60$ Hz. This could represent an electric motor, for example.

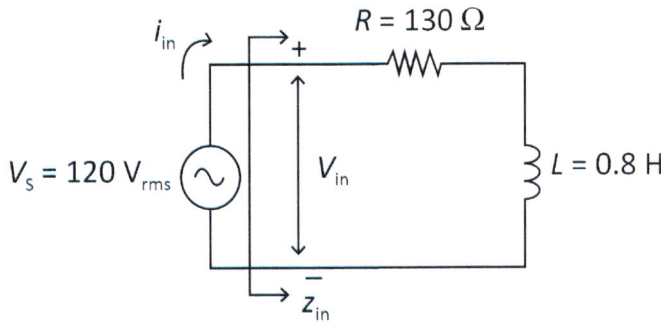

(a) What is the value of $X = X_L$? Include the units.

(b) What is the expression for $\mathbf{Z_{in}}$ including all numbers.

(c) At 60 Hz, what is the phase shift, ϕ, between the current, i_{in}, and voltage, V_{in}, of the load, $\mathbf{Z_{in}}$? Give your answer in both degrees and radians. Which is leading, current or voltage?

(d) Roughly hand sketch the relationship between sinusoidal $v(t)$ and $i(t)$ showing the phase in part (c). The axes will be volts and amps, respectively, and you don't need to put values. Concentrate on representing the phase accurately. Label each curve as V or I and show the phase difference between them. (You may choose to use a graphing program instead of hand sketching.)

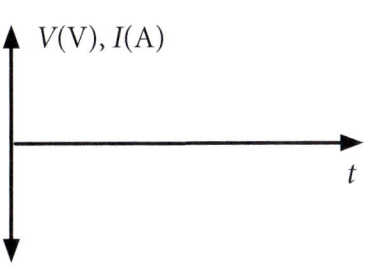

(e) Repeat parts (a), (b) and (c) for $f = 1$ kHz. What does the result tell you about how $\mathbf{Z_{in}}$ behaves with higher frequency? Does the load look "more inductive" at higher frequencies? Explain.

(f) Repeat part (d) for the results of part (e).

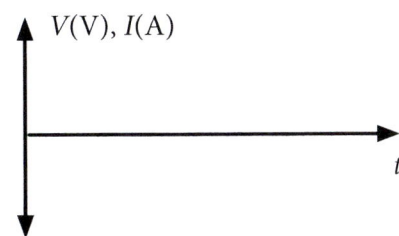

(g) Repeat parts (a), (b) and (c) for $f = 10$ Hz. Does the load look more or less "inductive" at lower frequencies? Explain.

(h) Repeat part (d) for the results of part (g).

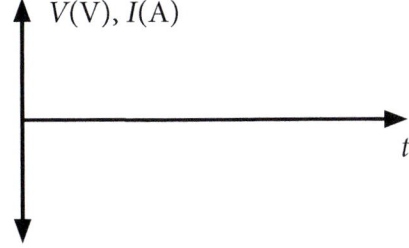

35. For an AC voltage, half the time the voltage is positive the other half it is negative, leading to an average of zero. Why then doesn't the average power consumed by a resistive load equal zero?

36. (a) What is the power factor, and what does it mean for it to be at 0? 1? In between?

(b) If voltage is lagging the current, which one should you add to bring them in phase, capacitance or inductance?

(c) Why do industrial power customers most-frequently add capacitance (often a whole bank of capacitors) to their load? How does this benefit power companies?

37. Why are commercial power customers charged a penalty for low power factors if reactive power is returned to the grid anyway?

38. Residential electric customers do not pay additional fees for low power factors. Would you prefer to be charged per kVAh or kWh?

39. (a) What is the power factor for a purely resistive load? Show the power triangle. (*Hint*: This is sort of a trick question, as are the following two.)

(b) What is the power factor for a purely capacitive load? Show the power triangle.

(c) What is the power factor for a purely inductive load? Show the power triangle.

40. This figure shows the power factor correction of a large inductive load using a capacitor. "You" are the power customer. Write one or two paragraphs that explain the various parts of the figure and what it says about power factor correction.

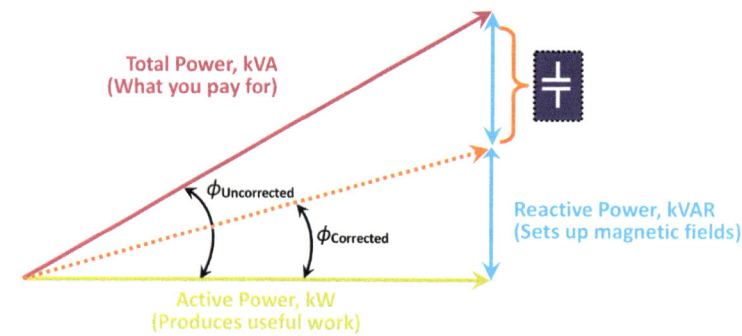

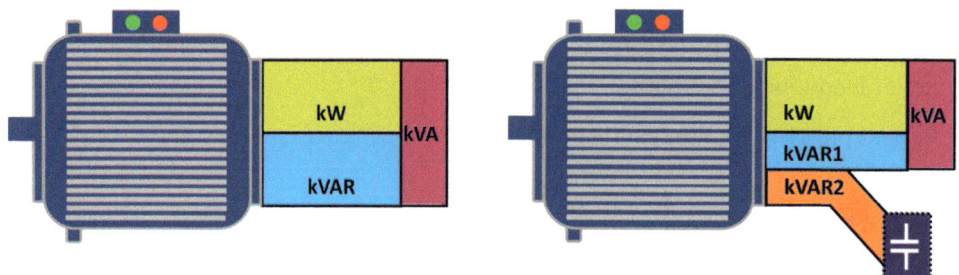

41. If the phase between the voltage and the current across some load is 45°, and the real power is 1 kW, what is the power factor, apparent power, and reactive power? Draw the power triangle. Be sure to use the correct units.

42. Consider $I(t) = 5\cos(\omega t)$ A and $V(t) = 3\cos(\omega t + \phi)$ V representing voltage and current, both at 60 Hz. Let $\phi = -60°$.

   (a) Use Desmos or another application to plot the two waves. Don't forget to convert between degrees and frequency to radians and radians/s. Label the waveforms on the plot.

   (b) Also, plot the power as a function of time on the same plot and label it.

   (c) Is this a capacitive or inductive impedance? How do you know? Show on the plot how you know.

   (d) What is the power factor of this load?

   (e) What is S, the apparent power in units of VA (always in RMS units)?

   (f) What is P, the active power?

   (g) What is Q, the reactive power?

   (h) Draw the power triangle for the previous parts. Label all relevant details, and draw to scale using a reasonably correct ϕ, with the sides having the right proportions.

43. Suppose you are the power engineer working for a large manufacturing plant. You are told that the power factor is too low, so you decide to add very-high-power resistors to your load to increase the power factor. Why is this not a good idea?

44. Your manufacturing plant uses a motor whose inductance is 0.19 H having a DC resistance $R = 8\ \Omega$. We are in the U.S., so the frequency is 60 Hz. In this case, $V = 480\ V_{rms}$. Do not forget to convert Hz to radians/second. Note that inductive reactance is $X = \omega L$ (and capacitive reactance is $X = -1/\omega C$).

   (a) What is the reactance, X_L, due to the inductance?

   (b) What is Z?

   (c) Calculate the current in A_{rms}.

   (d) What is ϕ, the angle between the voltage and the current?

   (e) What is the power factor?

   (f) Draw the power triangle including the values of S, Q and P. Use reasonably correct proportions.

Chapter 4

Introduction to Time and Frequency Domains: Filters, Waveforms, Frequency Components, and Resonance

4.1 Introduction and Overview

In this chapter, you will learn:

1. How inductors and capacitors affect the rate at which a circuit responds
2. What an electronic "filter" is
3. How frequency-dependent impedances enable filters
4. What "time domain" means and why it is important
5. How time and frequency are dual concepts
6. What "frequency domain" means and why it is important
7. What a "frequency spectrum analyzer" is and what it tells you
8. What resonance is and how it depends on impedance

4.2 Characteristic Time Behavior of Capacitors and Inductors with Resistors

This chapter introduces the concepts of **time domain** and **frequency domain**. Briefly, time and frequency domains are two related ways to represent and analyze signals or systems, which in our case include voltage and current waveforms, and how they are affected by circuits. The time domain is easier to grasp; it is exactly what you experienced in Chapters 2 and 3 when you looked at the outputs of the waveform generator and three-phase generator with the oscilloscope. Remember that the horizontal axis is time, and the trace on the oscilloscope is the voltage waveform, meaning how the voltage changes with time. Viewing waveforms on the oscilloscope and how they evolve passing through a circuit is an example of "being in the time domain."

The frequency domain is a subtler way to interpret functions of time, but it is pervasive in engineering and science. Later in this chapter, we will see that time-dependent waveforms can be thought of as collections of sinusoids of various frequencies. That frequency-dependent representation of the waveform is said to be "in the frequency domain." It will pay enormous dividends for you to understand these concepts now, since you will be using them extensively in later courses. Understanding the frequency domain is foundational to the world of engineering and science, and its importance cannot be overstated. We start with the time-dependent behavior of electrical circuits, apply that to electrical filter circuits and then move on to the frequency domain.

Introduction to Electrical Engineering

Gary H. Bernstein

Copyright © 2026 Jenny Stanford Publishing Pte. Ltd.

ISBN 978-981-5129-30-4 (Hardcover), 978-1-003-71345-6 (eBook)

www.jennystanford.com

DOI: 10.1201/9781003713456-4

In this chapter, we start to think about how voltages and currents change as a function of time. If a circuit comprises only voltage sources and resistors, the circuit properties will be exactly those of the sources. There will be no additional time dependence; whatever the sources do, the voltages and currents will all have the same waveform and timing (phase). Not so when there are capacitors and/or inductors in the circuit. They store energy, and it takes time for that energy to build up and be released. Therefore, there are lags in their voltage and current behavior. Depending on the waveform of the source, the resulting waveforms can look very different from those of the source. In Chapter 3 you saw that when the source waveform is sinusoidal, so also will be the voltages and currents. However, source waveforms can have various shapes, such as square and triangle waves, as produced by the waveform generator discussed in Chapter 2. In that case, the phase shifts due to the reactance of the circuit will change the shapes of the voltages and currents internal to the circuit.

Electrical circuits are a bit like water clocks, which date back perhaps 6000 years. Their principle is simple: water flows into a bucket at a constant rate, and the height of the water in the bucket either tells the time directly, or the weight drives a clock mechanism. Electrical circuits work on much the same principle. In the water analogy of circuit behavior, water is replaced by charge or current, and the water level is replaced by the voltage on a capacitor or current through an inductor. The currents and voltages within the circuit change with time. The charged capacitor (having electrical charges on the plates) or inductor (whose magnetic field charges by the current) underpin the behavior of all sorts of circuit behavior. These circuits include electronic filters, which is one of the main topics of this chapter.

4.2.1 Step Response and Time Constants for RC Circuits

To begin, it is useful to think more deeply about what it means for a capacitor to "charge." You might want to review this material in Chapter 1. In Section 1.4.3 you learned how current flows "through" a capacitor, and in doing so deposits charge on the plates. Consider the simplest capacitor: parallel metal plates separated by some insulating (i.e., dielectric) material. For this capacitor,

$$C = \frac{\varepsilon_r \varepsilon_0 A}{d}, \tag{1.20}$$

where ε_r and ε_0 are the dielectric constant of the material between the plates and the permittivity of free space, respectively, A is the area of the plates, and d is the distance between them. The magnitude of charge that collects on each plate (positive on one and negative on the other) is Q. Charge is proportional to the voltage, V, developed between them, with capacitance, C, the proportionality constant

$$Q = Cv_c \tag{1.19}$$

Since the amount of charge can change with time, voltage on a capacitor also changes with time as

$$i(t) = \frac{dQ(t)}{dt} \tag{4.1}$$

Current is the flow of charge and carries charge to the capacitor plates. Since current can change with time, we rewrite Eq. 1.1 as

$$i(t) = \frac{dQ(t)}{dt} \tag{4.2}$$

If we keep track of the current waveform and integrate it over time, we accumulate all of the charge that is on the plates of the capacitor (starting with the initial condition that $Q(t = 0) = 0$) as

$$Q(t) = \int i(t) dt \tag{4.3}$$

That charge on the plates causes an electric field, E, between the plates separated by a distance d, which integrates to the voltage on the capacitor,

$$v_c(t) = -E(t)d, \text{ so } E(t) = \frac{-v_c(t)}{d} \tag{1.22}$$

Plugging Eq. 1.19 into Eq. 4.2 we get the characteristic equation for a capacitor, namely

$$i_c(t) = C \frac{dv_c(t)}{dt} \tag{1.25}$$

Equation 1.25 tells us (among other things) that if we drive a constant current through a capacitor, its voltage waveform must be an increasing straight line, i.e., $v_C(t)$ increases linearly. This conjures up the image of water from a faucet filling a bucket at a constant rate (see Section 1.4.3.1), as in the water clock.

So, the charge on the plates changes with time as current flows, and voltage on a capacitor changes with charge. Can we change the charge arbitrarily quickly, jumping from one voltage to another in zero time, as in a step change? In mathematics, that is called "discontinuously." For the voltage to change abruptly, or discontinuously, Q would have to change abruptly, which would require an infinite current. You know that in real systems, current is limited by a lot of things. Even if you have heavy-duty wires and power supplies, the current cannot be greater than $i = v/R$, where R is all of the resistance in the current path; the resistance of the circuit limits the current, so charge can build up only so fast, and no faster. In the case of the water clock, the bucket can't fill up any faster than the rate at which water can be delivered. We expect, then, that resistance plays a major role in the rate at which the charge and voltage on a capacitor can change.

Let's consider the time behavior of the voltage across a capacitor in response to an abrupt, step change in input voltage. Consider a voltage source that increases abruptly or instantaneously from 0 to 1 V, as shown in Fig. 4.1. That is called a **unit step function**, denoted $v(t) = \boldsymbol{u(t)}$. (Step functions can describe changes in either voltage or current. The "step" refers to the time dependence.) The voltage unit step function $u(t)$ changes abruptly from 0 to 1 V at exactly $t = 0$. We can scale this to any voltage magnitude, V_S, as simply $v(t) = V_S u(t)$.

For later courses: In your physics and EE electromagnetics courses, you will learn that a changing electric field gives rise to displacement current. In a capacitor, the displacement current is almost entirely in the gap and is due to the changing voltage between the plates. This current looks as if charges are moving through the gap, although there actually aren't any charges in there. The displacement current keeps Kirchhoff's current law valid at the circuit level.

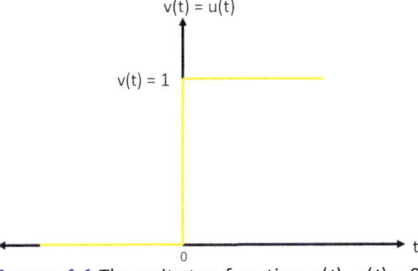

FIGURE 4.1 The unit step function $u(t)$. $u(t) = 0$ for $t < 0$ and $u(t) = 1$ for $t > 0$; $u(t)$ is undefined at exactly $t = 0$.

A mathematical step function, as shown in Fig. 4.2(a), can be realized by using a DC source of magnitude V_S, and a single-pole, single-throw switch. In Figs. 4.2(b) and 4.2(c), points B and B′ are ground (0), and point A is at a voltage V_S. The switch is connected to either B or A.

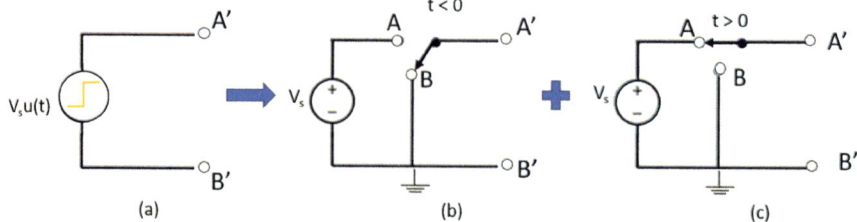

FIGURE 4.2 The mathematical voltage step function $V_S u(t)$, (a), can be represented as a switch that transfers a voltage that is (b) ground or 0 (switch contact at B) or (c) V_S (switch contact at A).

At the moment $t = 0$ that the switch changes position from B to A, the voltage between A′ and B′, $V_{A'B'}$, changes abruptly, as a step up, from $V_{A'B'} = 0$ to $V_{A'B'} = V_S$.

Figure 4.3 shows a step voltage of magnitude V_S input to a simple, first-order, **RC circuit**, which in this example is a series-connected resistor and capacitor. A **first-order** system has the equivalent of only one capacitor or inductor, and therefore responds in only one way as a function of time, namely exponentially. **Higher-order** circuits have combinations of more than one capacitor and inductor and respond in more-complicated ways; they are not discussed in this treatment of the step response.

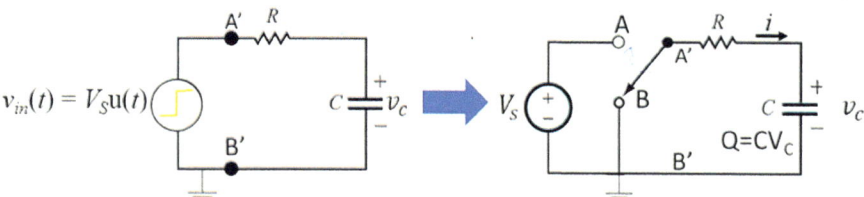

FIGURE 4.3 RC circuit with step-function input voltage realized using a switch. The switch moves from B ($V_{in} = 0$) to A ($V_{in} = V_S$) at $t = 0$.

For the reasons of energy transport and power limitations discussed above, capacitors and inductors require some amount of time to build up to a **steady-state**, or final, voltage or current waveform, respectively. The circuit behavior after a long time is called its **steady-state response**.

The voltages and currents change on the way to reaching their steady-state condition. This time-dependent behavior is called the **transient response**. The full behavior, including the transient and steady-state responses, due to a step-function input is called the **step response**. In general, the steady-state response due to an arbitrary repeating, or periodic, input can be some periodic waveform; however, the steady-state *step response* is always some DC voltage or current.

For $t < 0$, $v(t) = V_S u(t) = 0$. At the instant just before the step we call the time $t = 0^-$ (pronounced "zero minus"). Just after the step we call the time $t = 0^+$ (pronounced "zero plus," in case you needed to be told). When any ideal voltage source has a value of $v(t) = 0$, the source behaves exactly as a short circuit, or wire. How can you understand that? Well, a wire has no voltage drop but passes all the current through it that wants to flow. That is what a voltage source having zero volts does as well. When $v(t) = V_S$, that is exactly the same as an ideal DC supply or battery, as you already understand it.

You might at this very moment be asking, "Okay so far, but what is the voltage at *exactly* $t = 0$?" Well, we just don't talk about that. Seriously, the voltage is *undefined* at $t = 0$, meaning we can get as close to the discontinuity from the left, negative time, as we want, but just never get to $t = 0$. Also, we can get as close to the discontinuity from the right, positive time, as we want, but again, we never reach it. The time $t = 0$ is infinitely brief, and we simply don't know or care to discuss what the circuit does at exactly $t = 0$.

That said, what about the *real world*? (After all, you are learning to be an electrical engineer). In reality, no voltage or current changes instantaneously. Why, then, is the abrupt change in voltage at A′ in Fig. 4.2 allowed? That is because the length of wire from A to A′ is assumed to hold negligible charge – we model it has having zero capacitance. There is nothing that needs to charge for it to change voltage. No power is needed for the electric fields to readjust and the voltage to appear at A'. In reality, there is a tiny bit of capacitance; if we were concerned about things happening in really, really short times, this assumption would break down. When a switch is flipped, the voltage at A' can change between two discrete values very, very quickly, but never *actually* in "zero time." Voltages might change on a scale of nanoseconds, but we really don't care if we are interested in the behavior of the circuit in the range of, say, milliseconds. There is always some rise time due to the existence of tiny stray capacitances and inductances, but they are small.

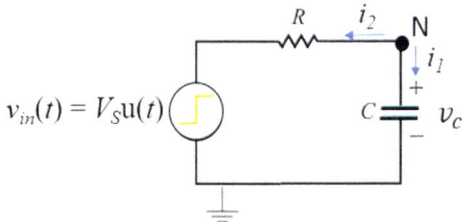

FIGURE 4.4 RC circuit with step-function input voltage where currents are summed at node N. This analysis leads to a differential equation whose solution is of the form $e^{-t/\tau}$, where $\tau = RC$ is the time constant.

In many real systems we ignore those very short transients at $t = 0$ and focus on the overall behavior relatively long after the switched voltage has very quickly settled down. To put it another way, here we assume here that the time scales of interest for the circuit behavior are much longer than the time scales of the equipment switching, but you are likely someday to be working on modern systems that operate at very short time scales, and such transients will not be ignored.

When we begin to analyze a circuit, we approach it in some pre-existing state. We have to make some assumptions about whether at $t = 0^-$ any currents are flowing, or any voltages already exist, i.e., there are charges already on the capacitors. These are called the **initial conditions**. Assume for our initial conditions that there is no charge on the capacitor plates at $t = 0^-$, so $v_C(t = 0^-) = 0$. That is assured in Fig. 4.3 by assuming that the switch has been at B for a very long time, so any charge on the capacitor will have long-ago leaked off through the resistor, and the capacitor has long-ago discharged.

We might be concerned with any voltage or current in this or any circuit, but we are at present concerned with the voltage across the capacitor. Using Eq. 1.25, $i_C(t) = C\frac{dv_c(t)}{dt}$, we derive a simple differential equation for $t > 0$ by summing the currents at node N in Fig. 4.4 as

$$i_1 + i_2 = 0 \tag{4.4}$$

which yields the differential equation

$$C\frac{dv_c(t)}{dt} + \frac{v_c(t) - V_s u(t)}{R} = 0 \tag{4.5}$$

You may not yet have had a course in differential equations and might benefit from a short explanation of what is going on here mathematically. A differential equation is like an algebraic equation, but the solution is a function rather than the value of a variable. The essential question asked by this differential equation is, "What function of time do you replace $V_C(t)$ with, so the time derivative satisfies Eq. 4.5 at every moment t?" Every differential equation has its own way of being solved, which is not the point of this discussion. Therefore, only a little bit will be said about it here, and the solution, meaning the function $v_C(t)$ that describes the capacitor voltage, will be provided to you. Using the algebra of basic calculus, and replacing $V_s u(t > 0)$ with V_s, we recast Eq. 4.5 as

$$\frac{dv_c(t)}{v_c(t) - V_s} = -\frac{dt}{RC} \tag{4.6}$$

In solving this, or any, differential equation we must account for the initial conditions. How do we apply the initial condition for $v_C(t > 0)$, i.e., $v_C(t = 0^+)$? Voltage across a capacitor cannot change instantaneously for all the reasons regarding energy and power discussed above. That requirement is called the **voltage continuity principle of capacitors**, namely that $v_C(t = 0^+) = v_C(t = 0^-)$. Again, we skip right over $t = 0$, since it is undefined. Nothing of interest to us happens there. Recall that we had assumed that the switch was at B for a long time before $t = 0$, so $v_C(t = 0^-) = v_C(t = 0^+) = 0$.

Solutions of this type of differential equation have solutions of the form e^{-x}. For a step input of $V_S u(t)$, it can be shown that $v_C(t > 0)$ increases from $V_C(t = 0^+) = 0$ as

$$v_c(t) = V_S(1 - e^{-t/\tau}) \tag{4.7}$$

where

$$\tau = RC \tag{4.8}$$

You can prove this to yourself by plugging Eq. 4.7 into Eq. 4.5 with $u(t) = 1$. If the equality holds, then it is the correct solution (and it does). Since $t = 0^+$ is infinitesimally close to $t = 0$, you may just plug in $t = 0$ when doing any calculations for the start of the change in input voltage. When $t = 0$, $v_C = 0$, as the initial condition demands.

Figure 4.5(a) shows not only a step up, but also an equal step down at $t = t_{S2}$, which combined is called a **pulse**. In a circuit, the end of the pulse occurs when, for example, the switch moves back from position A in Fig. 4.3 to position B. In a pulse, after some time spent at a higher voltage, the input voltage drops (back) to zero instantaneously. In the figure, $t = 0$ is denoted as $t = t_{S1}$ to signify the first of two switching events. The step response of Eq. 4.7 is shown after the rising edge of the step ($t = 0 = t_{S1}$) in Fig. 4.5(a); $v_C(t)$ will rise toward, and asymptotically approach, V_S. The rate of rise is characterized by $\tau = RC$ (Greek letter tau), the **time constant** of the RC combination. τ is important because it tells you a lot about how the circuit charges and discharges. Because current has dimensions of charge/time, or C/s, *time is built into electrical circuits*. τ is that amount of time that characterizes how quickly things change in a circuit. Note that $\tau = RC$ has dimensions of $(V/I) \times (Q/V) = \text{volts}/(\text{coul/s}) \times \text{coul/volt} = \text{s}$; in other words, $1\ \Omega \cdot F = 1$ s, i.e., one ohm-farad is 1 s.

Figure 4.5(a) shows that after inserting $t = \tau$, or one time constant's worth of time, into Eq. 4.7, $v_C(t)$ rises to $V_S(1 - 1/e) = 0.63\ V_S$. After 2τ, $v_C(t)$ increases to $V_S(1 - 1/e^2) = 0.865\ V_S$, and $0.95\ V_S$ after 3τ, etc. For example, for a $V_S = 5$ V step, if $R = 250\ \Omega$ and $C = 1\ \mu F$, it will take $t = \tau = 0.25$ ms for the capacitor to reach a voltage of $0.63\ V_S = 3.15$ V.

So far, we have ignored the voltage across the resistor, $v_R(t)$. Due to Kirchhoff's voltage law, for $t > 0$,

$$v_{in}(t) = v_c(t) + v_R(t) = V_s$$

or
$$v_R(t) = V_s - v_c(t) \tag{4.9}$$

which is also shown in Fig. 4.5(a). Eq. 4.9 can be stated as, "Whatever part of the input voltage that the capacitor takes causes less to be available for the resistor to take." So, when $v_C(t = 0^+) = 0$, it must be that $v_R(t = 0^+) = 5$ V. The full-time dependence of $v_R(t)$ can be found by plugging Eq. 4.7 into 4.9 to get

$$v_R(t) = V_s - v_c(t) = V_s - V_s\left(1 - e^{-\frac{t}{\tau}}\right) = V_s e^{-t/\tau} \tag{4.10}$$

At $t = 0^+ = t_{S1}$, the voltage across the resistor is the full source voltage, and then drops exponentially as $v_C(t)$ rises, as shown in Fig. 4.5(a). When $t \gg \tau$, i.e., after several time constants have passed, $v_C(t) = V_S$ (Eq. 4.7), which is just as if the capacitor were an open circuit – all of the input voltage is dropped across the capacitor; at the same time, the current is always $i = v/R$, so the current has dropped to 0 as well. We can use Eq. 1.25, $i_C(t) = C \dfrac{dv_c(t)}{dt}$, to make exactly the same point: when $v_C(t)$ is flat, as it is after it saturates at $v_C(t) = V_S$, then $\dfrac{dv_c(t)}{dt} = 0$, and therefore $i_C(t) = 0$ and $v_R(t) = 0$, which again, is due to the capacitor acting as an open circuit.

All of the above can be restated as "**A capacitor is an open circuit to DC.**" As far as the capacitor is concerned, until the end of the step happens at $t = t_{S2}$ in Fig. 4.5(a), the input $v_{in}(t) = V_S u(t)$ is simply a DC voltage. After waiting long enough for the transient charging phase to be over, i.e., waiting for a time that is several τ long, the capacitor responds to the DC voltage as if it were an open circuit.

After the end of the voltage input pulse, charge flows off the capacitor plates, and the voltage across the capacitor drops asymptotically toward zero as

$$v_c(t) = v_c(t = t_{S2}{}^+)e^{-(t-t_{S2})/\tau} \tag{4.11}$$

where $v_C(t = t_{S2}{}^+)$ is the voltage across the capacitor just after the input drops to zero. $v_C(t = t_{S2}{}^+)$ is the initial condition for $v_C(t)$ for $t > t_{S2}$. As drawn in Fig. 4.5(a), τ is short enough, and the pulse is long enough that $v_C(t)$ has saturated at its steady-state value of V_S by the time that the $v_{in}(t)$ drops. In this case, the initial condition for the decay phase of the pulse is $v_C(t = t_{S2}{}^+) = V_S$.

It is useful to think about the passive sign convention, PSC (Section 3.5.2.1), during charging and discharging. While the capacitor is charging, positive current flows into the positive terminal, so power and energy are flowing into the capacitor, and the energy is stored in the electric field. It is the same sort of thing as when a battery charges – positive current flows into the positive terminal, and the energy is stored in chemical bonds to be used later.

During the discharge cycle, for $t > t_{S2}$, the polarity of the voltage on the capacitor does not change because before and after t_{S2} the voltage continuity principle still holds, and $v_C(t = t_{S2}{}^+) = v_C(t = t_{S2}{}^-)$; $v_C(t)$ was positive just before the pulse drop and it remains positive just after. Now, according to the PSC, the capacitor, like a battery, is a source of power, i.e., it is an EMF; as the capacitor

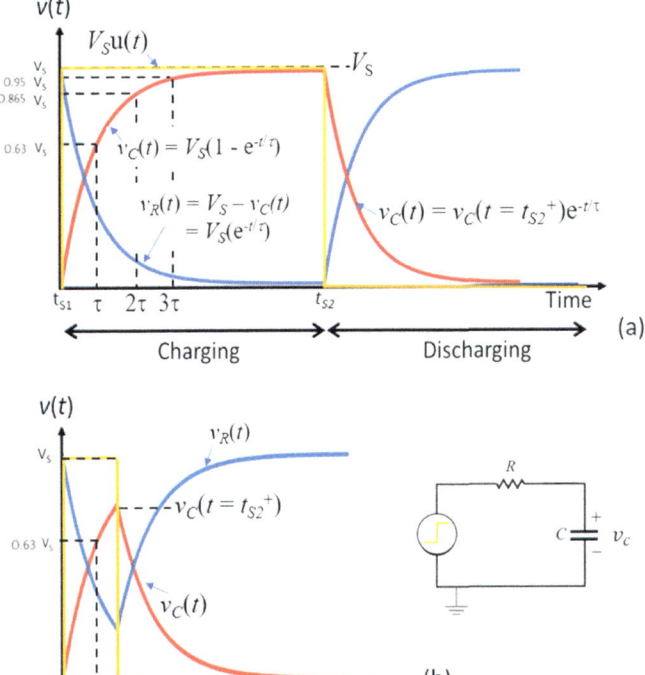

FIGURE 4.5 (a) Voltage across the capacitor of an RC circuit in response to a long voltage pulse. The capacitor voltage rises as $v_C(t) = V_S(1 - e^{-t/\tau})$ for several time constants τ, during which it reaches its steady-state value of V_S and saturates. When the pulse drops, the capacitor voltage drops as $v_C(t) = V_S e^{-t/\tau}$. The voltage drop across the resistor is whatever is left over from V_S that is not dropped across the capacitor, so $v_R(t) = V_S - v_C(t)$. (b) Same as (a) except that the pulse is less than 2τ long. When the pulse drops to 0, $v_C(t) = v_C(t = t_{S2}{}^+)$, and $v_C(t)$ begins to drop exponentially toward $v_C(t) = 0$.

discharges, positive current flows *out of* (current is negative at) the positive terminal, and the capacitor is a power *source*. This charging and discharging describes how supercapacitors are used in electric cars for regenerative braking (See sidebar in Section 1.4.3).

4.2.2 Step Responses due to Multiple Switching Events

Figure 4.5(b) shows a different and important situation in which the second switching event at t_{S2} is on the order of the length of τ. The first switching event is the application of $v_{in}(t)$ at $t = t_{S1} = 0$ and the second switching event is the drop in voltage at $t = t_{S2}$. In Fig. 4.5(b), t_{S2} is not much longer than τ; $v_C(t)$ has not been given a chance to reach its steady-state value, V_S, before the input abruptly changes again. At $t = t_{S2}^+$, the initial condition for the decay phase of the pulse, $v_C(t = t_{S2}^+)$, is less than V_S. In order to find $v_C(t)$ or $v_R(t)$ for $t > t_{S2}$, the differential Eq. 4.5 must be solved for different initial conditions for which $v_C(t = t_{S2}^+)$ depends on t_{S2}.

Figure 4.6 shows $v_C(t)$ in response to repeated switching with three different

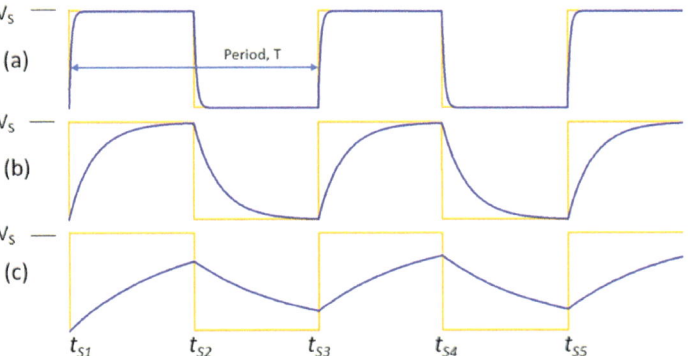

FIGURE 4.6 Evolution of the capacitor voltage waveform with time. The gold curve is the square wave and blue curve is the voltage on the capacitor. (a) For τ short compared to the square-wave period, T, $v_C(t)$ saturates at V_S each half cycle and looks somewhat like a square wave. (b) For τ on the order of T, as shown here, $v_C(t)$ barely saturates, and the waveform is a very distorted square wave. (c) For τ long compared to T, $v_C(t)$ does not have a chance to increase to V_S before discharging. As a result, the voltage across the capacitor begins to resemble a triangle wave. This plot was made using LTspice.

time constants, as compared to the switching period, T. This switching is equivalent to a square-wave input voltage. In Fig. 4.6(a), τ is short relative to the period, T, so $v_C(t)$ has sufficient time to saturate before $v_{in}(t)$ returns to 0. In this case, $v_C(t)$ still resembles a square wave. In b), τ is a larger fraction of T, and $v_C(t)$ barely saturates. Now, $v_C(t)$ is becoming a noticeably distorted version of a square wave. Finally, in c), τ is comparable in length to T, and $v_C(t)$ does not have enough time to rise all the way to V_S. Now the shape of the $v_C(t)$ waveform is beginning to resemble a triangle wave. As you can see, for small R or C (short τ), voltage waveforms can follow input voltage waveforms, but if either R or C is too large (long τ), voltage waveforms in the circuit can become distorted.

Unlike a square-wave, the switch in a circuit could change positions several times, having multiple different values of V_S and t_S, after each of which a new initial condition exists. Also, the circuit could have more than one physical switch that all change at the same or different times. Any combination of switching events is possible. The initial conditions for subsequent switching events at any $t = t_S$ are the voltage (or any circuit variable) at $t = t_{S2}^+$. In particular, every new initial condition for *capacitor voltage* $v_C(t = t_{S2}^+)$ just at the start of the next stage of the time evolution is the last value just before, which is $v_C(t = t_S^-)$ (the voltage continuity principle). In general, we will call the initial condition of *whatever variable that we wish to solve for* just after every switching event "$x(t_{S2}^+)$."

We also have to know after any switching event what the eventual, or steady-state, DC value would be if we wait a really long time, *whether or not we actually do wait that long* before a subsequent switching event. Let's call that value $x(t = \infty) = x_{SS}$. Now that we have a vocabulary for the times and values just after a switching event and after a long time, we can write the general solution to Eq. 4.6 for all cases as

$$x(t > t_s) = x_{SS} + [x(t_s^+) - x_{SS}]e^{-(t-t_s)/\tau} \tag{4.12}$$

which is the most general form of the step response. It accounts for any voltage or current in the circuit, *as long as it is a first-order system.* "Any voltage or current" can be that of either a resistor, capacitor, or inductor, for any magnitude and direction (up or down) of the switching event, after any length of time, as many times as you want to repeat the process. It indeed encompasses every condition that you can think of. Note that the quantity "$t - t_S$" is the amount of time that elapses after the switching event happens at $t = t_S$. Also, you must evaluate the value $x(t_S{}^+)$ just after the switching event. That is the initial condition for the next phase of the circuit evolution.

You may apply Eq. 4.12 to the single pulse in Fig. 4.5(b), where $v_C(t)$ doesn't have a chance to saturate after the first switching event, which happens at $t_S = 0$. In that case after the first step up at $t = 0$, $x(t_{S2}{}^+) = v_C(t = 0^+) = 0$. In order to apply Eq. 4.12 we need x_{SS1}. After many time constants would have passed, $v_C(t)$ would saturate at V_S, so $x_{SS1} = v_C(t = \infty) = V_S$. For the particular τ of the problem, we substitute $x(t_{S2}{}^+) = 0$ and $x_{SS1} = V_S$ into Eq. 4.12, and obtain

$$v_c(t)=V_s +[0-V_s]e^{-(t-0)/\tau} =V_s(1-e^{-t/\tau})$$

(4.13)

which happens to be Eq. 4.7 since it starts off with the same initial condition. Now the pulse drops at $t = t_{S2}$. The new initial condition uses Eq. 4.13 to obtain $x(t_{S2}{}^+) = v_C(t = t_{S2}{}^+)$. Finding $X_{SS2} = v_C(t = \infty)$ is straightforward – the capacitor eventually discharges to 0, so $X_{SS2} = 0$ and Eq. 4.12 becomes

$$v_c(t>t_{s2})=v_c(t=t_{s2}{}^+)e^{-(t-t_{s2}{}^+)/\tau}$$

(4.14)

This is the same as Eq. 4.11. Referring back to Fig. 4.6(c), you would have to apply Eq. 4.12 for every $t = t_S$ in order to find the behavior for every half-period of the square wave. Fortunately, there are circuit simulators for doing that. The figure was produced using LTspice, available at no cost from https://www.analog.com/en/design-center/design-tools-and-calculators/ltspice-simulator.html.

As an example of applying Eq. 4.12 to an RC circuit, consider that of Fig. 4.7. $R_1 = 1$ kΩ, $R_2 = 1.5$ kΩ, $R_3 = 2$ kΩ, and $C = 10$ μF. Switches 1 and 2 are at positions B and D for a long time before changing at $t_{S1} = 0$. Switch 1 moves from B to A, inserting V_1 into the circuit; simultaneously Switch 2 moves from D to E, removing V_2 from the circuit. For the second part of this example, Switch 1 moves back to B at $t_{S2} = 10$ ms. We want to know $v_C(t)$ for $0 < t < t_{S2} = 10$ ms and then $t > t_{S2}$. The solution is broken into two distinct procedures that are linked by $v_C(t = t_{S2} = 10$ ms$)$.

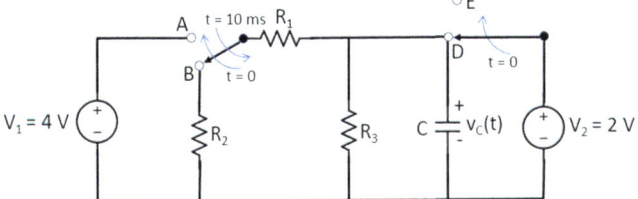

FIGURE 4.7 Example of multiple switching events in an RC circuit. Switch 1 moves from B to A at $t = 0$, and Switch 2 moves from D to E, also at $t = 0$. Switch 1 moves back to B at $t = 10$ ms.

Recasting Eq. 4.12 in terms of the capacitor voltage,

$$v_C(t)=v_{CSS1} +[v_C(t_{s1}{}^+)-v_{CSS1}]e^{-(t-t_{s1})/\tau}$$

(4.15)

We need to populate the general equation Eq. 4.12 with values for all of the variables. What is $v_C(t_{S1}{}^+)$? For the first switching event, $t_{S1} = 0$. First, we will set up the initial condition for $t < 0$. Since V_2 is in parallel with the capacitor, the capacitor has been fully charged after waiting a very long time and, due to the continuous voltage principle, $v_C(t_{S1}{}^+) = v_C(t = 0^-) = 2$ V.

Now we find $v_C(t = \infty) = v_{CSS1}$. After $t = 0$ when both switches have simultaneously changed positions, V_2 is removed from the circuit, and $V_1 = 4$ V is across R_1 in series with the parallel combination of R_3 and C. After waiting a really long time, the capacitor charges completely and acts like an open circuit, and $v_C(t = \infty) = V_{R3}$. With the capacitor basically out of the circuit, V_{R3} is just a voltage divider with R_1, and

$$v_{CSS1} = v_C(t = \infty) = V_{R3}(t = \infty) = 4 \times R_3 / (R_1 + R_3) = 8/3 \qquad (4.16)$$

Remember, we aren't actually going to wait that long in the problem, but we need to know what value the voltage is eventually heading toward, which determines its overall time dependence and shape.

All that remains to be found in order to use Eq. 4.15 is τ. You can see that when Switch 1 changes position, R_2 is removed from the circuit. This tells you right away that time constants $\tau = RC$ can change values with different switch positions. The general rule for obtaining τ is to find the equivalent resistance that participates in charging and discharging the capacitor. In order to do that, replace all voltage sources with short circuits, i.e., wires, and all current sources with open circuits. Current will flow the same way through the resistors whether charging or discharging the capacitor. In this case it is easy to see that if the capacitor discharges when V_1 is replaced by a wire, it goes through the parallel combination $R_{eq1} = R_1 || R_3 = 2/3$ kΩ. Hence, for $0 < t < 10$ ms, $\tau_1 = R_{eq1}C = 2/3$ kΩ × 10 μF = 6.7 ms. Now that we have $v_C(t_{S1}^+)$, v_{CSS1}, and τ_1, you may plug any time you want into the exponential term of Eq. 4.15 to obtain $v_C(t)$ from

$$\begin{aligned} v_C(t) &= \frac{8}{3} + \left[2 - \frac{8}{3} \right] e^{-(t-0)/6.7\,\text{ms}} \\ &= \frac{8}{3} - \frac{2}{3} e^{-t/6.7\,\text{ms}} \qquad\qquad (4.17) \\ &= 2.67 - 0.67 e^{-t/6.7\,\text{ms}}. \end{aligned}$$

Finding $v_C(t)$ first gives you the initial condition based on the continuous voltage principle, but if you know the initial and final states of any other variable, you may apply Eq. 4.12 with equal success. Again, Eq. 4.12 applies to any variable in a first-order circuit.

Now we solve for the circuit behavior for $t > t_{S2} = 10$ ms. The steps are the same, with some changes in the particulars. We plug $t = t_{S2} = 10$ ms into Eq. 4.17 to get the initial condition $v_C(t_{S2}^+) = v_C(t_{S2}^-)$, and we obtain $v_C(t_{S2}^+) = 2.52$ V.

Next we need the steady-state value for $v_{CSS2} = v_C(t = \infty)$. With Switch 1 at B, after a very long time the capacitor has completely discharged and $v_{CSS2} = v_C(t = \infty) = 0$. (If only every calculation could be that simple.)

Now we must find the new value of τ, which we call τ_2. Looking back from the capacitor, R_2 is back in the circuit and plays a role in how current flows while discharging. Now, $R_{eq2} = (R_1 + R_2)$ $|| (R_3) = (1$ kΩ $+ 1.5$ kΩ$) || (2$ kΩ$) = 1.11$ kΩ, which is different from R_{eq1} due to the new position of Switch 1. So, $\tau_2 = R_{eq2}C = 11.1$ ms. Again, you may plug all those values into Eq. 4.15 to obtain the time dependence $v_C(t > 10$ ms$)$. Let's randomly choose to find $v_C(t = 15$ ms$)$. Equation 4.15 becomes

$$v_C(t > t_{S2}) = v_{C_{SS2}} + [v_C(t_{S2}^+) - v_{C_{SS2}}] e^{-(t-t_{S2})/\tau_2} \qquad (4.18)$$

$$= 0 + [2.52 - 0] e^{-\frac{15\,\text{ms} - 10\,\text{ms}}{11.1\,\text{ms}}} = 1.60\,\text{V}$$

Using Eq. 4.18 you may find $v_C(t)$ for any time $t > 10$ ms or using the general Eq. 4.12 the value for any variable using the appropriate initial and final values, and τ_2.

Here is a summary of the steps used to solve step-response problems like the one above:

1. Cast Eq. 4.12 in the form of the variable whose time dependence you wish to know. For now, the circuit contains a capacitor, but the same steps will apply to circuits having an inductor.

2. Use your vast knowledge of circuit theory using whatever means possible to find the initial condition of the variable, $x(t_S^+)$, immediately *after* the switch. Warning: the continuous voltage principle applies only to the voltage across a capacitor.

3. Imagine that you wait forever: find the steady-state value of the variable at $t = \infty$, x_{SS}.

4. Apply the rules for replacing voltage and current sources and find the equivalent resistance R_{eq} that is in series with the capacitor. That will be some series and parallel combinations of the relevant resistors in the circuit. Not every resistor might play a part in the charging and discharging of the capacitor.

5. Plug all of those values into Eq. 4.12 and solve it for the time of interest.

6. If there is a second (or more) switching event, solve the equation in step 5 at t_{S2}. Use that as your initial condition, $x(t_{S2}^+)$ going forward.

7. Think through what the voltage that the capacitor will be charged to (or the current through the inductor) after a very, very long time, and use that for X_{SS2} in Eq. 4.12.

8. Find a new τ_2 based on the new positions of the switches.

9. Solve the new Eq. 4.12 for any time, using $t - t_{S2}$ in the exponential, that you want after the second switching event.

10. Repeat for every switching event thereafter.

4.2.3 Step Response and Time Constants for RL Circuits

Now, what about inductors? Let's replace the capacitor with an inductor, as shown in Fig. 4.8. What is the equivalent of charging a capacitor? It is the creation of magnetic flux due to current through the inductor. As a review, we saw that the energy in a capacitor

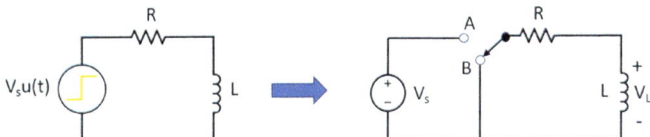

FIGURE 4.8 RL circuit with step-function input voltage realized using a switch. The switch moves from B ($v_{in}(t) = 0$) to A ($v_{in}(t) = V_s$) at $t = 0$.

was due to the electric field between the plates, which is due to the charge on the plates, which is proportional to the voltage across the plates. Therefore, we saw that the voltage across the plates cannot change discontinuously. In the case of the inductor, energy in the coil is stored in the magnetic field, which is due to the current through the inductor (Ampere's Law). We are, therefore, concerned with the current through the inductor, as we were concerned previously with the voltage across the capacitor. Current cannot change instantaneously between two values for the same reasons that the capacitor charge and voltage cannot, i.e., flux is stored energy. An instantaneous change in magnetic flux and the attending current would require infinite power, if even for a short time. Thus, whatever the current is at $t = 0^-$, it must be the same at $t = 0^+$. Therefore we have a similar principle to that of capacitors, the **current continuity principle of inductors**, namely that $i_L(t = 0^+) = i_L(t = 0^-)$.

Consider again the LR circuit shown in Fig. 4.8. Again, the step-input voltage can be achieved using a switch from B to A. We would like to know the time dependence of the current through the inductor, $i_L(t)$. Recall that the voltage across and inductor

$$v_L(t) = L \frac{di_L(t)}{dt} \tag{1.30}$$

is due to the *change* in current, and not its instantaneous value; in other words, no matter how large the current is, it must be changing in order for a voltage to appear.

We assume that the switch has been at position B for a very long time, the inductor is discharged, and no current is flowing in the RL circuit, $i_L(t = 0^-) = 0$. Because $i_L(t = 0^+) = i_L(t = 0^-)$, $i_L(t = 0^+) = 0$ just after the switch flips from B to A. Now, invoking Ohm's Law $v = iR$, $v_R(t = 0^+) = i_L(t = 0^+) \cdot R = 0$ V, so it must also be that $v_L(t = 0^+) = V_S$, i.e., all of the source voltage appears across the inductor and no current is (yet) flowing. In words, *the instant that the voltage input changes, all of the source voltage appears across the inductor and no current is yet flowing*. This is shown in Fig. 4.9 at $t = 0$.

Using the characteristic equation Eq. 1.30 for an ideal inductor $v_L(t) = L\dfrac{di_L(t)}{dt}$ as we did with the capacitor, we use Kirchhoff's current law to derive a differential equation, and then solve it with the initial condition to yield the time dependence of the current through the inductor as

$$i_L(t) = \frac{V_s}{R}(1 - e^{-t/\tau}) \tag{4.19}$$

where, for an RL circuit,

$$\tau = \frac{L}{R} \tag{4.20}$$

Also,

$$v_R(t) = i_L(t)R = V_s(1 - e^{-t/\tau}) \tag{4.21}$$

and

$$v_L(t) = V_s - v_R(t) = V_s e^{-t/\tau} \tag{4.22}$$

Equations 4.19, 4.21 and 4.22 are shown in Fig. 4.9. Plugging $t = 0$ into Eqs. 4.19, 4.21 and 4.22 yields $i_L(t = 0^+) = 0$, $v_L(t = 0^+) = V_S$, and $v_R(t = 0^+) = 0$, as expected.

Figure 4.9 shows $i_L(t)$, $v_L(t)$ and $v_R(t)$ due to a square-wave input with τ much shorter than the period, T. The behavior might be a bit less intuitive than that of the capacitor. First, the inductor voltage, $v_L(t = 0^+)$ jumps *instantaneously* to that of the input voltage. Whoa, doesn't that violate – well ... – *something*? No, it doesn't. The voltage across the inductor is not due to the increase in energy as it was with the capacitor. No power has to flow for the voltage to change, and no energy is transferred. The instantaneous

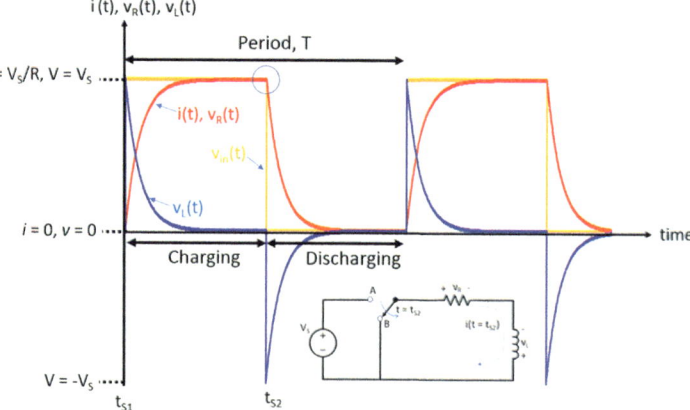

FIGURE 4.9 Step response of an RL circuit. The vertical axis represents both current and voltage. The current starts at zero at $t = t_{S1}$ and rises with time constant $\tau = L/R$ toward $x_{SS1} = i(t = \infty) = V_S/R$. After the switch at $t = t_{S2}$, $i_L(t)$ continues to flow in the same direction (indicated by the circle), but decays toward $x_{SS2} = 0$. At the same time, $v_L(t)$ flips from positive to negative according to the passive sign convention, since the current continues to flow in the same direction, but the inductor is now a power source.

change in $v_L(t)$ is, rather, due to electric fields from the source reaching the inductor, which happens very quickly. (For you cognoscenti, that is the same as saying that ideal wires have no capacitance or inductance. In reality, they do have a tiny bit, and there are very short delays due to the wires).

After a long time compared with τ, $i_L(t) = \frac{V_s}{R}(1 - e^{-t/\tau})$ approaches V_S/R, which implies that all of the input voltage will be dropped across the resistor and no voltage will be dropped across the inductor; the (ideal) inductor behaves just like a wire or short circuit. Briefly, **"An inductor is a short circuit to DC."** In a similar way as the capacitor in Fig. 4.5(a), as far as the inductor is concerned, until the end of the step happens at $t = t_{S2}$ in Fig. 4.9, the input $v_{in}(t) = V_S u(t)$ is simply a DC voltage. After waiting long enough for the transient charging phase to be over, i.e., waiting for a time that is several τ long, the inductor responds to the DC voltage as if it were a short circuit. We can use Eq. 1.30, $v_L(t) = L\frac{dI_L}{dt}$, to make exactly the same point: when $i_L(t)$ is flat, as it is after it saturates at $i_L(t) = V_S/R$, then $\frac{dI_L}{dt} = 0$, and therefore $v_L(t) = 0$.

Let's look in Fig. 4.9 at the decay phase of the $i_L(t)$ in response to the input voltage returning to zero. This happens when the switch in Fig. 4.8 returns to position B, which corresponds to $t = t_{S2}$ in Fig. 4.9. You can see that $i_L(t)$ starts off at its maximum value (indicated by the small circle) and starts to decrease when the pulse drops. Now, according to the current continuity principle of inductors, $i_L(t = t_{S2}^+) = i_L(t = t_{S2}^-)$.

This applies not only to the magnitude of the current, but also, the sign or direction of the current. This may seem obvious, but the implications are important. Refer to the inset circuit of Fig. 4.9. The current continues to flow in the same direction, but the circuit now comprises only the resistor and the inductor. The polarity of the voltages must be the same for both elements, which are effectively now in parallel with each other. Current flows in the same direction in the resistor, so its polarity is unchanged, but that *requires the polarity of $v_L(t)$ to change sign*! This polarity flip shown in Fig. 4.9 might be nonintuitive and take some time to get comfortable with.

We may also think of the polarity flip in terms of the passive sign convention as applied to the inductor. The current continues to flow in the same direction after the switch, but the polarity changes. Therefore, the inductor changes from a consumer of power to a source of power as the voltage across the inductor becomes a source of EMF (see the difference between voltage and EMF explanation in Section 1.4.4.1). The energy stored in the flux is used up as power is delivered to the resistor, which uses that power to create heat. In some device other than a resistor, it could be used to do other forms of work, but in this circuit, it is heat. If you recall the discussion in Chapter 3 about reactance, the energy stored in the inductive reactance was returned to the power lines for a power factor that was less than 1. The behavior here is the very same phenomenon, but for a step change rather than a steady-state sinusoidal voltage input.

There is yet another water analogy that is commonly used to describe the continuity principle of inductors. Figure 4.10 shows a water pump pumping water around a pipe, and that water drives a massive spinning flywheel. The pump is analogous to the power supply, water flow in the pipes is current in the wires, the viscosity of the water in the pipe is the resistance, and energy or momentum contained in the spinning flywheel is analogous to the magnetic flux. When the pump stops pumping it becomes like a plain piece of pipe (just like a voltage source of 0 volts acts like a wire). The flywheel's energy keeps it spinning, which continues to drive water around the water circuit in the same direction as it was already flowing. As the viscosity of the

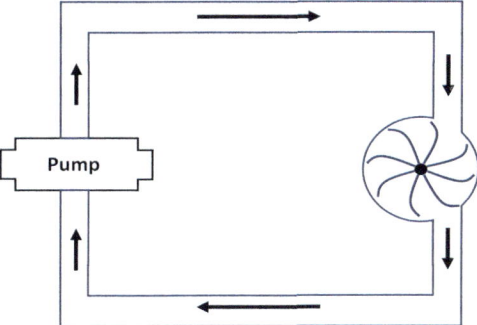

FIGURE 4.10 Water analogy of current flowing through an inductor. After the pump stops, the flywheel's momentum keeps the water flowing in the same direction.

water uses up the energy, the flywheel slows down and finally stops. This is the same as the behavior of the decay of current in the RL circuit.

Finding the step response of an RL circuit is just like that described for capacitors. You start out by writing Eq. 4.21 in terms of the inductor current as

$$i_L(t > t_s) = i_{L_{SS1}} + [i_L(t_{S1}^+) - i_{L_{SS1}}]e^{-(t-t_{S1})/\tau}. \tag{4.23}$$

Just as with the RC circuit, find the values for $i_L(t)$ at $t = t_{S1}^+$ and $t = \infty$ and τ, and apply those at every switching event. Again, Eq. 4.23 can be applied to the $v_L(t)$ or any other voltage or current behavior in the circuit.

One more thing about the time constant $\tau = L/R$. Note that 1 H/Ω = 1 s, i.e., 1 henry/ohm = 1 second. Suppose $L = 1$ µH and $R = 250$ Ω, which in magnitude are the same as the numbers we used for the capacitor. Now, $\tau = L/R = 4 \times 10^{-9}$ s = 4 ns. This is about 5 orders of magnitude shorter! This begs the question of why is it that for an RL circuit the time constant τ is *inversely proportional* to R ($\tau = L/R$) whereas it was *proportional* to R for a capacitor ($\tau = RC$)? Refer to the RC circuit in Fig. 4.7. R restricts current; a larger resistance slows the accumulation of charge on the capacitor plates, and therefore slows the increase in the voltage, and τ is larger. At long times, i.e., in **steady state**, the voltage on the capacitor approaches V_S.

What about the inductor? For a small series R, it takes longer to 'fill up' an inductor with a larger magnetic field. Here are the details: Refer to the RL circuit in Fig. 4.8. Let's take a rough approach to explain why lower R leads to a longer time constant. Inductors resist changes in current. Thus, the current starts at zero just after a voltage step. With zero current, $v_R = 0$ and the voltage is initially all across the inductor ($v_L(t = 0^+) = L[di(t)/dt]_{t=0}^+ = V_S$; $[di(t)/dt]_{t=0}^+ = V_S/L$). After a long time, the current increases to the steady state current, i.e., there is no more change in current and consequently no more voltage across the inductor. When that finally happens, the input voltage is all across the resistor, and the current after a long time is given by $i = V_S/R$; smaller R leads to larger final current. Physically, the magnetic field of the inductor is larger with larger current, so it takes longer to build up the larger field with smaller R, and τ is larger. Also, the final magnetic field is larger with larger L, so larger L also leads to longer rise times, and therefore a larger τ. Whew. Those are the details, but in short it takes longer to create the larger magnetic field for smaller resistance and larger inductance.

4.2.3.1   Frequency Dependence of Step Response

A hugely important part of nearly every area of electrical engineering is to analyze the behavior of an electrical system as a function of the frequency that it is operated at. This frequency dependence, or **frequency response**, shows up in virtually every treatment of just about everything you will do going forward. It begins as we ask how capacitors and inductors change their voltages and currents as the frequency of a sinusoidal input voltage changes.

Recall that a periodic waveform is characterized by the length of the period, *T*, or equivalently the rate at which it repeats, its frequency, where

$$f = 1/T \tag{2.16}$$

Figure 4.6 showed that for a given square-wave frequency, the time constant τ may or may not "fit" into the individual pulse widths. If τ is too long, then the current or voltage will not have a chance to completely rise and fall, as shown in Fig. 4.6(c). We extend that now to a similar situation, but instead of changing τ we keep τ constant and change the period, and hence the frequency, of the square wave. The result will have a similar feel to that of Fig. 4.6, but the implications of viewing the time dependence this way are enormous.

Figure 4.11 shows the rise and fall of the $v_C(t)$ of Figs. 4.3 and 4.5 as f increases (T decreases). The RC time constant of the circuit stays the same (the components don't change), but we vary the input frequency, just as you did with the square-wave generator in the Chapter 2 lab activities. At some frequency, Fig. 4.11(a), T is long enough relative to τ that $v_C(t)$ can easily saturate at V_S. As f increases, Fig. 4.11(b), T is closer to τ and $v_C(t)$ does not quite rise to the limit. As f further increases, Fig. 4.11(c), $v_C(t)$ rises a little bit, then falls a little bit, then rises a bit, etc. The amplitude of $v_C(t)$ as a periodic waveform is significantly decreased.

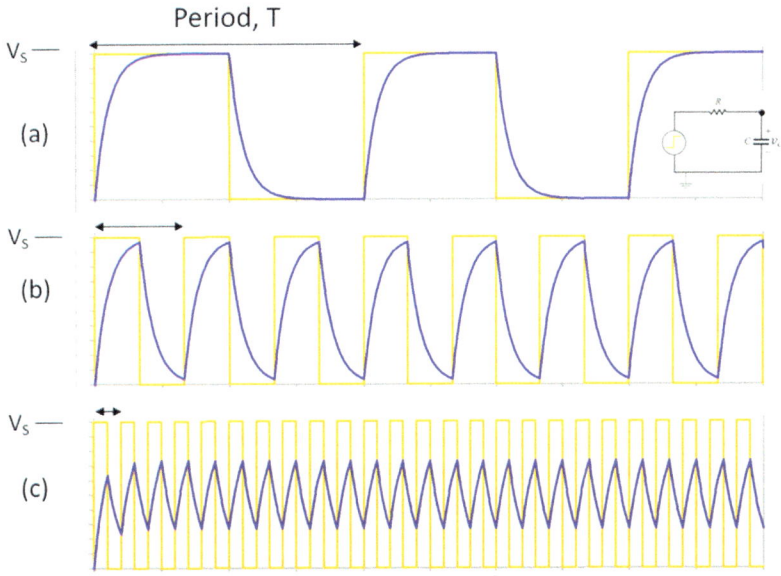

FIGURE 4.11 For the same $\tau = RC$ ($R = 1\,k\Omega$, $C = 1\,\mu f$), as the square-wave frequency increases from (a) to (c), the voltage across the capacitor (and current through the inductor) cannot keep up, and the amplitude decreases.

This is the essence of how an electrical filter works and is where all of the previous concepts about time constants and frequency come together. Let's extend the concept just presented from a square-wave excitation to a sine-wave excitation. You would be correct to imagine that the sharp edges will be smoothened, or even disappear. Figure 4.12 shows the same concept as Fig. 4.6, except that the square-wave input is now a sine wave. Keeping f constant and increasing τ from Figs. 4.12(a) to 4.12(c), you can see that the output is sinusoidal, and looks just like the input, so indeed, the sharp edges are gone. Also, the time behavior is the same – the response of $v_C(t)$ gets slower as τ increases, so $v_C(t)$ does not have time enough to rise and fall by very much, and its amplitude is decreased in (c) compared to (a).

Now, here is the big leap: Figure 4.13 shows $v_C(t)$ as a function of frequency for a given value of τ, but here it is in response to a sine-wave excitation rather than the square-wave excitation as in Fig. 4.11. Just as in Fig. 4.12, the response is a nice, clean sine-wave. Also, just as in Fig. 4.11, $v_C(t)$ cannot keep up as the frequency gets too high, and the amplitude decreases. Another way to say this is that the voltage across a capacitor decreases with frequency, or equivalently, the reactance of a capacitor decreases with frequency. Similarly, and not shown explicitly, the voltage across an inductor increases with frequency, i.e., the reactance of an inductor increases with frequency.

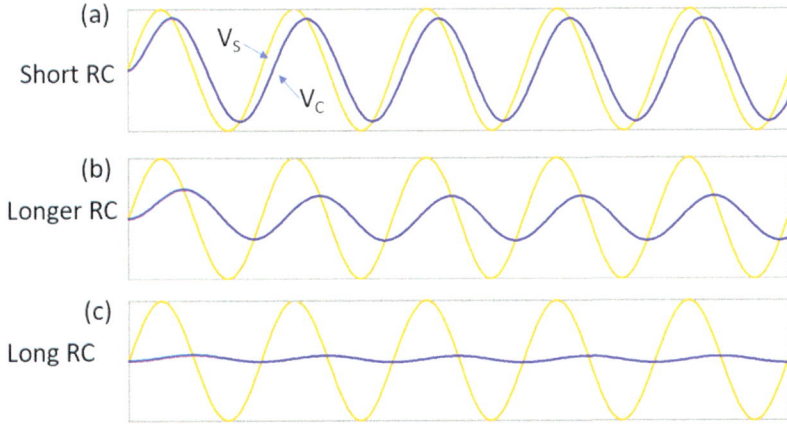

FIGURE 4.12 For the same frequency, as $\tau = RC$ increases from (a) to (c), the capacitor cannot charge up fast enough ($v_C(t) = Q(t)/C$) to follow the waveform. Therefore, the voltage on the capacitor cannot keep up. However, a sine wave input results in a sine-wave output. Not so with square wave.

It is critical to understand that the phenomenon is the same – τ limits the ability of the capacitor (and inductor) to charge and discharge within a certain amount of time relative to T. The big difference here is that when the input is a sinusoid, then the output is also a sinusoid, and there is no change to the actual shape. (The discerning student will notice that there is a phase shift that changes with frequency, which is due to the impedance discussed in Chapter 3 and again in Chapter 5). This completes

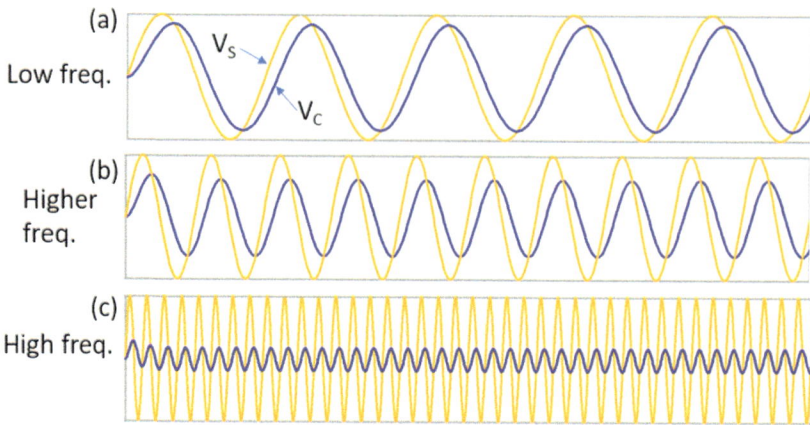

FIGURE 4.13 For the same $\tau = RC$, as frequency increases from (a) to (c), the capacitor cannot charge up fast enough to follow the waveform. Therefore, the voltage on the capacitor cannot keep up. However, a sine wave input results in a sine-wave output. Lower-frequency voltages 'appear' across the capacitor; higher frequencies do not.

the time-domain material of step-response of circuits, with the concept that underlying all of electrical filter behavior, to be discussed next, is the charging/discharging properties of the circuit due to a certain value of τ relative to $1/f$.

4.2.4 Frequency-Dependent Behavior of Reactive Components

Time has units of seconds (s) and frequency has units of cycles per second (Hz) (s^{-1}), and are thus inverses of each other. Time and frequency domains are intimately related. All of the step-response material covered in previous sections led us to Fig. 4.13: if a voltage or current cannot change very quickly (a time domain concept) then it cannot change repeatedly at a very high frequency (a frequency domain concept). Therefore, knowing the time-domain behavior of a circuit is equivalent to knowing its frequency-domain behavior. In this section, we discuss how voltages and currents in circuits with capacitors and inductors change with time, and how they obtain useful insights into their frequency domain behavior.

Filters are circuits that intentionally (or not) have some frequency dependence that is used to behave a certain way at certain frequencies. Circuits generally have lots of components, so it is useful to decide what voltage and what current at what component is the one we care about. That will be our **output**.

Consider the RC circuit that is a capacitor in series with a resistor, shown in Fig. 4.3. In order to understand its time-dependent behavior, we will begin by introducing the simple, but useful, notion of a circuit port. In common language, one use of the word "port" is an interface between two things, such as a port for boats (the interface between the ocean and land) or a place to plug in an electrical jack. In circuits, an **input port** is merely two nodes that connect the input source to the rest of the circuit. An **output port** is two nodes at which we consider the voltage or current to be of interest, to be measured or used to interface to some other circuit or component. In Fig. 4.3, the input port is between nodes A' and

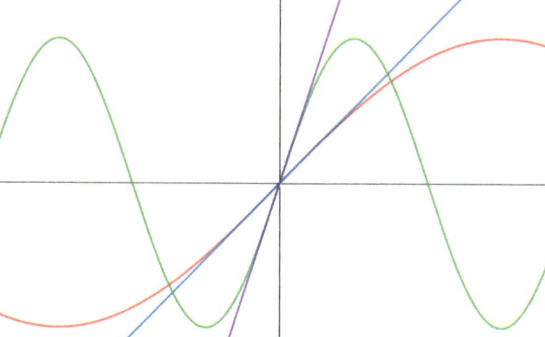

FIGURE 4.14 Graph of two sine waves having the same amplitude but different frequencies, and their tangent lines at $t = 0$. Note that the time derivative of the sine wave increases with frequency, i.e., the slope of the tangent increases with ω as shown in Eq. 4.24.

B′, where we have the input from the step-function voltage on the left, or the switch on the right. The voltage across any two nodes of our choice can be our output port. In Fig. 4.3, let's choose the nodes across the capacitor to be the output port, and thus call $v_C(t)$ our output voltage, $v_{out}(t)$.

Previous sections started with the time domain and motivated how τ = RC limits the ability of $v_C(t)$ to change quickly. Let's now try to understand the characteristic equation of a capacitor, $I = Cdv_C(t)/dt$, in terms of frequency. If the voltage applied to a capacitor is $v_C(t) = V_C\sin(\omega t)$ (such as the 60 Hz line voltage, or a sine wave from a waveform generator), then:

$$i_c(t) = Cdv_c(t)/dt = \omega CV_c\cos(\omega t) \tag{4.24}$$

which is proportional to ω, as illustrated in Fig. 4.14. This tells us that the amplitude of the current increases with frequency even when the amplitude of the voltage stays the same. Put another way, the voltage across the capacitor, V_C, must decrease with increasing frequency for the same current. Put yet another way, the reactance of the capacitor decreases as the frequency of the voltage across it increases, which was presented in Eq. 3.22.

Deeper understanding: The concepts presented in this chapter can be applied much more widely. Every electrical system has capacitance and inductance. In the case of transistors, discussed in Chapter 7, many of their individual components look like tiny capacitors, and the wires look like tiny inductors. Therefore, the analysis of modern transistor behavior must account for these tiny values at extremely high frequencies, up to hundreds of GHz. The small capacitances act like short circuits to the input, and the inductances act like open circuits, so the transistor no longer operates as-designed above some high frequency. This is critical to the behavior of computers, radar, communications systems, space systems, and much more.

So, why *physically* does the magnitude of the reactance of a capacitor, $|X_C| = 1/\omega C$, decrease with increasing frequency? Recall that an ideal voltage source, having no internal resistance, can produce whatever current the circuit asks of it. If an ideal AC sinusoidal source drives a capacitor directly, then the capacitor must be driven to the same voltage as the source *no matter what*. If the source voltage changes very rapidly, the capacitor must charge and discharge right along with that voltage, and to do so, the current must be very large to place a lot of charge on the capacitor in a very short time. Since lower reactance results in more current for a given voltage ($|V| = |I||Z| = |I|\sqrt{(R^2 + X^2)}$), this is consistent with $|X_C| = 1/\omega C$.

4.3 Electrical Filters

What is the general meaning of the word *filter*? A **filter** separates things based on some quality, like the size of rocks, pebbles, and sand. A sieve or metal screen lets the small particles (pebbles and sand) get through and blocks the larger stones. Likewise, we can make an electrical filter. Think of an electrical filter as a black box with a signal going in on one side (the input port) and a modified version coming out on the other side (the output port). The output is the filtered version of the input.

Many circuits are constructed in a sequence of stages in which different parts of the process happen; the load of each stage is the input to the next stage of electronics. Often, filters are one stage in the sequence of processing. In Fig. 4.15, V_{in} is input to

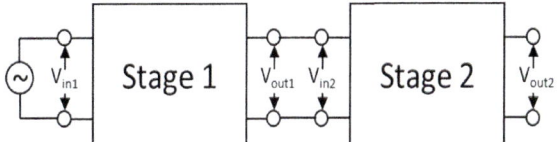

FIGURE 4.15 Representation of one stage of a circuit that passes its output to the input of a second stage of the circuit. The input port of Stage 1 is at V_{in1}, and the first-stage output port is at V_{out1}. V_{out1} passes directly to the input port of Stage 2, V_{in2}, whose output port is at V_{out2}.

Stage 1 resulting in an output from Stage 1 of V_{out1}. In turn, V_{out1} is passed as an input to, or **drives**, some other circuit **stage** (Stage 2 here), such as a stereo speaker (Chapter 5), a radio tuner (Chapter 6), music amplifier (Chapter 7), or rectifier stage of a power supply (Chapter 10), to name but a few. Even if you only observe V_{out1} with an oscilloscope, you must be aware that you are passing V_{out1} to the *input* of the oscilloscope, which can be thought of as the next stage of a larger circuit, which again, is the load for V_{out1}. So, as you saw in Chapter 2, the oscilloscope's input impedance could not generally be ignored because it became part of the overall circuit.

Figure 4.16 shows a black box representing a filter for electrical signals. In this diagram, the input port is on the left and the output port is on the right. Sinusoids of different frequencies will have different degrees of success at getting from the input to the output. The effect of the filter on low-frequency sinusoidal waveforms is shown at the top, where the output is the same amplitude as the input. The effect on high-frequency waveforms is shown at the bottom, where the output has lower amplitude than the input. Note that even though we are talking about the frequencies of sine waves, we are thinking of them as functions of time, so we are "in the time domain," and not frequency domain.

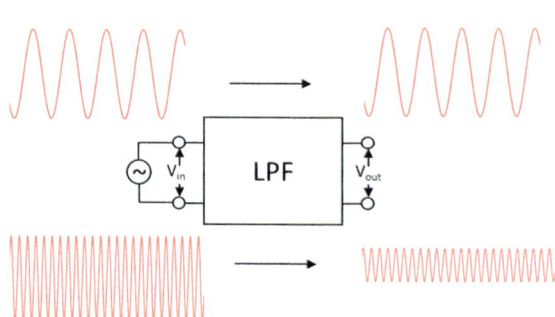

FIGURE 4.16 Representation of a filter that allows low frequencies to pass through but blocks high frequencies. It is presumed that V_{out} drives another circuit stage.

A signal is said to be **amplified** when its power or voltage amplitude is increased, and **attenuated** when its power or voltage amplitude is decreased. The ratio of the output to the input is called **gain**. In Fig. 4.16, for sinusoidal inputs of the same amplitude, the low-frequency signal makes it to the output port **unattenuated**, i.e., not decreased in amplitude, whereas the high-frequency signal is **attenuated**, i.e., decreased in amplitude, by the filter, and only a little of it shows up at the output port.

We consider only **passive filters**, which contain only passive elements (R, L, and C), and which do not have power gain greater than 1. That means that the filter does not transfer any more power to the output than it gets at the input. Furthermore, all of our filters for now are first-order circuits, and therefore are not able to increase the voltage. (You might ask, "What about transformers?" They can increase voltage, but they are passive because there is no *power* increase from input to output.)

All that said, what isn't a passive filter? **Active filters**, in which active components like transistors, are powered by an additional power supply and add complexity and functionality to the process of filtering. Active filters can amplify both voltage and power, and they can do so with various frequency dependencies. Transistors are presented in Chapter 7.

Think of the stages in Fig. 4.15 as subcircuits, any one of which might be a filter circuit like those to be discussed here. The function that describes the ratio of the output signal to the input signal of any stage of a circuit as a function of frequency is called the **frequency response** or the **transfer function**. A sieve separates sand and pebbles by the size of the holes, but what are the 'holes' for low or high frequencies in an electrical filter? The next section answers that question.

4.3.1   High-Pass and Low-Pass Filters

Figures 4.17 and 4.18 show the frequency response of a low-pass and high-pass filter, respectively. Electrical filters are described in terms of their frequency-dependent voltage gain, commonly called '$A_V(f)$,' or '$A_V(\omega)$.' When the frequency behavior of the filter is referred to as a "transfer function," it is most often called '$H(\omega)$.'

A **low-pass filter (LPF)**, such as the one whose frequency characteristics are shown by the solid line in Fig. 4.17, passes low frequencies to the output port and blocks high frequencies from getting

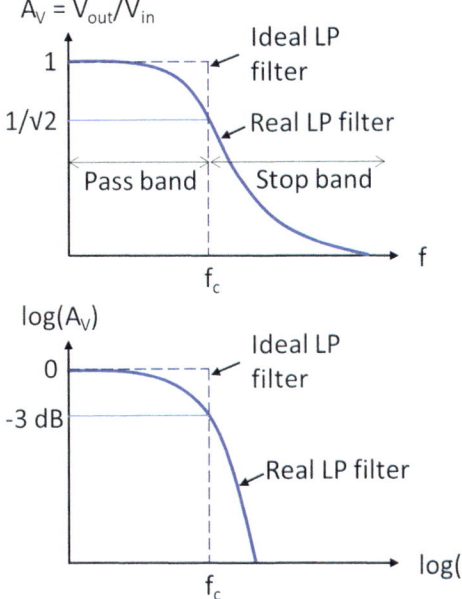

FIGURE 4.17 Graph of voltage gain of a low-pass filter as a function of frequency (i.e., its transfer function). Low frequencies are passed through while high frequencies are stopped. (top) Filter gain drawn on a linear-linear plot. (bottom) Gain drawn on a log-log plot.

through to the output port. The vertical axis plots the fraction of the input voltage that survives to the output port, and it is shown in the two common ways of making such plots, namely as a linear plot (top) and as a log plot (bottom). The frequency axis can also be plotted on either a linear or log scale. **A high-pass filter (HPF)**, such as the one shown in Fig. 4.18, passes high frequencies to the output and attenuates low frequencies.

An *ideal* low-pass filter, shown by the dashed lines in Fig. 4.17, would be very "sharp" – it would pass frequencies below some **cutoff frequency** f_c without causing any attenuation, and stop frequencies above f_c with perfect attenuation, i.e., their amplitudes are zero. An ideal *high-pass* filter, shown in Fig. 4.18, on the other hand ... oh, you can fill in the rest yourself. Ideal filters, like most other ideal electrical concepts, do not exist, but invoking them as approximations makes some ideas easier to understand.

Figure 4.17 also shows the frequency response of a real LPF (solid lines), which lets low-frequency signals through, but not the high-frequency, in varying amounts. Figure 4.18 shows the gain of a real HPF, which lets the high-frequencies through, but not the low-frequencies, again, in varying amounts. In practice, there are no ideal filters; in real filters, the transition from zero attention ($A = 1$) to full attenuation ($A = 0$) is always at least somewhat gradual.

When the input to the filter shown in Fig. 4.16 is a sinusoidal waveform at a frequency f, the gain, $A_V(f)$, is the ratio of the output voltage to the input voltage,

$$A_V(f) = \frac{V_{out}(f)}{V_{in}(f)} \qquad (4.25)$$

for that frequency. The collection of all values of $A_V(f)$ is a function of frequency and has some shape (as shown in Figs. 4.17 and 4.18 and many others after that); hence we write the description of the entire curve as $A_V(f)$ or $H(f)$. Note that you could express the input and output voltages as peak voltage, peak-to-peak amplitude, or RMS and you would get the same ratio. For a passive filter, the output voltage cannot be greater than the input voltage, so $V_{out}(f) \le V_{in}(f)$ and $A_V(f) \le 1$.

Following from Eq. 4.25 we see that

$$V_{out}(f) = A_V(f) \cdot V_{in}(f) \qquad (4.26)$$

At every frequency the output voltage is the input voltage times *the gain at that frequency*. In other words, the filter shapes the input frequencies by multiplying their magnitudes with the gain of the filter at every

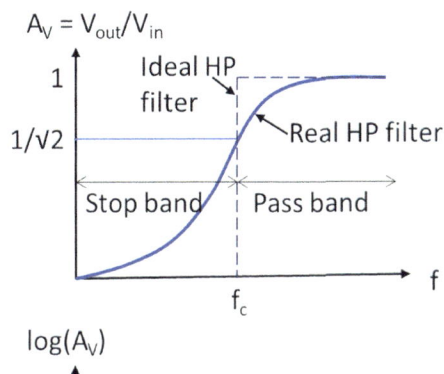

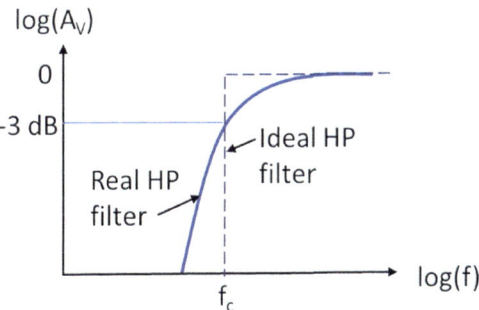

FIGURE 4.18 Graph of voltage gain of a high-pass filter as a function of frequency (i.e., its transfer function). High frequencies are passed through while low frequencies are stopped. (top) Filter gain drawn on a linear-linear plot. (bottom) Gain drawn on a log-log plot.

frequency. Both the signal, $V_{in}(f)$, and the system, $A_V(f)$ have a frequency dependence, and $V_{out}(f)$ is the product of those two dependencies.

Consider an electrical signal that is music. Using Fig. 4.17, top, as an example, very low frequencies (the low notes that are way to the left in the figure) would retain their original voltages $V_{in}(f)$ after passing through the filter since the gain is close to 1; very high frequencies (way to the right, or even off the plot to the right) would be totally attenuated even if their amplitudes were very high, since the gain is effectively zero. In between these extremes, the amplitude of the input at every frequency, whatever that might be, is multiplied by the gain of the filter at that frequency, which is between zero and one. This process yields a new amplitude that depends on how much of that frequency there is in the music and the amount that the filter attenuates that frequency. You already know this intuitively because this is what happens when you choose audio "equalizer" settings on your music player to give more or less bass, midrange, and treble frequencies. We will get much more into that later, especially in Chapter 5.

When the output of one filter or stage is input to a subsequent stage, as shown in Fig. 4.15, then the output voltage due to the combined stages is

$$V_{12}(f) = V_{in1}(f) \cdot A_{V1}(f) \cdot A_{V2}(f). \tag{4.27}$$

We can say this succinctly as

$$A_{V,tot}(f) = A_{V1}(f) \cdot A_{V2}(f) \text{ or}$$
$$H_{tot}(f) = H_1(f) \cdot H_2(f). \tag{4.28}$$

No matter what the purpose of a stage is, say as a filter, amplifier, or some other purpose, the combined effect of multiple stages is the product of the behavior of all of the stages. This is illustrated near the end of this chapter in the discussion on equalization.

A **band** is a contiguous range of frequencies. The frequencies for which a signal is passed relatively unattenuated are called the **pass band**, and the frequencies for which it is mostly attenuated are called the **stop band**. A common definition of the cutoff frequency, which separates the pass band from the stop band, is the **half-power point** at which $A_V = 1/\sqrt{2} = 0.707$, as shown in Figs. 4.17 and 4.18. (The astute reader will realize that power goes as voltage squared, so the "half-power point" corresponds to the frequency at which the voltage is $1/\sqrt{2}$ of its maximum value.) Because filters that closely approach ideal

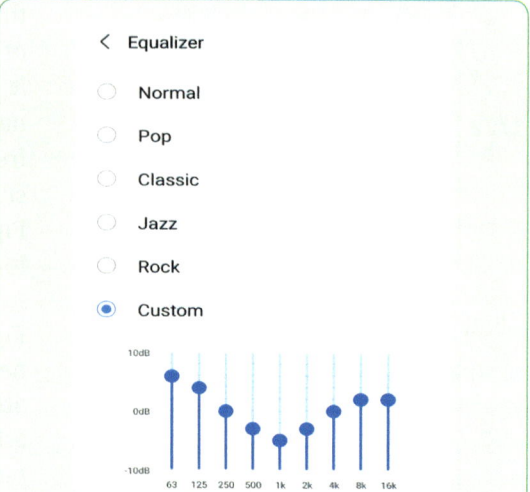

Concepts: You probably know how to intuitively use the equalizer functions on your music player. This setting is how I like to listen to music on my phone. Amplification is the same as multiplication by a certain gain, as discussed here. These settings amplify the lows and the highs and attenuate the midrange. This is usually done more-so to compensate for less-than-great earbuds than for an actual preference for this sound. Since poor earbuds don't reproduce lows very well, this amplifies them to make them more equal to the rest of the music, hence the term 'equalizer.' I like more highs likely because as I age I lose the ability to hear them.

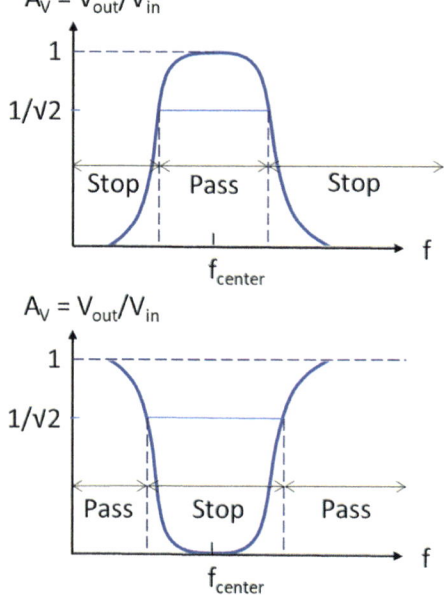

FIGURE 4.19 Frequency response of 'bandpass' filter (top) and 'notch' filter or 'band-stop filter' (bottom). The frequency in the middle is called the **center frequency**.

filters are often impractical, some slope of the pass and stop bands must be tolerated. Filter design is a field in itself, so you will learn only the basics in this chapter.

Passive filters can be constructed by combining inductors and/or capacitors to create a variety of filter behaviors, including **bandpass** and **notch** filters, whose frequency responses, or transfer functions, are shown in Fig. 4.19. Bandpass filters pass only a narrow range of frequencies and block the rest. Notch, or **bandstop**, filters pass all frequencies *except* for a narrow range. In the lab part of this chapter, you will use a **graphic equalizer** (in principle like the one on your phone as shown in the sidebar) that is a collection of 31 sophisticated filters configured in series, like the 2-stage filter of Fig. 4.15. Each of the equalizer filters is an adjustable bandpass or notch filter. Each filter can be set to either have no effect on the input, i.e., have a gain of 1 (be **flat**); amplify (boost) its narrow band of frequencies; or attenuate (cut) its band of frequencies. Outside its narrow band, the filter has no effect on the input, i.e., it has a gain of 1. The equalizer is discussed in more depth in Section 4.6.

4.3.2   Filters as Combinations of Frequency-Dependent Impedances

Passive filters are combinations of resistors, capacitors, and inductors, as noted above. Recall that the reactances of capacitors and inductors are frequency dependent (recall that $\omega = 2\pi f$):

$$X_L = \omega L \text{ and } X_C = -\frac{1}{\omega C} \tag{4.29}$$

(At the risk of some confusion, it is common to write '$-1/(\omega C)$' as '$-1/\omega C$,' as we will do going forward.) So, X_C *decreases* as frequency increases (goes as $1/f$) and X_L *increases* as frequency increases (goes as f). This is a clue that filters employing inductors and capacitors change their characteristics with frequency. This is, loosely speaking, where the 'holes' in the electrical filter come from.

A good way to understand how simple filters work is to think of them loosely as frequency-dependent voltage or current dividers. Just to be sure we are all on the same page, let's review voltage dividers and leave the review of current dividers to you. Refer to Fig. 4.20. If you have two resistors, R_A and R_B, in series, with a voltage V_{in} applied to the series combination, then the voltage across each resistor is its share of the total voltage: $V_A = V_{in} \times R_A /(R_A + R_B)$ and $V_B = V_{in} \times R_B/(R_A + R_B)$. The larger resistance grabs more of the applied voltage. In Chapter 2 you built a simple voltage divider and saw that the relationships held for both DC and AC signals. In fact, for resistors, the frequency that you used was completely arbitrary – the voltage divider would have worked the same way at any frequency because the circuit had no reactive elements.

For a deeper understanding: The word 'signal' implies that information is being conveyed. A sine wave does not convey much information by itself; however, a more complex signal, e.g., music, can be thought of as comprising a collection of sine waves of different frequencies, which is the subject of this chapter. So, I will use the term 'sine wave' and 'signal' interchangeably.

On a more physical level, the word 'signal' is intentionally vague. Is it a voltage? A current? Power flow? Is the filter placed in series or parallel with the next stage of the circuit? In fact, it could be any of these things. Often the 'signal' is a voltage that is dropped across either two nodes that are open, or across a load that is placed between those two nodes. If a voltage appears at just a node with no other circuit elements attached, then it is implied that another stage of the circuit will eventually go there as a load, as in Figs. 4.15 and 4.16 so that the voltage can be used for something. If Stage 1 of Figure 15 is a filter that acts like a large series resistance to an input at some frequency, then most of the input voltage will be dropped on that series resistance, and little or no voltage will be left to appear at the node. Therefore, that signal is 'blocked.'

Sometimes the signal is a current. Again, a large frequency-dependent resistance will block current. If the Stage 1 filter in Figure 15 is *a small* resistance in parallel with Stage 2, then, as a current divider, current will be shorted by the first stage and bypass the second stage. If it is *a large* parallel resistance, then current will be passed to the second stage.

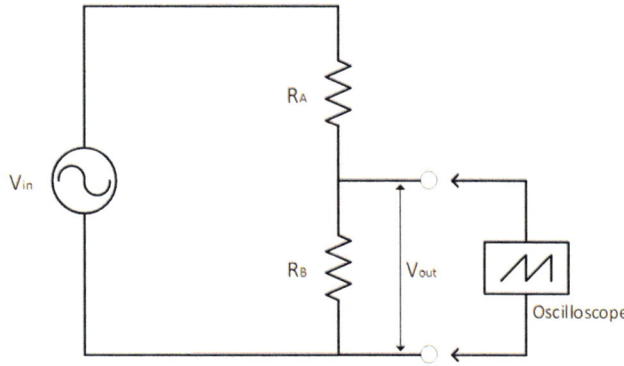

FIGURE 4.20 A simple voltage divider with an oscilloscope measuring the voltage across R_B.

The behavior of capacitors and inductors is not as simple as that of resistors. Whereas a resistor is described by a real number R, a capacitor or inductor is described by a reactance X. Any combination of resistors and reactances is described by an impedance $\mathbf{Z} = R + jX$ (Chapter 3). Because R and X are both real numbers, R is the real part and j·X is the imaginary part of the impedance. Again, an inductor has reactance $X_L = \omega L$ and a capacitor has reactance $X_C = -1/\omega C$. Recall also that the phase shift between voltage across, and current through, an impedance $\mathbf{Z}$ is $\phi = \arctan(X/R)$.

The 'j' is associated with the $\pi/2$ phase shift between the sinusoidal voltage and current in inductors and capacitors, as discussed below.

Impedances in series add just as resistors add in series:

$$\mathbf{Z}_{tot} = \mathbf{Z}_1 + \mathbf{Z}_2. \tag{4.30}$$

As with the addition of any complex numbers, the imaginary and real parts add together separately. Here are some examples. It is useful to recall that for resistances in series, $R_{tot} = R_1 + R_2$, and for resistances in parallel, $1/R_{tot} = 1/R_1 + 1/R_2$.

(a) The impedance of two inductances ($\mathbf{Z}_1 = j\omega L_1$, $\mathbf{Z}_2 = j\omega L_2$) in series yields $\mathbf{Z}_{tot} = \mathbf{Z}_1 + \mathbf{Z}_2 = j\omega(L_1 + L_2)$ $= j\omega L_{tot}$. Thus, we recover the formula for the addition of two inductances in series:

$$L_{tot} = L_1 + L_2. \tag{4.31}$$

Inductances add in *series* like resistances do in *series*.

(b) The impedance of two capacitances ($\mathbf{Z}_1 = -j/\omega C_1$, $\mathbf{Z}_2 = -j/\omega C_2$) in series yields $\mathbf{Z}_{tot} = \mathbf{Z}_1 + \mathbf{Z}_2 = -j/\omega C_1 - j/\omega C_2 = -(j/\omega)(1/C_1 + 1/C_2) = -(j/\omega)(1/C_{tot})$. Thus, we recover the formula for the addition of two capacitances in series:

$$\frac{1}{C_{tot}} = \frac{1}{C_1} + \frac{1}{C_2}. \tag{4.32}$$

Capacitances add in *series* like resistances do in *parallel*.

(c) Adding resistance R and inductive reactance ωL in series yields $\mathbf{Z}_{tot} = R + j\omega L$.

(d) Adding resistance R and capacitive reactance $-1/\omega C$ in series yields $\mathbf{Z}_{tot} = R - j/\omega C$.

(e) What about adding resistance R and reactances ωL and $-1/\omega C$? This is an important situation: $\mathbf{Z}_{tot} = R + j\omega L - j/\omega C$. Note that *it is possible for jωL and –j/ωC to cancel each other out!* This was the basis for increasing the power factor in Chapter 3. The condition where they cancel exactly is called **resonance** and leads to some very interesting behavior. Resonance is discussed near the end of this chapter and in greater depth in Chapter 5.

Figure 4.21 shows four examples of simple filter circuits. For each of the circuits we start by thinking of a single frequency. It will be useful to focus on just the magnitude of the reactance $|X|$ and compare that to the value of R in order to estimate how much voltage is dropped across the capacitor or inductor using the notion of a voltage divider. Note that $|X|$ is a positive number, and phase is ignored. $|X_L| = \omega L$ and $|X_C| = 1/\omega C$. We will analyze the behavior of these filters in terms of how large the reactances are compared to the resistances, and for now ignore the phase between the input and output voltages. This treatment will be rather hand-wavy, but will illustrate some key concepts. It will be refined to be more realistic and include phases across all the circuit components relative to the input voltage in Chapter 5.

First the hand-wavy stuff. As a consequence of Eq. 4.29, ideal capacitors have *high* reactance at *low* frequencies and *low* reactance *at high* frequencies. Ideal inductors behave the opposite way: inductors have *low* reactance at *low* frequencies and *high* reactance at *high* frequencies. Ideal resistors have no frequency dependence since they are purely real and have no imaginary term to their impedance. It is useful to consider this behavior in the limits of zero and infinitely high frequencies: at DC, i.e., 0 Hz, a capacitor is an open circuit, and inductor a short circuit (which was motivated in the previous discussion about step response after a long-time elapses). At very high frequency, the capacitor is a short circuit, and the inductor is an open circuit. *The sooner you memorize these basic properties the better.*

Here's the game: combine R, L and C components that have different frequency dependencies, and thus have different reactances, to create filters having different properties. Each of the four circuits in Fig. 4.21 is a simple low- or high-pass filter. Each is basically a voltage divider, where the output voltage, V_{out}, depends on the relative sizes of the resistance of the resistor and the frequency-dependent reactance of the capacitor or inductor, as in a voltage divider. Don't forget that V_{out} is wherever we define the output port to be. Each of the figures is drawn so that the output port is on the right. The output gets passed to the input of a subsequent stage, even if that stage is just an oscilloscope or digital multimeter. As another example, the filter could be the preamplifier of a stereo system, and its output could be the input to a power amplifier that drives the speakers. More on stereos in Chapter 5.

To understand the functionalities of the filters, we

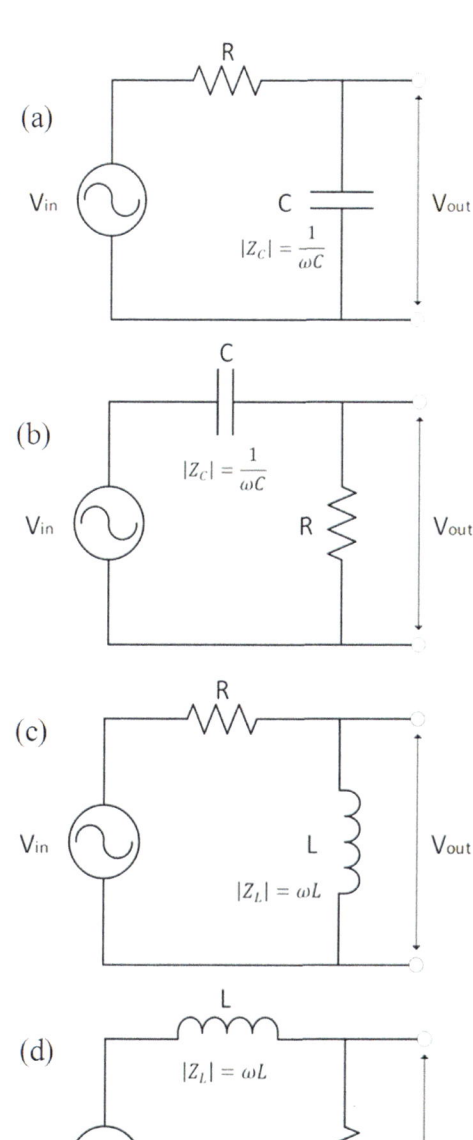

FIGURE 4.21 Four first-order filters that change the voltage at the output for different input frequencies. (a) LPF; (b) HPF; (c) HPF; (d) LPF.

investigate their behaviors at the limits of very low and very high frequency. Treating the "RC filter" of Fig. 4.21(a) as a voltage divider, the reactance of the capacitor, $|X_C|$, is much higher than R at sufficiently low frequency. Thinking of it as a voltage divider, most of the voltage is dropped across the capacitor, and V_{out} is high, i.e., essentially the same as the input voltage, so $V_{out} = V_{in}$, and $A_V = 1$. At very high frequency, $|X_C|$ is much lower than R, so V_{out} is low, essentially zero, so $A_V = 0$. Putting together the behavior at these limits tells us that the circuit of Fig. 4.21(a) is a low-pass filter, LPF. Similar logic explains the behavior of all of the circuits in Fig. 4.21.

Now the refined stuff. For both DC and AC current in a resistor, the relationship between current and voltage is a simple proportionality: $v = iR$. For a capacitor or inductor, the current and voltage are out of phase with one another, so that periodically the current is zero when the voltage is nonzero (so the resistance would seem to be infinite) or *vice versa* (for a resistance of zero). There is a way of recovering simple current–voltage relationships for nontrivial AC circuits. It involves, as you will have guessed from this and the preceding chapter, defining a complex-valued generalization of resistance, namely the impedance. It also involves defining complex-valued functions of time, called **phasors**, which are complex numbers representing voltage or current. Phasor representations of sinusoids is a powerful tool you will use in later EE courses and is the subject of Chapter 5.

Here we will give you a peek at what you can do with phasors (all complex quantities are represented in bold font). V and I are complex representations of the voltage and current waveforms and contain both magnitude and phase information. At this point, let's concern ourselves only with the magnitudes. For a pure sine wave, the phasor has a magnitude equal to the peak value of the wave: $|V| = V_p$, $|I| = I_p$. For power systems (Chapter 3), the magnitudes are usually assumed to be in RMS, i.e., $|V| = V_{RMS}$ and $|I| = I_{RMS}$. Whether you choose the peak or RMS values has no effect on mathematics. You just have to keep in mind which value you are using.

For a pure sinusoidal voltage and current acting on a complex impedance, the following resistor-like relationship of the magnitudes holds:

$$|V| = |I|\,|Z|. \tag{4.33}$$

Here $|Z|$ is the magnitude of the impedance (Z is not a phasor, but it *is* complex). For any Z of the form $R + jX$, $|Z| = \sqrt{R^2 + X^2}$ (see Eq. 3.24). We can use this to treat any of the filters in Fig. 4.21 using an AC version of the voltage divider relationship.

Suppose the output voltage of any filter in Fig. 4.21 drives a Stage 2 as in Fig. 4.15. Furthermore, suppose that Stage 2 has a very high **input impedance** – i.e., $|Z_2|$ is very large and draws very little current from Stage 1. Then, because no current leaks out of the Stage 1 filter, the two passive components of the filter share the same current. In other words, the current in the reactance and resistance for each of these simple filters is the same, i.e., $I_X = I_R = I$. Also, the total voltage across the series resistive and reactive components, with $Z_{tot} = R + jX$, is the same as the input voltage, $V_{in}(t)$, at every point in time, as must be the case due to Kirchhoff's voltage law.

In the "RC" and "RL" filters of Figs. 4.21(a) and (c), respectively, the as-drawn vertical components (the outputs) have reactances X and impedances jX. Using the relation of Eq. 4.33, we have $|V_{in}| = |I|$ $|Z_{tot}|$ and $|V_{out}| = |I||jX| = |I||X|$ (the j changes the phase, but not magnitude of reactances). Because the currents are the same, we can take the ratio of impedances to find the frequency-dependent transfer function:

$$H(f) = A_v = \frac{|V_{out}|}{|V_{in}|} = \frac{|I||jX|}{|I||Z_{tot}|} = \frac{|X|}{\sqrt{R^2 + X^2}}. \tag{4.34}$$

For the filter of 4.21(a), this works out to

$$A_v = 1/\sqrt{1 + \omega^2 R^2 C^2}. \tag{4.35}$$

(By now you are sophisticated enough not to care, or hardly even notice, if we use frequency f or radial frequency ω in our equations.) Plotting Eq. 4.35 as a function of frequency yields a plot similar the LPF plots shown in Fig. 4.17.

Swapping the positions of the resistor and reactive elements leads to a similar analysis for the RC filter of Fig. 4.21(b) and RL filter of Fig. 4.21(d) as

$$A_v = \frac{R}{\sqrt{R^2 + X^2}}. \tag{4.36}$$

This creates the opposite kind of filter, i.e., LPF instead of HPF and vice versa. For the RC filter of 4.21(b), this works out to the

$$A_V = 1 / \sqrt{1 + \frac{1}{\omega^2 R^2 C^2}}, \tag{4.37}$$

which then yields a plot similar to Fig. 4.18.

So how is the refined version of the two analyses different from the rough version? First, the rough version is only approximate except in the limits of very high and very low frequencies, when X is either very large or very small compared to R. Were X a true resistance, rather than reactance, the denominator in Eqs. 4.34 and 4.36 would be $R + X$, rather than $\sqrt{R^2 + X^2}$. The two versions give the same sort of behavior, but the correct value of the gain is obtained only by treating X as a reactance. The notion of the simple voltage divider using X as "sort of a resistance" is good only for identifying the type of filter. The refined version yields correct values of output voltage at every frequency. Also, even though we know there is a phase shift between input and output voltages, the gain A_V is only about the amplitudes of the waveforms – not their relative phases. The full phasor treatment in Chapter 5 provides more about the phase of V_{out}. Remember the warning in Section 3.3.1.1; when you see the word "phase," always ask, "The phase is between what and what?" Here, we mean the *phase is that of the time delay between the input and output sinusoids.*

The cutoff frequency $f_c = \omega_c/2\pi$ separating the pass- and stop-bands, as shown in Figs. 4.17 and 4.18, is defined to be where $A_V = 1/\sqrt{2}$. We can loosely regard frequencies above f_c as "high frequencies" and those below f_c as "low frequencies." This isn't very rigorous, but it gives us something to use when discussing the behavior of the filters. Using Eq. 4.35, you can derive that $\omega_c = 1/RC$. This is related to $\tau = RC$ as defined in Eq. 4.8:

$$\omega_c = \frac{1}{\tau} = \frac{1}{RC}. \tag{4.38}$$

So,

$$f_c = \frac{\omega_c}{2\pi} = \frac{1}{2\pi\tau} = \frac{1}{2\pi RC}, \tag{4.39}$$

where ω_c and f_c are the cutoff frequencies of the filters in Figs. 4.21(a) and 4.21(b). So, $\tau = 1/\omega; \tau \neq 1/f$! Keep that in mind!

Say $R = 1$ kΩ and $C = 1$ µF. Then, $RC = 0.001$ s, or $\tau = 1$ ms. τ is the time for the sinusoid to change by one radian, where, as you know, one period is 2π radians. The 1 ms time constant, τ, for this RC filter therefore corresponds to a cutoff frequency of 159 Hz, or 159 full cycles of 2π radians per cycle.

It is no coincidence that τ appears both here and in previous sections on step response. When the period $T = 1/f = 1/2\pi\omega$ becomes shorter than the time it takes for the capacitors in Figs. 4.21 (a) and (b) to charge, which is, roughly speaking, τ, then the voltage across the capacitor can no longer keep up. Since the output port in Fig. 4.21(a) is $v_C(t)$, the output voltage decreases as the input frequency increases, resulting in the LPF properties shown in Fig. 4.17.

The same basic ideas hold for a filter with R in series with an inductor of inductance L, as shown in Figs. 4.21(c) and 4.21(d). For an inductor, if $i_L(t) = I_0 \sin(\omega t)$ then

$$v_L(t) = L \, di_L(t)/dt = \omega L I_0 \cos(\omega t). \tag{4.40}$$

This equation tells us that the higher the frequency of the current (the higher the time derivative term) for the same amplitude of current, the higher the voltage is across the inductor. Put another way, the inductor looks increasingly 'AC-resistive' the higher the frequency of the current. Put yet another way, more voltage is needed to pass current at higher frequencies, which implies a larger reactance. The magnitude of the inductor's reactance is $|Z_L| = |X_L| = 2\pi fL$. This tells us that both higher frequency and higher inductance result in higher reactance.

Note that it is not the *value* of the instantaneous current that determines the voltage drop across the inductor, as it is in a resistor, but rather it is how *fast it changes*, i.e., the frequency, as illustrated in Fig. 4.14. The actual DC resistance of an inductor is that of the wire that is used to wind it, which is often only a few ohms (as you saw in Chapters 2 and 3). If you put a DC voltage across an inductor, it will behave like a resistance of typically a few to a few tens of ohms due to the windings. An *ideal* inductor is one having zero winding resistance. Even for an ideal inductor, which has zero DC resistance, you would need a high voltage across it to push even a small current that is oscillating at a high frequency. To the external circuit, then, the inductor at high frequencies exhibits a high impedance to the flow of current even as it presents a low DC resistance. This explains why an inductor in series is referred to as a **choke** when its purpose is to block high frequencies and let low frequencies through. (My how EEs use colorful language to describe technical ideas.)

Let's now consider a resistor in series with an (ideal) inductor, where the output port is across the inductor, as shown in Fig. 4.21(c). This configuration forms an HPF. At low frequencies the inductor has low reactance, and at high frequencies has high reactance. Again, you may roughly think "voltage divider": the voltage that appears at the output is low at low frequencies and high at high frequencies. For the filter of Fig. 4.21(c), doing the math works out to

$$A_v = 1/\sqrt{1 + \frac{R^2}{\omega^2 L^2}}. \tag{4.41}$$

Figure 4.21(d) is a LPF based on the same reasoning. For the RC filter of 4.21(d), this works out to

$$A_v = 1/\sqrt{1 + \frac{\omega^2 L^2}{R^2}}. \tag{4.42}$$

Plotting these as functions of frequency will reveal their natures as HPF and LPF, respectively. For an RL filter, $\tau = \frac{L}{R}$;

$$\omega_c = \frac{1}{\tau} = \frac{R}{L} \tag{4.43}$$

$$f_c = \frac{\omega_c}{2\pi} = \frac{1}{2\pi\tau} = \frac{R}{2\pi L} \tag{4.44}$$

where ω_c and f_c are the cutoff frequencies of the RL filters of Figs. 4.21(c) and 4.21(d).

4.3.3 Bandpass and Notch Filters

Figure 4.22 shows one of many possible ways to create a bandpass filter, BPF, introduced in Fig. 4.19. It is the combination of a low-pass and a high-pass filter configured in such a way that only a relatively narrow band of frequencies is allowed to pass to the output. Refer back to Eq. 4.27 for two filter stages. In Fig. 4.22, the first stage is a HPF, and the second stage is a LPF. The output of the first stage is input to the second stage. The HPF passes all frequencies greater than its cutoff frequency, which we call $f_{C\text{-HP}}$. Of all of those that reach the second stage, the subsequent LPF passes all of those frequencies that are less than its cutoff frequency, $f_{C\text{-LP}}$. All frequencies that fall between the two cutoff frequencies make it to the output. Based on Eq. 4.27,

$$V_{\text{out}}(f) = V_{\text{in}}(f) \cdot A_{V1}(f) \cdot A_{V2}(f). \quad (4.45)$$

Any form of a HPF and LPF will work, including those using inductors alone or combined with capacitors. Careful design is required, however, in order to avoid resonances, which is the topic of Section 4.4.

Regardless of the order of the filters, it is required that there be a range of frequencies that overlap both filters, which makes it to the output. For this, it is required that $f_{C\text{-HP}} < f_{C\text{-LP}}$. If instead, $f_{C\text{-LP}} < f_{C\text{-HP}}$, then there would be no frequencies that overlap, and no frequencies would get through.

Figure 4.23 shows one possible notch filter, also introduced in Fig. 4.19, in which there are two parallel paths for voltage frequency components to pass from input to output. In this example, the top path

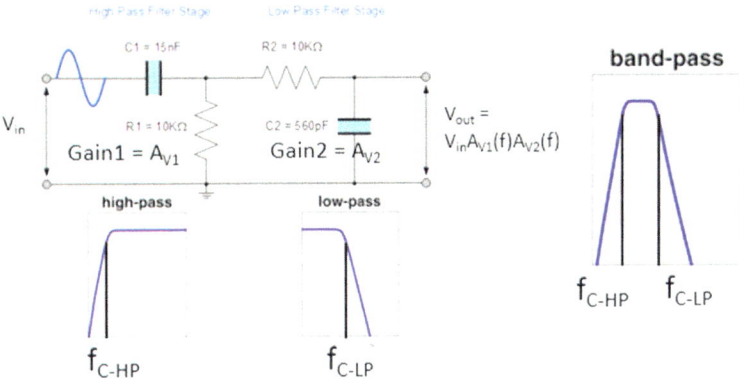

FIGURE 4.22 Circuit diagram of a bandpass filter using two first-order RC filters in series. The first stage is a HPF having gain $A_1(f)$ and the second stage is a LPF having gain of $A_2(f)$. The two filters combine so that only a narrow band passes through. Note that $f_{C\text{-HP}} < f_{C\text{-LP}}$ in order for this to operate correctly. If the cutoff frequencies are reversed, no frequencies will get through.

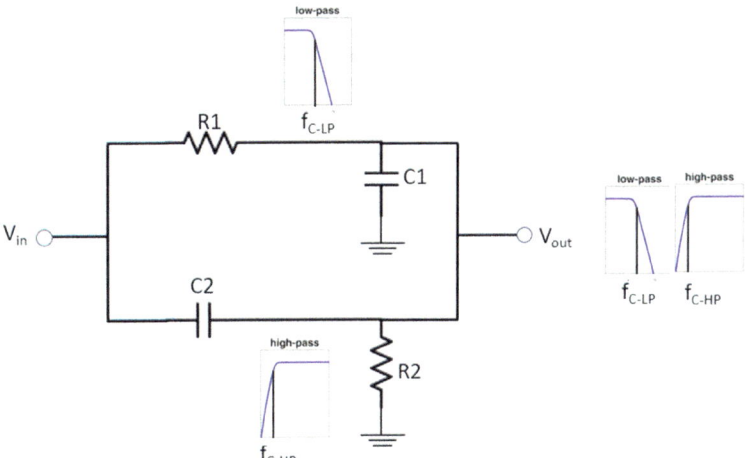

FIGURE 4.23 Circuit diagram of a notch filter using two first-order RC filters in parallel. The top stage is a LPF having gain $A_1(f)$ and the bottom stage is a HPF having gain of $A_2(f)$. The LPF passes all frequencies below $f_{C\text{-LP}}$ to get to the output, and the HPF passes all frequencies about $f_{C\text{-HP}}$ to get to the output. Therefore, it is required that $f_{C\text{-LP}} < f_{C\text{-HP}}$. Frequencies between $f_{C\text{-LP}}$ and $f_{C\text{-HP}}$ do not get through either of the parallel filter paths, so the result is a notch filter.

is a LPF and the bottom path is a HPF. Low frequencies pass along the top branch and high frequencies pass along the bottom branch. Any frequency between $f_{C\text{-LP}}$ and $f_{C\text{-HP}}$ will not have a branch along which it can pass. Therefore, it must be that $f_{C\text{-LP}} < f_{C\text{-HP}}$, and the notch of the notch filter is between these frequencies.

4.4 Frequency Domain

Recall that the **domain** of any function is the set of values on which the function is defined. A function maps points from the domain to another set of points called the **range**. So far in this course you have used the oscilloscope to plot a voltage waveform as a function of time. That, again, is called the **time domain**. Soon you will learn that for every voltage function (of time) in the time domain it is possible to derive a different voltage function, but now a *function of frequency*. This function is said to be in the **frequency domain**. You might be astute enough to ask, "Frequencies of what?" Well, that's what this section is largely about.

The two functions of time or frequency will seem to you to be very, very different, but in fact they contain the same information. That means, for example, that you can transform the time-domain function into its related frequency-domain function, and *vice versa*. It turns out that much can be known about the information conveyed by a waveform by viewing it in the frequency domain rather than in the (more-familiar) time-domain. Because these two functions are so closely related, one thinks of them as one thing that can be understood equally well in both the time and frequency domains, but they provide two different perspectives. Each is useful in different circumstances. Mastering this material at the level presented in this course will give you a firm foundation for mastering many complicated concepts in your upcoming courses, and is well worth the effort.

4.4.1 The Fourier Transform

To start this section, let's review some simple facts about sine and cosine functions, which are referred to as the "circular trig functions". They are fundamentally the same thing, but differ in time and phase, such that $\sin(\omega t) = \cos(\omega t - \pi/2)$; one is just a shifted version of the other. If you observe a sine or cosine function from a waveform generator on an oscilloscope, you can add delay so that it starts wherever you want it to on the screen. There is no distinction to whether it is called a sine or cosine function. Likewise, in most mathematical treatments of waveforms, it is arbitrary if one uses sines or cosines, or even a combination of them, to perform

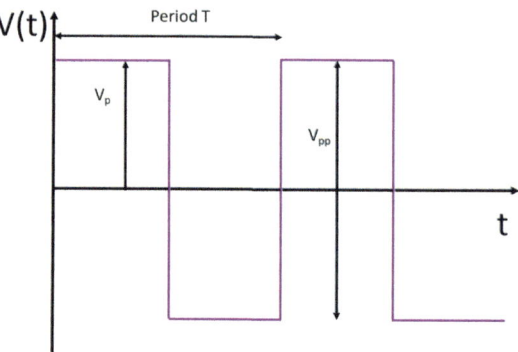

FIGURE 4.24 Square wave with period *T* represented in the time domain.

mathematical manipulations. In the forthcoming materials, we will just use the term "sine wave" to generically represent any sinusoidal waveform, like that seen on an oscilloscope. You should bear in mind that any equation can be written in a form to include any of the circular trig functions.

In Chapter 2 you saw that a periodic waveform (say, a square wave as shown in Fig. 4.24) repeats every amount of time, *T*, the **period**, where *T* is in seconds, s. Then, the **frequency** of that waveform is $1/T$, where $1/T$ can be expressed in "cycles per second," which is called **hertz** or Hz. For example, if the square wave repeats every 1 ms, then its frequency is 1 kHz (it repeats 1000 times per second, which is 1000 cycles per second). We call that the "frequency of the waveform."

You might notice that a sine wave that repeats 1000 times per second also has a frequency of 1 kHz, but intuit that describing both the square and sine waves as having a frequency of 1 kHz is inadequate to convey all there is to say about them. In fact, there is a lot more to the story. Around 1822, Jean Baptiste Joseph Fourier published his theory of how *a periodic function can be represented by the sum of (many) other periodic functions*, namely sines and cosines, much like a function can be represented by a Taylor series expansion.

Put another way, imagine that your square wave is a cake (this is not my own analogy). You might ask, "What are the ingredients of the cake? How much flour, how much egg, how much milk, vanilla, salt, sugar, etc." When a very experienced baker tastes a cake, she/he might be able to tell what ingredients are in there, and even be able to estimate how much of them there is. The 'ingredients' of the square wave are sine waves of many different frequencies, and the amplitudes of the sine waves are 'how much of each ingredient' there is. The function of frequency that represents the magnitudes of the sine waves that Fourier discovered is called the **Fourier transform**, FT. As part of the cake analogy, the cake would need lots of sugar to be very sweet. Likewise, if a square wave has a very high frequency, its FT will have to contain sine-wave frequency components that also have very high frequency. This section will help you to understand how to think of the FT, and how to interpret a specific FT of a time-domain waveform.

4.4.2 Decomposition, Frequency Components, and Superposition

At the heart of understanding the frequency domain is the fact that a periodic function (i.e., one that repeats), such as sine, triangle, or square waves, or any other periodic function, can be thought of as being made up of a 'mixture' of sines and cosines of different frequencies. By 'mixture', we mean a point-by-point (at every instant of time) addition of the values of the various sine or cosine functions with different frequencies. Note: this is true for functions that are not periodic as well, with some differences. This discussion is only about periodic signals.

Suppose you start to pick out the 'ingredients' (like the cake) of the triangle or square wave. The 'ingredients' are the sine waves at different frequencies, called **harmonics**, as shown in Figs. 4.25 and 4.26. These ingredients are also called **frequency components** since they are distinguished by their frequencies. (These frequency components must also be shifted in time, which is their phase. We will focus mostly on their frequencies in this chapter). We can say that the waveform can be **decomposed** into its frequency components by the Fourier transform.

If a signal is not periodic like the square wave, but is always changing, i.e., does not repeat (like, say, music), its F.T. is a continuous function of frequency components, without the discreteness that we see here. That case is specifically referred to as a Fourier transform. Fourier transforms of periodic functions are called **Fourier series** because of the discreteness of the individual frequency components. Every case presented in this chapter is a Fourier series. The general notion of the FT presented here will be developed more formally in your future courses.

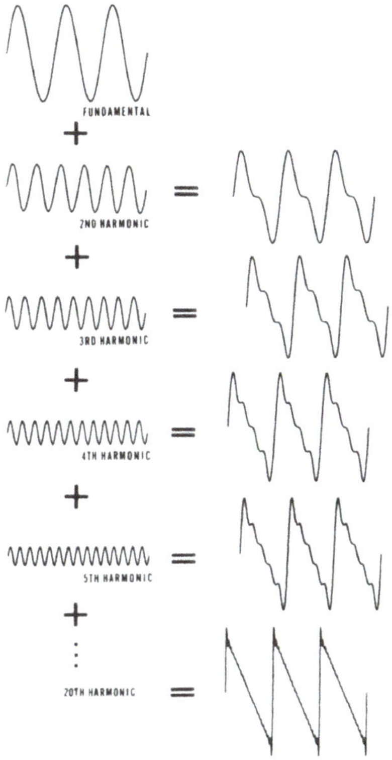

FIGURE 4.25 Summation of individual frequency harmonics to produce sawtooth waveform. Reproduced with permission of the World Soundscape Project, Simon Fraser University.

Let's 'bake' a square-wave cake, as if it were made from harmonic baking ingredients. If you had to choose one frequency to start building the 1 kHz square wave, you would probably pick a 1 kHz sine wave, which would be a good start. At least you would have something that repeats every 1 ms and has something vaguely like the desired shape – and it could become a square wave if you could create and sharpen up the corners. The other frequency components are the rest of the harmonics, where 1 kHz is the **fundamental frequency**, or **first harmonic**, or just "the fundamental." The fundamental is usually written as f_0, pronounced "f naught," even though it is the first harmonic. It is

Nomenclature: The word 'harmonic' shows up in many contexts in engineering and physics. In physics, an oscillator can be a 'simple harmonic oscillator.' In engineering, the various frequencies that make up a Fourier spectrum are called harmonics. Even in music, harmonic means that it sounds pleasant. What they all have in common is that they relate to sine waves.

also written as f_1, which is less confusing, but not as common. (I will use f_0 except where I want to be very explicit, as in this and the next paragraph.) $f_2 = 2 \times f_1 = 2 \times 1$ kHz is the **second harmonic**, and so on.

Note that often the amplitude of a particular harmonic is zero, meaning that you just don't need any of it. The **odd harmonics** are f_1, f_3, f_5, etc., which here are 1 kHz, 3 kHz, 5 kHz, etc., and the **even harmonics** are f_2, f_4, f_6, etc., which here are 2 kHz, 4 kHz, 6 kHz, etc. The square wave of Fig. 4.26 is made up of only the odd harmonics; all the even harmonics are absent, i.e., have zero amplitude. Together, all the harmonics taken as a collection of sine waves with their various frequencies make up the **frequency spectrum** of the waveform. Figures 4.25 and 4.26 show how a waveform is built up, or **synthesized**, by adding the appropriate amount of each sine wave at the appropriate frequencies. Thinking of the square wave as a collection of sine waves of various frequencies now puts you firmly in the frequency domain. I hope you are starting to

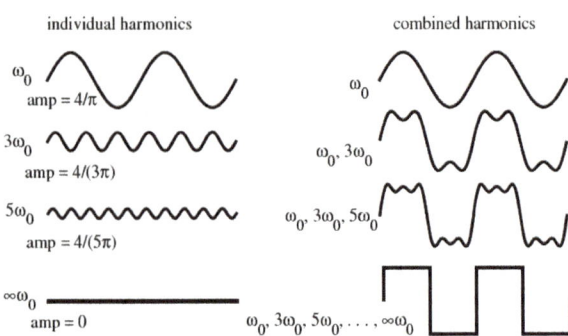

FIGURE 4.26 Summation of individual frequency harmonics to produce a square wave. Note that this figure uses **angular frequency**, $\omega = 2\pi f$ rad/s. Adapted with permission from Chidambaram, M., Saxena, N. (2018). *Relay Control System. In: Relay Tuning of PID Controllers. Advances in Industrial Control.* Springer, Singapore.

get it. As you gain experience, you will be able to intuit which frequency components, and how much of each, make up a waveform.

So far, we have seen that a periodic waveform, such as a square wave, can be decomposed using the Fourier transform into a collection of frequency components. When that square wave voltage is input to a circuit, such as a filter, that circuit performs some function on the waveform. We may choose to think of the input waveform not as a single time-domain function, but rather the collection of all of the frequency components, all applied independently to the circuit. *We may now analyze how that circuit acts on each of those individual sine waves of different frequencies, and later add those up to get the actual output of the circuit.*

A circuit or system is said to be **linear** if it responds to separate inputs as if they were acting alone – the output due to one input is not affected by the presence of another simultaneous input. (Linearity is discussed in more depth in Chapter 5). In this case, a linear circuit responds to each sine-wave frequency component individually, as given by Eq. 4.45, $V_{out}(f) = A_V(f) \cdot V_{in}(f)$. Since a waveform is composed of a collection of frequency components, the sum of the responses of the linear system to the individual frequency components add together to make up the response of the circuit to the original time-domain waveform. The property of a linear circuit to sum the individual responses due to separate inputs is called the **Law of Superposition**. Superposition is the opposite of decomposition. If a waveform is decomposed into its frequency components, and then all of those frequency components were added together by superposition, you would get back the original waveform. If each component were modified by a filter, then the superposition of the resulting frequency components would yield a different waveform that is the filtered version of the original waveform.

At the center of all of this is the assumption that the system or circuit is linear. If the output of the circuit were a "linear cake", it would be as if you baked all the ingredients *separately* and then put them together, and *you got the cake*. Since that isn't possible, baking a cake is a **nonlinear** process – the chemical reactions of some ingredients affect the behavior of the others during baking. Based on the discussion of linearity, we have to assume that in our analogy, we have "linear cakes" in which we can still taste the original ingredients.

4.4.3 Frequency Spectrum Analyzer

Fortunately, there is a tool that allows you to measure how much of each frequency component is contained in a waveform that you might encounter on a bench. This tool is called, appropriately, a

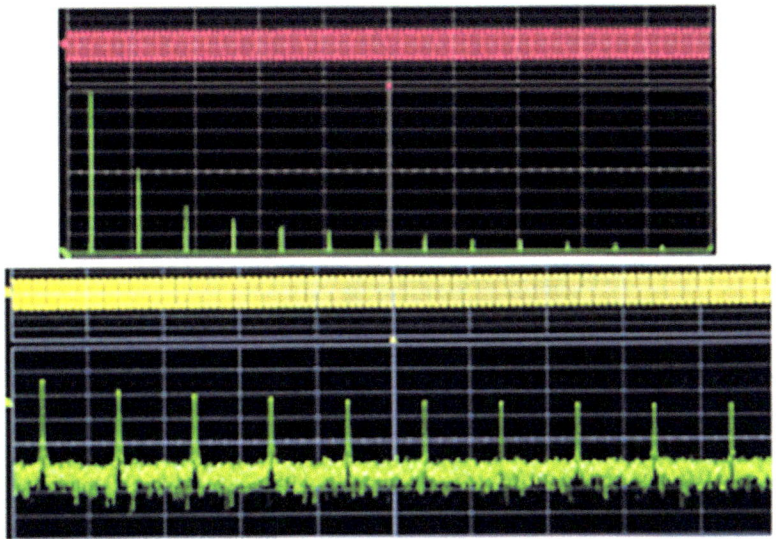

FIGURE 4.27 FFT of square wave as seen on the screen of an oscilloscope. The top plot is on a linear amplitude scale, and the bottom is a logarithmic scale.

frequency spectrum analyzer, or just **spectrum analyzer**. The spectrum analyzer displays on the vertical scale how much of each frequency component makes up the waveform. Dedicated spectrum analyzers are expensive. However, there is a simpler and cheaper way to get the frequency spectrum, but it relies on having at your disposal a means to capture the waveform to a computer and then apply mathematical algorithms to reveal the spectrum. This is often performed in a digital oscilloscope, like the one you use in the lab, as shown in Fig. 4.27.

After storing the voltage values of the time-domain waveform, the frequency components can be calculated using an algorithm called the **fast Fourier transform**, or **FFT**. The oscilloscopes you use in the lab have an FFT algorithm built in. It accesses the waveform values, calculates the FFT, and displays it as a frequency spectrum. This can be done so quickly that it appears to the user as if it is happening nearly in **real time**, i.e., as it is happening.

We refer to the frequency-domain representation of a waveform as its frequency spectrum, Fourier transform or Fourier series, but these terms have somewhat different meanings, so one must know what they are looking at when they see a plot with frequency components. Warning to avoid confusion: The Fourier transform and Fourier series are *mathematical* representations of the time-domain waveform in the frequency domain. They contain *both positive and negative frequencies and positive and negative values*. This explains why many plots in the frequency domain that you may see in other sources have such features. *The plot shown on the spectrum analyzer in Fig. 4.27 is better described as a* **frequency spectrum** *and has only positive frequencies and positive values* and ignores phase. It has elements in common with the Fourier transform and series but *is not the same*.

> *To be filed away for the future:* You will at some point be introduced to the Fourier transform in terms of both sines and cosines. *Adding a sine function to a cosine at the same frequency results in a sinusoid with a different amplitude and shifted phase.* So, it is just another way of saying the same things that you have seen here w.r.t. only sines but includes the phase. Store that fact away for later use in your Signals and Systems course.

It is often very useful to have the frequency spectrum of a waveform available. For example, it is very difficult to look at a time-domain waveform and see where problems in the form of external noise might be cropping in. The frequency spectrum could show extraneous peaks and provide important clues about where to start looking for the cause. Another example is to see how well a filter works by inputting various frequencies at once, as in a square wave, and seeing how the frequency spectrum responds. Yet another example is to measure how clean, or pure, a signal is. The power from the wall outlet should not, but could, contain harmonic components above 60 Hz, which could damage sensitive electronic equipment. Also, extra frequency components in a hi-fi audio system can degrade the quality of the music. This is related to harmonic distortion, discussed in the sidebar.

4.4.4 Phase of the Frequency Components

In this treatment we discuss the amplitudes of the frequency components but be aware that a mathematical Fourier transform specifies both an amplitude *and phase*, or time shift, for each of the sinusoidal frequency components. Both amplitude and phase are important so that the frequency components all add up correctly to make the waveform, as shown in Figs. 4.25 and 4.26.

We will not much discuss the phase of each frequency component in this introductory treatment, but you should be aware of it. What do we mean by "the phase of a frequency component?" Remember that any phase relationship must be between two waveforms of the *same frequency*. The time delay between them expressed as a fraction of a period in degrees or radians is the phase shift. When discussing the Fourier transform, the phase of every waveform is the time difference between the sinusoid

> *For deeper understanding:* This sidebar may at first seem esoteric, but in fact it reinforces some very important points about frequency spectra. If you are a 'hi-fi' stereo buff, you will undoubtedly run across 'total harmonic distortion' to describe the quality of an amplifier. It is worth discussing in the context of Fourier analysis. Figure 4.29 says that a pure sine wave has only one frequency component. However, *an imperfect sine wave MUST have other harmonics in order to make up that 'new' waveform that isn't a perfect sine wave.* By just looking at the waveform, your eyes may not be able to tell that it isn't perfect, but the spectrum will have harmonics in it. That is called 'harmonic distortion,' and adding up the power in all of the frequency components gives the 'total harmonic distortion,' THD. If an amplifier has a sine wave as an input, and the output is a perfect sine wave, then its THD will be 0%, which is basically impossible. My prized NAD 7600 has a THD of 0.03% at up to 150 watts of output power. This is a very low value for a lot of power, where 0.5% is more common. That means that my NAD reproduces the music faithfully with little distortion, even when played loudly.

having $\phi = 0$ (crossing through zero at $t = 0$ for sines) and the same sinusoid shifted by some phase. Put another way, the sinusoidal frequency components must be phase shifted away from $\phi = 0$ before they are added together in order to recover the correct time-domain waveform. Note that some transforms (including square and triangle waves) have a zero phase shift, so all of the frequency components cross through 0 at $t = 0$, but that is not generally the case.

4.4.5 Common Fourier Transform Pairs

There are a few functions whose time- and frequency-domain relationships, called their **Fourier-transform pairs**, you should be familiar with. Figures 4.28, 4.29 and 4.30 show both time and frequency domain representations of a few signals, where the height of the arrow or bar represents the magnitude of that particular frequency component. The arrow or bar is called a **Dirac delta function**, $\delta(f)$, and signifies that the function has a value only at a single point, which in this case is a single frequency (see sidebar below). (If you haven't already been introduced to delta functions in another course, don't worry – you will at some point.) $\delta(f)$ is a delta function at $f = 0$, and $\delta(f - f_0)$ is a delta function located at frequency f_0. The length of the arrow and the number next to it indicate the relative value of the delta function at that point. In this case, the delta function tells us the magnitude (e.g., V_p or V_{RMS}) of each frequency component contained in the waveform. The lengths of the arrows help us visualize the overall frequency-domain function. Delta functions themselves are important mathematical functions, with deep significance, and

> *Clarity:* Delta functions are very important in all of science and engineering, including EE. In this course we will use them to simply mean something that happens only at a single point on the domain. There is much to know about delta functions that you will learn in other courses.

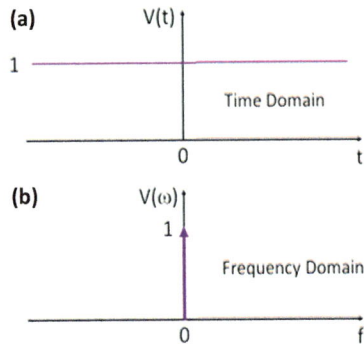

FIGURE 4.28 (a) Time and (b) frequency domain representations of a DC signal. The delta function is at 0 Hz and is the DC component.

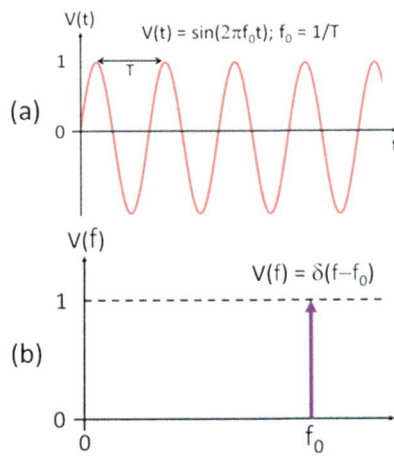

FIGURE 4.29 Time and frequency domain representations of a pure sine wave.

will be important in more-advanced courses. That said, here we use them only to indicate graphically how much of each frequency component we have at each frequency. The length of the delta-functions is chosen simply to reflect the magnitude of the frequency component but otherwise has no mathematical meaning. Also, arrows are typically chosen to represent the delta function, but even that isn't necessary, and simple bars or lines are often used instead.

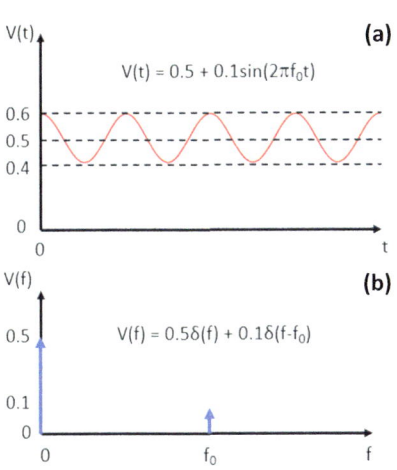

FIGURE 4.30 Time and frequency domain representations of a single frequency (f_0) sine wave with DC offset.

The dimensions of the frequency components of any function are the same as that of the time-domain function. If the time-domain function is $v(t)$, then the frequency components are voltage sinusoids, likewise for current or any other Fourier transform. After all, you add up a bunch of voltages to get a voltage waveform, so the dimensions of the spectral components must match those of the time-domain function.

4.4.5.1 DC Voltage

Our first example of a Fourier-transform pair is the simplest. Figure 4.28 shows both the time- and frequency-domain representations of a DC voltage. A DC voltage is constant in time – a flat line from $t = -\infty$ to $t = +\infty$. DC is, in fact, *a frequency component whose frequency is zero Hz*. Therefore, its frequency-domain representation is just a (voltage) delta function (arrow) at $f = 0$ Hz whose height represents the same value as the DC voltage itself. When combined with a more complicated waveform, the DC voltage is appropriately referred to as the "DC component" of the signal. You might wish to note that the RMS and peak values of a DC voltage are the same, regardless of which voltage representation you use.

4.4.5.2 Sine Wave of Frequency f_0

Here is another simple one: How many sine wave frequency components does it take to make up a sine wave? Ummm, one – right? So long as you pick the right one (at the right amplitude and frequency), it takes only one sine wave to 'make up' a sine wave (and you can't make up a sine wave from sine waves of other frequencies). Any impure sine wave must contain other sine-wave frequency components, as in the harmonic distortion mentioned in the sidebar.

If the function is a **pure** (i.e., perfect) sine wave with frequency f_0, then the only frequency component needed to form that waveform is a sine wave at f_0, i.e., $\sin(2\pi f_0 t)$. Figure 4.29 shows both the time domain representation and frequency spectrum of a pure sine wave. Mathematically, to form a true Fourier transform, we also need to add to it another component of negative amplitude and frequency at $-f_0$. The negative frequencies are a mathematical artifact that does not hinder our understanding of time-domain signals in the frequency domain. For now, they just come along for the ride, and we choose to not show them. As mentioned above, the frequency spectrum shown on a spectrum analyzer does not include negative frequencies. So, for our purposes, a sine wave is represented by a single delta function at f_0.

4.4.5.3   Sine Wave with a DC Voltage Offset

This is a simple example of the superposition of frequency components. Figure 4.30 shows both the time and frequency domain representations of a sine wave with a DC offset. We say that the AC signal is "offset by," "shifted by," or "riding on" a DC voltage. In the vernacular of the frequency domain, the total signal has a DC frequency component. Therefore, it takes two delta functions to describe it, namely one at zero frequency, the DC component, and one at the oscillation frequency. The total signal has a nonzero average voltage. Because the average of a sine wave is zero, the DC offset is the average value of the sum of the two frequency components. This fact played an important role in power transmission as discussed in Chapter 3.

When an oscilloscope is **DC coupled**, all of the frequency components of the signal, including the DC component, pass to the input of the amplifier. Oscilloscopes have an option to use a high-pass filter at the input from the probes in order to remove the DC part of the waveform if needed. Suppose that you wish to study the sine wave, or any equivalent AC waveform, of Fig. 4.30, except that instead of an amplitude of 0.1 V, as shown, the amplitude is 10 mV. In order to study the features of the AC signal, you would have to increase the sensitivity to, let's just say, 5 mV/div. In doing so, the 0.5 V DC component would be increased to a height of 0.5 V/5 mV/div = 100 divisions on the screen, which would require a screen to be about 1 meter tall in order to keep the AC waveform barely on the screen so that you could see it!

Since scopes are not built like that, it is necessary to do something to allow the AC components to be magnified on the screen while stopping the DC component. Well, a HPF is just the thing! Figure 4.30 shows the frequency characteristics of an ideal HPF filter overlaid on the spectrum that has both DC and AC components. Based on Eq. 4.26, where $A_V(f) = 0$ below the cutoff frequency f_C, such a filter would eliminate, or block, the DC component and allow only the AC components above f_C to pass.

Figure 4.31(a) shows the DC-coupled circuit, which allows both the DC and AC components of the signal to pass into the scope, which is represented by the resistance, R. Notice that in the circuit diagram, the DC source in series with the AC sources realizes the same signal as that represented in Fig. 4.30.

In Fig. 4.31(b) there is added a series capacitor that acts as a HPF along with the resistance of the scope, R. Now the scope input is **AC-coupled**, implying that only AC is allowed to pass to the oscilloscope. DC and AC coupling is a feature on every scope, and it is toggled by a button or a menu item. In order to know f_C, you would have to refer to the user manual of your scope, and it varies depending on the purpose for which the scope was built. You are likely to find cutoff frequencies from 10 Hz to 100 Hz, but much higher values are possible.

Figure 4.31(b) also shows the waveform of the AC components of the waveform after AC coupling, meaning after the DC voltage component is blocked. If you were to watch the waveform after pressing the "AC-coupling" button, which brings the capacitor into the circuit, the AC components would 'float' down to the DC average value line and would

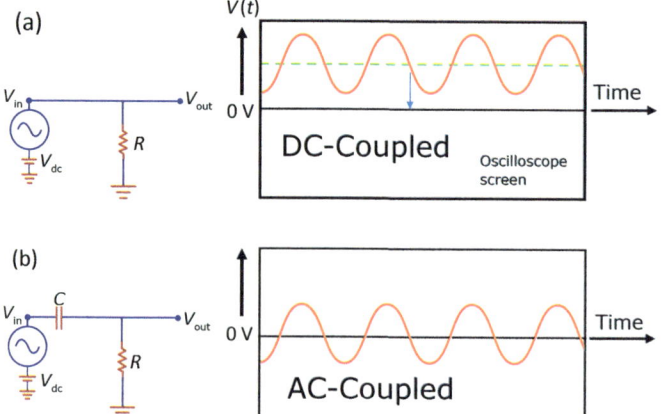

FIGURE 4.31 DC and AC coupling. The resistor, R, is the amplifier at the input of the oscilloscope. (a) DC coupling. When the input to the oscilloscope goes directly to the amplifier, both the AC and DC components are all present. (b) AC coupling. A series capacitor is a high-pass filter that blocks the DC component and lets the AC component get through to the amplifier. It is now free of the DC component and can be magnified on the screen.

still be very small on the screen. However, upon increasing the sensitivity to that 0.5 mV/div level, it would expand to fill the screen, and then you could study it.

4.4.5.4   Square Wave

More-complex waves, such as the sawtooth in Fig. 4.25 and the square wave in Figs. 4.24, require a larger set of frequency components to build them up. First, we will study the (periodic) square wave, as shown in Fig. 4.24 and here in Fig. 4.32, and later discuss triangle waves. The square wave needs many frequency components extending out to high frequencies to create, or synthesize, it sufficiently so that it looks like a good square wave with sharp corners, and only few oscillations at the edges. Square waves are made up of only odd harmonics, i.e., $\sin(n2\pi f_0 t) = \sin(n\omega_0 t)$, where n is odd. Figure 4.32(a) shows the square wave compared with the fundamental, or first harmonic. The frequency matches that of the square wave. Figure 4.32(b) shows a square wave built up from only three components, namely $n = 1$, 3 and 5. By the time the 11th harmonic is reached (which includes 6 nonzero components since all the even harmonics are zero), the plot starts to look like a fairly acceptable square wave. Finally, Fig. 4.32(d) shows the square wave synthesized by up to the 49th harmonic (25 components).

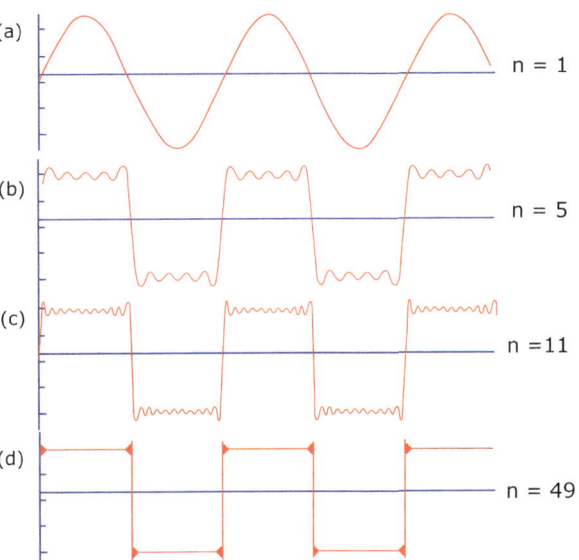

FIGURE 4.32 Gradual summation of harmonics to produce a square wave. The more harmonics added, the more the sum looks like a perfect square wave. The overshoot at the edges is called the "Gibb's phenomenon" and is an unavoidable consequence of the method of Fourier transforms, and it is not considered to be an error in the theory. The overshoot does not disappear with the addition of more harmonics.

Why only odd harmonics? The answer can be inferred from Fig. 4.32. The odd harmonics have an odd integer number of half waves fitting into one-half period of the square wave. Even harmonics would have an integer number of full waves fitting into each half wave. So what? Well, the even harmonics would be decreasing at the point where the square wave is increasing. Therefore, even harmonics would do more harm than good in our quest to build the perfect square wave from frequency components.

In order to synthesize the square wave, the harmonic components must be added point by point in time. Quantitatively, the sum of the voltages of each of the following harmonic components needed to create the square wave are:

$$f(t) = \frac{4}{\pi} \cdot \sum \frac{1}{n} \sin(n\omega t)$$
$$= \frac{4}{\pi}\left[\sin(\omega t) + \frac{1}{3}\sin(3\omega t) + \frac{1}{5}\sin(5\omega t) + \frac{1}{7}\sin(7\omega t) + \cdots\right], \qquad (4.46)$$

where $n = 1$ to ∞ and n is odd. In short, the amplitude of each harmonic component drops off as $1/n$, or equivalently, as $1/f$ since $f = nf_0$ and $1/n = f_0/f$.

Figure 4.33 is a rather busy plot that shows a comparison of the frequency components that make up both a 100 Hz unit square wave and 100 Hz unit triangle wave, to be discussed in the next section. The figure shows on a linear scale the relative voltages of several of the frequency components of the waveforms as delta functions, as has been done in the previous few sections; however, this is

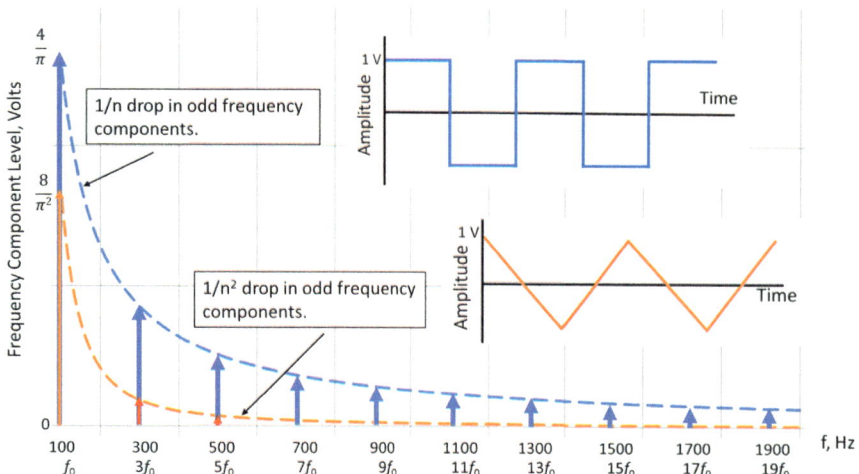

somewhat more complicated. The delta functions follow the amplitudes of the frequency components, which drop off as $1/n$ as prescribed by Eq. 4.46. The relative amplitudes of each harmonic follow the f_0/f line, which is included as an aid the eye. "Dropping off as $1/n$" means that the n-th harmonic is $1/n$ times the amplitude of the first harmonic. By the time the 19th harmonic is drawn, it is $1/19$ the amplitude of the first harmonic. That is still discernible on the plot, but not easily.

FIGURE 4.33 Time and frequency domain representations of a 100 Hz square wave and a triangle wave. The blue delta functions represent the odd harmonics summed to create the square wave and the red delta functions sum to create the triangle wave. The lengths of the arrows are proportional to that frequency component's voltage amplitude. Triangle waves have the same frequency components as the square wave, but they are smaller at the high frequencies, dropping as $1/f^2$ versus $1/f$ for the square wave.

4.4.5.5 Triangle Wave

Everything said about the square wave holds for the triangle wave, except for the ratios of the harmonics. The frequency components of a unit triangle wave for $n = 1$ to ∞, where n is odd, are:

$$f(t)=\frac{8}{\pi^2}\cdot\sum\frac{1}{n^2}\cos(n\omega t)=\frac{8}{\pi^2}\left[\cos(\omega t)+\frac{1}{9}\cos(3\omega t)+\frac{1}{25}\cos(5\omega t)+\frac{1}{49}\cos(7\omega t)+\cdots\right]. \qquad (4.47)$$

The magnitudes of the harmonics are different, but more importantly, the harmonics drop off as $1/n^2$ instead of the $1/n$ for the square wave. (Congratulations if you noticed that Eq. 4.47 is in terms of cosines rather than the sines of Eq. 4.46. This 90° phase shift is required because the phase of the triangle wave shown has a peak at $t = 0$. Cosines also peak at $t = 0$, so they are the basis for this spectrum.)

Figure 4.33 also shows the magnitudes of the harmonics of the 100 Hz triangle wave in comparison to those of the square wave. You can readily see that the high-frequency harmonics are much smaller as compared to those of the square wave. Just like the baker who can identify ingredients in a cake but not necessarily how much there is of each, we can make qualitative statements about differences between the two spectra. It makes intuitive sense that *fast changes in the time domain require high frequency components*. We can surmise that the spectrum of the triangle wave needs less of the higher harmonics as compared to those of a square wave because the sharp, fast changes at the edges of the square wave are absent in the triangle wave. Still, you need some of all the components to make a straight line as well as the sharp corners at the top.

Now the 19th harmonic is $(1/19)^2$ times less than the fundamental and is not even discernable on the plot! In fact, it is impossible to see the peaks after about the 9th harmonic. The solution to plotting very small and very large numbers on the same plot is to use logarithms. This is very commonly done in the form of the units of decibels, which are introduced in the next section.

4.4.6　Decibels

You have seen that after a few harmonics are represented in the square and triangle waves of Fig. 4.33, they become so small on the plot that their values are difficult to discern, especially for the triangle wave. Using logarithms helps to solve that problem by shrinking big numbers and exaggerating small numbers. Figure 4.34 shows the very same data as that on Fig. 4.33, but instead using logarithms of the ratios of voltages of the frequency components. In fact, it is very common to use logarithms for this reason, and also that it makes doing the arithmetic of calculating the gains of amplifiers and other things easier. This section is

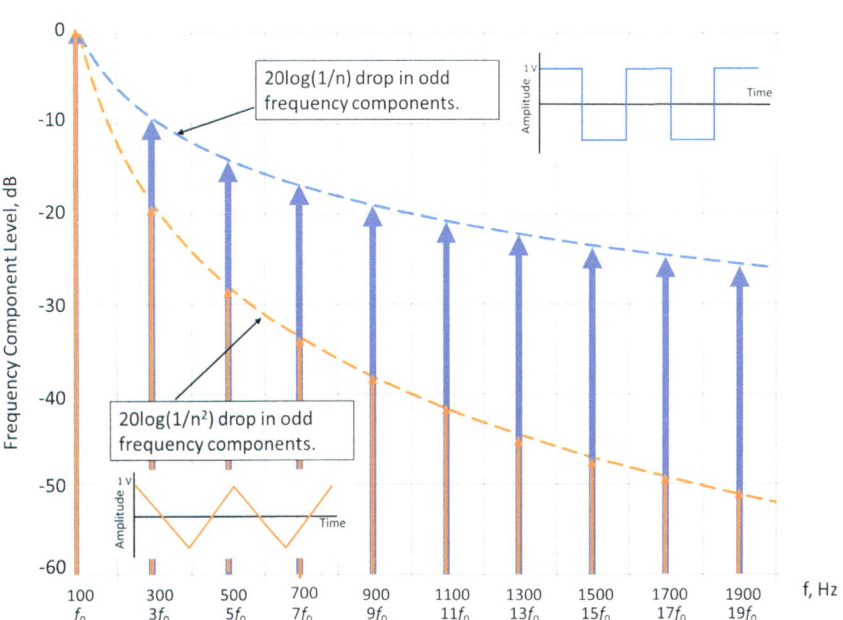

FIGURE 4.34 Time and frequency domain representations of the same 100 Hz square and triangle waves of Fig. 4.33 but shown in decibel units. The lowest dB value shown is completely arbitrary, so the lengths of the arrows have no actual meaning, unlike the arrows in Fig. 4.33.

intended to make you somewhat familiar and comfortable with using logarithmic units to display voltage and power values, which is very common, and worth the effort to master.

The frequency spectra of square and triangle waves using the logarithm of the ratio of the amplitudes of the harmonics to that of the fundamental, as provided in Eqs. 4.46 and 4.47, are shown in Fig. 4.34. The plot shows the logarithmic representation of the frequency components in terms of **decibels, dB**. The decibel is a dimensionless quantity that takes a little getting used to, and can be confusing, even for seasoned engineers. There are literally dozens of different types of decibels in use, but we will use only a few in this course. Because the argument of a logarithm is dimensionless, *any quantity expressed in dB is the log of some sort of ratio.* Your job is to first ask, "What is it a ratio of?" The definition of *voltage* ratios in dB is:

$$A_{V,\mathrm{dB}} = 20\log_{10}\left(V_2\big/V_1\right) \tag{4.48}$$

where (V_2/V_1) is the ratio of two voltage amplitudes. The ratio $\dfrac{V_2}{V_1}$ might represent the gain of an amplifier or a filter, i.e., the voltage (peak, peak-peak, or rms) at the output, V_2, relative to that at the input, V_1. A very specific example is the 'mute' button on a stereo receiver. No matter where you have the volume level set to, pressing the 'mute' button changes the amplifier voltage by –20 dB, i.e., makes it sound quieter due to a reduction in the amplifier gain. The amount of voltage attenuation caused by the mute button can be shown to be a factor of 10, as calculated by

$$-20\,\text{dB} = 20\log_{10}\left(\frac{V_2}{V_1}\right)$$

$$-1 = \log_{10}\left(\frac{V_2}{V_1}\right)$$

$$10^{-1} = \frac{V_2}{V_1} = 0.1 \tag{4.49}$$

In this example the ratio $\frac{V_2}{V_1}$ = 0.1 is that of the voltage output with the mute button depressed to the voltage with the button not depressed.

The concept of dB can be applied widely to any ratio. In the specific case of Fig. 4.34, we consider the ratio of the voltage of the harmonics to that of the first harmonic (the fundamental). In Fig. 4.34, the ratio is that of the amplitude of each harmonic divided by that of the fundamental. Looking back at Fig. 4.33, you can see that the amplitude of the fundamental was selected to provide the square and triangle waves having an amplitude of 1 V. However, in Fig. 4.34, that actual voltage disappears by taking the ratio. For example, dividing the first harmonic by itself provides a ratio of 1, which is equivalent, by Eq. 4.48, to 0 dB. By normalizing the harmonics to the fundamental this way, the information conveyed by the plot is the relative values of the frequency components for the shape of the waveform, independent of the actual voltage of the waveform.

Again, since log(1) = 0, the magnitude on the plot of the first harmonic is shown as '0 dB.' Since all harmonics have amplitudes less than that of the fundamental, and the logs of numbers less than 1 are negative, the amplitudes of the rest of the harmonics in dB are negative. Here is another example of negative dB: In Section 4.3.1 we pointed out that the cutoff frequency is the half-power point, which is when the output voltage has dropped by a factor of $1/\sqrt{2}$ (because $P = V^2/R$). That is the same as the voltage gain of $1/\sqrt{2}$. Applying Eq. 4.48, we get that $A_{V,\text{dB}} = 20\log_{10}(0.707) = -3$ dB. So, the output voltage is "down 3 dB at the cutoff frequency," as shown in Figs. 4.17 and 4.18, bottom. In short, negative dB values imply a ratio less than 1 and positive dB imply a ratio greater than 1. For an amplifier, the gain is the ratio of the output voltage to the input voltage, $\frac{V_2}{V_1}$, for which the output voltage is a scaled version of the input voltage. Applying Eq. 4.48, negative dB corresponds to a smaller voltage compared to the input voltage (gain < 1), which is attenuated. A larger output voltage compared to the input voltage (gain > 1) would be expressed in positive dB and is amplified.

As shown above, logarithms are often used to compress numbers that have a large range and would be inconvenient or impossible to view on a single plot. For example, if you need to plot 0.001 and 100 on the same linear plot, you would need a

Of interest: The process of building up a wave from its harmonic components is called **synthesis**. In the early days of electronic music, around 1965, engineers were experimenting with ways to create innovative sounds by building them up piece by piece, as described in the text. In the 1970s, musical recordings (vinyl records) of a commercial system called a Moog Synthesizer (after Robert Moog, its inventor) became very popular (see 'Switched-on Bach'). At first, the sounds did not resemble actual musical instruments. This led later to attempts to make the sounds resemble real musical instruments as closely as possible. Today, realistic electronic music is more likely to be a sample of an actual musical instrument played by a professional and recorded into a library. Individual notes are modified electronically and played back at different frequencies and volumes to create the appearance of a real musical instrument playing full passages.

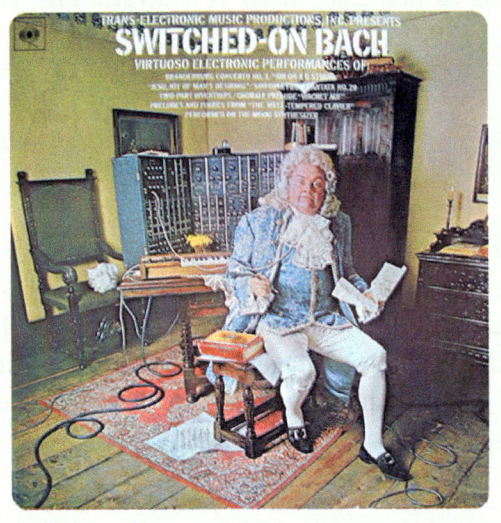

microscope to read the 0.001, if the printer or computer screen could even resolve it (which it could not). However, the logarithms of these two numbers are 2 and –3, respectively, which is easily resolved. If they were voltage gains, they would be 40 dB and –60 dB, respectively, which can easily be shown on the same plot. Keep in mind that the argument of the log must be positive - no negative gains allowed, but the gain can be greater or less than 1.

Because $\log(0) = -\infty$, the lowest value shown on a log plot is never zero, but rather some *arbitrary* value that the maker of the plot simply has decided to use. Really, it's just arbitrary; you can pick whatever lowest value you want. For the triangle wave of Fig. 4.34, the 19th harmonic is all the way down to about –52 dB (as compared to about –25 dB for the square wave) which is nearly 1/1000 of the amplitude of the fundamental. (You can calculate the value in dB exactly from the amplitude ratio of $1/19^2$.) That would be impossible to discern on the linear plot of Fig. 4.33. On the plot of Fig. 4.34 it *appears* that the 19th harmonic of the triangle wave is about 1/6 as large as the fundamental, which, again, is confusing if you aren't familiar with using dBs. In fact, the amplitude of the 19th harmonic is smaller than the fundamental by a factor of 1/19th or 0.056 or 5.6%. Notice that the horizontal axis does *not* correspond to zero voltage. The lower bound on the plot of Fig. 4.34 is *arbitrarily* chosen as –60 dB. It could also have been taken, say, as –100 dB, simply at the whim of the person making the graph. Then the arrows expressing the two harmonics would *seem* to be closer to the same size, but that would just be an illusion. (Caution: decibel values are smaller than they appear). Plotting numbers like 0.0001 on a linear scale compared to 1 is impossible but expressing it as –80 dB compared to 0 dB is not. So, using dB is very useful as long as you know how to interpret the plots.

Most digital oscilloscopes can display a frequency spectrum in many types of dB, implying that the $\frac{V_2}{V_1}$ term in Eq. 4.48 could, instead, be some other ratio of your choosing. One that you will use in this week's lab is 'dBmV'. For a voltage in units of dBmV, the denominator of $20\log_{10}\left(\frac{V_2}{V_1}\right)$ is taken as 1 mV, so all voltages are referenced to 1 mV. So, although dBmV is a log of a ratio, it actually provides a logarithmic representation of the voltage since the numerator is referenced to 1 mV. The use of dBmV (or its big brother, the dBV, which is the same thing referenced to 1 V) allows the comparison of voltages over a large range while maintaining actual voltage values, which can be calculated using Eq. 4.48.

4.4.7 Filters Applied to Signals

So far you have learned that signals contain a range of frequency components. You have learned about filter circuits, and how there can be more than one stage of filter. You should be clear that there are two distinct entities that have a frequency dependence, namely the signal and its Fourier spectrum, and the system, which in this chapter is the filter circuit. Both are frequency-dependent. Equation 4.26 says that $V_{out}(f) = A_V(f) \cdot V_{in}(f)$. In words, this tells us that the output can't contain any frequency components that aren't part of the original signal, $V_{in}(f)$. It also says that the output can't contain any frequencies for which the gain is zero (or very low). The only frequencies that can be found in the output are those that are contained at the input and are passed to the output by the filter.

Figure 4.35 shows the frequency spectrum of a clarinet playing the note C4, which is middle C on the piano. It has considerable frequency content at and above the fundamental frequency, all of which give the clarinet its distinctive sound. Each instrument has its own set of harmonics that are like its DNA, giving its own particular identity. We can imagine playing the sound of the clarinet through a very poor-quality amplifier, or even using a poor speaker that has a very restricted frequency range. Any of those can act as if it were a low-pass filter. Or if we want, we can simply imagine what an actual LPF would do to the sound. Because of Eq. 4.26, every frequency is individually multiplied by the gain of the system (amplifier, speaker, or whatever is in the signal path), to yield a new, distorted version of the frequency spectrum. If the spectrum were to look some-thing like that shown in Fig. 4.35, the frequency components above about 1.5 kHz would be significantly attenuated, including above about 3 kHz where they would be effectively eliminated. You might imagine that the sound might be recognizable as a clarinet, but would not sound very good, with most of the high-frequency content missing. All of that explanation is contained in the meaning of Eq. 4.26.

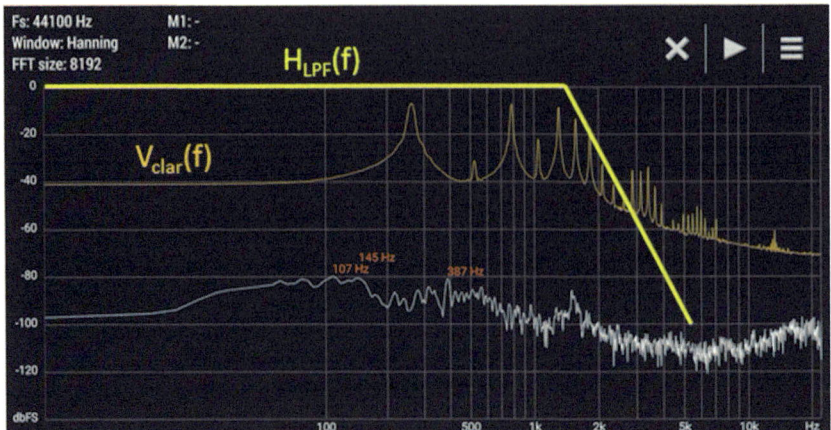

FIGURE 4.35 Frequency spectrum of a clarinet playing the note C4, one line below the staff (middle C on the piano), whose fundamental frequency is close to 261 Hz. There are many harmonics, which gives the clarinet its rich tone. If the waveform for that note is passed through a LPF, as shown schematically, many of the higher harmonics will be attenuated or even removed according to the gain of the filter at each frequency, and the tone will sound tinny. (This filter characteristic is not intended to represent a real filter.) (Software: Advanced Spectrum Analyzer PRO).

You might now better appreciate that music must pass through mechanical and electrical filters all the way from the source of the sound, the room, the microphone, the recording electronics, audio processing, and manufacturing, such as an old-style phonograph record or, more likely, as the digital file for online music. Then the playback process goes through the preamplifier, which includes the actual filters used to shape the desired tone of the music, followed by the amplifier and finally the speaker or earbuds. Great care must be used at every stage to pass all the frequency components faithfully in order to produce high-quality audio sound. Chapter 9 delves much more deeply into the digital recording and playback process.

4.4.8   Concluding Thoughts about Fourier Transforms in General

In closing this section about Fourier transforms, here is a broader perspective. The Fourier transform is a powerful tool that is applied widely throughout science and engineering. The domain of the FT of any function is the inverse of the domain of the original function. You have seen that in the case of time-dependent electrical signals, say voltage versus time, the Fourier transform is a function of inverse time, i.e., frequency (the ingredients of the cake). It is a plot of the voltage of all the sine-wave frequencies that are needed to add together to make up the original time-domain waveform. Engineering and science widely use FTs of functions whose domains are not time.

Here is one example of a Fourier transform of something other than a time-dependent waveform. In optics, images are light intensities as a function of position, say in mm; the FT of the image is represented in terms of the inverse of position, mm^{-1}, or "per mm." So, the domain of the FT of the optical image is not in units of "frequency" as we are using for our electrical signals but is instead called **spatial frequency**. Imagine a picture of several trees covered in leaves. The trunks of the tree are large and spaced far apart, so together they contain low spatial frequency terms in the FT of the picture. The leaves are small and closely spaced, and many leaves happen in a short distance, so they contain high spatial frequency information.

Just like a square wave in the time domain, a picture of a periodic 2-dimensional grid of lines has a Fourier transform that is composed of a pattern of delta functions. The analysis of an optical system in terms of frequency components is called **Fourier optics**. A delta function in Fourier optics is a bright spot on a new optical pattern that is the Fourier transform of the object. A great example of this can be found at https://www.thorlabs.com/newgrouppage9.cfm?objectgroup_id=11829.

Since an image is really a 2-D map of brightness, or light intensity, you would expect the FT of the image to also be a 2D map of intensity (remember, the units of the range of the FT are the same as that of the original function). The spots of light further out from the center are the higher-frequency components, with distance from the center having units of "per distance," since that is the inverse of "distance". More of a spatial frequency component is represented by a brighter spot. In the above example using a picture of trees with leaves, those spots further from the center would be caused by the leaves. If the leaves were bright, then their spots would be bright. The lower-frequency spots closer to the center would be caused by the large, widely spaced tree trunks. All of this is the essence of an important field of electrical engineering called **image processing**.

In summary, the FT of just about any collection of data yields insights into what is going on in the process that created the data, regardless of the units and their inverses in the frequency domain. Although we are mostly concerned here with time and frequency, the FT concepts may pop up in other courses that you will take in future semesters.

4.5 | Resonance

Before we begin a new topic, it might be useful for you to be reminded of some important things to keep in mind. After this review, we will start a discussion on circuits that use both inductors and capacitors.

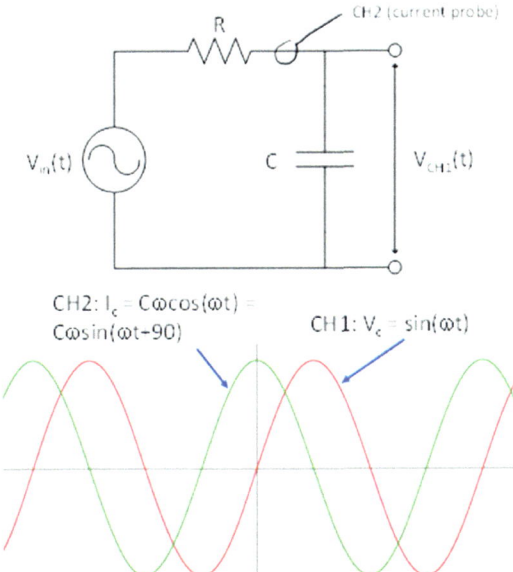

In Section 4.2.1 and Section 4.2.3 we saw that the mathematical relationship between currents and voltages in inductors and capacitors is $i_C(t) = C dv_C(t)/dt$ and $v_L(t) = L di_L(t)/dt$. It should be stressed that regardless of what other elements are in the full circuit, we are discussing specifically the *current through*, and the *voltage across*, a *single* element, namely an inductor or capacitor. (Recall: $d\sin(x)/dx = \cos(x)$; $d\cos(x)/dx = -\sin(x)$.) So, if $v_C(t) = \sin(\omega t)$ then $i_C(t) = C \cdot \omega \cos(\omega t)$, found simply by taking the derivative of cosine, with the chain rule. Likewise, if $i_L(t) = \sin(\omega t)$ then $v_L(t) = L \cdot \omega \cos(\omega t)$.

See Fig. 4.36 (top) for a simple circuit with a resistor and a capacitor driven by an ideal sinusoidal voltage source. If you view the voltage across a capacitor on CH1 of an oscilloscope and the current through it on CH2, it would look something like Fig. 4.36 (bottom). Note that the *i-v* relationship for the capacitor alone is $v_C(t) = \sin(\omega t)$ and $i_C(t) = C\omega\cos(\omega t) = C\omega\sin(\omega t + 90)$, so current leads voltage, as ELI the **ICE**man (Chapter 3) insists it should.

Note that it is common to use frequency in rad/s and the phase in degrees, the combination of which is called "mixed units." Hopefully it will be obvious when this is done and not cause any confusion. Note also that only for ideal capacitance and inductance are current and voltage out of phase by exactly 90°. In general,

FIGURE 4.36 Relationship between current and voltage of an ideal capacitor. (top) Schematic diagram of RC circuit illustrating the relationship between current through, and voltage across, the capacitor. Channel 1, CH1, is a probe placed across a capacitor to display voltage, $V_C(t)$ and CH2 is a current probe wrapped around the wire carrying current to the capacitor displaying $I_C(t)$. (bottom) Graph of the current through, and voltage across, the capacitor. As ELI the ICEman predicts, current leads the voltage. Note that only the voltage across and current through the capacitor are out of phase by 90 degrees. There will be a different phase between the current and voltage sources.

phase differences between voltages and currents due to an impedance, **Z**, within a circuit having a combination of *R*, *L* and *C* are more complicated; they depend on all of the components in the circuit that contribute to $Z = R + jX$. ϕ = arctan (X/R), which can be different from 90°. That is discussed at length in Chapter 5. For now, we focus only on the phase, with respect to the current, of the voltage across the capacitor and the inductor individually, and discuss the implications of those phase differences.

4.5.1 First- and Second-Order Circuits

The simple filters shown in Fig. 4.21 and in Fig. 4.36 contain only one resistor and one reactive component, i.e., inductor or capacitor. Circuits having only one reactive component are called **first-order circuits**. (If you can combine two inductors together, or two capacitors together, in series or parallel to form a single component, that counts only as one. No cheating!) The input to the filters is an AC source that is positive for half of the cycle and negative for the other half. Depending on which filter we are discussing, i.e., low-pass or high-pass, they let either low or high frequencies through, respectively, but not both. For part of each cycle, energy is stored in the capacitor as an electric field, and for an inductor, as a magnetic field. During each full cycle, the component stores energy and releases that energy, which is either absorbed by the resistor or returned to the source (as discussed in Chapter 3).

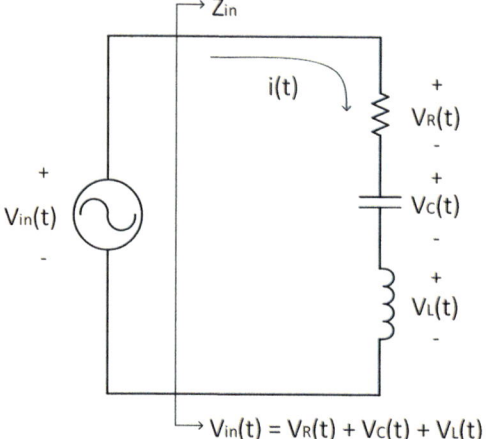

FIGURE 4.37 Series RLC circuit driven by a sinusoidal voltage source. The bracket labeled 'Z_{in}' indicates that everything to the right of it can be lumped into a single impedance value that is called Z_{in}. As is always the case with Kirchhoff's voltage law, at every moment in time, the voltages across the three load components must sum to that of the input.

A circuit that contains two reactive elements, i.e., both a capacitor and an inductor, is called a **second-order circuit**. Figure 4.37 shows a series RLC second-order circuit. The large bracket labeled 'Z_{in}' means that everything to the right of it can be collected together as one impedance called Z_{in}. The same current passes through all the series components in Fig. 4.37. The loop equation (Kirchhoff's voltage law) is:

$$v_{in}(t) = v_R(t) + v_C(t) + v_L(t). \tag{4.50}$$

Noting that for an inductor $v_L(t) = L\,di(t)/dt$, and for a capacitor, $i(t) = C\,dv_C(t)/dt$ or $v_C(t) = (1/C)\int i(t)\,dt$, we get:

$$v_{in}(t) = i(t)R + \frac{1}{C}\int i(t)\,dt + L\frac{di(t)}{dt}. \tag{4.51}$$

Taking the derivative of each side, we find that:

$$\frac{dv_{in}(t)}{dt} = \frac{1}{C}i(t) + R\frac{di(t)}{dt} + L\frac{d^2i(t)}{dt^2}. \tag{4.52}$$

Note that this is a second-order differential equation (hence the term *second-order circuit*) that would have to be solved to find the voltages and currents. Don't worry – this is not a differential equations course, and we won't try to solve it here. An easier method is to use phasors, as presented in Chapter 5.

4.5.2 Resonance in a Series RLC Circuit

The second-order circuit behaves very differently from a first-order circuit having only one reactive element. (This will look somewhat familiar to you, as this is the basis of power factor correction in Chapter 3.) Let's say that the resistor is our "load," meaning it is the component of the circuit that we need to drive with current to get some kind of output, such as a stereo speaker, motor, or the like. You can think of a speaker as a resistance in series with some extra capacitance and inductance. Some of it is there on purpose and some of it is there as an unavoidable result of the wiring. (Chapter 5 will go into this in considerable depth.) The point is that many circuits have all three components, R, L and C, in them, and the behavior is somewhat complicated.

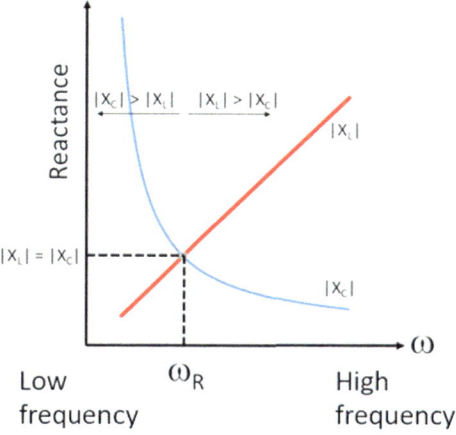

FIGURE 4.38 Plot of magnitudes of capacitive reactance, $|X_C| = 1/\omega C$, and inductive reactance, $|X_L| = \omega L$, as a function of frequency. The reactances have opposite signs, so their voltages are out of phase by 180°. When they are equal in magnitude at the resonant frequency, their voltages cancel perfectly.

Refer to Fig. 4.38. At low frequencies, the inductor has a low impedance ($|Z| = |X| = \omega L$ is low) and, at very high frequencies, the inductor has a high impedance and acts almost like an open circuit. At very low frequencies, the capacitor impedance is very high ($|Z| = |X| = 1/\omega C$ is high) and it acts like an open circuit. At high frequencies it is very low and acts like a short circuit. Therefore, the total impedance of series R, L and C is high at both ends of the frequency range because one of the series components will have a high impedance; the current out of the source is, therefore, small, as shown in Fig. 4.39. At some intermediate frequency the impedance of each reactive element is lower, and the current increases as the frequency approaches ω_R, as shown in Fig. 4.39. However, that is not the whole story, as explained next.

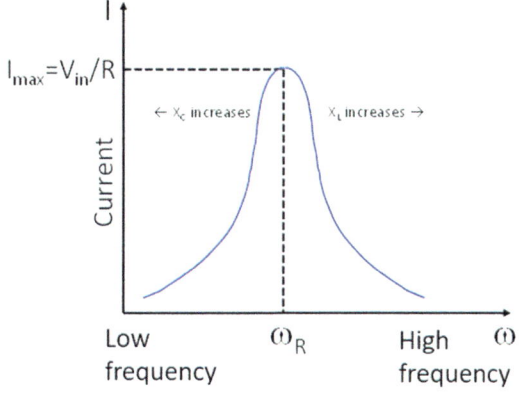

FIGURE 4.39 Current in a series RLC circuit as a function of frequency. At low frequencies the capacitive AC reactance dominates and at high frequencies the inductive AC reactance dominates. At the resonant frequency, the voltages across the inductor and capacitor cancel each other, and the only voltage drop is across the resistor, so the current is a maximum.

The previous paragraph does not tell the whole story because it does not include the phases of the voltages across all of the components. It is important to understand that because of ELI the ICEman, for the series circuit in which the current is the same for all of the circuit components at any time, the voltages across L and C are 180° out of phase with each other, i.e., they are the 'opposite of each other.' This happens because $v_L(t)$ leads the current by 90° and $v_C(t)$ lags the current by 90°. This implies that the voltages are 180° out of phase, i.e., their voltages are always of opposite polarity, as shown in Fig. 4.40 (top). This further implies that the total voltage across the two of them combined is less than either of them alone. At some frequency, called the **resonant frequency**, ω_R, the magnitudes of the reactances of the inductor and capacitor are equal, $|X_L| = |X_C|$, as shown in Fig. 4.38, while they have opposite signs, i.e., $X_L = -X_C$. This situation is called **resonance**. Because the reactive elements have opposite voltages (they are 180° out of phase) we then have the condition that $v_C(t) + v_L(t) = 0$, as shown in Fig. 4.40 (bottom). At ω_R, therefore, all of the input voltage appears across just the resistor.

The most current that can possibly flow in the circuit at any frequency is $i_{max}(t) = v_{in}(t)/R$, which happens at resonance as shown in Fig. 4.39, as if there were no reactances and therefore no other voltage drops. At any frequency other than ω_R, the magnitudes of the reactances are not equal, so some voltage is dropped across the series combination of L and C. At frequencies "off resonance," the current is less than it is at resonance, again as shown in Fig. 4.39.

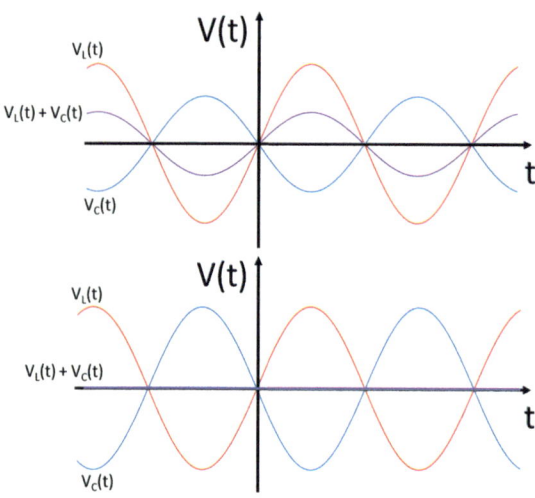

FIGURE 4.40 (Top) Graph of voltages across the capacitor and inductor that are out of phase by 180° but have unequal magnitudes. Although they are out of phase, they do not completely cancel. (Bottom) Graph of voltages that are of equal amplitude at resonance and cancel perfectly.

4.5.3 What, Physically, Is Resonance?

Resonant circuits are at the heart of communications circuits of all kinds, including analog radio tuners (like your AM radio), which respond only to a narrow range of frequencies. Resonant circuits are discussed more in Chapter 5, and radios are discussed in Chapter 6.

It is worth taking some time to get a sense of the concept of resonance. Imagine a child sitting on a playground swing. You might recall from a physics course that the natural frequency of a pendulum is $\omega = \sqrt{(g/l)}$, where g is acceleration of gravity and l is the length of the pendulum. When a pendulum swings, energy is converted from stored potential energy at the peak of the arc to kinetic energy at the bottom, and back to potential at the other peak. If the child on the swing (pendulum) is raised to a certain height and let go, the child will swing with the same period, but decreasing amplitude, as energy is lost through damping due to air resistance and friction with the rope. The role of the adult is to replace that energy at every period to keep the child swinging at a constant amplitude (and to keep the child happy). The adult knows intuitively when the swing is at its maximum height and how to add just the right amount of energy to push the swing, thus precisely replacing the energy lost during the previous cycle.

> This baby swing has a motor that keeps it swinging back and forth. The motor replaces the energy lost to damping in the system.

Now imagine that a tired parent (with a degree in electrical or computer engineering) programs a robot to swing the child in the parent's place. The robot is programmed to push with the same frequency as the swing's oscillation. If the robot were perfect, it would push on the swing at the very top of its cycle and put just enough energy into the swing to replace energy lost to damping in the previous cycle. Let's imagine instead that the frequency of the robot is not set correctly. The robot might start out pushing the swing at just the right moment, but over time gets out of sync with the swing's oscillation. Sometimes it gives just the right push, sometimes it misses, and sometimes it may even slow down the swing's oscillation by pushing while it is still coming toward the robot. On average, the swing will just stop swinging. Not a happy child.

In true engineering fashion, the parent reprograms the robot to push only when the child is exactly at the top of the arc near the robot. The robot pushes exactly enough to counteract the damping loss from each swing, and the child is happy. This is dangerous business, though, because if the robot pushed a little too hard, the swing might gain more energy than is lost at each cycle, and the child would eventually be pushed too high and might be injured. (The lesson for parents is that it is better to push the child yourself rather than trust that to a robot – for now, anyway.)

Resonance in the RLC circuit works like the swing. Instead of conversion between kinetic and potential energy for the swing, energy flows out of the field of one reactive device and into the field of the other one. When $|X_L| = |X_C|$, i.e., $1/\omega C = \omega L$, the magnitude of their impedances is identical, and the frequency is the resonant frequency, ω_R. Since the current through both is the same, the voltage across both will be the same, but out of phase by 180° (Fig. 4.40 (bottom). However, with a resistor in the circuit, as the current flows, some of that energy is converted to heat, as the swing loses energy to air resistance, etc.

Let's leave the resistor out for now and have just L and C in series. If the input voltage were set to zero (which is the same as a short circuit) and you charged up the capacitor, thus storing energy in its electric field, and then let it completely discharge, it would do so as current through the inductor, which would store all of that energy in its magnetic field. After the energy had been converted to magnetic energy, the magnetic field would then decrease as current flows out of the inductor and charges up the capacitor. In this ideal case, i.e., with no damping due to resistance, the energy flows back and forth between L and C – forever. That would be the equivalent of the child swinging with no damping, i.e., no frictional losses, in which case the parent could just sit and watch while doing WhatsApp on their phone (*not advised*).

On a real swing, the child would stop swinging due to losses in the swing system if the parent stopped pushing. Likewise, resistors provide 'friction' in electrical circuits, that is, they heat up and dissipate energy. If we include resistance due to the wires of the inductor and capacitor (or add a resistor to the circuit, as we have in Fig. 4.37), energy would be absorbed during each cycle as current passes through it ($v = iR$; $P = i^2R$), and the oscillations would die out, but the frequency would remain nearly the same as it decays (under the right conditions), just as it would for the swing as it slows down to stop. The waveform of such a **damped sinusoid** is shown in Fig. 4.41.

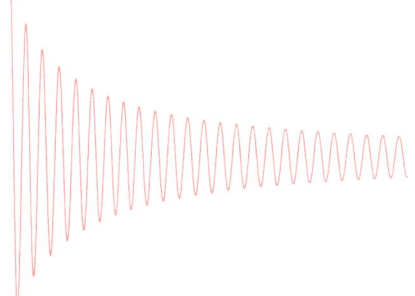

FIGURE 4.41 Damped sinusoid typical of energy loss in a resonant circuit.

Let's consider a nearly ideal LC circuit, i.e., one that has little resistance. When we add an AC source that provides a sinusoidal voltage, it drives current through the series LC pair. At resonance, $|X_R| = |X_L|$. Assuming that R is low (because that could change things a little bit, which we will ignore), and solving $1/\omega_R C = \omega_R L$ for ω_R, we get that $\omega_R^2 LC = 1$, or:

$$\omega_R = \frac{1}{\sqrt{LC}}. \tag{4.53}$$

When the frequency of the source is ω_R (as if the robot were pushing the swing at exactly the right time every time the swing gets to the top of its arc), it replaces the lost energy at every cycle. *Interestingly, the voltages across the inductor and capacitor at resonance can be quite large, even much larger than the input voltage, but their sum is zero at resonance. This does not violate Kirchhoff's voltage law since the voltages still sum to zero around the loop.* You will see this for yourself in the activities of Chapter 5.

If the source voltage frequency is far from ω_R, its oscillations will not cause the LC pair to resonate, and the pair will gain little energy from the source. The voltage from the AC source will be either higher (hence, charging) or lower (hence, discharging) than the voltage of the LC pair at random parts of the charging cycle, just like the robot pushing at the wrong time, and on average will transfer little energy to the pair. In summary, both views of the resonant behavior of the system are equally valid – viewing the circuit elements as impedances that are equal and opposite to each other summing to zero, or as resonant elements passing energy back and forth to each other. The series RLC circuit is only one example of a second-order circuit, and we have discussed its properties. Other configurations will also be resonant, but the resonant frequency will be a different function of its component values, i.e., $\omega_R = 1/\sqrt{(LC)}$ will not necessarily apply and will be more complicated.

4.6   The Graphic Equalizer

The **graphic equalizer** (**EQ**) (Fig. 4.42) is not a piece of electronic test equipment, but is, rather, a piece of stereo equipment used by hi-fi enthusiasts, or professional audio gear used in recording and live concerts. The "EQ" is introduced here as a real-world example of filters and will be used in the activities to make certain filter behavior available to you.

You might know that stereos have "bass" and "treble" tone controls. You might even know that some have a "mid-range" tone control that brings out voices from the music. Normal tone controls can be adjusted to either amplify (increase) or attenuate (decrease) the loudness of music that is played within that narrow band of frequencies. Now imagine that you have separate tone controls for 'really really low bass,' 'really low bass,' 'low bass,' 'bass,' 'higher bass,' ... 'low high notes,' 'middle high notes,' 'higher notes,' etc. The graphic equalizer does exactly this, and it is characterized by its number of bands.

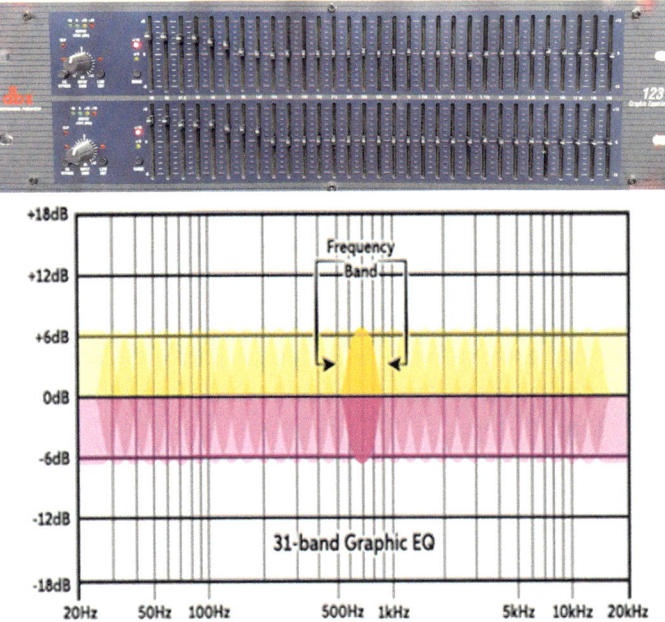

You will use a 31-band EQ (dbx 1231 from Harman International, Fig. 4.42, top) on your bench for this chapter's lab to demonstrate some properties of the frequency spectra that make up the waveforms available on your function generator. Each vertical slider controls the gain of an amplifier whose bandwidth is very narrow. The gain due to the position of the sliders is represented in dB relative to the zero position. As stated above, voltage gain in decibels is expressed as $G = 20\log(A_V)$ where A_V is the ratio of the voltage out of the EQ to that at the input to the EQ, i.e., $A_V = V_{\text{out}}/V_{\text{in}}$. Gain is 0 dB for an unchanged signal, i.e., is a gain of $A_V = 1$.

In Fig. 4.42, the vertical sliders in the middle and high frequencies (to the right) are shown as set to zero gain, which is the middle of the vertical range. This allows the signal to pass through to the output unaffected, i.e., "changed by 0 dB." When the sliders are in the upper half of the range, as shown for the low frequencies to the left, those frequencies

FIGURE 4.42 (Top) dbx 2231 dual 31-band graphic equalizer. Each slider controls a narrow band of frequencies. (Bottom) Frequency spectrum of all bands in a 31-band EQ. Notice that the bands overlap each other. Each band covers 1/3 of an **octave**, where an octave is a factor of two in frequency. When all the bands are at the same level, all the gains contribute to a flat frequency spectrum.

are amplified (positive dB), and when they are in the lower half, the frequencies are attenuated (negative dB, not shown). You may have already used this function in your phones to adjust the sound of your music, as mentioned at the beginning of this chapter. Pushing a slider up, to positive dB, i.e., greater than 0 dB, makes that narrow band of frequencies louder; the output voltage in that range of frequencies within the music is amplified. Pushing a slider down makes that band of frequencies quieter, i.e., the signal is attenuated, and gain less than 1 in dB is negative. For this particular EQ, pushing the sliders all the way up provides a gain of +15 dB and all the way down produces a gain of –15 dB.

Each slider controls a narrow range, or **band**, of frequencies that overlaps to some extent with the neighboring band, as shown in Fig. 4.42, bottom. For example, the '1K' slider will have some effect on frequencies between about 800 Hz to 1.25 kHz. If you step back from an EQ and take in the entire row, the positions of the sliders make a graph of the **equalization** of the audio signal. Strictly speaking, the term "equalization" is used when the goal is to make all frequencies of equal amplitude, so that the output sound most exactly matches the original source of music. However, most often an equalizer is used to change the frequencies to suit individual tastes in music. In this chapter's lab activity, you will use the full function of the EQ as a filter, either high-pass, low-pass, or band-pass, to investigate the frequency components of waveforms from the waveform generator.

Here is an example of equalization in the strictest sense. Starting in about 1954, the Recording Industry Association of America, RIAA, standardized the RIAA equalization curve for phonograph records. (You know- those large, round, flat, black spinning things that people used to listen to all the time? Perhaps you've seen them?) You see, to match the properties of human hearing, low notes in music have higher amplitudes than do the high notes (Fig. 4.43, top). Thus, making the music itself automatically undergoes a form of equalization. However, in order to play these larger amplitudes, the grooves in the record would have to be physically wider and would thus take up more physical space on the phonograph record. (This stuff is explained more in Chapter 9.) In order to fit more music on records, recording engineers attenuate the bass notes by the equalization curve shown in Fig. 4.43, bottom, dashed line, to shrink the physical width of the grooves. When played back, the stereo's phono preamplifier circuit provides the opposite equalization; it electronically boosts the low notes according to the curve shown at the top, solid line, of Fig. 4.43 in order to counteract the original equalization and recreate the original frequency spectrum of the live music. For one reason or another, nearly every type of recorded music is equalized in some way. Now you understand that you build your own equalization curve for your earbuds whenever you change the properties of the sound on your

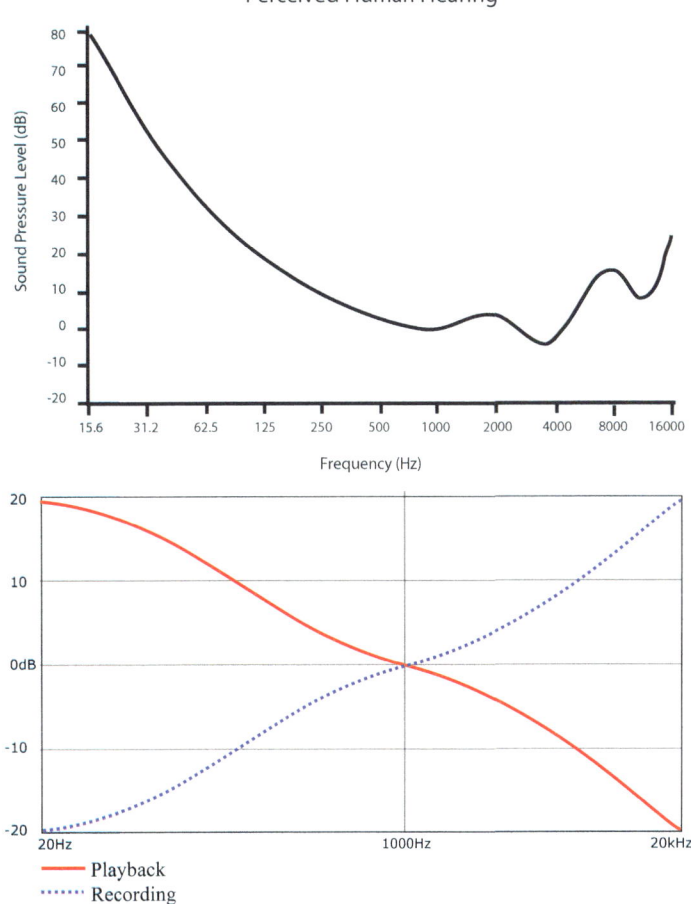

FIGURE 4.43 (Top) Perceived equal-loudness contour of human hearing. (Bottom) The RIAA equalization curve adopted in 1954. This equalization is encoded into every modern phonograph record.

phone or MP3 player (as if anyone still uses those) by moving the sliders of the built-in (software) equalizer.

So, we come to the end of this exposition on time and frequency domains. You should really try to grasp the basic ideas because it will make your formal introductions to this material in later courses much easier. You will then be able to focus on the mathematical details rather than on the concepts, which can be difficult if you are trying to get them both at the same time. Also, this material is a portal into important and advanced science and engineering concepts including such varied subjects as electronic circuits of all types, the way bridges sway, how telescopes work, the way electrons move through crystals, and how electron microscopes can image single atoms. If you can obtain an appreciation for this material, you will be among the somewhat elite group of people who have entered the realm of advanced science and mathematics.

4.7 | Lab 4 Activities

Bring wired earbuds or headphones to this lab as well as a music source. Also don't forget to bring a thumb drive to store your scope screenshots. If you don't have these, they will be provided for you.

1. **Choosing RC values for f_c of about 1 kHz.** Refer to the schematic in Fig. 4.44. Look in your parts kit and find an approximately 0.005 µF (5 nF) capacitor. You will want to fairly accurately know the actual value. Turn on the Keysight U1733C LCR meter. (Fig. 4.45). Attach the capacitor to the leads. Press **Frequency** and select **1 kHz**. Press **ZLCR** and select **C**, indicating that it is measuring capacitance. Record the capacitance and calculate the resistance for a cutoff frequency ($f_c = 1/(2\pi RC)$) of 1 kHz. Connect long wires from the breadboard wire kit to the leads on the JET RS-200 resistance substitution box and use it as your resistor. That box allows you to select a wide range of resistances. Use the LCR meter to measure the resistance by changing **ZLCR** to **R** at 1 kHz as before. (Or, if you prefer, use your DMM ohmmeter). Make changes to get close to your desired **R**. Build the circuit using the breadboard. What capacitance did you measure and what resistance value did you select? What value of f_c did you calculate? Is this circuit an LPF or HPF? Explain.

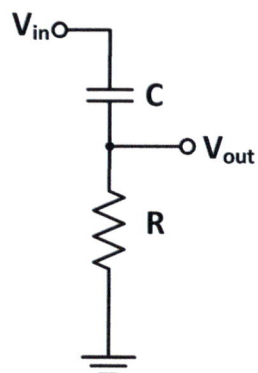

FIGURE 4.44 (Steps 1, 2, 3). RC filter circuit #1.

2. **Sweeping the frequency to observe HPF behavior.** Set the output of the Rigol waveform generator to 1 kHz at 10 V_{pp}. Connect the signal from the output of the waveform generator WG) to V_{in} of the circuit in Fig. 4.44. Connect the ground from the WG to the grounded lead of the resistor. Place a BNC 'tee' on the output of the WG and connect it to CH1 of the scope with a BNC cable. Use a scope probe to connect the voltage across the resistor, V_{out}, to CH2. You can put the ground connection of the probe to the same ground point as that of the generator. If you were to connect it elsewhere you would "ground out" one of the circuit elements. Set the trigger to CH1 and see if the trace is stable.

You could leave the probe ground disconnected, but that might introduce interference (aka "noise.") Try it both ways. When noise is a problem, since many environments

FIGURE 4.45 (Step 1) Keysight U1733C LCR meter.

are electrically noisy, the trace can be erratic due to triggering on the noise instead of the signal. One way to improve it is to sync the trigger as follows (do this whether or not you need it now): Connect the WG CH1 **Sync** output to the **Ext** input on the scope. Press the **Trigger Setup** button and under **Source** select **All** and **Ext**. That should calm things down a bit.

For this part you will need the sweep function of the waveform generator (explained below). This provides a sine wave output whose frequency increases and/or decreases between two values over a certain amount of time.

Display both channels, V_{in} and V_{out}, so that the whole waveforms are visible on the screen. Make sure that both channels have **Averaging** turned off, i.e., the scope waveform is not averaged over multiple traces, as it would interfere with this measurement.

To use the sweep function on the WG: Press **Sweep**, located between **Mod** and **Burst** on the front panel. Choose **Linear** sweep. **SwpTime** selects the duration of the sweep. **Return Time** selects the time it takes the waveform to reset to the starting frequency, like a reverse sweep. **Start** and **End** change the range of the frequencies swept. The rest of the parameters will not be used in this lab and should not be changed. Set the **Start** value to 100 Hz and the **End** value to 2 kHz. Set **SwpTime** = 3 s and **Return** = 0 s.

Set the horizontal time base to 1 ms/div. If you don't see a large change in amplitude due to your chosen RC cutoff frequency, expand the range of frequencies until you do. Also, the sweep time is not critical. Choose a smaller or larger value that lets you really see what is going on. The same with the time base. Change all of these until you get a good sense of what a HPF does. Compare V_{out} (CH2) with V_{in} (CH1).

Recall that the capacitor looks like a lower resistance as the frequency increases, so the voltage is dropped more on the resistor, and V_{out} increases as frequency increases. You should see the amplitude of V_{out} (across the resistor) start at a low voltage at lower frequency and increase to V_{in} at higher frequencies. Describe what happens to the amplitude of the voltage across the resistor as you increase the frequency from low frequency to high frequency. Thinking back to internal resistance of the WG from Chapter 2 (R_{int} = 50 Ω), and based on your choice of R, did you expect V_{in} change during the sweep? Why or why not?

3. **Plotting HPF behavior.** Turn off **Sweep** on the WG. Set the **Averaging** to 3 sweeps (on the scope) for both channels. Using individual frequencies, use cursors to measure V_{out} (CH2). (If there is noise, the measurement function can give you errors). Record 10 values as a function of frequency, spanning the range from low to high over which the output changes considerably. Include f_c = 1 kHz as one of your frequencies. Later, for your lab report, make a linear-linear plot of the output voltage as a function of input frequency. Show zero and V_{in} volts as the lower and upper limits. Does it look like a high-pass filter? What is the ratio of V_{out}/V_{in} (i.e., the gain) at f_c = 1 kHz? Is it (close to) 0.707 as it should be? Qualitatively explain the results you observe in terms of the reactance of the capacitor as a function of frequency.

4. **Changing a HPF to a LPF.** Now switch the positions of the resistor and the capacitor as shown in Fig. 4.46. Set the sweep frequency of the WG to go from 100 Hz to 5 kHz. Repeat Step 2. Turn off averaging. Explain why the circuit is now an LPF. How does the voltage across the capacitor, V_{out}, change with frequency?

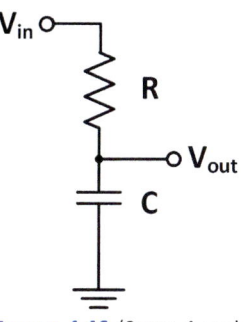

FIGURE 4.46 (Steps 4 and 5) RC filter circuit #2.

5. **Plotting LPF behavior.** Take data for ten frequencies as you did in Step 3 and make a plot in your lab report. Show your plot in your lab report.

Does it look like an LPF? What is the gain at f_c? Again, explain the results you observe in terms of the reactance of the capacitor as a function of frequency. Compare the results in Step 3.

6. **Observing phase shift due to a reactance.** In this next step you will look at the phase shift between the input and output voltages. You should see the phase change as the frequency increases. However, you will need to do this by sweeping the frequency manually. You already know that the phase between current and voltage is related to the reactance, but for now just note that Chapter 5 will explain the phase shift in detail. For now, just observe that the phase does, in fact, depend on frequency.

Turn off **Sweep**. Set the frequency to 100 Hz and use the control knob to manually sweep the frequency upwards to a few kHz. As you change the frequency, also change the time base so that you can see the phase. As expected, the amplitude of V_{out} decreases as frequency increases, but now you should watch the phase between V_{in} and V_{out}. Note qualitatively the phase shift as shown in Fig. 4.47 at various frequencies, in particular how the phase moves between 0° and 90°. State what happens in general terms. You do not need to understand it in detail. Note that this dependence of phase versus frequency is due to the reactance of the capacitor. In this chapter we used the magnitude of the reactance only in reference to the amplitude of the output of the filter while ignoring the phase relationships. Here you can see that the phase also changes.

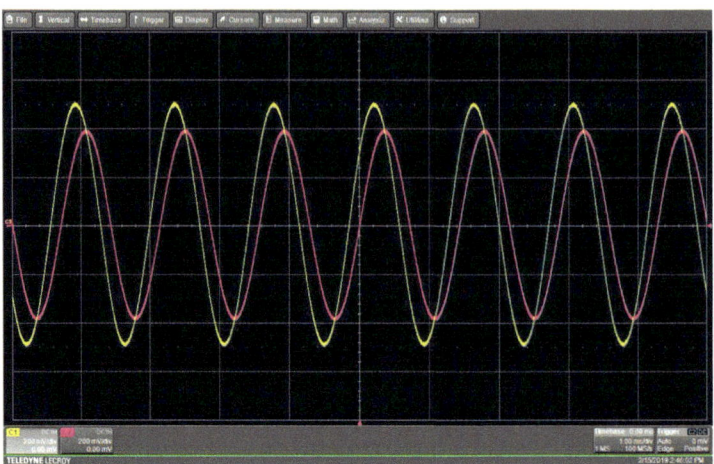

FIGURE 4.47 (Step 6) Phase shift between the input voltage and the voltage across the capacitor.

Skip this paragraph during the lab. Something to think about *later* that explains the phase behavior at low and high frequencies: *At low frequencies,* X_C is large, so the input voltage drops mostly across the capacitor. You would then expect the phase of the input voltage and the capacitor voltage to be nearly the same. *At high frequencies,* X_C is small and nearly all of the input voltage is dropped across the resistor. When this happens, and because $I = V/R$ with no phase shift, the current is nearly in phase with the input voltage. Recall that voltage lags current by 90° for a capacitor, ICE. So, since V_{in} is in phase with the current, and V_C lags the current by 90°, then V_C lags V_{in} by 90°.

7. **Observing a *fast* rise time.** In this and the next few sections you will observe the charging of the capacitor and compare the rise time to the period of a square wave. Use the LPF circuit of Fig. 4.46 where $f_c = 1/(2\pi RC)$. Change the resistor to provide a cutoff frequency that is about 10× higher than 1 kHz. Decreasing R decreases the RC time constant, τ. What is your new RC time constant? Switch the input from a sine wave to a square wave at 1 kHz. What is the period, T? Make the vertical scale the same for

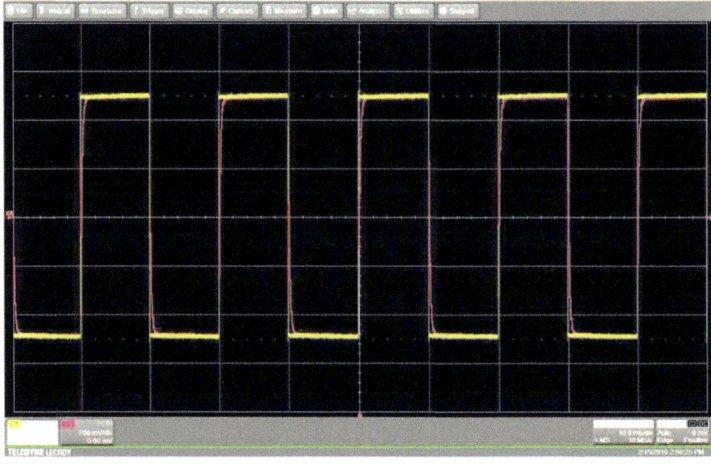

FIGURE 4.48 (Step 7) Output of the low-pass filter showing fast rise and fall times relative to the long period of the square wave.

both channels. Save a screenshot comparing the voltage waveform across the capacitor, V_{out}, to the input square wave, V_{in}. How does RC compare with the period, T? The rise time of this RC circuit is very short compared to the period, so the square wave at V_{out} is still pretty square (Fig. 4.48). The voltage across the capacitor, V_{out}, appears to rise nearly as quickly as the input because the charging time, i.e., the time constant, is short compared to the period.

8. **Increasing the frequency for a given τ.** Increase the frequency from 1 kHz while expanding the time base (horizontal scale) until the capacitor voltage can no longer charge fast enough to saturate at the input voltage. Save a screenshot. What happens to the capacitor voltage waveform as you increase the frequency? What waveform does it start to look like? What is τ compared to this shorter period ($f = 1/T$)? Explain what is happening in terms of charging and discharging the capacitor.

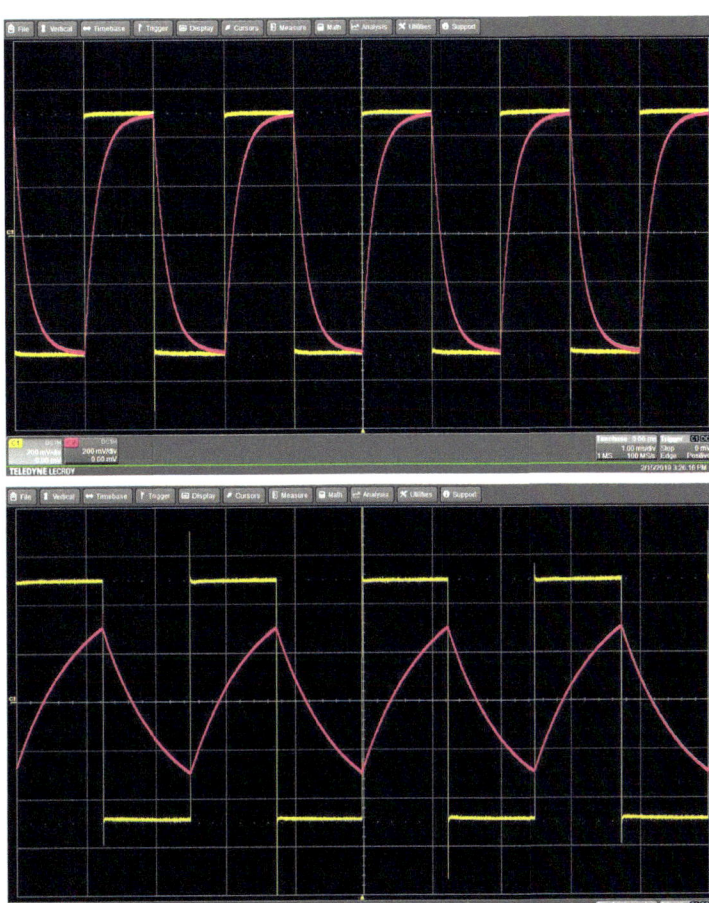

FIGURE 4.49 (Steps 8 and 9) Output of the low-pass filter showing slower rise and fall times relative to the shorter period of the square wave. Top, 2 ms, 500 Hz, and bottom, 200 μs, 2 kHz.

9. **Observing a *slower* rise time.** Reset the frequency to 1 kHz. Increase the resistance ($\tau = RC$) so that the rise time is a substantial fraction of the period of the square wave, as shown in Fig. 4.49, top. What value of resistance are you using, and what is τ? Save a screenshot. In this case, τ has been increased, so now the waveform takes a significant fraction of the period to rise to the peak voltage of the square wave. Just as you did in the previous step, explain why the waveform is no longer a square wave.

10. **Observing a *slow* rise time.** Again, increase τ so that the waveform looks roughly like a sawtooth, as shown in Fig. 4.49, bottom. What value of resistance are you using, and what is τ? Save a screenshot. For your lab report, explain the distortion in terms of charging

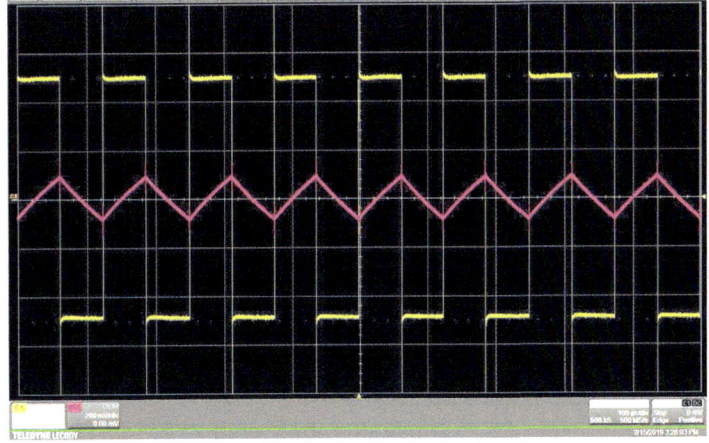

FIGURE 4.50 (Step 10) Output of the low-pass filter showing very slow rise and fall times relative to the much shorter period (100 μs, 8 kHz) of the square wave.

and discharging the capacitor and τ. Steps 22 and 23 will relate the new shape to the frequency components at the output of the LPF.

11. **Observing a very slow rise time.** For the last time, increase τ yet again until the sawtooth gets very small in amplitude compared to the input square wave, as shown in Fig. 4.50. Save a screenshot. Now the rise time is so slow compared to the period that it barely has a chance to rise. Can you reason out what must happen in the frequency domain to make this happen? You can wait to answer this question until after Steps 22 and 23 if you don't know the answer.

12. **Setting up the FSA to observe the FS of a sine wave.** Now to the frequency spectrum. You will learn to use the frequency spectrum analyzer (FSA) to examine the frequency components of your time-domain waveforms. A frequency spectrum (FS) is calculated by the fast Fourier transform (FFT) algorithm using the data that is the waveform stored in the oscilloscope memory but *uses only the data that you can see on the screen.*

Disconnect the probe from the circuit and move it out of the way. Create a sine wave of 10 V_{pp} at 1 kHz. Make sure that the waveform does not extend beyond the top or bottom of the window because the frequency spectrum analyzer, FSA, operates only on data that is visible on the screen. Change the time base to show many periods.

Push the **Spectrum** button on the panel (lower right button). The window splits into two sections. The top is the waveform and the bottom is the frequency spectrum (FS) of that waveform. To change the waveform on the top panel, you must select the channel first. Selecting the **SpecAn** tab allows you to make changes to the spectrum panel.

Now select the **SpecAn** tab to access the FSA settings. Choose the **Start Stop** option under the **Frequency & Span** box and choose a **Start Freq.** of 0.0 kHz and **Stop Freq** of 5 kHz. You may change these later. Press **Output** and select V_{RMS}. *Sometimes you have to toggle between the various Output choices before the one you want is activated. Watch out.*

Select a **Mode** of **Average** to make a cleaner-looking spectrum. Averaging 3 traces is a good number. After you make changes, it will now take a few seconds to compute the average. Pressing the **Vertical** button places the FS in the panel, but the vertical size is usually small. Adjust the vertical scale and **Vertical** offset if you need to in order to change the full height of the largest FS peak. If you change the vertical scale, you will need to adjust the vertical offset. *The vertical offset knob is very delicate and must be rotated very slowly. It takes patience. Warning: this is potentially time consuming, so be patient and rotate knobs slowly.*

In order to avoid having to read grid lines or cursors, display a data table of all of the peaks: Press **Peaks/Markers**, then **Peaks**, **Show Table**, and **Show Frequency**. Do not set the center frequency to any peak. Ignore that feature. Compare to Fig. 4.51. You will refine your spectrum in the next step. As usual, take a screenshot.

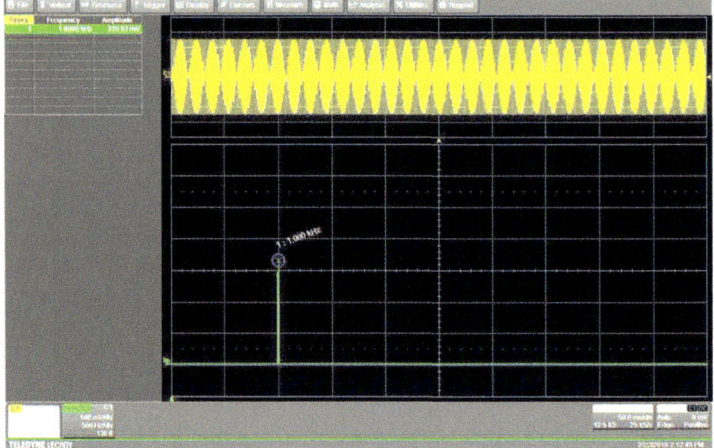

FIGURE 4.51 (Steps 12 and 13) FFT of a 1 kHz sine wave. There is only one peak in the spectrum of a sine wave because the FFT of all waveforms is made up of sine waves, and only one sine wave is needed to "describe itself" in the frequency domain. Notice how smooth the plot is where there are no frequency components.

13. **Choosing many periods of the waveform and relating the peak of the FS to it.** Be aware that the LeCroy scope ignores data that is not displayed on the screen. Also, the peaks of the FFT get sharper when the FFT has many periods to calculate from. Select CH1 and change the time base so that you get many periods of the sine wave in the top panel. It might be easier to just turn off the FSA and do this strictly on CH1. When you go back to the FSA, more periods show up on the top screen and the time-domain trace becomes impossible to view in detail. That is not a problem for the FSA because all the points are stored in memory, whether or not they are discernable on the screen. If you choose such a long time base that you cannot see details of the waveform, or the waveform looks distorted (see Fig. 4.51), that will *not* cause errors in the FS. In general, more periods of the waveform are better, but too many can take uncomfortably long to capture the waveform. You will get a sense of this through practice.

   Press CH1 and change the time base to get just a few and then many periods of the sine wave. Capture a couple of screenshots demonstrating the effect on the shape of the FS peak. Describe what you observe in the FS as you change the time base from showing only a few periods to many, many periods. The peak should have sharpened as you displayed more periods.

14. **Meaning of the value of the FS peak.** Determine the peak height. The peak should have a height of the RMS value of your 10 V_{pp} waveform, i.e., 3.5 V_{RMS}. In your notebook, show that 10 V_{pp} is 3.5 V_{RMS}. Why is the height of the peak in the FS the same as V_{RMS} from the WG?

15. **Relating the FS in dB to the waveform.** In this step you will learn about viewing the spectrum in decibels, dB, rather than the linear mode of V_{RMS}. Measurements in decibels are more complicated. *You must review the notion of decibels in the chapter. If you haven't read it, you have no chance of understanding these steps.* As a brief reminder, decibels are based on the logarithm of a *ratio*. A voltage expressed in dBmV is that voltage referenced to (i.e., divided by) one mV and is equal to V_{dBmV} = 20 log ($V_{RMS}/0.001$ V_{RMS}). For example, 5 V_{pp} = 2.5 V_p = 1.773 V_{RMS}; the related value in dBmV

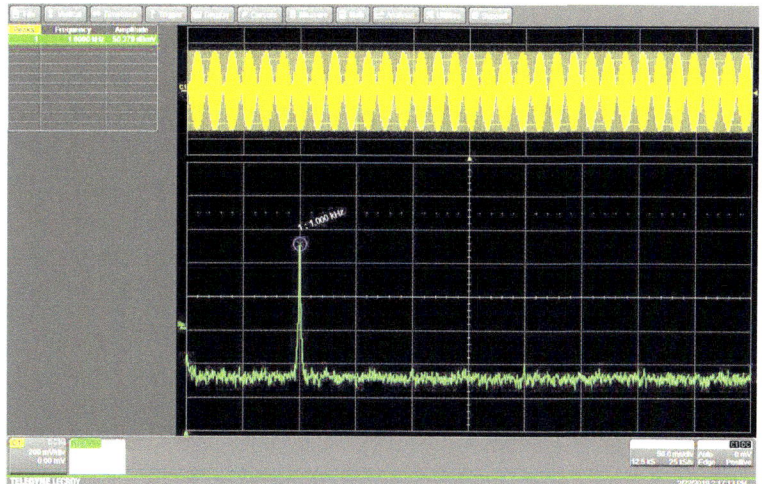

FIGURE 4.52 (Steps 15 and 16) FFT of a 1 kHz sine wave in dBmV. The "grass" seen in the figure is the background noise, which is always present in some form or other. It is the nature of displaying voltages in dB that large differences in voltage are made measurable by enhancing very small voltages. This is why the background noise that was not visible when viewing the FFT in linear mode, i.e., V_{RMS}, is now visible. The number of dB of the noise is called the **noise floor**.

is V_{dBmV} = 20log(1.773 $V_{RMS}/0.001$ V_{RMS}) dBmV = 65 dBmV. Set the **Scale** to **dBmV** by pressing the **Output** box and selecting **dBmV**. Compare to Fig. 4.52. What value does the peak measure in dBmV? What do you calculate that it should be?

16. **Calculating voltage in dBmV and interpreting the noise floor.** Calculate the voltage level as expressed in dBmV that you expect for your sine wave (as demonstrated in Step 12). Using the measurement in dBmV, confirm your calculation. Read the caption to Fig. 4.52 to learn about the **noise floor**. You might notice that the table and the display using the dB scale show several small peaks rising out of the noise floor because logarithms enhance small numbers, as discussed in the chapter. A perfect sine wave would have only one peak. Any imperfections show up as extra frequency components. You may ignore all of those since they are actually very

small in amplitude and are effectively nonexistent when viewed in V_{RMS}. However, lots of extra taller peaks might mean that the sine wave suffers from significant harmonic distortion, i.e., it is a poorly formed sine wave, or there is some interference creeping into the measurement. Turn on the cursors and move them around. Notice that they read out in the **SpecAn** box in dB. Determine the noise floor in dB. What is the dBmV difference between the noise floor and the peak? That difference in dB is related to the ratio of their voltages.

(a) Converting back from dBmV to voltages and calculating a ratio, how many times larger is the peak compared to the noise floor?

(b) Another way to do this is to calculate with dBs: The ratio of two voltages in dB is the difference in dB between the two different dB measurements, regardless of the type of dB (dBmV, dBV or something else). For example, if the peak value is +25 dB and the noise floor is –15 dB, the difference is 40 dB, which corresponds to a ratio of $10^{40/20} = 100\times$. Do it both ways to try this out.

17. **Calculating voltage in dBV.** You can also express a voltage as dBV, which is the voltage referenced to 1 V, rather than 1 mV as in dBmV. The units of dBV are found from the same calculation as dBmV, but referenced instead to 1 V. Set the **Scale** to **dBV** by pressing the **Output** box and selecting **dBV**. Use the cursors again to measure the peak and the noise floor. What are the peak and noise floor values in dBV? How have those numbers changed? Now re-calculate the ratio of the peak to noise floor. Did it change? Why or why not?

Change the WG output from 10 V_{pp} to 1 V_{pp}. Now you would expect to get a negative dBV reading because your waveform is 0.35 V_{RMS}, which is less than 1 V_{RMS}. Measure the peak in dBV and calculate the RMS value of the peak from that. Does your calculation agree with the output from the WG?

You should understand that the reason the same voltage can have either a negative or positive value in dB is that it depends on the 'flavor of dB' you choose. It is due simply to the nature of the log of a ratio that is less than or greater than 1. For example, your 0.35 V_{RMS} has a positive dBmV value but a negative dBV value due simply to the size of the ratio inside the log term.

18. **Demonstrating amplifier gain in dB.** In the following steps you will use an audio graphic equalizer (EQ) as a versatile filter to investigate the frequency spectra of various waveforms. Refer to Fig. 4.53. The EQ has two banks of 31 *sliders*, where each bank, top and bottom, is a different channel, usually left and right in a stereo. Each slider controls the output of a narrow range of frequencies, called a **band**. The range of frequencies affected by each slider is shown in Fig. 4.42, bottom. Sliding the control up makes the output voltage at the center of that band of frequencies larger (i.e., amplifies it), by up to +15 dB (5.6 times larger), and down makes it smaller (i.e., attenuates it), by up to –15 dB (5.6 times smaller). Frequencies outside the band are not affected. In this case the units of dB represent the gain, or change in amplitude, of the output voltage with respect to the input voltage of the EQ.

FIGURE 4.53 (Step 16) dbx 2231 dual 31-band graphic equalizer.

Set the output of the WG to 1 V$_{pp}$ since that is in the range that audio signals, called **line levels**, into the EQ are expected to be. Again, use a 1 kHz sine wave. The EQ is stereo: one channel is the top front and back, and one channel is the bottom, front and back. For this lab, two channels are not needed, so just use the top front and back controls and jacks. Plug ¼-inch phono-to-BNC adapters (shown in Fig. 4.54 and found in the Audio box) into the input and output jacks at the back of the EQ. Use a BNC "tee" at the output of the WG to split the signal to both the input of the EQ and CH1 of the scope.

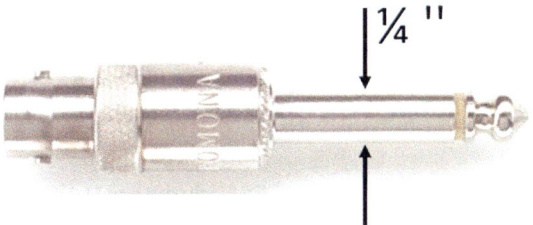

FIGURE 4.54 (Step 16) ¼ inch phono to BNC adapter.

Attach the output of the EQ to CH2 of the scope. Now you can compare the signal that goes into the EQ, CH1, and what the waveform looks like after you adjust the various frequency components at the output of the EQ, CH2. Make sure that all sliders are set to 0 dB, i.e., **flat**, across the spectrum. The spectrum is "flat" when none of the frequencies in the entire spectrum are affected by any gain or attenuation. However, the **Input Gain** knob adds gain (amplifies or attenuates) across the entire spectrum equally, which does not change the "flatness" of the spectrum.

Turn off the FSA. Using cursors if the signal is noisy, compare the RMS voltage at the output to the input for **Input Gain** (the knob on the left side) set to 0 dB. Now set **Input Gain** to +6 dB; measure and calculate the voltage you expect to measure. Do your calculation and measurement agree? Repeat for –6 dB. Calculate and show that +6 dB and –6 dB represent voltage gains of 2 and ½, respectively. Reset **Input Gain** back to 0 dB.

19. **Observing the effects of sliders on frequencies in other bands – the EQ does not create frequencies.** The order of the following steps is important. You are still using the same 1 V$_{pp}$, 1 kHz sine wave from the WG as input to the EQ. Set **Input1** of the FSA to CH2, the output of the EQ. Set the **Scale Output** to dBV. **Start** at 0 kHz and **Stop** at 5 kHz. Set **Averages** to 1. Select CH2 and set the scale to 1 V/div to make the waveform small in the upper panel. Select the FSA and press the **Vertical** button to bring the FSA trace onto the screen (its height will correspond to the amplitude shown on CH2).

Set all sliders as close to 0 dB (centered) as possible, where they have a click stop (the **detent**), i.e., flat. Move the sliders up and down one at a time starting from the lowest frequency and observe the output of the EQ in the time domain (waveform) and the frequency domain (FSA) at the same time. Be sure to set the slider back to zero each time. Note each slider changes the frequency content only within its own range of frequencies, which is not particularly narrow. There is no need to write down the result of every move of the controls until something interesting happens. What do you observe happens to the sine wave and its FS at frequencies when you move sliders that are far away from 1 kHz? Near 1 kHz? At 1 kHz? This result demonstrates that *the EQ does not 'create' any new frequencies, but merely boosts or cuts what is already there* as a frequency component of the input signal. (Note that new frequencies would be created if the EQ distorted the sine wave, which it does not.) You have also seen that the band of each slider overlaps frequencies in nearby bands, as suggested by Fig. 4.42, bottom. Explain how what you observe supports these assertions.

20. **Gain in dB affected by the sliders.** Make sure you are observing the frequency spectrum in units of V$_{RMS}$. The range of the sliders when the red light is lit is +/− 15 dB. Change the setting of the 1 k slider on the EQ by 15 dB down (all the way to the bottom, i.e., −15 dB) from the zero position as well as 15 dB up (+15 dB). Do the changes to the amplitudes in dBV on the FS correspond to changes in the amplitudes in dB on CH2?

'Up', or amplification greater than 1, is called **boost** and 'down,' or attenuation, is called **cut**. Set the 1 kHz slider back to 0 dB. For each of the sliders to either side of 1 kHz, push them one at a time all the way to the top and measure the new height of the FS peak. The gain is the difference in dB between the flat and boosted outputs. For example, if the output reads –9 dB when flat and –2 dB when boosted, the change is +7 dB, which corresponds to an amplification having gain of $10^{7/20}$ = 2.2 times. What is the maximum gain for each of the two *adjacent* sliders?

21. **Observing the FS of a square wave.** Now things get more complicated as you move from a simple sine wave to more-complicated waveforms. Generate a 1 V_{pp}, 1.11 kHz square wave with a duty cycle of 50% and pass it through the EQ with all bands set back to flat. Press EQ **BYPASS**. The input is unaffected by the EQ, and the output should be essentially identical. Now turn off the EQ **BYPASS** so that the square wave is going through the EQ. There might be a very slight difference, but this will not affect your results. Any sliders that are not exactly at the detent can affect the shape of the output square wave, so double check them.

Note that the frequency of 1.11 kHz is chosen because the 9th harmonic occurs at 10 kHz, which is an adjustable band on the EQ, and will be important later. Turn **BYPASS** on. Select averaging of 1 and time base of 10 ms/div. Switch to the FSA. Set the bandwidth from 0 kHz to 15 kHz, and select **Output** to be V_{RMS}. (You might have to toggle from V_{dBV} to V_{RMS} to get it to change from dB to V_{RMS}.) Look at the spectrum of the output from the EQ (CH2). It should look like Fig. 4.55 where only the harmonics are visible with no noise in between. Save a screenshot.

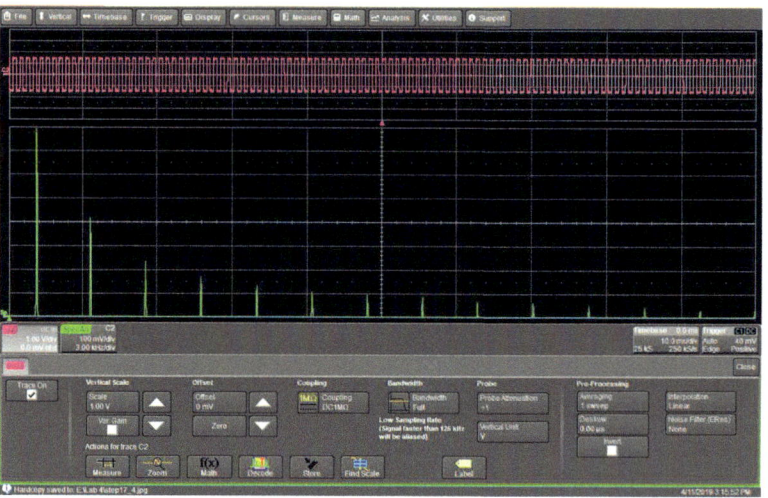

FIGURE 4.55 (Step 21) FFT of a 1.11 kHz square wave in V_{RMS}. Frequency components drop off as 1/*f* as discussed in the chapter, and only odd harmonics are present.

Switch the FSA to dBV. Now you will see lots of small peaks between the harmonics because the square wave is not perfect. Since they are about "50 dB down," they are in fact very small compared to the harmonics. Save another screenshot.

22. **Comparing the FS of square and triangle waves.** Turn **BYPASS** on. Use V_{RMS}. Now toggle the function on the WG between square and ramp several times while observing the FS. Can you see that the harmonics of the triangle wave drop off faster, as $1/n^2$ for a triangle wave than they do for the square wave, $1/n$, as described in the reading? Save a screenshot of each spectrum for comparison. Discuss in the lab report.

23. **Passing a square wave through a LPF.** Here, you will use the EQ as an LPF in order to reduce the high-frequency harmonics and turn the square wave into a crude triangle wave, as shown in Fig. 4.56. Set the sliders to flat. Observe the output in both time and frequency domains. Set the FS to a bandwidth of 0 kHz to 15 kHz in V_{RMS} and the time base to 20 ms/div. At this setting the peaks will be a bit broadened, but the spectrum will still be observable, and will refresh quickly. Record the peak values for the fundamental and the first harmonic. The ratio should be something like 1/3, because the harmonics drop of as $1/n$ as discussed in the chapter. You will compare that ratio to the case after you apply the LPF.

Now bring up just CH2 so you can get a good look at the waveform. Starting from the right, one at a time lower each of the high-frequency sliders to their lowest positions and leave them there. (On the dbx 1231 EQ, the lowest position is – 15 dB relative to 0 dB as indicated by the lit LED on the left of the sliders). Combined, this turns the EQ into a (poor) LPF. What happens to the waveform as you lower each band?

By the time you get to the 2.5 kHz band, the waveform should look roughly like a triangle wave, and your FS should look like that of Fig. 4.56. Did you see that it begins to look like a triangle wave after just a few bands are reduced? Turn on the FSA under the same conditions as before. What happened to the FS? Does the new FS resemble the Fourier transform of a triangle wave as seen in Step 20? Again, find the ratio of the fundamental to the first harmonic. Is the ratio still about 1/3 or is it smaller, to better reflect that of a triangle wave, as discussed in the chapter? Now the harmonics drop off more like $1/n^2$, as it should for a triangle wave.

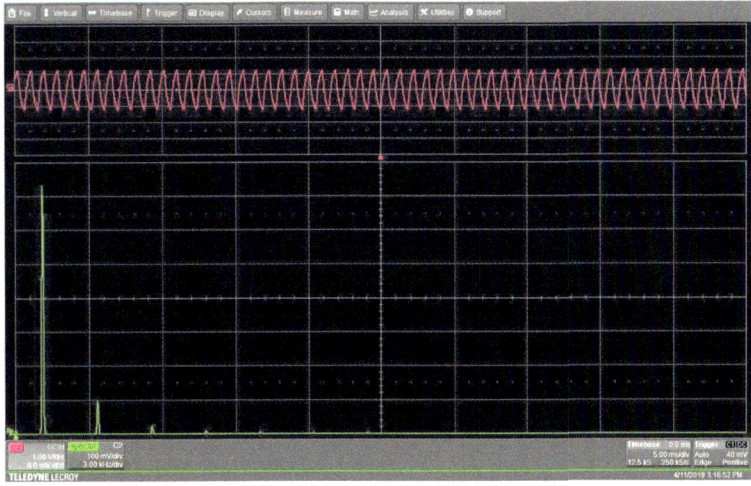

FIGURE 4.56 (Step 23) FFT of a 1.11 kHz square wave in mV that has passed through the EQ acting as a low-pass filter. The high-frequency components are attenuated, and the waveform has become much like a triangle wave. Now the harmonics drop off more like $1/f^2$, as it should for a triangle wave.

Can you describe how the shape of the waveform is related to the new frequency components? Think back to how the shape of the square wave changed as you took out more and more of the high frequencies. Note that it is the same reasoning and answer as Steps 10 and 11.

24. **Passing a square wave through an HPF.** Leave all the FSA settings alone. Set all of the bands back to flat. It is important that they all be on the detent stop. Now use the EQ as a HPF as you remove the low frequency components one at a time, starting from 20 Hz through 2.5 kHz. Turn off the FSA to get a good look at the waveform as you do this. What shape do you approach as you remove the low frequencies, leaving only the high frequencies? This waveform will be new to you it has not been shown anywhere yet. This is not very pretty but can still be explained in terms of the resulting frequency components as you explained the shape of the square wave after the LPF.

Set the sliders back to flat. Turn on the FSA and repeat while watching the changes in

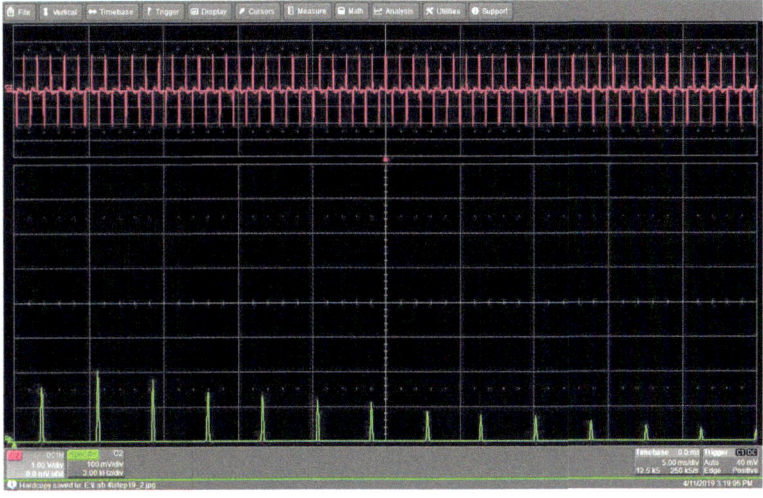

FIGURE 4.57 (Step 24) FFT of a 1.11 kHz square wave in V_{RMS} that has passed through the EQ acting as a high-pass filter. The amplitudes of the low-frequency components are reduced, and the waveform has become much like a series of spikes.

the FS. See Fig. 4.57. Think of the flat parts of the square wave as low frequencies and the edges of the square wave as high frequencies. In your report, explain qualitatively why removing the low frequencies makes the shape that it does.

25. **Amplifying and attenuating selected frequencies within the square wave spectrum.** Set all the sliders back to flat. Turn off the FSA for now and just look at the square wave. Add (i.e., slide up) some of the gain of the 10 kHz frequency band, which is the 9th harmonic. See Fig. 4.58, which shows the 9th harmonic amplified (as well as those nearby). Do you see the oscillations that show up on top of the square wave? *This is equivalent to amplifying (mostly) that frequency component from all of those that make up the square wave. Remember, the EQ does not create frequency components – it can only amplify ones that are already within the spectrum, which we showed in Step 19.*

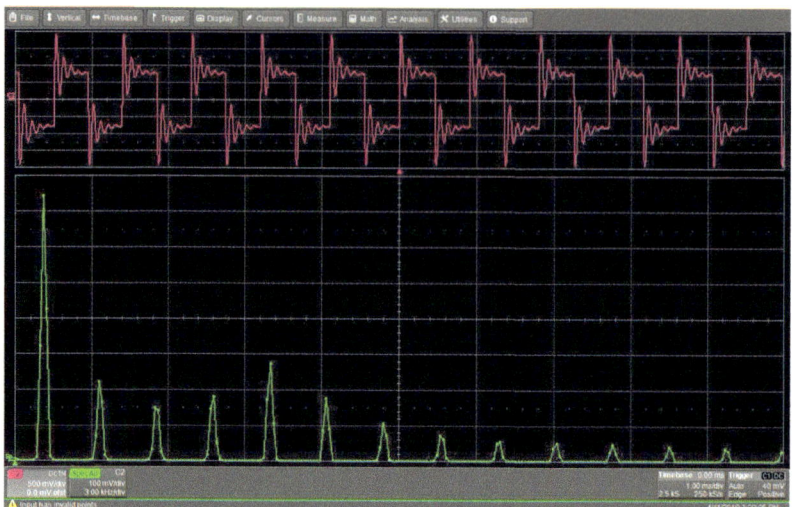

FIGURE 4.58 (Step 25) FFT of a 1.11 kHz square wave in dBV (**Linear** mode) that has passed through the EQ and has only the 9th harmonic, i.e., 10 kHz, increased. The shape of the waveform reflects that amplified harmonic. Also, the spectrum shows a larger peak for that harmonic. (Note that the width of the peaks is large because here only a few periods are used to calculate the FFT.)

Using the cursors, measure the frequency of that quasi-sinusoidal artifact that is imposed on the square wave. Does it correspond approximately to the frequency that you have changed, i.e., 10 kHz?

Look at the FS as you again increase the 10 kHz band. You should see that the 10 kHz peak is increased in the FS. Each slider is broad-band, so nearby peaks increase with it, but the one that is most increased is that of the 9th harmonic at 10 kHz. Note that the precise shape of the output waveform is affected by phase shifts due to the filters in the EQ, and strictly speaking the EQ does not remove only one frequency component, but this demonstration still illustrates the basic idea of frequency spectra. Take a moment to play around with this by selecting other slider frequencies and seeing what happens. The frequency and amplitude of the oscillations will conform closely with the band that you choose.

26. **Relating sounds to frequency spectrum of sine and square waves.**

CAUTION FOR NEXT STEPS:

SET THE OUTPUT OF THE WAVEFORM GENERATOR TO 10 mV$_{pp}$.

DO NOT HAVE EAR BUDS IN YOUR EARS BEFORE YOU CHECK CAREFULLY FOR LOUDNESS LEVEL.

PUT EARBUDS INTO YOUR EARS ONLY AFTER YOU BRING THE EARBUDS SLOWLY TOWARD YOUR HEAD.

This demonstration translates the results from Step 18 and onward into audible sounds to relate the results to music. When you are done you should be able to compare what you hear to what you saw in previous steps.

Special note: Earbuds are often wireless, being connected to the music source by a Bluetooth connection. This activity is intended to be done using the older-style 1/8-inch stereo jack, without the third connection for a microphone. These can be purchased at very little cost. The sound quality for music will be sub-optimum, but they will work fine for these demonstrations.

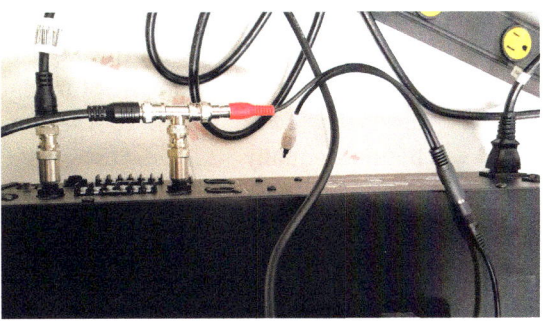

FIGURE 4.59 (Step 26) Setup for listening to one output and also viewing the waveform on the scope.

Remove all earbuds and headphones from your ears. On the EQ, turn **Input Gain** CCW, all the way down to –12 dB, to start with. Refer to Fig. 4.59 to see how to set up the output so that you can hear it with your earbuds and view it on the scope: One side of the tee goes to the scope (CH2), and the other has a BNC to RCA adapter. RCA connectors are common in home audio equipment, and sound out from TVs. Plug one channel of the stereo RCA-to-1/8-inch jack adapter into the RCA jack. The other side allows you to plug in the 1/8-inch jack from earbuds. Plug in the earbuds. Output from the WG a 1.1 kHz, 30 mV$_{pp}$ sine wave.

Because it is only one jack, you will hear sound from only one side. Bring the earbuds toward your head making sure that the volume is not excessive – **do not insert the earbuds yet**. Slowly increase the output voltage from the generator as necessary to obtain a **barely audible** tone. **Do not make it loud enough to cause any discomfort.** You may need as much as several hundred mV depending on your earbuds, but *do not make it any louder than you need to*. You can use **Input Gain** on the EQ as a fine volume control. Starting at a low voltage, **you may now *slowly* put the earbuds into your ears.** Decrease the volume as necessary. These pure tones are annoying and are louder than they might seem at first. **Stop listening as soon as you have addressed the questions.**

Basically repeat step 19, but only for a few sliders: Move the EQ sliders up and down for frequencies well above and below 1 kHz and then at 1 kHz. Describe what you hear. Relate the sound to what you did in Step 19. You should hear no changes other than an increase or decrease in volume.

Look only at CH2. Change the input to a square wave, which now has a fundamental frequency of 1.11 kHz and significant harmonics. Again, start at only 30 mV$_{pp}$. Describe the sound. How do you relate the sound to the frequency spectrum of the square wave? Decrease the sliders together at 8 kHz, 10 kHz and 12.5 kHz all the way down. What does that do to the sound? Set them back to flat. Now do the same for 800 Hz, 1 kHz and 1.25 kHz.

Repeat this step so far while observing the FS.

Describe how the sound changes with each frequency change. Relate those changes in sound to the modified frequency spectra of the 1.11 kHz square wave that you just observed. Can you hear those changes to the waveform? This is the foundation of signal processing, but here we are doing it in an analog rather than digitally, which is the subject of Chapter 9.

27. **Listening to music with equalization.** *Take the earbuds out of your ears.* Now use music from your music source (e.g., cell phone or provided CD player) as input to the equalizer. Turn the volume on the music source all the way down. Use two cables with RCA-to-¼ inch adapters

FIGURE 4.60 (Step 27) Setup for listening to music through the EQ.

to connect the source to the EQ using both channels, four plugs, for stereo. (See Fig. 4.60). Turn the gain control on the EQ all the way CCW. Now use headphones or earbuds, and turn the volume up very slowly. Starting at a low volume, **you may now *slowly* put the earbuds into your ears.** Listen to the music as you change the frequency components on both channels equally. Describe how well the EQ works in shaping the sound of your music.

28. **More music filtering–Karaoke?** Use a song with vocals. Try to remove the vocals by dropping out the higher midrange frequencies. Typically, these might be frequencies in the range of 0.8–4 kHz. Start at a low to medium volume and only drop frequencies out – don't make them louder. Can you block out the vocals and still keep the background music? This is a really bad way of doing Karaoke. In fact, it really is done in cheap Karaoke, and is called 'vocal suppression.' Why do you think it is a bad way? Read about Karaoke on Wikipedia to learn about several clever ways to remove the vocals from recordings.

Congratulations! You are now much more knowledgeable about the frequency domain. Try to keep these lessons in mind for the rest of this course and as you go through the next years of your engineering and science courses. Doing so will serve you well.

Problems

1. Use the basic equations of the definition of current ($I = dQ/dt$), current through a resistor ($V = IR$), charge on a capacitor ($Q = CV$), and the induced voltage on an inductor ($V = LdI/dt$) to convince yourself that the units of RC and L/R are actually seconds (time). That fact alone suggests that electronic components can be used in time-dependent circuit operation.

2. The switch in the following closes at $t = 0$, and the capacitor is initially charged to 18 V.

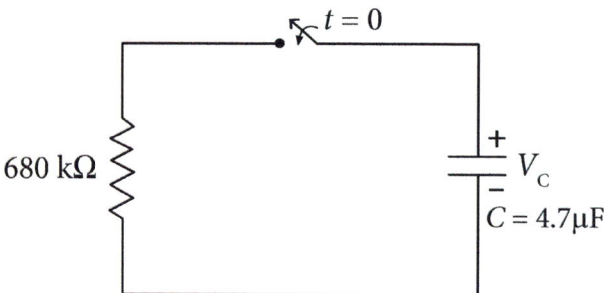

(a) What is τ?

(b) Write the expression for $V_C(t > 0)$.

(c) Roughly hand sketch $V_C(t > 0)$. Include values for the initial voltage and τ.

(d) What is $V_C(t)$ at $t = 5$ s after the switch closes?

(e) Use the time derivative of your answer to (b) to write an expression for, and then calculate, the slope of $V_C(t = 0^+)$.

(f) **The answer to this question tells us that the tangent line at the initial value of any changing variable crosses the final value at a time τ later. That line can be used as a guide for drawing plots more accurately.**

At what time, t, does the tangent line found in (e) cross the time axis for the final value of $V_C(t) = 0$? What is that time in terms of τ? (You must recall the equation for a straight line.)

(g) Draw $V_C(t > 0)$ again, this time including the tangent line starting at $t = 0$ and crossing the time axis. Include all of the values found in the previous parts of this question.

3. The switch shown in the figure is open for a long time for $t < 0$ and closes at $t = 0$.

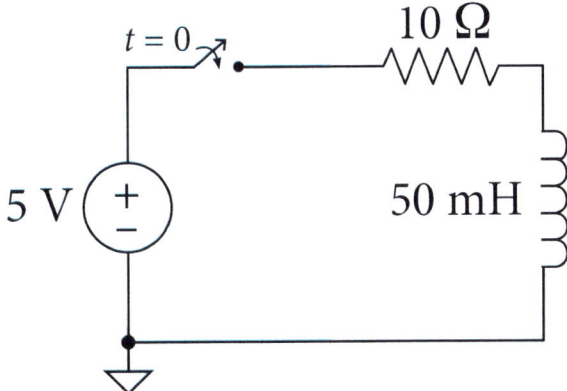

(a) What is the time constant?

(b) What is the current after a long time after the switch is closed?

(c) What is the current at 2.5 ms after the switch was closed?

(d) Calculate the voltage across the resistor and the inductor at $t = 2.5$ ms.

4. The capacitor in the figure is initially uncharged ($V_C(t = 0) = 0$ V). The switch closes at $t = 0$.

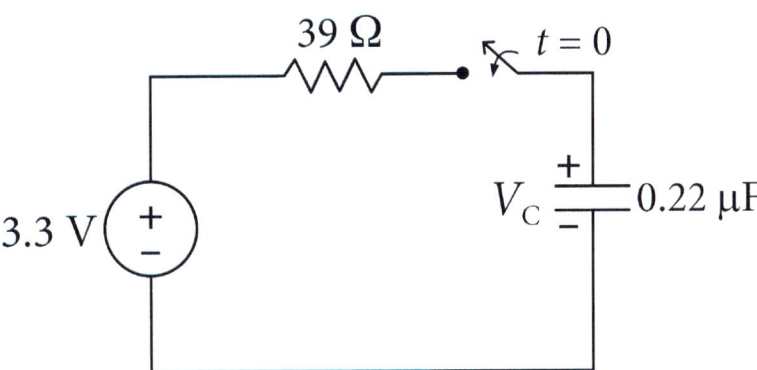

(a) What is τ?

(b) Write the expression for $V_C(t > 0)$ including the values provided.

(c) Sketch $V_C(t > 0)$. Include values for τ, $V_C(t = 0^+)$ and $V_C(t = \infty)$.

(d) What is $V_C(t)$ at $t = 15$ μs after the switch closes?

(e) Use the time derivative of your answer to (b) to write an expression for, and then calculate, the slope of $V_C(t = 0^+)$.

(f) At what time, t, does the tangent line at $t = 0^+$ found in (e) cross the asymptote of the final value of $V_C(t)$? What is that time in terms of τ? (Use the equation for a straight line.)

(g) Draw $V_C(t > 0)$ again, this time including the tangent line starting at $t = 0$. Label the time at which the tangent line crosses the horizontal asymptote.

5. The switch in the figure is open for a long time for $t < 0$ s and closes at $t = 0$ s.

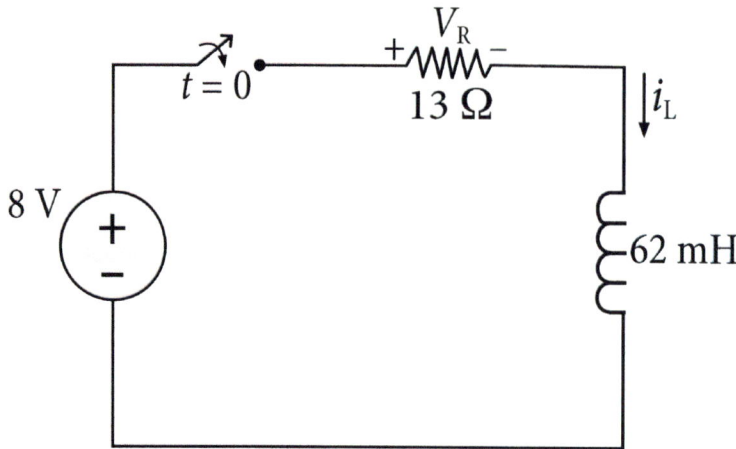

(a) What is the value of the time constant τ?

(b) What is the current just before the switch closes, $i_L(t = 0^-)$?

(c) What is the current just after the switch closes, $i_L(t = 0^+)$?

(d) What is the current a long time after the switch closes, $i_L(t = \infty)$?

(e) Use the general form of the step response and fill in the values to obtain an expression for $i_L(t > 0) \cdot x(t) = x_f + [x(t_0^+) - x_f]e^{-(t-t_0)/t}$. You already have everything you need.

(f) Sketch $i_L(t > 0)$.

(g) Calculate $i_L(t = 2.35$ ms$)$

(h) Calculate $V_R(t = 2.35$ ms$)$

(i) Calculate $V_L(t = 2.35$ ms$)$.

6. The switch in the figure is at A for a long time for $t < 0$. Calculate the current $i_L(t > 0)$ using the general solution and plot it.

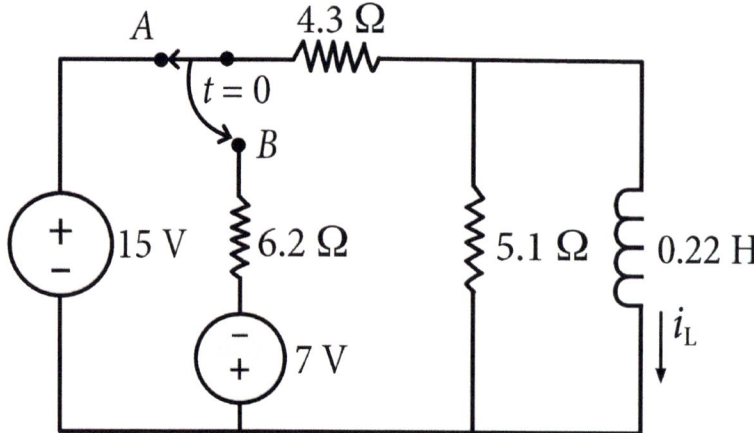

7. S1 in the figure is at A for a long time and switches to B at $t = 0$. Then, S2 switches from C to D at $t = 2$ s. **Find and plot $V_C(t)$ from 0 to 5 seconds.**

Hint: There are two stages to this problem. Each stage has its own time constant. In stage 1, the capacitor charges up for 2 s under the influence of the 12 V supply, and then does something (could be charge or discharge) after 2 s under the influence of the 15 V source. Each stage of the problem has its own initial and final conditions that must go into the general solution for that stage.

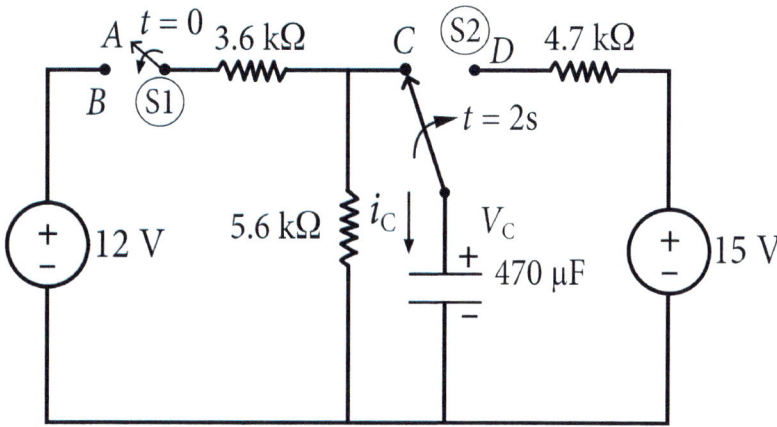

8. The switch in the figure is at A for a long time before $t = 0$. At $t = 0$ it moves from A to B.

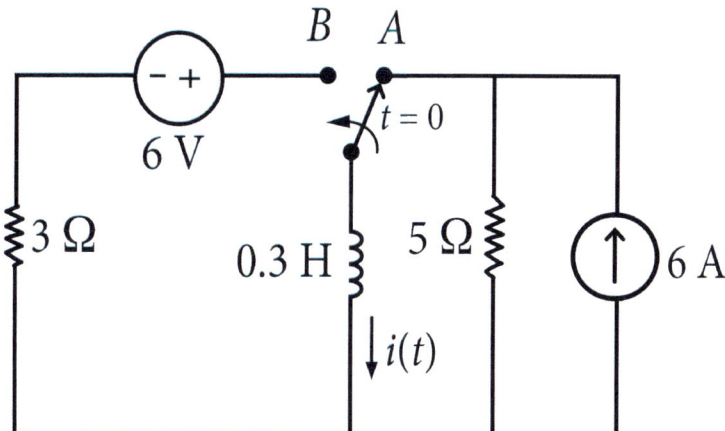

(a) What is the value of the current $i(t)$ at $t = 0.1$ s?

(b) At $t = 0.1$ s the switch moves back from B to A. What is the value of the current $i(t)$ at $t = 0.15$ s?

(c) Draw the current versus time from $t = 0$ to $t = 0.15$ s. Show the values of current at each switching event.

9. (a) What is the gain in decibels (dB) for a voltage **increase** by a factor of 70?

(b) What is the gain in decibels (dB) for a voltage **attenuation** by a factor of 70?

(c) What gain in dB corresponds to a voltage amplification by a factor of 4?

(d) What voltage gain corresponds to +15 dB?

(e) What voltage gain corresponds to –15 dB?

(f) What **voltage** (NOT GAIN) corresponds to +80 dBmV?

(g) What voltage corresponds to –100 dBmV?

(h) What voltage corresponds to +60 dBV?

(i) What voltage corresponds to – 60 dBV?

(j) How many dBs is a power gain by a factor of 70?

(k) How many dBs is a power gain of 4 times?

10. What is 2.4 V in (a) dBmV and (b) dBV? (c) Can you express 2.4 V in voltage gain dB, as in the previous question? Why or why not?

11. When the output voltage of a circuit or amplifier is the same as the input voltage, what is that gain in dB?

12. (a) For the circuit shown in the figure, the input is V_i = 2 mV and the voltage gains are A_{V1} = 2, A_{V2} = 3, A_{V3} = 10 and A_{V4} = 5. What is the total gain of the amplifier, and what is the output voltage in volts?

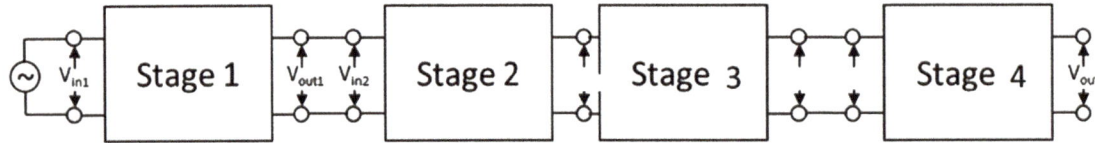

(b) In dB, what is the voltage gain of each stage above, and what is the total gain of the amplifier as calculated by conversion of the total gain into dB?

(c) Show that the total gain of the four stages in dB as found above is the sum of the individual gains of each stage in dB.

(d) What are the input and output voltages in dBmV?

(e) What are the input and output voltages in dBV?

13. (a) Draw a low-pass RL filter.

(b) What value of R do you need in order to have a cutoff frequency of 2 MHz if L = 0.07 mH?

14. The figure shows a filter:

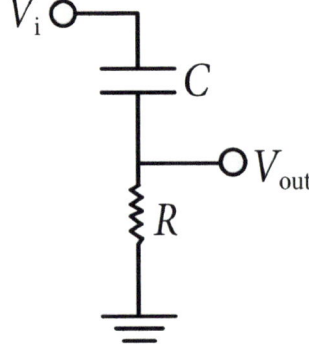

(a) Is the capacitor like a short or an open at very high frequencies?

(b) Explain in terms of capacitive reactance.

(c) Is the capacitor like a short or an open at very low frequencies?

(d) Explain in terms of capacitive reactance.

(e) Is this a low-pass or a high-pass filter?

(f) Explain why.

15. (a) For $R = 1$ kohm, what capacitance is needed to make an RC filter with a cutoff frequency f_c of 37 kHz?

(b) Does f_c depend on whether the filter is a high-pass or low-pass filter? Draw the transfer functions of both HPF and LPF showing where f_c is on each, including the numerical value from part (a).

16. (a) You just built a power supply (see Chapter 7) that has very low ripple, which means that the output is almost perfectly smooth but has some periodic variations in voltage over time. The DC voltage is 10 V, and the ripple has a peak-to-peak voltage of 1 mV. You want to see the waveform of the ripple on your oscilloscope. Do you use AC coupling or DC coupling? Why?

(b) Write out the steps in setting up the oscilloscope that you use to observe the ripple well enough to make voltage measurements on it.

(c) Suppose the lowest frequency component of an AC signal that you would be able to observe in AC coupling mode is 10 Hz. What is the value of the blocking capacitor internal to the oscilloscope? *Hint*: you must recall the input resistance of the oscilloscope.

17. Consider the input to an oscilloscope using a 1× probe, set to AC coupling mode. The internal blocking capacitor is 1 nF.

Draw the circuit including the capacitance and resistance values. Show graphically what frequencies will be visible on the oscilloscope trace. Your answer will be only qualitative in nature, but must make the relevant points about the behavior of the filter. Specifically: Draw the filter characteristics of the circuit indicating the value of the cutoff frequency. To do so, you must know the input resistance of standard oscilloscopes.

On your filter characteristic graph, show where the following frequencies lie, and state in words qualitatively (e.g. extremely, a lot, considerably, a little, a bit, not at all, etc.) how much each frequency will be attenuated at: 30 Hz, 120 Hz, 180 Hz, 1.5 kHz.

18. Identify each type of filter as LP, HP, bandpass or notch. Indicate approximately where the cutoff frequencies are.

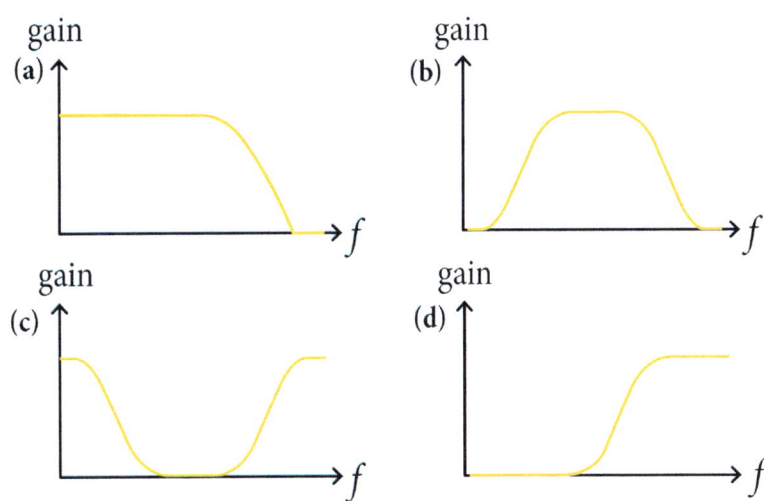

19. For each of the filters in the figures, state if it is a LPF or HPF, and say how you know.

(a)

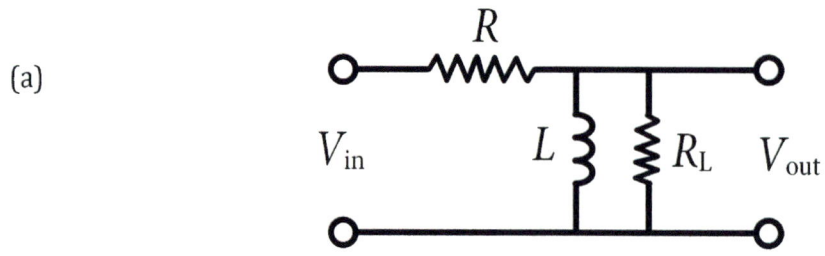

(b)

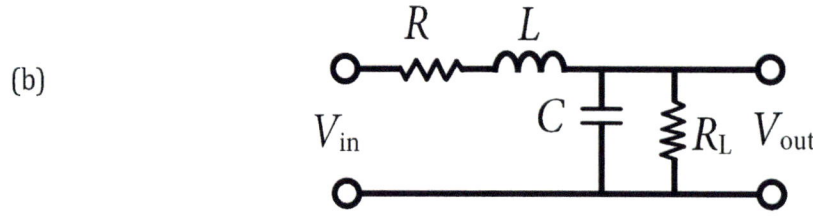

(c)

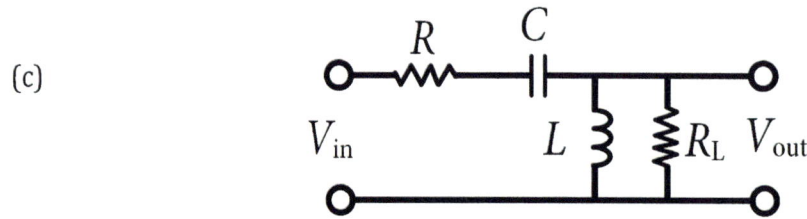

20. Answer the following questions about the circuit shown in the figure:

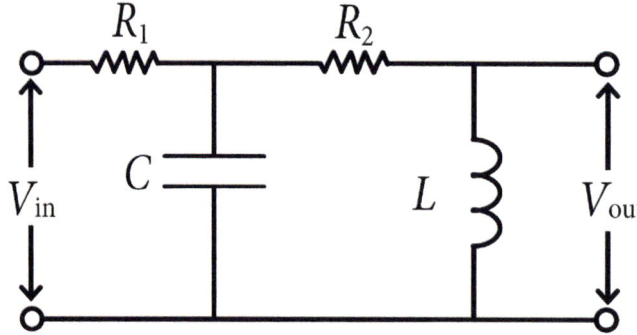

(a) What kind of filter is the combination of R_1 and C? Select one: *HPF, LPF, bandpass filter, notch filter.*

(b) What kind of filter is the combination of R_2 and L? Select one: *HPF, LPF, bandpass filter, notch filter.*

(c) What kind of filter is the overall circuit? Select one: *HPF, LPF, bandpass filter, notch filter.*

(d) What must be the relationship among R_1 with C and R_2 and L for this circuit to function properly as your selection in part (c)? Your answer must be explained using equations and a figure of the gains of the two separate filters versus frequency.

21. Consider this bandpass filter shown in the figure.

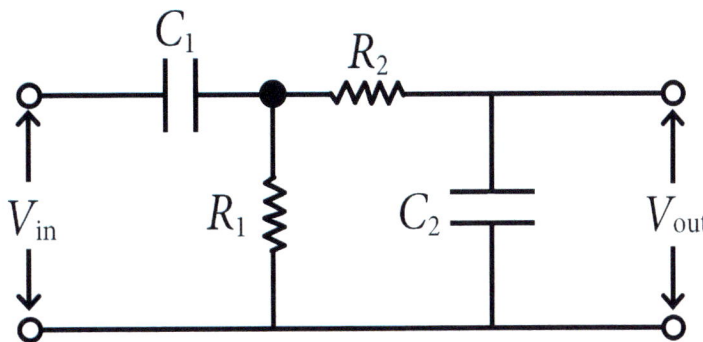

(a) Explain how it works. Let C_1 = 0.10 microfarad (μF) and R_1 = 20 kohm. Also, let C_2 = 2 nanofarad (nF) and R_2 = 5 kohm. Draw the filter response labeling all of the relevant frequencies.

(b) What is V_{out} if the two sets of values of R_1 and C_1 are switched with R_2 and C_2? Draw that filter response. Explain how it works.

(c) Which set of device parameters are acceptable for its operation (choose a line from (a) to (d))? Capacitances are in farads and resistances are in ohm. All but one answer are wrong for some reason. Choose the only correct answer.

	C_1	R_1	C_2	R_2
(a)	4×10^{-6}	500	4×10^{-6}	700
(b)	9×10^{-6}	700	4×10^{-6}	200
(c)	4×10^{-6}	500	3×10^{-5}	400
(d)	4×10^{-6}	300	9×10^{-5}	500

22. Consider the filter circuit shown in the figure:

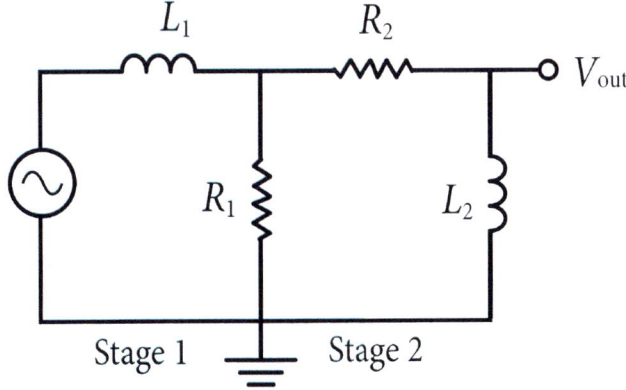

(a) What kind of filter is stage 1?

(b) What kind of filter is stage 2?

(c) What kind of filter is the entire circuit?

(d) Does it matter what the values of L_1, R_1, R_2 and L_2 are? Will it work no matter what values you choose? Draw a picture that supports your explanation.

23. Consider the LPF and HPF shown in the figure:

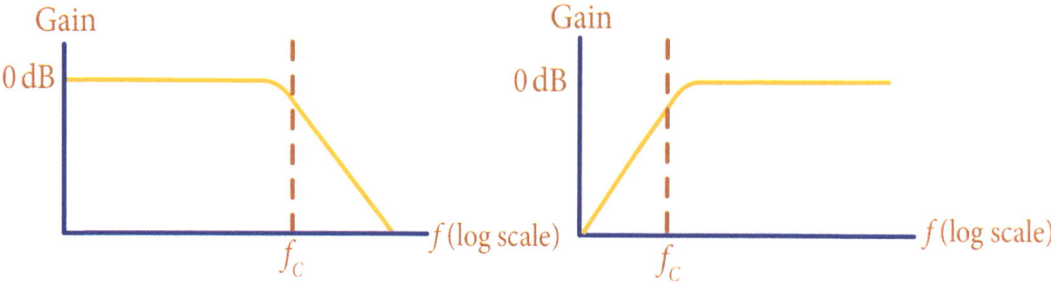

Now consider the following frequency spectrum of a signal that is input to the filters.

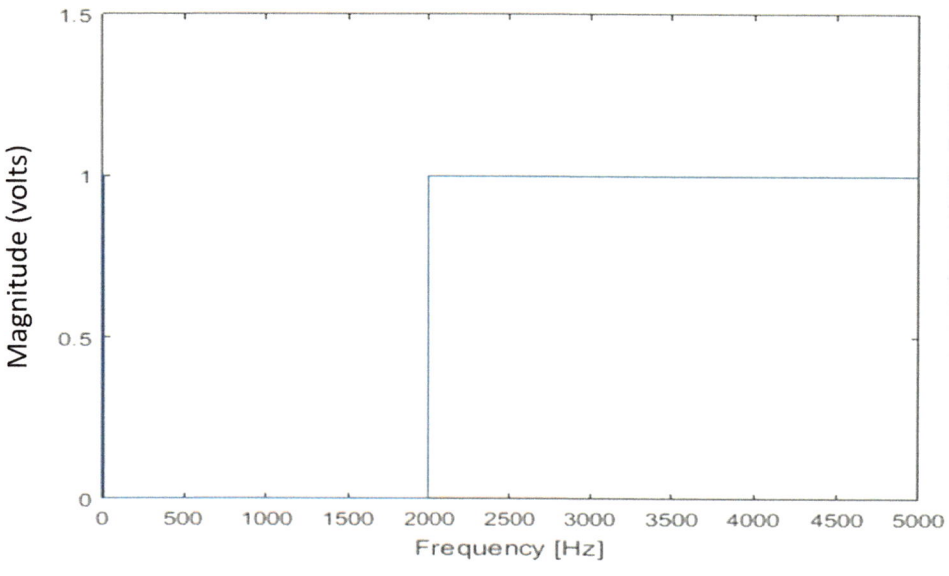

This frequency spectrum might be referred to as "flat above 2 kHz." It might be the spectrum of an audio signal coming from the output of a microphone that is picking up the sound of a small waterfall. (Why small? Because it has only higher frequencies in it.)

The output of a filter is the input frequency spectrum multiplied by the gain of the filter. Drawing right over the filter plots provided below, sketch the spectrum of the outputs of both filters under the following conditions. Other than f_c, you can make your own choice about the frequencies on the horizontal scale since they are not explicitly stated. Note that the plots are logarithmic on both gain and frequency.

(a) f_C of the filters is 1000 Hz.

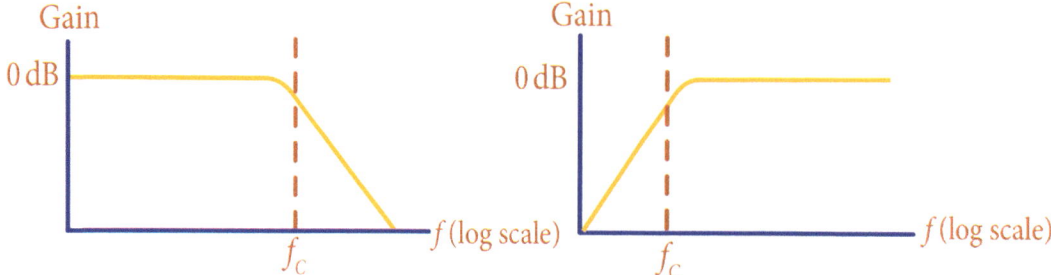

(b) f_C of the filters is 3000 Hz.

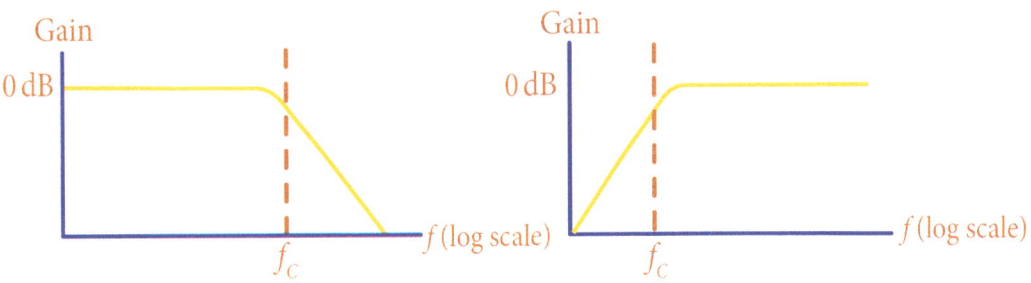

24. Hand sketch how you would set up a graphic equalizer in the lab to make it behave as a bandpass filter between 100 Hz and 5 kHz.

25. The figure shows the spectrum of a clarinet playing the note A4 at 440 Hz:

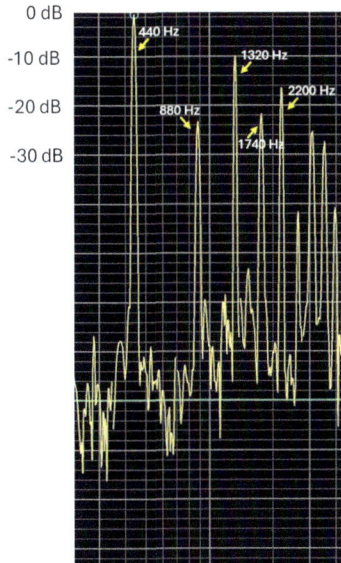

The amplitude of the spectrum is in dB, where the largest peak is normalized to 0 dB. Assume that 0 dB represents a voltage magnitude of 3 V from an audio amplifier. The vertical scale is 10 dB per large division. *Hint*: Carefully estimate the number of dB for each of the frequency components and convert that voltage into the voltage magnitude using the formula that converts dB to voltage ratios. Calculate:

(a) The voltage magnitude of the 880 Hz component

(b) The voltage magnitude of the 1320 Hz component

(c) The voltage magnitude of the 1740 Hz component

(d) The voltage magnitude of the 2200 Hz component

26. The plots below show the spectrum of a tuba playing a note with a filter drawn over it. For each plot, (a) through (f), describe in words what you would hear at the output of the filter.

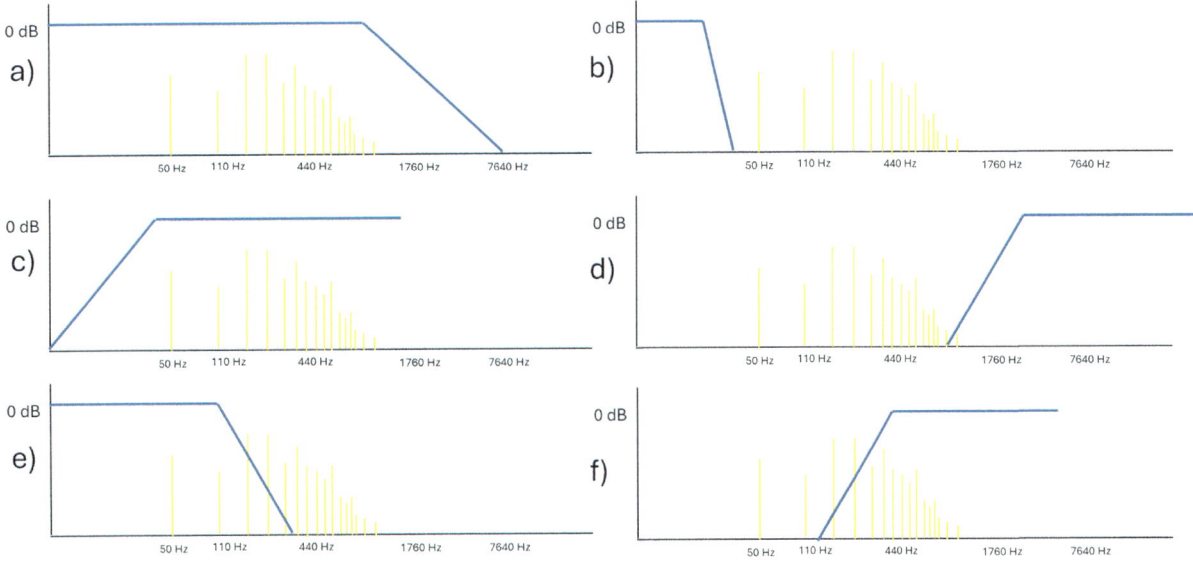

27. Download any free phone application that displays a frequency spectrum of sounds in the microphone. Search apps under "frequency spectrum analyzer". Try whistling different notes or placing it near an electric motor or fan, or other source of noise. Show what you observe by sketching it, or better yet, do a screenshot and print it. Describe the shape compared to what you hear.

28. Consider a square wave of 50% duty cycle, amplitude 5 V, frequency of 100 Hz. Refer to Eqs. 4.46 and 4.47.

    (a) What is the fundamental frequency?

    (b) What is the voltage of the fundamental frequency component in both peak and RMS units?

    (c) What is the first harmonic frequency?

    (d) What is the voltage of the first harmonic frequency component in both peak and RMS units?

    (e) What is the second harmonic frequency?

    (f) What is the voltage of the second harmonic frequency component in both peak and RMS units?

    (g) What is the third harmonic frequency?

    (h) What is the voltage of the third harmonic frequency component in both peak and RMS units?

    Repeat the above for a triangle wave. Note that a triangle wave is not a sawtooth. It has the same slope on the rising and falling edges.

    (i) What is the fundamental frequency?

    (j) What is the voltage of the fundamental frequency component in both peak and RMS units?

    (k) What is the first harmonic frequency?

    (l) What is the voltage of the first harmonic frequency component in both peak and RMS units?

    (m) What is the second harmonic frequency?

    (n) What is the voltage of the second harmonic frequency component in both peak and RMS units?

    (o) What is the third harmonic frequency?

    (p) What is the voltage of the third harmonic frequency component in both peak and RMS units?

29. Sketch the frequency domain representation of the following function. Plot frequency in Hz. The delta functions should represent RMS, rather than peak, voltages. Don't forget to convert the peak voltages to RMS (Note: DC voltages are the same as their RMS values.)

$$V(t)=6+5\sin(2\pi \cdot 3000 \cdot t)+3\sin\left(\frac{t}{0.03\text{ms}}\right)$$

30. For the following function:

$$V(t) = 5 + 2\sin(2\pi \times 2\text{ kHz} \times t + \pi/2) + 3\cos(1.5 \times 10^4 \times t + \pi/4),$$

plot (a) the time domain representation using a graphing calculator (such as Desmos), and (b) the frequency domain representation, plotted by hand. Be sure that the time-domain plot of the function shows the waveform clearly. Also, your frequency spectrum should have units of V_{RMS} and Hz. Note that a DC voltage is the same as its RMS value.

31. Consider a square wave having 50% duty cycle at 300 Hz. Its frequency spectrum is shown in the figure:

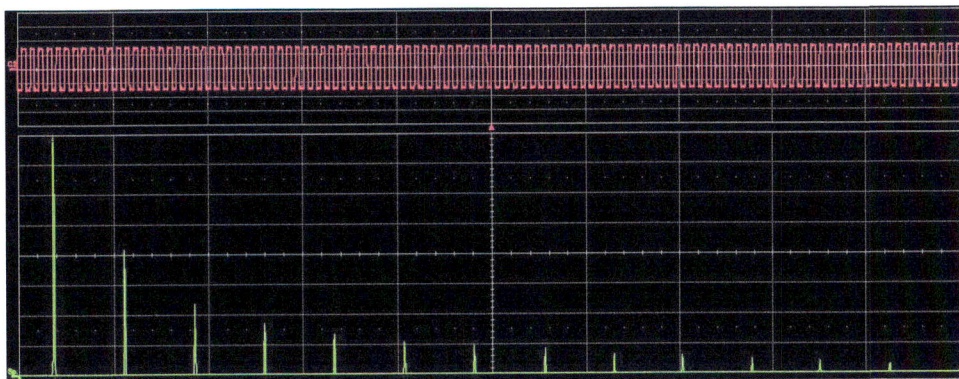

You pass it through an ideal low-pass filter whose frequency spectrum is shown in the following. It has a cutoff frequency of 2 kHz.

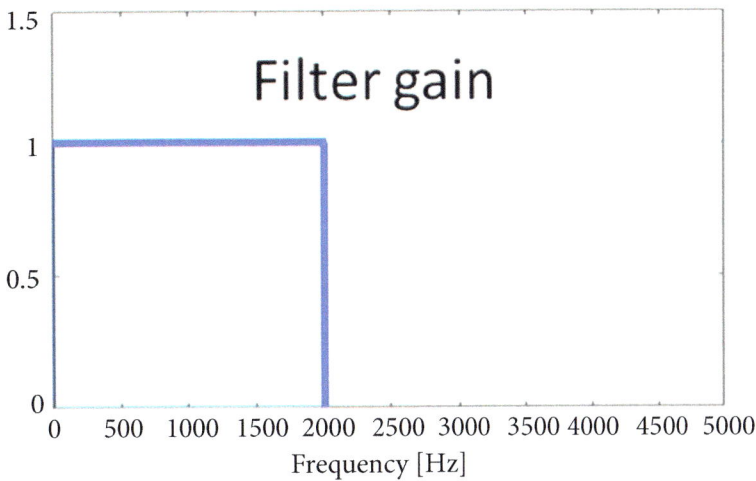

Draw the frequency spectrum at the output of the filter. You must label the frequencies that you include in your drawing. Draw it to scale as presented to you in the first figure.

32. (a) On a linear (not log) scale, draw the frequency spectrum of a square wave having a duty cycle of 50%, an amplitude of 1 V, and a frequency of 1 kHz. Show five harmonics, and label their amplitudes and frequencies.

(b) Repeat part (a) for a triangle wave of the same amplitude and frequency. Draw it to scale relative to the previous part.

(c) In words, discuss the similarities and differences between the two frequency spectra from above.

(d) Draw the square-wave spectrum from (a) after it passes through an ideal LPF having a cutoff frequency of 6 kHz.

33. See the frequency spectrum of a square wave in Eq. 4.46. Per the following instructions, use a graphing calculator (such as Desmos) to add up the first 5 nonzero frequency components (up to the 9th harmonic) of a square wave having zero DC component, a frequency of 1 kHz, 50% duty cycle, and amplitude of 8 V.

    Show one picture each for the inclusion of each additional frequency component. As each harmonic is added, the sum should look more and more like a square wave with a $V_p = 8$ V.

    Finally, include a last plot that shows all the individual frequency components and the resulting square wave on a single plot. (It will be very dense, but let's see what it looks like anyway.)

    Your solution should include six separate plots.

34. The frequency spectrum of a triangle wave is given in Eq. 4.47 where ω represents the fundamental (radial) frequency of a square wave having a duty cycle of 50%, and $n = 1$ to ∞:

$$f(t) = \frac{8}{\pi^2} \cdot \sum \frac{1}{n^2} \cos(n\omega t) = \frac{8}{\pi^2}\left[\cos(\omega t) + \frac{1}{9}\cos(3\omega t) + \frac{1}{25}\cos(5\omega t) + \frac{1}{49}\cos(7\omega t) + \cdots\right]$$

    Assume that $f = \omega/2\pi = 11.1$ kHz.

    (a) Use Desmos, Matlab or your favorite software to plot the sum of frequency components five times on the same plot, starting with just the fundamental frequency ($n = 1$) and then adding an additional odd frequency component each time.

    (b) Another equation that seems to be different is as follows. Note that it is based on sines instead of cosines:

$$x_{triangle}(t) = \frac{8}{\pi^2} \sum_{k=0}^{\infty} (-1)^k \frac{\sin\left(2\pi(2k+1)ft\right)}{(2k+1)^2}$$

$$= \frac{8}{\pi^2}\left(\sin(2\pi ft) - \frac{1}{9}\sin(6\pi ft) + \frac{1}{25}\sin(10\pi ft) - \cdots\right)$$

    Repeat part (a) using this equation. What is the difference between these two triangle functions?

35. Imagine that you need to produce a triangle wave but all you have available is a square wave. You figure out, however, that you can use a filter to transform this square wave into a triangle wave. Recall that the frequency components of a square wave drop off as $1/n$ and those of a triangle wave drop off as $1/n^2$. Do you use a low-pass filter or a high-pass filter? Explain your answer.

36. Consider the RLC series circuit in the figure having values of $R = 10\ \Omega$, $L = 5$ mH, and $C = 10\ \mu$F.

    (a) What is its resonant frequency?

    (b) What is X_L at resonance? Include the sign.

    (c) What is X_C at resonance? Include the sign.

    (d) At resonance, what is the total value of the impedance $Z = R + jX$? Don't forget this is a series circuit, and the total impedance is the sum of the three impedances, including the signs.

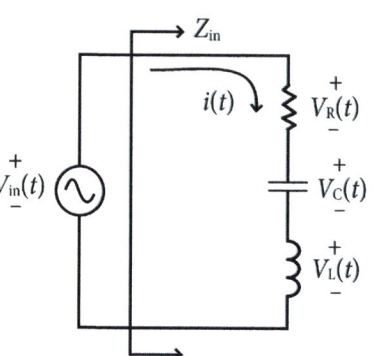

(e) Hand sketch the current as function of frequency from 100 Hz to 10 kHz. Use $V_{in,p}$ = 1 V. Include the resonant frequency and the peak current. Everything else will be approximate. The current should be on a linear scale, and the frequency may be on either a linear or log scale. Log is preferred, but not required for this rough drawing.

37. You have a series resonant circuit with L = 75 μH, C = 60 μF, and R = 8 Ω.

(a) What is the resonant frequency? Express your answer in both rad/s and kHz.

(b) Calculate the inductive and capacitive reactances at the resonant frequency, and show your work. Remember that the units of reactance is ohms, just like resistance, and capacitive reactance is negative.

(c) Redraw here by hand the basic information in Fig. 4.38 and explain what it is about. Relate your answer back to the previous question.

38. (a) What does the sweep function on the Rigol waveform generator do? Why might you want to use it in this lab?

(b) Say you input a frequency-swept sinusoid to the EQ. The sinusoid is swept from 100 Hz to 20 kHz over the span of 4 seconds. All of the EQ sliders are at 0 dB except for the one slider that controls frequencies around 5 kHz, which is pushed all the way up. Also, the time base is very slow – it also takes 4 seconds to scan across the whole screen, so position corresponds to frequency of the input. Describe in words and a rough sketch what you would expect to see as the trace (time domain) on the oscilloscope at the output of the EQ. (The vertical gain is low enough that all of the trace fits on the screen.) Explain what is happening.

Chapter 5

Phase and Phasors

5.1 Introduction and Overview

This chapter deals with how complex arithmetic and phasors are used to model linear circuits under sinusoidal steady-state conditions. These and related terms will be discussed throughout the chapter. At the end of the reading and lab activity, you will know about:

1. The power of mathematics in modeling circuit behavior
2. Properties of sinusoids and complex-number arithmetic
3. What phasors are and their different mathematical representations
4. Impedance, resistance, and reactance
5. How L, C and R behave in circuits
6. What is meant by "linearity"
7. Second-order resonant circuits
8. Stereo speakers
9. Decibel representations of voltage and power
10. Speaker crossover circuits

A general engineering concept: The term "model" is used throughout the science and engineering community and deserves some discussion. You may have assembled plastic models of airplanes or cars, or built model rockets, or drawn pictures of buildings. Architects build scale models and make extensive computer renderings of buildings as part of their design process. Anything that helps you envision a complicated object that is impractical to build and work with could be called a model. It may not have occurred to you that if you write equations to represent something, which could be loop and node equations of a circuit (as just one example), you are constructing a mathematical model that stands in for the real circuit. Once you have the mathematical model, you are then free to "experiment" with it much more easily than if you had to build a new circuit each time. Insofar as the model accurately describes the circuit, you may reasonably expect that the circuit will behave as predicted. However, you need to be sure of your constraints, such as the frequency range that the circuit will operate at. For example, simple equations might fail at high frequencies; microwave circuits, operating in the several GHz range (such as cell phones or satellite communications), require different models and mathematical methods than circuits operating in the audio frequency range (kHz). By the end of this chapter, you will, for example, understand that very small capacitances might have very small effects at low frequencies, but very large effects at high frequencies, and therefore must be included in high-frequency models.

5.1.1 The Use of Models in Electrical Design

Thomas Edison was one of the greatest inventors of the modern age. His contributions to technology changed our way of life. His most significant inventions include the first practical light bulb and the first practical way to record and distribute music, and he made huge contributions to the electrical

An Introduction to Electrical Engineering with Lab Activities
Gary H. Bernstein
Copyright © 2026 Jenny Stanford Publishing Pte. Ltd.
ISBN 978-981-5129-30-4 (Hardcover), 978-1-003-71345-6 (eBook)
www.jennystanford.com
DOI: 10.1201/9781003713456-5

grid (although Tesla won out in the end), and to motion pictures (movies). However, he was not particularly progressive when it came to his development process. He preferred to use empirical methods (i.e., trying things out) rather than at-that-time newer mathematical methods. His intransigence tended to lengthen the process of product development.

Modern engineers often use mathematical **models** of their systems with which they can do calculations, and they 'experiment' by varying different parameters in order to predict the results before expending effort on building a real system. Electrical engineers are doing this when they write equations that describe the operation of electrical circuits. Those equations are a model that will predict how that circuit will behave after they build it. To that end, it is extremely important that engineers, including EEs, gain a solid mathematical foundation as part of their educations. In this chapter, you will learn some mathematical techniques that help to analyze circuits that have L, C, and R components, as discussed regarding the complex loads of Chapter 3.

This course is intended to be heavier on the basic concepts and as light on the mathematics as possible, but no lighter than is necessary. This chapter is the most heavily mathematical in this course, but it will be accessible to you, and perhaps also fun and interesting.

Near the end of Chapter 4, you read, "In general, phase differences between voltages and currents within a circuit having a combination of *R*, *L*, and *C* are more complicated; they depend on all of the components in the circuit that contribute to $\mathbf{Z} = R + jX$. φ = arctan (X/R), which can be different from 90°. That is discussed at length in Chapter 5." And here we are. Since any periodic waveform can be represented as a sum of sine and cosine waves, we can analyze a circuit in terms of *sinusoidal* inputs that have a particular amplitude, frequency, and phase. The relationships between current and voltage of inductors and capacitors are frequency-dependent, so we must know how to determine how the currents, voltages, and phases among them depend on frequency.

5.2 Representations of Functions

Let's pull together some concepts that you will need to be comfortable with going forward. See Fig. 5.1 to see that sine and cosine have the identical shape but are shifted in phase by 90°. As you saw in Chapter 4, a periodic voltage or current waveform can be represented by a Fourier series of sine and cosine functions (collectively referred to as circular functions, or sinusoids). In fact, this reliance on trig functions is so strong that I am tempted to call EE 'sine-wave engineering' instead. That may be hyperbole, so let's just say that a lot of what you will do in EE has sinusoids 'under the hood,' as in Fourier analysis discussed in Chapter 4.

Fortunately, you are familiar with sines and cosines from trigonometry, and they are easy to work with and plot. Trig identities have some simple properties of addition and multiplication that help immensely. Also, they can be represented in another form, called

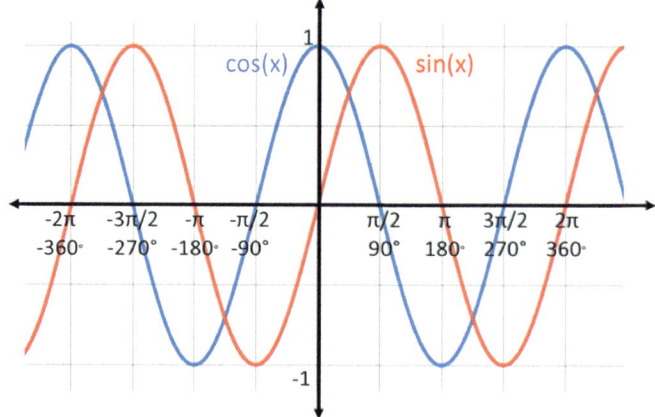

FIGURE 5.1 Sine and cosine functions. The shape of sine and cosine are the same, but cosine is phase shifted by $\pi/2$ radians. You can convert sine to cosine by adding a phase shift of 90° ($\pi/2$ radians). So, you can write $\cos(x) = \sin(x + 90°)$ or $\sin(x) = \cos(x - 90°)$. We say that cosine **leads** sine by 90° since it crosses the horizontal axis before sine does (by a quarter of a period). It would not be strictly wrong to say that cosine **lags** sine by 270°, but it is customary to restrict leading and lagging angles to the range between −180° and 180°.

phasors, that makes doing the math easier than even the trig identities. The idea is this: To simplify the math, time-dependent sinusoidal signals can be transformed from the time-domain into a type of frequency domain called the **phasor domain** or **phasor space**. Then, AC circuits can be analyzed in the phasor domain, and finally transformed back into the time domain to get the response. This chapter introduces phasors in some detail.

It is important to gain familiarity with how sinusoidal functions are used in EE because mathematical methods used in later courses will very quickly mask the nature of their underlying structure. In this chapter you will be introduced to some of those tools with the goal of seeing that they really have trig functions at their heart, even though it might not always be obvious. Everything presented in this chapter is valid for sinusoidal inputs and linear circuits. Linearity is discussed in Section 5.6.

5.2.1   Another Look at Sinusoids

Since sines and cosines will be so important in this chapter, as well as the next three years of EE course material and beyond, we should take a little time to review basic properties of sinusoids. Some of you may find this section to be a bit elementary and prefer to skim through it, but be aware that it is necessary to know what is presented here.

Figure 5.2 shows the definition of $\sin(\theta)$ and $\cos(\theta)$, where θ is the Greek letter theta. Any letter would do for the argument, but θ is traditionally used in trigonometry, so we will use it here as well. Later we will switch to ϕ, since that is more traditional in EE circuit theory.

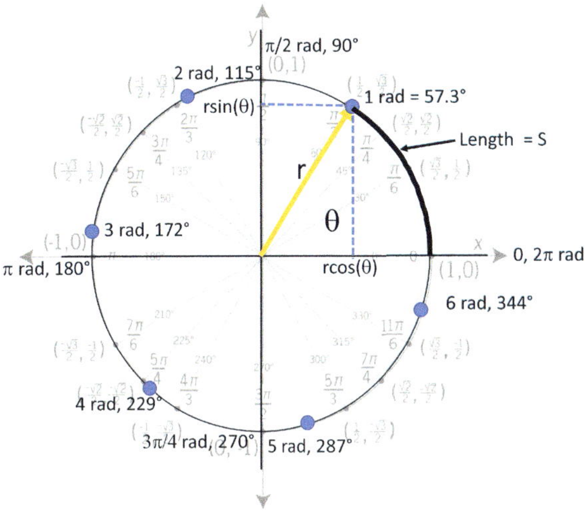

An angle θ can be expressed in degrees or radians. You can think of the circle as having 360° in one revolution, or you can think of it as having a circumference of 2π times its radius. Recall that the definition of the angle, θ, in *radians* is the ratio of the arc length on the circle, S, to the radius, r, i.e.:

FIGURE 5.2 The unit circle (i.e., $|r| = 1$) showing the values of $\cos(\theta)$, $\sin(\theta)$, and the relationship between radians and degrees. (The brackets of $|r|$ means "the length of r.")

$$\theta(\text{radians}) = \frac{S}{r}, \text{ or } S = r\theta. \tag{5.1}$$

"Degrees" is a measure of "how many 360ths of a circle you have gone around" and "radians" (abb. 'rad') is a measure of "how many radii you have gone around" the circumference of the circle. 360° is equivalent to 2π (about 6.3) radians. It is easy to convert between degrees and radians using

"angle in degrees" = "angle in radians" $\times$ 180/π. (5.2)

Just as it is useful to have an intuition about sizes in cm or inches without having a ruler handy, so it is true for degrees and radians. Figure 5.2 shows a circle with each integer radian of angle marked around the

Of interest only: Believe it or not, the use of π in trigonometry is under attack. Serious people are suggesting that 2π be replaced by a symbol, τ (tau) where τ, with units of "tau-radians," suggests one *turn*, cycle, or period of oscillation. For example, using τ, one quarter around a circle or a period is not $\pi/2$, but rather $\tau/4$, suggesting that it is, in fact, ¼ of a period. See, for example, https://www.youtube.com/watch?v=83ofi_L6eAo&t=4s/. Also, you might watch this video for a nice review of the unit circle.

circle along with the equivalent angle in degrees and important angles in radians. Also shown are the coordinates of *x* and *y* on the unit circle for each angle.

A note about "dimensions" and "units." Dimensions are measurable quantities without regard to how much of it we are dealing with, including length, mass, time, and charge. Units are simply the numbers we use to describe how much of each dimension we have, e.g., cm and inches, seconds and minutes, etc. Because we represent angles as ratios of two things having the same dimensions ($\theta = S/r$), they are actually *dimensionless*. We choose to refer to angular measure in *units* of "radians" or "degrees," but that is only the fraction of the circumference around the circle, and fractions are ratios, and hence dimensionless. This distinction actually matters because the argument of trig functions must always be dimensionless.

In Fig. 5.1 it is seen that the sine and cosine functions are identical except that they are displaced from each other by a phase shift of 90° (or $\pi/2$ rad). This is worth exploring because the same concepts will arise in the discussion of phasors. Imagine that you project the shadow of a point on a spinning disk

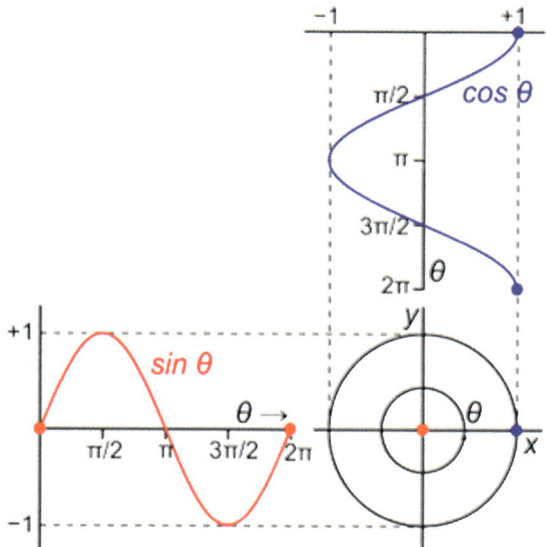

FIGURE 5.3 Screenshot from animation showing how the projections of the rotating radius sweeps out both sine and cosine functions, depending on which axis is used to plot the projection of the line as it rotates. Later, this will be central to the concept of phasors. *Source*: https://commons.wikimedia.org/wiki/File:Sine_and_cosine_animation.gif.

onto the *x* and *y* axes. That would more formally be called "the projection of the point on the axes." Plotting the position of the shadows on the axes as a function of time will trace out circular functions. This is demonstrated in the GIF shown in Fig. 5.3.

What determines if the shadow makes a sine or cosine? If you shine the light from the right side instead of the front of the spinning disk, you would get exactly the same shape of the position as a function of time, but the two traces would be displaced 90° ($\pi/2$ rad) apart from each other, as generated in Fig. 5.3 and plotted in Fig. 5.1. This relationship is expressed as:

$$\cos(\theta) = \sin\left(\theta + \frac{\pi}{2}\right) \text{ and} \tag{5.3}$$

$$\sin(\theta) = \cos\left(\theta - \frac{\pi}{2}\right), \tag{5.4}$$

which means that sine and cosine are the same function with only a phase shift of $\pi/2$, as shown in Fig. 5.1. (Recall that $\sin(\theta - \pi/2)$ means to plot sin (θ), but shifted to the right, or lagging, by $\pi/2$ rad.) By now, you might be thinking, "Hey, please stick to either degrees or radians!" Typically, science and math treatments use radians, and engineering uses degrees, so both units are appropriate, and you should intuit each of them equally. This text will use both, but mostly radians in equations and degrees when describing the functions. It is just more common to think in terms of degrees than radians.

As shown in Fig. 5.2, the argument of a trig function is the angle between the hypotenuse of the triangle that contains the radius of the unit circle and its projection onto the *x*-axis. As the angle increases, i.e., sweeps around the circle, the projection of the line on the *x*-axis sweeps out the cosine function and simultaneously the projection on the *y*-axis sweeps out the sine function, as discussed above and shown in the gif of Fig. 5.3. You should take some time to have a look at this animation because we will be referring to it later in the phasors discussion when we talk about "taking the real part" (which will be explained later). You can see that, as mentioned above, the projections along each axis sweep out curves of identical shape (sinusoids), but they are offset from each other along the θ axis by a phase shift of $\pi/2$.

Now we will think of the horizontal axis in Fig. 5.1 as time rather than θ because θ is a function of time, $\theta(t)$, and will increase with time as

$$\theta = \omega t + \phi, \tag{5.5}$$

where ω is the **angular frequency**, i.e., in rad/s and φ is the phase shift for the sinusoid at $t = 0$. This corresponds to the radius rotating counterclockwise (CCW) with time.

Now we will relate that rotating radius as a sinusoidal function of time. See Fig. 5.4 which shows two identical sinusoids that differ only in a phase shift of $\phi = \pi/4$. As mentioned in Section 3.3.1.1, when you see the word "phase" you should always ask "phase between what and what?" In this case, φ is the phase between the sinusoid rising through $y = 0$ when $t = 0$ and the same, shifted sinusoid rising through $y = 0$ at some other time. That difference in time, converted to an angle, is ϕ.

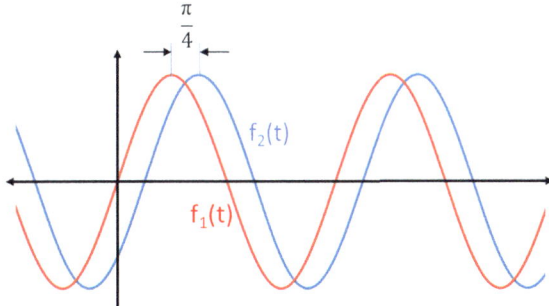

FIGURE 5.4 Two sinusoids of the form $f(t) = \sin(\omega t + \phi)$ that are phase-shifted by $\pi/4$. $f_1(t) = \sin(\omega t)$ and $f_2(t) = \sin(\omega t - \pi/4)$. $f_2(t = 0) = \sin(-\pi/4)$. Later we will represent these functions with the phase shifts in the form of complex vectors.

Consider the function $f(t) = \sin(\omega t)$. Since ω has units of radians per second, then multiplying by t yields the (dimensionless) radians. In short, using $\theta(t) = \omega t$ means that $\theta(t)$ is increasing with time, which is the same as the radius spinning around like the seconds hand on a clock (but, by convention, backwards). So, when you see $\sin(\omega t)$, keep in mind that ωt is the number of radians that the radius has moved through after t seconds, as it is sweeping around the circle. $\omega = \omega t$ can increase without bound, but we don't care – after ωt has reached 2π radians, it just keeps on rotating around (modulo 2π).

5.2.2 Real, Imaginary, and Complex Numbers

Central to our development of the kind of math that is very heavily used by EEs (and much of science and engineering as well) is the use of imaginary numbers. I would wager that although you know (even from elementary school) that $\sqrt{-1}$ is called "i," and you can do some simple arithmetic with i, you probably don't yet realize the broad utility of this extension of real numbers. Imaginary numbers used in phasors might at first seem complicated, but are simply a way to introduce an angle, or phase, into a periodic function (much more on that later). Just get it out of your head right now that there is something 'mysterious' about imaginary numbers – again, they just let us represent phases in a compact way. So, we now have two types of numbers: the conventional ones that we have used up until now, which are technically called **real** numbers (although they are no more real than imaginary numbers are imaginary), like '5';

> *Know this:* This text uses bold font, e.g., **A**, to represent a complex number. Also, $|A| = A$ are two ways of representing the magnitude of the complex number. If a number is not bolded, it is a real number.

and, we have imaginary numbers, like 'i7.' Here is a small wrinkle in the discussion: Because currents are so often represented as the letter '*i*,' electrical engineers usually use '*j*' to represent $\sqrt{-1}$. So, '*j*' it is from now on. A short review of properties of *j*:

$$j = \sqrt{-1}, \text{ so } j \times j = j^2 = -1;$$
$$j \times j \times j = j^3 = -j;$$
$$\text{and } j \times j \times j \times j = j^4 = 1. \tag{5.6}$$

Now that we have the '*j*-thing' out of the way, we can refer to any number as "complex" if it contains a part that is real and a part that is imaginary, which can be written in what is called the **rectangular form** (there are three other forms) as:

$$\boldsymbol{A} = a + jb. \tag{5.7}$$

Consider $\boldsymbol{A} = 5 + j7$; the **real part**, *a*, is '5' and the **imaginary part**, *b*, is '7.' In print, vectors are designated by **bold** font. However, in writing in your notebook or report, that is difficult to do, so just put a "bar" or a "caret" (^) over the name to indicate that it is a vector.

Let's get something else out of the way. Suppose we have a really complicated complex number, for example $\boldsymbol{A} = (5 + j7)/(3 + j12)^2 + (9 + j8)/(4 + j2)$. If you take the time to work out all the algebra, you can write any complex number in the rectangular form $\boldsymbol{A} = a + jb$. (You will get plenty of practice in this chapter doing complex arithmetic, so be patient.) Put another way, every complex number has one real part and one imaginary part, no matter how complicated it may appear. This will become clear in the upcoming section about the complex plane. By the way, in this example, $\boldsymbol{A} = 2.59 + j0.64$.

5.3 Introduction to Phasors

A **phasor** is a complex number used to represent sinusoids, having both magnitude and phase. In Chapter 3 we dealt with sinusoidal voltage $v(t)$ and current $i(t)$, and the complex impedance $\boldsymbol{Z} = R + jX$. In this chapter, $v(t)$ and $i(t)$ will be transformed into complex quantities $\boldsymbol{V}$ and $\boldsymbol{I}$, analyzed using complex arithmetic (which may be new to you) and then transformed back into time-dependent sinusoids. We will call the ones that change in time, i.e., $\boldsymbol{V}$ and $\boldsymbol{I}$, phasors. The impedance, $\boldsymbol{Z}$, which does not change in time, is just a constant, although it is a complex quantity. The arithmetic will be similar for all of them. **Sinusoidal steady-state, SSS, circuit analysis**, assumes that the input to a system is sinusoidal, and that all transients (as presented in Chapter 4) have died out. Phasors discussed in this chapter are used only in sinusoidal steady-state circuit analysis.

Trigonometric identities can be used to describe the behavior of any sinusoidal behavior. However, trig identities are easy only when a couple of terms are combined since the algebra gets complicated when many terms are included. As you will see in this chapter, the use of phasors adds an amazingly "elegant" (i.e., simple) way to manipulate combinations of sine waves, and also affords an intuitive way to visualize voltages and currents in SSS analysis.

Again, $\boldsymbol{A}$ denotes complex numbers that are *constants*, i.e., do not represent functions of time (including impedances, $\boldsymbol{Z}$). You may think of a time-independent number as just a constant, like '8' or '8 + j6', and a phasor as a new way to express a sinusoidal function, like $4\sin(\omega t + \pi/4)$. Time-dependent voltages and currents, $v(t)$ and $i(t)$, represented as phasors, $\boldsymbol{V}$ and $\boldsymbol{I}$, and complex impedances, like **Z**, have much in common, so often there is not much point in distinguishing them from each other mathematically. However, sometimes the difference is crucial, since **Z** is not a function of time, but rather it is a constant used to relate **V** to **I** in a way that is similar to $v(t) = i(t)R$. First there is a fair bit of more math background to cover, and then we will get back to voltages, currents, and impedances.

Now we discuss some math that you will use here and in many of your later courses. Complex numbers can take the form $A = Ae^{j\theta}$, where e is the mathematical constant called "Euler's number," whose value is 2.718.... Like π, it is irrational, so the digits go on forever without settling into any infinitely repeating pattern. Around the year 1740, Leonhard Euler discovered what is now called "Euler's equation," namely that:

$$e^{j\theta} = \cos(\theta) + j\sin(\theta). \tag{5.8}$$

As described in *e: The Story of a Number*, by Eli Maor, Euler discovered this by writing an infinite series for e^x. Substituting $j\theta$ for x and collecting odd and even powers of j, out popped his eponymous formula.

When we substitute $\theta = \omega t$ into Euler's equation, we get the important phasor formula of:

$$e^{j\omega t} = \cos(\omega t) + j\sin(\omega t). \tag{5.9}$$

($e^{j\omega t}$ is sometimes written as "cis(ωt)," but not here.) Referring to Eq. 5.7, we see that the real part of Eq. 5.9 is $a = \cos(\omega t)$ and imaginary part is $b = \sin(\omega t)$. As discussed above, using ωt instead of θ means that *the angle of the argument increases with time*, or the radius of a circle *rotates* in the counter-clockwise direction, like the behavior shown in Fig. 5.3. Now, instead of having a complex constant we have a time-varying phasor. So, you can see that phasors are related to time-varying sine and cosine functions.

The exponential version of a phasor that represents a sinusoid, $Ae^{j\omega t}$, may look foreign to you; one goal of this chapter is to demystify this mathematics so that you are comfortable with it here and as you go forward in future courses. Phasors can represent a variety of physical quantities, as long as they vary with a sinusoidal dependence. In this course we transform sinusoidal steady-state AC voltages and currents into their phasor forms as

$$\boldsymbol{V} = |\boldsymbol{V}|e^{j\omega t} \text{ and } \boldsymbol{I} = |\boldsymbol{I}|e^{j\omega t}. \tag{5.10}$$

Whenever you see complex exponentials that vary in time, as in Eq. 5.10, always remember that they fundamentally represent sinusoids. Phasors, like other vectors, have a magnitude designated as, e.g., $|\boldsymbol{V}|$. We can write $|\boldsymbol{V}| = V$ and $|\boldsymbol{I}| = I$, so we can write Eq. 5.10 using another notation as

$$\boldsymbol{V} = Ve^{j\omega t} \text{ and } \boldsymbol{I} = Ie^{j\omega t}. \tag{5.11}$$

Phasors are *phundamental* (I couldn't resist). Complex exponentials show up whenever steady-state sinusoidal or wave behavior of a system needs to be solved for, such as: analyzing sinusoidal steady-state linear circuit behavior; power transmission on the grid; optics and lasers to describe the wave nature of photons; the density of an electron in space (the electron wave function in quantum mechanics) in semiconductor devices; the design of high-frequency circuits; and in communications to represent electromagnetic waves propagating through space in radio waves. So you see, phasors are *really* important! It is to your advantage to learn as much as you can about them now.

5.3.1 The Complex Plane

Complex numbers can be represented by *vectors* having their vector components of the real and imaginary parts when placed on the **complex plane**. We denote a vector in the complex plane in the rectangular form as $A = a + jb$, where, again, a is the real part and b is the imaginary part. See Fig. 5.5. The horizontal component of the vector, along the **real axis**, is the real part; the vertical component, along the **imaginary axis**, is the imaginary part. You can think of this plane as a two-dimensional vector space, just like x and y components of vectors in Cartesian coordinates. The unit vector on the real axis is simply '1' and the unit vector on the imaginary axis is simply 'j.' Put another way, numbers along the real axis are just (real) numbers, but numbers along the imaginary axis have the form '$j\#$'.

The rectangular form of a complex number shows the coordinates explicitly as a along the horizontal axis and jb along the vertical axis. Impedance, Z, is one example of a complex vector, or complex constant. Vectors in the complex plane are *also* used to represent sinusoidal, time-dependent behavior, in which case the vector is called a phasor. Phasors are treated mathematically as any complex vector would be, but phasors have additional properties. For now, we will discuss complex vectors in general and later focus on what constitutes a phasor versus a complex vector constant.

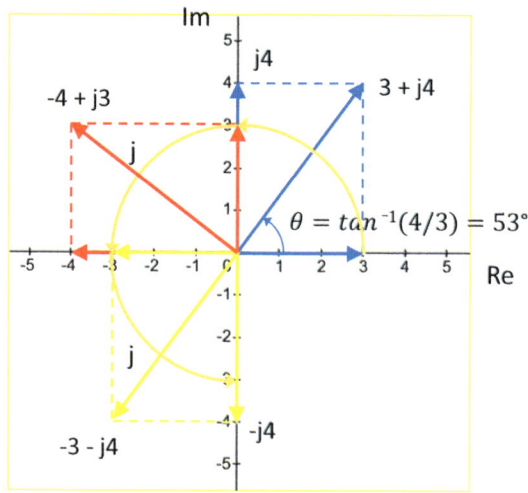

FIGURE 5.5 Examples of phasors in the complex plane. Multiplying Z by j causes a CCW rotation of 90°.

Deeper understanding: Multiplying a vector Z by j is equivalent to rotation CCW of a complex number by $\pi/2$ in the complex plane, i.e., adding 90°. If Z is real, you may think of multiplying by j as flipping a number *halfway around to being negative*. That might not make much sense for a real number, but it does make sense when the j causes a sinusoid to take on an extra phase of 90°, which is halfway toward 180°, which is the equivalent of multiplying the function by –1.

Figure 5.5 shows the following complex vectors written in the rectangular form: $A = 3$; $A = j4$; $A = 3 + j4$; $A = -3 - j4$; $A = j \times j4 = -4$; $A = j \times j \times j4 = -j4$; and others. Be sure that you understand where each value of A lies on the complex plane. As you know, the length of the vector $A = a + jb$ is

$$|A| = A = \sqrt{a^2 + b^2} \qquad (5.12)$$

just as with any vector on the Cartesian plane. For example, $A = 3 + j4$, $|A| = |3 + j4| = \sqrt{9 + 16} = 5$. The angle that A makes from the horizontal, i.e., real, axis is

$$\theta = \arctan(b/a). \qquad (5.13)$$

which in this case is $\theta = \arctan(4/3) = 53°$. $|A|$ is called the **magnitude** or **modulus** of A, and θ is called the **argument**, or **arg**, of A, written arg(A). Recall that

$$\tan(\theta) = \sin(\theta)/\cos(\theta), \tag{5.14}$$

or side opposite/side adjacent of a triangle. The arctan, or tan^{-1}, function is the inverse of tangent and returns the angle between –90 and 90°.

Multiplying any phasor or vector **A** by a vector constant, which we will call **A₁**, results in a new vector $A' = A \cdot A_1$ whose length is $|A'| = |A||A_1|$ and whose angle, $\theta_{A'}$, is the sum of the two angles $\theta_A + \theta_{A1}$. In other words, $|A_1|$ scales the length of **A** and simultaneously rotates **A** by an additional angle θ_{A1}. For $A = j$, the scaling is simply a factor of 1 (the length of j is $|1j| = 1$) and the rotation is that of the angle of j, or 90° (arctan(1/0) = 90°). This was demonstrated above and in Fig. 5.5, where you see that multiplication by j rotated the vector by 90°. *Scaling and rotation is a general property of multiplication by a complex constant.*

Note that the real part of any phasor or vector **A** is just the length of its projection along the horizontal axis (found by multiplying the magnitude of the vector by cos(θ)), and the imaginary part is the length of its projection along the vertical axis (found by multiplying the magnitude of the vector by sin(θ)). Hence, we can write **A** as a version of the rectangular form, called the **trigonometric form**, as

> *Of interest:* You can now see that $e^{j0} = e^0 = 1$; $e^{j\pi/2} = j$; and $e^{j\pi} = -1$. It turns out, then, that $e^{j\pi} + 1 = 0$. This last relationship has received considerable attention from math junkies because it relates the numbers e, j, π, 1, and 0. (See http://www.bbc.com/earth/story/20160120-the-most-beautiful-equation-is-eulers-identity).

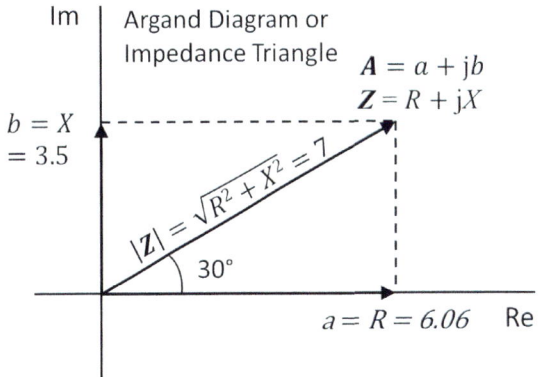

FIGURE 5.6 Argand diagram, or impedance triangle, for **A** $= a + jb$. The rectangular, trigonometric, exponential, and polar forms of the complex number **A**, where $A = Z = R + jX = 6.06 + j3.5 = 7(\cos(30) + j\sin(30)) = 7e^{j30} = 7\angle30$.

$$A = A\cos(\theta) + jA\sin(\theta), \tag{5.15}$$

again where $\theta = \arctan(b/a)$. The rectangular form of the complex number, given in Eq. 5.7, $A = a + jb$, is just the trivial extension of Eq. 5.15 in which $A\cos(\theta)$ and $A\sin(\theta)$ are calculated.

Using Euler's equation, Eq. 5.15 leads directly to the same form as that of Eq. 5.8, namely that $A = Ae^{j\theta}$. This is called the **exponential form** of **A**. Again, it may seem strange to see the imaginary number 'j' in the exponent but demystifying that is one of the main goals of this chapter.

$Ae^{j\theta}$, in turn, can be represented by a shorthand notation called the **polar form**:

$$Ae^{j\theta} = A\angle\theta, \tag{5.16}$$

spoken as "*A* at an angle θ" or just "*A* at θ." Reminder: here, *A* (not bold) is the magnitude of the vector **A**, and θ is the angle of **A** CCW from the real axis. The polar form isn't something different – it's just a shorthand notation for the exponential form that makes it faster and easier to write and (later) manipulate products and quotients of vectors.

Like Fig. 5.5, Fig. 5.6 shows the relationship among the real and imaginary components of a complex number $A = a + jb$. This simple complex vector diagram is called the **Argand diagram** (named for Jean-Robert Argand) when **A** is a generalized complex number. (Life can be so confusing. Recall that the argument of a complex number is arg(**A**). Well, "arg" could derive from "Argand", but no, it stands for "argument.")

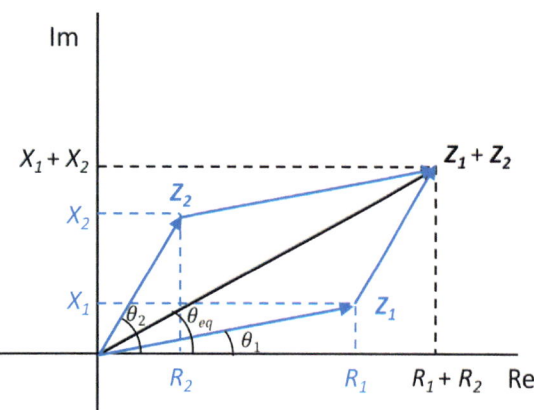

FIGURE 5.7 Addition of two complex vectors in the complex plane.

Know this: Let's collect and review some information, and present an example. You can do complex math instantly on any modern calculator, Matlab, or many other platforms. Even the Google search box will calculate it for you. (As an undergrad in the late 1970s, I had to slog through all of this with barely a 4-function calculator.) In order to get a sense of the arithmetic, it is important practice to do these calculations yourself using a calculator. Here is an example of the four forms using a previous expression:

Example:

$A = (5 + j7)/(3 + j12)^2 + (9 + j8)/(4 + j2)$

Rectangular form: $A = 2.59 + j0.64$

$|A| = 2.67;$
$\theta = \arg(A) = \arctan(\text{imaginary/real}) = 0.24 \text{ rad}$

Trigonometric form:

$A = 2.67(\cos(0.24) + j\sin(0.24))$ (in radians)

Exponential form: $A = 2.67e^{j0.24}$

Polar form: $A = r\angle\theta = 2.67\angle0.24$

Every form has its place and utility, so you need to know all of them and how they are related to each other.

When that complex number is impedance, or specifically $A = Z = R + jX$, the Argand diagram is called the **impedance triangle**. The impedance triangle should remind you of the power triangle of Chapter 3. There, the angle and magnitudes of the sides of the power triangle had the same relationships as the impedance triangle here. As an example of using the Argand diagram/impedance triangle, let's consider a complex number that is an impedance Z, where $\theta = 30°$, and $|Z| = Z = 7 \, \Omega$. (Recall that $\cos(30) = 0.866$ and $\sin(30) = 0.5$.) Figure 5.6 shows $Z = 7e^{j30} = 7\angle30 = R + jX = 7\cos(30) + j7\sin(30) = 6.06 + j3.5$. In order, those are the exponential, polar, trigonometric and rectangular forms of Z.

Adding two complex numbers in rectangular form is easy. Figure 5.7 shows that phasors and complex vectors add just like any vectors do. Each component of the Argand diagram adds independently of the other component, i.e., the real parts add together, and the imaginary parts add together. Suppose we want to add two impedances in series as

$$Z_1 = R_1 + jX_1 \text{ and } Z_2 = R_2 + jX_2. \tag{5.17}$$

Then the equivalent impedance vector is

$$\begin{aligned} Z_{eq} &= Z_1 + Z_2 \\ &= (R_1 + R_2) + j(X_1 + X_2) \\ &= R_{eq} + jX_{eq}. \end{aligned} \tag{5.18}$$

Since $\tan(\theta) = \sin(\theta)/\cos(\theta)$, or side opposite/side adjacent of the unit triangle, $\tan(\theta) = X/R$, and

$$\theta_{eq} = \arctan(X_{eq}/R_{eq}). \tag{5.19}$$

If the impedances (or any complex numbers) are in the exponential form of $Z_1 = Z_1e^{j\theta_1}$ and $Z_2 = Z_2e^{j\theta_2}$, or the polar form $Z_1 = Z_1 \angle \theta_1$ and $Z_2 = Z_2 \angle \theta_2$, then *you have to convert them to trigonometric or rectangular form before you add them together.* This process may be time consuming, depending on how much practice you've had, and what computational tools you have available.

5.3.2 Multiplying and Dividing Complex Numbers

We write an impedance as $Z = R + jX$, where the real part R may or may not be just the value of some resistor, and the imaginary part X might not be just the reactance of a circuit component. In general,

R and *X* are expressions containing ω, *R*, *L* and *C*. The units of **Z**, *R* and *X* are all ohms. Impedances, **Z**, add in series and parallel the same way that you have so far treated resistances in series and parallel. The equivalent impedance of two series impedances is

$$\mathbf{Z_{eq}} = \mathbf{Z_1} + \mathbf{Z_2} \tag{5.20}$$

and in parallel it is

$$\frac{1}{\mathbf{Z}_{eq}} = \frac{1}{\mathbf{Z}_1} + \frac{1}{\mathbf{Z}_2} \tag{5.21}$$

$$\mathbf{Z}_{eq} = \frac{\mathbf{Z}_1 \mathbf{Z}_2}{\mathbf{Z}_1 + \mathbf{Z}_2} \tag{5.22}$$

Clearly, multiplication and division are needed to calculate parallel impedances. Multiplication and division of complex numbers is tedious in the rectangular form and trivial in the polar form. Multiplying two complex numbers in the rectangular form is done by brute force:

$$\mathbf{Z_1} \cdot \mathbf{Z_2} = (R_1 + jX_1)(R_2 + jX_2) = (R_1 R_2 - X_1 X_2) + j(X_1 R_2 + X_2 R_1). \tag{5.23}$$

The result of this multiplication is another complex number. For impedances, $\mathbf{Z_1} \cdot \mathbf{Z_2}$ has units of Ω^2. After dividing by $\mathbf{Z_1} + \mathbf{Z_2}$, $\mathbf{Z}_{eq}$ has units of Ω. So, what *about* division?

The goal in division is to clear all the *j*'s from the denominator so that they appear only in the numerator. Division by a complex number, $\mathbf{Z_1}/\mathbf{Z_2} = (R_1 + jX_1)/(R_2 + jX_2)$, is done by first multiplying the numerator and denominator by the **complex conjugate** of the denominator. The complex conjugate of a complex number is found simply by flipping the sign of the imaginary part. The complex conjugate of $\mathbf{M} = a + jb$ is denoted with an asterisk as $\mathbf{M}^* = (a + jb)^* = a - jb$. So,

$$\mathbf{MM}^* = (a + jb)(a + jb)^* = (a + jb)(a - jb) = a^2 + b^2 = |\mathbf{M}|^2, \tag{5.24}$$

which is *real*. Put another way, *multiplying a complex number by its complex conjugate yields a real number*. So, by multiplying the top and bottom of $\mathbf{Z_1}/\mathbf{Z_2}$ by the complex conjugate of the denominator, i.e., $(\mathbf{Z_1} \cdot \mathbf{Z_2}^*)/(\mathbf{Z_2} \cdot \mathbf{Z_2}^*)$, we turn the denominator into a real number, and then we have only multiplication to perform in the numerator:

$$\frac{\mathbf{Z_1}}{\mathbf{Z_2}} = \frac{\mathbf{Z_1}\mathbf{Z_2}^*}{\mathbf{Z_2}\mathbf{Z_2}^*} = \frac{(R_1 + jX_1)(R_2 - jX_2)}{(R_2 + jX_2)(R_2 - jX_2)} = \frac{(R_1 + jX_1)(R_2 - jX_2)}{(R_2^2 + X_2^2)}. \tag{5.25}$$

This may seem like a bit of math, and it can get a bit long (but fun to do). Fortunately, multiplication and division using the exponential and polar forms is much simpler. Writing $\mathbf{Z_1} = Z_1 e^{j\theta_1}$ and $\mathbf{Z_2} = Z_2 e^{j\theta_2}$, just as multiplying any exponential we get

$$\mathbf{Z_1} \mathbf{Z_2} = Z_1 e^{j\theta_1} Z_2 e^{j\theta_2}) = Z_1 Z_2 \, e^{j(\theta_1 + \theta_2)}. \tag{5.26}$$

Using the even-simpler polar form we write $Z_1 \angle \theta_1 \; Z_2 \angle \theta_2 = Z_1 Z_2 \angle (\theta_1 + \theta_2)$. It's just that simple, *as long as you already have the exponential or polar form.*

Division is equally simple using the exponential and polar forms:

$$\frac{Z_1}{Z_2} = \frac{Z_1 e^{j\theta_1}}{Z_2 e^{j\theta_2}} = \frac{Z_1 \angle \theta_1}{Z_2 \angle \theta_2} = \frac{Z_1}{Z_2} \angle(\theta_1 - \theta_2) \qquad (5.27)$$

Equation 5.26 makes it clear that multiplying the complex number, or vector in the complex plane, Z_1 by Z_2 scales the modulus Z_1 by Z_2 and rotates Z_1 CCW by (the angle) $\arg(Z_2) = \theta_2$. Division (Eq. 5.27) does the same sort of thing except that Z_1 is divided by Z_2, and the resulting vector is Z_1 rotated *clockwise* (CW) by (the angle) $\arg(Z_2)$.

5.3.3 Rotating Phasors

So far, we have reviewed simple trig concepts, imaginary and complex numbers, vectors on the complex plane, introduced Euler's formula, shown that there are four different ways to represent complex numbers or vectors, and demonstrated some simple arithmetic using complex numbers. Now we start to put all that together into some useful methods for circuit analysis to calculate sinusoidal currents and voltages and their phase relationships. Our next discussion pertains to vectors in the complex plane whose *angles change with time*, and which can be related to sinusoids. If the angle of a vector in the complex plane is time-dependent, i.e., $\theta = \omega t$, then the vector will rotate around the origin like a spinning hand on a clock (but where increasing time goes in a CCW direction).

As mentioned, Euler's formula for vector constants is of the exponential form $e^{j\theta}$, but phasors, which are time-dependent, are of the form $e^{j\omega t}$. The difference is that when $\theta = \omega t + \phi$ (units: radians/sec × sec = radians), the angle θ is increasing with time. We use A to represent a generalized complex number, and Z for impedance. Z might depend on frequency (as in X_L and X_C), but it is a constant in time, so it is not a phasor. However, voltages and currents change in time, so they *can be* represented as phasors.

Let's represent a voltage phasor in the trigonometric form:

$$V(t) = V \cos(\omega t + \phi) + jV \sin(\omega t + \phi). \qquad (5.28)$$

The magnitude of the phasor, (not-bold) V, could represent a peak, peak-to-peak, or RMS amplitude. You just have to pick one at the start of the problem. Let's assume peak amplitudes here to be more consistent with conventional math. As a phasor, $V(t)$ rotates around the circle, as in Fig. 5.5. Do not get θ and ϕ confused in the context of sinusoids and phasors: θ is the angle of rotation as t increases; ϕ is the angle at the start of that rotation when $t = 0$. ϕ is a constant phase relative to $t = 0$.

So far, we have implied that a phasor is a time-dependent vector, or a rotating vector, on the complex plane. This is not strictly true, and requires some clarification. To be precise, a **phasor** is generally defined to be a *fixed* vector on the complex plane that *would* rotate as t increases, but it is *fixed at time $t = 0$*, i.e., $e^{j(\omega t + \phi)} = e^{j\phi}$. A complex vector whose angle *changes* with time, i.e., $e^{j(\omega t + \phi)}$ spins CCW with ωt, is often called a **rotating phasor**. This is an important distinction as we go forward.

To summarize: If the origin of a vector on the complex plane is an impedance, it is *not* a phasor. If the origin of a vector on the complex plane is a sinusoidal AC current or voltage, then it is a phasor. A *phasor* is a "snapshot" of a *rotating phasor* at $t = 0$. When the rotating phasor is frozen at $t = 0$, it becomes a stationary vector on the complex plane, but that is only because *the rotation is temporarily ignored*. The rotation of the phasor is restored at the step when we transform back into the time domain to obtain the AC sinusoidal behavior. So you see, there are subtle but important distinctions amongst a complex vector constant, a phasor, and a rotating phasor.

Here is an important fact regarding the addition of rotating phasors. In fact, it is the foundation for all phasor diagrams and mathematics: *The addition of any two (perfect) sinusoids of the same frequency, regardless of their phase or amplitudes, results in a new (perfect) sinusoid having a new phase and amplitude but the same frequency.* This is neither obvious nor intuitive, so it is worth calling out specifically and discussing.

For example, $V(t) = 5\cos(100t + \pi/3) + 2\cos(100t - \pi/10)$ (where φ and ω are in rad and rad/s, respectively) is the sum of two sinusoids of different amplitude and phase, but *same frequency*. Figure 5.8(top) shows that the sum yields a *perfect sinusoid* having a third phase and amplitude, but *same frequency*. Maybe you would have guessed that without resorting to trig identities, but probably not! The waveforms shown in Fig. 5.8(top) are examples of the time-domain functions that phasor diagrams and calculations represent. It's because it is so much more cumbersome to do all of that with the waveforms or trig identities that phasor arithmetic becomes so important.

Let's use phasors to do the example computation represented by the waveforms in Fig. 5.8(top) in order to show how much simpler it is. This is done by **transforming each time-dependent sinusoid into a phasor, adding them, and then transforming back to the time domain by *taking the real part*.** For example, let's add two sinusoidal voltages in series (it could be currents in parallel, and it could be more than two) where

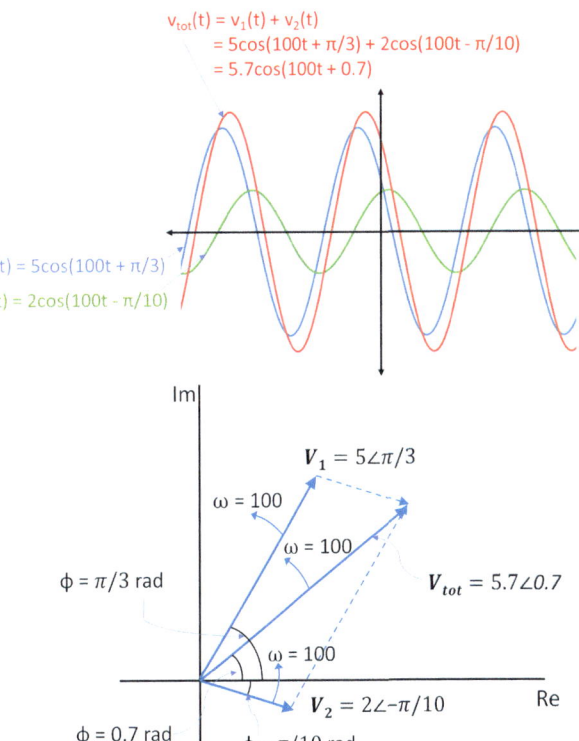

FIGURE 5.8 (Top) Plot of the addition of $\mathbf{V_{tot}}$ = $5\cos(100t + \pi/3)$ + $2\cos(100t - \pi/10)$ demonstrating that the addition of any two sinusoids of the same frequency results in a third sinusoid of the same frequency. The two terms might represent two voltages added in series or two currents added in parallel. (Bottom) Phasor representation of the same calculation. The v_{tot} plot in part (a) is actually two curves that lie on top of each other showing that the calculation done here and in the text are in agreement.

The top and bottom figures convey exactly the same information, but obviously look very different. Each has its place, but it is easier to do calculations with phasor notation. The waveforms in (Top) would be seen on an oscilloscope, and the phasors in (Bottom) would be used to analyze a circuit or some other system of sinusoidal behavior. The arrows are a reminder that the phasors are actually rotating as sinusoids, but the calculation is done at $t = 0$, as shown.

$$v_{\text{tot}}(t) = v_1(t) + v_2(t) = 5\cos\left(100t + \frac{\pi}{3}\right) + 2\cos\left(100t - \frac{\pi}{10}\right). \tag{5.29}$$

As stated above, the resulting $v_{\text{tot}}(t)$ will have the same frequency at some magnitude and some phase relative to the two component voltages.

Let's test it out. The two components of $v_{\text{tot}}(t)$ are transformed as rotating phasors to

$$\mathbf{V_{tot}} = [5\cos(100t + \pi/3) + 5j\sin(100t + \pi/3)]$$
$$+ [2\cos(100t - \pi/10) + 2j\sin(100t - \pi/10)], \tag{5.30}$$

which, for completeness, can also be written in exponential form as:

$$V_{\text{tot}} = 5e^{j(100t + \pi/3)} + 2e^{(j100t - \pi/10)}$$
$$= 5e^{\pi/3}e^{j100t} + 2e^{-(\pi/10)}e^{j100t}$$
$$= [5e^{\pi/3} + 2e^{-(\pi/10)}]e^{j100t}. \tag{5.31}$$

The fact that the sinusoids are rotating phasors, as suggested in Fig. 5.8(bottom), does not help or play any role in the vector analysis. All of the rotating phasors rotate together since they are all of the same frequency. (A good illustration of how phasors rotate together can be found at https://en.wikipedia.org/wiki/Phasor#/media/File: Sumafasores.gif.) Therefore, their relationships to each other do not change, even as they all spin around (CCW). Hence, the rotation is not relevant to the analysis, and it can be dropped from the calculation by setting $t = 0$. This results in stationary phasors, or vectors on the complex plane, also shown in Fig. 5.8(bottom), but without the curved arrows.

Later we will put the time dependence back in to transform back from phasor space to recover the time domain behavior.

Follow these steps as a general guideline for working with phasors. This example is specifically about adding two waveforms, but it can be generalized to other problems.

1. Transform the sinusoids from time domain to phasor domain, as in Eq. 5.30.

2. Set $t = 0$ to eliminate the time dependence, i.e., change the rotating phasors into stationary, non-rotating phasors:

$$V_{\text{tot}} = \left[5\cos\left(\frac{\pi}{3}\right) + j5\sin\left(\frac{\pi}{3}\right)\right] + \left[2\cos\left(-\frac{\pi}{10}\right) + j2\sin\left(-\frac{\pi}{10}\right)\right] \tag{5.32}$$

These are the two component vectors shown in Fig. 5.8(bottom).

3. To add phasors, collect the real and imaginary terms and calculate their values in the form $A = a + jb$:

$$V_{\text{tot}} = \left[5\cos\left(\frac{\pi}{3}\right) + 2\cos\left(-\frac{\pi}{10}\right)\right] + j\left[5\sin\left(\frac{\pi}{3}\right) + 2\sin\left(-\frac{\pi}{10}\right)\right] = 4.4 + j3.7, \tag{5.33}$$

which is also a (non-rotating) phasor. This is the resultant vector shown in Fig. 5.8(bottom).

4. Now you can find the magnitude and angle of the resultant vector. The magnitude is $|V_{\text{tot}}| = V_{\text{tot}} = \sqrt{4.4^2 + 3.7^2} = 5.7$ V; the angle ϕ is $\arctan(3.7/4.4) = 0.7$ rad $= 40.2°$. In general, it is a good idea to confirm for yourself that this is the correct quadrant before accepting what a calculator spits out, especially when negative arguments of arctan are involved. You should see from Eq. 5.33 that $4.4 + j3.7$ lies in the first quadrant.

5. Convert the result into whatever phasor form helps make solving the problem easiest. In the exponential form $V_{\text{tot}} = 5.7e^{0.7}$; in the polar form $V_{\text{tot}} = 5.7\angle 0.7$; in the trigonometric form; $V_{\text{tot}} = 5.7\cos(0.7) + j5.7\cos(0.7)$; and in the rectangular form, $V_{\text{tot}} = 4.4 + j3.7$.

6. Once you have the phasor form of whatever answer you are looking for (here it was the sum of two voltages in series), transform the result back to the time domain. You return the time dependence to get a rotating phasor by simply replacing the time dependence back into the phasor in the trigonometric form.

$$V_{\text{tot}} = 5.7\cos(100t + 0.7) + j5.7\cos(100t + 0.7). \tag{5.34}$$

7. The final step to getting fully back into the time domain from the phasor domain is to simply *delete the imaginary term*. This is called "**taking the real part**." (This may not make intuitive sense; it will be discussed in depth later.) Converting back to the time domain we get:

$$v_{\text{tot}}(t) = 5.7\cos(100t + 0.7). \tag{5.35}$$

Study Fig. 5.8 very carefully as it encapsulates the foundations of phasors. You can see in Fig. 5.8 (bottom) that this is the same result as that found by adding the two sine waves in the time domain.

In summary, the addition of sinusoidal waveforms represented as rotating phasors assumes that they are all at the same frequency, so combining their vectors results in a vector of some new magnitude and phase that is also a rotating vector at the same frequency, as shown in Fig. 5.8(bottom). This underlying fact, demonstrated by the previous calculation, is what makes the manipulation of phasors so powerful.

It is also worth noting that *multiplying* sinusoids (for example in power calculations as $P = V \cdot I$ as in Chapter 3) results in *sums and differences of frequencies*, so the resultant phasors would rotate at different frequencies, and are generally not analyzed using phasor diagrams unless all of the phasors are, say, power and are rotating at the same frequency, which is the case for power factor and the power triangle discussed in Chapter 3.

5.4 Real and Complex Impedances

In steady-state sinusoidal analysis, voltages and currents are represented as phasors V and I, respectively, and obey Ohm's law $V = I \cdot Z$. As suggested by the calculation performed in the previous section, we will transform sinusoidal voltages and currents into phasors instead of attempting to manipulate the sinusoidal functions themselves. We will do the calculations in phasor space and then transform back to the time domain to get our solutions.

Every electronic device has a relationship between the voltage across it and the current through it. Here we limit ourselves to passive, linear, two-terminal devices, namely resistors, inductors, and capacitors. A **linear** device is one that drops proportionately more voltage as the current through it increases, e.g., twice the current corresponds to twice the voltage. A semiconductor diode (Chapter 7) is an example of a nonlinear device because its current increases exponentially with the voltage across it. Linearity is discussed more deeply in Section 5.6.

Back to impedances. The impedance, Z, of a device or circuit (recall the black box of Fig. 3.38)

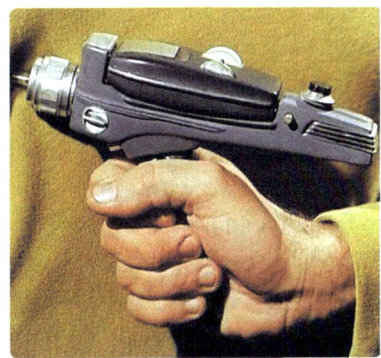

https://www.pinterest.com/pin/343892121529212358/

Just for fun: "Phasor" is short for "phase vector." It is clever how the Star Trek series, from the very beginning, used the "phaser" as their weapon of choice. This is clever on so many levels. The phaser weapon is portrayed as emitting what looks like a beam of light, whereas a later program reveals that it is a stream of fictitious subatomic particles. Buck Rogers was a comic-book hero of the 1930's. He was in a coma and woke up in the 25th century where he had access to a "ray gun." Although the type of ray wasn't exactly identified, it could have been a laser. Although such weapons were only science fiction, they are now science fact. There really do now exist "directed energy" weapons that use a powerful laser to put a huge amount of energy into a very precise location, which can immobilize or destroy a target. Such systems are now deployed on ships and even ground-based mobile platforms. The mathematics of light propagation relies heavily on the use of phasors.

controls the amount of current flowing through it for a given voltage across it, as well as the phase between that current and voltage. Z is in general a complex constant, which means that it can be due to a resistor, inductor or capacitor, or a circuit that is a combination of those elements. An impedance will "impede" the sinusoidal current according to the magnitude of the impedance, $|Z|$, as $|V| = |I|\cdot|Z|$ or $|I| = |V|/|Z|$ where V and I are phasors.

The relationship $|V| = |I|\cdot|Z|$ gives the relationship amongst the magnitudes of voltage, current and impedance, but not the phase between the voltage and current. The current and voltage due to a complex impedance are not in phase, as they are with a simple resistor. For that we need to use the phasor form $V = I\cdot Z$. Ohm's Law for phasors states that multiplying a phasor, I, by a complex number, Z, results in another phasor, V, whose angle is that of I rotated CCW by the angle of Z. (It is worth pointing out that although we are working with complex numbers, the units are still volts, amps, and ohms for both the real and imaginary parts.) To put it another way, it is the impedance that imparts the magnitudes of the current and voltage, as well as the phase shift between the voltage and current. That phase shift is the phase of the impedance! For the specific case of a resistor, Z is real, the imaginary part is zero, and the phase imparted by Z to the current to obtain the phase of the voltage is zero, so voltage and current are in phase. For the case of inductors or capacitors, Z is purely imaginary, which on the complex plane has a phase of $\pm90°$, so the phase imparted by Z to I to get V is $\pm90°$. In general, current and voltage due to Z are separated by $\arg(Z)$.

5.4.1 Real Impedances

Of course, you know that $v(t) = i(t)R$. Still, this deserves some more discussion because we will expand our view of all of the terms in Ohm's Law. We will start with this simple equation and then make it more complex (pun intended).

You know that if you apply 10 V_{DC} to a 20 ohm resistor, then the current through the resistor must be 0.5 A_{DC}. Now let's change the power supply from a DC source to a sinusoidal AC source having peak voltage of 10 V_p with a frequency of, say, 1 kHz. Now the current will have a value of 0.5 A_p and will also be sinusoidal since $v(t) = i(t)R$ applies at every moment in time.

We can arbitrarily represent the current as $i(t) = I_p\cos(\omega t + \phi)$. Here, ω = 1000 Hz × 2π rad/s/ Hz = 6283 rad/s. What is ϕ? That is included for completeness, but without another sinusoid to compare it to, or any particular time that we call $t = 0$, we can choose the phase of current to be any arbitrary value we like (refer back to Fig. 5.4). So, the easy thing to do is to let $\phi = 0$ and $i(t) = I_p\cos(\omega t)$. Doing so simply means that the plot of $i(t)$ crosses through $i(t = 0) = I_p$. $i(t)$ would shift along the time axis with some other choice of ϕ.

Voltage and current through a resistor is actually a very special case since the phase shift between them is zero. The waveform for $i(t)$ will be exactly the same as that of $v(t)$, scaled by R of course. Put another way, if one channel of the oscilloscope displays $v(t)$ and another channel displays $i(t)$, as in Fig. 5.9, and if you change the scales properly, then the two waveforms can be made to lay exactly over each other (not shown). Even if the amplitudes were not the same as displayed, they would both cross zero at the same times. Put another way, *there is no phase shift between the voltage across the resistor and the current through the resistor.*

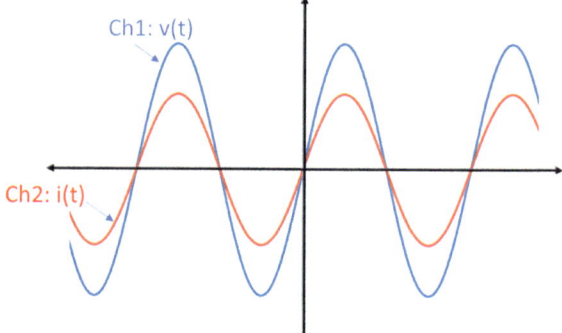

FIGURE 5.9 Two channels of an oscilloscope showing two waveforms (here they are $\sin(\omega t)$) that are in phase. They both cross zero at the same times.

We can represent a resistive load as $Z = R + j0$. We say that the impedance is "purely resistive" or "purely real" or just "real." Since the reactance, X, is zero, there is no phase shift between the current and the voltage displayed on the scope. As stated in Chapter 3, complex numbers provide a way to include phase in the analysis; a resistor doesn't have an imaginary component and doesn't add any phase shift to the temporal relationship between its voltage and current. For $X \neq 0$ transforming $v(t) = i(t) R$ into phasors in the form $V = I \cdot Z$, yields a nonzero phase shift between V and I.

5.4.2 Complex Impedances

The following has been covered in Chapters 3 and 4, but here is a review. For an *ideal* inductor,

$$Z_L = 0 + jX_L = j\omega L \tag{5.36}$$

and for an *ideal* capacitor,

$$Z_C = 0 + jX_C = -\frac{j}{\omega C} \tag{5.37}$$

The impedance is "purely reactive" for ideal inductors and capacitors since the real part is zero. Since $X_L = \omega L$ and $X_C = -1/\omega C$, it is obvious that the impedances depend on the frequency at which current is pushed through the device. For an inductor, X_L increases with frequency, i.e., it takes a higher voltage to get current to go through an inductor as frequency increases. It is the opposite for a capacitor; higher-frequency currents pass more easily through a capacitor, i.e., drop less voltage.

Just like resistors, any parallel or series combination of impedances can be combined into an equivalent impedance expressed as $Z_{eq} = R + jX$. If X is positive, Z is said to be "inductive" and if X negative, Z is said to be "capacitive." The 'j' in front of the reactance is the 'cause' of the phase shift between current and voltage in $V = I \cdot Z$. As discussed above, including the jX term causes rotation by $\arg(Z) = \arctan(X/R)$ in the complex plane, which becomes the phase between V and I on the complex plane, and is, therefore, the phase shift between $v(t)$ and $i(t)$ in the time domain.

5.4.3 Parasitic *R*, *C*, and *L*

Why is the word "ideal" italicized with regards to Eqs. 5.36 and 5.37? Because no device is only capacitive, inductive, or resistive. Every device has some of everything, and how much that matters depends on the frequency that the device is driven at. Also, by the way, there are only those three linear, passive device properties: resistance, inductance and capacitance, because those properties are ultimately determined from the relative phase between the voltages and currents. (There is a fourth, proposed, device called a "memristor," but it has altogether different properties.)

Any unwanted or unintended circuit element that arises as part of the physical nature of a

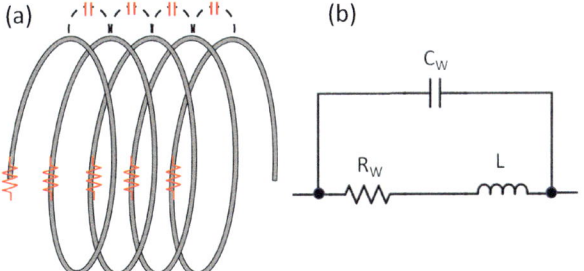

FIGURE 5.10 (a) Drawing of a wound inductor showing how how parasitic resistance and capacitance arise between the windings. (b) One equivalent circuit of a non-ideal inductor in which the parasitic resistance is in series with the inductance, and the parasitic capacitance is in parallel with both of them. At fairly high frequencies the parallel capacitive reactance is low, so current leaks past. the inductor.

device is referred to as **parasitic**, because it usually has a deleterious effect on the intended circuit behavior. Combined, all of this extra behavior is due to the **parasitics**. Let's consider, for example, the parasitics of an inductor. Each winding of an inductor has some parasitic resistance in the wires, and parasitic capacitance between the adjacent windings, as shown in Fig. 5.10.

Recall from Eq. 1.6 that $R = \rho l / A$, where ρ is the resistivity of the wire, l is the length of the wire, and A is the cross-sectional area. There is virtually no escaping this parasitic resistance (except for superconducting wires). When measured at DC frequency with an ohmmeter, this DC resistance is referred to as the **DCR**. In fact, the parasitic resistance is more complicated than just the DCR of the wire that makes up the inductor. The value of R in $Z = R + jX$ changes with frequency. In the case of using the device at some frequency other than DC, the term DCR is replaced with the term **ESR**, which stands for **equivalent series resistance**. It is better to think of ESR as the real part of the total impedance, also represented by R. But now, it isn't really a resistance that you can locate and measure with an ohmmeter, but rather is caused by a combination of the DCR, inductance, and parasitic capacitance of the inductor. ESR can be due to many combined effects, including the wire resistance, how current flows in the cross-section of wires at various frequencies (called the **skin effect**), core losses, as discussed in Chapter 1, and more. ESR can either increase or decrease as frequency increases. DCR and ESR apply as well to capacitors due to properties of the dielectric material's frequency response.

In circuit models, the parasitic resistance can be modeled as either in series with the inductor, as shown in Fig. 5.10, or in parallel. The choice depends on many factors, including frequency range. Both models can be used and compared to see which best describes the physical behavior as frequency changes. For those of you who are really interested, Appendix 2 delves much more deeply into this.

Referring to Fig. 5.10, you will notice that there are many small "parasitic capacitors" between the windings, and all that capacitance can be modeled as being in parallel with the inductor. At a high frequency (say, 1 GHz, but 'high' is always relative to the application and the device), the inductance might be sufficiently large that all of the current at that frequency is stopped, but the parallel capacitance might look like a low impedance, so high-frequency current might flow right 'through' the inductor – as if it were nearly a short circuit!

The lesson for the practicing engineer is that one must be careful to choose devices with the right properties at the operating frequencies for, say, cell phones, radar, or microwave communications. There are two lessons here: the goodness of a model of a circuit element depends on the frequency range, and increasingly sophisticated models lead to more accurate results over larger frequency ranges. Can you see now why the breadboards introduced in Chapter 2 are not particularly good for prototyping high-frequency circuits? If you want to, go back to the exploded view of the breadboard, and see how it looks like a huge group of capacitors! Every one of those parasitic capacitors is part of the circuit. Yikes!

So, circuit elements are not always as they appear, as there can be more to them than just their intended device properties. Here are some common rules of thumb. We usually ignore the resistance and inductance of capacitors. We often ignore the interwinding capacitance of inductors but must include it in those cases where higher frequencies demand it. We frequently have to include the resistance of inductor windings because the wire can be resistive enough to affect the circuit operation, e.g., from a few ohms to a few tens of ohm. Finally, we usually ignore the inductance and capacitance of resistors. Understand that any of these assumptions may be different depending on the type of device (e.g., wire-wound versus carbon-paste resistors) level of accuracy that you need from your model, and the relative values of R, ωL and $1/\omega C$ at the frequencies of operation.

5.5 Relationship between *V* and *I* for Complex Impedances

Figure 5.11(top) shows an ideal inductor driven by an AC source

$$v_{in}(t) = V_m \cos(\omega t). \tag{5.38}$$

What is the current through the inductor? We already know that $v_L(t) = L\dfrac{di(t)}{dt}$, so

$$\frac{di(t)}{dt} = \frac{V_m}{L}\cos(\omega t). \tag{5.39}$$

Because

$$\int \cos(\omega t)dt = \frac{1}{\omega}\sin(\omega t), \tag{5.40}$$

$$i(t) = \frac{V_m}{\omega L}\sin(\omega t). \tag{5.41}$$

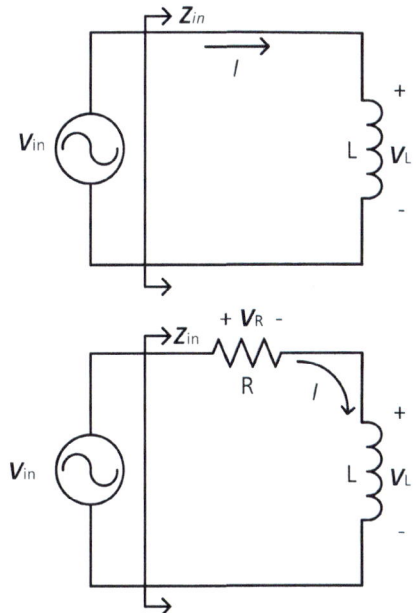

Since $\cos(\omega t)$ and $\sin(\omega t)$ are out of phase by 90°, so are the voltage and current.

If we display current and voltage on an oscilloscope using two different channels, we would see the two sinusoids are 90° out of phase, like Fig. 5.1. Also note that the factor of **ω** in Eq. 5.41 implies that for a given amplitude of input voltage, as frequency increases, the current through the inductor decreases, which is the basis of **$Z_L = j\omega L$**.

Figure 5.11(bottom) shows a real inductor driven by an AC source, where R might represent, or model, the parasitic resistance of a real inductor measured at some frequency or be an actual resistor in series with an ideal inductor. In Fig. 5.11(top), there are only two circuit elements (source and inductor), so the voltage across the inductor is exactly the same as the source voltage; in Fig. 5.11(bottom), the source voltage is split in some non-obvious way across the resistor and the inductor.

For Fig. 5.11, we write using phasors that **V_{in} = $I \cdot Z_{in}$** and **I** = **V_{in}/Z_{in}**. The notation "**Z_{in}**," pronounced "Z_{in}," implies that the impedance is that "looking into" the circuit from the voltage source. Recall that in Section 5.2.2, referring to Eq. 5.7, it was stated that "you can write any complex number in the rectangular form $A = a + jb$." The large bracket with arrows in the figure means that

FIGURE 5.11 (Top) Ideal inductor driven by AC source. (Bottom) Inductor including the real part of the impedance, driven by AC source. The impedance "seen" by the source is labeled "Z_{in}".

you should think of everything to the right as a *total combined impedance reduced to a single complex quantity* **Z_{in}** = $R_{in} + jX_{in}$, no matter how many components there are in the circuit that make up **Z_{in}**. In Fig. 5.11(top), **Z_{in}** = $j\omega L$, and in Fig. 5.11(bottom) **Z_{in}** = $R + j\omega L$. If there were more components to the right of the source, say inductors, capacitors, and resistors in series and parallel, they would all be combined (using complex arithmetic) to make up a different **Z_{in}** = $R_{in} + jX_{in}$.

For example, suppose there was a capacitor in parallel with the inductor in Fig. 5.11(bottom). Now,

$$Z_{in} = R + X_L \parallel X_C = R + \left(\frac{X_L X_C}{X_L + X_C} \right)$$

$$= R - \left(j\omega L \cdot \frac{j}{\omega C} \right) \Big/ \left(j\omega L - \frac{j}{\omega C} \right)$$

$$= R - j\omega L / (\omega^2 L C - 1)$$

(5.42)

which is in the form of $Z_{in} = R_{in} + jX_{in}$.

Figure 5.12(top) is the phasor diagram for a voltage source having an arbitrary phase ϕ driving an ideal inductor, as in Fig. 5.11(top). The phasor V_{in} is drawn for time $t = 0$, and reflects its initial phase, ϕ. The inset in the upper left is the complex vector diagram for the impedance of a pure inductance. Note that impedances are complex numbers, and therefore *can* be placed on the complex plane. However, Z_{in} is *not* a phasor – it is a constant for a given frequency. It is *not* sinusoidal, does *not* change with time, and does *not* rotate. For that reason, it is not drawn on these phasor diagrams, but it is not uncommon to see voltages, currents and impedances placed all together on phasor diagrams. Of course, they all have different units so their lengths can only be used to compare like quantities on the diagram. Specifically, there is no meaning to the length of a voltage phasor relative to a current phasor or an impedance. However, the lengths of two voltage phasors on the same diagram should be indicative of their relative values.

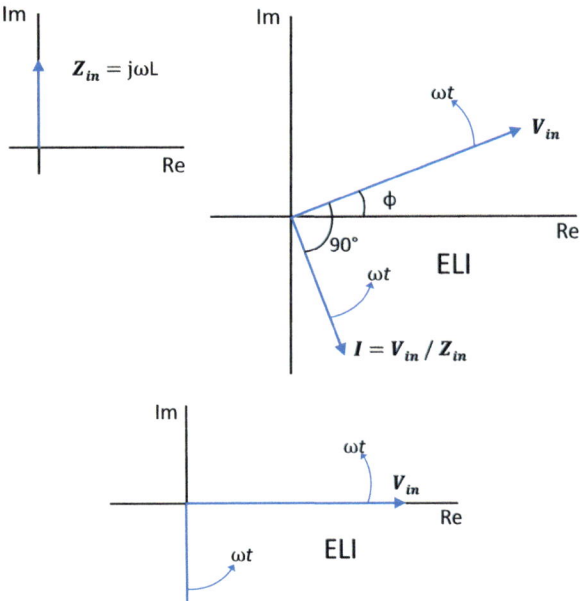

FIGURE 5.12 (Top) Phasor diagrams related to current and voltage (voltage having arbitrary phase ϕ) for a pure inductance, with the impedance diagram of the inductor in the inset. (Bottom) Same as a) except that the voltage is taken as the reference having $\phi = 0$.

Figure 5.12(top) shows that the phase shift between V and I due to a pure inductance is 90°, as in Fig. 5.11(top). Since the angle of Z (arg(Z)) in the inset is 90°:

$$I = \frac{V_{in}}{Z_{in}} = \frac{V_{in} \angle \phi}{Z_{in} \angle 90} = \frac{V_{in}}{Z_{in}} \angle (\phi - 90°).$$

(5.43)

What does a negative phase shift mean on a phasor diagram? The phasors are actually rotating CCW, so in time the current passes a particular angle after the voltage. The angle arg(V_{in}) is ϕ and arg(I) is ϕ–90°. The current is lagging, which agrees with ELI in which current lags voltage in an inductor. So, everything is copacetic.

Figure 5.12(top) includes arrows that indicate that V and I rotate at an angular frequency of ω in the CCW direction, which is generally understood on a phasor diagram, but not usually shown explicitly as it is here. Note that the inset impedance diagram does not have a rotating arrow because (again) impedances are not phasors and do not rotate – they are just complex constants. This figure includes an arbitrary voltage phase of ϕ. Figure 5.12(bottom) shows the same relationship as Fig. 5.12(top), except with the voltage phase arbitrarily chosen as $\phi = 0$. That is the only difference between Figs. 5.12(top) and 5.12(bottom).

We will first use the cumbersome rectangular form for V and Z to solve for $i(t)$ in Fig. 5.12(top), and then use the polar form, which can sometimes be easier. We will follow the seven steps outlined in Section 5.3.3. To use phasors for circuit analysis, current and voltage are first transformed into phasor notation. Calculations using Z and $V = I \cdot Z$ can be performed using any standard phasor form but must then be cast in the trigonometric form as $A = A_m\cos(\phi) + jA_m\sin(\phi)$. Once that is finished, to get the answer as a function of time, the last step of the calculation is to transform back into the time domain by replacing the time dependence (putting the ωt back in) as $A = A_m\cos(\omega t + \phi) + jA_m\sin(\omega t + \phi)$, and taking the real part $A_m\cos(\omega t + \phi)$ of the quantity, V or I, that you are solving for. This simply means that we discard the imaginary term. This will be demonstrated in the following sections.

Let's derive the current due to a sinusoidal voltage source at frequency ω for the circuit of Fig. 5.11(a). There is a long way and a short way to look at the process. First the long way, using the steps to solving phasor problems outlined above. To begin, we convert the time-dependent, sinusoidal voltage

$$v(t) = V_m \cos(\omega t + \phi) \tag{5.44}$$

into a phasor by adding a complex term, and then removing the time dependence:

$$V = V_m \cos(\phi) + jV_m\sin(\phi). \tag{5.45}$$

We get current by dividing voltage by the impedance, Z.

$$V = I \cdot Z = I \cdot j\omega L \tag{5.46}$$

$$I = \frac{V}{Z} = \frac{V}{j\omega L} \tag{5.47}$$

$$I = \frac{1}{j\omega L} [V_m\cos(\phi) + jV_m\sin(\phi)] = \frac{1}{\omega L} [V_m\sin(\phi) - jV_m\cos(\phi)]. \tag{5.48}$$

Finally, discarding the imaginary term, or taking the real part, and replacing the time dependence we get:

$$i(t) = \frac{V_m}{\omega L} \sin(\omega t + \phi), \tag{5.49}$$

which is the same result as Eq. 5.41, as expected (except that here we use $\phi \neq 0$). Note that this particular problem is very basic. If Z were more complicated, say $Z = 4 + j5$, then there would have been considerably more arithmetic needed to get from Eq. 5.47 to Eq. 5.48.

It might have occurred to you to remark about this process, "We just *choose* (where italics are read as derisive disbelief) to add an imaginary term, and then later throw away all that complex stuff and keep the real part? Huh? How can we just *choose* (again read indignantly) to add terms, do lots of math, and then toss out a whole bunch of stuff that we used to do the calculation? What's with all *that*?"

The rationale for simply **taking the real part** is often confusing for beginners, so an explanation is provided in Section 5.6. For now, just know that the last step of the process is simply to take the real part of the resulting phasor. The time-domain current is the real part of the phasor representation of the current and is written $i(t) = \textbf{Re}\{\textbf{\textit{I}}\}$, which in this case is:

$$i(t) = \text{Re}\{\textbf{\textit{I}}\} \frac{V_m}{\omega L} \sin(\omega t + \phi). \tag{5.50}$$

The resulting current is (a) sinusoidal with frequency ω, as expected; (b) is a sine function, which lags the cosine voltage (ELI); (c) is phase shifted by 90° (from $\cos(\omega t + \phi)$ to $\sin(\omega t + \phi)$), which we expect for a purely reactive load; and d) decreases with ω, which we expect for an inductor. Neat, huh? You know, phasors are pretty amazing!

5.5.1 Using the Polar Form of Phasors

Now the short way to do the previous calculation. The polar form of phasors is sometimes quicker to use than the rectangular form. Keeping in mind that $|\textbf{\textit{V}}| = V$ and $|\textbf{\textit{Z}}| = Z$, we can solve for $\textbf{\textit{I}}$ in polar form as (note the use of mixed units in the following equations):

$$\textbf{\textit{I}} = \frac{|\textbf{\textit{V}}|\angle\phi}{|\textbf{\textit{Z}}|\angle 90°} = \frac{V\angle\phi}{Z\angle 90°} = \frac{V}{Z}\angle(\phi - 90°) = \frac{V}{\omega L}\angle(\phi - 90°). \tag{5.51}$$

Compared to Eq. 5.48, this was relatively easy, especially for a more-complicated value of Z after calculating the magnitude and angle of $\textbf{\textit{Z}}$, i.e., $|\textbf{\textit{Z}}|$ and $\arg(\textbf{\textit{Z}})$. This problem is basically finished at this point, but we must still transform back to the time domain. Rewriting Eq. 5.51 in exponential form,

$$\frac{V}{Z}\angle(\phi - 90°) = \frac{V}{\omega L}e^{(\phi-90)}. \tag{5.52}$$

After restoring the time dependence $e^{j\omega t}$, we get:

$$\frac{V}{\omega L}e^{(\omega t + \phi - 90)} = \frac{V}{\omega L}\cos(\omega t + \phi - 90°) + j\frac{V}{\omega L}\sin(\omega t + \phi - 90°). \tag{5.53}$$

Taking the real part of the current phasor,

$$i(t) = \text{Re}\{\textbf{\textit{I}}\} = \frac{V}{\omega L}\cos(\omega t + \phi - 90°). \tag{5.54}$$

Comparing to the form of Eq. 5.49, this can be written as

$$i(t) = \frac{V}{\omega L}\sin(\omega t + \phi). \tag{5.55}$$

This is the same result found using the trigonometric form throughout.

Let's summarize the steps to solving for current in $\textbf{\textit{V}} = \textbf{\textit{I}} \cdot \textbf{\textit{Z}}$ but keep only what we need to get the job done. Let's do the abbreviated version of finding the current through the inductor. To start, you must have already determined $\textbf{\textit{V}}$, $\textbf{\textit{Z}}$ and $\arg(\textbf{\textit{Z}})$.

1. Write $v(t) = V_m \cos(\omega t + \phi)$ in phasor form directly as $V \angle \phi$, and $\mathbf{Z}$ as $Z \angle 90°$.

2. Write $\mathbf{I} = \dfrac{V \angle \phi}{Z \angle 90°} = \dfrac{V}{\omega L} \angle (\phi - 90°)$ (or whatever you are solving for).

3. Transform directly to the time domain by inserting ωt:

$$i(t) = \frac{V}{\omega L} \cos(\omega t + \phi - 90°)$$

$$= \frac{V}{\omega L} \sin(\omega t + \phi).$$

As before, if $\mathbf{Z} = R + jX$ is more complicated than in this example, you will need to solve for Z and $\arg(\mathbf{Z})$ to do step 1. So, although you have seen all the intermediate steps, nothing stops you from just skipping to the important ones, as demonstrated here. With a bit of practice, it is a short process.

More about phase: As previously stated, (non-rotating) phasor $e^{j\phi}$ at $t = 0$ corresponds to the rotating phasor $e^{j(\omega t + \phi)}$. Phase is always relative between two sinusoids or two points in time – the phase of one sinusoid is meaningless. Figure 5.13 relates ϕ back to the underlying sinusoidal meaning of phasors. For $t = 0$ and $\phi = 0$, a phasor is coincident with the horizontal axis on the phasor diagram, as shown by phasor $\mathbf{A}$. Adding a phase ϕ is equivalent to rotating a phasor

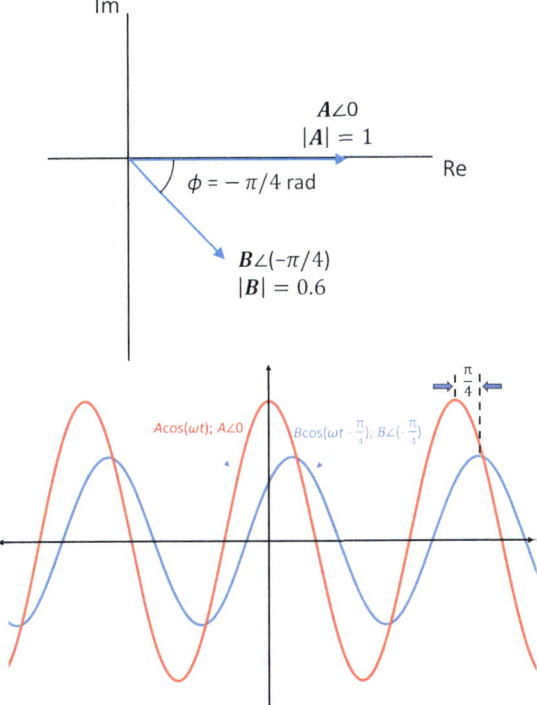

FIGURE 5.13 (Top) Phasor diagram for two phasors that are shifted in phase by $\pi/4$ radians from each other. They could be two voltages, two currents, or a voltage and a current. Two cosine functions that are apart. (Bottom) After transforming back to the time domain, we see that the phasor diagram in (Top) corresponds to two sinusoids of the associated magnitudes separated by $\pi/4$ radians.

by ϕ, and shifting the sinusoid by that phase along the time axis, as shown by phasor $\mathbf{B}$, which has a phase relative to $t = 0$, and to $\mathbf{A}$, of $\phi = -\pi/4$. A negative phase for $\mathbf{B}$ means that $\mathbf{B}$ lags $\mathbf{A}$. This is intuitive because as both $\mathbf{A}$ and $\mathbf{B}$ rotate CCW, $\mathbf{B}$ reaches every angle after $\mathbf{A}$ does. Figure 5.13 (Bottom) is the same information contained in the top part of the figure, but in the time domain. The amplitudes of the cosine functions correspond to the amplitudes shown at the top, and the time difference between $\mathbf{A}$ and $\mathbf{B}$ corresponds to the phase difference shown at the top.

5.5.2 Phasor Diagrams at Different Frequencies

Figure 5.14 shows the total impedance, $\mathbf{Z_{in}} = R + j\omega L$ for two frequencies, $\omega_1 > \omega_2$. (Here we assume that R is independent of frequency, but as discussed below, that isn't always the case.) The magnitude of the complex vector $\mathbf{Z_{in}}$ is

$$|\mathbf{Z_{in}}| = Z = \sqrt{R^2 + \omega^2 L^2}, \tag{5.56}$$

which increases with ω because of the contribution of $X_L = \omega L$. Now,

$$I = \frac{V_{in}}{Z_{in}} = \frac{V \angle 0}{Z \angle \phi} = \frac{V}{Z} \angle - \phi, \tag{5.57}$$

where $\omega = \arctan(\omega L/R)$. As shown in Fig. 5.14, as ω increases, the inductive reactance X_L increases as ωL, and even while R stays constant, the phase angle, ϕ, of the impedance approaches 90°. At low frequencies, X_L approaches zero, ϕ approaches 0°, as shown in Fig. 5.14.

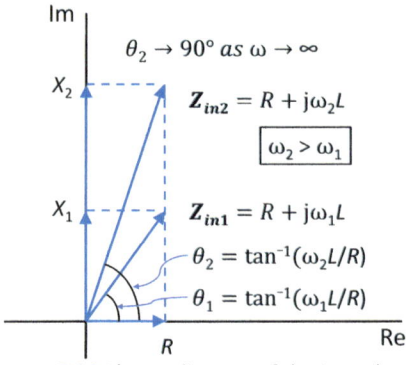

FIGURE 5.14 Phasor diagram of the impedance of a resistance in series with an ideal inductor at two different frequencies. The magnitude of Z_{in} increases as frequency increases.

For the inductor at high frequencies, you can say that "the imaginary part dominates the real part of the impedance." At very high ω, X_L is so large that R is vanishingly small in comparison, in which case $\mathbf{Z_{in}}$ looks nearly purely inductive and ϕ is nearly 90°. This is reflected in the fact that $\arctan(\omega L/R) = \arctan(\infty) = 90°$.

Figure 15.15 shows the phasor diagram for the current due to the inductance as frequency changes and $\mathbf{Z_{in}}$ changes, as in Fig. 5.14. Because $\mathbf{I} = \mathbf{V_{in}}/\mathbf{Z_{in}} = (V/Z)\angle{-\phi}$, we see that $\mathbf{I}$ lags $\mathbf{V_{in}}$ by ϕ, and $|\mathbf{I}| = V/Z = V/\omega L$, decreases as ω increases. So, both the phase and magnitude of $\mathbf{I}$ with respect to $\mathbf{V}$ are affected as frequency changes.

Figure 5.11(Bottom) is the same circuit as that in 5.11(Top), but includes some resistance, R. The resistance R in Fig. 5.11 might be due to a **discrete** resistor ("discrete" means "separate") in the circuit. Also, a real inductor has some internal resistance, ESR, which is effectively in series with the purely reactive inductance. There might be other circuit components that are part of $\mathbf{Z_{in}}$. In order to model $\mathbf{Z_{in}}$ as accurately as possible, the real part of $\mathbf{Z_{in}}$, R, could include the resistive effects of ESR, wires, series resistors, internal resistance of supplies, and more. It is up to you, the engineer, to decide what level of accuracy you need the model to be. For example, if wire resistance is 0.0001 Ω, and a discrete resistor in series is 5.6 kΩ, then you can safely ignore wire resistance. If ESR increases to 500 Ω at some frequency, you should probably leave that in the model of $\mathbf{Z_{in}}$. As component values change with frequency, so do impedance, current, voltage and phase angle. All of this can be represented on phasor diagrams as in Figs. 5.14 and 5.15, but both R and X are functions of frequency, so the magnitude and angle of $\mathbf{Z_{in}}$ change with frequency.

5.5.3 Capacitive Reactance

So far, all of the examples have been about inductors, for which $X_L = \omega L$. For capacitors, the reactance is $X_C = -1/\omega C$. As noted above, we typically ignore the resistance of a capacitor, so we would not generally see a parasitic resistor included in the model as we did in Fig. 5.11. However, resistors appear routinely, of course, in series with capacitors, as we saw in the LPF and HPF of Chapter 4, so the same sort of circuit and phasor properties apply equally.

The difference for capacitive impedances is that the reactance is negative, so it points down rather than up. Figure 5.16(a) shows just the impedance or reactance of an ideal capacitor, and Fig. 5.16(b) shows the total impedance, Z_{in}, of a resistance in series with the capacitor at two different frequencies. What does a negative reactance mean in Fig. 5.16(a)? For an inductor, the positive reactance was the cause of the voltage leading the current. We know this qualitatively because of ELI, and quantitatively because:

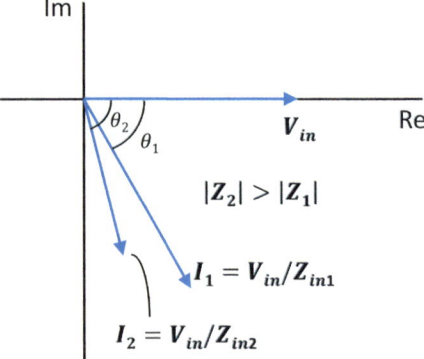

FIGURE 5.15 Phasor diagram of the impedance of a resistor in series with an ideal inductor at two different frequencies.

$$V_L = |I_L| \angle \phi \cdot |Z_L| \angle + 90° = |I_L| \cdot |Z_L| \angle (\phi + 90°). \qquad (5.58)$$

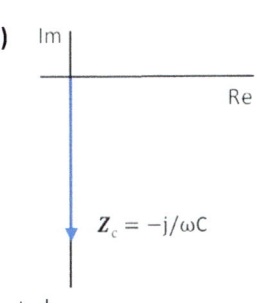

The phase of the inductor impedance, arg(Z_L), contributes 90° to the current phase to get the voltage phase, which as a result *leads* the current by 90° (ELI). For a capacitor, because the reactance $X_C = -1/\omega C$ is negative, and $Z_C = -j/\omega C$, arg(Z_C) = −90°, and

$$V_C = |I_C| \angle \phi \cdot |Z_C| \angle -90° = |I_C| \cdot |Z_C| \angle (\phi - 90°). \qquad (5.59)$$

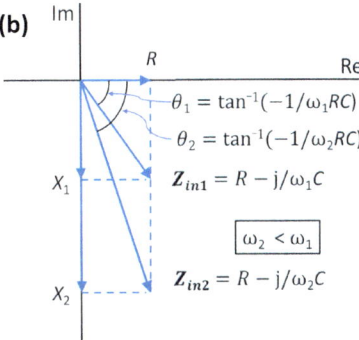

The phase arg(Z_C) subtracts 90° from the current phase to get the voltage phase, which as a result *lags* the current by 90° (ICE).

Here is an example. Suppose that the impedance Z comprises a resistance in series with a capacitor, so that $Z = R - j/\omega C$. For $R = 100\ \Omega$, $\omega = 377$ rad/s and $C = 10\ \mu F$, $Z = 100 - j265\ \Omega$. Let the magnitude of I, $I = 2$ A$_p$. Let the phase of the current be $\phi = 0$ to simplify the calculation. What is the voltage across Z and the phase of V relative to I?

To use the impedance, we calculate its magnitude and phase as $Z = \sqrt{100^2 + 265^2} = 283\ \Omega$ and $\phi = \arctan(X_C/R) = \arctan(-265/100) = -69°$. Now,

FIGURE 5.16 (a) Impedance diagram for ideal capacitor. Current leads voltage for a capacitor because $V_C = I_C Z_C$, and the phase contributed by Z_C is −90°. (b) Z_{in} for a resistance in series with an ideal capacitor at two different frequencies. The magnitude of Z_{in} decreases as frequency increases.

$$V = I \cdot Z = I \angle 0° \times Z \angle \phi = 2 \angle 0° \times 283 \angle -69° = 566 \angle -69°\ V_p. \qquad (5.60)$$

This means that for whatever the impedance represents, a voltage $V = 566$ V$_p$ across Z lags the current I of 2 A$_p$ through Z by 69°. The current could have been specified as peak, peak-to-peak or RMS. The calculations are all correct as long as you are consistent throughout.

5.6 Linearity

The notion of linearity is an important engineering and science concept, so you should understand it. Here we discuss how phasors work in linear systems, but linearity applies throughout science and engineering.

As posed above, "How is it that we can just *choose* to represent a sinusoidal function as a phasor, do our calculations in the complex domain, and then just *choose* to throw away the imaginary part (i.e., take the real part) to get our solution in the time domain?" This section explains that seemingly mysterious process.

Phasors have been presented as a transformation from the time-domain waveform into the phasor domain. We saw in Chapter 4 that a Fourier transform changes a function of time into a function of frequency. We can do certain things to the waveform more easily as it is represented in the frequency domain, and then transform back into the time domain, in which the original function has been changed. We can, say, enter the frequency domain of a square wave, remove the high-frequency components, and then in the time domain see that the square wave has been transformed into a triangle wave. (You did this in the Chapter 4 activities.) Phasors are something like that. We take the time-domain functions, transform them into phasors, work in the complex phasor space, and then take the result and transform it back to the time domain by *throwing away the complex part and keeping only the real part.* Adding the complex dimension allows the inclusion of phase information.

But *why* can we do this? Phasor theory applies only to **linear** circuits, which means that the circuits respond to various inputs, *x* and *y*, as:

$$\text{(a) } f(Ax) = Af(x) \tag{5.61}$$

and:

$$\text{(b) } f(x+y) = f(x) + f(y). \tag{5.62}$$

Often these two conditions are written as one equation:

$$f(Ax + By) = Af(x) + Bf(y). \tag{5.63}$$

An example of condition (a) is a linear amplifier, for which if the voltage input to the amplifier is made twice as large, then the output will also be twice as large, and not be distorted in any way. The output is a scaled version of the input. Or, if you stretch a spring (but not too much), the amount it stretches will be proportional to how much force you apply to it – twice as much force results in twice the stretching distance.

Condition (b) tells us that multiple inputs to a circuit do not interfere with each other; the output will be the sum of all of them as if they were all acted on individually. You might think of it this way: when you start two songs playing at the same time on your computer with two different applications, they can both play through the speakers at the same time. The output (sound) is the linear sum of the two inputs (audio tracks). Each song sounds just as clear as if the other one wasn't playing, if you could mentally separate them out. If the system were not linear, the output would be some kind of distorted version of the two separate songs. Fourier analysis of waveforms assumes linearity, as a linear circuit responds to each sinusoidal frequency component independently, as if it were the only one; the output is then the sum of the responses to each frequency component. If the system were nonlinear, then frequency components would 'mix' with each other, and the output would be distorted, containing new frequency components that were not present in the original signal.

Now we can ask, "What is the output of our linear circuit due to an input $V\cos(\omega t + \phi)$?" Because of linearity, we can form a phasor by adding a term to the real input that is 90° out of phase, i.e., the imaginary component, which is sometimes called the **quadrature component**. Due to linearity, the system reacts independently to the real and imaginary parts of the input. When we are finished with our calculation, we get rid of the part of the result that is due only to the extra, quadrature, term, which we have already called "taking the real part."

Let's refine this point a little further. A voltage $V\cos(\omega t + \phi)$ can be converted to a phasor by adding the quadrature term, resulting in:

$$\mathbf{V} = V\cos(\omega t + \phi) + jV\sin(\omega t + \phi). \tag{5.64}$$

Now that we have transformed our time-domain voltage signal into a phasor, our linear system acts on both the real and imaginary parts to create some output for both parts. Translating that to the math of linearity, where "Output" is a linear function of some "input" $V\cos(\omega t + \phi) + jV\sin(\omega t + \phi)$:

$$\text{Output}[V\cos(\omega t + \phi) + jV\sin(\omega t + \phi)] = \text{Output}[V\cos(\omega t + \phi)] + \text{Output}[jV\sin(\omega t + \phi)] \quad (5.65)$$

When we are finished doing our phasor calculation, we throw away the imaginary part, i.e., "take the real part," leaving the result due only to our real input. During the intermediary process of having both the real and imaginary terms we work with phasors that provide the benefit of simplicity and graphical interpretation.

Referring to Fig. 5.34, as the phasor rotates, its projection on the vertical, or imaginary, axis, is the sine function and its projection on the horizontal, or real, axis is a cosine function. "Taking the real part" of the result simply means eliminating the sinusoidal part of the output phasor on the vertical axis, which is due only to the imaginary part that we added to transform us to phasor space in the first place, and keeping the sinusoidal function that comes out of the horizontal axis, which is due to our actual original input. Again, this is what is shown at https://en.wikipedia.org/wiki/Phasor. (Note that sometimes you will see treatments that keep the vertical, or imaginary component. That is the same result as taking the real part, but lagging by 90°, so it portrays the same qualitative result.)

Mathematically one can find the real part of a complex number by adding the complex conjugate of the number and then dividing by 2. Complex conjugates are denoted by a raised asterisk and are denoted by adding the word "star." The complex conjugate of $Z = R + jX$ is $Z^* = R - jX$ and is called "Z-star." Adding the complex conjugates Z and Z^* and dividing by 2 yields the real part, R.

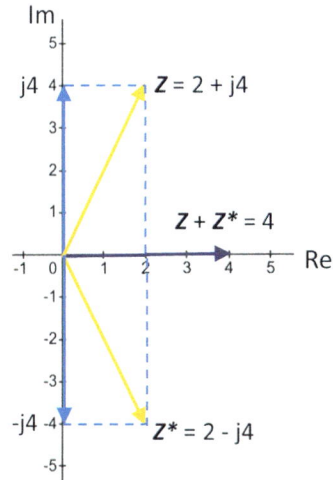

FIGURE 5.17 Graphical representation of adding a complex vector and its complex conjugate to take the real part, shown before dividing by 2. Note how the imaginary parts cancel.

$$\frac{1}{2}(Z + Z^*) = \frac{1}{2}(R + jX + R - jX) = \text{Re}\{Z\} = R \quad (5.66)$$

The imaginary part cancels, and upon dividing by 2 one recovers the real part, R. This is shown graphically in Fig. 5.17. Adding the complex conjugate is like adding the negative of the imaginary component, so they cancel, but with an extra R component that has to be divided away by the '2.' As for phasors, the complex conjugate of a phasor, let's call it V, rotates in the opposite direction through the $e^{-j\omega t}$ term. As the two rotate in opposite directions, their imaginary parts are always of equal magnitude and opposite sign, and cancel. The real parts add together, and are therefore represented twice in the sum, which must be divided by 2 to recover finally the real part of the time-dependent waveform. The arithmetic of this process is

$$\frac{1}{2}[\text{Re}\{V\} + j\text{Im}\{V\} + \text{Re}\{V\} - j\text{Im}\{V\}] = \text{Re}\{V\} = V\cos(\omega t + \phi), \quad (5.67)$$

where $\text{Im}\{V\}$ is the imaginary part of V.

5.7   Second-Order Resonant Circuits

The concept of resonance was discussed at the end of Chapter 4. Section 4.5 introduced resonance in a series, second-order RLC circuit, whereas here we approach it more quantitatively, using phasors to describe the same behavior. Figure 5.18 shows the second-order series circuit from Fig. 4.37. We *cannot* simply say that $Z_{in} = R + 1/\omega C + \omega L$! That would completely ignore the phases of the voltages that result from the series current that they all share! Instead, we must use complex impedances of the three components to obtain

$$Z_{in} = R + jX_C + jX_L = R - \frac{j}{\omega C} + j\omega L. \qquad (5.68)$$

Now we have correctly included the ±90° phase shifts between the current and voltages of the capacitor and inductor.

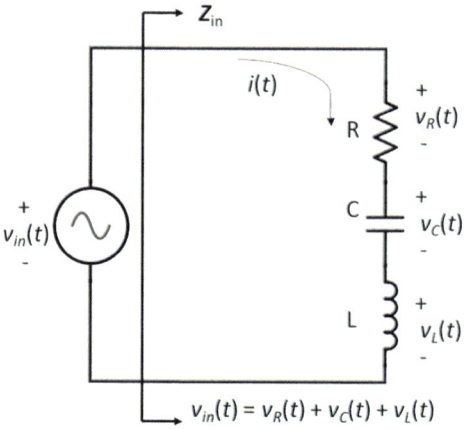

$v_{in}(t) = v_R(t) + v_C(t) + v_L(t)$

FIGURE 5.18 Circuit of 4.37 showing a series LCR circuit driven by an ideal AC voltage source. The current is the same through all components, but the voltages are not in phase with each other. The impedance is R at the resonant frequency.

Figure 5.19 (Top) shows the individual impedance vectors in the complex plane. The critical aspect of the diagram of the impedances is that the inductive and capacitive reactances point opposite to each other. Since $V = I \cdot Z$, for the same series current, the voltages across the capacitor and inductor are of opposite signs, so they tend to cancel each other out. This was described in Section 4.5.1 but will be extended here.

Let's take this one step at a time. Suppose ω were very large. Then the capacitive reactance $X_C = -1/\omega C$ would be very small, the inductive reactance $X_L = \omega L$ would be very large, and the total impedance would be almost the same as $X_L = \omega L$, as shown in Fig. 5.14 for the pure inductance at large ω. For very small ω, $-1/\omega C$ would be very large and ωL would be small, so the same situation applies with the opposite phase relationship between the current and voltage across Z_{in}. This is shown in Fig. 5.16(b).

The resonant frequency, ω_R (often written as ω_0), is that at which the impedance of any second-order circuit is either a maximum or minimum. Figure 5.19 (Middle) shows that as frequency increases, X_L increases while X_C decreases. Keeping in mind that they have opposite signs, the total reactance is zero at resonance. Resonance occurs for this series RCL circuit where $|X_L| = |X_C|$. Changing either L or C shifts the point where the curves cross each other, and therefore changes the resonant frequency.

As introduced in Chapter 4 for the series LCR circuit:

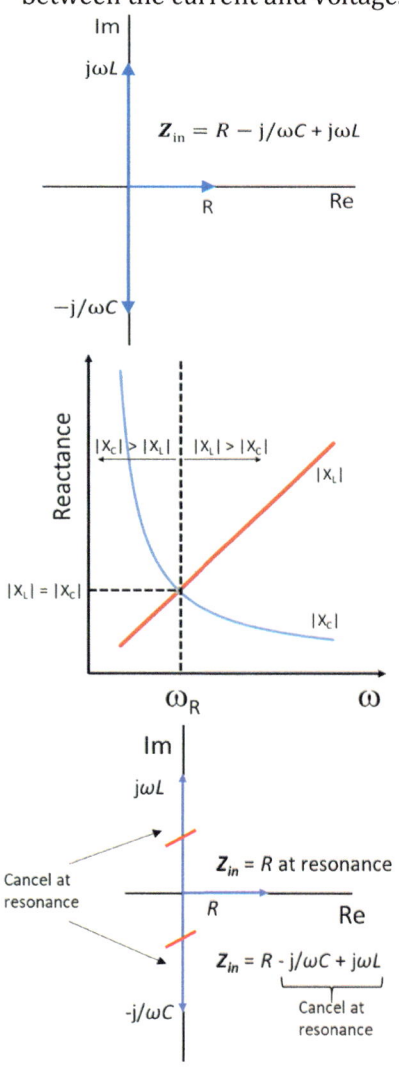

FIGURE 5.19 (Top) Impedance diagram showing the individual components of $Z_{in} = R + jX_C + jX_L = R - j/\omega C + j\omega L$. (Middle) The magnitudes of X_L and X_C are equal at resonance. (Bottom) The same phasor diagram as in (Middle) except at the resonance condition of $\omega = \omega_R$ at which the capacitive and inductive reactances cancel exactly.

$$\omega_R L = \frac{1}{\omega_R C}, \text{ or } \omega_R = \frac{1}{\sqrt{LC}} \left(\text{or, } f_R = \frac{1}{2\pi\sqrt{LC}} \right). \qquad (5.69)$$

As shown in Fig. 5.19 (Bottom), the lengths of the vectors are now exactly the same, and since the phases are opposite, the two vertical vectors perfectly cancel each other out. So, is $Z_{in} = R + jX_C + jX_L = R - j/\omega C + j\omega L$ now zero? No, only the imaginary part is zero, and all that is left now for the total impedance is the real

To enhance your basic understanding of impedance in second-order circuits: We have discussed $\mathbf{Z} = R + jX$ at length, but it would be easy to oversimplify what that equation means. The real part of the impedance, R, of a second-order circuit (for example any of the three filters in Fig. 5.19) is not necessarily simply a combination of resistances, and likewise X is not just due to the circuit's various reactances. In fact, resistance, capacitance, inductance and frequency may appear in the terms that represent the real and imaginary parts. How is that possible? Remember that the energy 'sloshes' around amongst the various reactive components. In doing so, it passes partly through resistances. Resistive energy losses depend on how much 'sloshing' is going on at any frequency. So, you would expect that the real part, which describes that energy loss, would depend on the resistor values, the reactances in which the energy is stored and released, and of course the frequency of the sloshing. Therefore, do not be surprised that solving circuits for their total impedances can lead to pretty complicated expressions for both R and X.

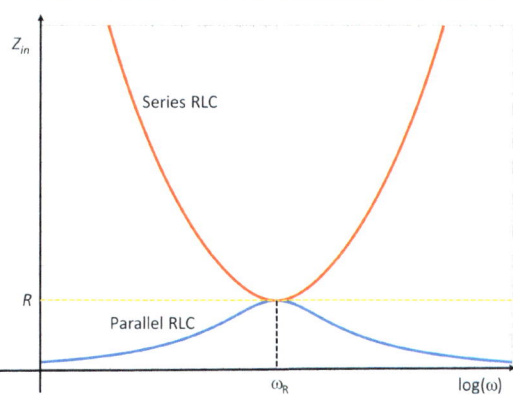

FIGURE 5.20 Resonance in series and parallel LCR circuits. At resonance, the impedance is that of the resistor.

part R, and the current is as large as it possibly can be, but limited by R to $v_{in}(t)/R$. Figure 5.20 shows the dependence of the magnitude of the impedance $\mathbf{Z_{in}}$ for the series RLC circuit of Fig. 5.18 on the logarithm of frequency. As can be seen, at resonance, the capacitive and inductive reactances cancel, leaving only the resistance.

Is the resonant frequency the same for all second-order circuits? No. You can calculate ω_R by finding where the total reactance goes to zero, but it will be different depending on the circuit.

What would happen if R were very low, as if the windings had very little resistance (which is a practical scenario)? In that case, energy would slosh back and forth between magnetic and electric energy and dissipate very little energy in R during each cycle. Since the total reactance is zero at resonance, the current can get very large when there are small losses due to R. Just because the total reactance is zero does not necessarily mean that the individual voltages across the individual impedances $|\mathbf{Z_L}| = \omega L$ and $|\mathbf{Z_C}| = -1/\omega C$ are themselves small. In fact, the large current $i(t)$ flowing through those finite impedances can lead to quite large instantaneous voltages across them, $i(t)|\mathbf{Z_L}|$ and $i(t)|\mathbf{Z_C}|$, but they are out of phase with each other by 180°, which, again, is why their voltages exactly cancel at resonance. You will see in the lab activities that the voltages across the inductor and capacitor can themselves be quite large for a relatively small input voltage, as long as R is small.

Let's review: (i) Voltage leads current by 90° in and inductor (ELI), and (ii) voltage lags current by 90° in a capacitor (ICE). Furthermore (iii) the current in the series LCR circuit is the same for all the components. Therefore, $v_L(t)$ and $v_C(t)$ are always out of phase by 180°. At frequencies lower than ω_R the capacitive reactance is larger and at frequencies higher than ω_R the inductive reactance is larger, as seen in Fig. 5.19 (Middle). It is only at resonance that their agnitudes are equal and the inductor and capacitor voltages cancel exactly.

We have until now discussed only a *series* LC combination of reactances. What about *parallel* combinations, as shown in Fig. 5.21? Just as was shown for resistances and conductances in Chapter 2, we can define the inverse of the impedance: $1/\mathbf{Z} = \mathbf{Y}$ is called **admittance** and is written $\mathbf{Y} = G + jB$, where G is the **conductance** and B is the **susceptance**. The units for $\mathbf{Y}$, G and B are all siemens. If you work out the math, for $\mathbf{Z} = R + jX$, you will find that $G = R/|\mathbf{Z}|^2$ and $B = -X/|\mathbf{Z}|^2$. Many problems of parallel current paths are simplified by writing parallel impedances as admittances, and we will use it here.

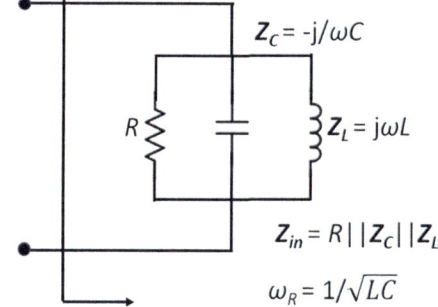

FIGURE 5.21 Parallel LCR circuit. The impedance is infinite at the resonant frequency.

Current chooses the path that has the least impedance, or highest admittance. Since for an inductor, $Z_L = X_L = j\omega L$, and

$$Y_L = 1/j\omega L = -j/\omega L. \tag{5.70}$$

So, the inductive admittance has the same form as the capacitive impedance. In short, inductors conduct (we could say 'admit') less current at high frequencies. For a capacitor,

$$Y_C = 1/(-j/\omega C) = -\omega C/j = j\omega C. \tag{5.71}$$

The capacitive admittance has the same form as the inductive impedance. We can say that capacitors conduct more current at high frequencies. That all said, we can add the two admittances in parallel and get the equivalent admittance

$$Y_{eq} = Y_L + Y_C = j\omega C - j/\omega L. \tag{5.72}$$

At low frequencies, current will flow most easily through the inductor, and at high frequencies it will flow most easily through the capacitor. So, at very low and very high frequencies, the parallel RLC combination is like a short circuit. When $Y_{eq} = 0$, no current flows through the inductor or capacitor. Put another way, at the resonant frequency, the parallel combination will look like zero admittance or infinite impedance, and again $\omega_R = \frac{1}{\sqrt{LC}}$. It remains as an exercise to write the expression for the total impedance of the parallel L and C combination, and show that its magnitude is infinite at $\omega_R = \frac{1}{\sqrt{LC}}$. When at resonance the LC combination is acting like an open circuit, the resistor is the only current path available, and $Z_{in} = R$. Away from ω_R, either or both the inductor and the capacitor offer another current path in parallel, and the magnitude of the impedance is less than R alone.

In both series and parallel LCR combinations, at resonance the energy sloshes between the electric field in the capacitor and the magnetic field in (and around) the inductor. Recall from Lab 3 that only the real part of an impedance dissipates energy – reactances only store and release energy. With little resistance, the energy could slosh back and forth for a long time before dissipating in the resistance. We use the **quality factor**, or **Q factor**, or **Q**, to express the ratio of energy stored to energy dissipated per radian (i.e., $1/2\pi$ of a cycle) by the real part of the impedance. A quality factor can be assigned to individual inductors or capacitors, or to a whole circuit. It makes sense that more reactance for some amount of resistance leads to higher Q values. For an inductor alone:

$$Q_L = \frac{X_L}{R_L} = \frac{\omega_L}{R_L}, \tag{5.73}$$

where R_L is the ESR the inductor, and

$$Q_C = \frac{X_C}{R_C} = \frac{1}{\omega C R_C}, \tag{5.74}$$

where R_C is the ESR of the capacitor due to leads and other internal components. Because ESR is complicated and frequency-dependent, the Q of inductors and capacitors is also a complicated function of frequency, and does not increase or decrease with frequency without bound.

It is possible to calculate the Q of a circuit comprising more than one element, each having its own Q factor. The rule for combining Q_s is the same as that of resistors in parallel and capacitors in series, namely the reciprocal sum.

In the series LCR circuit, the capacitor adds little ESR, so Q for the circuit tends to be close to or about the same as that of the inductor, including total DCR and ESR. At resonance, $\omega_R = 1/\sqrt{(LC)}$, so:

$$\omega_R L = \frac{1}{\omega_R C}, \text{ and} \tag{5.75}$$

$$Q = \frac{\omega_R L}{R} = \frac{1}{\omega_R RC} = \frac{1}{R}\sqrt{\frac{L}{C}}, \tag{5.76}$$

where R is the real part of the impedance of the full LCR circuit.

Figure 5.22 shows a security device commonly attached to clothes in department stores to discourage theft. This particular device is a resonant circuit, called a **tank circuit**, formed from just a coil connected to a capacitor. Both elements are inexpensive sheets of metal foil on paper. That's it. For these applications, these devices are called "radio frequency electronic article surveillance," or **EAS**, tags. These are a type of radio frequency identification, or **RFID**, tag. When an EAS tag passes through the security gates at the exit of a store, an alarm goes off.

Here is how it works. A radio frequency signal at the resonant frequency of the tank circuit, at a few MHz, is transmitted between gates at the exit of the store. As the EAS tag passes through the security gates it resonates and absorbs energy from the field. Some of that energy is reradiated, much like a mirror reflects light. A receiver sees a change in the power and phase of the RF energy when the EAS tag goes through the gates, which sets off an alarm.

FIGURE 5.22 Case and inside of an EAS tag. The tank circuit comprises just a coil and a capacitor.

5.8 Example of Second-Order Filters: Speaker Crossover Networks

In order to fully appreciate and understand this section, you must understand the material in Chapter 4 related to filters, including high-pass filters, HPF, low-pass filters, LPF, and bandpass filters, BPF. If you are unsure of this material, you should take some time to review it and fill in whatever gaps you might have.

In Chapter 4 you saw that a filter is an electrical circuit that allows current or voltage at only certain frequencies at an input port to be passed to an output port. A LPF has a stop band at high frequencies, i.e., stops the high frequencies from going through the circuit from the input port to the output port, and has a pass band at the low frequencies, i.e., passes the low frequencies to the output port. A HPF does the opposite of an LPF, of course. A BPF has a pass band at frequencies between some upper and lower cutoff frequencies, and thus stops those lower and higher. What is called 'high' and 'low' in any filter depends on the values of its components. In an audio circuit, 'low' might be below 50 Hz; in a microwave-communications system, 'low' might be below 10 GHz. Still, the concepts and principles of filters are the same.

One application of filters that you might have some personal experience with (whether you know it or not) is that of crossover networks found in "loudspeakers," or just "speakers." Let's define a few terms with the help of Fig. 5.23(a). The **speaker** is the cabinet and everything in it; the music is produced by separate **drivers**, which are the individual transducers that convert electrical energy into sound energy. (Although it might be confusing, we speak of **driving** the drivers to produce the sound. The origin of "driver" to describe the speaker elements presumably stems from their action of driving the air pressure to make sound.) The speaker driver was introduced in Chapter 1. The "woofer" is the largest driver and produces the lowest sound frequencies. If the speaker is a "three-way" speaker, then it has a midrange and tweeter as well. The midrange handles, guess what, the mid-audio frequencies, and the tweeter produces the highest audible frequencies.

Many of you spend your day with music delivered to your ears by earbuds or headphones. Generally, those do not contain filters inside the earpieces, but rather just pass on the full

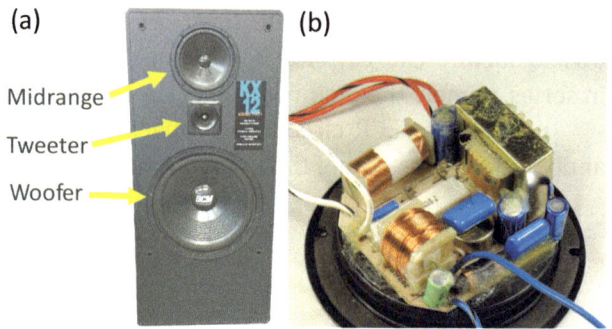

(a)   (b)

Midrange
Tweeter
Woofer

FIGURE 5.23 (a) Speaker cabinet and drivers. (b) Crossover network inside the speaker cabinet, which connects to the wire terminals. The crossover separates the audio frequencies into bands for each of the drivers. For interest: The word tweeter likely derives from the high-frequency sound made by small birds. Midrange drivers used to be called "squawkers," presumably after the sound of larger birds. Somehow, "squawking" doesn't sound like a good thing for a speaker to do. The term "woofer?" A woof is the bark of a large dog, hence, lower frequencies.

bandwidth of the music signal because there is only one driver inside the earpiece. However, large speakers (Fig. 5.23) often have multiple drivers. The frequency range of the bare drivers overlap. You don't want more than one speaker to generate sound that overlaps within a given frequency range; this would lead to poor sound quality. Therefore, filters are needed like traffic cops to send only certain frequencies to certain drivers (Drivers... traffic cop. Get it?). The collection of filters is called the **crossover network**, or just **crossover**. Crossovers also protect the drivers from operating outside their intended range, and possibly being damaged. Later we will use complex impedances to analyze some speaker crossover filters.

Optional reading for those interested in hi-fi stereos: The audio response of a speaker is due to many factors, including the drivers, crossover design, and cabinet design including size, materials, location of drivers, and much more. Also, a flat response is not necessarily what many listeners prefer. For example, listeners of rock music may want a heavier bass response, whereas classical may want flatter response in the midrange at the expense of bass.

Speaker design must account for the characteristics of the cabinet, in which some frequencies may be more resonant than others, and hence louder, and the drivers themselves may have peaks and dips at various frequencies. Sound reproduction is very much a matter of taste, and an art.

One other important aspect of speaker design is that of the phase of the sound produced by each driver. At the crossover frequency, two drivers are attempting to produce the same sound, which must be accounted for so that they do not interfere. Also, at any position in the room, two speakers, left and right, contribute to the sound. Audiophiles highly value the "soundstage" of music listening in which the source appears to be localized in space between the speakers. Achieving good soundstage can be a trade-off with achieving a wide frequency response. This fact accounts for the vast range of cabinet designs found among high-end speakers.

Speaker drivers are chosen so that without any extra electrical filtering, the bare woofer and midrange frequency responses overlap, as do the midrange and tweeter. That bare frequency response is further modified by the crossover filter network. The role of a crossover network, like that shown in Fig. 5.23(b), is to level out the contributions to the music from each of the drivers so that the acoustic frequency response of the entire speaker is uniform, or "flat," across the audio range, i.e., from 20 Hz to 20 kHz. The frequency response of a circuit could be a frequency-dependent current, voltage, or power, or in the case of the speaker the sound pressure at some distance away from it - whatever characteristic we happen to be most interested in. Note in Fig. 5.24 that the woofer and midrange frequencies "cross over" each other at about 350 Hz, and the mid and tweeter frequencies cross over at about 2500 Hz. Those are the **crossover frequencies** in this example. We want the volume from each speaker to be the same at the crossover frequency, so at crossover, the power from each driver should be half of the total power needed to make the total response from the speaker flat. (This is called a "constant-power crossover.") The cross-over frequency is set by the choice of crossover filter component values. If the power is greater than half of the total power, the frequencies at and near the crossover frequency will stand out too much, and if it is less, then those frequencies will be attenuated; in either case, the response will not be flat.

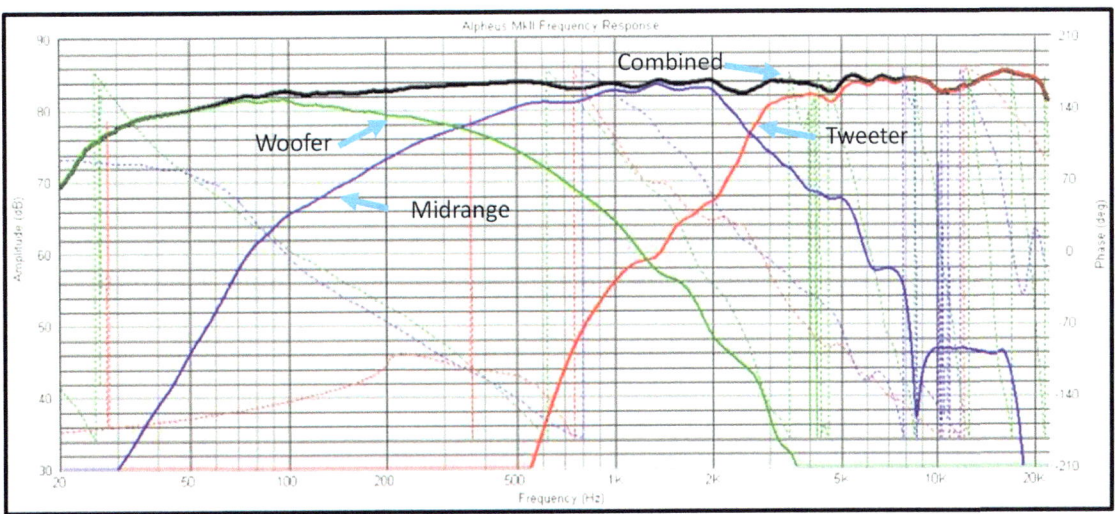

FIGURE 5.24 Loudness, or sound pressure, of a speaker, in audio decibels. 0 dB is the sound pressure at the threshold of human hearing. The graph shows the individual response curves for the woofer, midrange and tweeter, and the total sound coming from the speaker. The designer chooses an overall shape to the response curve, which in this case is fairly "flat."

5.8.1 Crossover Network Design Example

Figure 5.25 shows just one possible three-way crossover network. The optional sidebar explains that there are many more considerations than just the crossover network, and a lot more than just the simple notion of output power goes into the crossover design, but that is all we will work with here.

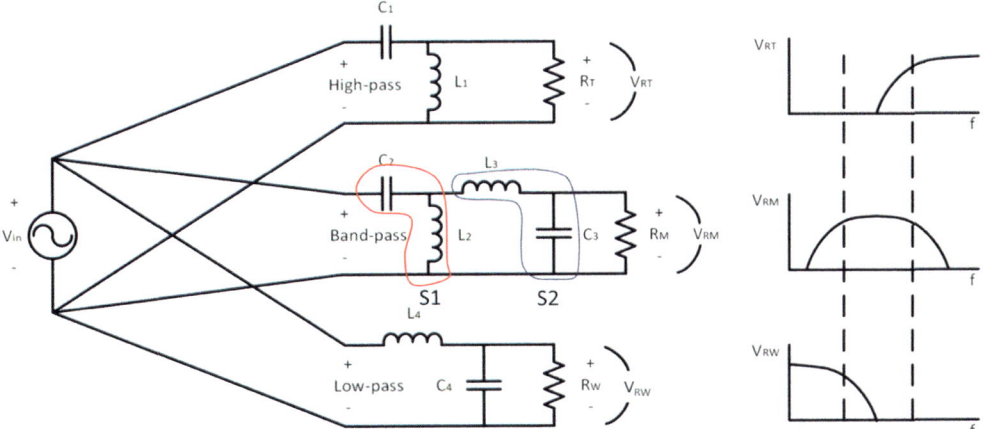

FIGURE 5.25 Parallel second-order three-way speaker crossover network. From the top are the tweeter, midrange, and woofer. The input voltage from the amplifier feeds all three filters in parallel. The graphs at the right show the desired voltage response versus frequency for each driver.

We are making the following simplifying assumptions about our hypothetical speakers: The woofer, midrange and tweeter produce the same amount of sound for a given power input (i.e., they are equally efficient) over a wide frequency range. Also, our drivers and cabinets do not add any variations that need to be accounted for by the crossovers. Therefore, we are designing our crossovers for only one thing, namely a flat power response from three resistors that represent the drivers, each having the typical value of 8 ohm. Finally, our design completely ignores any effects of the phase of the driver response, which is also important in high-quality speaker design.

A plot of the gain of a circuit such as a filter or amplifier as a function of frequency is called a **Bode plot** (pronounced "bo-dee"). The voltage or power gain is the ratio of voltage or power output to the voltage or power input. There are two parts to a Bode plot, namely the log of the gain in dB and the phase shift between the input and the output. These are usually plotted as functions of the log of frequency. In this chapter, we will focus only on the gain, and not phase. Figure 5.24 is a Bode plot of the sound pressure, in dB, as a function of log of frequency that was found by placing a calibrated microphone in front of a speaker and sweeping the frequency. Since we won't be doing that, we will settle for knowing the voltage, and therefore power, delivered to the drivers across the audio frequency spectrum covering 3 orders of magnitude, or **decades**, from 20 Hz to 20 kHz.

5.8.2   More About Decibels

The language of circuit analysis is often that of decibels, dB, as introduced in Chapter 4. Because power expressed in dB is important in the next sections, and it tends to be confusing for beginners, we will now continue and extend the discussion from Chapter 4.

Let's discuss how decibels are calculated for voltage and power. Any quantity expressed in dB must be done so as a ratio of two quantities having the same dimensions, since the argument of the logarithm must be dimensionless. The gain of any system (e.g., a filter or an amplifier) is the ratio of the output to the input, be that power, voltage or current. We will limit the discussion to power and voltage, since current and voltage are related by a constant, R. Recall that $P = V^2/R$. Importantly, power is related to the voltage raised to the power of 2. So, for a given voltage gain, A_V, the power gain is A_V^2. You would be correct to ask, "But what about the value of R? Is that the same for the input and output?" The answer would be "Generally no, but we often assume that it is." Let's start with power gain in dB (which is consistent with the original definition of decibels):

$$A_{P,\text{dB}} = 10 \log\left(\frac{P_{\text{out}}}{P_{\text{in}}}\right). \tag{5.77}$$

Because $P_{\text{in}} = V_{\text{in}}^2/R_{\text{in}}$ and $P_{\text{out}} = V_{\text{out}}^2/R_{\text{out}}$, this relationship becomes:

$$A_{P,\text{dB}} = 10 \log\left(\frac{V_{\text{out}}^2}{R_{\text{out}}} \div \frac{V_{\text{in}}^2}{R_{\text{in}}}\right) = 10 \log\left(\frac{V_{\text{out}}^2 R_{\text{in}}}{V_{\text{in}}^2 R_{\text{out}}}\right) = 10 \log\left(\frac{V_{\text{out}}^2}{V_{\text{in}}^2}\right) + 10 \log\left(\frac{R_{\text{in}}}{R_{\text{out}}}\right), \tag{5.78}$$

where R_{in} is the resistance seen by V_{in} looking into the system, and R_{out} is the voltage across the load R_{out} at the output of the system. Normally, in simple discussions, the difference in these two resistances is not considered. By taking the exponent '2' outside the logarithm [$\log(x^n) = n\log(x)$], and assuming that $R_{\text{in}} = R_{\text{out}}$, this relationship becomes the familiar expression for voltage gain in dB from Eq. 4.48,

$$A_{P,\text{dB}} = 10 \log\left(\frac{V_{\text{out}}^2}{V_{\text{in}}^2}\right) = 20 \log\left(\frac{V_{\text{out}}}{V_{\text{in}}}\right) = A_{V,\text{dB}}. \tag{5.79}$$

This tells us that for a given system, whether we use the decibel expression for voltage or power, we get the same number of dB, as discussed below.

The assumption that $R_{in} = R_{out}$ is a good approximation in the world of radio communications where resistances are standardized. If necessary, accounting for the difference in input and output voltages and resistances yields

$$A_{P,dB} = 10 \log \left(\frac{V_{out}^2}{V_{in}^2} \right) + 10 \log \left(\frac{R_{in}}{R_{out}} \right) \tag{5.80}$$

In practice, power in dB is almost always expressed using Eq. 5.79. Assuming that, we make the following observations:

1. The system has the same gain in dB for voltage and power. For example, if the power has gain of $A_P = 2$, then the voltage has gain $A_V = \sqrt{2}$, and then $A_{V,dB} = 20 \log (\sqrt{2}) = A_{P,dB} = 10 \log (2) = 3$ dB. So, power dB and voltage dB are the same, here 3 dB, for the same system, but the ratios of V_{out}/V_{in} and P_{out}/P_{in} are different.

   The cutoff frequency of a filter, f_C, is defined as that frequency for which the power is decreased by a factor of 2, or put another way, the power gain is $A_P = \frac{1}{2}$. That frequency is said to be the **half-power point**. It is also said to be "three dB down." If you are talking about voltage gain, it is still the half-power point, and calculation of dB for both power and voltage is −3 dB. It's just that the −3 dB half-power point corresponds to a voltage ratio of $\sqrt{\frac{1}{2}} = 0.707$. You can see that in Figs. 4.17, 4.18 and 4.19. This is a confusing point for a lot of students.

2. Gains greater than 1 and their inverses, gains less than 1, e.g. 2 and ½, convert to the same magnitude in dB, but gains greater than 1 are positive and gains less than 1 are negative. For example, if $A_P = 2$ then $A_{P,dB} = 3$ dB, but if $A_P = \frac{1}{2}$ then $A_{P,dB} = -3$ dB. This is useful to keep in mind.

3. Consider a system that has two stages where the first stage passes its output to the input of the following stage, as in Fig. 4.15. Suppose that the first stage has gain A_1 and A_{dB1} and the second stage has gain A_2 and A_{dB2}. Then, the total gain is $A_1 \times A_2$ corresponding in dB to $A_{dB1} + A_{dB2}$; multiplying gains is the same as adding dB. [$\log(a \cdot b) = \log(a) + \log(b)$]. So, it is said that besides making it easier to compare very large and very small numbers, using dB makes it easier to calculate the gain of several amplifier stages in series because you just have to add dB rather than multiply ratios.

5.8.3 Crossover Network as Example of Second-Order Filters

One goal of some speaker designs is to make the frequency response flat across the audio spectrum. "Flat" means that the same power input at all frequencies results in the same sound power. Put another way, flat means that all of the music frequency spectrum sounds the same volume. For example, if the midrange power is too low (less air pressure), singers will sound as if they are far away from the microphone. If the high frequencies are louder than the rest of the spectrum, the cymbals will stand out too much, or acoustic guitar strings will sound very bright. The crossover network is designed to filter the frequency spectrum to account for the properties of each driver, the cabinet and other speaker properties to make an overall flat spectrum from the lowest to highest audio frequencies.

Without consideration of other factors, we choose R, L and C such that the LPF/BFP and BPF/HPF filter-pairs have their half-power points, i.e., −3 dB, at the same frequency so that the two drivers combined yield the same power at all of their overlapping frequencies. The frequencies, f_{CL} and f_{CH}, are the low and high crossover frequencies, respectively, as illustrated in Fig. 5.26. The slope of the gain as it changes with frequency in a Bode plot is its **rolloff**. The gain of first-order filters drops

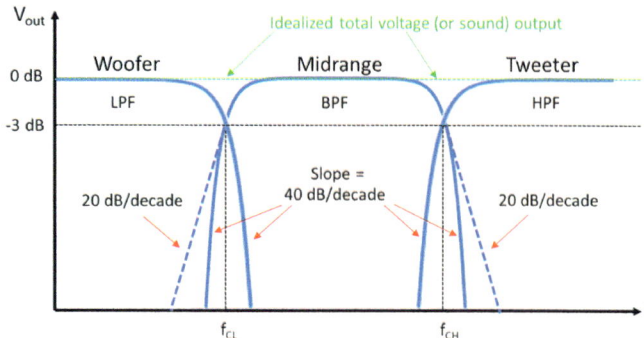

FIGURE 5.26 Idealized output voltage of woofer, midrange, and tweeter. For a perfectly flat network, the voltages would add to 0 dB for all frequencies. The components are selected so that two adjacent curves cross at a gain of -3 dB (half power) at the selected crossover frequency.

off with a slope, i.e., 'has a rolloff of,' or 'rolls off at,' 20 dB/decade, and the rolloff of second-order filters is steeper at 40 dB/decade. *Note:* A **decade** is a factor of ten in frequency, e.g., 20 Hz to 200 Hz. An **octave**, in both engineering and music, is a factor of two in frequency, e.g., 440 Hz to 880 Hz (A4, or concert A, to one octave up, A5). A steeper rolloff implies that the drivers exhibit less frequency overlap, so the music comes from two drivers over less of the music spectrum. (It is left as an exercise to convert dB/decade to dB/octave.)

Let's analyze how each of the three filters in Fig. 5.25 functions. We will discuss the voltage across each of the loads, i.e., drivers, keeping in mind that all three filters are driven by the same input voltage in parallel, and power into each driver load is $V_{\text{driver}}^2/R_{\text{driver}}$.

Most speaker drivers have the same nominal resistance of 8 Ω – the actual value varies from driver to driver, and over the frequency range. Assuming a driver resistance of 8 Ω, is the equivalent resistance of the three drivers all driven together in parallel equal to 8/3 Ω? It would be if they were all driven in parallel at all frequencies, but any particular frequency component of the music is passed to only one driver, so the resistance of most speakers is still generally rated at 8 Ω across the entire musical spectrum. If two speakers are run in parallel, say in separate rooms of a house, then their combination is typically rated around 4 Ω, since each frequency of the signal drives two 8 Ω speakers in parallel.

Let's now analyze each of the separate, second-order crossover filters, which are driven by the power amplifier and whose loads are one of the speaker drivers. Refer to Fig. 5.25.

1. **High-pass filter to the tweeter** (top of Fig. 5.25): The HPF comprises C_1 and L_1 (abbreviated here as "C_1/L_1") driving the tweeter as a load. Low-frequency currents are nearly stopped by C_1 ($Z_C = 1/\omega C$), and what little does pass through sees a short circuit at L_1 ($Z_L = \omega L$), so most of the current passes through, or is shunted, through L_1; no current passes through the tweeter, the voltage drop on the tweeter, $v_{RT} = i_{RT}R_T$, is nearly zero, and the voltage gain is, let's say, –60 dB.

Higher-frequency current is somewhat passed through C_1, whose impedance decreases with increasing frequency, and is somewhat stopped by L_1, whose impedance increases with frequency, so current is somewhat shunted through R_T; v_{RT} is now, say, –15 dB.

Finally, very high-frequency current passes easily through C_1, which now looks almost like a short circuit, and is nearly completely stopped by L_1, which now looks almost like an open circuit, so current from V_{in} is passed almost unimpeded to R_T, and the voltage gain is now 0 dB. Note that C_1/L_1

Optional for hi-fi enthusiasts: Notice in Fig. 5.25 that one set of wire connections feeds all three crossover filters. In some high-end speakers there are separate input jacks to the woofer filter and to the mid/tweeter filters. This is to allow the use of two separate amplifiers to power the two sets of drivers. This is called 'bi-amping'. The purpose of bi-amping is supposedly to relieve the burden on a single amplifier to simultaneously drive both the high-power bass and lower-power mids and highs. Some enthusiasts think that most amplifiers cannot do that without distortion. The vast majority of listeners don't do this, and simply connect, or 'strap,' the inputs together using one amplifier, so the circuit looks as it does in Fig. 5.25. There are others, though, who swear it sounds better to separate the wires going to the separate jacks, but use one amplifier. That is called 'bi-wiring.' This topic comes under the category of 'things about hi-fi that are more likely to be psychological than real.'

work together to stop low-frequency signals from reaching R_T and to pass high frequencies to R_T, and the effect of changing frequencies is more pronounced than if either C_1 or L_1 were used alone, as in a first-order filter. That is to say, a second-order filter has a sharper roll-off than a first-order filter, as shown in Fig. 5.26.

We can use complex impedances to show the behavior described above. Let's calculate the impedance looking in from the amplifier to the tweeter as a function of frequency. The capacitor is in series with the parallel combination of the inductor and the tweeter (resistor). The circuits have both series and parallel parts. Recall that impedances add in series and parallel the same way as resistances, and "||" means "in parallel with." For series circuits, impedances add as $\mathbf{Z_{tot}} = \mathbf{Z_1} + \mathbf{Z_2}$, which is generally straightforward. For parallel circuits, $\mathbf{Z_{tot}} = \mathbf{Z_1}||\mathbf{Z_2} = \mathbf{Z_1}\mathbf{Z_2}/(\mathbf{Z_1} + \mathbf{Z_2})$, which is not always quick to calculate, but is doable. So,

$$\mathbf{Z_{in}} = \frac{-j}{\omega C} + j\omega L \,||\, R_T \tag{5.81}$$

$$= \frac{-j}{\omega C} + \frac{R_T \cdot j\omega L}{R_T + j\omega L} \frac{(R_T - j\omega L)}{(R_T - j\omega L)} \tag{5.82}$$

$$= \frac{-j}{\omega C} + \frac{R_T \omega^2 L^2 + j\omega L R_T^2}{R_T^2 + \omega^2 L^2} \tag{5.83}$$

$$= \frac{R_T \omega^2 L^2}{R_T^2 + \omega^2 L^2} + j\left[\frac{\omega L R_T^2}{R_T^2 + \omega^2 L^2} - \frac{1}{\omega C}\right] \tag{5.84}$$

We can predict the behavior of the tweeter crossover circuit by asking what $\mathbf{Z_{in}}$ is as the frequency goes from DC to some very large value. For $\omega \to 0$, $\mathbf{Z_{in}} \to -j\infty$, so the reactance is very high and no current flows to the tweeter. When $\omega \to \infty$, then $\mathbf{Z_{in}} \to R_T$, which is exactly what you want it to do so that only the high-frequency components of the music come through the tweeter.

2. **Low-pass filter to the woofer** (bottom of Fig. 5.25): The LPF L_4/C_4 drives the woofer as the load, and works exactly the same way as the HPF, except that the roles of the inductor and capacitor are reversed. Try to work that out for yourself.

3. **Bandpass filter to the midrange** (middle of Fig. 5.25 and Fig. 5.27): This requires some discussion but is based on the same operations as the LPF and HPF. The BPF drives the midrange as its load. Notice that C_2/L_2 form a HPF. This is the first stage of the filter, S1. The second stage, S2, is L_3/C_3, which form a LPF. Notice that L_3, C_3 and R_M are the load for the HPF C_2/L_2, just as R_T is the load for the tweeter's HPF. Therefore, the high frequencies get through and are passed to the second-stage L_3/C_3 LPF. In turn, those high frequencies can get to the midrange driver *only if the cutoff frequency for the LPF C_3/L_3, f_{CL}, is higher than the cutoff frequency of the HPF C_2/L_2, f_{CH},* as shown in Fig. 5.27.

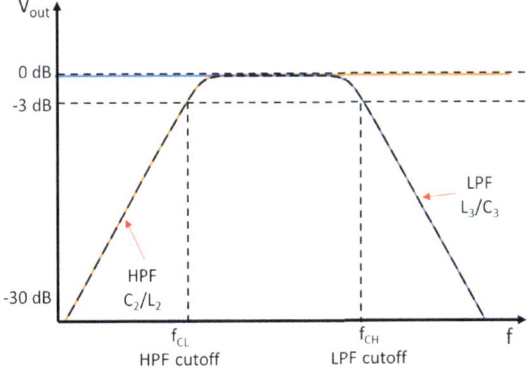

FIGURE 5.27 Explanation of BPF. The BPF is really an LPF and HPF in series. Note that the cutoff frequency for the HPF is lower than that of the LPF, so all 'high' frequencies are passed to the LPF, of which only the lower frequencies get through to the midrange driver.

5.8.3.1 Calculating Impedance of Crossover Networks Using a Computer

We can analyze the three crossover filters using complex impedances as in Eqs. 5.81–5.84, but it is especially tedious for the BPF. This is a great place to point out that computational programs such as Matlab, Wolfram, etc., or the better hand calculators can handle complex math, saving a lot of pencil lead.

Let's use the LPF with the woofer as the load (Fig. 5.25, bottom) as an example of how to easily do a calculation of impedances, voltages, and currents with a computer. We can define

$$Z_R = R_W;$$

$$Z_{C4} = \frac{-j}{\omega C_4};$$

$$Z_{L4} = j\omega L_4;$$

$$Z_P = R_W \| Z_{C4} = \frac{R_W Z_{C4}}{R_W + Z_{C4}};$$

$$\text{and } Z_{in} = Z_{L4} + Z_P. \tag{5.85}$$

Software packages like Matlab treat all of the complex math internally, so you need only write equations with complex numbers, and not do all of the associated algebra, as in Eqs. 5.81–5.84. Using those factors, the voltage divider equation used to get the voltage across R_W is:

$$V_{RW} = \frac{V_{in} Z_P}{Z_{in}}. \tag{5.86}$$

Of course, all calculations are done as phasors since phase makes all the difference to the driver voltage.

Once all the filters and driver voltages have been determined, and the same input signal goes to all of them, you can set out to match the voltages, and therefore power, to all of the drivers. Since the drivers all have the same nominal resistance, i.e., $R_T = R_M = R_W = 8\ \Omega$, you can match the voltages and know that you are matching the powers. Doing the same for the other two filters, BPF and HPF, you can plot the voltage or power at the driver and change the component values to find any particular crossover frequency for your speaker design. Finally, you can add the power output of all three drivers to get the behavior of the entire speaker and adjust component values to make the response flat, or have other characteristics. Without too much time invested, you can write Matlab (or other) code to simulate the entire 3-way crossover network. That software would simply use repeated applications of relationships like Eq. 5.81.

Note: In this week's lab activity, you will be running the SeriesRLCTool.m Matlab code provided to you. Be sure to have Matlab running on your computers and bring it with you to lab ready to go. Download the software before you get to lab. Also, you should at the very least read the first step before coming to lab. It will save you time in the lab.

5.9   Lab 5 Activities

About answers to my questions: For many of the steps I have answered in the explanations the very questions I am asking. It is okay to just give back what I wrote to show that you read and understood. This way I know that your brain was engaged while doing the activities.

Note: The use of angles θ and ϕ mean the same thing throughout this section.

1. **Getting to know the LCR meter.** You will find on Canvas a document called "LCR Meter Interpretation." You are not required to read it, but if you are very, very motivated, it will explain this first set of activities in much more detail. The following is a short synopsis that you must read.

FIGURE 5.28 (Step 1) Keysight U1733C LCR meter display and control panel.

Here, you will become more familiar with the Agilent U1733C LCR meter (Fig. 5.28). The LCR meter attempts to tell you the component value of inductors, capacitors and resistors. They are black boxes as far as the meter is concerned; all the meter knows internally is the current and voltage at its terminals, and the phase angle between them. Put another way, the meter only knows the impedance of what is between its terminals; it does not know why there is that impedance. Since devices always have some inductance, resistance and capacitance, the contributions of the various reactive components, including the parasitics, can change with frequency. The measured impedance could be any combination of R, L and C, but the meter's job is to report a single value of R, L or C to you.

The device being tested is often cleverly called the "device under test," or **DUT**. What the meter reports about the DUT depends on the chosen model, either series or parallel (Fig. 5.29), and the frequency that you use for the measurement. In fact, the series model is just a statement of $Z = R + jX$. The parallel model is a statement about **admittance**, $Y = 1/Z$. The R value that the meter provides isn't necessarily a resistance but is, rather, the real part of the impedance or admittance. At low frequencies it usually does correspond to an actual resistance or conductance embedded in the device.

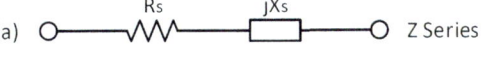

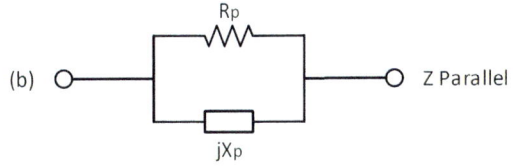

$$R_s \neq R_p; \ X_s \neq X_p$$

FIGURE 5.29 (Step 1) Series and parallel impedance models. Note that the real and imaginary parts in (a) are not equal to those in (b), but the combination of each provides the same measured value of Z.

The rule of thumb for choosing series or parallel models, i.e., Z or Y, to best express the dominant value L or C is that 'low' impedances should be measured with the series model and 'high' impedances should use the parallel model. The model converts the impedance into series and parallel resistances and reactances, which hopefully provide a useful value for L or C.

The test frequency must *not* be near the **self-resonant frequency** (SRF) of the DUT. That is the frequency at which the DUT's parasitic reactance (e.g., capacitance for an inductor) equals the device's low-frequency reactance, and the device, itself, resonates. When that happens, energy sloshes around inside the device itself between electric and magnetic fields. That messes up the measurement completely because the device becomes either an open circuit or a short circuit to an external measurement. The second requirement is that the test frequency produces an impedance that falls within the measurement limits of the meter. Sometimes the current is just too small to measure, and the results will be either unstable or no number will be displayed. Look at Table 5.1.

Table 5.1 Series/parallel model guidelines

Frequency (kHz)	Capacitors		Inductors	
	Use series model when	Use parallel model when	Use series model when	Use parallel model when
0.1	>16 µF	<16 µF	<160 mH	>160 mH
1	>1.6 µF	<1.6 µF	<16 mH	>16 mH
10	>160 nF	<160 nF	<1.6 mH	>1.6 mH
100	>16 nF	<16 nF	<160 µH	>160 µH
1000	>1.6 nF	<1.6 nF	<16 µH	>16 µH

2. **Measuring the DCR and ESR of an inductor.** Choose the large hand-wound inductor. All inductors have some resistance due to the wires that make it up. Here you will measure its DC resistance, **DCR**, using the Fluke DMM. Some care is necessary to remove the effect of the leads. Clip the leads together and press **REL**, which zeros out the meter. Now insert the inductor and measure its DCR without the effect of the leads. Make a note of it.

Turn on the LCR meter. Hold down **P<–>S** for 1 s to toggle between series and parallel models, and make sure it is set to the series model (the resistor in series with the inductor). Toggle **Freq** to choose 100 Hz. Toggle **ZLCR** to select the real part of the impedance, R. Now, insert the short alligator clip leads on the LCR meter, clip them together and press **ΔNull**, which zeroes it out. Attach the inductor and measure its real part, *R*. Toggle **ZLCR** to measure the inductance, *L* at 100 Hz. Record the value in your lab notebook.

At all frequencies, the real part of the impedance is called the **equivalent series resistance ESR**, here denoted '*R*.' *R* is the real part of the series model corresponding to $Z = R + jX$. Note that the DCR and ESR values can be very different, but are very close at low frequency. For an inductor, the DCR is just the resistance of the wire, but the ESR is the real part of the impedance, which may be very different from the DCR depending on frequency. The ESR is due to energy sloshing between the inductance and capacitance that flows through the DCR, giving a real part that depends on frequency, wire resistance, capacitance, and inductance of the **DUT**. Increase the frequency through all five available settings and note how the ESR, i.e., *R*, changes. How well do the DCR measured with the DMM and real part of **Z** measured with the LCR meter agree at 100 Hz? In general, what happens at the higher frequencies? What explains higher values of *R* as frequency increases?

3. **Observing the effect of the core on inductance.** You learned in Chapter 1 about relative permeability and magnetic fields. Now you will observe the effect of adding a ferrite core having a high relative permeability, μ_r, to the inductor. You will see that inductance can be increased considerably with a high-permeability core. The magnetic core from the AM radio antenna in Lab 1 has a relative permeability of about 1000 to 2000, which is much greater that of air alone, i.e. 1. You will use this as a simple inductor core. Set **Freq** to 100 Hz. Starting from several inches away, watch the increase in inductance and record the inductance when the end of the ferrite a) just touches the edge of the coil and b) when the ferrite is totally inserted (symmetric on both ends). In only a couple of lines, briefly describe qualitatively in terms of μ_r what you

did and what happened. Your answer should include "magnetic field lines" and "concentrated" in them. This is the same explanation as that of the antenna in Chapter 1.

4. **Learning about inductor nonidealities.** You will now see that a large inductance can be achieved with a small form factor (i.e., shape), much like an electrolytic capacitor having a high dielectric constant, when it has many windings and a high-permeability core. Choose the inductor that has black tape and "503" printed on it. It is about the size as the tip of your pinky finger. Use the short alligator clip leads on the meter, or remove the clip leads and plug the inductor directly into the '+' and '-' slots on the meter. Fill in the following table for different frequencies and both series and parallel models. Read the quality factor, Q, and the phase, $\boldsymbol{\theta}$, for the inductor from the meter by toggling *DQθ*.

Before you start filling in the table, measure the DCR of the inductor with the DMM and record that for later comparison.

Filling in the table below: Start at 100 Hz and increase the frequency but omit 120 Hz. Circle the values that correspond to the recommended model in Table 5.1.

You will find that many of the values you get seem like nonsense or are unstable. Record those anyway. This happens at frequencies for which: (a) the device is self-resonant, (b) the parasitic reactance dominates the total reactance, or (c) real or imaginary parts of the impedance are close to or are outside the usable range of the meter. One of the goals of this exercise is for you to see how nonideal a real inductor can actually be, and see how limited the frequency range might be that a real device adheres to its rated value.

Frequency	L_{meas} (mH)		R_{meas} (Ω)		$|Z|_{meas}$	θ_{meas}	Q_{meas}
kHz	Series	Parallel	Series	Parallel		Degrees	

Answer these questions **after** the lab session. Answering these questions will bring some sense to what may at first seem like a nonsensical jumble of meaningless numbers. Keep your answers short.

(a) Which measurements of L come closest to the 50.3 mH rated value of the inductor? (b) How do the R_{meas} values correspond to the DCR that you got measured? Read the box in the chapter in Section 5.7 to increase your understanding of reactance. (c) As you increase the frequency, can you see that the real part of the inductor's impedance is not just the DCR, but can be greater due to 'sloshing' of energy? Explain in your own words after reading the box. (d) How well did your data correspond to Table 5.1's recommendations? (e) In general, what was the trend of Z_{meas} with increasing frequency? (f) Does this make qualitative sense? (g) What simple equation explains it? (Hint – look at the equation review in the next step.) (h) In general, what was the trend of θ_{meas} with increasing frequency? (i) Does this make qualitative sense? (j) What simple equation explains it? (Hint – look at the equation review in the next step.) (k) At what frequencies do both the series and parallel models allow good measurements? This means that at those frequencies, the meter currents are within the measurable range. (l) Finally, what might explain the decrease of θ_{meas} and Q_{meas} at 100 kHz? Think resonance.

5. **Practicing some inductor impedance calculations and plots.** Fill in the table below as follows. This step illustrates some of the calculations done internally by the LCR meter. Figure 5.10 showed that inductor windings form a series of capacitors through which current can pass when the capacitive reactance is low, or capacitive admittance is high. Note that LCR measurements of inductance are unreliable at higher frequencies for which parasitic capacitive reactance is on the order of and lower than the parallel inductive reactance. This is exactly the same thing as a current divider, but with some reactive phase relationships. When the frequency is high enough that capacitance starts to change things, it is difficult to interpret the readings. Also, at either very low or very high frequencies, reactances might be so large that the meter cannot measure the associated very low currents. Therefore, we limit our analysis to frequencies at which the model allows an interpretation of **Z**.

Recall that for the series model, $\mathbf{Z}_L = R + jX_L = R + j\omega L$ where $\omega = 2\pi f$; $|\mathbf{Z}| = \sqrt{(R^2 + X^2)}$; $\theta = \arctan(X_L/R)$; $Q = \omega L/R = X_L/R = 2\pi f L/R$. For 100 Hz and 1 kHz only, use the series model for L and write the expression for **Z**, calculate $|\mathbf{Z}|$, θ, and Q, fill in the table below, and compare all to the table in Step 4.

Frequency	L_{meas}	R_{meas}	$X_L = (2\pi f)L_{meas}$	$|Z|_{calc}$	θ_{calc}	Q_{calc}
kHz	Series	Series	Ω	Ω	Degrees	

For each frequency, *hand-draw* a complex impedance diagram (right now in your notebook) that shows the relationship of X_L, R and θ. The drawings do not have to be perfect, but *be neat* and draw them somewhat to scale. Do NOT waste your time trying to be more accurate than just using your eye and best judgement. How well do your calculated values match the measured ones in Step 4? Do you see that the meter is just doing simple impedance (and admittance) measurements and reporting different forms of the results to you?

6. **Measuring capacitor parameters.** You have been provided a brown mylar capacitor. It is about the size of a thick postage stamp and has a value of about 5.0 nF. Fill in the following table for the capacitor. Note that R may not come up on the display, which indicates $R = 0$.

Frequency	C_{meas}		R_{meas}		$|Z|_{meas}$	θ_{meas}	Q_{meas}
kHz	Series	Parallel	Series	Parallel		Degrees	

Answer these questions after the lab session. Notice that the measurement of capacitance is much more uniform across frequencies than was the inductor. Why do you think that could be the case? There could be two different reasons. Answer in terms of the parasitic inductance of the capacitor as well as the amount of capacitive reactance presented by this particular value of capacitance (namely 5 nF). How well did your data correspond to Table 5.1's recommendations?

7. **Practicing some capacitor impedance calculations and plots.** Recall that for the series model, $Z_C = R + jX_C = R - j/\omega C$; where $\omega = 2\pi f$; $|Z| = \sqrt{(R^2 + X^2)}$; $\theta = \arctan(X_C/R)$. Also, for the series model, $Q_C = 1/\omega RC = X_C/R$. Fill in the table below. For 10 kHz only, use the series model for C and write the expression for Z, calculate $|Z|$, θ, and Q, and compare all to the table in Step 6.

Frequency	C_{meas}	R_{meas}	$X_C = -1/(2\pi f C_{meas})$	$|Z|_{calc}$	θ_{calc}	Q_{calc}
kHz	Series	Series	Ω	Ω	Degrees	

Hand-draw a complex impedance diagram (right now in your notebook) that shows the relationship of R, X_C and θ at 10 kHz. Don't forget that X_C is negative, i.e., points down in the diagram. How well do your calculated values match the measured ones in Step 6? Do you see that the meter is just doing simple impedance (and admittance) measurements and reporting different forms of the results to you?

8. **Setting up the Rigol power amplifier.** Your circuit can load down the WG if the input impedance of the circuit is too low. (The output resistance of the WG is 50 ohm.) To prevent this, you will use the Rigol PA-1011 power amplifier (PA) to drive your circuit. The PA has a voltage gain of 1, but has a relatively low output resistance, also called internal resistance, of just a few ohm. As such, the output voltage of the PA will not sag as much as it would if you drove the circuit with the WG alone, as discussed in Chapter 2.

The PA does not have an on/off switch; instead you just plug in the power supply to turn it on. First, *unplug* the power supply from the back of the PA; power off the WG; connect the USB cable from the PA to the WG; insert the power cord into the electrical outlet and the plug the power supply into the PA (display lights should come on); power on the WG. Press **Utility** on the WG; scroll to PA **Setup**; set **Gain** to 1X and **Switch** to ON. Press **Utility** again to exit. The **Output** and **Link** lights should be green.

Connect the output of the WG to the input of the amplifier and use the PA output as the input to the circuit that you will build in Step 9. The menus on the WG can be used to turn the PA output on and off and to toggle the PA gain between 1X and 10X output gains. Make sure it is set to 1X. When an amplifier has a voltage gain of 1 and provides more power than the input supply, it is called a **buffer amplifier**. That is how you are using the PA today. If you have trouble getting the PA to function correctly, tell your TA.

9. **Using the WG and oscilloscope as an LCR meter.** In this step you will do the same kind of measurement that the LCR meter does internally. Build the LR LPF circuit of Fig. 5.30 on a breadboard using the 50 mH inductor of Step 4 and a resistance R_S of about 1 kΩ. Measure the value of the resistor accurately using the DMM and use that value in calculations.

You will now measure the phase between the input voltage and the input current. This is the phase of Z_{in}, as $V_{in} = I_{in} Z_{in}$. $Z_{in} = R_{tot} + j\omega L$, where

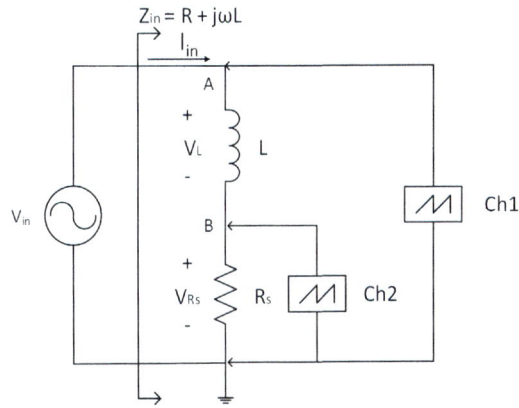

FIGURE 5.30 (Step 9) LC circuit used as low-pass filter.

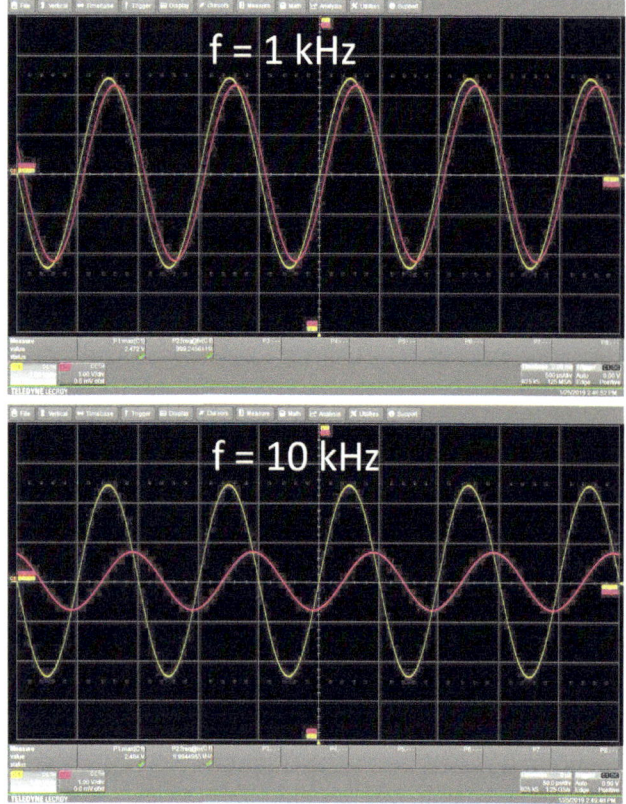

FIGURE 5.31 (Step 9 and 10) From top to bottom, input voltage CH1 and current CH2 at 1 kHz and 10 kHz, respectively. Note that the current decreases with increased frequency, and the voltage leads the current according to ELI for an inductor, with the phase increasing with frequency as the inductive reactance increases.

R_{tot} is the entire real part of the series model of the input impedance, namely the resistance R_s plus the real part of the inductor impedance at the operating frequency.

You know from Chapter 4 that as ω increases, $|V_L|$ will increase and $|V_R|$ will decrease due to the voltage divider effect. In this week's lab you will focus more on phase than you did in previous labs. The phase between the input current, I_{in}, and input voltage, V_{in} (same as V_A at A) will approach 90° as the inductive reactance dominates the total impedance $\mathbf{Z_{in}}$ at higher frequencies. You will see all of this for yourself in the following steps.

Set the WG to 5 V_{pp} at 1 kHz and connect the WG output to the input of the PA. Use the output of the PA, which we call V_{in}, as the input to the circuit. Use a 10X scope probe for point A to ground on CH1 and a 1X scope probe for point B to ground on CH2. Set P1 to measure V_A and P2 to measure V_B.

Remember from Lab 2 that voltage across a resistor is a proxy for (i.e., represents) the current because $V = IR$. The resistor used this way is called a **shunt resistor**. Note that the voltage across the resistor, V_B, is in phase with its current, so the phase of CH2 is the same as that of the current; the phase of CH1 is the phase of V_{in}.

Again, the phase of an impedance is the phase between the current into that impedance and the voltage across that impedance. Set P3 to display the phase φ between CH1 and CH2, which is the same as the phase of $\mathbf{Z_{in}}$.

For 1 kHz, display several periods on the screen and make sure that the waveform is completely on the screen (as always, the computer accesses only data that is visible). You should get a result that looks like Fig. 5.31, top. Does the phase correspond to ELI, as it should for an inductor?

Record P3, the phase angle θ of the impedance $\mathbf{Z_{in}}$. Capture a screen shot that clearly shows the phase. Record the amplitude of V_{in}, $|V_{in}|$, from CH1. Record the amplitude of I_{in}, $|I_{in}|$, from V_{RS} on CH2 ($V_{RS} = I_{in}R_S$).

Now you will do some calculations: (1) You can get $|\mathbf{Z_{in}}|$ from $|V| = |I_{in}||\mathbf{Z_{in}}|$. The polar form of the impedance is $\mathbf{Z_{in}} = |\mathbf{Z_{in}}|\angle\theta$. Right now, in your notebook, write the polar form of $\mathbf{Z_{in}}$ with the numbers inserted and neatly hand draw the complex impedance diagram for $\mathbf{Z_{in}}$ for 1 kHz. (2) Calculate R_{tot} from R_S plus the real part of the inductor impedance at 1 kHz; (3) Calculate R, the real part of $\mathbf{Z_{in}}$, from the polar plot ($R = |\mathbf{Z_{in}}|\cos\theta$.) (4) How does R compare to R_{tot}? Ideally, they are identical. (5) Get X_L from $|\mathbf{Z}|\sin\theta$. (6) Compare to $X_L = 2\pi fL$. Ideally, they are identical.

10. **Observing phase shift due to a larger inductive reactance.** Increase the frequency to 10 kHz. Now the inductive reactance is ten times larger; that increase shifts the angle and decreases the current. You should get a result that looks like Fig. 5.31, bottom. Repeat the previous steps:

Record P3, the phase angle ϕ of the impedance $\mathbf{Z_{in}}$. Capture a screen shot that clearly shows the phase. Record the amplitude of V_{in}, $|V_{in}|$, from CH1. Record the amplitude of I_{in}, $|I_{in}|$, from V_{RS} on CH2 ($V_{RS} = I_{in}R_S$), or what you read off of P2 if you are displaying it directly.

Now you will do some calculations: (1) You can get $|\mathbf{Z_{in}}|$ from $|V| = |I_{in}||\mathbf{Z_{in}}|$. The polar form of the impedance is $\mathbf{Z_{in}} = |\mathbf{Z_{in}}|\angle\theta$. Right now, in your notebook, write the polar form of $\mathbf{Z_{in}}$ with the numbers inserted and neatly hand draw the complex impedance diagram for $\mathbf{Z_{in}}$ for 10 kHz. (2) Calculate R_{tot} from R_S plus the real part of the inductor impedance at 10 kHz; (3) Calculate R, the real part of $\mathbf{Z_{in}}$, from the polar plot ($R = |\mathbf{Z_{in}}|\cos\theta$.) (4) How does R compare to R_{tot}? Ideally, they are identical. (5) Get X_L from $|\mathbf{Z}|\sin\theta$. (6) Compare to $X_L = 2\pi fL$. Ideally, they are identical. (7) In your report, describe the differences between the two impedance plots at the two different frequencies.

11. **Comparing your measurements with the same measurements using the LCR meter.** Turn off the output from the WG. Disconnect all of the leads from the circuit coming from the WG and probes. Connect and use the LCR meter at 10 kHz to measure the impedance with the leads connected from A to ground using the series model. (1) Compare R from the meter with R_{tot} and R from the previous step. (2) Compare θ from the meter with the value of θ you got in the previous step. (3) Compare Z from the meter with the value of $|Z_{in}|$ you got from the previous step. Do you see now what the LCR meter does internally? It is just what you did, but with a lot less equipment. Also, you may realize that as the current gets very small due to a combination of frequency and component values and the associated voltages get very hard to measure, the LCR meter cannot provide a number. This accounts for some of the bad readings you got in the first few steps.

12. **Setting up the circuit to experiment with series resonance.** For this next activity, you will sweep the WG frequency and observe resonance in a series LCR circuit. Resonance happens at that frequency $\omega = \omega_R$ (remember that $\omega = 2\pi f$) for which the reactances $X_L = \omega L$ and $X_C = -1/\omega C$ cancel each other out, $|\mathbf{Z}| = |R + j\omega L - 1/j\omega C|$ becomes a minimum, and the current becomes a maximum. At resonance, the current is limited only by the real part of the series LCR circuit impedance.

Make sure that averaging on the scope channels is off, i.e., set to 1 scan. At resonance, potentially very large voltages can appear across inductors and capacitors, and we don't want them to be too, too large. Set the WG for an output voltage from the PA of about 1 V_{pp} (or 0.5 V_p). It doesn't have to be exact. Just be sure to record its value.

Build the circuit shown in Fig. 5.32 using the output of the PA as V_{in} as you did previously. Choose the same inductor and capacitor you analyzed in the first several steps ($L \sim 0.05$ H and $C \sim 5$ nF). R_S is now 10 Ω. Measure the 10 Ω resistor accurately with the DMM minus the lead resistance as you did previously. Using $v = iR$ you will convert the resistor voltage to current, and also monitor the phase of the current. Most of the time, $V_{RS} = V_C$ on CH2 will be quite small, but at resonance much more of the input voltage will be across R_S, so it will be easy to measure. You are done with the DMM. If you wish you may put away the leads and turn it off.

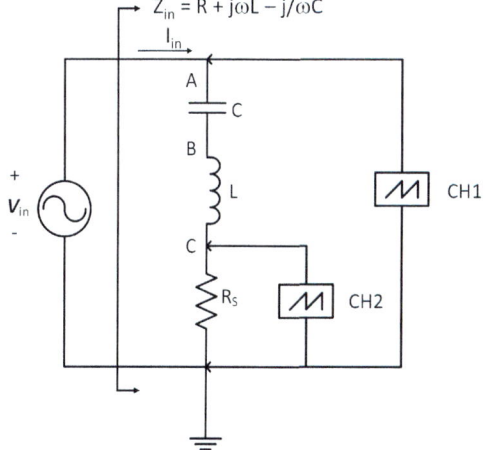

FIGURE 5.32 (Steps 12–16) Resonant circuit with shunt resistor.

13. **Running a Matlab resonance simulation.** Before you do the experiment, run the Matlab simulation called "SeriesRLCTool.m." From the zip file on Sakai, the m file is in the folder labeled "SeriesRLCTool." In the following, use the values that you measured with the LCR meter at 10 kHz. Of course it would be better to use the actual values at each frequency, but this is the best we can do with our simulator and our LCR meter. Since your resonant frequency is close to 10 kHz, you will use the values of L_{Meas}, R_{Meas} for the inductor, and C_{Meas} all measured by the LCR meter at 10 kHz. Also, add in the resistance of the shunt resistor, i.e., ~10 ohm, and the internal resistance of the PA, i.e., 2 ohm. (So $R = R_{shunt} + R_{meas10kHz} + R_{PA}$.) The circuit simulator uses an input voltage of 1 V_{pp}.

Enter the total resistance R as well as L and C into the boxes at the bottom of the sliders. Lock in the new values and run the simulation by clicking anywhere in any of the boxes. (You can, if you wish, simulate how the circuit would behave for other values by using the sliders.) In the upper right corner of the Matlab graphical box, drop down the selection "voltage." You will see all of the possible voltages plotted. The resonant frequency of the second order series circuit is $\omega_R = 2\pi f_R = 1/\sqrt{(LC)}$, which is the frequency of the voltage peaks. You may select to see the current, I, as a function of frequency, from which you estimate the peak frequency. Note that the simulator lumps all of the resistances into one element, whereas in your real circuit the effective resistances are distributed in the various components. Print your results to include in your lab report. Simulate the resonant frequency from your measured values. What f_R and I did you simulate? Record the Q value that you predict from the simulation. You will need it later.

14. **Observing resonance without measuring the resonant frequency.** Now let's use the circuit. Note: if you have problems, first suspect the trigger settings. Also, if the display does not update quickly, but there are delays between traces, press the **Timebase** tab in the lower right and set the **Delay** to zero. You want to check that regularly as it tends to change with settings.

Choose an input sine wave of 1 V_{pp}. (Don't forget that this is an amplitude of 0.5 V_p.) Press **Sweep**; from **SwpType** choose **Linear**. Set **SwpTime** to 10 s; **Return Time** 0 ms; **Start** frequency 10 Hz; **End** frequency 20 kHz. The sweep starts when you press **Output1**. Wait until you set up the scope before you do that.

The oscilloscope: Select vertical gain of CH1 to be 1 V/div with a 10X probe (because some voltages will be large); select CH2 to be 20 mV/div with the 1X probe (because some voltages will be small). (The 1X probe has a yellow slide switch – slide it to the 1X position.) Set the time base to about 50 μs/div. Check the timebase delay – set it to zero. Trigger on CH1 with a trigger level of about 200 mV. Set the trigger to **Auto** and turn on **Output1**. Watch what happens to the current (voltage on CH2) as the input frequency sweeps a few times across the frequency range and describe it in your lab book. You should see the source voltage (CH1) change period with time. The current (CH2) should be small at the start of the sweep, but momentarily increase and then decrease due to resonance.

15. **Observing resonance and measuring resonant frequency.** The goal of this step is to quantify the circuit response as a function of frequency. A peak in the current like that shown in Fig. 5.33 occurs at the resonant frequency, as you just saw.

Turn off the output from the WG. Reset the sweep time to 2 s and the time base to 200 ms/div; turn the small **Horizontal Delay** knob until the triangle, the trigger point, at the bottom of the screen is at the left edge of the screen. Make sure the **Timebase Delay** is zero.

Now you are set to sweep the frequency over 20 kHz and display the current at every frequency on one horizontal sweep of the screen. Since you want the horizontal position to represent

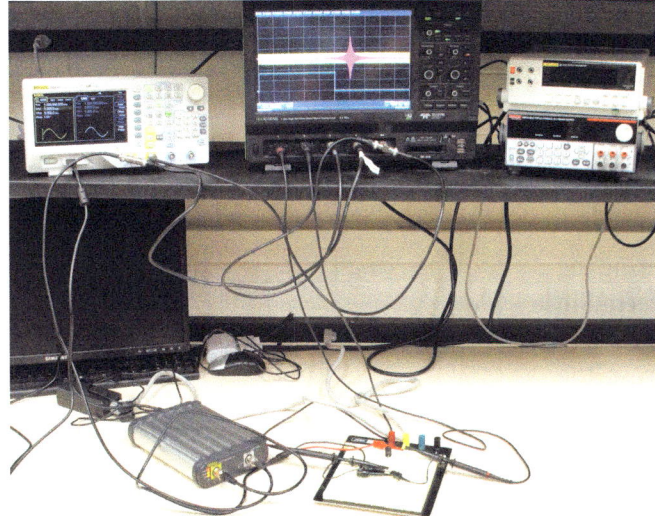

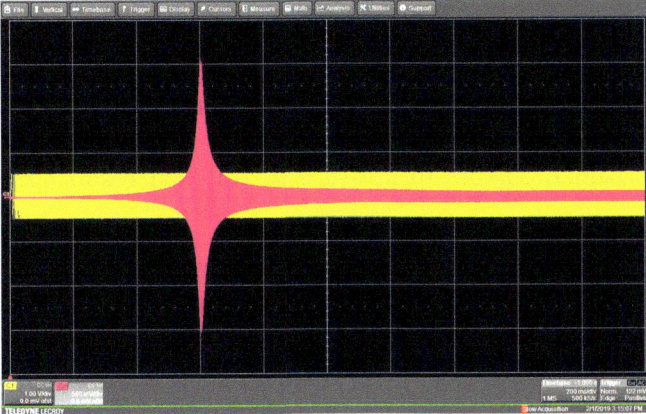

FIGURE 5.33 (Step 15) (Top) Bench setup and (bottom) screen shot of resonance. This shape will be the same whether you are displaying V_L, V_C or V_R, i.e., the current. The sharpness of the peak will depend on the Q of your circuit, and may not look exactly like this example. The top picture shows the trigger that starts the sweep, which you will not do here.

frequency, the sweep must begin at the same time as the oscilloscope trace. To do this, you will sync the oscilloscope to start the trace when the sweep begins. Connect the **Sync** output from the WG to the external trigger input of the scope marked **EXT**. Select **Trigger Setup** and **Source**, then **Other**, then **Ext**.

On the WG, scroll down with the arrow at the bottom of the column of buttons and change **Source** to **Int**. Select the **Auto** trigger on the scope. Turn on **Output1**. View CH1 (sweep output from the PA) and CH2 (current, or voltage across the 10 Ω shunt resistor).

Your screen should look like Fig. 5.33. By moving the trigger point to the left edge of the screen, you have set up the horizontal scope sweep to take 2 full seconds (200 ms/div X 10 div = 2 s) to traverse once from left to right corresponding exactly to the 2 s of the frequency sweep time. When appropriately triggered at the start of the sweep, the left side of the screen represents 10 Hz and the right side represents 20 kHz. Since 10 Hz is nearly zero, we will just call it "zero Hz." Each of the 10 horizontal divisions now represents 2 kHz of frequency on a linear sweep. Based on this, you can read off directly from the screen the frequency at which resonance happens. You don't have to measure it exactly because you will do that in the next few steps. What frequency is your resonance happening at? Does this fairly closely agree with the $f_R = 1/2\pi\sqrt{(LC)}$ simulation and calculation based on L and C values measured near f_R (i.e., at 10 kHz) from Steps 4 and 7?

16. **Observing change from capacitive to inductive reactance.** Turn off the sweep function and turn on the sine wave function so that you can input your choice of frequency. Set the trigger source to CH1 and use **Auto** triggering. Observe (don't measure) the phase between the current (CH2) and the input voltage (CH1) just below resonance, at resonance, and just above resonance. You should see the phase of the current change from leading to lagging the input voltage as you increase the frequency from below f_R to above f_R. Comment on why it behaves that way.

Now you will do a very slow frequency scan to measure f_R more precisely: Set the digit that is controlled by the parameter-select wheel to the one's place (last digit before the decimal). (You may have already noticed that the Rigol WG switches decimal point and commas). Increase the frequency while watching the phase and amplitude of the current. At f_R the current will be a maximum as you have already seen and the phase shown on P3 will be very close to zero. What phase between current and voltage does resonance correspond to? What is the precise value of f_R that you read from the WG display? How close is that to your derived value of $f_R = 1/2\pi\sqrt{(LC)}$ that you get from the values measured with the LCR meter at 10 kHz?

17. **Observing voltage across the inductor at resonance.** Put aside the 1× probe and use three 10× probes for these steps, since you will be measuring fairly large voltages, and the 10× probe enables the measurement of larger voltages (because it attenuates the voltage at the input of the scope by 10×).

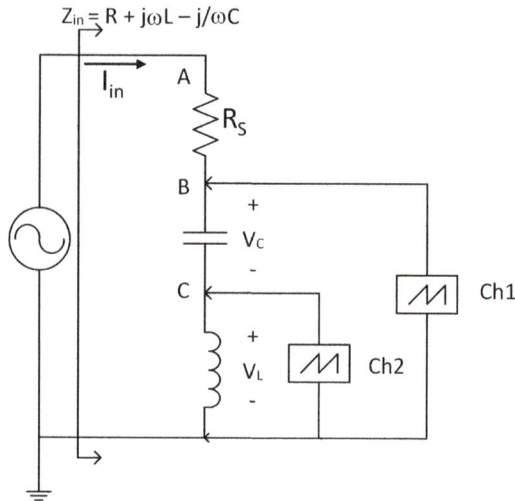

$$Z_{in} = R + j\omega L - j/\omega C$$

FIGURE 5.34 (Steps 17 and 18) Resonant circuit reconfigured to measure voltages across the capacitor and inductor.

Build the LCR circuit shown in Fig. 5.34 using the same components. Place the probes as directed in the figure. In this configuration of probe placement, CH2, at point C, displays the voltage across the inductor as a function of frequency. CH1, at point B, will be used to observe the voltage across the capacitor. Display the input voltage, at point A, on CH4. You will soon see that the inductor and capacitor voltages can get quite large.

Turn on CH2 and CH4 and turn off CH1 for now. Change the vertical gain of CH2 to 10 V/div. Repeat the previous step sweeping the frequency slowly across f_R. You will see that the voltage across the inductor at resonance is much greater than the 0.5 V_p input voltage. Use the horizontal cursors to measure V_L (CH2) at resonance. At resonance, how many times larger is V_L than V_{in}?

18. **Observing the voltage across the capacitor and the phase between the inductor and capacitor voltages.** In order to view the voltage across the capacitor, V_C, create a function F1 = CH1 − CH2 = V_C. Select **Auto** trigger; Press the **Horizontal Delay** knob when CH2 is selected and then do it again when F1 is selected. Reselect CH2; change the horizontal scale until both of the sine waves are easily visible. If the traces are in separate windows, press **Display**, **Display Setup**, **Single** to put them both in the same window, as shown in Fig. 5.35. Select the approximate resonant frequency and change the frequency slowly until the amplitudes are maximized, i.e., you are exactly at resonance. You should see that CH2, which is V_L, and F1, which is V_C, are out of phase by 180° and are of equal amplitude, so that they exactly cancel. Save that screenshot. It should look like Fig. 5.35. Briefly explain why V_L and V_C are out of phase and of equal amplitude, and therefore cancel at resonance. At resonance, where is all of the input voltage showing up in the load impedance? This is a tricky question. Don't forget about the DCR of the inductor.

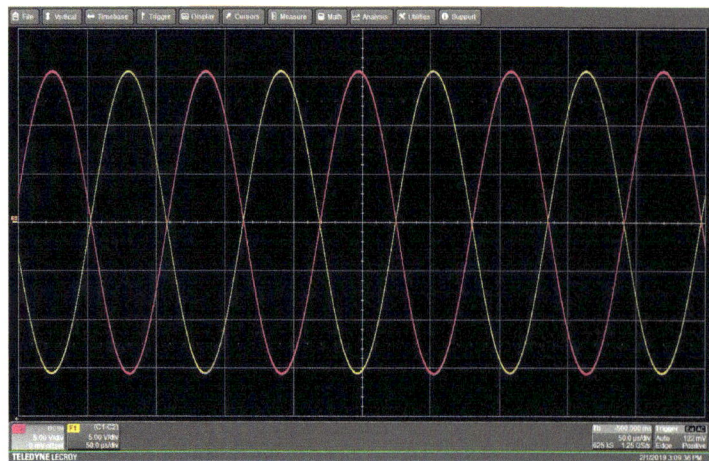

FIGURE 5.35 (Step 18) Voltages across the inductor and capacitor at resonance showing that they are both large (here, about 16 V_p) and out of phase by 180°.

Go to a frequency around $0.7f_R$ and compare the two voltages. Which voltage is larger? What is the phase now? Do the same for about $1.3f_R$. Which voltage is larger? What is the phase now? Explain the voltage relationships for each frequency. Also, note that the current that flows through both the capacitor and inductor

are in series. In terms of ELI and ICE, briefly explain why the two voltages are always 180° out of phase.

19. **Estimating the Q of a resonant circuit.** This is your first determination of Q of the resonant circuit. The Q of series devices can be calculated the following way: $1/Q_{tot} = 1/Q_1 + 1/Q_2$, just like resistors in parallel and capacitors in series. Therefore, $Q_{tot} = Q_1 Q_2/(Q_1 + Q_2)$. You can think of your circuit as the capacitor having its value of Q_C in series with the inductor/resistance combination having another value of Q_L. Ignoring the 10 Ω shunt resistor, use the values of Q_L and Q_C that you obtained for 10 kHz (corresponding to f_R) from Steps 4 and 6 to get Q_{tot} for this circuit. Compare the calculated value of Q_{tot} to the one that the Matlab program predicted in Step 13.

20. **Directly measuring the Q of a resonant circuit.** In this step, you will measure Q of your circuit directly. This is usually how Q is determined. Figure 5.36 shows how Q is extracted from the current versus frequency curve like what you got in Step 15. Recall that 3 dB down in voltage is the same as a factor of 0.707, which is the "half power point." You need the center frequency and the lower and upper half-power frequencies. You will find those next. Q for any resonator, not just today's circuit, is found as $Q = f_R/(f_2 - f_1)$ $= f_R/\Delta f$.

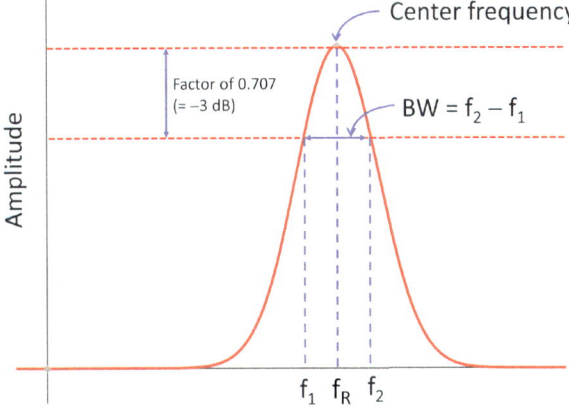

FIGURE 5.36 (Step 20) Definition of center frequency, f_R, lower frequency, f_1, and higher frequency, f_2, used to calculate actual Q of a resonant circuit.

Rebuild the circuit of Fig. 5.32. Also, use the X1 probe again for CH2. You don't need any other probes for this measurement. Display the voltage across the resistor, V_R, (CH2) (remember, this is the current). Set the time base to 100 ms/div so all you get is a big block of color. You don't need to see the wave explicitly – just the amplitude. Enter approximately the resonant frequency, f_R, using the keypad and then select the one's digit of resolution in the frequency. Set up the horizontal cursors to measure the peak of the waveform. Sweep the frequency up and down while using the cursor as a reference to maximize and measure V_{Rmax}. The frequency at which the waveform has maximum amplitude is the true f_R, just as you saw in Step 15. Record the precise center frequency f_R from the display on the WG.

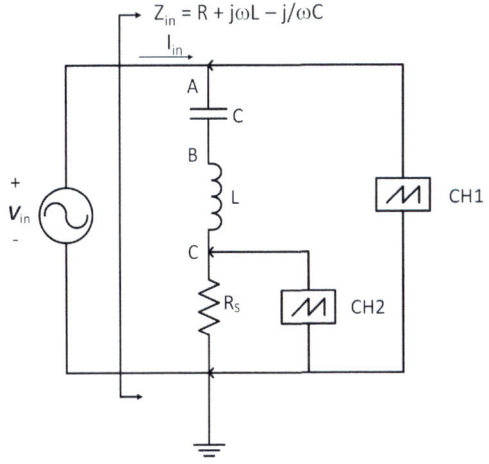

FIGURE 5.32 Repeated.

Calculate the half-power voltage $V_{hp} = 0.707 \times V_{Rmax}$. Now, as a reference, move the cursor down so that it reads the same as V_{hp}. Decrease the frequency using the one's digit of resolution until the peak of the waveform closely matches the cursor level so that $V_R = V_{hp}$; this frequency is f_1. Without moving the cursor, increase the frequency past f_R until again $V_R = V_{hp}$. This new frequency is f_2. Calculate $Q = f_R/(f_2 - f_1) = f_R/\Delta f$. How well does your direct measurement of Q at resonance match the calculated values from the previous step? What might cause any differences? There is no right answer to this very last question.

Good work! Save and check your data and go home!

5.9.1   Appendix 2: Models, Modeling, and Using an LCR Meter

This appendix is about using the Keysight U1733C LCR meter (Fig. A2.1) to determine passive component values. In particular, we will be concerned with series and parallel models of inductors and capacitors as reported by the meter. The meter provides several pieces of information about your **device under test (DUT)** and requires a fair bit of interpretation. An excellent resource for understanding impedance measurements can be found in the Keysight Impedance Measurement Handbook, which can be downloaded from https://www.keysight.com/us/en/assets/7018-06840/application-notes/5950-3000.pdf. This appendix refers to figures shown in that document, and as such, should be downloaded for reference as you read through this appendix.

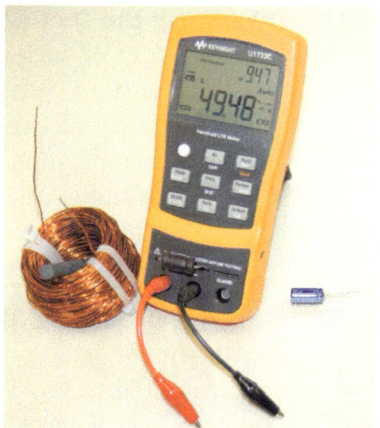

FIGURE A2.1 Keysight U1733C with hand-wound inductor and core, and electrolytic capacitor.

Let's begin by discussing the terms *model* and *modeling* as they pertain to electrical circuit analysis. So far, we have been analyzing circuit diagrams as if each component were ideal. By "ideal," we mean that if it is labeled, for example, "50 mH inductor," then it will *be* a 50 mH inductor, and *only that*, no matter how you use it in a circuit. In real life that is often not even approximately true.

Electrical engineers, and those in other technical fields, often build mathematical **models**, or representations, of their systems before they commit to building them. This saves considerable time and effort, but is useful only if the equations that make up their models can accurately predict physical behavior. One way to improve electrical circuit **modeling** is to replace individual components whose behavior is not ideal by simple circuits consisting of other devices that are themselves taken as ideal. These subcircuits are called **equivalent circuit models**. This is common practice in the use of commercial modeling tools such as SPICE, VHDL, and many more.

When we pick up a component, say a "50 mH" inductor, we might expect that when we measure its inductance, a measurement will simply yield the **nominal** value of "50 mH." However, it is nowhere near that simple. The inductor will work as 'expected' only over a certain frequency range. Outside of that, it could behave in some non-obvious way. The behavior of any device will change over a wide enough range of frequencies, and it is important to know this behavior in order to confirm that a device will function as intended in a real circuit.

Chapter 5 tells us that what we would like to think of as an inductor actually has embedded in it some **parasitic**, or **stray**, capacitance and resistance along with its inductance. The simplest of these is the DC resistance, **DCR**, of the wire in an inductor. We might have some intuition about the size and effects of these parasitic elements, but it is impossible to predict exactly their behavior over a wide range of frequencies. An actual measurement is needed in order to make decisions about how to best represent, or model, the device in a circuit.

A device model for a passive element can be any combination of passive components, L, C and R. You could combine them in any way that provides the same overall impedance from the calculation that the measurement provides. For example, the circuits shown in Fig. A2.2 are relatively simple models for a real (in the physical sense) capacitor and inductor. In Fig. A2.2a, showing a device model of a real capacitor, the small, stray inductance L_S due to the leads in series with the capacitor is an open circuit at sufficiently high frequencies, also at which the ideal capacitor at the heart of the model is a short circuit. Therefore, at very high frequencies the inductance dominates the behavior. In a similar way, in Fig. A2.2a showing a model for a real inductor, the stray capacitance C_P that shunts (is in parallel with) the inductor becomes a short circuit at very high frequencies, effectively eliminating the ideal inductor at high frequencies. These stray reactances clearly affect the component behaviors in real circuits.

As an illustration of how a device might behave very differently than expected, suppose you are at a remote base station at the Antarctic, and a 100 pF capacitor in your 5 GHz weather radar station has failed. You look in your spare parts bin and find only a 5 nH inductor. You might conclude that you are out of luck, but thankfully you understand stray capacitance. You throw it on your super-duper LCR meter that can do measurements at 5 GHz and find that at (only) that frequency the parasitic capacitance is so large that this 'inductor' is actually swamped by the stray capacitance, and in fact has the same impedance as a 100 pF capacitor. In fact, it may as well be a 100 pF capacitor at that frequency. You pop it into your radar system and see an oncoming storm in time to save the base station. It doesn't physically look like a capacitor, but at that frequency it is acting like one. In this case, you are the hero. The point here is that a model of the inductor as a capacitor may be valid at 5 GHz, but likely is not valid at other frequencies.

In order to design a system that works over a range of frequencies, engineers need equivalent

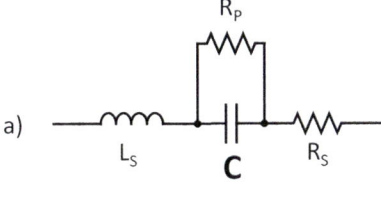
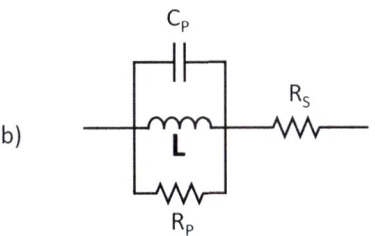

FIGURE A2.2 Simple physical models for capacitors and inductors. (a) Ideal capacitor embedded in basic model of device parasitics and (b) ideal inductor with its device parasitics.

circuit models that are sophisticated enough that the modeled impedance matches the measured impedance adequately over the needed frequency range. Doing so often starts with some physical insight into what electromagnetic phenomena may be present internal to the device e.g., wire-to-wire capacitances, wire inductance and resistance, etc., but that intuition is never perfect. Figure A2.3 is a somewhat more complicated model, this one of a device called a **varactor diode**. A varactor is a type of diode whose capacitance changes significantly

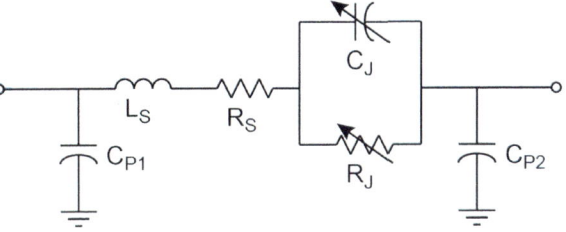

FIGURE A2.3 Equivalent circuit model for a varactor diode. All of the devices in the model except for C_J are parasitic devices that must be included to predict its behavior over a range of frequencies. Figure courtesy of P. Fay.

with the amount of reverse bias applied to it (see Chapter 7 for more information about diodes), and can be used as a variable capacitor rather than the diode that it actually is. The AC behavior changes depending on some external DC voltage, and this AC model assumes that the amplitude of the AC signal (the **perturbation**) is small, so it is called a **small-signal model**. Varactor diodes are used in all kinds of modern devices for tuning and filtering, including radios and TVs. Clearly this device could have significant deviations in behavior if used over a wide frequency range. We see that shunt capacitors C_{P1} and C_{P2} will effectively short out the diode to ground at high frequencies, and the series inductance, even though it is small, will also limit its high-frequency operation. Getting all of this sort of thing right is the full-time job of many electrical engineers.

The purpose of this appendix is not to discuss details of circuit design tools, but rather to illuminate what the LCR meter is telling us and how that data is used to develop very simple equivalent-circuit models that can go into design tools. The Keysight U1733C LCR meter provides parameters that summarize the properties of DUTs in terms of two simple device models, namely *series* resistance and reactance (inductance or capacitance) or *parallel* resistance and reactance.

An LCR meter tells the user how it behaves, in terms of its impedance only, at specific frequencies. The DUT is a black box as far as the meter is concerned; the meter does a measurement by which it knows only *the magnitudes of the voltage across the device and the current through it, and the phase angle between them*. Based on these values, the meter *attempts* to report the component value of inductors and capacitors. Put another way, the meter only knows the impedance, **Z**, of what is between its terminals; it does not know *why* there is that impedance. The meter does not know any details of the physical device and is, therefore, ignorant of what might or not be parasitic elements. If the phase if the voltage leads the current, ELI, then the meter will report a positive value of inductance when the meter is set to measure "L." If we set the meter to measure capacitance when the DUT is inductive, the meter will report a *negative* value of *C* whose magnitude of impedance will correspond to that of the inductor. If the phase angle is lagging, then ICE applies, and positive values of *C*, or negative values of *L*, will be reported.

The U1733C LCR meter knows the impedance only at the available test frequencies, which are limited to a few values, namely 100 Hz, 120 Hz, 1 kHz, 10 kHz and 100 kHz. It cannot know the behavior at other frequencies. The goal of selecting a good model is generally to be able to use it in a circuit design; a complicated equivalent circuit model may be more representative of the behavior, i.e., impedance, *over a larger range of frequencies*, and therefore be more useful. Figure 1–14 of the Handbook shows impedance behavior for a real capacitor (left side of middle) and an inductor (right side of middle) on a log-log plot over a wide frequency range. It can be seen that the behaviors are overall somewhat complicated. Three regions of the frequency range will be discussed below, namely, the linear behavior at lower frequencies, the peak (or dip) in each plot, and the linear region at higher frequencies.

Let's get into more detail about what the LCR meter tells us. Devices always have some inductance, resistance and capacitance. Therefore, the reactive contributions, $j\omega L$ and $-j/\omega C$, to the total impedance **Z** due to the various components, including the parasitics, change with frequency. The measured impedance could be any combination of resistances, inductances and capacitances, but the meter will only report a single value of R as well as L *or* C to you, as in the real and imaginary parts of a single complex number.

The Keysight LCR meter offers two simple models from which you can choose, namely a resistance in series with the reactance, or a resistance in parallel with the reactance, as shown in Fig. A2.4. In the first case, A2.4a, R_S is the real part of the impedance and X_S is the imaginary part. In the second case, A2.4b, a resistance is in parallel with the reactance, but they are not the real and imaginary parts of the impedance, as discussed next.

Remember, the meter is aware of only one value of **Z**. In that sense, the two models are *exactly equivalent*. The series model can be converted to the parallel model, and vice versa – *there is only one Z at any one frequency*.

So, are these two models useful? In some sense, the series and parallel models are not models at all – rather they are just representations of the impedance (series

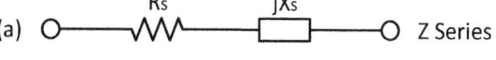

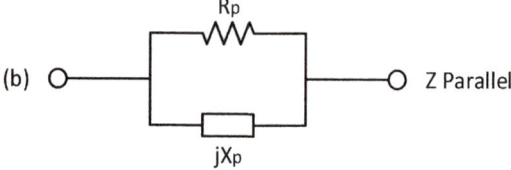

$R_S \neq R_P; \ X_S \neq X_P$

FIGURE A2.4 (a) Series and (b) parallel impedance models. Note that each of these total impedances is a complex number on the complex plane, as discussed in Chapter 5. Note that the resistances and reactances are not equal to each other, but the combination of each provides the same measured value of **Z**.

model) and **admittance**, the inverse of impedance (1/**Z**), (parallel model) in the complex plane. They are not all that useful in device modeling compared to something like that of Fig. A2.3. The meter reports the impedance (or admittance) of the DUT at only a handful of frequencies. More sophisticated instruments measure impedance across a continuous span of frequencies, which can be very useful for sophisticated circuit modeling tasks. Such a measurement would look like the middle of Fig. 1–14 of the Handbook. However, the purpose of *this* meter is more to give a *snapshot* of a device's behavior at a few frequencies across a broad range. Unfortunately, this is like looking at the world through a soda straw at only five places, meaning that it is difficult to get a complete picture of the device's behavior over a wide range of frequencies. In a way, it would be a little like picking out the Eiffel Tower in a photograph from a small area of steel girders. More sophisticated tools can do these measurements over a continuous range of frequencies, and show the entire picture at once, but they are also much more expensive and complicated to use.

How do engineers use simple LCR meters in practice? One use is to check devices against their **data sheets**, or **specifications**, or 'spec sheets.' Choosing a frequency and comparing to the expected values can reveal defective devices. Another, more sophisticated, purpose is to build a mathematical model of the device that can extend across a range of frequencies in order to attempt to replicate at least part of the impedance measurement shown in Fig. 1–14. To do this, a series of equations is developed with values at each measured frequency, which can be used to predict the device behavior over a more useable, wider frequency range. This can often work well enough for circuit modeling in the absence of having continuous device behavior over a wide range of frequencies.

As you know, impedance is $Z = R_S + jX_S$. This equation corresponds exactly to the series circuit model shown in Fig. A2.3. Using the parallel model shown in the figure is often easier when using the inverses of resistance and reactance, which can then be added together. The inverse of Z is Y, called the **admittance** is

$$Y = G + jB, \tag{A2.1}$$

where $G = 1/R_P$ is **conductance** and B is **susceptance**; $jB = 1/(jX_P) = -j/X_P$. (Recall that X_L is positive and X_C is negative.) Both Z and Y can be written as complex numbers with a real part and an imaginary part, but those real and imaginary components are not the same and do not have the same units: both $Z = R_S + jX_S$ and $Y = G + jB$ are used in circuit modeling.

The relationships between R_S and R_P and as well as X_S and X_P can be found by doing some algebra. Both models represent the same measured impedance, i.e., magnitude and phase angle. Setting both impedance models to be equivalent, and slogging through the algebra, we get:

$$Z = R_S + jX_S = R_P||jX_P = \frac{R_p X J_p^2}{R_p^2 + X_p^2} + j\frac{R_p^2 X_p}{R_p^2 + X_p^2}. \tag{A2.2}$$

From this, we extract the series resistance in terms of the parallel model parameters as:

$$R_S = \frac{R_p X_p^2}{R_p^2 + X_p^2} \tag{A2.3}$$

and series reactance in terms of the parallel model as:

$$X_S = \frac{R_p^2 X_p}{R_p^2 + X_p^2} \tag{A2.4}$$

Similarly, since both models represent the same Z,

$$Z^{-1} = Y = G + jB = \frac{1}{R_p} - j\frac{1}{X_p}. \tag{A2.5}$$

After more algebra, we see that the parallel resistance in terms of the series model is:

$$R_P = \frac{R_s^2 + X_s^2}{R_s} \qquad (A2.6)$$

and the parallel reactance in terms of the series model is:

$$X_P = \frac{R_s^2 + X_s^2}{X_s}. \qquad (A2.7)$$

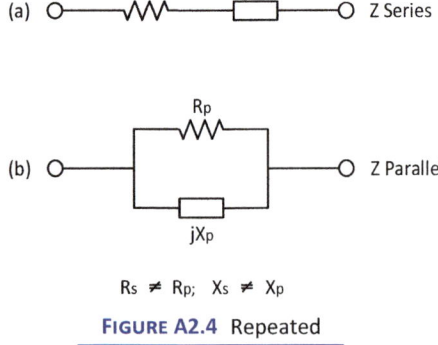

FIGURE A2.4 Repeated

You may not be as comfortable thinking in terms of parallel resistances and reactances as you are with series ones, so let's look at the equations and get a general idea of how they are related. We will try to gain some physical intuition about these equations. Let's think of 'small R' as nearly a short circuit and 'large R' as nearly an open circuit. Large R_S impedes current from flowing through the ideal X_S and small R_P shunts current around the ideal X_P. Small R_S has little or no effect on measuring the ideal X_S because R_S drops relatively very little voltage. Similarly, large R_P has little or no effect on measuring the ideal X_P because relatively very little current flows through R_P.

Thinking as we did above, if R_P is small, it shorts out the ideal DUT in the model, so X_P contributes less to the overall device impedance (in extreme cases, this may make it difficult for the meter to accurately measure X_P). Similarly, an R_S that is much larger than X_S may make it difficult for the meter to accurately determine X_S.

When selecting the series model for, say, an inductor, the meter reports a real part R_S and inductance, from $X_S = \omega L_S$, that make the model match the measured impedance. In this case, R_S is called the **equivalent series resistance, ESR.** When selecting the parallel model, the meter reports a real part R_P and reactive part, from $X_P = \omega L_P$, that make the model match the measured impedance.

It is important to note that R_S, the ESR, feels physical because it is natural to think of wire resistance in an inductor or capacitor, or some other series resistance as being a real thing. In fact, thinking this way is the beginning of mentally forming a model from our intuition about the physical device. It might occur to you to ask, "Is the ESR the same as the DCR? Why does ESR depend on frequency, and how can it change if it is just the DCR of the wire?" Well, they are the same at DC, but the ESR in general it isn't just the DCR of the wires. In fact, as the frequency changes and parasitic capacitance or inductance kicks in, there are some elements of resonance in which energy sloshes around inside the device between the capacitive and inductive reactances. Every slosh (each period) of the current causes losses through the device's DCR, so the total losses increase, and the ESR changes. Therefore, DCR is just the series resistance of the wires, but the ESR depends on the DCR as well as the frequency, capacitance and inductance.

So we might also ask, "Is R_P a physical thing? Is current really leaking around the inductor?" Sometimes, yes, but in the specific case of an inductor made with resistive wire, R_P is mostly just the transformed value of R_S and does not represent an actual leakage path around the ideal device. Yes, there are sometimes actual leakage paths that should be put in as shunt resistors for many devices. Since the component is a black box and being limited to these simple models, you may think of both R_S and R_P as just mathematical parameters that make **Z** agree with the impedance read by the meter.

We might think that each model would report the same value of L so that $L_S = L_P$, but as the math above shows, this is not always the case. Which one is the 'true' L? L_S or L_P? In other words, which model best represents the internal *ideal* device? In a way this is a philosophical question; the impedance is what it is for a complex interplay of physical reasons. Since the meter only knows the impedance at discrete frequencies, it will report reactances that correspond to the measured impedances. A rule of thumb is that the parallel model works best for devices that give a relatively high $|Z|$ at the measurement frequency, and the series model works best for devices that give a relatively low $|Z|$ at the measurement frequency, as shown in Fig. 1-15 of the Handbook.

One way that meters of this type are used is to provide a quick check of component values. For this type of a use, having the meter quickly (and approximately) identify the likely value of the "intrinsic" device element can be helpful. Figure 1–15 of the Handbook shows some practical frequency ranges over which series and parallel models work best, and are in fact assumed for the U1733C meter. The decision about which reported value of L to trust as best representing the internal ideal DUT depends on the measured $|Z|$. The series model is best for low $|Z|$ and the parallel model is best for high $|Z|$. Between these two limits, both models usually give about the same value of X, and therefore of L or C.

As you also know from Chapters 4 and 5, combinations of capacitances and inductances have resonant frequencies, ω_R. An inductor always has some parasitic C in parallel and a capacitor always has a parasitic L in series. Combining the values of L and C results in a **self-resonant frequency** $\omega_R \sim 1/\sqrt{LC}$ (which is exact only for zero R) for which $|Z|$ can be either very low for a series LC equivalent circuit or very high for a parallel LC equivalent circuit.

The plots in the middle of Fig. 1–14 shows on a log-log plot how $|Z|$ varies with frequency for the simple models of the capacitor and inductor. Each graph shows at lower frequencies a region that is linear because the ideal reactance dominates the device behavior; impedance changes linearly on the log-log plot with frequency for a constant value of L or C. Also, the graphs show the self-resonance phenomenon as a peak or dip in the plots of $|Z|$. Increasing the measurement frequency will often bring the measurement near or past the resonance peak. If you are especially astute you might realize that because the meter calculates the reported value of L or C based only on Z, it will report 'wrong' values near ω_R. As the meter approaches frequencies closer to the resonance peak it reports a higher value of C because larger capacitances cause a lower $|Z|$; likewise, for an inductor it will report a higher value of L because larger inductances lead to larger $|Z|$. The meter isn't smart – it just knows what it can see (the impedance at the measurement frequency) and nothing more.

What happens when the measurement frequency is so high that $|Z|$ is beyond the resonance peak? Figure 1–14 shows that $|Z|$ goes in the other direction, increasing or decreasing with frequency for capacitance and inductance, respectively. In that case, the parasitic reactance has completely taken over, and the opposite reactance (L changes to C and vice versa) takes over. (This is what saved the Antarctic base station from the effects of the storm.)

One final note about the utility of the LCR meter. Sometimes the meter will not be able to report either the series or the parallel model parameters even though it can report the other one. This is not so mysterious: In such cases, it simply does not have the digits of resolution to display both model parameters even though the impedance stored internally is valid. Could you transform it yourself from one model to the other? Yes, using the equations derived in this appendix. Be careful, though, to follow the rules of significant figures in the arithmetic.

This appendix has been a deeper dive into what the LCR meter is doing internally and how you can interpret the readings that it provides. Hopefully, it is not so mysterious anymore and you will have some confidence in the measurements you get. Good luck moving forward!

Problems

1. Show the following complex numbers on the complex plane as phasors:

   $A = 3 + j6$

   $B = jA$

   $C = j \cdot jA$

   $D = j \cdot j \cdot jA$

2. Consider the expression for an impedance: $Z = (5 + j7)^2/(7 + j4) - (3 + j7)/(5 - j7)$.

   (a) Reduce Z to a single expression in the form of $R + jX$.

   (b) What is the admittance in the form of $Y = G + jB$? Be sure to show the values of G and B.

3. Consider the following circuit:

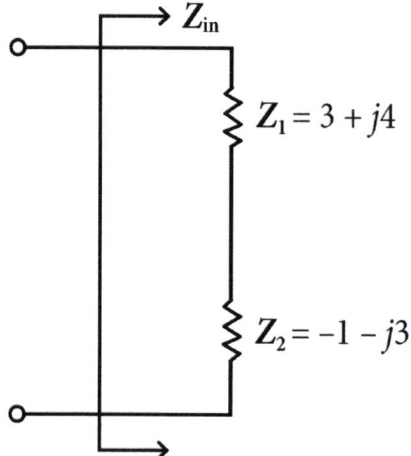

   (a) What is Z_{in} in the form $R + jX$?

   (b) Draw a complex impedance plot that shows the relationship among Z_1, Z_2 and Z_{in}.

4. Derive an expression for Z_{in} for the circuit in the given diagram. It must be in a simplified form of $Z_{in} = a + jb$. In your answer, "j"s must be in the numerators and not the denominators.

   Is this a HPF, LPF or BPF?

   Show the limits of Z_{in} for low and high ω to support your answer.

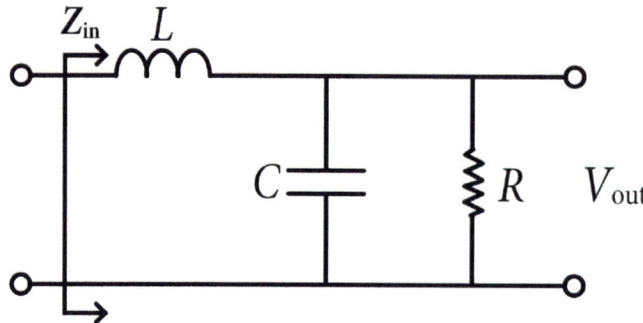

5. Derive an expression for $\mathbf{Z_{in}}$ for the circuit in the given diagram. It must be in a simplified form of $\mathbf{Z_{in}} = a + jb$. In your answer, "j"s must be in the numerators and not the denominators.

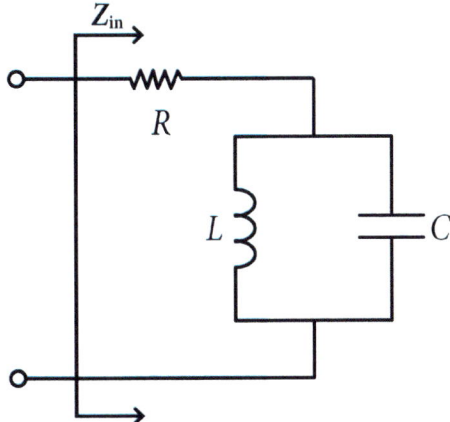

6. Consider the high-pass filter of Fig. 5.25 that filters the music going to the tweeter. Derive the frequency-dependent input impedance of the filter and tweeter in the form $\mathbf{Z_{in}} = R + jX$. Convince yourself that the tweeter gets only the high frequency parts of the music.

7. Discrete circuit elements can exhibit non-ideal characteristics due to parasitic reactance at high frequencies. You are given the high frequency equivalent circuit for a capacitor in the following diagram, which includes the inductance of the lead wires.

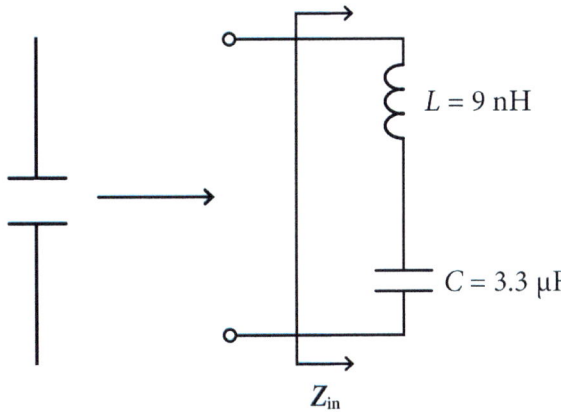

(a) Fill in the table of X_L, X_C, $\mathbf{Z_{in}}$ and $|\mathbf{Z_{in}}|$ for three frequencies: 1 kHz, 1 MHz, and 1 GHz.

	1 kHz	1 MHz	1 GHz
$X_L = \omega L$			
$X_C = \dfrac{-1}{\omega C}$			
$\mathbf{Z_{in}}$			
$|\mathbf{Z_{in}}|$			

(b) Plot $\mathbf{Z_{in}}$ on a phasor diagram for frequencies of 1 kHz, 1 MHz, and 1 GHz. The values do not have to be exact, but be sure that it shows the trends of your answers. Include the values that you calculated. Using the diagram, comment on how the total impedance changes as frequency increases.

8. (a) Express the complex constant $A = 8 + j9$ in all four complex forms, namely, rectangular, exponential, trigonometric and polar forms.

   (b) Draw the vector A on the complex plane. Show all relevant values related to the four forms.

9. Write the waveform $I(t) = 4\cos(2\pi 75t - \pi/3)$ in the four phasor forms. (Note that all angles are expressed in radians.) All expressions must be simplified. Hint: the first four answers can be obtained by simple inspection of the waveform with minor computation.

   (a) Trigonometric

   (b) Rectangular

   (c) Exponential

   (d) Polar

   (e) Draw the phasor diagram showing all the above forms including vectors, magnitudes and angles.

   (f) Draw the time-domain representation of the waveform including magnitude and angle. Also include the time at which the waveform reaches its first peak.

10. This exercise demonstrates that adding any sinusoids of the same frequency, regardless of their amplitudes and phases, results in a sinusoid at the same frequency with some new amplitude and phase. This would be an example of, say, adding the voltages across two series impedances, or adding three phases of a three-phase power system. The fact that the sum is a sinusoid at the same frequency explains why we can add phasors together and get a third phasor that is at the same frequency, and why we can do all our calculations at $t = 0$, i.e., the phasors all rotate together.

    The following time-domain functions are in radians:

    $A(t) = 6\cos(377t + \pi/4)$; $B(t) = 5\cos(377t - \pi/2)$; and $C(t) = A(t) + B(t)$.

    (a) Convert $A(t)$ and $B(t)$ into phasors A and B in the trigonometric and rectangular forms.

    (b) Calculate C from $A + B$. Express C in all four forms in the following order: Rectangular, Polar, trigonometric, and exponential.

    (c) Express the phasor C in the time domain as $C(t)$.

    (d) Carefully draw the phasors A and B in the complex plane. Show that graphically adding them together gives C. Note that the phasor diagram conveys the same information as that of the sinusoids in the next part.

    (e) Use Desmos or your favorite plotting program to plot $A(t)$, $B(t)$ and $C(t)$ separately. Also, plot $A(t) + B(t)$. You should see from this plot that $A(t) + B(t)$ gives you the same waveform as $C(t)$ that you calculated in the previous part. If it does not, you have made a mistake somewhere.

11. Consider the following diagram:

For $Z = 50 + j70$ and voltage $V = 5\cos(377t)$, calculate the current and express it as a time-domain function $I(t)$.

12. A current $I(t) = 4\cos(377t + \pi/4)$ amps flows through an impedance Z. Z is the series combination of a 300 Ω resistor and a 7 microfarad capacitor.

Note the argument of $I(t)$ is radians. $\omega = 377$ rad/s corresponds to $f = 60$ Hz.

(a) Draw a circuit diagram that shows current flowing into Z including the resistor and capacitor.

(b) What is Z in rectangular form?

(c) What is Z in polar form (use radians)?

(d) What is the phasor representation of $I(t)$ in polar form?

(e) What is the polar form of the phasor representation of the voltage V across Z?

(f) Draw a phasor diagram that includes I and V. Be sure to indicate clearly their magnitudes and angles.

(g) What are the time-domain representations of the current and of the voltage across Z? Your expressions should be in radians.

(h) Make a plot (Desmos or your favorite software) of the current and voltage waveforms.

To plot the voltage and current to be approximately the same size on the same graph, you will have to scale them accordingly so they both fit on the plot. That is not a problem because they are in different units anyway. Be sure to plot only a few periods in the horizontal direction to make the phase clearly visible.

Be sure to include the functions that you are plotting along with the plot, as Desmos allows you to do with the panel on the left side of the screen.

Write on the plot all the relevant parameters of your answer.

13. (a) What kind of filter is this? Calculate the transfer function $A(\omega) = V_{out}(\omega)/V_{in}(\omega) = a + jb$. Leave $A(\omega)$ in terms of ω; you will add values at a particular frequency in the next part. Note that $A(\omega)$, $V_{in}(\omega)$ and $V_{out}(\omega)$ are complex. Your answer should be in the form $A(\omega) = a + jb$. Hint: use the voltage divider equation with complex quantities. The input $V_{in}(t)$ is expressed in radians.

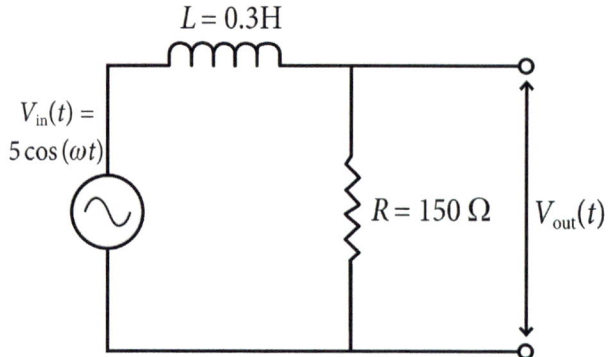

(b) Plot (using Desmos or any other plotting program you like) both the magnitude and phase angle of the transfer function as a function of frequency. Plot the vertical axis on a linear scale and the horizontal axis (frequency) on a log scale. Remember, the phase of the transfer function is the shift in the phase of the output relative to the phase of the input. It is not a phase shift between voltage and current. (Always ask, "Phase of what relative to what?")

(c) Cast $V_{in}(t)$ as a phasor $\mathbf{V_{in}}$ in the polar form. Hint: you can do this by inspection in one line.

(d) Calculate the magnitude and phase of the gain $\mathbf{A}$ at ω = 500 rad/s. Show that the equation works out to what you got on your plot. What is the significance of ω = 500 rad/s?

(e) At ω = 500 rad/s, what is the polar form of $\mathbf{V_{out}}$? Hint: use the equation you derived in the previous question, but use the polar form for both $\mathbf{V_{in}}$ and $\mathbf{A}$.

(f) Convert $\mathbf{V_{out}}$ at ω = 500 rad/s back to the time domain. Hint: This can also be done in one line by inspection.

14. Look at Fig. 5.36, which shows the amplitude of current, or voltage across the resistor, in the circuit of Fig. 5.32 as a function of excitation frequency. The resonance of the circuit is characterized by its quality factor, Q. For a center frequency of 25 kHz, a lower –3 dB point of 23.5 kHz and an upper –3 dB point of 26.5 kHz, what is the Q of the circuit?

15. Explain Fig. 5.33, bottom and how it depends on Fig. 5.32. What does the peaked trace on the oscilloscope show?

16. Assume that the inductor and capacitor in Fig. 5.32 are ideal, i.e., they have zero DCR and ESR. The following questions pertain to Fig. 5.33. All your answers should correspond to your actual lab instructions.

(a) What is the yellow trace?

(b) What is the pink curve?

(c) $|V_L|$, $|V_C|$, and $|V_R|$ are the magnitudes of the voltages across the inductor, capacitor and resistor, respectively. Pick one correct answer. Choose (v) if none is correct.

   At the resonant frequency:

   (i) $|V_L| = |V_C| = 0$

   (ii) $|V_R| = 0$

   (iii) $|V_L| = |V_C|$ and both are small

   (iv) $|V_L| = |V_C|$ and both are large

   (v) None of the above is correct.

(d) Consider the regions of the data of Fig. 5.33 to the left of the peak, exactly at the peak, and to the right of the peak. Pick one correct answer: Z_{in} at the left, center and right of the peak are, respectively:

   (i) All resistive

   (ii) All capacitive

   (iii) All inductive

   (iv) Capacitive, resistive, inductive

   (v) Inductive, resistive, capacitive

17. We say that the Fourier transform is a linear operator, meaning that calculating the Fourier transform of the sum of two scaled time-domain functions is:

$$\mathcal{F}[af(t) + bg(t)] = \mathcal{F}[af(t)] + \mathcal{F}[bg(t)] = a\mathcal{F}[f(t)] + b\mathcal{F}[g(t)] = aF(\omega) + bG(\omega).$$

Here, $\mathcal{F}$ is the operator that transforms a time-domain function into its Fourier transform; $f(t)$ and $g(t)$ are two time-domain functions; a and b are constants, and can be thought of as gain when multiplied by a time-domain function; and $F(\omega)$ and $G(\omega)$ are the Fourier transforms of $f(t)$ and $g(t)$, respectively.

(a) Focus on the linearity part. State in words and diagrams what it means for the Fourier transform to be 'linear'. Draw diagrams of the time and frequency domain functions that show the linear relationship presented above.

(b) The following is the definition of the Fourier transform. Show that it is linear. Note: this is easier than you might expect.

The Fourier transform, $\mathcal{F}$:

$$\mathcal{F}[f(t)] = F(\omega) = \int_{-\infty}^{\infty} e^{-i\omega t} f(t)dt$$

Chapter 6

Wireless Communications

6.1 Introduction and Overview

In this chapter, you will learn about:

1. Radio waves
2. How radio waves propagate through space
3. Modulation of radio waves
4. Amplitude modulation
5. Waveforms of the baseband and modulated signals
6. Bandwidth of the baseband and modulated signals

6.2 Radio Waves

This chapter discusses the underlying technology behind something that we take for granted nearly every day – audio, video or data by over-the-air television, AM/FM radios used in cars, satellite television, walkie-talkies, air traffic control radar, cell phones, Bluetooth®, Wi-Fi, satellite TV, tire-pressure sensors, automobile collision-avoidance radar and countless others. It is so common that we might think of it as old technology, and it does go back to the late 1800s, but it is also a modern and evolving area of research and development.

A complete radio link comprises the following, as roughly shown in Fig. 6.1:

- An electronic source of information, such as an analog microphone or a streaming device that uses digital bits for a movie or audio broadcast (Chapter 9), produces a signal.
- A transmitter that transforms the signal into an electromagnetic form that can travel through space.

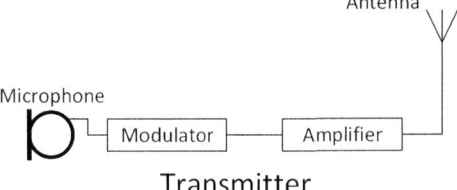

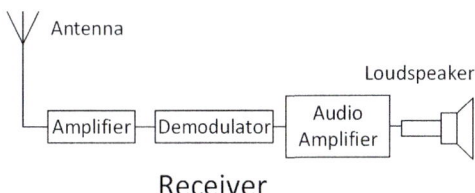

FIGURE 6.1 Radio link block diagram.

An Introduction to Electrical Engineering with Lab Activities
Gary H. Bernstein
Copyright © 2026 Jenny Stanford Publishing Pte. Ltd.
ISBN 978-981-5129-30-4 (Hardcover), 978-1-003-71345-6 (eBook)
www.jennystanford.com
DOI: 10.1201/9781003713456-6

- An antenna that launches the radio wave into space by converting energy in an electrical current into energy in an electromagnetic wave.
- A medium for transmission, which here we assume is **free space** (i.e., not a wire or cable).
- An antenna that converts the electromagnetic wave back into current.
- A receiver that converts, or **demodulates**, the signal back into a form that we can process, which could be, for example, sound or data.

This week's lab activities will include every one of these elements, but focus mostly on the processes of modulation and demodulation. The study of antennas is very important, but will not be the focus of this week's activity. Antenna theory is usually taught in upper-level courses on waveguides and antennas.

6.2.1 Transmission of Electromagnetic Waves and Maxwell's Equations

You may already know that light is the oscillation of **electromagnetic** (EM) energy as electric and magnetic fields coupled in such a way that they support each other and **propagate** (i.e., spread or broadcast). You may, however, not already know that radio waves and light are exactly the same EM phenomenon, except for the frequency of oscillation, which can vary by an extremely large factor. A typical frequency for **amplitude modulation**, or **AM**, radio, commonly found in most car and household radios, is about 10^6 Hz (or 1 MHz). (Later, we will discuss why it is called *amplitude* modulation). The frequency of oscillation of green light is about a billion times higher than that of AM radio, but it is *exactly the same stuff.* The frequency of X-rays is higher yet by another 100 to 1000 times.

FIGURE 6.2 Water wave propagation in a bowl caused by a dropped marble.

When you think of a traveling wave, you may imagine a water wave moving across a still pond from the spot at which a stone hits the surface (see Fig. 6.2 for a mini-example). Peaks and valleys of water height move away from the spot in all directions. Figure 6.3 illustrates how an electromagnetic wave propagates through space. The wave comprises electric and magnetic fields having peaks at the same physical location with orientations that are perpendicular to each other. Figure 6.3 shows a 'snapshot' of the magnitudes and directions of the electric and magnetic fields as a function of

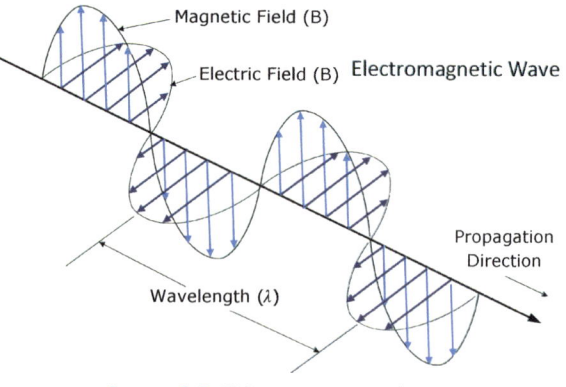

FIGURE 6.3 EM wave propagation.

position along the direction of propagation. In reality, the wave is 'sliding' forward, but the drawing shows it frozen in time and space. **Maxwell's equations** tell us that this form of energy is necessary for the propagation of an EM wave, meaning that it propels itself forward through free space, e.g., from a radio transmitter to the AM radio in your car, or in this week's lab, on your benchtop. (Or for that matter, radio-frequency waves or light that travels billions of light-years across deep space.)

Maxwell's equations (in differential form) for waves traveling through free space are as follows:

$$\nabla \cdot \mathbf{E} = \frac{\rho}{\varepsilon_0} \tag{6.1}$$

$$\nabla \cdot \mathbf{H} = 0 \tag{6.2}$$

$$\nabla \times \mathbf{E} = -\mu_0 \frac{\partial \mathbf{H}}{\partial t} \tag{6.3}$$

$$\nabla \times \mathbf{H} = J + \varepsilon_0 \frac{\partial \mathbf{E}}{\partial t}, \tag{6.4}$$

James Clerk Maxwell developed his eponymous equations around 1861. These four short mathematical statements describe the behavior of electromagnetic interactions. They are the starting point of just about every treatment of the dynamics of electric and magnetic fields. Their significance cannot be overstated.

where $\mathbf{E}$ is the electric field, $\mathbf{H}$ is the magnetic field, ρ ("rho") is the density of electric charge, $\mathbf{J}$ is current density, and μ_0 ("mu nought") and ε_0 ("epsilon nought") are constants called the **permeability of free space** and the **permittivity of free space**, respectively. The symbol ∂ is *not* a Greek letter. It is pronounced "partial," and $\partial \mathbf{E}/\partial t$ ("the partial of E with respect to t") can be treated like the time derivative of $\mathbf{E}$ for our purposes. You may have seen this in a multivariable calculus course, but not likely at this stage of your education. Sometimes ∂ is called "del," but it isn't common for reasons coming right up, so let's not do that here.

In the equations, the upside-down capital delta is the derivative in three dimensions. ∇ (also not Greek), is always called the **del operator**. ∇ is a *derivative vector* and is equal to $(\partial/\partial x, \partial/\partial y, \partial/\partial z)$, which is the three-dimensional derivative in the x, y and z directions. These operators are pronounced "the partial with respect to x of ...." etc. In Eqs. 6.1 to 6.4, the small dot ($\cdot$) is the **dot product**, and the small cross ($\times$) is the **cross product**. You will see the dot and cross products in future multivariable calculus and physics courses. Although this new math may seem daunting, all four equations can be understood in simple ways. In order to fully understand the dot and cross products, you will have to wait for your course in multivariable calculus.

The *first* of Maxwell's equations, Eq. 6.1, $\nabla \cdot \mathbf{E} = \rho/\varepsilon_0$, says simply that electric charge, ρ, is a **source** of electric field. Electric field lines, as shown in Fig. 6.4, are quantified by electric flux, which you may think of as how *many* lines are shown in the picture. Flux *density* is how *close* together the lines are. Electric field lines start at **sources** of electric flux (positive charges) and end at **sinks** of electric flux (negative charges). The expression "$\nabla \cdot \mathbf{E}$," pronounced "del dot E," is called "the **divergence** of E" (also written as "div **E**"). This is appropriate because the field lines diverge, or spread

File away for later courses: Electromagnetic fields are traveling waves, where the electric and magnetic fields are expressed as complex exponentials, or phasors, just as presented in Chapter 5. The form of a traveling wave is $Ae^{i(\omega t - kx)}$, where k is a constant called the "wave number" or "spatial frequency." As the name implies, the spatial frequency describes how many periods of the oscillation happen in a given distance. A large k value (many wavelengths in a certain distance) corresponds to a short wavelength. Recall that the phase in a rotating phasor is $\omega t + \phi$. At a particular phase in the argument of a sinusoid, the function is at the same value – let's just call that the peak of the sinusoid. In the traveling wave, ϕ is a function of position due to the kx term. After a time, t, the peak shifts by a distance x to the right such that the entire wave has moved. So now, the wave isn't just sinusoidal in time (at some point in space), it is also sinusoidal in position (at some fixed time), like that of Fig. 6.3. Calculations of traveling waves use the phasor form to keep track of the phase relationships as waves are added together and/or travel. As in Chapter 5, to get the actual electric and magnetic fields, the imaginary part is removed and the real part represents the solution.

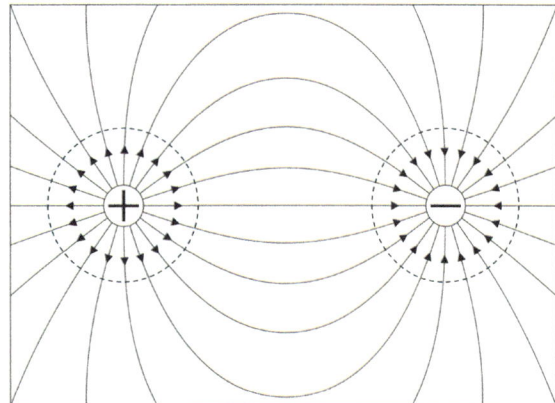

FIGURE 6.4 Electric field lines beginning on positive charge and ending on negative charge.

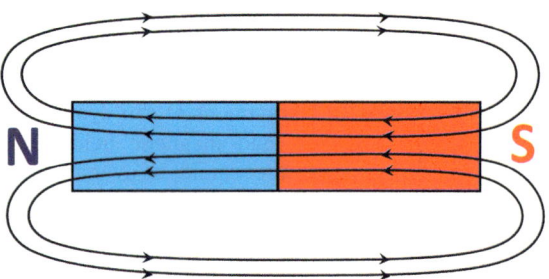

FIGURE 6.5 Magnetic field lines circling around and through the magnet. There are no magnetic monopoles.

out, from the charge. This first equation is equivalent to the statement that "the amount of electric flux coming out of a region, shown by the dashed circle, is proportional to the amount of positive charge within the region." Doubling the amount of charge inside the circle would double the number of lines, or total flux, coming out of the circle. The situation is the same for negative charge, but the lines go into the circle, and it is therefore called a sink.

The *second* of Maxwell's equations, Eq. 6.2, $\nabla \cdot \mathbf{H} = 0$, tells us that "there is no source or sink of magnetic field lines," as there is for electric fields. In other words, field lines must flow *through* any point, and not to or from a point. This statement reduces to simply that "there are no magnetic monopoles." This means that there do not exist any individual magnetic bits like there are electrons and protons. Because of that, magnetic field lines can't begin or end anywhere, but rather *always form closed loops*. All of the magnetic flux that leaves the north pole of a magnet must circle back to the north pole of the magnet through the south pole, as shown in Fig. 6.5. This is true for all magnets including small ones such as atoms, and large ones such as the Earth. Therefore, the poles of a magnet are neither sources nor sinks of magnetic flux. In contrast, we can imagine a single electrically charged particle, say a proton, floating in space with field lines emanating from it. That situation cannot exist for magnetic field lines.

The *third* equation, Eq. 6.3, $\nabla \times \mathbf{E} = -\mu_0 \partial \mathbf{H}/\partial t$, is more complicated. As you can see, it has a cross product as well as a time derivative (scary). The expression "$\nabla \times \mathbf{E}$" can be pronounced "del cross E" or the "**curl** of **E**." Figure 6.6 and its caption offer a way to visualize the curl of a function.

Ignoring the details of the curl, the presence of the time derivative of the magnetic field tells us that "a changing magnetic field generates an electric field." If the magnetic field were to remain constant, then its time derivative would be zero, and it would generate no electric field, even if **H** were huge.

The curl of **E**, $\nabla \times \mathbf{E}$, rather than just **E**, tells us something about the direction of the electric field. $\nabla \times \mathbf{E}$ is a vector that is the spatial derivative of **E** and whose direction is perpendicular to **E**. Also note that the time derivative of a vector, as in $-\partial \mathbf{H}/\partial t$, is also a vector, but more simply, it is in the same direction as **H**. Therefore, since $\nabla \times \mathbf{E}$ is in the same direction as **H** and $-\partial \mathbf{H}/\partial t$, and **E** is perpendicular to $\nabla \times \mathbf{E}$, then **E** must be perpendicular to **H**. This agrees with Fig. 6.3.

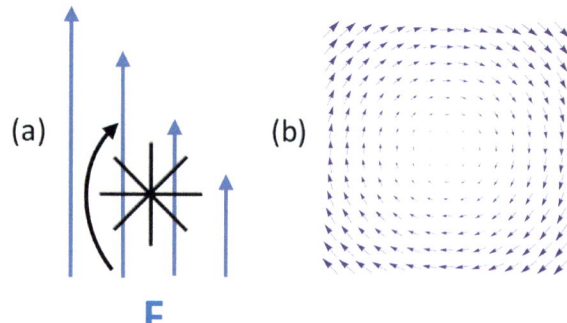

FIGURE 6.6 Visualization of the curl of a vector field, $\nabla \times \mathbf{F}$. (a) Think of the vector field as flowing water. Place a paddle wheel into the water flow. The axle of the paddle wheel is perpendicular to the page. The larger field (or flow), represented by the longer arrows, will produce more force on the wheel causing it to rotate. Use a right-hand rule to obtain the direction of the curl. The cross product is always perpendicular to the field lines, or along the axle direction. Curl the fingers in the direction of the rotation of the paddle wheel. The axle is always perpendicular to the field **F**, as is $\nabla \times \mathbf{F}$. The direction of the curl is that in which the thumb points. In this case, the positive direction of $\nabla \times \mathbf{F}$ is into the page. (b) Another way to visualize $\nabla \times \mathbf{F}$ is by observing the rotation, or *curl*, of the field around a point. Such fields have nonzero curl whose value at each point can be visualized using the method presented in (a).

Figure 6.3 shows that the extrema (peaks) of the electric and magnetic fields coincide. Let's see if we can intuit that based on what we know so far. The third equation, $\nabla \times \mathbf{E} = -\mu_0 \partial \mathbf{H}/\partial t$, tells us that a spatial derivative equals a time derivative. If both are maximum at the same point in space and at the same instant of time (think *snapshot*, which stops time), then they will be in phase in both space and time, rather than be separated by a phase shift along the direction of propagation.

Look closely at the maximum of the magnetic field in Fig. 6.3. At that point as the wave slides to the right, the time derivative at that point is low – it changes more slowly than when it crosses, say, through zero, where the time derivative is a maximum. Where the two maxima coincide in space and time, $\nabla \times \mathbf{E}$ is also a minimum because the electric field changes the least along the direction of propagation. Therefore, we see that both the spatial derivative in the form of $\nabla \times \mathbf{E}$ and time derivative in the form of $-\partial \mathbf{H}/\partial t$ are both maximum and minimum at the same place and time, so the electric and magnetic fields are in phase, as shown. (Caveat: The peaks in electromagnetic waves can be separated when propagating through certain materials, i.e., not free space.)

The third Maxwell equation in one dimension reduces to **Faraday's Law**, EMF = $-d\Phi/dt$. Ignoring any field directions, it states that a changing magnetic field causes an electric field. We already saw in Chapters 1 and 3 that an EMF or voltage develops across an inductor or a transformer secondary coil due to changing magnetic field (which is caused by changing currents according to Ampere's law). Here, we see that just floating out in space (no physical coil needed), if a magnetic field changes, it is accompanied by an electric field. This third equation tells us, furthermore, that if the magnetic field at a point in space changes sinusoidally in time, then the electric field also changes sinusoidally in time ($d \sin(t)/dt = \cos(t)$); this is central to electromagnetic wave propagation.

The *fourth* Maxwell equation, Eq. 6.4, $\nabla \times \mathbf{H} = \mathbf{J} + \varepsilon_0 \partial \mathbf{E}/\partial t$, is yet more complicated. It has two terms

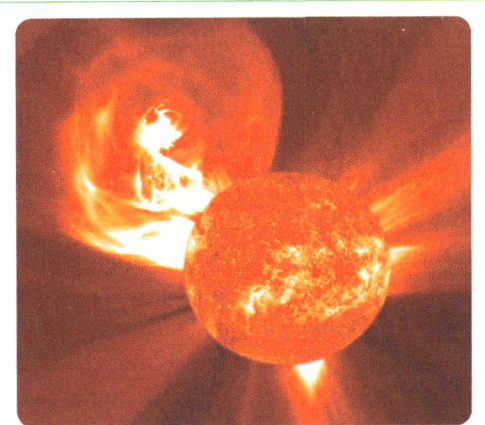

on the right. The first term $\mathbf{J}$ is current density in amperes per unit area, and represents the physical movement of charge, like that in a wire, assuming that there is any charge to move. This term involving $\mathbf{J}$ says that a constant current leads to a magnetic field; and the term $\nabla \times \mathbf{H}$ says that the magnetic field wraps, or curls, around the current. You already know that about Ampere's Law, as presented in Chapters 1 and 3. We can ignore this first term for an EM wave traveling in space since there are generally no charged particles in a traveling electromagnetic wave through free space. Sometimes there are traveling waves *and* charged particles, such as radio waves bouncing off of

---

Magnetic Field (B)

Electric Field (B) Electromagnetic Wave

Propagation Direction

Wavelength (λ)

FIGURE 6.3 Repeated.

Of possible interest: Did you know that coronal mass ejections that reach Earth can cause geomagnetic storms that can knock out satellites and even the power grid? How can that be? Faraday's Law, of course. Those billions of tons of charged particles that reach the Earth represent a current, and through Ampere's Law create a magnetic field. That field interacts with the Earth's magnetic field, causing it to change wildly. The huge $d\Phi/dt$ causes huge EMFs to form throughout the Earth that are so large they can generate massive currents in power lines, overloading them and causing widespread damage. This has already happened to some degree, but the 'big one' of modern times has not yet happened, and is just a matter of time before the Earth takes a direct hit from the Sun. Google 'Carrington Event' and prepare to be shocked.

the ionosphere or charged particles coming from the Sun to the earth that cause auroras. Charged particles in space, discussed in the sidebar about coronal mass ejections, are often referred to as a **plasma**, which is discussed in Chapter 8. Here we ignore the **J** term in Eq. 6.4 because we are discussing only traveling EM waves.

After removing the **J** term, we see that the fourth equation becomes $\nabla \times \mathbf{H} = \varepsilon_0 \partial \mathbf{E}/\partial t$, which looks very much like the third equation. Now we see that a sinusoidal electric field causes a sinusoidal magnetic field in the same way that the third equation told us that a sinusoidal magnetic field causes a sinusoidal electric field. Also, it is apparent that **H** is perpendicular to **E** for the same reasons discussed in the third equation.

The term $\varepsilon_0 \partial \mathbf{E}/\partial t$ must have the same units as **J**, i.e., current density, and is called the **displacement current**. Inside of a capacitor, or in free space, the displacement current causes a magnetic field *as if charge current were actually flowing* (but it isn't). Displacement current is as 'real' as charge current in the sense that both of them cause a magnetic field. Of course, all of this can be analyzed more mathematically, but you won't be ready for it until you take a full course in electromagnetism.

Finally, what does all of this have to do with light? Well, the T-shirt in Fig. 6.7 says it all. Light, and radio waves, propagate according to Maxwell's Equations. These equations can be encapsulated by stating that "a changing electric field causes a magnetic field, and that changing magnetic field, in turn, causes a changing electric field." Without absorption, this can go on *ad infinitum* as light propagates through space. With the naked eye (on a dark night) you can see the Andromeda galaxy, which is 2.5 million light years away; those photons make it all the way to you because of this repeated oscillation. There is no absorption of the fields in perfectly empty space,

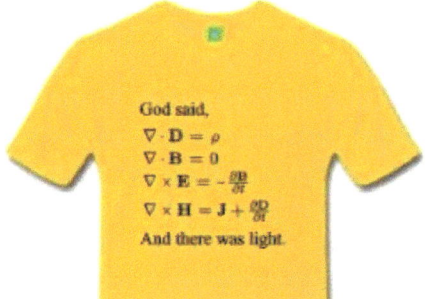

FIGURE 6.7 Maxwell's-equations T-shirt that tells the world how proud you are to be a nerd.

so EM oscillations can go on for infinite time and distance unless some physical object absorbs the energy in the fields. When we talk about the spectrum of sunlight in Chapter 8, we will see that molecules in our atmosphere absorb light at various wavelengths, so not all sunlight reaches the ground.

6.2.2 The Electromagnetic Spectrum

Now you know that light and radio waves are traveling waves that propagate by electric and magnetic fields oscillating at some frequency, and travel at the speed of light, represented by c. The speed of light, c, is different in space, water, glass, etc., but all EM waves travel at the speed of light in a given medium. In free space, the speed of radio waves and light is very nearly 3×10^8 m/s, or 186,000 mi/s. It is slower in any other medium. The velocity, v, frequency, f, and wavelength, λ (lambda), of any wave, be it an EM wave or any other, are related by

$$\lambda f = v. \tag{6.5}$$

The range of oscillation frequencies and wavelengths of EM waves is vast. Figures 6.8 and 6.9 show two versions of the electromagnetic spectrum with a few differences. Some frequencies cannot penetrate Earth's atmosphere, and are therefore not useful for terrestrial free-space communications. **Extremely low frequencies**, called **ELF**, go down even to 3 Hz with extremely long wavelengths. Applications of ELF at various frequencies are submarine communications, probing the Earth's crust, and radio communications with deep mines.

THE ELECTROMAGNETIC SPECTRUM

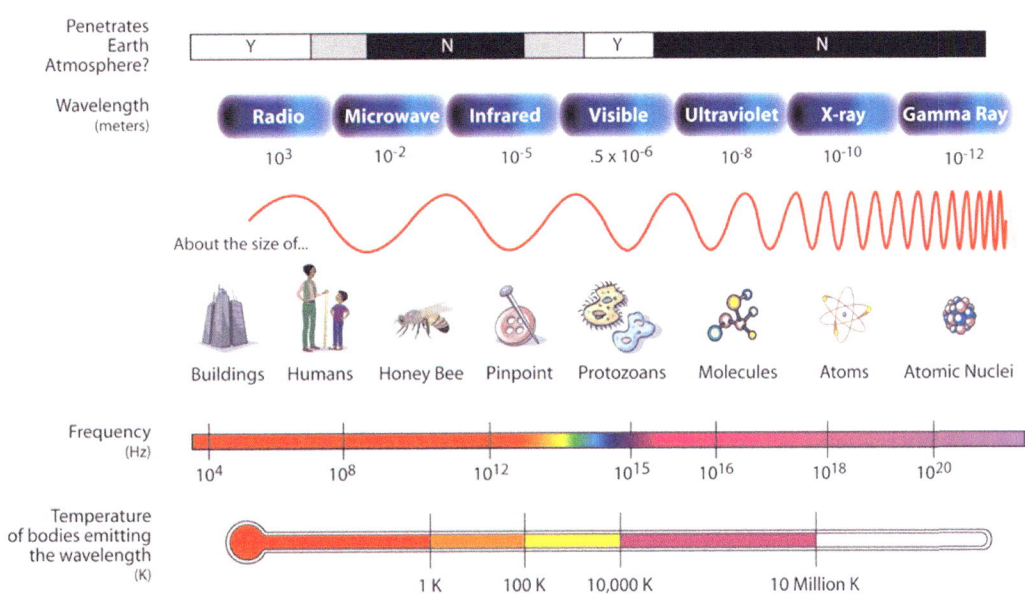

FIGURE 6.8 Electromagnetic spectrum-atmospheric penetration, wavelength, frequency and temperature of emitter. *Source*: http://mynasadata.larc.nasa.gov/science-processes/electromagnetic-diagram/.

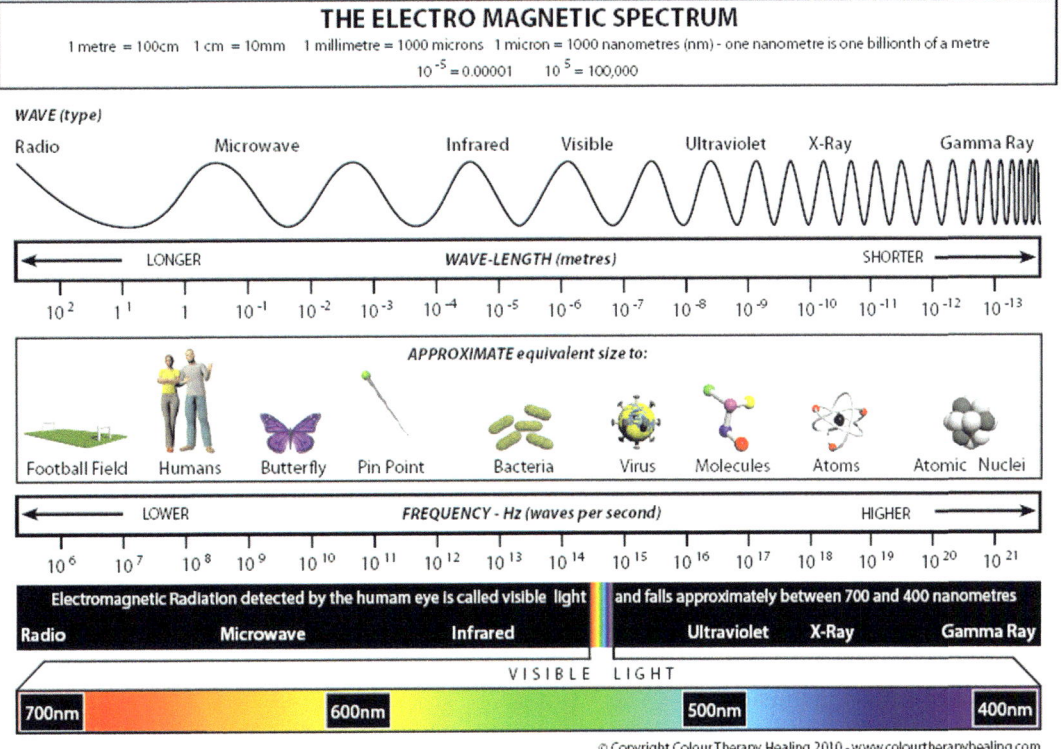

FIGURE 6.9 Electromagnetic spectrum: emphasis on visible light. Reproduced with permission from Color Therapy Healing.

The upper range of EM frequencies that modern electronics can directly create is under constant expansion. By this I mean an electronic circuit with moving electrons generating an EM wave. There are other ways to generate high frequencies, such as light, that use properties of nature and special materials, which we will see in Chapter 8 involving light-emitting diodes. In the 1990s the upper limit was about 250 GHz, and today it is about 1000 GHz, or 1 THz (teraHertz). The fundamentals of semiconductor devices that function at ever-higher frequencies is the subject of future courses that you might choose to take.

For fun: The furthest-known galaxy MoM-z14 was discovered in 2025 by the James Webb Space Telescope, JWST, and is 33.6 billion light years away. The wavelength used to detect it was in the near-infrared spectrum at about 1.5 micron. Use either Fig. 6.8 or 6.9 to calculate how many oscillations of electric and magnetic fields these photons make by the time it reaches the JWST from there.

6.3 Antennas and Amplitude Modulation

This chapter's lab activity is based on one of the simplest forms of radio transmission, or **wireless communications,** called **amplitude modulation**, or **AM radio**. This section describes how information is carried by radio waves, and specifically by the instantaneous strength, or amplitude, of the signal.

How does an antenna work? Antennas come in many shapes and varieties, but let's just assume it is a straight piece of wire. A signal source like your waveform generator causes currents, i.e., moving electrons, to oscillate back and forth on the wire.

Figure 6.1 shows schematically an antenna at the output of a receiver and input of a receiver. Even though that is schematic, it does capture the basic shape of an antenna, which in this case resembles a **dipole antenna**. Dipole antennas are made of a straight piece of wire that is driven at one end and ends at an open circuit at the other end, or it can be driven at the center. Until now, we have assumed that charge does not build up anywhere in a circuit except for the plates of a capacitor. Antennas are different because you must include the capacitance of the wire in your thinking. Even though the antenna is an open circuit to the generator, it has capacitance along its length, which is called a **distributed** capacitance. Due to this capacitance, as AC currents flow along it, charge builds up at various points along its length. Hence, somewhat unexpectedly, currents can flow on an antenna even though it is an open circuit at its ends.

These currents generate magnetic and electric fields that then launch into space as an EM traveling wave, according to Maxwell's equations. In order to launch a wave into space, an antenna must be the right size to fit the wave into it so that it achieves the largest possible fields and can emit a wave into space with the highest efficiency. Like resonant vibrations on a string, low frequencies need big antennas to launch long waves, and high frequencies need small antennas to launch shorter wavelengths. Think about low notes from an acoustic bass and high notes from a ukulele. You don't get low notes from a ukelele.

The **half-wave dipole antenna** has a total length equal to half of the wavelength of the most-efficient frequency transmitted. Frequencies away from the center frequency are not as efficiently transmitted, but it works well enough over some range of frequencies. A receiving antenna works in reverse: the electric field of the EM wave causes currents to flow on the antenna; the receiver circuit detects those currents and converts the high-frequency wave to lower frequencies for further processing, such as playing music.

Regarding antennas, we are always concerned with what signal frequencies we wish to transmit. **Bandwidth** is defined as the span of frequencies in any analog signal. The bandwidth of human hearing is generally taken to be 20 Hz to 20 kHz, although nearly everyone loses the high frequencies as they age (I certainly have). We can take our 'information' to be human audio, such as voice, a symphony or rock music. The frequencies of the information are called the **baseband** frequencies. It is simply not practical to launch a wave whose frequency corresponds to baseband audio information because the wavelengths would be so long. (I will let you figure out the wavelength and the antenna size you would need.)

This problem of not being able to transmit information at the low baseband frequencies is solved by shifting the baseband to much higher frequencies using a technique called **modulation**, or **mixing**, and then launching the signal from a shorter antenna into space. The job of transmission is relegated to the high-frequency **carrier** wave, and the information content is relegated to the baseband **modulation frequency** that "rides" on the carrier. This is shown in Fig. 6.10. The AM radio that you built receives the combined, or **modulated**, high-frequency wave and **demodulates** it by removing the carrier and recovering the information content (e.g., music) for us to listen to. Any reference to frequencies common to radio transmission, or high frequencies in general, can be called simply **radio frequency**, or **RF**.

Let's discuss modulation. Modulation is the change of some characteristic, be that amplitude, frequency, phase or some other property of a wave, usually repetitively. Mathematically, amplitude modulation is **multiplication**. In AM radio, we **multiply** the carrier wave, which is a high-frequency sinusoid at say 1 MHz, by our baseband signal, i.e., we modulate the amplitude of the carrier by the voice or music that is at the baseband frequency range. Figure 6.10(c) shows an amplitude-modulated signal. Its shape is due to the baseband sinusoid (a) multiplying, or "riding on,"

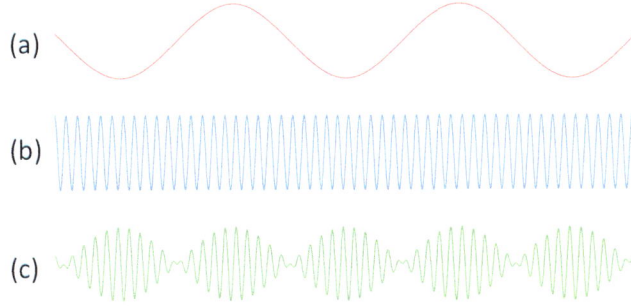

FIGURE 6.10 The basic notion of amplitude modulation. (a) Baseband information, or modulating signal. This sinusoidal waveform in the audio range would be a simple tone, like a flute sound; (b) carrier wave at a high frequency transmitted by an antenna; (c) amplitude-modulated waveform showing the envelope caused by the flute tone. After demodulation, the radio would play the sound of the flute tone. The waveforms of Fig. 6.11 are improved for undistorted signal transmission.

Thinking like an EE: You may think of any signal as a voltage that is a function of time, and can describe that signal as a simple function, $f(t)$. So, we have jumped now from the realm of physics, i.e., what causes radio waves, what is the nature of an electric and magnetic field, etc., to that of mathematics. You are certainly used to using math to describe physics, but it is useful to point out explicitly that whatever the mathematics predicts, presuming it is done correctly, will turn out to be a fact when you look at the resulting physics. So, if we describe the carrier as $C(t)$, and mathematically we think something will happen to $C(t)$ if we do something to that waveform, then we expect that the real wave that results will have that exact property. The strength of this concept has driven the study of physics and engineering throughout modern times, and is the reason why math is so closely tied to engineering and physics.

the high-frequency carrier sinusoid (b). The outline of the modulated carrier is called the **envelope**. The shape of the envelope is exactly the same as that of the baseband, or modulating, signal.

In commercial AM radio in the United States, the carrier is between 530 kHz and 1710 kHz, depending on where you are, and in **shortwave** radio it is between about 2.3 MHz and 26.1 MHz. Another term for what we call simply "AM" is **medium wave**. The use of "medium" to describe "amplitude modulation" is somewhat misleading in that it is "medium" compared with what? Nowadays, radio broadcasting happens over much higher frequencies, so the term, although still used for commercial AM radio, is a bit outdated.

Refer to Figs. 6.10 and 6.11 during the following explanation. The carrier signal is described by:

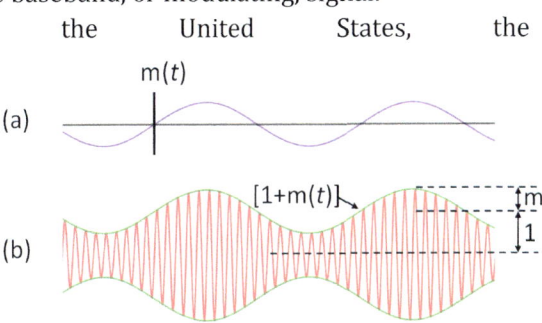

FIGURE 6.11 Practical amplitude modulation waveforms. (a) Baseband signal $m(t)$ of amplitude m; (b) carrier wave modulated by $[1 + m(t)]$. In this example, $m = 0.5$.

Of interest: Some musical instruments, such as the flute or modern singers, strive to create a 'tremolo' effect that is essentially amplitude modulation by controlling the speed of the airstream, creating fast variations in volume, whereas violins or other stringed instruments wiggle their fingers slightly up and down a string, an effect called 'vibrato,' which is essentially frequency modulation. These terms are often confused in practice, and both are usually just called 'vibrato.'

$$C(t) = C\sin(2\pi f_c t) \tag{6.6}$$

where f_c is the frequency of the carrier, e.g., 1 MHz, and C is the amplitude of the carrier signal, shown in Fig. 6.10b. To get the transmitted AM radio signal, $y(t)$, we multiply the carrier function $C(t)$ by a **gain** factor $M(t)$:

$$y(t) = M(t) \cdot C(t) \tag{6.7}$$

where:

$$M(t) = 1 + m(t). \tag{6.8}$$

The function $m(t)$ is the voice or music information we want to transmit. In actual AM radio broadcasting, $m(t)$ has an amplitude less than 1, so that $1 + m(t)$ never goes negative (the reason for which will be discussed below). We can say that we "modulate $C(t)$ by," or "mix $C(t)$ with," a modulating signal $m(t)$. Because $C(t)$ is a high enough frequency to interact well with an antenna, it is $C(t)$ that launches the low-frequency music, $m(t)$, into space from the antenna.

Let's assume that the modulating signal is a pure sine wave, (again, shifted up by a constant so that it never goes negative), i.e.,

$$M(t) = 1 + m(t) = 1 + m\sin(\omega_m t). \tag{6.9}$$

$M(t)$ has the shape of a sine wave of some amplitude m, and angular frequency ω_m, but shifted up by the constant value '1.' $M(t)$ and $m(t)$ may be voltages or currents in a transmitter amplifier. Furthermore, the carrier is also a sine wave, $C(t) = C\sin(\omega_c t)$. Remember that in AM radio, the carrier is a high-frequency sine wave (say, 1 MHz) and the modulating signal could be as simple as just an audio-frequency sine wave (say, 1 kHz), i.e., $\omega_c \gg \omega_m$, or more complicated, like rock-band music. For a 1 kHz sine-wave, the modulating signal is, therefore:

$$M(t) = 1 + m\sin(2\pi \cdot 1000 \cdot t). \tag{6.10}$$

So far, we have described the transmission of a pure, 1 kHz sine wave, which is firmly within the audible range. What does a pure sinusoid *sound* like? Recall from Chapter 4 that a pure sine wave has no harmonics other than its own fundamental frequency. To me, it sounds like an annoying, featureless tone. You might think of it as a simple note played on a tin whistle without any form of tremolo or vibrato. Since it contains no harmonics it is uninteresting from a musical point of view. (In this lab, you will hear, and be annoyed by, some pure sine waves.)

Now we have the modulated RF signal as

$$y(t) = [1 + m\sin(\omega_m t)]\,C\sin(\omega_c t). \tag{6.11}$$

Figure 6.11 shows (a) the time-domain modulating signal, $m(t)$ (say, music or a 1 kHz sine wave) and (b) the resulting modulated carrier, $y(t)$. The modulated carrier looks much like the modulating signal, but the envelope is 'filled up' with the high-frequency carrier. Note that the edges of the envelope

For deeper understanding: Note that for $m = 0$, the carrier is just a pure sine wave with amplitude C. That carrier is there, but the sound playing on that radio station is silence. If you tune into an AM radio station when nobody is talking, you hear silence without the background static that you normally hear when tuned between stations. Intuitively you know you have a station and wait for someone to start talking. Now you know why that happens. During this silence, the broadcaster is still sending a 'flat' carrier wave and your radio is playing just the silence. Some types of communications radio stop the carrier during the silent moments in order to conserve power. This would not be good for broadcast radio transmission of music and conversation. What would you hear?

have the same overall shape as the modulating signal. This waveform is different from that of Fig. 6.10 because here the baseband signal is shifted up by a DC offset of magnitude 1 so that the modulation never goes negative. That is critical to undistorted AM radio transmission, as discussed below.

Recall from what you have learned about harmonic content of a signal that anything that isn't a perfect, or pure, sine wave must contain, or be composed of, more than one frequency component. It has extra frequencies in the frequency domain that are needed to make up that waveform. In the case of amplitude modulation, the modulated signal does not look like a perfect sine wave and therefore has additional frequency content. The frequency spectrum is different from both $\sin(\omega_m t)$ and $\sin(\omega_c t)$, which can be derived mathematically, as follows.

Recall the trigonometric identity:

$$\sin(\alpha)\sin(\beta) = (1/2)[\cos(\alpha - \beta) + \cos(\alpha + \beta)]. \tag{6.12}$$

(Recall that we used this equation (Eq. 3.30) to explain the DC offset in the real part of power in Chapter 3 when we computed $V \cdot I$ where each was a sine wave of the same frequency). Using Eq. 6.11, multiplying through and using Eq. 6.12, we get:

$$y(t) = C\sin(\omega_c t) + \frac{1}{2}Cm[\cos(\omega_m t - \omega_c t) + \cos(\omega_m t + \omega_c t)]. \tag{6.13}$$

We see that modulation in Eq. 6.11 causes two *new* frequency components of modulated spectrum, namely, $[\omega_c - \omega_m]$ and $[\omega_c + \omega_m]$, as shown experimentally in Fig. 6.12. Each of these sinusoidal functions is a frequency component of the spectrum of the amplitude-modulated wave. (The sine and cosine are essentially the same, but differ in phase. The distinction is not important here.)

The '1' term in Eq. (6.11) is a level shift that leaves the carrier signal unaffected. Also, since ω_m is small compared to ω_c, both $[\omega_c + \omega_m]$ and $[\omega_c - \omega_m]$ are close to ω_c. We call these two frequency components the **upper and lower sidebands**, respectively. For our example, the carrier frequency is at 1 MHz, and the signal, i.e.,

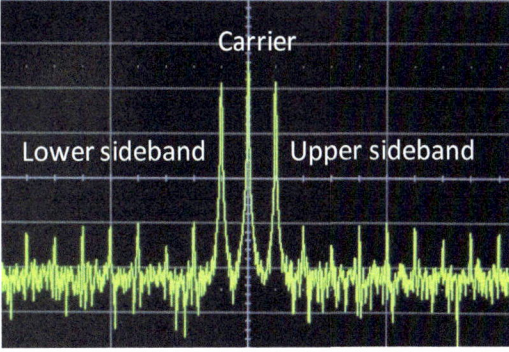

FIGURE 6.12 Modulated signal and sidebands in the frequency domain.

the modulating frequency, is 1 kHz, so the lower sideband is at 999 kHz, and the upper sideband is at 1001 kHz. In practice, with music for example, the modulating signal would be more complicated than a single tone, and would carry multiple frequencies, as we shall soon see.

If m is less than C, as it should be for undistorted AM radio transmission, then, as shown in Fig. 6.12, the sidebands are not as large as the carrier frequency component. Depending on the relative sizes of m and C, there is more or less of the sidebands compared to the carrier in the total harmonic content. Put another way, if there is not much modulation, i.e., m is small, then the music will be quiet (low amplitude) and the sidebands will be small compared with C. If m is large, i.e., there is a lot of modulation, then the music will be loud and the sidebands will be large, approaching or even exceeding the size of the carrier in the spectrum. We define a **modulation index**,

$$h = 100\% \times \frac{m}{C}. \tag{6.14}$$

For $m < C$, we have a waveform like those shown in Figs. 6.13(a) and 6.13(b). None of the envelopes of the waveform reaches zero. In that case, there is a moderate amount of power in the frequency components at the sidebands. The carrier frequency component does not change. You will see this effect in this week's lab when you use the frequency spectrum analyzer to observe the harmonic content of your modulated sine waves.

When $m = C$, Fig. 6.13(c), we have 100% modulation. Now the envelope just reaches zero. In that case, there is a lot of power in the sidebands compared to the carrier. For modulation index > 100%, Fig. 6.13(d), $m > C$. This condition is called **overmodulation**. It may be advantageous for radio stations or amateur radio operators to broadcast with $h > 100\%$ because the signal would be received at a louder volume and can be received farther away, since more power is in the signal.

FIGURE 6.13 Four different values of modulation index, m: (a) 50%; (b) 80%; (c) 100%; and (d) 150%.

Overmodulation is against government **Federal Communications Commission (FCC)** regulations. Abrupt changes in modulation, i.e., distorted baseband and carrier signals, can lead to frequency harmonics that reside outside of the sidebands and can bleed into "neighboring" radio stations (i.e., those with adjacent carrier frequencies), a phenomenon known as **splatter**. In this chapter's lab activities, those harmonics will be difficult to see, so it won't be obvious that it is happening. If you wish to learn more, download and read this document: https://www. nrscstandards.org/standards-and-guidelines/ documents/references/am-mod-overmod-1986.pdf?. By the time you finish the activities, you will be able to understand a good deal of this report.

Now, let's assume that we are broadcasting not just a single 1 kHz tone, but a range of

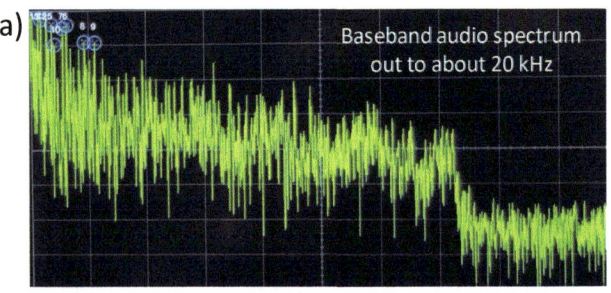

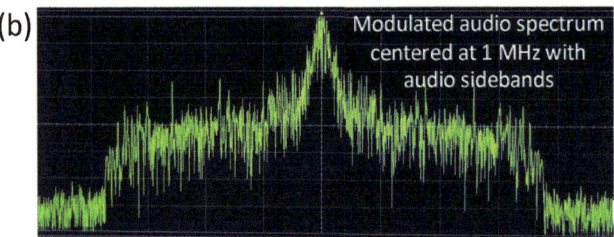

FIGURE 6.14 (a) Baseband audio signal from about 20 Hz to 20 kHz; (b) Modulated audio spectrum centered at the carrier frequency and spreading out by 20 kHz in both lower and upper sidebands.

frequencies over the full baseband, as you would have in voice or music. If the baseband range were the full audio spectrum of 20–20 kHz (which is in fact *not* the case for commercial AM radio), then we would have a 20 kHz bandwidth of modulating frequency, as shown in Fig. 6.14(a). In this case, the AM sidebands stick out from the sides of the carrier by about 20 kHz each for a total broadcast bandwidth of 40 kHz, as shown in Fig. 6.14(b).

Although negative frequencies in a Fourier spectrum are a mathematical artifact of the frequency content of signals, they do in fact translate into the lower and upper sidebands like those in Fig. 6.14(b). That results in a total broadcast bandwidth, from the lower edge of the lower sideband to the upper edge of the upper sideband, of 40 kHz as measured on the spectrum analyzer. Since there is a limited total range of AM frequencies that are allowed to be broadcast, more bandwidth devoted to each channel results in fewer available channels on the radio dial. This is a tradeoff between quality and quantity of commercial radio stations made available by FCC regulations.

Historical: The original AM 5 kHz bandwidth was set when the recording industry was in its infancy, and records hardly had more bandwidth than that. For that reason, AM radio is mostly used now for talk shows. FM radio came around in its modern form in the early 1950's to transmit better musical quality. If you want to know, "What does music sound like with a 10 kHz bandwidth?" just turn on an AM radio, listen to some music, and then switch to FM ('frequency modulation') radio for a comparison. FM has a bandwidth of 15 kHz, which is much closer to the maximum that most people can hear. The wider bandwidth is enabled by much-higher carrier frequencies (from 87.5 MHz to 108 MHz) that can be separated into many more FM stations.

For music, it isn't economical to use up a lot of bandwidth for sounds that people can't hear or appreciate, or are not necessary. Early audio recordings had very low bandwidth, so not much was needed in broadcasting. For that reason, commercial AM radio used to be limited by the FCC to about 5 kHz of baseband bandwidth (i.e., the audio bandwidth) in the United States until the 1980s. In order to compete with FM broadcasting, which has higher bandwidth, and therefore sounds better, it was briefly increased to 15 kHz, but has since been set at 10.2 kHz. Because of the two sidebands, the total bandwidth of the broadcast spectrum for a single channel is twice the audio bandwidth, namely, 10 kHz, 30 kHz and 20.4 kHz in the above discussion, respectively. The limitation to a total broadcast bandwidth of 20.4 kHz is one reason why AM radio does not sound as good as many other sources of music.

6.4 AM Demodulation

Demodulation is required because speakers don't work at RF, and if they did, we wouldn't be able to hear it, and if we could, we probably wouldn't understand it. Therefore, we need to bring the RF signal down to the audio range. In the earliest days of radio, demodulation was done with what was called a **crystal radio** set. A schematic for a very simple crystal radio is shown in Fig. 6.15. The crystal was, in fact, an actual mineral crystal (often galena, but there are many others) that had the property of passing current in only one direction, a process called **rectification**, when a thin wire was pressed against it; the wire was called a "cat's whisker." This rectification is exactly the same property provided by today's manufactured semiconductor diodes. (Diodes are discussed in Chapter 7.) In the figure, the diode symbol (the triangle with line) represents either a modern diode or a crystal of yesteryear.

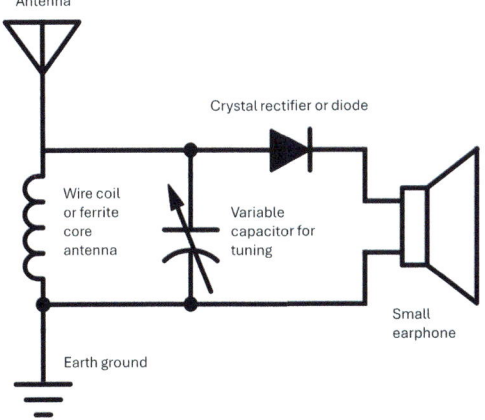

FIGURE 6.15 Crystal radio set schematic.

Here is how a very simple crystal radio set, like that of Fig. 6.15, works to demodulate an RF AM radio signal so that it can be heard at audio frequency. (You can build one yourself from parts or kits.) Electrons on a long antenna are pushed back and forth (oscillating current) due to the oscillating EM fields of *all* of the frequencies of *all* of the AM radio stations to which it is exposed, including the one station that you want to listen to. The trick here is to isolate the one station you want to hear. That current is in series with a parallel inductor and capacitor combination, called an LC **tank circuit**. A parallel tank circuit has a *high impedance* at its resonant frequency. That frequency is the same as that of the series resonant circuit discussed in Chapter 5, namely $\omega_0 = \sqrt{1/RC}$.

Only one of the carrier frequencies (let's say WSBT at 960 kHz) is at the resonant frequency of the tank circuit. The resonant frequency is chosen by you by changing the capacitance of the variable (tuning) capacitor, as mentioned in Chapter 1 about the AM radio kit. Of all the carrier frequencies oscillating on the antenna, only at the resonant, tuned, frequency is the tank circuit pretty much an open circuit. Now the current on the antenna at the resonant frequency cannot pass through the tank circuit, and is shunted toward the diode and the load, which is the earphone. That RF current is rectified by the diode, meaning that it flows in only one direction. Don't forget that the signal has the low-frequency modulation riding on the high-frequency carrier. This circuit is as simple as it gets: the earpiece responds only to the low-frequency baseband signal; it acts like a low-pass filter because its mass cannot respond to the high carrier frequency. Hence, the listener hears only the low-frequency baseband signal, which is the end result of the demodulation.

If the output goes to an amplifier as the load rather than an earpiece, the carrier frequency is not limited by the physical motion of the earpiece, as in the crystal radio; rather, demodulation is aided by placing a capacitor in parallel with the load, as shown in Fig. 6.16. The capacitor

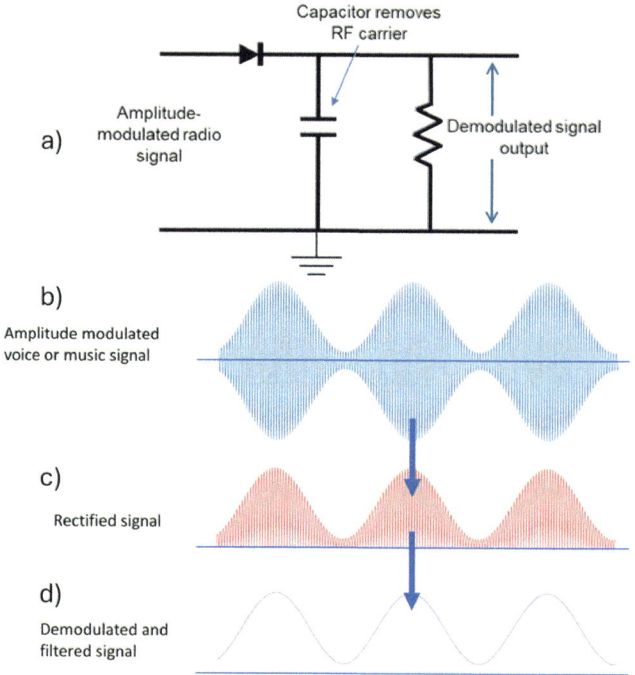

FIGURE 6.16 (a) AM demodulator that includes a capacitor. This is essentially the same as the crystal radio circuit with headset, but here the load is more general, and could be an audio amplifier. (b) The voice or music is amplitude modulated. (c) The diode stops current from flowing in the negative direction, and thus converts the AC RF signal to DC, i.e., current flows in only one direction. (d) The capacitor is a low-pass filter to the load (Chapter 4), so the high frequencies are shorted past the load, low frequencies are passed to the output, and only the low-frequency envelope remains.

stores charge during each half cycle, fills in the valleys while discharging, and helps to smoothen out the fast carrier ripples in the signal, thus eliminating the carrier but keeping the baseband. Put another way, when placed in parallel with the load resistor, the capacitor forms a low-pass filter, and the carrier frequency is shorted to ground, so only the baseband appears at the output. This will be analyzed at length in Chapter 10, as the same type of filter is used in power supplies.

What does the diode do? The diode causes the current in the earphone to go in only one direction. If current flowed in both directions, the diaphragm of the earphone in Fig. 6.15 (the part that goes back and forth to make the sound) would rapidly, at the carrier frequency, see a positive force followed quickly by a negative force of almost the exact same magnitude. Therefore, it would not be able to respond to any of the force, and there would be an average zero net motion. Hence, no overall sound would result. With the diode, each pulse in the same direction contributes to the overall baseband movement of the diaphragm. In a similar way, the capacitor in Fig. 6.16 would not charge to a DC voltage between peaks, but would instead average to zero, and there would be no DC value of the modulation, again resulting in no sound.

6.5 About the AM Radio Kit

The radio that you built in Lab 1 is much more sophisticated than a vintage crystal radio. It uses something called **superheterodyning**, invented in 1918 by Edwin Armstrong, for the demodulation circuit. A superheterodyne receiver internally uses **heterodyning**, which is the same technique as *modulation* – it multiplies the signal received by the antenna by a sinusoid at some frequency that mixes the desired signal down to a fixed **intermediate frequency** (IF), which is well above the baseband. In the case of AM radio, it is typically 455 kHz. In fact, "super" in "superheterodyning" means "supersonic." The variable tuning capacitor does the same task as that of the crystal radio, i.e., selects the station that you want to listen to, but instead controls the frequency of the superheterodyne circuit.

The intermediate frequency falls within the bandwidth of a fixed, bandpass filter of high quality having a steep rolloff on each side (see Chapter 4). As you, the listener, change that internal modulation frequency, the carrier and sidebands of different carrier frequencies, i.e., different radio stations, are made (using Eq. 6.12) to fall within that IF filter's bandwidth. The rectifying circuit is optimized at that one intermediate frequency, and the reception is made overall much better quality in terms of **selectivity** of a particular frequency (radio station), as well as the **sensitivity** of the receiver, i.e., its ability to detect weak stations. Your radio kit uses a Rectron MK484 integrated circuit (IC) that does all of the heterodyning and rectifying.

The AM-780K radio kit uses a ferrite core antenna (see Chapter 1) instead of a very long wire antenna. In the long-wire or metal antenna, the electric field of the EM wave pushes the electrons and forces the small signal current to flow in the tank circuit. A ferrite core antenna is more like a transformer in which the primary is the external EM wave – it produces a voltage in response to the changing magnetic field. The core antenna senses the magnetic field component of the EM wave, and not the electric field component, and produces a voltage (or equivalently, current) by Faraday's Law that causes oscillations in the tank circuit. The ferrite core antenna can work as just a loop of wire, which it is, but the magnetic field that threads the coil is increased by the ferrite material in the core, just as any transformer's efficiency, or inductor's inductance, is increased by a magnetic core. The ferrite core antenna in small radios is actually several having different inductances. A ferrite core may not be the very best antenna, but it works well enough for portable radios to offer a wide range of radio frequencies, and takes up very little space.

6.6 Other Forms of Modulation

There are many forms of modulation besides AM that exploit changing either frequency, phase or sideband frequency content of the total signal, among others. Next to AM, the most common modulation method is **frequency modulation**, or **FM**. Most drivers have their car radio presets set to FM stations rather than AM stations because FM radio has a larger bandwidth and less noise. In other words, it sounds a lot better than AM.

As the name implies, the FM transmitter signal modulates the frequency of the carrier,

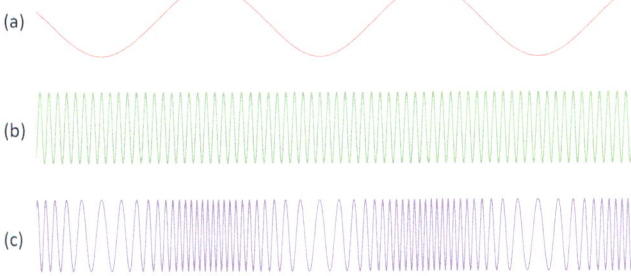

FIGURE 6.17 Frequency Modulation (FM). (a) Baseband modulating signal; (b) carrier; (c) frequency-modulated carrier. Note that the amplitude of an FM signal is constant but the frequency changes with the baseband signal. The detector circuitry converts frequency to voltage to recover the baseband signal.

rather than the amplitude. The carrier frequency changes around some center frequency *at the frequency of the modulation*, as shown in Fig. 6.17, rather than the amplitude at a fixed carrier frequency, as in AM. The amount above and below the carrier frequency that it is modulated is called the **frequency deviation**. Let's consider a 1 kHz tone again. The center frequency of the carrier, or the carrier frequency, might be, say, 101.5 MHz. The 1 kHz tone changes that carrier frequency above and below 101.5 MHz by some frequency deviation at a rate of 1 kHz, so the carrier frequency itself is modulated. The amplitude of the carrier does not change, but louder music causes *larger carrier-frequency deviation*. The most that it is allowed by the FCC to deviate is 75 kHz to one side. Also, extra bandwidth is allotted to carrying a stereo signal, which is beyond the scope of this chapter. Because of the need for two-channel signals, i.e., stereo, the total broadcast bandwidth of an FM station is 256 kHz. FM radio stations are allowed a baseband bandwidth of about 15 kHz, versus 10.2 kHz for AM radio. Such a large bandwidth (compared to AM radio) is made possible by placing the frequencies of FM radio stations in a much wider frequency range, namely from 87.5 MHz to 108 MHz. That 20.5 MHz bandwidth is compared to the AM band, which is only about 1.2 MHz wide.

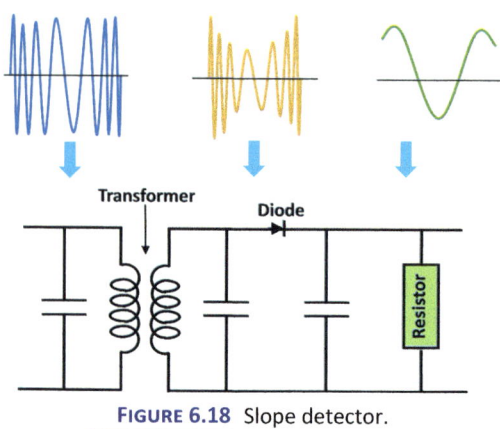

FIGURE 6.18 Slope detector.

One of many types of FM demodulators is called a **slope detector** (Fig. 6.18). The **discriminator** is a filter that has a gain that is frequency dependent, so the amplitude at the output depends on the instantaneous frequency. Hence, the FM is converted internally to AM in the sense that the amplitude of the carrier now depends on the baseband frequency. The resulting AM signal can be demodulated in the conventional way, using superheterodyning, which as in AM strips away the modulation leaving the baseband.

AM radios make great lightning detectors because lightning produces electromagnetic radiation over a wide range of frequencies that overlap the radio spectrum. The frequency components of the lightning add to those of AM radio, so a loud crackle is heard during a lightning event. By tuning into a weak radio station, or a space on the dial with no station at all, the crackle of far-off lightning can be clearly heard, even when no storm is in sight. This can be a useful tool when traveling cross-country by car, or just to impress your friends!

However, there is no component of the lightning that shifts any of the higher frequency components of FM signals, so FM radio does not produce the crackle sound due to the lightning, and is much quieter generally in the presence of external noise sources. Also, as an FM station becomes weaker, say from driving on the highway, it tends to remain clear until it starts to drop out and create noise due to the low signal, much like a cell phone has dropouts. For these reasons of noise immunity, as well as using smaller antennas for the higher frequencies, FM is the choice of public service radios, such as fire and police bands.

Historical: Hedy Lamarr was a film actress from about 1930 to 1957. At one point she dated technology mogul Howard Hughes, who was willing to share technical guidance and information about weapons systems. She became interested in technology to such an extent that she co-invented, with composer George Antheil, an important communications technology called **frequency hopping**, which does what it sounds like. Frequency hopping avoids **jamming** by making it difficult for an opponent to find it and respond to it before it changes frequency. She was so far ahead of her time that her patent U.S. 2,292,387 from 1942 was finally adopted for defense systems in 1962.

Another form of frequency modulation is **frequency shift keying**, **FSK**, in which the frequency changes discretely between two frequencies in order to transmit digital information. **Phase shift modulation**, **PSM**, encodes digital data in the relative phases of signals. **Single-sideband modulation**, **SSM**, uses a filter that removes one of the sidebands that arises due to amplitude modulation. All of the information is contained in one sideband, but both are needed to generate the original signal. Therefore, in order to save bandwidth, one of the sidebands is removed by filtering during broadcast, and the receiver reconstructs it by modulating the remaining sideband. The two sidebands are now back in place and used to demodulate the carrier.

Congratulations again! You have finished this chapter about the nature of electromagnetic waves and radio broadcasting. Hopefully you will now have greater insights and appreciation for your car and home radios whenever you tune in for news or music.

6.7   Lab 6 Activities

Because of your superior taste in music, you've just been hired by a major media corporation to broadcast on your own AM radio station. You have been given complete creative control over the content (within FCC regulations), but due to budgetary constraints, you must broadcast your station using only the equipment found in the EE undergrad lab. In this lab, you will use the concepts of amplitude modulation/demodulation discussed above to broadcast a real (but low-power) AM radio station that can be received by your AM radio.

1. **Creating the carrier.** First, you're going to generate a high-frequency carrier signal that can be transmitted by an antenna attached to the output of the waveform generator. That is, you're going to generate the carrier frequency that you can tune to using your radio.

   Connect the output of the function generator to the input of the scope using a coax cable. View a 10 V_{pp} sine wave with frequency at 1 MHz on the scope and also note its frequency using the frequency spectrum analyzer (FSA). As shown in Fig. 6.19, is it a clean, narrow delta function in the frequency domain? Don't forget that the quality of the spectrum on the FSA depends on how many periods are displayed in the time domain waveform; more is better. It is okay if you have so many periods displayed that the details cannot be discerned, as shown.

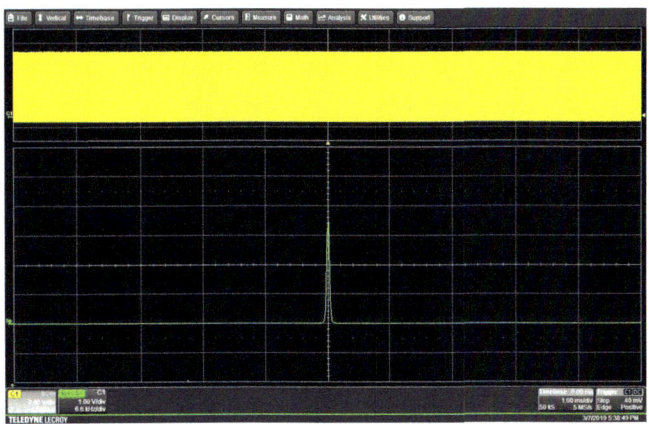

FIGURE 6.19 (Step 1) Carrier sine wave and its Fourier spectrum showing the 1 MHz peak in the frequency domain. The "solid" bar across the top is actually a sine wave whose period is too small to be resolved by the screen resolution.

2. **Modulating the carrier.** Now, you need to modulate the carrier signal in the lower audio frequency range that can be heard by human ears. Later in the lab you will modulate the carrier with music (and you can play 'DJ'), but for now you will demonstrate this with a simple sine wave.

   Press **MOD** on the waveform generator to activate and set the modulation. **Source** should be **Int.** Set the modulation frequency, **AM Freq**, to 1 kHz. **DSSC** should be off. Set **AM Depth** to 70%. The "depth" of the modulation signal is the modulation index, h, of the modulated wave.

The scope will have trouble syncing the modulated signal, so for triggering, you need to use the **Sync** output from the generator to the **Ext** input of the scope as you did in Lab 5, Step 15 when you synced the resonance peak. Now you can expand the horizontal time base to do measurements of both the carrier and modulating signals, as shown in Fig. 6.20.

Using the cursors, confirm on the scope that the modulation frequency and the carrier frequency are what you set them to. Save the original scope trace and the expanded one to show both frequencies for your lab report.

3. **Changing modulation depth.** Change **Depth** to 30%, 70%, 100% and 120% and observe how the signal in the time domain changes. As you increase

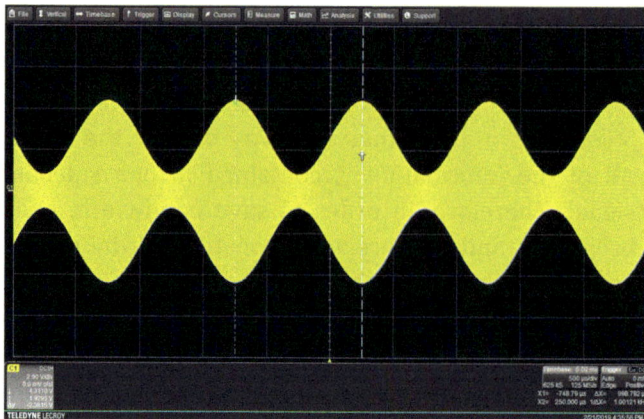

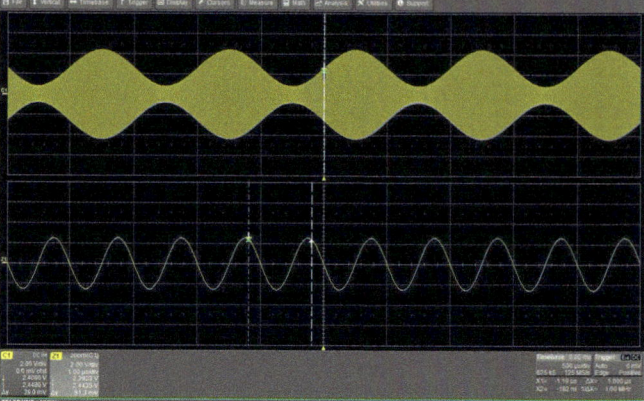

FIGURE 6.20 (Step 2) (Top) Amplitude modulation with a 1 MHz carrier and 1 kHz sine wave signal, including measurement cursors. (Bottom) Modulated carrier and zoomed carrier. Cursors are shown measuring the carrier frequency.

FIGURE 6.21 (Step 3) Amplitude modulation with modulation depth, *h*, of (from top) 30%, 70%, 100% and 120%. Composite figure from several traces.

h, the envelope of the wave should get deeper into the carrier, and then distort, as shown in Fig. 6.21. Save some waveforms for your lab report. What is happening when *h* > 100%?

4. **Setting up the FSA.** Set your modulating frequency to 1 kHz for this step. Turn on the spectrum analyzer. Set **Center Freq** to 1 MHz. Set **Freq. Span** to 50 kHz. Set **Scale** to V_{RMS}. Turn on **Peaks** under **Peaks/Marker**.

5. **Observing the FS.** Remember, the FSA provides sharper peaks when more periods of the waveform are displayed at the top of the screen by increasing the time base setting to more time per division. Make sure that you have many periods of the modulated signal on the time waveform at the top window. In fact, you should select the time base to be so long that all of the periods on the time waveform may even blend together. Don't worry, the memory is storing all of values even if the screen cannot display all of them. Also, you cannot get a spectrum if your time-domain waveform at the top of the screen has data that is above or below the window,

so you may have to lower the vertical gain to get it all inside the window. If you use **Normal** triggering, there may be some fluctuations, but the FS is still good. If you want it to be more stable, press **Single** every time you make a change to refresh the waveform and the FS.

Capture the spectrum. It should look like Fig. 6.22. Note that the center peak, which is the largest, is due to the carrier. The two side peaks are the upper and lower sidebands,

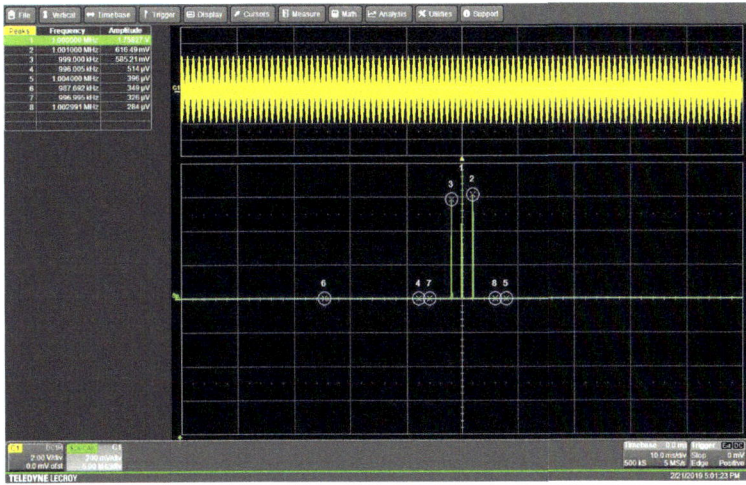

FIGURE 6.22 (Step 5) Frequency spectrum of AM signal. **Peaks** are turned on. Sidebands are clearly visible and identified in the table in the upper left corner.

which are symmetric and due to the modulating signal, or envelope. The bandwidth of the signal includes the sidebands from 999 kHz to 1001 kHz, so is a total of 2 kHz. In general, the bandwidth of the signal is twice the maximum frequency of the modulating signal. If it were music, as you will use later, it could be as much as 2 × 20 kHz, or 40 kHz.

Once you have a sharp spectrum, change the modulation index for all of the same values of h that you used in Step 3 and see how the spectrum changes. Notice that as depth increases, the sidebands increase in height as predicted by the modulation equations given in Eq. 6.13. Discuss in your post-lab report.

6. **Powering and checking your AM radio.** In order to use your AM radio today, you will power it with the benchtop power supply. Use the connectors with clips on the end to attach to the battery terminals. Set the voltage to 9 V. Be sure to put positive to positive, etc.

Before you test your very own radio station, make sure your radio is working properly by tuning in a local, strong station. Somehow you will need to identify that station and its carrier frequency. Having a commercial AM radio handy will be helpful. How does the label on the tuning dial of your radio match that carrier frequency? Can you hear any other radio stations? (At some point, you might use your radio late at night in an open location and see how many stations you can receive. Welcome to the world of "DXing," although this radio is not sensitive enough to do serious DXing.)

7. **Setting up your broadcast antenna.** Now that you've generated a modulated signal, you're going to broadcast a pure sine wave to your AM radio over your carrier. A pure sine wave sounds like a featureless tone, much like a flute without any vibrato, or a plain whistle.

Turn off the generator. Attach a BNC tee connection at the generator output. At the tee, add a BNC cable going to an adapter and a black flexible antenna at the end

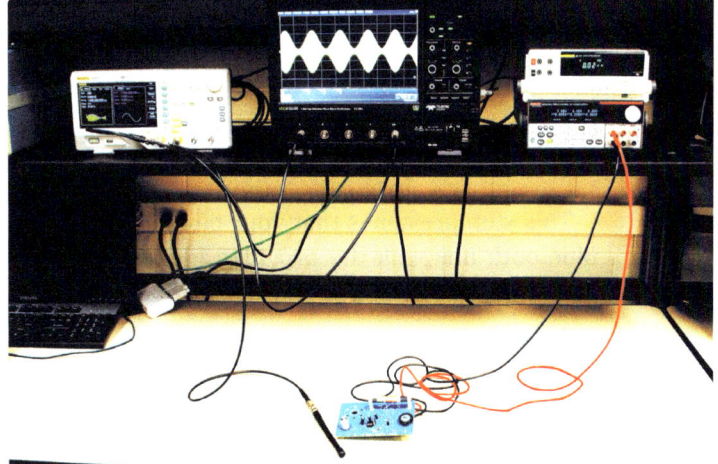

FIGURE 6.23 (Step 7) Setup for your AM radio broadcast station.

of it. Turn on the generator and reset the frequency to 800 kHz with the same voltages as before. Set the modulation frequency to 1 kHz. Choose a suitably high value of *h*. Your setup should look now like Fig. 6.23.

8. **Tuning the radio and hearing changes in depth.** Now you will tune in to your radio station to hear the sine wave you are broadcasting. It is important to find the 'sweet spot' for the distance from the antenna to your radio. If it is too close, you will hear the tone over the entire tuning range; if it is too far, you will not hear the tone at all. You need to place the antenna at a distance such that only one spot on the dial allows you to hear it. That spot will be correctly tuned to the carrier frequency. That distance may be around 1 to 2 feet, but will depend heavily on the setup of your antenna.

Turn your tuning capacitor until you hear a strong 1 kHz tone with the loudness (volume) up, and then turn the volume down in order to preserve your sanity and that of your fellow students. Change the modulation frequency (e.g. from 800 Hz to 1.2 kHz) and retune the radio to prove that you are listening to *your* radio station. Does the tone sound clear?

Change the modulation index, or depth, while you are listening to the tone. Does the volume of the tone change with *h*? Can you see why the broadcaster would like to use the largest legal modulation index? Can you also see that *h* = 0 is silence? Discuss what you observed.

Also, listen to your tone as you increase *h* from less than 100% to greater than 100%. Does the tone change? Would you call that "distortion?" That is one reason that broadcasters do not overmodulate their signals.

9. **Testing your radio's receiving frequencies.** Next, you will determine the end limits of your reception frequencies. Modern AM radios in the United States can tune from 530 kHz to 1700 kHz, but without additional adjustments, your radio will likely tune over a very different range. That is what you want to determine.

Set *h* = 70%. Turn your dial to the lowest frequency. Adjust the carrier to obtain the maximum volume of the tone to determine the actual frequency at that end of the dial setting. You won't find a precise setting of the carrier frequency that plays the tone loudest, but do your best to estimate where that point is. That is the lowest radio frequency that you can tune to. You might find that this is above or below the range of commercial AM broadcasts in the United States.

Now, set your dial to the highest frequency and repeat what you did for low frequency. What are the limits to your frequency reception? How do they compare with AM radio broadcasts in the United States?

10. **Listening to music and seeing its waveform and spectrum.** Now, instead of a boring sine wave, you're going to modulate using an actual music signal from your phone/media, or other (provided) device. However, first you're going to see what a music waveform generated by your phone looks like without any modulation.

This step presumes that your phone or music player has an audio/music jack. If it has only a Bluetooth link, link the provided Bluetooth adapter to your phone and follow the instructions for the audio output from the adapter.

Take the output of your phone/MP3 player/media/Bluetooth device using the following sequence of connectors. The purpose of this setup is to amplify the audio from the player while simultaneously allowing two people to listen using earbuds or a headset. To summarize Fig. 6.24, the music source splits to up to two sets of earbuds for listening and an amplifier whose output drives the oscilloscope and the modulator input on the waveform generator.

a. Media player out using 1/8 inch stereo plug with long splitter.

b. Into one side of the splitter plug in the second, short splitter.

c. Into the other side of the long splitter, plug in the long 1/8 inch stereo extension cable.

d. The extension cable goes to the input of the amplifier and later to the waveform generator at the modulation input. The **GND** pin on the amplifier goes to the bottom right pin on the audio-jack board, labeled **COM**, at the end of the rows closest to the input jack. For mono (not stereo), one of the amplifier signal input pins, **IN1** or **IN2**, connects to one of the two middle pins on either side that have **R** or **L** in between them.

e. The amplifier is powered by an AC adaptor. (From Wikipedia: Other common names include plug pack, plug-in adapter, adapter block, domestic mains adapter, line power adapter, wall wart, power brick, and power adapter. ☺)

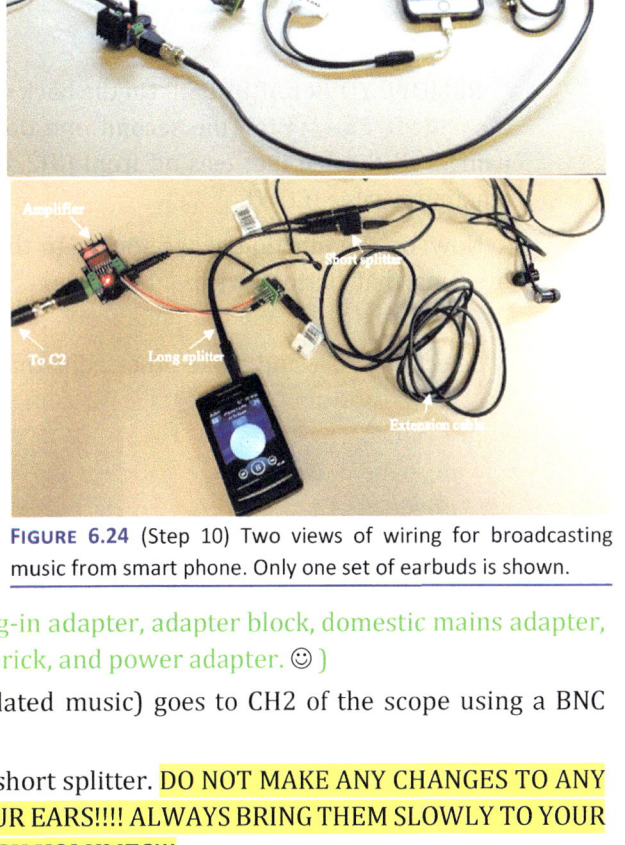

FIGURE 6.24 (Step 10) Two views of wiring for broadcasting music from smart phone. Only one set of earbuds is shown.

f. The output of the amplifier (unmodulated music) goes to CH2 of the scope using a BNC cable, as shown.

g. Two sets of earbuds can plug into the short splitter. DO NOT MAKE ANY CHANGES TO ANY SETTINGS WITH THE EARBUDS IN YOUR EARS!!!! ALWAYS BRING THEM SLOWLY TO YOUR HEAD WHILE BEING CAUTIOUS OF HIGH VOLUMES!!!

Now you can see the music on the scope (output of the amplifier, CH2) and listen to it at the same time. You may have to adjust the gain of the amplifier (volume control) or the volume of your music source.

Note: At this point, turn off your modulated output and your bench light. Both can cause noise on the cables.

Turn the volume down on your music player. Pick a song. Turn the amplifier gain all the way CCW (maximum gain). Set the FSA source to CH2. Set the limits of the FSA to range from 0 Hz to 30 kHz. View CH2 and its spectrum. If necessary, increase the volume out of your player. ***Don't listen if it is too loud for your ears.*** You should be able to hear and see the waveform change and relate

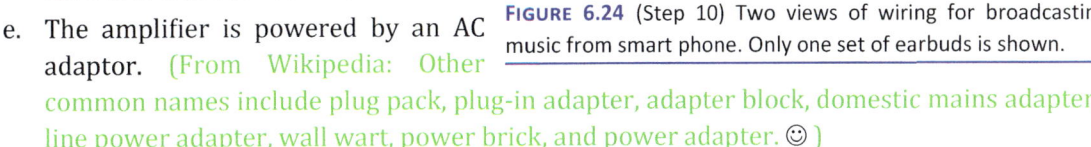

FIGURE 6.25 (Step 10) Example of waveform of music and associated frequency spectrum. In this example, the music spectrum drops off around 20 kHz.

the changes to the music. As the music changes, the spectrum should also change. Look at the waveform in the time and frequency domains, as shown in Fig. 6.25. Describe what you see. Save waveforms using the **Stop** button to freeze images.

11. **Viewing modulated music.** Now you will use this audio signal to modulate the carrier wave on the function generator. This will be working like an actual AM radio station that is broadcasting music.

    REMOVE YOUR EARBUDS!!! On the back of the function generator, there is a BNC input labeled **Ch.1 MOD/FSK/TRIG** (the second one down). Plug the audio amplifier output into this port using a BNC cable that tees off from CH2, and make sure the gain on the amplifier is set to its maximum value (CCW).

    Now, on the function generator, go to the **MOD** menu, select **Source** and change it to **Ext** by pressing that button twice. This modulates the carrier signal using the audio signal you just plugged into the back of the generator.

    Select $h = 70\%$ and make other adjustments using the volume control on your music source. Now you are modulating the carrier frequency on the waveform generator using the audio signal. The output of the waveform generator should still be connected to the scope as before (CH1). Turn off the FSA to save space on the screen and look at the time domain waveform of the modulated signal. Describe what you see. View both CH1 and CH2 at the same time to see the unmodulated (CH2) and modulated (CH1) signals simultaneously. Again, press **Stop** to get a good image of the waveform.

    NOTE: REMOVE YOUR EARBUDS FROM YOUR EARS!!! YOU MAY HAVE TO INCREASE THE VOLUME OF YOUR MEDIA DEVICE TO SEE GOOD MODULATION.

12. **Changing the modulation.** Change the amplitude of the modulation, m, by changing the volume control of the music player. What happens?

    To prove to yourself that the modulated signal is the same as the unmodulated one, change the vertical scale in fine amounts by pressing the **Vertical** button and make small changes to both CH1 and CH2 until they are

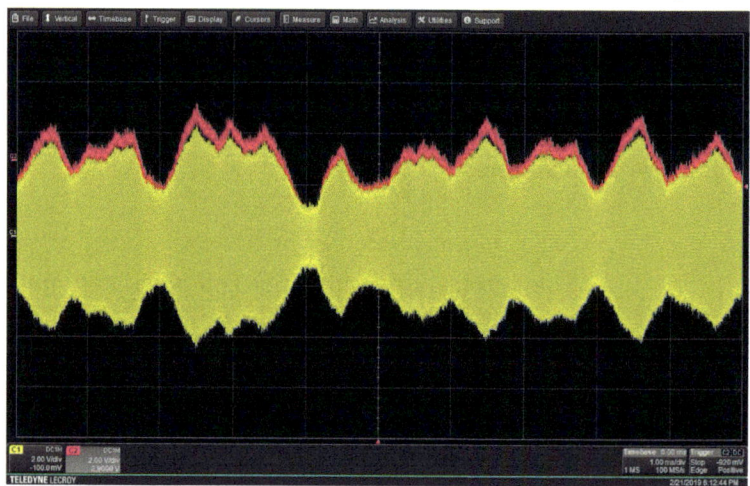

FIGURE 6.26 (Step 12) Audio signal both unmodulated and modulated. Their traces are overlapped to show that the AM signal is a modulated replica of the original waveform.

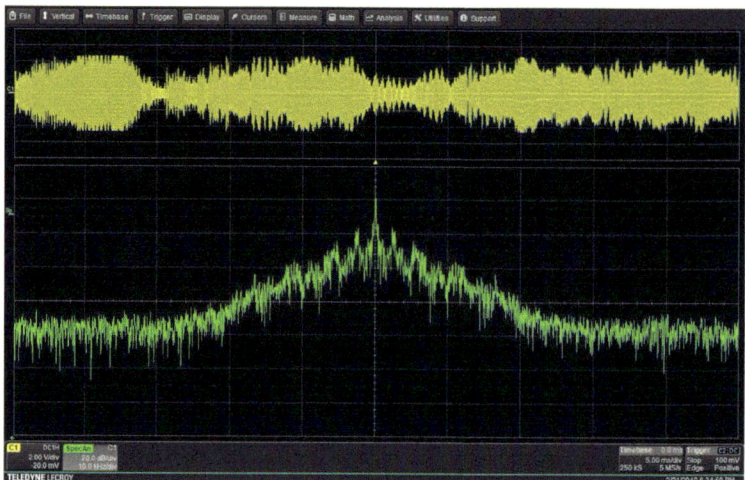

FIGURE 6.27 (Step 13) Modulated audio signal in both time and frequency domains. Notice that the spectrum is symmetric around the carrier frequency of 1 MHz.

very nearly the same amplitude. Then change the offsets so that the unmodulated wave overlaps the edge of the modulated wave, as shown in Fig. 6.26. They should be nearly identical. This demonstrates that the envelope of the modulated carrier is the music.

13. **Observing sidebands due to music.** Set the carrier frequency to 1 MHz. Look at the frequency spectrum of the modulated music in logarithmic mode. Use a bandwidth of 80 kHz, which is 40 kHz on each side. About half of the spectrum will be music and the rest just background noise. Use *h* of 100% in order to really accentuate the sidebands. See Fig. 6.27. Can you see the sidebands on each side of the carrier? What frequency do they extend out to on each side? Are they symmetric as they should be? Do you understand that those sidebands are due to the spectrum of the music that is being broadcast by AM radio?

14. **Changing *h* with music.** Decrease *h* in increments. What happens to the time- and frequency-domain traces? Now change *h* to zero. What happens to the sidebands? Why?

15. **Broadcasting through the air.** Finally, you are ready to bring your radio station "on the air" by broadcasting your music at the carrier frequency using an antenna as you did in Step 7. Turn on the output of the generator so that the antenna is now broadcasting through the air. Tune your radio to listen to the music on your radio the same way you did with the sine wave. You are now an AM broadcast radio station. How is the quality? Does it sound as clear as the commercial AM radio station that you can receive? (It could sound much better, actually, since reception inside the building may not be good at all!)

16. **Move your radio away from your antenna.** Are you violating FCC regulations as provided below?

    Unlicensed operation on the AM and FM radio broadcast bands is permitted for some extremely low-powered devices covered under Part 15 of the FCC's rules. On FM frequencies, these devices are limited to an effective service range of approximately 200 feet (61 meters). See 47 CFR (Code of Federal Regulations) Section 15.239, and the *July 24, 1991 Public Notice*. On the AM broadcast band, these devices are limited to an effective service range of approximately 200 feet (61 meters). See 47 CFR Sections 15.207, 15.209, 15.219, and 15.221. These devices must accept any interference caused by any other operation, which may further limit the effective service range.

    If you are violating FCC regulations, arrest yourself and report to the nearest police station. Or as an alternative, turn down the output voltage to the antenna so that you don't broadcast as far. Your choice. (But, it probably won't broadcast more than a couple of feet or so, so no worries.)

17. **Observing the received modulated music.** You will now see on the scope what your AM radio is actually picking up over the air from your radio station. Your radio should be turned off for this step (only).

    TURN OFF THE POWER TO THE RADIO. IN THE FOLLOWING, BE CAREFUL NOT TO SHORT CIRCUIT YOUR IC. LOOK CAREFULLY TO BE SURE THAT THE PROBE HOOK IS NOT TOUCHING TWO WIRES.

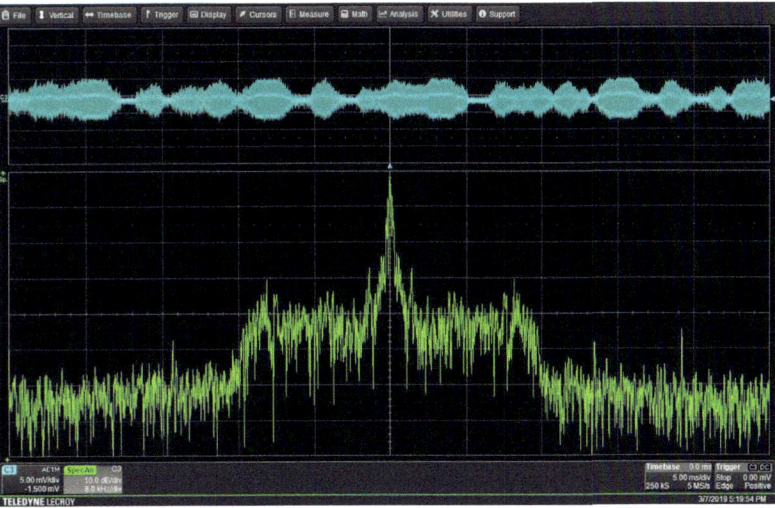

FIGURE 6.28 (Step 17) Waveform and frequency spectrum of modulated music at the antenna of the AM radio.

For this step, set the trigger to **Normal**, rather than external sync from the function generator, as you did in Step 2. Attach a 10× scope probe on CH4 to the input of the three-terminal IC1 on the AM radio. Why 10×? Because the 10 MΩ of the 10× probe will load down a very weak signal less than the 1 MΩ of the 1× probe will. If you are looking at your radio with the volume knob in the upper right-hand corner, the input pin is the leftmost pin on IC1. Put the probe ground to the negative terminal of the battery. Make sure you are not shorting out any two wires.

The power to the radio is still turned off. Broadcast your music and move the antenna close to the radio. Turn on CH4 to see the radio signal and also CH1 to see what you are broadcasting. Set your tuner dial to the same position that you found before for your 800 kHz carrier with the same settings that you used in Step 11. You may need to fine-tune it to get a strong signal. Set your time base to 5 ms/div to see the modulated music. You should see the modulated signal being picked up by the antenna and tuned by the tank circuit, and it should look much like the broadcast signal. View the signal in both time and frequency domains, like that shown in Fig. 6.28. Describe it in terms of what you saw in Step 11.

18. **Observing the received demodulated music.** Disconnect the probe and move it to the demodulated signal from the output of IC1 on the radio. The output is the topmost pin on IC1. Now turn on the power to the radio, turn on the radio and tune it to your radio station. The demodulated output is the envelope of the actual music that is amplified by the audio amplifier (op-amp) and drives the speaker. You may have to invert the signal to get it to match the original music.

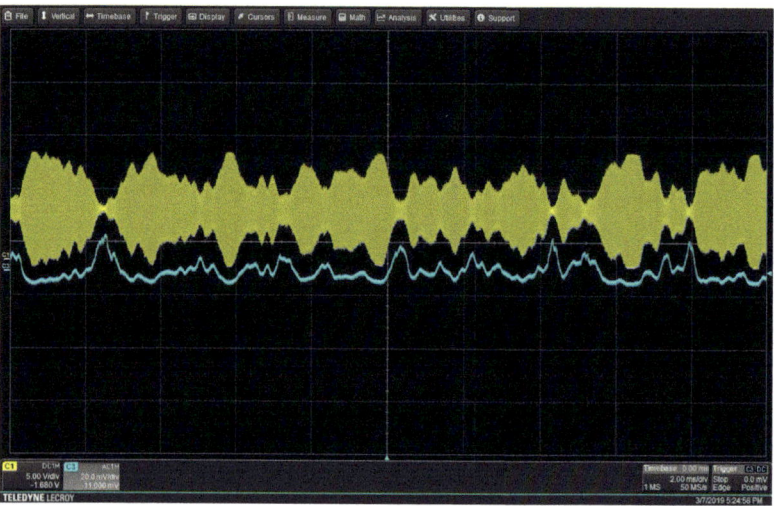

FIGURE 6.29 (Step 18) Waveform of demodulated music at the output of the IC1 chip on the AM radio. Superimposed on it is the AM signal from the waveform generator.

Describe this signal in the time domain. Compare to the previous version of the unmodulated music in Step 10, which should still be on CH2. Does it look pretty much the same? Compare what you see to Fig. 6.29.

19. **Observing frequency-modulated sine wave.** Now we will very briefly introduce frequency modulation (FM). In FM, the carrier *frequency* rather than its amplitude changes with the music waveform. You don't have an FM radio to play with, so you will just look at what the modulated signal looks like. You will go back to using a sine wave, or **tone**, to modulate the FM signal.

Keep the carrier frequency at 800 kHz. Using the **MOD** menu, reset the modulation source to **Int**. Change **Type** to FM. Change **FM Freq** to 1 Hz (yes, that is one Hz). Press **Deviation** and set **FMDev** to 35 kHz. Looking again at CH1 (the modulated signal), can you see that the frequency is changing at the rate of once per second? The sine wave is 'breathing.' By that, I mean that it is expanding and contracting horizontally, which is actually the frequency changing with the same frequency as the sine wave, which is once per second. That is frequency modulation.

20. **Exploring FM.** Play around with FM any way you want. Put in your lab report what you did. Also in your lab report, be sure to summarize what you did and comment on what you observed. What did you do to play around? Did everything work as you thought it should, or were there surprises and/or difficulties?

Congratulations on a job well done! You have built your own radio (Lab 1) and your own broadcast radio station. You have learned about modulation and now have a basic understanding of how radio waves travel through space, interact with an antenna, are received by a radio, and are converted to sound. Good work!

Problems

1. Write out the correct Maxwell's equation. Draw a picture that demonstrates each one.

   (a) Which of Maxwell's equations states that the total electric flux through a closed surface is proportional to the charge enclosed within the corresponding volume?

   (b) Which of Maxwell's equations states that the total magnetic flux through a closed surface is zero, implying the absence of magnetic monopoles?

   (c) Which of Maxwell's equations states that a time-varying magnetic field induces an electric field?

   (d) Which of Maxwell's equations states that a magnetic field is generated by electric currents and time-varying electric fields? Draw a diagram illustrating this concept, including a wire carrying current and a capacitor with a changing electric field.

2. Using the units of volts (V), coulombs (C), and meters (m), show that the Maxwell equation $\nabla \bullet \mathbf{E} = \dfrac{\rho}{\varepsilon_0}$ is consistent in its units. Note that the del operator is a spatial derivative, so its units are m^{-1}; the following quantities can be expressed using the following units:

$$\mathbf{E} \Rightarrow \frac{\mathrm{V}}{\mathrm{m}}, \quad \nabla \bullet \mathbf{E} \Rightarrow \frac{\mathrm{V}}{\mathrm{m}^2}.$$

3. The Sun's photosphere is the layer from which sunlight is emitted. Notably, the spectrum of sunlight peaks in the green region of the visible spectrum. Referring to Fig. 6.8, estimate the approximate temperature of the photosphere. First, make your best estimate based on the peak wavelength, then verify the actual value using an internet search.

4. (a) Use Fig. 6.8 or 6.9 and information in the chapter to approximate the wavelength in meters pertaining to the following applications of electromagnetic radiation.

   (b) To look at the internal health of your teeth at a dentist's office.

   (c) To observe the external structure of a plant cell in a microscope.

   (d) To listen to music on an FM radio in your car.

   (e) To listen to news on an AM radio in your car.

   (f) To detect supernovas using extremely high-energy electromagnetic waves.

5. A half-wave dipole antenna has a length equal to half the wavelength of the signal it is designed to receive. If we were to use a half-wave antenna to communicate with a submarine at the extremely low frequency (ELF) of 76 Hz, how long would the antenna need to be in miles, assuming a free-space wavelength? (Note: Real ELF communication antennas used until 2004 were over 10 miles long and operated at extremely low data rates.)

6. What would a listener hear from an AM-radio broadcast if the modulation index is 0%? Pick one answer and explain.

   (a) silence, (b) a regular broadcast with no distortion, (c) a regular broadcast with some distortion.

7. Questions about modulation: State whether the following are true or false.

   (a) An AM radio signal received by the antenna is converted into an audio frequency in the radio through a process called "modulation."

   (b) Radio waves are composed of either electric or magnetic fields, but not both.

   (c) Maxwell's equations are needed to describe how sidebands appear in the frequency spectrum of an AM radio signal.

   (d) For commercial broadcast AM radio, the baseband bandwidth (music, for example) might reasonably be, for example, 10 kHz with a carrier frequency of 40 kHz.

   (e) In frequency modulation, the baseband signal (music, for example) changes the frequency of the carrier.

   (f) Frequency modulation creates an "envelope" on the carrier that resembles the baseband signal.

   (g) If $h = 0\%$ then the AM sidebands would be absent from the frequency spectrum.

   (h) $h > 100\%$ results in overmodulation.

   (i) $h = 100\%$ is the case for broadcasting total silence.

   (j) The two AM sidebands for music broadcasting have the same bandwidth as each other.

8. (a) Briefly describe (~1 sentence each) the parts that make up a radio link.

   (b) Briefly describe how each of the above parts is represented in the lab.

9. (a) Using your preferred graphing software (a free one you can use is https://www.desmos. com/) make a plot of an AM signal with a 900 kHz carrier wave, a 20 kHz baseband sine wave, and a modulation index of 50%. Assume that $C = 1$. Write down the equation you plotted. Repeat with modulation indices of (b) 100%, then (c) 150%. (d) For $h = 100\%$, in the AM spectrum, what is the amplitude of the sideband relative to the carrier? (e) For $h = 100\%$, sketch the frequency spectrum. Show correct frequencies and relative amplitudes.

10. Consider a modulated signal made up of a 1 Vp, 1 MHz carrier wave with a 0.25 Vp, 1 kHz modulating sine wave. (a) Sketch its representation in the frequency domain and show the calculated frequencies and amplitudes of the spectral components. (b) What is h for this signal?

11. (a) Draw and describe the properties of a parallel LC circuit at resonance as discussed in Fig. 6.15, and compare to the series LC circuit that you studied in Lab 5. Discuss in terms of currents and voltages, resonant frequencies, and impedances.

(b) Describe how tuning is done with an LC tank circuit, as shown in Fig. 6.15. This question is not about the diode and earpiece – just the coil and variable capacitor. Explain what happens to the radio stations that are far from the resonant frequency and at the resonant frequency.

12. Suppose you have built your own crystal radio, as shown in Fig. 6.15. (a) If the inductor has a value of 4 mH, what value of capacitance do you need to tune to Chicago's WGN radio, AM 720 kHz? (b) What device discussed in Lab 1 would you use to do that?

13. In FM radio transmission:

(a) An increase in the amplitude of the baseband audio (listening volume) is converted into a change in what property of the transmitted signal?

(b) What about the frequency of the baseband audio, such as the notes from a tuba or piccolo? How is this represented in the transmitted signal?

Chapter 7

Electronic Materials, Semiconductors, and Integrated Circuits

7.1   Introduction and Overview

In this chapter, you will learn:

1. The relationship between materials and electronic devices
2. How charges move through materials
3. What are conductors, insulators, and semiconductors
4. What a diode is and why it is important
5. What a transistor is and how it is like both a light switch and a light dimmer
6. What the basic types of transistors are
7. What an amplifier is
8. What an operational amplifier is
9. What an integrated circuit is and how it is made

7.2   Electronic Materials

At the dawn of civilization, humans recognized that if they hacked away at trees they could shape them into useful things, such as sticks for huts. Likewise, it was discovered that mud could be dried in forms to make bricks. Eventually, new materials were developed, such as concrete (early 1800s), that were much stronger than the dried bricks, as they were engineered to have certain properties that were better than those that occur naturally. Electronic devices have undergone the same sort of evolution, but instead of over millennia it has happened over the past 100 years, and much more so over the past 60 years. And, it is still advancing at a rapid pace.

As an example of a simple electronic material, metals were studied for hundreds of years for their ability to

> *Deeper understanding.* In contrast to insulators, metals, by definition, are highly conductive. They get their distinctive properties from their electronic bonding structure, called 'metallic bonds,' which is that each metal atom contributes several electrons to a 'sea' of electrons. This sea of electrons all work together with the positively charged nuclei to hold the metal together in the solid form. Due to having so many 'mobile' electrons available that can conduct current, metals typically have low resistivity (high conductivity). Also, they are shiny, since the reflection of light depends on the interaction of the electromagnetic light wave with electrons that can respond and re-radiate the light. Also, they tend to conduct heat well for the same reason that they conduct electricity – the large population of mobile electrons can carry heat. So, the properties of metals depend directly on the availability of electrons that can move through the material.

An Introduction to Electrical Engineering with Lab Activities
Gary H. Bernstein
Copyright © 2026 Jenny Stanford Publishing Pte. Ltd.
ISBN 978-981-5129-30-4 (Hardcover), 978-1-003-71345-6 (eBook)
www.jennystanford.com
DOI: 10.1201/9781003713456-7

carry electrical current, and it was found that silver carried the most current for the same voltage and wire shape. The intensive property of a material to carry current is called its **conductivity**. The inverse of conductivity is **resistivity**, which is the intensive property of material to resist the flow of current. As you might expect, metals generally have high conductivity, and are called **conductors**. Many materials, such as rubber and glass, have very low conductivity, i.e., high resistivity, and are called **insulators**.

Since silver is expensive, copper, which is only slightly less conductive, is the go-to material for electrical wiring. It doesn't take a lot of fancy technology to make (elemental) copper wire (Figure 7.1) – just mine and extract the copper, refine it, melt it

FIGURE 7.1 Copper wire on a spool.

into ingots, and draw it into wires. (We'll ignore the outer insulation, which has undergone significant improvements over hundreds of years.) Copper wire is an example, then, of an electronic material that is useful basically as nature gave it to us. In this module, we will discuss highly engineered devices that take advantage of the unique electronic properties of a class of materials called **semiconductors**.

7.3   Electrical Properties of Materials

Metals conduct current by the flow of (negatively charged) electrons. You know from your chemistry courses that atoms have a nucleus surrounded by electrons, with the electrons residing at various energy levels from the **core shells** out to the **outer shells**, or **valence electrons**. Atoms in a metal are arranged such that the outer-shell, i.e., valence, electrons are free to move off of their "parent" atom and wander away to visit other atoms. This results in what is often called a "sea" of *conduction* electrons in the metal (see Figure 7.2). None of the electrons in the inner, filled, shells can move off of their parent atoms, and so do not participate in electrical conduction.

The resistivity and conductivity of a material depend on both how many electrons are available to move, i.e., the (conduction) **electron concentration**, and how easily they move in the material, a property called **mobility**. The electrical resistance of a chunk of material, which could be a simple carbon resistor for example (see Fig. 7.3), depends on its material characteristics, and is:

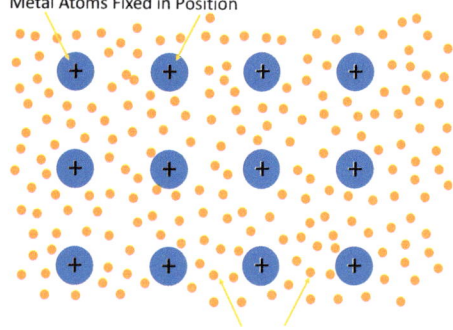

Metal Atoms Fixed in Position

"Sea" of Mobile Conduction Electrons

FIGURE 7.2 'Sea' of conduction electrons in a metal.

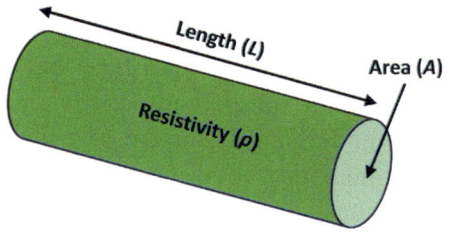

$$\text{Resistance}\,(R) = \rho \frac{L}{A}$$

FIGURE 7.3 Resistor and resistance equation.

$$R = \frac{\rho L}{A} = \frac{L}{\sigma A}, \qquad (7.1)$$

where

$$\sigma = \frac{1}{\rho}. \qquad (7.2)$$

R is the total resistance of the piece of material in units of ohms, ρ (Greek "rho") is the resistivity of the material, typically in units of Ω-cm, L and A are the length and cross-sectional area of a piece of the material, respectively, and σ is the conductivity of the material in units of siemens/cm, S/cm or $(\Omega\text{-cm})^{-1}$. The equation says that for a given type of material, a longer piece has more resistance, and one with a larger cross-sectional area has less resistance. The value ρ tells us that the resistance of a device depends on what kind of material it is made of. The higher the resistivity of the material, the higher the resistance.

A water analogy for a resistor is that of a pipe full of gravel (sort of like Fig. 7.4). The aspect ratio of the pipe is given by L/A, and the amount of gravel in it restricts the water flow and represents its resistivity, ρ. Although most electronic devices are more complicated than resistors, it is generally true that the longer the device, the more resistive it is, and the thicker it is, the less resistive it is. The inverse dependence on area is a simple matter of more parallel paths for **charge carriers** to take to pass current for a given voltage drop across the device. (We use the term "charge carrier" because later we will see that it isn't only electrons that carry current in electronic devices). The decreasing current for longer devices is due to the following two facts. For a given voltage, V, the electric field, E, decreases for some voltage dropped over some distance:

FIGURE 7.4 Gravel in a pipe acts like a "resistor" to the flow of water.

$$E = \frac{V}{L}; \ (V = E \cdot L). \tag{7.3}$$

Also, the force on an electron in an electric field is:

$$F = e \cdot E, \tag{7.4}$$

where e is the charge of an electron ($e = 1.6 \times 10^{-19}$ C). So, for a given voltage drop, say across a resistor, the electric field E decreases as length L increases; lower electric field doesn't push the charge carriers as strongly, so they don't move as quickly, so there is less current. Keeping those relationships in mind, it is possible to intuit that $R = \frac{\rho L}{A}$. In summary, we see that we can use materials that occur naturally or are simple to manufacture, like metals, carbon, plastic, glass, rubber, and others, to make useful electronic devices, namely wires, resistors and insulators.

Well, electronics would be pretty limited with just simple conductors and insulators. Just like the advent of concrete, engineers have sought to take advantage of the widest variety of both naturally occurring materials and engineered materials. Classes of materials include conductors (like metals) and insulators (like glass, rubber and plastics); **semiconductors** (like silicon, which is one topic in this module); **ferromagnetic** materials (like iron and many iron alloys) along with other classes of **magnetic** materials; and **piezoelectric** materials, such as lead zirconate titanate (PZT). Piezoelectric materials are very useful in that they change their shape with applied voltage or, inversely, produce a voltage if they are squeezed or vibrate. They are used in, for example, acoustic guitar pickups (Fig. 7.5), inkjet printers, ultrasonic imagers such as echocardiographs or other ultrasound machines, and many other applications.

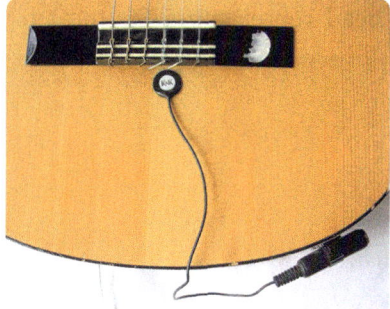

FIGURE 7.5 Some acoustic guitar pickups use piezoelectric materials. *Source*: https://commons.wikimedia.org/wiki/File:Piezoelectric_pickup1.jpg.

Most of these things don't just happen naturally, sitting in the ground waiting to be plucked out. Most materials used in electronics must be engineered, refined, and processed in complicated ways to have special electrical properties. Although that activity can be called "materials science," electrical engineers building electronic devices must also be familiar with this area.

7.4 Semiconductors

Here we focus on the class of materials called **semiconductors** because it forms the basis for that most interesting and important of devices called the **transistor**, which has reshaped human existence (more on transistors later). Semiconductor atoms are arranged in a regular **lattice** that forms a **crystal**, like that shown in Fig. 7.6. All of their electrical properties depend on the details of the regular lattice of atoms in the crystal structure. The importance of the crystalline properties of many classes of materials cannot be overstated. It is fundamental to electronic devices.

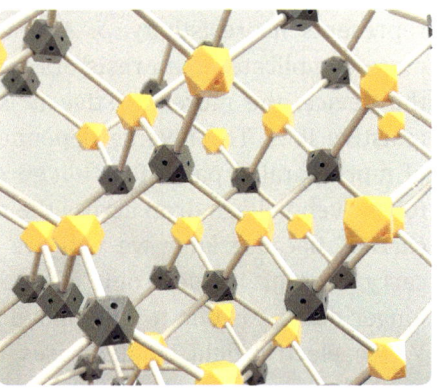

FIGURE 7.6 Model of a semiconductor crystal in which atoms are spaced at regular intervals. The sticks are covalent bonds, and the polyhedral beads are atoms. This crystal is the zinc blende structure showing two types of atoms, e.g., gallium and arsenic, bonded together in a regular pattern forming gallium arsenide. When every atom is silicon, this model represents the silicon crystal lattice.

Many electronic functions that are performed by semiconductors today were done previously with **vacuum tubes**, shown in Fig. 7.7, which comprised electrodes in glass bulbs that got really hot and used a lot of power. Inside the vacuum tube was just that – a vacuum, so that the electrons could fly through space without hitting air molecules and scattering, and the hot electrodes would not burn up. Because electrons flow through semiconductor materials in modern devices, and not a vacuum, the field of semiconductor-based electronics is called **solid-state electronics**.

You will most often see a semiconductor defined as "a material that has a conductivity somewhere between a good conductor, like copper, and an insulator, like glass." Even though that is not wrong, it is about the *least* useful thing you can say about semiconductors. So, what is the *most* useful thing about semiconductors? *It is that inside of a semiconductor, current can be carried by both negative charge carriers* (i.e., *electrons*) *and by positive charge carriers, which in semiconductors are called holes.*

Silicon is the most common and important semiconductor material (and there are dozens of them), so let's use it as an example. Si is a Group IV element on the periodic table, which means that the outer shell of Si has four valence electrons. Figure 7.6 shows that each Si atom (when every bead represents a Si atom) is bonded to four other Si atoms. In the crystalline state at very low (cryogenic) temperatures, all four valence electrons of an atom participate in bonding with its four nearest neighbors. Since each bond has a contribution from two atoms, this effectively totals 8 electrons in the filled outer shell of every Si atom. Because (at such low temperatures) the outer shell is full, i.e., there are no missing bonds for an electron to move over to, none of the valence electrons is free to move, and therefore cannot carry current. It is like a checkerboard where every space has a checker on it; there is no room to move the checkers across the board.

FIGURE 7.7 Vacuum tubes must be very hot in order to operate. They may use an excessive amount of power, are relatively slow, and are unreliable, but their glow is pretty in the dark.

Based on this, Si should be a perfect insulator, which it is when it is very cold, but generally it is not, for reasons described next. You will see that temperature plays a huge role in how semiconductors behave.

7.4.1 Energy Bands

Electrons are subject to the rules of atomic-sized things, generally called **quantum mechanics**. Quantum mechanics is a mathematically challenging topic that is covered in upper-level undergraduate and early graduate-level courses, so we will not go into it much here. I will tell you, though, that understanding your Physics II course and then your electromagnetics fields and waves course will give you a strong background to understanding how electrons also behave as waves. You will find that the interference of the electron waves as they move through the crystal, much like radio waves in an antenna or light waves reflecting from a soap bubble, gives rise to all of the important electronic behavior in solid-state devices.

Electrons in the outer shells of the atoms, here pure Si with no other types of impurity atoms, are collected into two energy bands, shown in the energy-band diagram of Fig. 7.8. **Energy-band diagrams** (or just 'band diagrams') plot energy vertically and position horizontally. The energy of the electrons increases with vertical height on energy-band diagrams. Figure

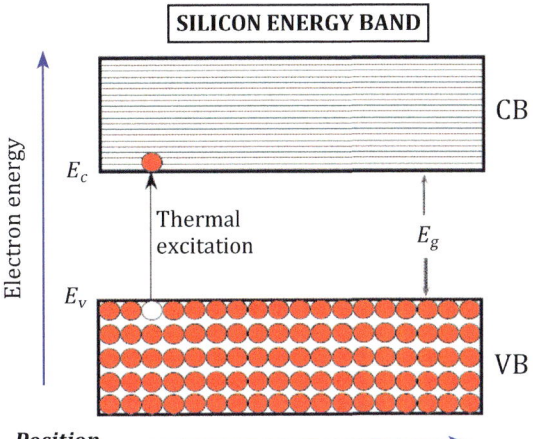

FIGURE 7.8 Band diagram showing valence band and conduction band in silicon. The bandgap energy is denoted as E_g. The diagram shows thermal energy creating an electron-hole pair. The vertical axis is energy and the horizontal axis is position. For this picture specifically, the free electron (in the conduction band, CB) is co-located with the hole in the valence band, VB, that it has left behind. It hasn't yet moved away from it. *Source*: https:// physics.stackexchange.com/.

7.8 shows two bands – the **conduction band** above and the **valence band** below. Electrons that are free to move, i.e., mobile electrons, are energetically in the conduction band, and the electrons that hold the crystal together, the **bonding electrons**, or **outer-shell electrons**, or **valence electrons**, are energetically in the valence band. As Fig. 7.8 shows, the usual case is that the valence band is almost completely filled with valence electrons, and there are very few conduction electrons in the conduction band. Much more will be said about this.

What does energy plotted vertically mean? Pretty much the same thing as if it were the potential energy of books placed on a bookshelf – the ones closer to the bottom shelf have the least energy and ones stacked up higher have more energy. For the books, the energy is potential energy, and the book has moved up on the shelf to gain that potential energy. In the energy bands, the energy above the top of the valence band and at the bottom of the conduction band is potential energy, and energy above the bottom of the conduction band is kinetic energy. Electrons at the bottom of the conduction have barely broken free of the bonds and are free to *start* moving in an electric field. The electrons higher up in the conduction band are moving more quickly than those toward the bottom of the band. One more thing to keep in mind: the horizontal axis is position, so in Fig. 7.8, the electron in the conduction band is shown physically at the same place as where it was in the valence band just as it broke free from its bond and has not yet moved away. The energy of the bandgap is the energy that it needed to break out of the crystal lattice and become free to move, which it will do in an electric field.

The valence electrons that participate in crystal bonding, as in Fig. 7.6, are in the valence band. They literally hold the crystal together. Without those, the Si crystal would literally evaporate into a gas. For very cold temperatures in pure Si, all of the bonding electrons are sitting at the lowest energy they can. As an analogy of inner shells of an atom, i.e., those with energy less than the valence electrons, think of a lower bookshelf that is totally crammed with books – there is no room to fit any more books, and you can't move any of them around without removing any first.

Similarly, electrons in a filled valence band cannot conduct current. Every way in which an electron can exist in the valence band, called an **electronic state**, is in use, or **occupied**, by an electron. According to quantum mechanics, no two electrons can occupy the same state. This is called the **Pauli Exclusion Principle**, and it is the reason that you cannot pass your hand through your desk – the electrons cannot occupy the same state as those in the desk. In a very rough sense that implies that none of them can move through the crystal because if an electron wants to move to another location, it would have to be in a state at another location that is already occupied, and that isn't allowed. Later we will talk about electrons that are absent from the valence band, leaving unoccupied states, which *does* allow current to flow, just as taking a few books off of the lower shelf allows the others to be moved around. (This picture of valence electrons will be refined and expanded in a real semiconductors course.)

Heat energy, or **thermal energy**, makes it possible for devices to function; in a solid, heat energy is mostly in the form of atomic vibrations, what I like to call 'jiggling.' When atoms jiggle, so do their electrons. At sufficiently high temperature, just by pure chance a very few of the valence electrons will break free of their parent atoms, as shown in Fig. 7.8. We say that the valence electron is **excited**. (As you should also be to be learning about this extremely important area of science and engineering!) Mobile electrons, regardless of how they were created, are referred to as **free electrons**, or **conduction electrons**. Free electrons can move through the crystal and are said to be "in the **conduction band**." Literally, this means that the valence electrons have broken away from their parent atoms and can now wander around inside the crystal. That same idea, but stated in terms of energy bands, is that they move to higher energy in the conduction band and can now conduct current.

The energy needed for an electron to just barely escape from the bonding state in the valence band to the conduction state, and hence carry current, in the conduction band is called the **bandgap energy**. The bandgap energy is one of the most important parameters for all types of semiconductor materials. Very analogously, if you were to leap out of a deep trench in the ground and land outside the trench, the depth of the trench would be the bandgap for you. The same is true of an electron – it needs a certain energy to just barely break out of the atom and become free to wander throughout the material. If the electron is then actually moving, its kinetic energy is represented by its vertical position on the conduction band diagram. The 'wandering' of the electron is its horizontal position on the conduction band diagram.

Don't take Fig. 7.8 too literally – there aren't some kind of actual bands separated by a band gap. You will not see a band diagram, no matter how powerful a microscope you use to look at the atoms. Bands are *models* for the energies of electrons at various places in the crystal. It is convenient to use the band model in order to make predictions about the behavior of electronic devices. Figure 7.8 is a *simple* band diagram. Band diagrams of real devices are more complicated than that of Fig. 7.8, and do an amazingly good job of predicting real device behavior. You will see slightly more complicated band diagrams later when we discuss the diode.

> *Important:* What do we mean by the word 'electron' used in a semiconductor? In this field of electronic materials, it actually has two distinct meanings. In one sense, it is the valence electrons that are contained in the outer, filled shell of the atoms in the lattice. Those electrons do *not* participate in carrying current. The other sense of the word electron is the 'conduction' electrons that have broken free of the valence band and are now free to move through the lattice, and *do* carry current. In most semiconductor textbooks, 'electron' refers to the conduction electrons, whereas 'valence electrons' is used to denote those electrons that do not contribute to the flow of current.

7.4.2 Thermal Energy

Heat is energy. A chunk of material has a total amount of heat, or thermal, energy in it, so it is an extensive quantity. Temperature is a measure of how vigorously the atoms are vibrating and is an intensive quantity. Obviously, the hotter a material is, the more thermal energy a bulk quantity of it contains. That all said, we need to be thinking of temperature at the atomic level. Atoms are more active and vibrate more around their lattice sites at higher temperature. A bunch of atoms vibrating together in the form of a wave in a solid is called a **phonon**. (Collectively, when a wave at a football game happens, the whole wave is like a human phonon! We might be tempted to call that a "humon" ☺). A phonon is an extended jiggle of the atoms. The phonons *are* the thermal energy! Cold temperature, no heat, no extended jiggles, no phonons, no electrons in the conduction band.

Every way in which an atom can vibrate, or store energy in the form of motion, is called a **degree of freedom**. Each degree of freedom of an atom or molecule has an amount of thermal energy of $\frac{kT}{2}$, where k is Boltzmann's constant (1.38 × 10^{-23} J/K, joules per Kelvin, or 8.62 × 10^{-5} J/eV, where "eV" is explained below) and T is the absolute temperature in "Kelvin" or sometimes "Kelvin." So, kT is the fundamental quantity of thermal energy that affects the electrons in a crystal, and you will see "kT" appear in nearly every equation related to the electronic behavior of semiconductors.

Kelvin units are basically the same as degrees Celsius, except that zero Kelvin is the state of no thermal energy, what is called **absolute zero**. Zero degrees Celsius is an arbitrary number based on how much we all like ice in our drinks. All temperatures used in discussing semiconductors are about how much total thermal energy there is, so they are relative to 'no energy,' or absolute zero. The use of Celsius and Fahrenheit degrees in equations is practically meaningless – all temperatures used in semiconductor physics and devices are expressed in Kelvin. For reference: the units of Celsius and Kelvin are the same amount of thermal energy. In other words, a difference in temperature of 1° Celsius is the same as 1 Kelvin. Also, 0° Celsius is the same as 273.15 Kelvin. (It is considered incorrect to say "degrees Kelvin." Why? Because Kelvin is the unit of temperature that is directly related to total thermal energy. Aside from the constant k, saying 'Kelvin' is like saying how much thermal energy something has.)

All of the atoms vibrating, i.e., phonons, with relatively low energy can occasionally, and rarely, come together all at once to impart a lot of energy to a bonded, i.e., valence, electron. Then all of that thermal energy at one place and at the same time can knock the valence electron loose from a bond and free it to move around, i.e., jump up to the conduction band, and therefore carry current. That is shown crudely in Fig. 7.8.

In order to imagine this process, think of an ocean on a very windy day. There are lots of big waves. Every once in a while you will see some water spray out of the surface and fly up much higher than the surrounding waves. That phenomenon is due to the statistical probability that occasionally some waves will *crash together* and impart more energy to the water, launching the water spray upwards. You can imagine that the water will spray higher when it is windier. In this analogy, the water spray height is the height of the electrons above the bottom of the conduction band (refer to Fig. 7.8), and the wind speed is the thermal energy, or temperature, of the semiconductor material.

The same thing happens when many low-energy thermal phonons converge at a point, interact with a valence electron and give it a lot of energy. An electron gains all that energy and is knocked loose from the bond, and is free to roam around, carrying current, i.e., it is a conduction electron. The probability of that happening is really, really, really small, but there are a lot, lot, lot of electrons in the crystal, and the phonons vibrate very, very, very frequently. So, at reasonable temperatures there can be an appreciable number of thermally generated conduction electrons, just like you are likely to see some water spray if you look out over the ocean. This all depends on the semiconductor material, temperature and bandgap.

The probability of finding electrons at some energy (relative to some reference energy that needs to be defined) decreases exponentially with energy (analogous to height for the water spray.) This is expressed as

$$P \sim e^{-E/kT} \tag{7.5}$$

where P is related to the probability of finding the electron at some energy above the reference energy. If the material is very cold, and T is small, then P gets very close to zero. If the material is warm, and T is high (it is very windy), then P approaches 1, meaning that the probability is much higher, i.e., you are more likely to find electrons way up in the conduction band. Statistics like that of Eq. 7.5 that drop off exponentially with energy are called **Maxwell–Boltzmann statistics**. Just for completeness – when you take a full course in semiconductors, you will find that the reference energy is called the "Fermi level" and you will see the term "Fermi–Dirac statistics," which applies to the electrons at lower energies just at the very edge above the conduction band. Those details are very important, but not for this simple introduction.

So, now we have some free electrons in the conduction band. Free electrons accelerate due to electric fields (see Chapter 1). The energy gained by an electron is related by:

$$U = e \cdot V, \tag{7.6}$$

where U is the energy that one electron gains by falling through (i.e., accelerated by) an electric field due to a voltage difference V, and (the magnitude of) $e = 1.602 \times 10^{-19}$ coulombs is the charge of a single electron (just as often you will see q for the electric charge). In this equation, the energy is in units of joules, and is a very small number indeed. Electrons are really tiny, and it is awkward to discuss their energies in joules, because you would have factors of 10^{-19} or so everywhere. That is simplified by converting to **electronvolts**, **eV**, in which those factors go away. (eV is pronounced "ee-vee".) The term "volts" refers to the external *electrical potentials*, as in power supplies, and "electronvolts" refers to the *energy* that the charge carrier gains *due to the voltage V*. As an example, an electron that falls through 1.1 V gains an energy of 1.76×10^{-19} J, but, and so much more simply, it gains 1.1 eV. In fact, the abbreviation for electronvolts, eV, is also the equation for electron volts, eV! Easy-peasy. So, voltage is not energy, but electronvolts *is* energy. You may use the voltage in an equation, but then have to multiply by e to get energy in joules or just use eV units in an equation without the *e*. We use eV as the energy unit mostly when we are discussing what happens to an electron in a material, because of those annoying extra factors.

For a deeper understanding: An aside about 'energy.' You have been using that word from the first days of your physics classes, and seen it a lot in chemistry and in this course. One simple definition of "energy" is "the ability to do work." Great. But what is it? This question is very deep, and fundamental to physics, but here I try to give you something that you can use to help wrap your head around it. First you must think of a 'system,' by which I mean all of the things that play a role in the problem you are trying to understand. For example, if you are discussing an atom, you don't care about the moon, but if you were talking about the gravity on Earth, you might need to include the moon in your system. I think of energy as a kind of 'stuff' that is built into a chosen system. You get only a certain amount of energy stuff in the system, which is present when it is created, unless you choose to add more energy to your system, or let some energy escape. That energy stuff can change forms depending on what it is related to, such as an electric field or the thermal motion of atoms. You may choose to add more energy to your system by, say, shining light on it, or heating it up. In a semiconductor, some valence electrons can absorb heat or light energy of at least E_g, and in so doing move up into the conduction band. That extra energy now exists (partially) as potential energy within the electric fields of the crystal plus kinetic energy of the electrons as they move. The actual change in energy can be found only by using quantum mechanics, which is not covered in this course.

As for the thermal energy, kT, we know that Boltzmann's constant, $k = 1.38 \times 10^{-23}$ J/K. To convert from joules to its voltage equivalent, divide by e, according to Eq. 7.6. In that case kT is expressed in eV when a value for $k = 1.38 \times 10^{-23}/1.602 \times 10^{-19} = 8.61 \times 10^{-5}$ eV/K is used. The equivalent value of eV is often written as "kT/e" or "kT/q", where k is assumed to be in the form of J/K.

Thermal energy, kT, is also an energy (duh) and at the atomic level is also best expressed in units of eV. The most important example is kT at **room temperature**, **R.T.**, which is by definition often taken as 300 K, but this varies. 300 K is just so nice and round! (300 K is an uncomfortably warm room). The thermal energy at R.T. is 0.026 eV = 26 meV (milli-electronvolts). That is the energy that an electron gains falling through only 26 mV. Most voltages that describe the internal workings of semiconductor devices are much larger than 26 mV, as you will see. As temperatures change, kT changes, but voltages applied to the devices do not. Put another way, as temperature increases, atomic jiggling gets worse, electrons are banged around more, and gain higher energies compared to those energies that are due to the applied voltages. Therefore, device behavior, meaning how currents flow at given voltages, can change with temperature. For this reason, temperature is an ever-present consideration in semiconductor devices.

7.4.3 Holes as Carriers of Positive Charge

In Si, E_g is 1.12 eV, which is 43 times the thermal energy at room temperature. 1.12 eV is vastly more energy than thermal energy collisions can *typically* provide; you need 43 phonons to collide at the same place and time, which is very, very, very unlikely. Statistically, very few electrons (1 in every 10^{12}) break free of the valence band and jump to the conduction band. That is suggested in Fig. 7.8 by showing only one of many valence electrons being thermally excited to the conduction band (since it is impossible to draw 1 in 10^{12} of anything).

When an electron is so extremely lucky as to be freed from being a valence electron, it becomes a conduction electron and moves away from its parent atom. (I draw no analogies here between electrons and people!) In doing so, it leaves behind a (positive) proton in the parent nucleus that is no longer paired with the (negative) valence electron. The missing bond is therefore a site of positive charge and is called a **hole**. The term "hole" is descriptive in that it is like a tiny hole in the continuous 'fabric' of the filled valence band. The electron and hole are collectively referred to as an **electron–hole pair**, or **EHP**.

Valence electrons can now take advantage of the new space created by the removal of the electron, and can now move around to the missing sites. In this way, holes move around as valence electrons shift positions. The moving empty bonds give the *appearance* that the positively charged empty bonding site (hole) is moving through the crystal when it is actually the other valence electrons that are moving in the opposite direction. This complicated charge motion behaves simply like a moving positive charge. Thus, *holes behave just like positive charges that are free to move through the valence band just as free electrons move through the conduction band.*

FIGURE 7.9 Bubbles in water are a good analogy for holes in semiconductor devices. Though the water (valence electrons) is the actual material moving downwards, it appears as though the bubbles (holes) are moving upwards. The water and the air move in opposite directions in the presence of the field, which here is gravity, but in an electronic device is the electric field.

A simple analogy for a hole is a bubble of air in a glass of water (Fig. 7.9). As air moves upwards in a bottle of water, it looks like something substantial is moving upward, but it is really the downward movement of water. The net result of the downward-moving water to the viewer is the upward movement of the air. The air is 'missing water,' which we might think of as positive 'air charge,' or as missing negative 'water charge.' The 'air' in the semiconductor is the holes, which are exposed positive charges in the nucleus due to missing valence electrons. Even though all that 'negatively charged water' is moving downwards, it is effectively the same as 'positively charged air' moving upwards. Factoid: The lowest energy for holes in the valence band is at the top, as drawn in Fig. 7.8. Energetic holes, i.e., holes that are moving quickly, are lower down in the valence band. You will get a much more complete understanding of holes in an introductory semiconductors course.

An electric field causes *all* electrons that *can* move to move in the same direction, be they in the valence or conduction bands. Again, in a semiconductor, some of the (negative) valence electrons are moving, but all we care about is the apparent motion of the (background) positive charge, which is opposite to the motion of the valence electrons that are moving. Both negative conduction electrons and positive holes carry current. Recall that positive current is the direction that positive charges move. Metals carry current *only* by electrons (there are no holes), so the electron motion is opposite to the direction of positive current in metal wires. In semiconductors, there are both conduction electrons and holes, which move in *opposite* directions in an electric field but carry current in the *same* direction. In short, both electrons and holes contribute to current flow in semiconductors.

7.4.4 Dopant Atoms

At room temperature, in pure Si there is a carrier concentration of only about $n = 10^{10}$ electrons per cubic cm (cm^{-3}) and an equal $p = 10^{10}$ holes cm^{-3}, which is about one in every five trillion electrons in the material. It is called **intrinsic** when the concentrations are the same, i.e., $n = p$. **Undoped**, i.e., perfectly pure, semiconductor material has an equal number of electrons and holes ($n = p$) and is clearly intrinsic. That isn't a very interesting situation for two reasons: the electrons and holes would carry very little current, and it would just basically be a resistor with a high resistivity. We need more useful properties to make diodes, transistors, and other devices.

Since so far in our discussion one excited valence electron results in one hole and one conduction electron, how can an imbalance in numbers happen? We can change the number and balance of electrons and holes through a process called **doping** (*not* what athletes sometimes do illegally). Doping is a critical part of the semiconductor manufacturing process. The type and number of charge carriers can be changed by adding tiny amounts of **impurities** (i.e., atoms that are not Si) that act as **dopant atoms**, which contribute either electrons or holes. Without dopant atoms, only thermal energy can create electron–hole pairs, i.e., one electron for each hole. Since they are created in pairs, they are equal in number. Since it takes a lot of energy to promote the valence electrons across the bandgap, there are very, very few of them as compared with **doped** material, i.e., material with some dopant atoms mixed in with the Si atoms. A doped semiconductor is called **n-type** when it has more conduction electrons than holes, and **p-type** when it has more holes than electrons.

Dopant atoms in a crystal comprise only from about 1 in 10^7 Si atoms to about 1 in 10^3, but even those small fractions are enough to drastically change the electrical properties of the semiconductor. Using many different methods, dopant atoms can be injected into the semiconductor crystal lattice, replacing the Si atoms at the lattice sites. Although it is a huge industry, and very important technologically, semiconductor device manufacturing is barely touched on in this chapter. You are encouraged to learn more on your own.

How do dopant atoms create electrons or holes? Recall that Si is group IV, and each atom shares two valence electrons bound to each of its four neighbors to make a total of 8 electrons needed to fill the outer shell (the valence band). **Donors** replace Si atoms and contribute (**donate**) extra conduction electrons, making the material n-type. Donors (in Si) are members of group V, which have five electrons in their valence shell; these include phosphorus and arsenic. Five valence electrons are one more than is needed to complete the 8 bonds of the valence band using one electron from each of the four neighbors. That last electron from the donor is only weakly bound to the donor atom, so it needs very little thermal energy to break free. This is called the **ionization energy**. Figure 7.10 shows how donors and acceptors are represented on a band diagram. The dopant atoms are shown as separate short lines to signify that they sit alone by themselves, far away from the other dopant atoms.

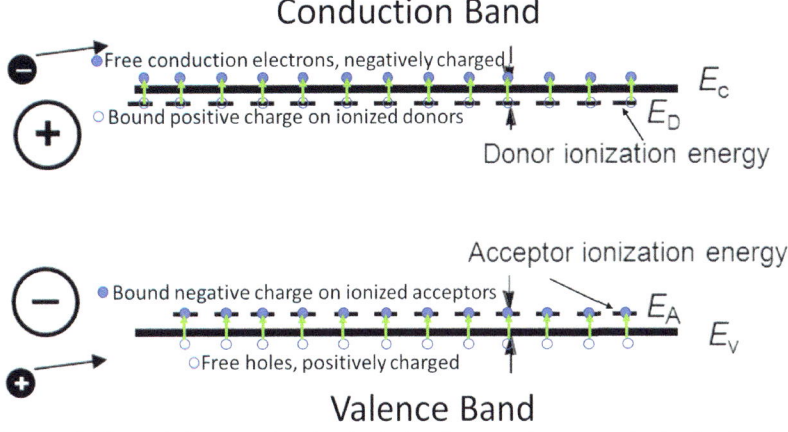

FIGURE 7.10 Band diagram showing both the donor and acceptor ionization levels. Filled circles are negative charge and empty circles are positive charge.

Remember that the vertical scale is energy. The non-ionized dopant energies are placed very close to the band signifying that it takes very little thermal energy to free the extra electron (donors) or accept an electron (acceptors, below). The ionization energy of a donor atom is (at room temperature) less than $kT = 0.026$ eV, and much, much less than $E_g = 1.12$ eV, which you know is the energy needed for an electron to go directly from the valence band to the conduction band. Therefore, at room temperature, effectively every donor atom adds a conduction electron to the material, as shown in Fig. 7.10. The total density of electrons can be millions of times more than the intrinsic concentration of electrons, which are due to thermal energy without dopants, as discussed above.

Not required: One day in about 2010 or so, I was walking across campus with a visitor who said to me, "You know, your name actually means 'electron.'" I replied, "Huh?" He explained to me that Bernstein means amber (literally, burned (bern) stone (stein)) in German (which I, of course, already knew). He also told me that the word 'electron' is derived from the Greek word for amber, 'elektron,' (which I did not know) because static electricity was first studied by rubbing animal fur on pieces of amber. So, 'Bernstein' is actually 'electron' in some real sense. How cool is that?

Holes are a bit more complicated: **Acceptors** in Si are from group III; they include aluminum and boron and have only three valence electrons. When they bond to the four surrounding Si atoms, they come up short by one electron and cannot fill the 8 required bonds. (Remind yourself of the bubble analogy of hole motion. The following is the physical explanation of how that happens in the p-type material.)

Remember that the Si crystal is **charge neutral**, i.e., there are exactly the same number of electrons in total as there are protons in all the nuclei. Also, the dopant atoms are charge neutral before they gain or lose an electron and become ionized. The missing bond at the non-ionized acceptor atom does not in itself represent any charge (i.e., it is charge neutral) because there is also one less proton in the acceptor nucleus to go with it.

Thermal energy gives a tiny bit of energy, the acceptor's ionization energy, to an electron in the valence band, and it is excited enough to jump to an acceptor atom, again, as shown in Fig. 7.10. Figure 7.10 shows electrons in the valence band gaining a tiny bit of thermal energy and jumping onto acceptor atoms, making them ionized. The acceptors **accept** electrons, and hence get their name. Those extra electrons on acceptors cannot participate in current flow as they are stuck in place.

Now think about the valence band. When a valence electron jumps up to the empty bonding site giving its electron to the acceptor, *that* missing valence electron leaves behind a missing bond with a *net positive charge on its parent Si atom* from that unpaired proton. As valence electrons move from position to position, the empty, positively charged site in the valence band moves in the opposite way as the valence electrons and behaves just like a positive charge moving through the crystal – a hole!

Hole current flows by the motion of valence electrons from atom to atom so that the missing positively charged hole in the valence band is free to move around just like a free electron moves throughout the crystal, but with a positive charge. To summarize, those missing bonds in the Si lattice are positively charged holes, which can be treated as positive charge carriers just as we treat the negative conduction electrons.

It is the ability to make n-type and p-type materials that makes possible the 400-billion-transistor modern microprocessor that gives the power to your newest, greatest laptop computer, or is at the heart of supercomputers that

> *Safety:* Ions can move rather easily in a liquid, which is why it is dangerous to use electrical devices while taking a bath or shower. Pure water is actually quite resistive, but also difficult to come by. House water is very conductive due to dissolved minerals. If water touches line voltage (120 V_{AC}), ions can conduct lots of current to your body, through your skin and through your heart, which could be fatal. That is why ground fault interrupters in homes, discussed in Chapter 3, are so important to protect human life. Blame the mobile ions for that.

render 3-dimensional graphics for blockbuster movies, predict the weather, perform simulations of solar activity, or discover new drugs. Other semiconductor devices can operate with thousands of volts to control the smart power grid, or convert light into breathtaking images of galaxies billions of light-years from Earth, as in the James Webb Space Telescope. Yet more applications of semiconductors include the light-emitting diodes that now dominate the television and lighting industries, providing bright light at a small fraction of the energy and cost of older technologies. Chapter 8 goes into more detail about the uses of semiconductors for lighting and energy production.

7.4.5 Charge Carriers and Semiconductor Diodes

Now we discuss the most basic semiconductor device, which is called the **diode**. You already saw a diode in the first lab and soldered it into your transistor radio. The diode is kind of a magic resistor. It is a two-terminal device like a resistor, but it has a very different relationship between current and voltage. In a resistor, if you double the voltage across it, the current doubles, i.e., it is **linear**. Not so for a diode, which is highly **nonlinear**. The diode is *nonlinear* because the current does not increase as the same factor as the voltage across it. (Linearity is a very important concept and is explained in depth in Section 5.6). When current flows, it increases *exponentially* ith the voltage across it (again, nonlinearly), so it increases very quickly with small changes in voltage as shown in Fig. 7.11. Also, diodes allow current to flow in only one direction. (Well, sort of. This will be expanded on below). For this reason, diodes are sometimes referred to as **rectifiers**. At various times we will be discussing independently the significance of the exponential dependence of current on voltage, and the rectification properties.

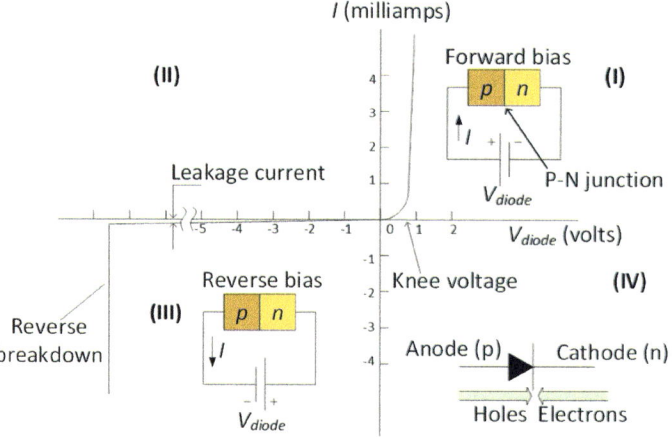

FIGURE 7.11 Characteristic (*I–V*) curve for a typical p–n junction diode. Quadrant I is the forward-bias region where the diode is "on" and conducts current. Quadrant II does not happen in normal use. Quadrant III is the reverse-bias region. There, the voltage is large, but very little current flows, so the diode is "off." When the voltage gets large enough, the diode "breaks down" and lots of current flows. This damages the device only if the diode gets too hot. Quadrant IV is not used for a normal diode, but is used for a solar cell when light shines on it (Chapter 8).

> *Perspective.* Some advice for you: in every course going forward, when you hear the terms 'linear' and 'nonlinear,' pay special attention because there are many important implications of any system that stem from those properties. The resistor and diode comparison is just one example.

Both are very important to how diodes are used in circuits. Chapter 10 goes into detail about how diode rectifiers are used to make power supplies.

Using water as an analogy for current, the diode as a rectifier acts like a water check valve (Fig. 7.12). In a check valve, water flows easily in the direction in which a diaphragm can open, but no water flows in the opposite direction in which the diaphragm is forced shut. Diodes don't have diaphragms of course, but the directional flow of current is similar.

Diode behavior is due to the nature of semiconductors. As already stated, semiconductors come in two flavors – n-type and p-type. In n-type semiconductors there are vastly more electrons than holes, so current is carried by negative charge. In p-type semiconductors, there are vastly more holes than electrons, so current is carried by positive charge. A diode is composed of a piece of n-type material, the **n-side**, that is in contact with a region of p-type material, the **p-side**. The interface between the two is called a **p-n junction**, as shown in the upper right corner of Fig. 7.11 and expanded in 7.13. Figure 7.11 also shows the **characteristic curve**, or **current-versus-voltage curve**, or ***I–V* curve**, which is a plot of the current through a device as a function of voltage across it. Remember, for a resistor, the *I–V* curve was just a straight line. We will get back to the shape of the diode *I–V* curve after discussing what physically happens inside the diode.

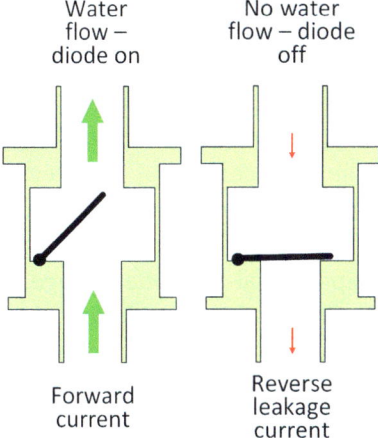

FIGURE 7.12 This valve passes fluid in only one direction. Likewise, a diode passes current in only one direction.

The circuit schematic symbol for the diode is shown in the lower right corner of Fig. 7.11. The wide end of the triangle is the p-type material and is the **anode**. The line at the point of the triangle is n-type material and is the **cathode**. When current flows, positive holes leave the anode and flow toward the cathode. Simultaneously, negative conduction electrons leave the cathode and flow toward the anode. When a positive V_{diode} is applied from the anode to the cathode such that current flows from left to right through the diode, it is said to be **forward-biased**. Significant current flows in forward bias. When V_{diode} is reversed, made negative, then the diode is **reverse-biased** and only a very tiny reverse-bias current flows (until the reverse-bias voltage becomes relatively large). This small reverse current is referred to as **leakage current**.

Figure 7.13(a) is a simple picture of the carriers in forward bias, and Fig. 7.13(b) shows the same thing but in more detail with the energy bands. In forward bias, holes are pushed away from the p-type material across the p-n junction into the n-type material, and electrons are pushed across the junction into the p-type material. Electrons flowing to the left in the figure and holes flowing to the right both carry current in the *same direction*, which here is to the right.

The motion of electrons and holes across the p-n junction is called **carrier injection**. Where the electrons and holes overlap on each side of the junction, electrons fall from the conduction band, **filling** holes in the valence band, a phenomenon called **recombination**. When a conduction electron and hole recombine, both together become a valence electron, which does not carry any current.

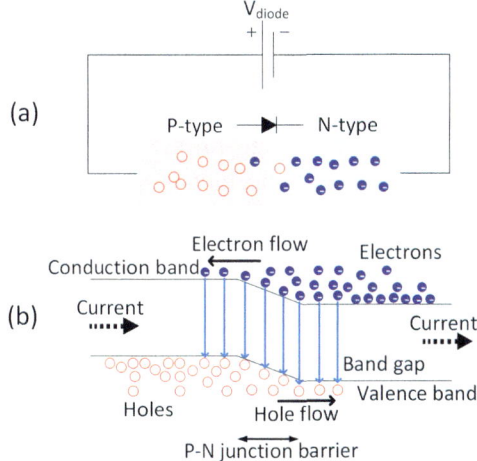

FIGURE 7.13 Two views of electrons and holes in forward bias. Electron energy is up and hole energy is down. (a) Simple view showing applied voltage, V_{diode}, electron and hole populations, and p-n junction. (b) Detailed view showing electron and hole populations decreasing with increasing energy (upwards in conduction band, downwards in the valence band), lowering of p-n junction barrier due to V_{diode}, carrier injection with overlapping electron and hole populations, recombination, and current direction.

(Again, only conduction electrons and holes carry current, not valence electrons.) Both the conduction electron and hole have been eliminated from the carrier populations.

So the electrons and holes disappear near the junction? How does that cause current to flow? The semiconductor diode must be connected to the external circuit in order to work properly. The connection happens at metal **contacts** on the outer surfaces of the diode. Both holes and electrons come in from the metal contacts to replace the electrons and holes that recombine. If that didn't happen, the diode would run out of carriers, and current could not flow for more than a brief moment. The replacement of the electrons and holes from the contacts is the flow of current in the circuit outside of the diode.

Let's look at carrier injection in more detail in order to better understand why the *I–V* curve is exponential in diode voltage. In band diagrams, electron energy increases upward and hole energy increases downward from the band edges. Most electrons and holes are sitting at low energy, near the edges of the bands. Electrons are mostly at the low energies at the bottom of the conduction band near its edge and holes are mostly at low energy near the top of the valence band, near its edge. (Remember, you would expect the air bubbles in water to go to the top of the glass, and likewise holes float up to the top of the valence band.) There are no carriers inside the bandgap, between the valence and conduction bands. The donors and acceptors do not participate in current flow. Their jobs were done when they gave their electrons and holes to the lattice.

There are exponentially fewer electrons and holes high into their respective bands (up in the conduction band and down in the valence band), which is crudely illustrated in Fig. 7.13(b). Remember, this is because thermal energy is imparted to the electrons according to Maxwell-Boltzmann statistics, so they are *exponentially* less likely to be higher in the bands, that is, the concentration of electrons and holes goes as $e^{-E/kT}$, where E is the energy into the conduction band for electrons or into the valence band for holes. At relatively high energies, high up in the bands, i.e., much greater than kT, the carriers are like a 'mist' of water vapor – there are very few carriers. Why "much greater than kT"? Because kT is the measure of the effect of the temperature of the material. After a few kT of energy above the band edges, the probability of finding an electron or hole has decreased a *lot*.

As already presented, current in forward bias depends *exponentially* on the voltage applied to the diode. Why exponentially? *The p-n junction is like a wall* – it is a **barrier** that keeps the holes and electrons separated on their respective sides of the junction. *The applied voltage lowers this barrier.* The applied voltage lowers the wall, allowing the highest-energy holes to move one way across the barrier and the highest-energy electrons to move the other way across the barrier. No current flows at zero bias ($V_{diode} = 0$) because the barrier (the wall) is all the way up. As the barrier is lowered ($V_{diode} > 0$), exponentially more electrons and holes are **injected** across the barrier. Think of this as opening a window in a room full of bees, where there are only a few bees near the ceiling. In this metaphor, the window opens from the ceiling toward the floor so the first bees to fly out are the ones near the ceiling. Likewise, those carriers (both electrons and holes) in the mist high up in their respective bands are the first ones to get injected across the junction barrier when a small voltage is applied. Because the mists of electrons and holes have few carriers, becoming denser lower down toward the bands, at first the current is very low, but increases exponentially as voltage increases, i.e., as the barrier is lowered, letting exponentially more carriers (both electrons and holes) across to the other side of the p-n junction. This is illustrated in Fig. 7.13(b).

Now, what about reverse bias? If the opposite polarity is applied with the positive battery terminal applied to the n-type side and the negative terminal applied to the p-type side, the diode is reverse-biased. In this case, the barrier height *increases*, and charge carriers are pushed *away* from the p-n junction; there is *no carrier injection*, and *very little current flows* in this opposite direction. Congratulations if you have followed this discussion of how the carriers in a diode lead to exponential current in only one direction! It isn't easy.

7.4.6 The Diode *I–V* Curve

For complicated reasons there is still a very tiny current, called **leakage current**, as shown in Fig. 7.11. If this were an analogy to the water check valve, it would be water leaking past the closed diaphragm. The actual cause of the leakage current is beyond the scope of this introductory treatment, but here is a simple explanation. EHPs are generated by thermal processes, which we talked about. Those electrons and holes flow out of the reverse-biased junction in the reverse direction to that shown in Fig. 7.13. There are very, very few of these EHPs created near the narrow junction, so there isn't much current flowing. This is the "leakage current," so named as if something were leaking. I suppose it is – kind of.

At room temperature, the entire *I–V* characteristic, except for the breakdown region shown in Fig. 7.11, is described by (here, *q* is used instead of *e* to avoid confusion)

$$I = I_0(e^{qV/kT} - 1) = I_0(e^{V/(kT/q)} - 1) = I_0(e^{(V_{diode}/0.026)} - 1), \tag{7.7}$$

where V_{diode} is the voltage across the diode and I_0 is the tiny reverse-bias or leakage current (nano- to picoamps). Let's discuss the exponent in Eq. 7.7. Because exponentials are dimensionless and qV is in units of electronvolts, so also must be kT. Moving the 'q' to the denominator leaves the numerator in volts and converts the denominator to volts as well. Therefore, the ratio is expressed easily, as shown in the last expression for room temperature, as the ratio of the diode voltage, in volts, to the magnitude of the thermal energy, as expressed in electronvolts. For example, a forward bias of 0.3 V across the diode is compared to kT in volts, so at room temperature $e^{V_{diode}/(kT/q)} = e^{0.3/0.026} = e^{11.5}$. This factor, when multiplied by a few pA that is I_0, does not result in all that much current. It takes more voltage than 0.3 V to 'turn on' a Si diode.

Equation 7.7 describes the shape of the current–voltage, *I–V*, characteristic in Fig. 7.11 for both positive and negative voltages (but not breakdown, discussed next). Note that at large and positive V_{diode} (0.5 V < V_{diode} < 0.7 V), the exponential term is very large, and large forward current results. In reverse bias, where V_{diode} is negative and only slightly larger in magnitude than 0.026, the exponential term becomes very small, the '–1' term dominates, and the tiny reverse leakage current, I_0, results. So that equation more-or-less describes the behavior in both forward and reverse bias.

At some large, negative V_{diode}, the current rapidly increases way beyond the leakage current. This regime is called **reverse breakdown**. Reverse breakdown occurs abruptly at a voltage designated V_{br}, and can be (but isn't always) steeper than forward bias, as indicated in Fig. 7.11. One cause of the breakdown is as follows. The high electric field of reverse bias rips a few electrons out of the valence band, promoting them to the conduction band. These few electrons are moving very fast; when they collide with more valence electrons, they create yet more EHPs and conduction electrons. Once the process starts, a lot of electrons and holes are available to collide, causing even more electrons to be ripped out of the valence band, and so on. This process is called **avalanche multiplication**, reminiscent of a little bit of snow on a mountain cascading into a full-blown avalanche. (The other cause of breakdown is quantum mechanical tunneling. Quantum mechanical processes are important in all semiconductor devices. Don't worry, we won't get into quantum mechanics in this course.)

Although the term *breakdown* suggests a destructive event, a diode can operate continuously in breakdown as long as the power dissipated by the diode, given by

$$P_{diode} = V_{br}I \tag{7.8}$$

does not cause the generated heat to exceed the amount that the diode and its package can safely dissipate to its surroundings. Equation 7.8 tacitly assumes that the breakdown voltage, V_{br}, is constant regardless of the current flowing in it, as suggested by Fig. 7.11. Diodes that break down due to avalanche multiplication are called **Zener diodes**. We will see in Chapter 10 that some circuits use the breakdown regime of a Zener diode as a source of constant voltage that allows variable current. In that case, the Zener diode is used as a **voltage regulator**.

Because of their unique *I–V* curves and carrier transport properties in both forward and reverse bias, diodes are used in all kinds of applications. Where current must flow only in one direction, as in a battery charger, it is used simply as a rectifier. Diodes can be used as voltage regulators in power supplies, as discussed above. **Light-emitting diodes** are used for home and industrial lighting (this and Chapter 8). Diodes in the form of **solar cells** can generate electricity (Chapter 8). The applications are endless. As a building block, p-n junctions are part of all transistors (discussed below) and just about every other semiconductor device. As such, it will be crucial to understand their basic operation and properties in order to understand their role in these devices. You will see p-n junctions and diodes more in future courses on semiconductor materials and devices and electronics.

The diode *I–V* equation says that current increases exponentially with the forward voltage across the diode. If you put a power supply across the diode, measure the current, and *very gradually* increase the voltage (so you don't burn out the diode by mistakenly letting the current get too large), you will see that the current stays very small until you can measure a reasonable amount of current. That voltage at which the diode "turns on" is called the **threshold voltage, turn-on voltage, cut-in voltage, knee voltage, diode voltage,** or **diode drop**.

All of these terms for the voltage at which the diode current is appreciable are misnomers. There is current at every forward voltage whether or not you can see it on an *I–V* plot, and it is always increasing exponentially with increased voltage. In fact, we can *arbitrarily* decide that currents below a few tenths of a mA levels don't really matter all that much, and currents above about 1 mA do matter, or perhaps use some other criterion. In other words, there is nothing precise about the turn-on voltage except for what everyone has agreed is a reasonable current to 'notice.' Generally, with respect to common circuitry (whatever that means) we have little current out to about 0.5 V to 0.7 V, after which the current gets very large very quickly, although it was increasing exponentially all the time, but was too small to appreciate. Frequently, when a diode is conducting in forward bias, it is just assumed that the voltage drop is something like 0.7 V. Or, if the circuit uses very little current you might assume it is 0.5 V. Most of the time this small difference in voltage doesn't make much difference in the circuit analysis. If it does, then you should use the actual *I–V* curve to determine how much current is actually flowing at any given voltage.

In Eq. 7.7, the reverse leakage current I_0 is multiplied by the exponential term. I_0 is very small, typically on the order of 10^{-12} to 10^{-9} A. Diode current calculated by Eq. 7.7 is low at voltages less than about 0.5 V because current is the product of I_0 and the exponential term, which is also relatively small. The exponential term has a long way to increase before the diode current gets as large as even 1 mA. (Note that all of the above numbers are just rough approximations of real diode parameters.) The bandgap E_g and doping concentrations are the major factors in the turn-on voltage, and varies among different materials. Larger-bandgap semiconductors have higher barriers and therefore fewer carriers high up near the top of the barrier, so it takes more forward voltage to get appreciable current flow. For comparison, Si has a bandgap of 1.12 eV. Germanium (Ge) has a bandgap of 0.66 eV, and hence a Ge diode has a lower turn-on voltage (about 0.2 V to 0.3 V). Gallium arsenide (GaAs) has a bandgap of 1.43 eV, so a GaAs diode turns on at a larger voltage (about 1.2 V). All of these numbers can vary slightly as reported in different sources. In this week's lab activity, you will observe the diode *I–V* curve for both Si and **light-emitting diodes, LEDs**. LEDs give off light when an electron and hole recombine. The bandgaps of LED materials are very large in order to produce photons that we can see as different colors. The larger bandgap increases their turn-on voltages.

7.4.6.1   Creating a Diode *I–V* Curve on an Oscilloscope

The current-voltage characteristic, or *I–V* curve, shows the current through the device as a function of the DC voltage across it. Every device is described by its *I–V* curve (and its band diagram explains why the *I–V* curve is what it is). As you know, the *I–V* curve of a simple resistor is a straight line whose slope is $1/R$. If you go on to study semiconductor devices and circuits, it is vital to understand what "the current through the device as a function of the DC voltage across it" actually means. This is fundamental to understanding the significance of *I–V* curves for all sorts of semiconductor devices.

A circuit has various currents in loops and voltages across nodes. When the circuit is operating, each current and voltage is the result of interactions among the circuit elements. *However, every device MUST obey its own I–V curve, since the curve represents the only way that the device can exist internally.* The *I–V* curves must be satisfied for every device in the circuit at all times. For example, if you know that the voltage across a resistor is such-and-such, then the current MUST obey $V = IR$. Or, if the voltage across a diode is in the reverse-bias direction, then the diode acts effectively as an open circuit. It simply must be so because the carriers inside those devices behave that way.

So, what *is* an *I–V* curve? It tells you the current through the device if you absolutely know the voltage across its terminals. Think of it as being very local to the device: the current through *it* as a function of voltage across *it*, no matter what is happening anywhere else in the circuit. For an individual device such as a Si diode, you could get an *I–V* curve by putting the leads from a DC power supply across the diode and then observing the current as you ramp up the voltage. If you plot current as read by an ammeter versus the output of the power supply, you are guaranteed that you will trace out the *I–V* curve. Don't forget that a power supply can source an 'infinite' amount of current; if you are sloppy and accidentally increase the output voltage to 1 V_{DC}, the exponentially large current would be intolerably large, and you will surely burn out the device! (Put 1 V into Eq. 7.7 and see what you get!) A resistor in series with a diode limits the current, protecting it from burning out. Figure 1.30 shows various types of diodes including large, heat-sinked ones on the left side of the figure. Perhaps if the one on the far left were bolted to a huge chunk of metal that has chilled water running through it then that diode could tolerate the large currents and power dissipation that happens for *maybe* a 0.8 V_{DC} drop, but certainly the ones on the right would not survive!

Conversely, you could use a current source to force a known current through the diode and measure the voltage across it with a voltmeter. At least doing it this way you are in control of the current, and hence power dissipation, so you are much less likely to damage the diode. Or, if you are just observing the *I–V* curve of a device in a circuit, you can put an ammeter in series with it, thus knowing the current, and simply observe the voltage that happens naturally during operation. (That is what you will do in the lab.) In each of these cases, if current versus voltage is plotted, then you will observe a good *I–V* curve for the device because the current and voltage simply must maintain their relationship. That is what you will learn more about in this section.

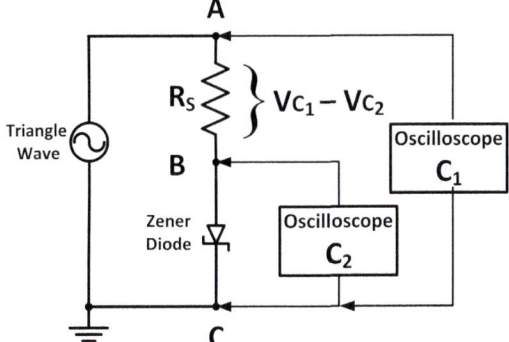

FIGURE 7.14 Circuit used to create *I–V* characteristic of a diode on an oscilloscope.

In the activities, you will build a circuit like that in Fig. 7.14 to display the *I–V* characteristics of a diode, as described by Eq. 7.7 and shown in Fig. 7.11. You are already used to using an oscilloscope to show a voltage as a function of time. However, the horizontal axis can be set proportional to *another voltage* rather than time. This feature, called an **X-Y plot** or **X-Y mode**, is standard on all oscilloscopes regardless of age or sophistication. In this way you can plot $I = I(V)$, or current as a function of voltage.

In using the circuit shown in Fig. 7.14 in the activities, you will do a few new things with your oscilloscope – first without the benefit of the current probe, and later with it. The horizontal axis for the *I–V* curve, as in Fig. 7.11, is the voltage across the diode, as measured by scope CH2. Because oscilloscopes only measure voltages and not currents, you will use the voltage across a resistor to measure the current, as you did in Labs 2 and 5. But what about the current probe? Doesn't the scope measure current with that? The current probe translates the current value it measures into a voltage to be read by the oscilloscope. A circuit that converts current to voltage is called a **transimpedance amplifier**.

You will plot the voltage across the series resistor (R_s) on the y-axis in order to represent the current as $I = V_{RS}/R_s$. A resistor used to measure current is called a **shunt resistor**. You are mostly concerned with the shape of the curve, but since you know R_s, you can easily calculate the current.

Measuring current with the current probe is, of course, possible, but is so simple as to be almost cheating, and you won't always have a current probe available. More importantly, a shunt resistor does a better job of measuring small currents because the voltages that are developed at small currents are well within the sensitivity range of the oscilloscope input preamplifiers.

Don't forget that all the voltage probes are referenced to the same ground. That means that if you (wrongly) put one probe and its ground wire across the shunt resistor from A to B, and the other probe and its ground wire across the diode from B to C, then the two ground wires, being the very same electrical point, will short across the diode, and the circuit won't work. It would be exactly as if you just put a wire across the diode from B to C. That 'wire' is the ground. To measure the voltage across the shunt resistor you will do a **differential measurement** as follows. You will take the total voltage from A to C on CH1 of the scope and subtract from it the voltage from B to C on CH2 to get the voltage from CH1 to CH2, or $V_{AC} - V_{BC} = V_{AB} = V_{CH1} - V_{CH2}$. This is your measurement of the current using the shunt resistor.

As already stated, all oscilloscopes are able to use the horizontal axis to display voltage instead of time. This produces an X-Y plot that can be used to show how one voltage depends on another voltage. In our case, the x-axis is the voltage across the diode, which comes in on CH2. The y-axis is the voltage across the resistor, $V_{CH1} - V_{CH2}$, which represents the current through the resistor and diode. Thus, the X-Y plot displays the diode *I–V* curve.

7.5 Transistors and Electric Switches

As important as diodes are, they are not the stars of the semiconductor show. That honor goes to the **transistor**. Transistors are made up of parts that are fundamentally diodes, but are so close together that their behaviors overlap, and new and extremely important behaviors emerge. As we move forward we will build up a model of a transistor from simple electromechanical parts, each of which is important in its own right. So, before we talk more about the transistor, let's start simply – really simply!

Let's start building our transistor model with a wall switch, like the kind you flip on when you enter a room, since in important fundamental ways, a wall switch and a transistor are similar. A schematic symbol and the inside of a switch are shown in Fig. 7.15. A switch is about as simple an electronic device as can

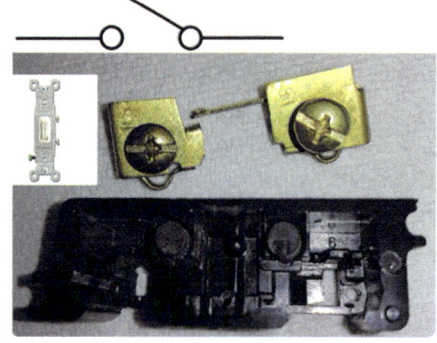

FIGURE 7.15 (top) Schematic symbol of open switch and (bottom) innards of a simple toggle wall switch. The plastic switch presses the metal bar against the metal tab on the left. The metal is thick enough to safely carry enough current for a ceiling light fixture or wall outlet.

be imagined, just slightly more complex than a wire – it is a temporarily broken wire. When the switch is flipped to the 'off' position, a physical contact between pieces of metal is broken. This creates an open circuit, so, for example, a ceiling light fixture is separated from the source of electrical power and the light turns off. When the light switch is moved into the 'on' position, the metal parts physically touch and current flows. It is now just a wire having negligible resistance. It is just that simple.

7.5.1 Inductors, Solenoids and Relays

Continuing to build our transistor analogy, we go next to a variation of the switch called the **relay**. A relay is an *electrically controlled switch* based on a simple device called a **solenoid**.

First the solenoid part. A solenoid is a minor variation of an inductor. It is like a robotic finger – instead of a person flipping a switch, the solenoid does it for you. Recall that a basic inductor is a hollow coil of wire. If a ferromagnetic metal rod is placed just outside the coil, the magnetic field due to current flow magnetizes the rod, which is pulled into the interior space of the coil (Fig. 7.16). The rod in Fig. 7.16 is called the **plunger**. If the plunger compresses

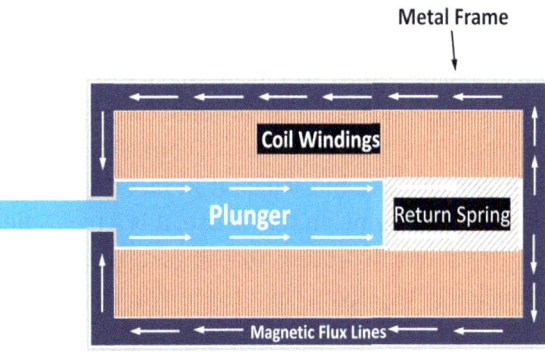

FIGURE 7.16 Solenoid. A current in the windings creates a strong magnetic field that pulls the plunger into the coil. A spring pushes it back.

a spring, it will return to its original position when the current is removed. In effect, the solenoid does for the plunger what your finger does for the switch in Fig. 7.15. Solenoids are used in applications in which electrical circuits are used to actuate mechanical systems, such as remote-controlled deadbolts on doors, manufacturing equipment, automated valves, and many more. (YouTube has many videos of solenoid-controlled pianos.)

*Adding an electrical switch to a solenoid makes it a **relay***, as shown in Fig. 7.17. The particular solenoid shown in Fig. 7.17 is a little bit different from what was described above. In this embodiment, the magnetic coil is an electromagnet and attracts a metal plate, but the plate does not enter the solenoid, it just gets pulled toward it. The plate can be connected to multiple switches. If exactly one switch just opens and closes, the relay is called **single-pole, single-throw**. If one switch has a contact in each position, up and down, like that shown in Fig. 7.17(a), it is called a **single-pole, double-throw relay**. Figure 7.17(b) shows a relay having four poles.

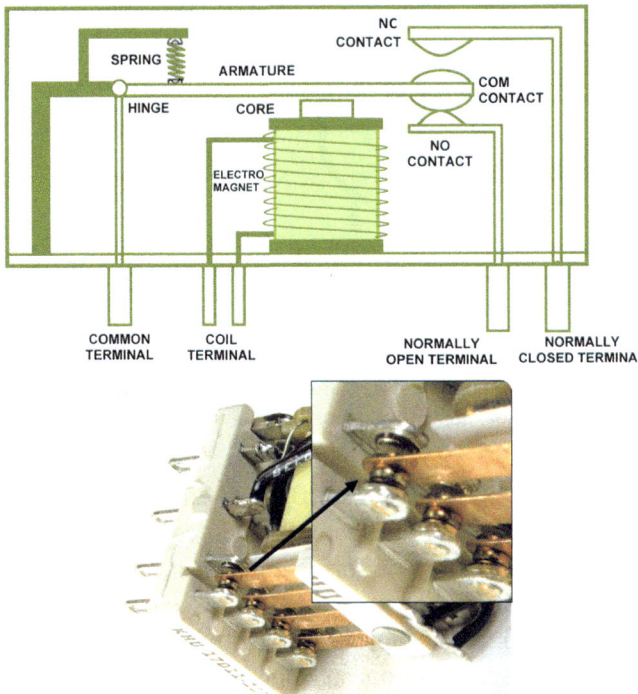

FIGURE 7.17 Relay. A current through the coil terminals creates a magnetic field that pulls down the **armature**. (a) Single-pole double-throw relay shown in the energized state, i.e., current flowing in the coil. The two contacts are called **normally open (NO)** and **normally closed (NC)**. The word "normally" refers to the state when the coil is not energized. (b) Quadruple-pole double-throw relay.

Each of the four switches has two positions, so this is a quadruple-pole, double-throw relay. I think you get the idea.

We are getting much closer now to a model of the transistor. Relays are so named because they relay the signal that goes into the coil to another switch. They are used to control relatively large amounts of current or power at the contacts using relatively small currents or power into the coil. Whenever you hear a piece of electronic gear make a clicking sound inside, it is almost certainly a relay doing it. If you ever press a power button and then hear a click (either loud or soft), you have activated a relay. That relay, then, carries the power to operate the system.

The starter on a gas-powered automobile needs tens to hundreds of amps to turn the engine and start the car. Do you think you want that much current passing through a switch inside the dashboard? Those would be big wires and it would be a big switch – large and expensive. Instead, a key switch carries only a small current that controls a relay that in turn carries the large starter current. The relay in a car is (confusingly) referred to as the "solenoid," as if it did not actually control a larger current, but which it in fact does. The solenoid is also responsible for moving a gear into place to turn the engine. In that sense, it is a true solenoid.

It is worth noting that this discussion of starters is applicable only until all cars are electric vehicles (EVs). Even then there will be a lot of relays involved. It is projected that by 2030, something like three quarters of all car sales will be electric or hybrid (electric plus gas) (rmi.org). Imagine how many EEs will be employed in the electric car industry!

7.5.2 Transistors Overview

Finally, we come to the actual **transistor**. A transistor is a three-terminal semiconductor device, unlike a semiconductor diode, which has two terminals. Transistors act like really tiny relays, without the coil and contacts. *Transistors control a flow of current through two terminals using a current or voltage on a third terminal, much like the relay.*

There are two major flavors of transistor, namely **unipolar** and **bipolar**. The unipolar device has just one type of carrier flowing through it, namely either electrons or holes, but not both at the same time. A bipolar transistor depends on the flow of both electrons and holes at the same time. The most common transistor is a unipolar device called the **metal oxide semiconductor field effect transistor**, or **MOSFET**. The MOSFET is by

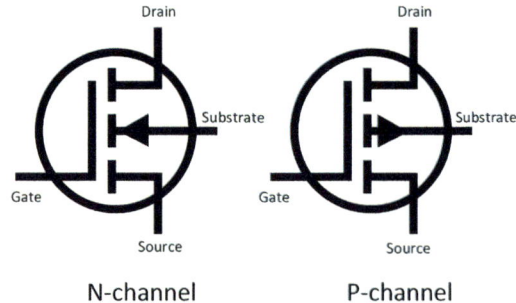

N-channel MOSFET   **P-channel MOSFET**

FIGURE 7.18 MOSFET schematic symbols. MOSFETS are unipolar devices because either electrons (N-Channel) or holes (P-Channel) move from source to drain, but not both.

far the most important electronic device (or perhaps any device) ever made, since it forms the basis for nearly all modern digital electronic devices, such as computers and anything else with a microprocessor (of which the list goes on endlessly...TVs, cell phones, video games, tablets, cars, planes, cameras, satellites, refrigerators, toasters, greeting cards, etc.).

Bipolar devices were the first commercial transistors, and are still important, but today are far less common than MOSFETs. The most common of these is called the **bipolar junction transistor**, or **BJT**. Symbols for the MOSFET and BJT are shown in Figs. 7.18 and 7.19, respectively.

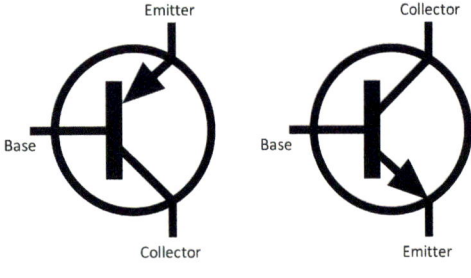

PNP BJT   **NPN BJT**

FIGURE 7.19 BJT schematic symbols. BJTs are bipolar devices because both electrons and holes move through the devices.

Here is what transistors have in common with relays: When a small voltage or current is applied to the gate (MOSFET) or the base (BJT) of a transistor, a relatively large current can flow between the other contacts, either drain and source (MOSFET), or collector and emitter (BJT). We assume here that negligible current flows into the gate or base terminal of the transistor. This isn't exactly true, but isn't important here, and will be clarified in your upcoming electronics courses.

7.5.3 Analog and Digital Information

At this point, we digress to understand the distinction between **analog** and **digital** data. This is only a brief introduction to support the two modes in which transistors can be used. A much deeper dive is the subject of Chapter 9.

Figure 7.20 shows an analog signal. The analog signal changes continuously, as did the music you saw on the oscilloscope in Labs 4 and 6. Analog signals may seem a bit archaic to you in our digital universe... phonograph records, 8-track and cassette tapes, old, deep and heavy TVs having

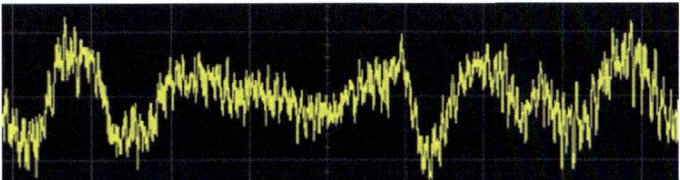

FIGURE 7.20 Analog signal (here music) changing continuously in time.

cathode-ray picture tubes, and more. In all of those, the signals that were used to create the music or the picture were continuous, smoothly varying voltages.

Digital data is the language of digital processing systems of all types, such as computers, cell phones, digital cameras, music players, Blu-Ray video disks, digital television, etc. A digital signal is a series of 1's and 0's that can represent just about any form of information – numbers, sounds, images, etc. However, the world around you is analog. When you hear music, you are obviously not doing so with 1's and 0's, as in the digital world, but rather are experiencing a continuously varying pressure wave in the air that conveys frequency and volume. For a computer to understand anything in its language of 1's and 0's, analog information must be converted to digital data, and for you to understand it, it must be converted back to an analog signal. Figure 7.21 shows both digital and analog signals. Transistors can be operated in a continuous analog fashion, such as in a stereo amplifier that creates a large current to drive stereo speakers. Or, transistors can work as tiny switches, as discussed above, to represent **binary digits**, or **bits**, in a computer chip.

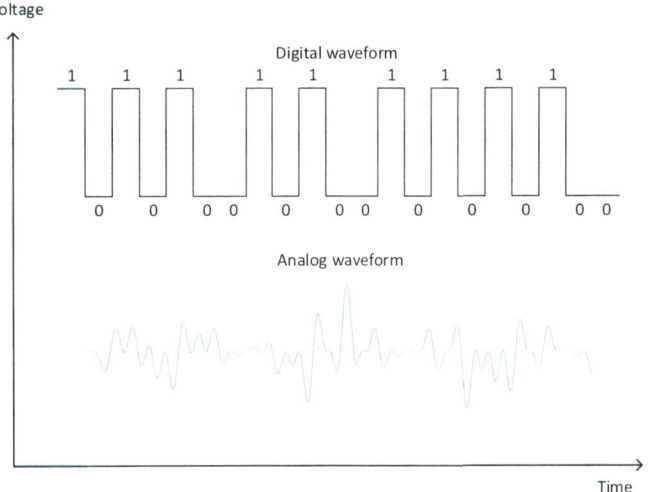

FIGURE 7.21 A digital and analog signal. The analog signal changes continuously whereas the digital signal has only two voltage levels, representing 1's and 0's. There is no implied relationship between these two waveforms.

7.5.4 Transistor Circuits

Let's get back to transistors. Again, a good analogy for the transistor is the relay; as in a relay, an external signal to the coil of a relay actuates the relay and allows current to pass through the secondary contacts to other relays, just like the transistors. In fact, the first electronic computers were made with relays well before transistors were invented (Fig. 7.22). Alan Turing's machine depicted in the movie *The Imitation Game* was built from relays.

In the **digital** mode, a transistor is like a light switch; the input to the transistor at the gate or the base, depending on the transistor type, is the equivalent of your finger on the switch flipping it either on or off. That input signal can turn a transistor on or off, where current either flows or does not flow through the other two contacts, just like a switch. In each case of switch up or down, or input voltage or current on or off, the transistor conveys either a digital 1 or 0. The current that flows can then go to other gates of other transistors (this is the digital logic built into the computer architecture) to turn them on and off, and so on, until, billions of transistors later, you have a supercomputer.

FIGURE 7.22 The IBM Harvard Mark 1 computer used relays instead of transistors. Its calculations were used for the Manhattan Project in 1944.

Safety: You may be familiar with audio amplifiers in which a tiny voltage and current, say from the audio output of a microphone or a computer is amplified to such a high voltage and current that huge speakers can blast enough sound to fill a stadium at high volume. Wear ear plugs at rock concerts. These massive amplifiers have damaged many eardrums. Keep in mind that the speakers are loud enough to be very loud at the back of the arena, so if you are up close, you very possibly could suffer hearing loss.

Transistors were first used in the analog mode, originally for telephony by the Bell Telephone System in the early 1950s, and then later for simple digital circuits, starting in the early to mid-1960s. The idea of analog operation is more like a light *dimmer* switch (as opposed to the on/off light switch) where the voltage input to the gate or base of the transistor is like your finger on the dimmer, and the output is the light being anywhere from fully off to fully on, at any brightness in between. In that sense, it is analog because it can be changed continuously, and not just full-off and full-on as in digital mode.

When the output of an analog transistor circuit is an exact replica of the input, but either larger or smaller, the transistor is working as an **amplifier**. Such an amplifier is said to be a linear amplifier because the output is proportional to the input. (If it were distorted, then it would be nonlinear.) The ratio of the output (voltage, current, power) to the input is called the **gain** of the amplifier. Gain can be greater than, equal to, or less than 1. Most amplifiers make signals larger, so the gain is most often greater than 1.

Just for fun: You can go here (https://www.youtube.com/watch?v=gsNaR6FRuO0) to listen to what digital data used to sound like when it was transmitted over a telephone for internet access using an archaic modem.

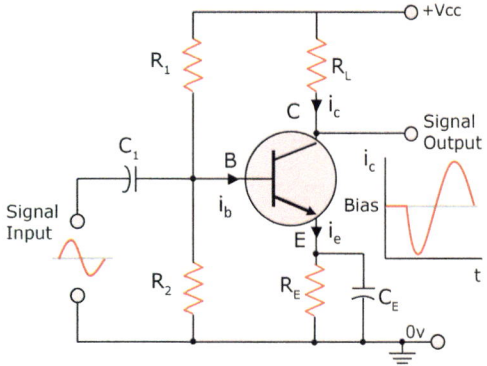

FIGURE 7.23 BJT inverting amplifier having gain greater than 1.

Amplifier circuits can take many forms, but a simple one based on a BJT is shown in Fig. 7.23. This amplifier is called an **inverting amplifier** because, subtracting the DC voltage offset, the polarity of the output is the negative of the input, i.e., it is **inverted**. In the figure, E is the **emitter**, B is the **base**, and C is the **collector**. The amplifier of Fig. 7.23 is inverting with a gain greater than 1; the input to the BJT is at the base and the output is at the collector. Figure 7.23 shows the output of the inverting amplifier with a gain greater than 1 and the output taken at the collector.

Here is a short explanation of how the single-transistor BJT amplifier of Fig. 7.23 works. You will analyze similar circuits in more detail in an electronics course. Chapter 4 told you that capacitors have high impedance at low frequencies, including DC, and low impedance, i.e., act like a simple wire, if the frequency is high enough. In this circuit, the capacitors are high-pass filters that block DC current and let through the AC signal. For sufficiently high frequencies, the information, such as music, passes right through them as if they were just a piece of wire. Therefore, the AC signal, labelled "signal input," is connected, AC-wise, directly to the base of the BJT. Low-frequencies, particularly DC, are blocked by the capacitors. Using a capacitor in this way is called **AC-coupling**, and the input signal is **AC-coupled**. With C_1 in series with the input, the DC voltage at B due to the voltage divider of R_1 and R_2, is blocked from the input.

Central to BJT operation is that they pass much more current through the transistor from collector (top of transistor) to emitter (bottom) than comes into the base (left of transistor). If this were a relay, it would be as if there were, say, 1 mA through the coil (input) used to close the switch, and 100 mA flowing through the relay switch (output). The ratio of the output current to the input current is, in general, called the **current gain**. In BJTs, the ratio of the collector current to the base current is called the **common emitter current gain**, which is also called simply β (**beta**). β is a large number, like one hundred to several hundred. Current into the base is i_b, current into the collector is i_c, and current out the emitter is i_e. A β–times larger current, i_c, goes into the collector at C than goes into the base at B, i_b. The whole transistor can be treated as one node, so currents into and out of the BJT must sum to zero. Hence, the emitter current is

$$i_e = i_b + i_c, \tag{7.9}$$

where i_b and i_c are positive into the transistor and i_e is positive out of the transistor. Since i_b is very small, i_c is approximately equal to i_e. To summarize, a small current into the base causes a large current to flow from the collector to the emitter right through the transistor.

Note that i_c must pass through R_L to get into the collector. For a small change in i_b, or Δi_b, (the Greek delta Δ often means "change in") there is a β-times-larger change in i_c, or:

$$\Delta i_c = \beta \cdot \Delta i_b. \tag{7.10}$$

The current gain, A_i, of the output to the input is:

$$A_i = \Delta i_c / \Delta i_b = \beta. \tag{7.11}$$

The voltage gain of the whole amplifier, A_v, is related to the currents multiplied by the resistances that the currents go through. The signal current passes through parallel and series combinations of devices including the base of the BJT, but the collector current goes directly through R_L. Because i_c is β-times larger than i_b, the change in output voltage due to changes in V_{RL} is greater than the change in input voltage, so the voltage gain is also greater than 1.

Notice that when the input voltage increases, i_b increases, i_c increases, the voltage drop across R_L increases, and the output voltage *decreases*, i.e., it *gets closer to ground*. For this reason, the output voltage changes in the opposite direction as the input voltage changes, and the circuit is an **inverting amplifier.** Figure 7.23 shows the output signal as being 180° out of phase with the input signal, as it should be for an inverting amplifier. We will not derive the actual voltage gain of the circuit, as you will in later courses, but in lab you will build the circuit, measure the gain, and then amplify music from your music player (and listen for distortions). What are R_1 and R_2 for? Those are called **bias resistors**, and as a voltage divider, they set the voltage at the base, and hence the current going into the base, without any signal.

Now let's look at MOSFETs. Figure 7.24 is a cross-section of a MOSFET. MOSFETs come in two basic flavors, namely n-channel and p-channel, which describes whether current inside the MOSFET is carried by electrons or by holes, respectively (but not both, as this is a unipolar device). We limit this discussion to n-channel devices, n-MOSFETs, only because the voltages are all positive and we avoid a bunch of minus signs.

As shown in Fig. 7.24, the input to an n-MOSFET is a capacitor at the **gate** (G) formed by a metal

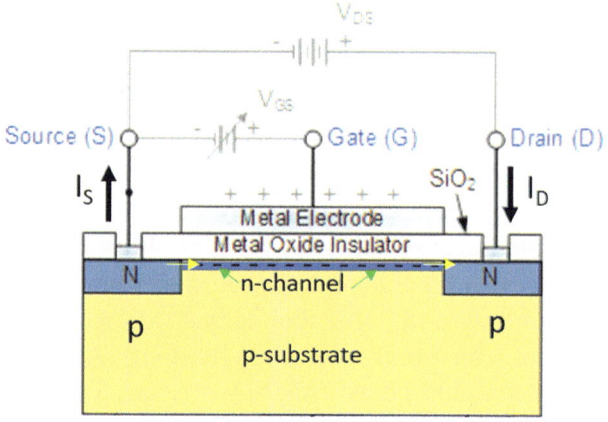

FIGURE 7.24 Cross-section of an n-channel MOSFET.

layer, which is the gate electrode, an oxide layer, the **gate oxide**, which is the capacitor's dielectric, and the p-type substrate, which forms the other electrode. The term **substrate** is reserved for the bulk piece of material that the transistor is built on. There are p-n junctions at the **source** (S) and **drain** (D). Both the gate-to-source voltage (**V_{GS}**) and drain-to-source voltage (**V_{DS}**) are positive when the MOSFET is operating. Note that the semiconductor under the gate is p-type Si, and the source and drain are both n-type Si. For a positive V_{DS}, the drain p-n junction diode is reverse-biased. That is a good thing, or current could flow into the substrate and then into the forward-biased source junction, effectively short-circuiting the entire transistor. Because of this reverse bias, no current flows through the drain until you do something to the gate voltage, as discussed next.

When the gate electrode is biased positively w.r.t. the source, and substrate, electrons accumulate on the negative plate just as you would expect in any capacitor. These electrons get there by entering through the S contact and go the bottom side of the gate oxide, the surface that is inside the MOSFET below the gate oxide. Electrons on the bottom surface of the oxide create a conductive layer called the **channel**. The channel electrons, shown in Fig. 7.24, connect the drain to the source with some value of resistance, and this connection allows electrons to flow from source to drain. As the names imply, when the MOSFET is conducting drain-to-source current, I_{DS}, electrons enter at the source and exit at the drain, as shown by the yellow arrows. Since the direction of current is opposite that to the motion of the electrons, the source current, I_S, leaves the MOSFET at the source contact and drain current, I_D, enters the MOSFET at the drain contact.

It might bother you that electrons can now flow against the direction of the reverse bias of the drain contact, where only the tiny leakage current can normally flow. It should. In a nutshell, those electrons from the source that reach the reverse-biased drain p-n junction are exactly like those that participate in leakage current, except that there are a lot more of them, so they *do* actually flow through the reverse-biased junction and out the drain contact. Without showing you any of the details of band diagrams, imagine that the drain is a waterfall, and those channel electrons fall over the waterfall and end up in the drain contact. This is not a trivial detail in the overall story. The details of this somewhat complicated phenomenon is explained fully in a course on semiconductor materials and devices.

Continuing, do electrons enter the MOSFET at the gate? Well, what do you already know about capacitors and how current flows through them? The gate is a very tiny capacitor, so DC current does not flow in, and only a very small amount of AC current flows 'through' the capacitor to the substrate. We will ignore that and just say that current does not flow into the gate. (Okay, I can't help myself: The AC current through the gate must be accounted for in accurate device simulations. Also, in modern devices, the gate oxide is so thin that a tiny amount of DC current *does* flow through. Although it is small for a single MOSFET, on a large integrated circuit the total of all that current is problematic, and a huge amount of effort is spent minimizing the gate leakage current. Again, we ignore it.)

So, how does a MOSFET act as a switch? Below a certain V_{GS}, there are no electrons in the channel, and I_{DS} is zero. Above a certain value of V_{GS}, called the **threshold voltage**, V_{th}, the channel begins to fill with electrons, acting like a resistor that connects the source to the drain, so current can flow due to V_{DS}. As V_{GS} increases, more electrons collect in the channel, lowering the channel resistance, so yet more current flows. In a very real sense, a MOSFET

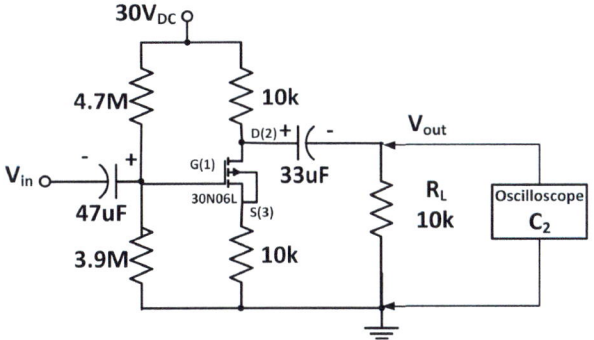

FIGURE 7.25 Analog amplifier using an n-MOSFET.

is an electrically controlled resistor. Hence, modulating the gate voltage modulates the drain current. That is the analog action of the MOSFET.

In very simple terms, if V_{GS} changes between 0 V and some value above V_{th}, it turns I_{DS} on and off, representing digital bits. This I_{DS} can charge other gates in other transistors, allowing digital circuits called **logic gates** to be implemented. Do this a billion times and you have a powerful microprocessor. The really great thing about MOSFET digital logic gates is that after they are switched, all of the capacitive gates charge up and essentially no more current flows. So, MOSFET-based logic uses much, much less power than the older BJT-based digital logic since at any point in time, only a small fraction of the transistors is being charged. Current flows continuously in BJTs using power to do so, so today's powerful microprocessors would be impossible if only BJT technology were available.

Let's repeat the analog amplifier exercise using an n-MOSFET instead of a BJT, as shown in Fig. 7.25. It actually works the same way as the BJT amplifier, but now instead of modulating the current into the base of a BJT, we change the voltage on the gate of the MOSFET. The drain current, I_{DS}, changes as the gate voltage changes, so the voltage across R_3 changes, thus modulating the output voltage. Everything else about its operation is exactly the same as the BJT amplifier, assuming that we choose the resistor values of R_1 and R_2 correctly to bias the voltage on the gate.

7.6 Operational Amplifiers

The simple transistor amplifier discussed above has many uses but could have shortcomings for advanced applications like high-fidelity music, accurate control of measurement equipment, or high-power or high-frequency operation. Therefore, there are many variations of transistor amplifiers using multiple transistors. The **operational amplifier**, or "op-amp," is one of the most basic and common circuits there is, and is available as an integrated circuit (IC), or "chip," costing from just a few pennies to several dollars, depending on the variety. An op-amp is used in your transistor-radio circuit to amplify the sound to drive the speaker.

The long-serving (since 1968), cheap and ubiquitous 741 op-amp IC is shown in Fig. 7.26. The pinout of the package is shown. The chip itself is inside the package and connected to the pins. The amplifier is shown as a triangle. Note that pin 1 is identified by a small dot that is either printed or embossed on the package. The pin at the minus sign (pin 2) is called the **inverting input** and the

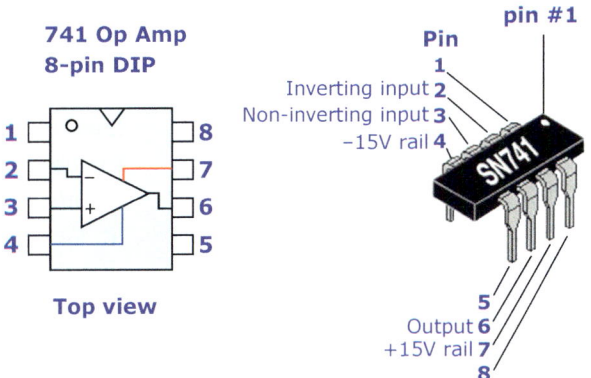

FIGURE 7.26 Schematic symbol (left) and physical packaging (right) of an SN741 op-amp.

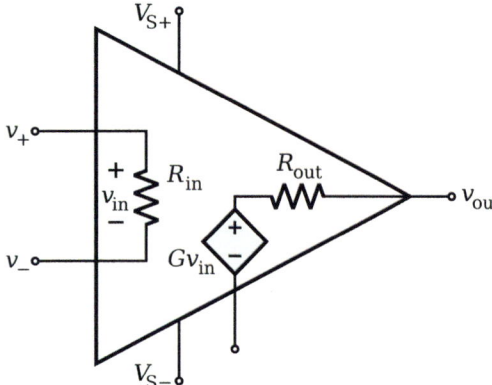

FIGURE 7.27 Model of the operation of an op-amp, as shown here without any external circuit components. Models are simplified versions of more complicated things. When interpreting how it works, just go with the simplicity of the model, and you won't be far from the actual operation.

pin at the plus sign (pin 3) is called the **non-inverting input**. The input voltage signal to the op-amp is across the inverting and non-inverting inputs. Some 741 op-amps come four to a larger package with more pins, and those are called "quad op-amps."

The symbol for an op-amp, shown in Figs. 7.26, 7.27, and 7.28, is based on a sideways triangle that is also generally used to represent an amplifier. An op-amp circuit internally uses several stages of amplification, meaning that the output of one stage is the input to the next stage. Just as we multiplied transfer functions in Chapter 4, the overall gain of the op-amp is the product of the gains of the individual gain stages. For example, three stages each having voltage gain 10^2 would have a total voltage gain, "G" shown in Fig. 7.27, of 10^6. This overall capability is made possible by using on the order of 20 transistors.

The components shown in Fig. 7.27 are just models for the more complicated actual circuit comprising about 30 resistors and transistors. We write circuit equations based on these simple models so that we don't have to analyze the full circuit. Some models seem so simple that your first reaction might be that you aren't really understanding. Many models are in fact, very, very simple. Here, you just model the behavior of the input of the circuit between non-inverting and inverting inputs as a resistor, R_{in}. What could be easier? Well, an open circuit would be easier. (To first order, the input to a MOSFET is often modeled as an open circuit). When I was an undergraduate EE student, it took me some time to just trust the simplicity of the models that I was given; I imagined that they were more complicated, and I must have been missing something. Nope. Just go with the models.

The diamond-shaped component in Fig. 7.27 labeled $G \cdot v_{in}$ is a **voltage-dependent voltage source**, meaning that it is like a power supply whose output voltage depends on another voltage somewhere else, which in this case is v_{in}. For the op-amp shown in Fig. 7.27 with no external components, the open-circuit output voltage is:

Deeper understanding: The next time you are at an oscilloscope, look at the signal with the input open. The trace will be a flat line, but when you turn up the sensitivity to, say, 10 mV/div, you will see how much electrical noise there is. There are many sources of electrical noise, including just the random motion of electrons in the devices. This is called Johnson noise. *The study of noise is very important to every area of electrical engineering. 'Noise' that comes from the environment, such as a nearby elevator or a fan motor, is better referred to as 'interference,' since 'noise' is generally reserved for random currents and voltages that result from the fundamental processes of electron motion in solids.*

$$v_{out} = G \cdot v_{in} = G \cdot (V^+ - V^-) . \qquad (7.12)$$

You might imagine that you provide in input voltage and the op-amp amplifies that input voltage by 10^6. That is called the **open-loop gain**, and the op-amp is being run "open-loop." That large amount of voltage gain, on the order of 10^6, is not very useful for amplification by itself because input voltages would have to be impractically small, on the order of microvolts, μV, to get a practical output voltage of only a few volts. Also, small amounts of electrical **noise** (random variations of the voltage) at the input, which are also typically in the μV range, would lead to a lot of noise at the output, after being amplified by a million times. (Noise is discussed in Chapter 9). Instead, op-amps are used in the **closed-loop** mode that uses feedback with all that gain to give the circuits useful properties with an overall moderate voltage gain that is in the same range as that of a simple amplifier, e.g., 10 to 100. **Feedback** refers to taking a voltage at the output and connecting it back to the input in such a way that the circuit has improved properties, as you will see next. The op-amp provides a great introduction to feedback.

Let's examine this point in more depth using the amplifier shown in Fig. 7.27, which is the op-amp and no other components. The voltages V^- and V^+ are referenced to the local ground. The op-amp is a **differential amplifier**, meaning that the output voltage is the internal gain, G, times the voltage *difference* between the two inputs, as in Eq. 7.12.

If v_{out} is reasonably small, say 5 V, then after dividing v_{out} by 10^6, you can see that $(V^+ - V^-)$ must be very small (in this case, only 5 μV). In the real world, without taking special precautions, that 5 μV might be lost in the electrical noise and environmental interferences that are commonly present. Therefore, an internal gain of 10^6 isn't in itself useful for

amplification by the op-amp. It is really only useful in the context of throwing much of it away to get good linear amplifier performance, as happens when feedback is used. We will write the equations to analyze an op-amp-based amplifier shortly.

Refer to Fig. 7.28. Working from the assumption that $(V^+ - V^-)$ must be very small, and that V^+ is set to ground, or zero volts, then V^- must also be very, very close to 0 V, again, on the order of μV. Since the operating voltages in the circuit are on the order of volts, we don't care much about a few μV, (what's a few μV among friends?) so we consider both V^+ *and* V^- to be at zero for the sake of calculations using relatively large numbers. V^- is not exactly zero volts, of course, because then there would be no actual input signal to the op-amp, but it close enough to allow us to write simple circuit equations that treat it as zero volts. This situation, in which V^- is nearly 0 V is called **virtual ground**.

In Fig. 7.28, an op-amp and two external resistors are used to make an inverting amplifier having moderate gain, much like the BJT amplifier discussed above. The inverting amplifier circuit is based around an op-amp using additional components to provide feedback from the output to the input, and to define the function of the overall circuit. This figure uses $V_{in,c}$ and $R_{in,c}$ as input circuit voltage and input circuit series resistance to distinguish them from V_{in} and R_{in} of the op-amp itself, shown in Fig. 7.27.

Another assumption of op-amps is that their input resistance R_{in} in Fig. 7.27 is very large, essentially infinite. We say that op-amps have "high input resistance," meaning that the

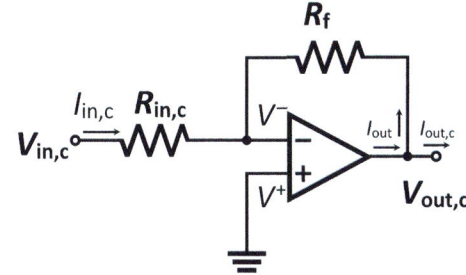

FIGURE 7.28 An op-amp circuit with external components that determine the overall gain and function of the amplifier. This circuit is an inverting amplifier.

voltage at V^- experiences, or "sees," a very high input resistance with respect to V^+. Because of that, and also that the differential voltage at the input is very small, the current into the op-amp is negligible, essentially zero.

Note that $V_{out,c}$ is not connected to anything. What good is a circuit that doesn't go to anything? It isn't expressly shown, but the assumption is that some load is connected between $V_{out,c}$ and ground, but it is a sufficiently large resistance that it doesn't affect the circuit analysis. If we did include that, we would take into account voltage drops due to the output resistance of the amplifier, as presented in Chapters 2, 4 and 5. Now that we have made the preceding assumptions, we can write simple loop and node equations to analyze the circuit behavior.

Just a quick note: R_f has subscript "f" because the voltage V_{out} is "fed back" to the input. So, "f" stands for 'feedback.' Again, $V^- \approx V^+ = 0$ because G is very large and V^+ is grounded. Let's get started as we refer to Figs. 7.27 and 7.28.

$$I_{in,c} = (V_{in,c} - 0)/R_{in}; \quad \text{(virtual ground)} \tag{7.13}$$

$$I_{out} = (V_{out,c} - 0)/R_f. \quad \text{(virtual ground)} \tag{7.14}$$

The two currents entering the V^- node sum to zero because the current into the op-amp is assumed to be zero:

$$I_{in,c} = -I_{out}, \quad \text{(KCL with infinitely large } R_{in}) \tag{7.15}$$

and we get:

$$\frac{V_{in,c}}{R_{in}} = -\frac{V_{out,c}}{R_f} \tag{7.16}$$

The voltage gain of the full circuit, A_v, is, then

$$A_v = \frac{V_{out,c}}{V_{in,c}} = -\frac{R_f}{R_{in}}. \tag{7.17}$$

Note that A_v is the gain for the full circuit, while G is the gain for the op-amp alone. The output waveform is a larger version of the input waveform by a factor of "A_v," and the minus sign means that the waveform is inverted, or upside-down, relative to the input. Thus, this circuit is an inverting, linear, voltage amplifier having gain that is set by the resistors, as given in Eq. 7.17. If you ever need a simple and effective amplifier, just grab a 741 op-amp and some resistors.

So, an op-amp internally has a huge voltage gain, G, but that can all be boiled down in the external circuit to a manageable voltage gain, A_v, that can be easily set by the simple choice of resistances. Op-amps are good for way more than just being amplifiers. They can be used to add analog signals together in summing circuits, can be used as

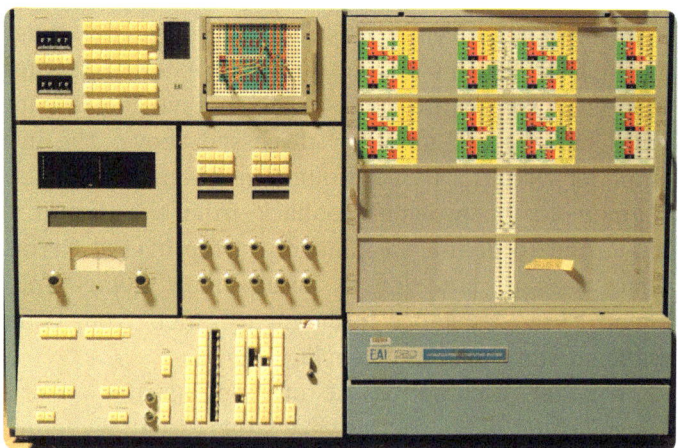

FIGURE 7.29 An old analog computer, EAI 580, introduced in 1957. This particular one was at the University of Notre Dame, and in 1995 was given to Prof. Dennis Bernstein at the University of Michigan who donated it to the Rhode Island Computer Museum. If interested, see here for some ND history: https://www.ricomputermuseum.org/collections-gallery/analog-computers/eai-580. Reproduced with permission from David Fischer (https://www.cca.org/).

filters of all types, and can compare two voltages, just to name a few applications.

Note, the input waveform is a function, $v(t)$, just like any mathematical function you are comfortable working with in your math classes. For example, sine, square and triangle waves are functions of time. If a capacitor replaces R_f, the circuit is called an **integrator**, whose output is the integral of the input. If a capacitor instead replaces $R_{in,c}$, the circuit is called a **differentiator**, whose output is the derivative of the input. Such functionality is the basis of an older type of computer called an **analog computer** (Fig. 7.29) that could model, and therefore solve, differential equations for controlling things like jet planes, factories, etc. Although analog computers, *per se*, are no longer used, the principles of these analog circuits are sometimes used in signal processing for audio processing, control systems, communications, and some specialized neural networks (for computer artificial intelligence).

7.7 Integrated Circuit Manufacturing

So far, we have discussed electronic materials, semiconductors, diodes, transistors and op-amps. Although the star of the semiconductor show is the transistor, the blockbuster semiconductor movie is the **ultra-large-scale integrated circuit, ULSIC**, or simply **IC**, or "chip." So far, we have mostly discussed separate, or **discrete**, transistors, and the small IC op-amp. Since transistors are made of semiconductors, one might once have reasonably asked, "Can we build more than one transistor on a single chunk of silicon?" Well, if you had asked that in 1958, you might have won a Nobel Prize, like Jack Kilby did, for doing just that. His work led to the development of integrated circuits. ICs have been evolving continuously ever since. Today

FIGURE 7.30 Packaged IC's (along with other components) soldered onto a printed circuit board.

there are many types of devices that can be built on an IC, including many varieties of transistors besides the MOSFET and BJT already discussed, as well as resistors, diodes, capacitors, inductors, and much more.

Figure 7.30 shows a printed circuit board with several black rectangular squares, which are the IC **packages** (you cannot see the ICs inside the packages). ICs have been developed for just about every application, including microprocessors, or computer chips, as well as amplifiers, power controllers, radios, etc. Figure 7.31 shows an IC in a (de-lidded) package that allows communication with the chip through very small wires, called **wire bonds**. Wire bonds are made of gold, aluminum or copper, and are on the order of about ¼ the diameter of a human hair, i.e., about 25 microns (1 **mil**, or thousandth of an inch). A somewhat more modern approach to packaging of ICs is the use of small malleable balls, called solder balls, that allow the chip to be flipped upside down and connected through hundreds of balls to contacts on the package. This is called the **flip-chip** process. There are literally dozens, if not hundreds, of proprietary ways of combining chips with packages. Wire bonds were the first, and is still quite common.

Manufacturing of ICs is an extremely expensive and complex process, but is done in such massive quantities that often the resulting chips are very inexpensive. That's how an op-amp with tens of devices

FIGURE 7.31 IC package. (Top) Microprocessor in package and (bottom) expanded view of small "wire bonds."

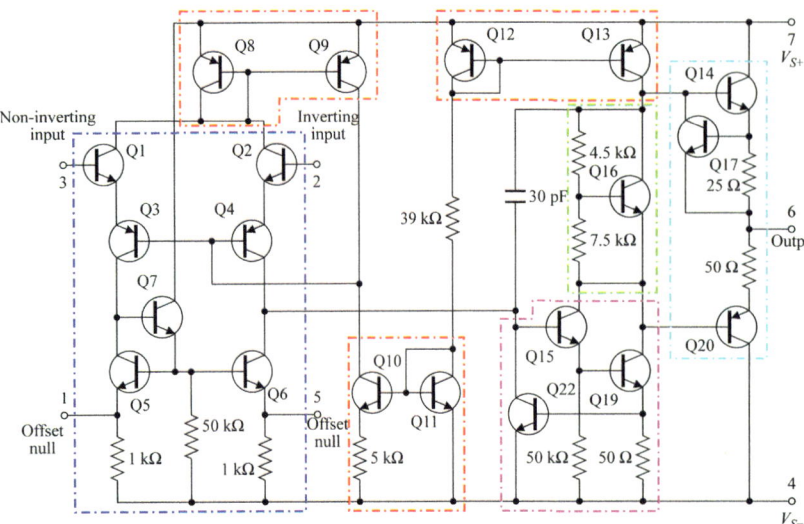

FIGURE 7.32 The schematic for SN741 op-amp contains multiple transistors (though still very few in the grand scheme of semiconductor devices). *Source*: https://commons.wikimedia.org/wiki/File:OpAmpTransistorLevel_Colored_Labeled.svg.

on the chip, as well as other much more complex circuits, can sell for all of a few pennies. The schematic in Fig. 7.32 shows the circuitry inside the 741 op-amp. An op-amp IC is an example of **small-scale integration**, meaning that it has relatively very few transistors. Nevertheless, if you tried to breadboard an op-amp, it would take up a considerable amount of space. Instead, it takes up a negligible fraction of a square millimeter on a silicon substrate. Modern ICs may have billions of transistors on them, with the most on a single chip being about 208 billion as of 2025 (NVIDIA's Blackwell B100 GPU). This is an almost unbelievable claim, but is nevertheless true. It is certain that within a few years, even these numbers will be less impressive as technology continues to evolve.

Figure 7.33 is a scanning electron microscope (SEM) image of part of an IC. This image shows relatively large features compared with today's most advanced technology but makes it clear that stuff on IC chips is small. The width of the region shown in the image is about 2 mm. For comparison, a human hair is about 100 microns, or 0.1 mm, in diameter. The features sizes shown on this chip are the highest level of metallization, and therefore are very large compared to the 0.005 micron (5 nm or 50 Angstrom) feature sizes of today's most advanced chips. Figure 7.34 is an optical microscope image (much lower magnification showing a large part of a chip) in which features are so small that almost no individual features like those shown in the SEM image can be discerned, but large functional blocks are visible. For a modern IC of about 1 cm on a side, almost nothing in Fig. 7.33 would be visible! In fact, the chip would have to be magnified to about the size of two football fields before the smallest features would be visible to the naked eye!

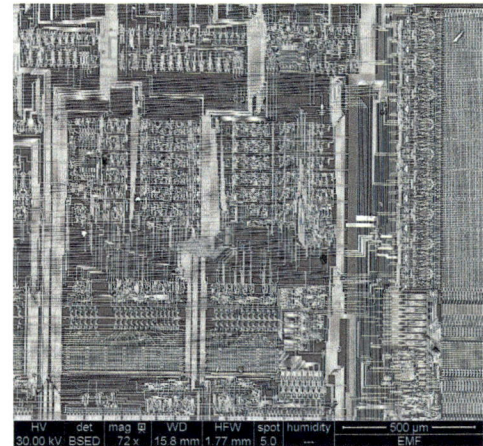

FIGURE 7.33 Electron microscope image of integrated circuit showing interconnect wiring. Reproduced with permission from https://wp.unil.ch/emf/examples-in-sem-2/.

How are such small features made on integrated circuits? The study of this field can occupy many entire careers, but here is a brief overview. IC chips are small (say, 1 cm^2) rectangles of semiconductor, usually Si, but often other materials. A single IC starts out as one small area on a large **wafer** that looks like a thin, flat dinner plate (Fig. 7.35). Larger wafers afford greater economies of scale – most processes are done to the whole wafer at once, and processing a larger wafer hardly costs more money. This lowers the cost per chip substantially. Today, the largest wafers in production are 300 mm (12 inches) in diameter.

Changing over to using larger wafers is a big, big deal for manufacturers. Such a change renders nearly all existing manufacturing equipment obsolete for the newest chips, and requires many billions

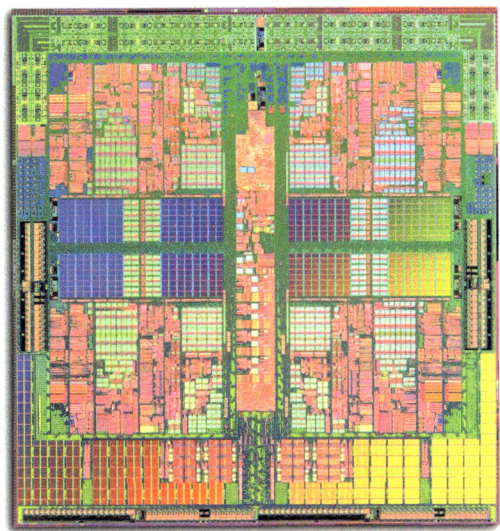

FIGURE 7.34 Optical-microscope image of a modern IC, showing the large functional blocks on the chip. *Source*: https://www.amd.com/us-en/Corporate/VirtualPressRoom/0,,51_104_572_573%5E15191~118854,00.html.

FIGURE 7.35 A Si wafer. Each rectangle is an IC prior to being cut out of the wafer.

of dollars of investment for a *single* manufacturing plant. Because a small group of companies who design and build new fabrication tools would have to invest many years and many billions of dollars on development, the most advanced semiconductor companies worldwide must agree on any changes in wafer size. A change happens only every couple of decades or so, and 300 mm wafers came out in 2002. Although the chip industry has considered increasing wafer size to 450 mm (18 inches) in diameter, this is almost certainly *not* going to happen within the next 20 years (private communication, Dr. Chris Mack.) One major issue is that the lithography step, discussed below, is done several dozen times on a wafer, and can be done only one chip at a time, so the economies of scale for larger wafers would not be as great as they have been in the past.

ICs are made in multi-billion-dollar manufacturing plants called **semiconductor foundries** or **wafer fabs**. Leading companies doing this in the United States include Intel, Texas Instruments, Micron, Global Foundries, TSMC, On Semiconductor, Samsung and others. (As of 2024, large-scale expansion or new fabs in the U.S. are planned by the top chip manufacturers, totaling some $500 billion over the next decade.) Wafers are processed in **cleanrooms** (Fig. 7.36) in which dust is nearly entirely eliminated, because even a tiny speck of dust is huge compared to features on a chip, and even one defect can ruin an entire circuit. Hundreds of processing steps are carried out in the cleanrooms. When the entire process is complete, the chips are tested on the wafer, and a saw or laser is used to cut the wafer up and separate out all the good, working chips that then go into the small packages shown above. Chip packages are then mounted onto printed circuit boards, as discussed in Chapter 1 and shown in Fig. 7.30.

How can features be made at such small size scales, as small as 5 nm? That is only about 45 atoms wide! Very small patterns are made on glass using fine beams of electrons in a process called **electron-beam lithography**. This is done much like patterning is done with an Etch-a-Sketch toy. The patterns made by the electron beam on the glass **photomasks** or **reticles** are a few times larger than the final optical patterns to be printed on the wafers.

The patterns on the glass reticle are projected using light of very short wavelengths onto the wafer surface in a process called **optical lithography**. Optical lithography is performed using highly sophisticated optical and mechanical systems, called **wafer steppers**, as shown in

FIGURE 7.36 Cleanroom in a semiconductor foundry.

Fig. 7.37. Wafer steppers are like optical microscopes used in reverse; rather than projecting something very small to a large screen, as would be the normal mode for a microscope, wafer steppers project the larger reticle pattern onto a smaller area. This somewhat reduces the extreme difficulty of making the small glass patterns on the reticles. In fact, reticles are extremely complex, with a set of them for a full IC costing upwards of $20M. Wafer steppers cost $50M to well over $100M *each*, with perhaps a *dozen* in use in a single foundry. The very newest advancement in this area uses light having extremely short wavelengths – called "extreme ultraviolet" light, whose wavelength is 13.5 nm, as compared to previous state- of-the-art "deep UV" (DUV) commonly used in industry now, whose wave-length is 193 nm. EUV tools can cost almost $400M each! That's for only one tool

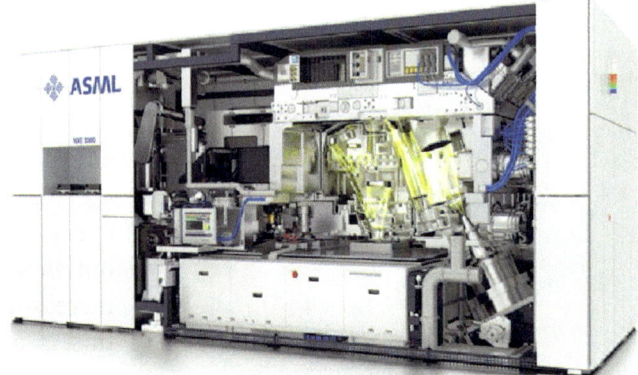

FIGURE 7.37 Modern DUV wafer stepper with the light path colorized over the photograph. A system like this can print features on integrated circuits that are 10,000 times smaller than a human hair. *Source*: ASML.

in a wafer fab, where the most advanced, high-volume fabs can use up to 30 of these, for an investment in just this one step of the manufacturing process of over $10B. This is only one component of the total cost of building an advanced fab at more than $20B.

Each optical lithography step sets the stage for a subsequent step, either the removal of material (called **etching**), or **deposition** of material. The shape and size are defined by the optical lithography step. In all, there are some 50 optical lithography steps required by a complete IC, and more than 500 total separate steps in the fabrication of a semiconductor wafer over several weeks from start to finish.

Great economies of scale are used to push more than 4M wafer starts per year through a wafer foundry, with each wafer having up to two thousand separate ICs on it. Of the individual chips that are started in the fab, at first only about 50% of those work well enough to sell. That fraction is known as the **yield**. Yields improve over time and approach about 95%, while manufacturing costs decrease. The most expensive of the advanced chips might sell for up to $2k. All of the relevant numbers are highly variable, but a new chip foundry can recoup its cost in 5 to 10 years.

One additional note: On August 9, 2022, the $280 billion CHIPS Act, which seeks to rebuild the U.S. microelectronics industry, was signed. As of 2023, the United States produces about 12% of the world's integrated circuits, which isn't very impressive considering that almost all of the technology came from the United States. This is a serious economic and security issue as the United States is dependent on other nations for the chips that go into most commercial and military systems. Although the results of the CHIPS Act will not be evident for several years, a significant increase in training in semiconductor manufacturing has begun, and several new fabs will come online starting in a few years. EEs will be in greater demand in the area of chip manufacturing and semiconductor technology in general.

7.8   Lab 7 Activities

1. **Building your circuit.** Set up the circuit in Fig. 7.38. In order to save some clutter at your circuit, use a BNC "tee" at the source output with a BNC (coax) cable directly to CH1 of the scope. That saves on using an extra probe from cluttering up your circuit. Use the glass diode with orange and black band (1N5230). This is a silicon Zener diode. Use R_s = 680 ohms. The resistor protects the diode by limiting the current. Put the black band (cathode) toward ground and

the orange band (anode) toward the resistor. Set up the scope probes as shown in Fig. 7.38. It is recommended to attach wires from the probes to the breadboard rather than attaching the probe directly to the circuit components. This will help you avoid pulling leads out of the breadboard.

Here is a review to help you understand and interpret today's lab. (Refer to Fig. 7.11, repeated.) When the input voltage is positive, the diode is *"forward-biased,"* and if the voltage is larger than the turn-on (or knee) voltage, V_t, of the diode, a current passes through it. V_t for a Si diode is around 0.6 V to 0.7 V. When the input voltage is negative, the diode is *"reverse-biased"* and the diode does not allow current to pass (ignoring breakdown).

Beyond V_t, the slope of the *I–V* curve is steep. Therefore, the diode voltage changes (ΔV) only a little bit even as the current changes (ΔI) significantly.

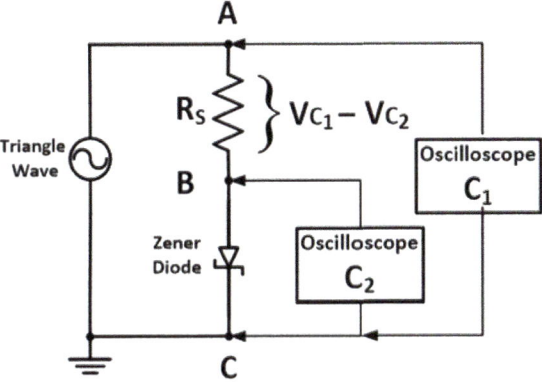

FIGURE 7.38 (Step 1) Circuit used to view *I–V* characteristic of a diode.

2. **Observing V_t of the Si diode.** Start with a *triangle* wave with V_{in} = 1 V_{pp} at 1 kHz. View CH1 and CH2 simultaneously. Be sure the scale is the same V/div on both channels. Also, be sure to press the **Vertical** knob to set the vertical offset to zero. For V_{in} below about 0.5 V or so, the diode voltage is below the threshold, as seen in Fig. 7.11, and almost no current flows.

You should see that almost all of V_{in} appears across the diode, which is acting electrically like an open circuit. Thinking like a voltage divider, why does all the voltage appear across the diode?

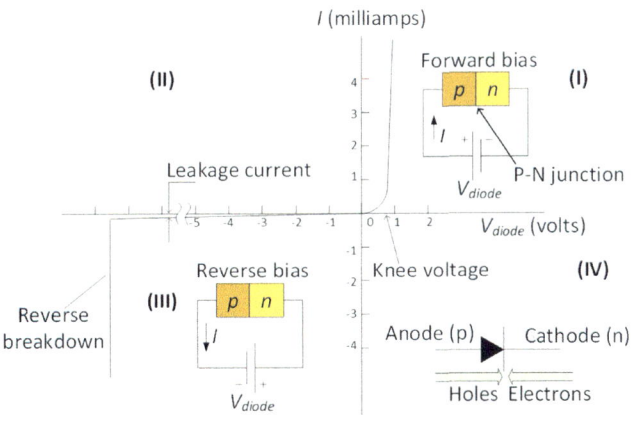

FIGURE 7.11 Repeated.

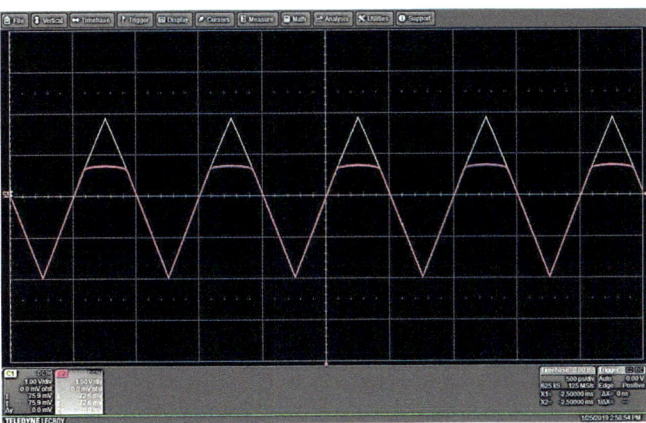

FIGURE 7.39 (Step 2) Waveforms of input voltage and clipped diode voltage.

Continue to increase V_{in} slowly. At some voltage, the input (CH1) will exceed the diode turn-on voltage (CH2). At what voltage does that happen? Continue increasing V_{in}. The voltage across the diode plateaus (i.e., maxes out and stays approximately constant) in the forward direction while the diode voltage less than that and in the reverse direction is the same as the input voltage, as seen in Fig. 7.39. What is the plateau voltage? How is that voltage related to V_t? As V_{in} increases beyond the turn-on voltage of about 0.6 V (in one direction), the diode begins to conduct current, and its forward voltage is limited to about V_t. That is called **clipping**. Again, this is the effect of Fig. 7.11 that shows that current increases very quickly just at the turn-on voltage in the forward bias. As a result, there is a relatively large range of currents for which the diode voltage hardly changes; the diode voltage now looks like a truncated triangle wave, i.e., the positive peaks are clipped, as shown in Fig. 7.39. Explain why the two curves overlap for all voltages less than the turn-on voltage, and in the reverse direction.

Now increase V_{in} to 16 V_{pp}, or 8 V_p negative. See Fig. 7.40. (Make sure you have the scale, volts/division, high enough to see the features.) Now the negative peaks in the diode voltage start to clip at V_{br} (the breakdown voltage), which is around 4 V. This is because the diode is undergoing reverse breakdown (again, see Fig. 7.11), and now current is flowing through the diode and the resistor, but in the reverse direction.

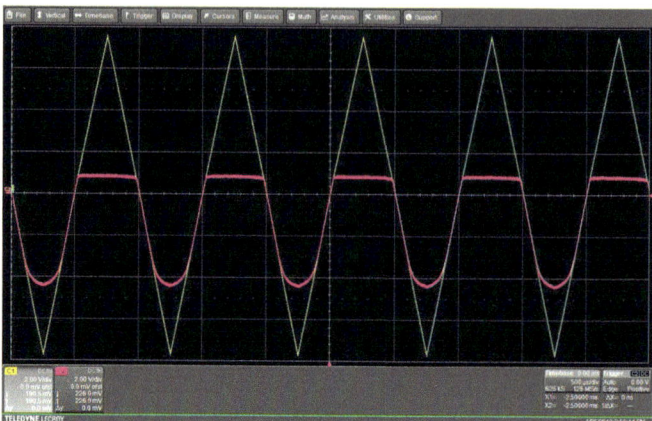

FIGURE 7.40 (Step 2) Waveform for diode voltage clipped in the forward and reverse voltage directions.

Measure the reverse breakdown voltage accurately at its maximum value. Just as in forward bias, the current changes over a wide range while the voltage hardly changes. Compare the shapes of the diode voltage curve in forward bias and reverse breakdown. Based on that, which regime has the higher dI/dV slope? Also, explain why the two curves overlap for the voltages that are between the turn-on voltage in the forward direction and the breakdown voltage in the reverse direction, i.e., over most of the purple curve. Turn off the WG and leave everything as it is.

3. **Observing voltage across two back-to-back Si diodes.** Now add a second Zener diode in series between the first one and ground, with the black band adjacent to the first diode's. These diodes are now "back-to-back." At low voltages, they will not pass any forward current because one of them is always reverse biased. But, they can still break down at higher AC voltages, one or the other, but not both at once. Why can't both diodes be in breakdown at the same time?

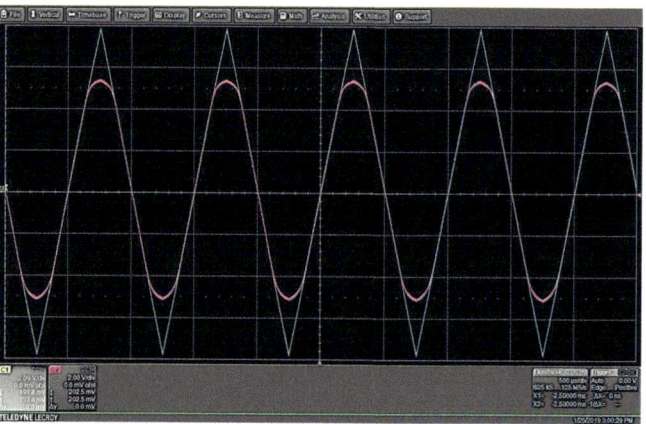

FIGURE 7.41 (Step 3) Waveform for two back-to-back Zener diodes, clipped in breakdown in each direction.

Now, CH2 should show both the positive and negative peaks clipped at around 5 V, as shown in Fig. 7.41. Using what you know about when current flows through the Zener diode, explain why the purple curve is symmetric in response to the input voltage.

Measure the breakdown voltage in both directions and take their average. Are they exactly the same? If not, by how many volts do they differ? They won't be exactly the same, but they will be close. Compare the breakdown voltage from Step 2 with the average breakdown voltage you just measured. Explain the difference between the breakdown voltages in this and the previous step in terms of the voltage drops of both diodes in series, where one is in breakdown and the other is in forward conduction.

4. **Introducing the light-emitting diode.** In this step, you will visually confirm the current passing through a diode! A **light-emitting diode**, or **LED**, is a diode that gives off light when current passes through it. In steps 2 and 3, you learned that a diode allows current to pass only in one direction, i.e., when forward biased. In forward bias, the LED diode glows! V_t will be larger because the material is not silicon – closer to 2 V whereas V_t for a Si diode is around 0.7 V. The LED material has a larger bandgap, which increases V_t. When the LED is reverse biased, i.e., the input voltage is negative, the LED does not conduct current, and does not emit light.

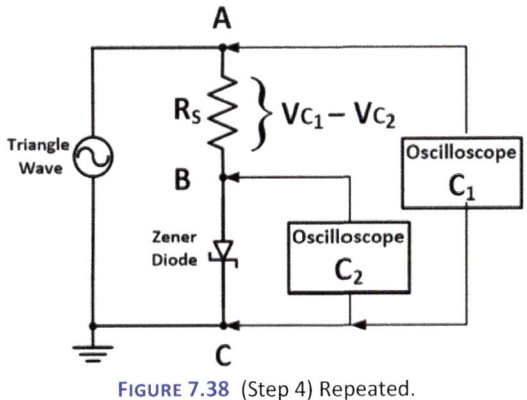

FIGURE 7.38 (Step 4) Repeated.

   Use the LED in the same configuration as Step 2, Fig. 7.38. The Zener had a black band on one end (cathode) that you connected to the ground, but your LED lacks that indicator. However, you can figure out the cathode of the LED by the length of its pins. The shorter pin is the cathode, and goes to ground. If you insert it backwards, this step will still work, but your waveforms will be inverted.

   Now the input frequency must be much lower in order to see the LED blink on and off. If the input frequency is high, say 1 kHz as in Steps 1 and 2, then the LED turns on and off with this high frequency, but your eye is not able to detect it at such a high rate. So, you will test the LED with a very low frequency for the input waveform. Set the frequency of the input triangular waveform to 10 Hz on the waveform generator.

   The following is similar to what you did in Step 2. Start with V_{in} = 1 V_{pp} (which is only 0.5 V_p) on the waveform generator and slowly increase it. While you look at the waveforms, i.e., the input voltage and the voltage across the LED, on the scope, notice when the LED starts to blink. At what peak voltage, V_p, does the LED barely turn on? This is one way to measure the LED turn-on voltage, V_t. Increase the input voltage gradually up to 8 V_{pp} while watching the LED brightness. Describe how the brightness of the LED changes as a function of the input voltage peak.

   At the negative peak of the input the LED is reverse-biased so, as you may expect, no current passes through the LED and so it should be turned off. How about the positive voltage? Observe the plateau voltage of the diode on the scope. What is V_t measured this way? Does it more or less agree with the voltage when you started to observe light flashing?

   Very slowly increase the frequency beyond 10 Hz. At what frequency can you no longer discern the blinking of the LED? If you are interested, later have a look at https://en.wikipedia.org/wiki/Crab_Pulsar for an interesting story about the limit of ability to discern flashing light.

5. **Displaying current through the diode.** You will employ the following instructions for your HDO-4104 LeCroy oscilloscope to do the differential measurement of resistor voltage and then display that voltage (multiplied by a proportionality constant to give the diode current) versus the voltage across the diode. In short, you will create your own diode *I–V* curve.

   Go back again to the circuit of Fig. 7.38, Step 1; i.e., just replace the LED with the silicon Zener diode. Set V_{in} to 20 V_{pp} at 100 Hz. CH1 measures the voltage across the whole circuit, i.e., sum of the voltages over the resistor and the diode. The CH2 probe measures the voltage across the diode directly. Therefore, in this setup, the voltage across the resistor is the difference between the CH1 and CH2 measurements.

In order to obtain the difference between CH1 and CH2, use the F1 Math function. To do this: Select **Math** at the top of the screen, **F1 Setup**. The trace window of **F1 Math** function appears. Select **Source 1** and **C1** and then **Source 2** and **C2**. Press **Operator 1, Basic Math** and **Difference**. To see all three traces on one plot, press **Display** at the top of the screen and **Single Grid**. Now, F1 is set up to show the voltage across the resistor, which again is proportional to current through both the resistor and diode.

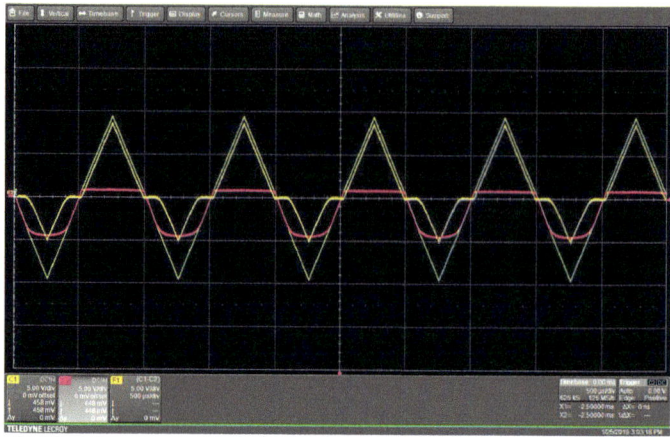

FIGURE 7.42 (Step 5) Waveforms of CH1, input voltage (yellow), CH2, diode voltage (pink), and F1 = CH1 − CH2, current (also yellow, lower magnitude).

Now, CH1 shows the voltage across the whole circuit, i.e., V_{in}. CH2 is the voltage across the diode, and F1 is the voltage across the resistor (current). Show all the corresponding waveforms on the scope, like that shown in Fig. 7.42.

6. **Scaling the resistor voltage trace to amperes.** The current through the diode and resistor are the same. Multiplying F1 by $1/R$ gives the current. To do this we use the F2 Math function. Show the waveform and values for the *current* by using the F2 Math function of the scope. To do this: Press **Math** and **F2 Setup, Operator 1, All Functions, Rescale, Source 1, Math (or All), F1.** Now, the F2 Math function is chosen to be a rescaled version of F1. The value of rescaling can be set in the right section of the F2 trace window. Press

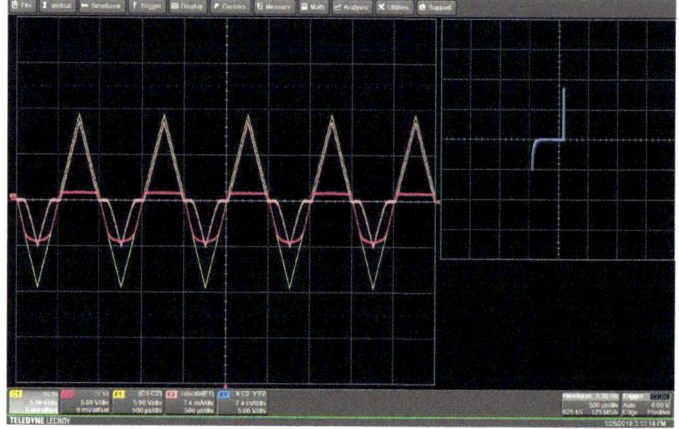

FIGURE 7.43 (Steps 6 and 7) Waveforms of CH1, CH2, F2, and X-Y. The diode characteristic is shown at the right.

First multiply by and enter the value of rescaling. For our case, this is equal to $1/R$, where R is the value of the resistor in your circuit (680 Ω). Then add 0 V. Choose **Override units** and enter the output units as "A" for amps. It automatically reads in mA. You can see that in Fig. 7.43. You should now have the very same curve as previously, but the units for the current will be in mA/div.

7. **Displaying an X-Y plot of a diode characteristic.** In this step you will observe the current-voltage, or *I–V*, characteristic (curve) of the Zener diode. The *I–V* curve is a plot of the current through the diode as a function of the voltage across the diode, as shown in Fig. 7.11 and explained in Sections 7.4.5 and 7.4.6. Also, review Section 7.4.6.1 if you haven't read it carefully. Oscilloscopes always offer a mode called "X-Y" that replaces the time base with a voltage source, and therefore plots one voltage versus another voltage. You know that the voltage across a resistor is proportional to the current through it. That will be the 'Y' in your 'X-Y' plot. The 'X' will be the voltage across the diode.

Now you have the current through the diode on F2 and the voltage across it on CH2. Next plot F2 versus CH2 in the X-Y mode to observe the characteristic curve of the diode. To do this: Press **Display** and **XY Grid**. A trace window appears. Also, instead of the normal display of the scope, a new square window appears in which you are going to plot the characteristic curve. In the above trace window, press **Input X** and **CH2**. This defines the x-axis of the XY display. You want your x-axis to be the voltage across the diode that is measured in CH2. To define the y-axis, press **Input Y** and **F2**.

In the trace window of X-Y display, at the left, there are three buttons that you can use to select the display mode of the scope (and are also available under **Display** at the top of the screen):

- **Single Grid** gives the common display of the scope, the measured waveform versus time appears.
- **XY Grid** gives the XY grid on the scope display.
- **XYSingle Grid** gives two windows side-by-side. The left one is the waveform and the right one is the XY-mode. Choose this mode.

Select V_{in} = 12 V_{pp}. Figure 7.43 shows the entire screen, including the diode characteristic that you should be viewing.

Zooming the curve on **XY Single**:

- **Horizontal zoom:** In order to zoom your characteristic curve in the horizontal direction, first you should identify which waveform is assigned to the x-axis of the XY-display. In our case, it is the voltage across the diode, which is CH2. So, activate the trace of CH2 by pressing the C2 box just above the trace window. If there is no such box above the trace window, the CH2 channel is turned off. Just turn it on by pressing the CH2 button on the scope panel. Now, by turning the **V/div** knob on the scope panel you can zoom in and out of the horizontal axis of XY-display. Also, note that, by turning this knob, the CH2 waveform in the other window is also being zoomed in and out at the same time, but in the vertical direction.
- **Vertical zoom:** In order to zoom your XY characteristic curve in the vertical direction, first recognize which waveform is assigned to the vertical direction (y-axis) of the XY-display. In your experiment, it is the diode current. You measured this current by the F2 Math function. So, simply activate the trace of F2 by pressing the F2 box just above the trace window. If there is no such box, it is deactivated. Under the Math menu on top of the screen, select **F2 Setup**. The F2 trace will be activated and the corresponding waveform appears on the scope. Now, by turning the **V/div** knob on the scope panel you can zoom in and out the vertical axis of XY-display. Again, it is interesting to notice that by doing this, the F2 waveform in the other window is also being zoomed in and out at the same time, in the vertical direction.

Notice – Very important!!! If the corresponding waveform assigned to one of the axes in the XY-display is zoomed outside of the screen on the waveform-vs-time display of the scope, it will not be displayed on the XY-curve! If your X-Y plot looks bad, check for this first!

Comment on any issues you had setting it up and show your results. Using the XY box at the bottom of the screen, estimate on the XY plot what the turn-on and breakdown voltages are. Note that you have made your own curve tracer like the one you will use in Step 10.

8. **Displaying the *I–V* curve of an LED.** Now, plot the characteristic curve of the LED in the same manner as in step 8 that you did for the Zener diode. Just as before, replace the Zener diode in your setup of Step 5 with the LED. You may safely increase V_{in} up to 20 V_{pp}. What do you observe on the XY grid? Compare it with the curve you observed for the silicon Zener diode. What differences are there between the two? Also, do they have the same V_t? Can you measure V_{br}? What does that tell you about V_{br}?

9. **Observing the limitations of a current probe for low currents.** Repeat this experiment with the LED but using the current probe instead. You don't need to disconnect any wires or change any settings except for one. Plug the current probe into CH4 and instead of using F2 for the input Y, use CH4. Degauss and zero the probe. Wrap the current probe around any wire in series with the resistor and diode. You should get a diode characteristic plot that is similar to the one in the previous step. Comment on your success. Is one better than the other? Which method gives a worse *I–V* curve? Why? Remove the current probe and carefully put it back in its storage box.

10. **Using a dedicated curve tracer.** Next you will use a dedicated **curve tracer** to observe the diode *I–V* characteristics. The curve tracer does all the work of the oscilloscope that you did, and can also show more complicated traces under a variety of conditions. Additionally, it can be used for transistors. The curve tracers are labeled "Tektronix 571." These are older tools but are still very useful for measuring the properties of common semiconductor devices.

FIGURE 7.44 (Step 10) Tek 571 curve tracer. The diode is inserted into the clips as shown at the upper right.

Turn on the curve tracer at the lower right of the front panel. Slide the plastic cover up, if it isn't already up. Slide the Zener diode into the two connectors labeled **DIODE** as shown in Fig. 7.44. When inserting devices, make sure that they are in contact with the small metal clips inside the plastic outer housing. The diode should have the black band to the left. Press **Menu.** Pressing the arrow keys up, down, left and right selects the line and the parameters. Input the following parameters for the Zener diode in forward bias:

Function: Acquisition

Type: DIODE

V_a max: 1 V

I_a max: 10 mA

R_{load} = 0.25 ohm

P_{max} = 0.1 W

Now press **START**. You should get a characteristic like that shown in Fig. 7.45. This is the forward-biased current region of the curve. Approximately what do you get for a turn-on voltage of the diode? Take a picture of the display for your report, or draw it in your notebook. Compare your result with that of Step 7.

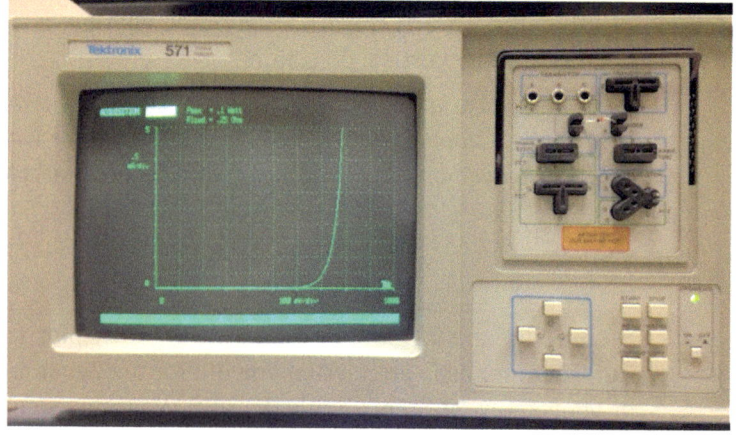

FIGURE 7.45 (Step 10) Forward-biased diode I–V characteristic displayed on the curve tracer.

11. **Measuring the breakdown voltage of a Si diode.** Now turn the diode around so that the black band is to the right. Press **Menu** to return to the **Settings** page. Change V_a max to 10 V and I_a max to 20 mA. Press **START**. This is the reverse-biased voltage part of the curve. What voltage does the diode break down at? How does the reverse curve compare to that which you found in Step 7?

    Does the curve look steep? That is the point of a Zener diode – it can support a wide range of current for a very narrow range of voltages in the reverse breakdown region. If this diode is in a circuit, it can regulate the voltage across anything that is in parallel with it to a very narrow voltage range near the Zener voltage. That is the subject of Chapter 10.

12. **Measuring the turn-on and breakdown voltages of an LED.** Repeat Steps 10 and 11 with the LED. Note that the plastic lid must be closed for voltages in this range. You should see that the breakdown voltage of the LED is close to 50 V and the turn-on voltage is greater than 2 V. *Do NOT try to defeat the safety feature.* Figure out for yourself what parameters you need to change. Compare your results with what you got in step 8. Answer all the same questions.

13. **Building a BJT amplifier.** (If instructed, skip to step 19 and follow Steps 19–24 for the MOSFET amplifier. All the steps are the same except for the details of the amplifier.) Next you will build a simple bipolar junction transistor (BJT) amplifier. The transistor will be of the 2N2222A "NPN" type, as shown in Fig. 7.46. After determining the part number, use the figure to determine the pin configuration. Label the pins in your notebook before building your circuit in order to avoid mistakes.

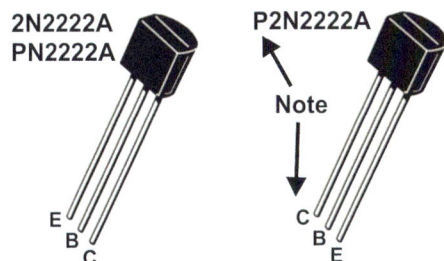

FIGURE 7.46 (Step 13) Pinout and schematic diagram of two NPN BJTs.

You will build the simple BJT amplifier shown in Fig. 7.47. Don't worry that you don't know the details of how this circuit works, since you will get that in your upcoming electronics course. For now, you should just have some fun practicing to build and test a real and useful circuit. Today you will get practice injecting a signal at the input and viewing a voltage at the output. This transistor circuit should have voltage gain (i.e., $A_V = V_o/V_i$) having a magnitude greater than 1. Since this is an inverting amplifier, the gain is negative.

Breadboard the BJT amplifier shown in Fig. 7.47. V_{in} is the signal from the waveform generator. I will wait here for you to finish. Take your time and double check everything. You don't want to make a mistake.....

Done?

Now, use CH1 for V_{in} and CH2 for V_{out}. For V_{in}, use a sine wave at 1 kHz and V_{in} = 0.5 V_{pp}. In order

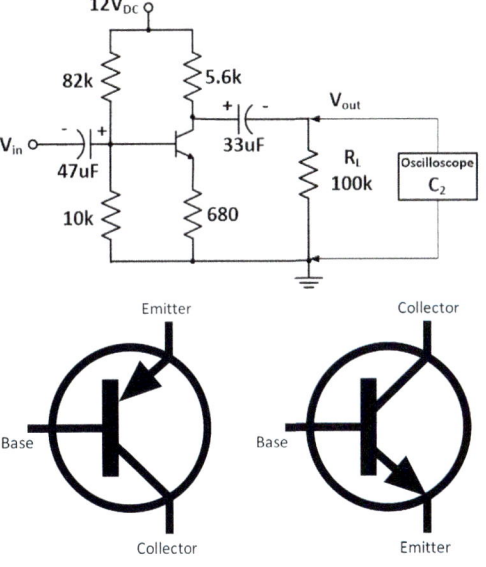

FIGURE 7.47 (Step 13) Schematic of simple BJT amplifier and BJT symbols.

to save some clutter at your circuit, use a BNC "tee" at the source output with a BNC (coax) cable directly to CH1 of the scope. That saves on using an extra probe from cluttering up your circuit. See Fig. 7.47. Set the DC power supply voltage to 12 VDC. In case you don't remember: Press **CH1**; **V-Set**; 12; **Enter**. When you are ready to power up your circuit, press **Output On/Off.**

14. **Observing gain and phase shift from input to output.** Set the sensitivities (V/div) of CH1 and CH2 of the scope to allow both waveforms to be clearly evaluated. Viewing both the input and output on the scope at the same time, you should see something like the traces of Fig. 7.48. What is the gain of your amplifier? Observe the phase between the input and the output. Can you see why this is called an "inverting amplifier?" Explain.

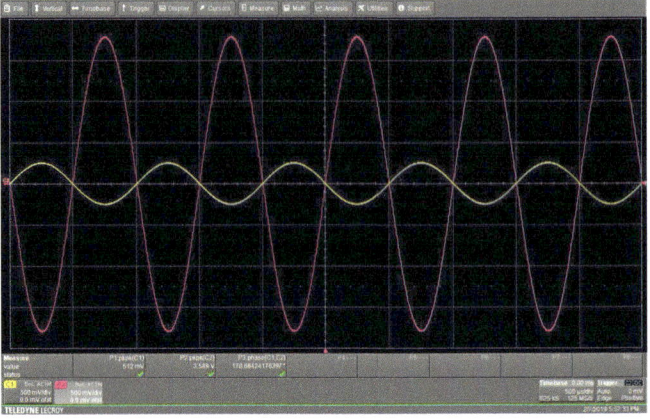

FIGURE 7.48 (Step 14) Waveform of input and output of the BJT amplifier. The output is larger than the input, so the gain is greater than 1. Also, it is evident that the two signals are out of phase by 180°. Measurements of peak voltages and phase are in the lower part of the screen shot.

15. **Checking distortion of sine wave from a waveform generator.** Look at the frequency spectrum (FS) of the input waveform from 0 kHz to 10 kHz (CH1). See Fig. 7.49. Don't forget that the best result for a spectrum on the scope is achieved by showing so many periods in the time domain that all of the features blend together, i.e., you want to display very many, say 100, periods in order to get a good FFT spectrum. Set the spectrum scale to dBmV. Make sure **Peaks** is selected under **Peaks and Markers.** For a perfect sine wave there should be only one frequency component. Is that what you observe? Remember that dB readings are based on RMS values. Convert the peak dBmV value to V_{rms} and then to V_p. Does it agree with the peak voltage as measured on CH1?

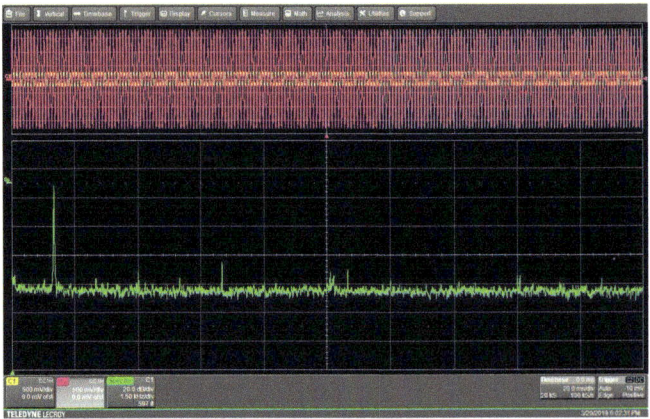

FIGURE 7.49 (Step 15) FFT of input sine wave at 0.5 V_{pp} and 1 kHz.

16. **Checking distortion of sine wave from a BJT amplifier.** Now, look at the FS of the *output*. (See Fig. 7.50). How does that FS compare to that of the input?

    For a perfect sine wave there should be one frequency, but there may be some small harmonics. Are there?

    How many "dB down" is the second harmonic? By that, we mean how many dBs difference is there between the fundamental and the second harmonic? That number in dB is the ratio of the first harmonic (the fundamental)

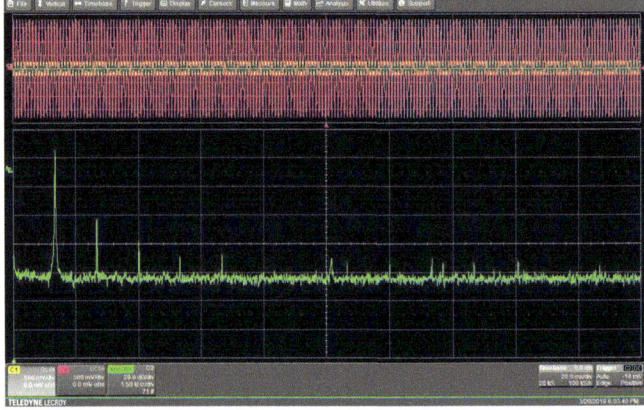

FIGURE 7.50 (Step 16) FFT of output sine wave at 0.5 V_{pp} and 1 kHz. If the harmonics are still very low, then you have made a clean, linear amplifier.

to the second harmonic. You can calculate the ratio of the two frequency components of your not-so-perfect sine wave using: difference in dB = 20log (ratio of first harmonic to second harmonic). How many times smaller is the second harmonic compared to the first? If the second harmonic is much smaller, then you have a very "linear" amplifier. Comment on any additional peaks. Do you think you could just look at the output sinusoid and determine "how distorted it is" without using the spectrum analyzer? Comment.

17. **Observing harmonics generated by distorted amplifier output.** Now, increase V_{in} until V_{out} clips, i.e., the amplifier output is "saturated." This means that the sine wave at the output becomes flat at the top and bottom. The output will be clipped after V_{in} gets to about 1.5 V_{pp}. Increase V_{in} to 2 V_{pp}. What is the FS of the output now? (See Fig. 7.51.) Are there more harmonics and are they larger? Does it remind you of the FS of a square wave from Chapter 4? What does the clipped output have in common with a square wave in the time domain? Repeat the calculation of the

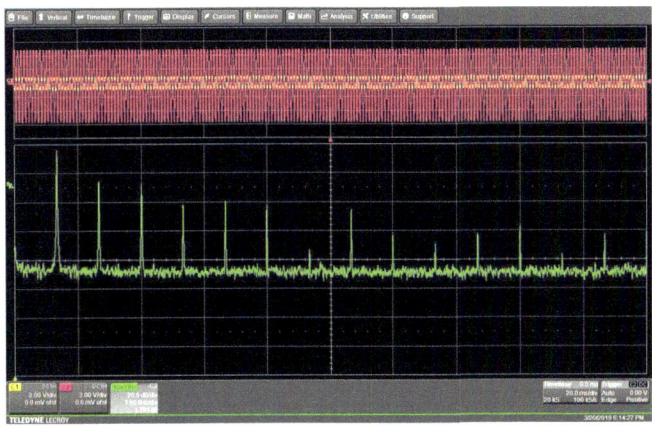

FIGURE 7.51 (Step 17) FFT of output sine wave at V_{in} = 1.8 V_{pp}. The significantly higher harmonics indicate that there is a considerable amount of distortion in the output.

previous step to find the ratio of the first harmonic to the second harmonic. Is it much smaller now? Do you still have a linear amplifier when the input voltage is so large that clipping occurs?

18. **Measuring frequency response of BJT amplifier.** What do you guess is the frequency response of the amplifier, i.e., how high of a frequency do you think it can operate at? The frequency at which the gain, A, of an amplifier drops to 1 is called the "transition frequency," denoted f_T. Let's test your intuition – write down in your notebook your guess of f_T. No points off for wild guesses.

Reset your input voltage to that of Step 13. Increase the frequency until the output voltage of the sine wave is the same as the original input voltage. It is recommended that rather than view the trace of the signal, you read the "measure" function of the voltages. Be sure that the traces are inside the upper and lower limits in order to get proper readings. Approximately what is the transition frequency, f_T, of your amplifier? Note that f_T for your amplifier may actually depend more on the breadboard stray capacitance than it does on the BJT operation. f_T of the transistor is actually somewhere between 100 MHz and 300 MHz.

Do not tear down your BJT amplifier. You will need it later. Use other components to build the next circuit.

19. **Building a MOSFET amplifier.** (If instructed, you might build the BJT amplifier instead of this one with a MOSFET, Steps 13 to 18. All the steps are the same except for the details of the amplifier). Next you will build a simple MOSFET amplifier. The MOSFET is the FQP30N06L n-channel MOSFET as shown in Fig. 7.52. Use the figure to determine the pin configuration. Label the pins in your notebook before building your circuit in order to avoid mistakes.

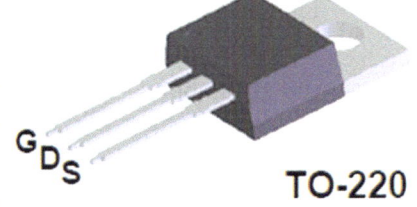

FIGURE 7.52 (Step 19) Pinout and schematic diagram of the FQP30N06L n-channel MOSFET.

You will build the simple MOSFET amplifier shown in Fig. 7.53. Don't worry that you don't know the details of how this circuit works, since you will get that in your upcoming electronics course. For now, you should just have some fun practicing to build and test a real circuit that could be used for all kinds of useful things. Today you will get practice injecting a signal at the input and viewing a voltage at the output. This transistor circuit should have voltage gain (i.e., $A_V = V_o/V_i$) > 1.

Breadboard the MOSFET amplifier shown in Fig. 7.53. V_{in} (V2 in the diagram) is the signal from the waveform generator. I will wait here for you to finish. Take your time and double check everything. You don't want to make a mistake.....

Done?

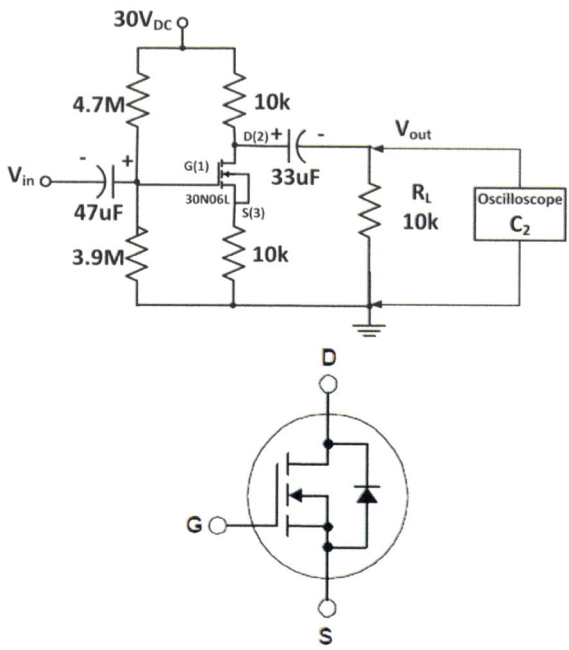

FIGURE 7.53 (Step 19) Schematic of simple MOSFET amplifier and MOSFET symbol.

Now, on your scope, use CH1 for V_{in} and CH2 for V_{out}. For V_{in}, use a sine wave at 1 kHz and V_{in} = 0.1 V_{pp}. In order to save some clutter at your circuit, use a BNC "tee" at the source output with a BNC (coax) cable directly to CH1 of the scope. That saves on using an extra probe from cluttering up your circuit. See Fig. 7.53. Set the DC power supply voltage, V1, to 30 V_{DC}. In case you don't remember: Press **CH1**; **V-Set**; 30; **Enter**. When you are ready to power up your circuit, press **Output On/Off**. What voltage gain, $A_V = V_{out}/V_{in}$, do you measure?

20. **Observing gain and phase shift from input to output.** Set the sensitivities (V/div) of CH1 and CH2 of the scope to allow both waveforms to be clearly evaluated. Viewing both the input and output on the scope at the same time, you should see something like the traces of Fig. 7.54. What is the gain of your amplifier? Observe the phase between the input and the output. Can you see why this is called an "inverting amplifier?" Explain.

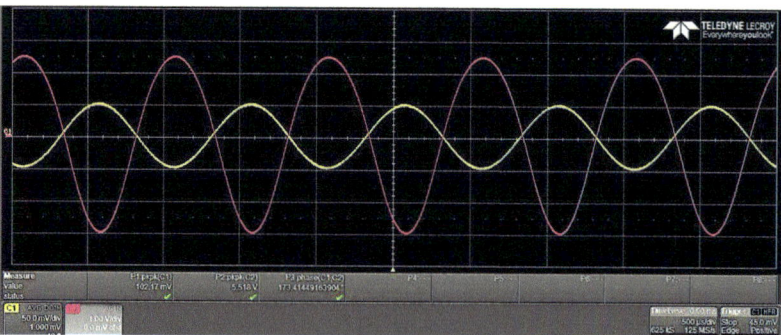

FIGURE 7.54 (Step 20) Waveform of input and output of the MOSFET amplifier. The output is larger than the input, so the gain is greater than 1. Also, it is evident that the two signals are out of phase by 180°. Measurements of peak voltages and phase are in the lower part of the screen shot.

21. **Checking distortion of sine wave from a waveform generator.** Look at the frequency spectrum (FS) of the input waveform from 0 to 10 kHz (CH1). See Fig. 7.55. Don't forget that the best result for a spectrum on the scope is achieved by showing so many periods in the time domain

that all of the features blend together, i.e., you want to display very many, say 100, periods in order to get a good FFT spectrum. Set the spectrum scale to dBmV. Make sure **Peaks** is selected under **Peaks and Markers.** For a perfect sine wave there should be only one frequency component. Is that what you observe? Remember that dB readings are based on RMS values. Convert the peak dBmV value to V_{rms} and then to V_p. Does it agree with the peak voltage as measured on CH1?

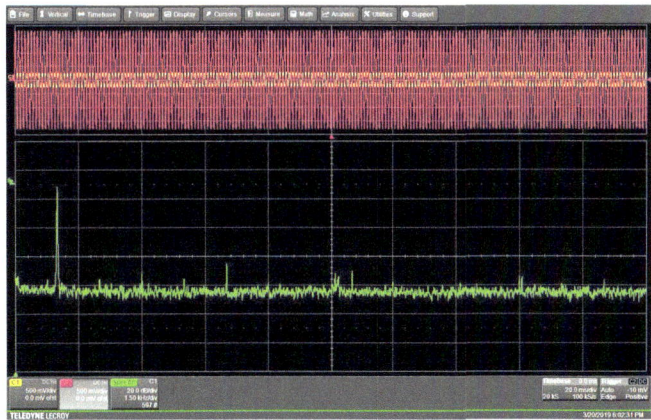

FIGURE 7.55 (Step 21) FFT of input sine wave at 0.5 V_{pp} and 1 kHz.

22. **Checking distortion of sine wave from a MOSFET amplifier.** Now, look at the FS of the *output.* (See Fig. 7.56.) How does that FS compare to that of the input?

For a perfect sine wave there should be one frequency, but there may be some small harmonics. Are there? In fact, MOSFET amplifiers are not all that linear – they are less linear than BJT amplifiers. If you built the BJT amplifier, compare your results to those shown in Step 16, Fig. 7.50. Qualitatively, what is the difference between the spectrum in Fig. 7.50 to your spectrum?

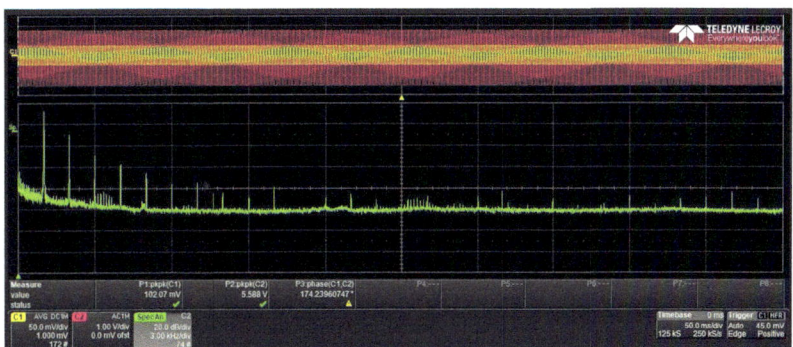

FIGURE 7.56 (Step 22) FFT of output sine wave at 0.5 V_{pp} and 1 kHz. If the harmonics are still very low, then you have made a clean, linear amplifier.

How many "dB down" is the second harmonic? By that, we mean how many dBs difference is there between the fundamental and the second harmonic? That number in dB is the ratio of the first harmonic (the fundamental) to the second harmonic. You can calculate the ratio of the two frequency components of your not-so-perfect sine wave using: difference in dB = 20 log (ratio of first harmonic to second harmonic). How many times smaller is the second harmonic compared to the first? If the second harmonic is much smaller, then you have a very "linear" amplifier. Again, the MOSFET amp isn't very linear. Comment on any additional peaks. Do you think you could just look at the output sinusoid and determine "how distorted it is" without using the spectrum analyzer? Comment.

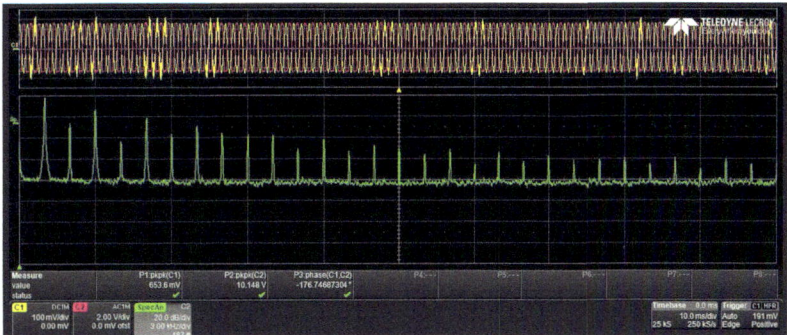

FIGURE 7.57 (Step 23) FFT of output sine wave at V_{in}=1.8 V_{pp}. The significantly higher harmonics indicate that there is a considerable amount of distortion in the output.

23. **Observing harmonics generated by distorted amplifier output.** Now, increase V_{in} until V_{out} clips, i.e., the amplifier output is "saturated." This means that the sine wave at the output becomes flat at the top and bottom. The output will be somewhat clipped after V_{in} gets to about 0.2 V_{pp}. Increase V_{in} to 0.5 V_{pp} where clipping is clear. What is the FS of the output now? (See Fig. 7.57.) Are there more harmonics and are they larger? Does it remind you of the FS of a square wave? What does the clipped output have in common with a square wave in the time domain? Repeat the calculation of the previous step to find the ratio of the first harmonic to the second harmonic. Is it much smaller now? Do you still have a linear amplifier when the input voltage is so large that clipping occurs?

24. **Measuring frequency response of MOSFET amplifier.** What do you guess is the frequency response of the amplifier, i.e., how high of a frequency do you think it can operate at? The frequency at which the gain, A, of an amplifier drops to 1 is called the "transition frequency," denoted f_T. Let's test your intuition – write down in your notebook your guess of f_T. No points off for wild guesses.

Increase the frequency until the output voltage of the sine wave is the same as the original input voltage. It is recommended that rather than view the trace of the signal, you read the "measure" function of the voltages. Be sure that the traces are inside the upper and lower limits in order to get proper readings. Approximately what is the transition frequency, f_T, of your amplifier? Note that f_T for your amplifier may actually depend more on the breadboard stray capacitance than it does on the MOSFET operation. f_T of the transistor is actually somewhere between about 25 MHz and 50 MHz. Generally, MOSFETs are slower than BJTs.

Do not tear down your MOSFET amplifier. You will need it later. Use other components to build the next circuit.

25. **Building op-amp inverting amplifier.** Now you will experiment with an op-amp-based inverting amplifier, as discussed in the chapter. The pinout diagram of the 741 op-amp is shown in Fig. 7.58. This op-amp is commonly made in a **dual inline package (DIP)**. Pin 1 is shown relative to the only notch on the surface of the op-amp package. Insert it into the breadboard straddling two sets of rows, as shown in Figs. 7.59 and 7.60. Using that pinout, build the simple op-amp circuit shown in Fig. 7.61. Be sure to measure and record the values of the resistances. You will repeat several of the experiments you did with the BJT amplifier. Therefore, many fewer details are provided in the following steps.

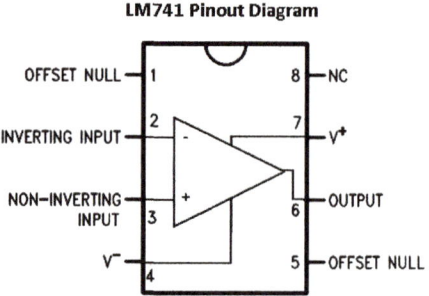

LM741 Pinout Diagram

FIGURE 7.58 (Step 25) Pinout of 741 op-amp DIP package.

FIGURE 7.59 (Step 25) ICs are inserted so they straddle rows of five tie points.

FIGURE 7.60 (Step 25) DIP package inserted on breadboard. Connections to pins are made through the remaining four available tie points.

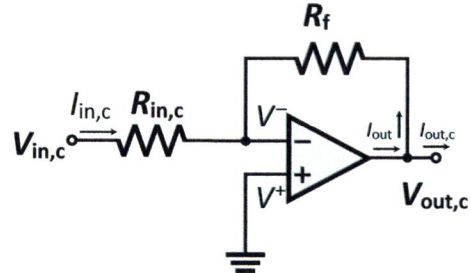

FIGURE 7.61 (Step 25) Schematic diagram of your op-amp amplifier with component values.

Draw Fig. 7.61 in your notebook and label all the pins before you start. Use R_{in} = 680 Ω and R_f = 5.6 kΩ. Use +5 V_{DC} and –5 V_{DC} for your supply voltages and an input frequency of 1 kHz. Connect the two supplies together in series to get a bipolar supply voltage, attaching ground in the center as you have done in previous labs. For convenience, a review is found at the end of this lab section. Connect the common (ground) to pin 3 of the op-amp, which is the non-inverting input.

26. **Testing the op-amp inverting amplifier.** Using a 10X probe, look at V_{in} on CH1 and V_{out} on CH2 of the scope. Start with a small voltage (say 0.1 V_{pp}) for V_{in} and increase it if necessary without causing the output to saturate, i.e., flatten out. What is the gain of your amplifier? Confirm that it is an inverting amplifier. Does the gain agree with the op-amp gain Eq. 7.17: $A_v = V_{out}/V_{in} = -R_f/R_{in}$?

27. **Changing the gain.** Change one of the resistor values with something comparable, but different value. What is your new gain, and does it agree with the equation?

28. **Finding the transition frequency.** In the linear regime, increase the input frequency and determine the transition frequency of your op-amp circuit as you did with the BJT amplifier in Step 18. What is it? Is it higher or lower than that of your BJT amplifier? Many variations of 741 op-amp ICs have f_T of about 500 MHz.

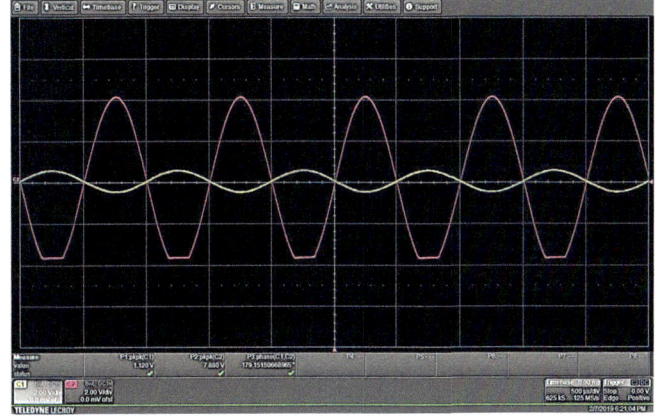

29. **Finding the saturation input voltage.** Using 1 kHz, increase your input voltage until the output is saturated. (See Fig. 7.62.) What input voltage causes that? Turn the input back down again to get back to the linear regime of your amplifier before doing the next step.

FIGURE 7.62 (Step 29) Waveform of input and just-saturated output of the op-amp amplifier. The output is larger than the input, as it should be because $R_f > R_{in}$. Also, because the input signal is fed back to the inverting input of the op-amp, the output is inverted and the phase shift is 180°.

30. **Observing spectrum of distorted sine wave.** Look at the frequency spectrum of the output below and well into saturation. Does it have extra frequency components below saturation? How about after saturation? Show both plots. Explain the behavior in terms of Steps 16 and 17 or Steps 22 and 23.

31. **Setting up the BJT or MOSFET amplifier for music.** Here you will listen to music from your BJT or MOSFET amplifier and judge how it sounds.

Go back to your BJT or MOSFET circuit, which you have not disassembled. Add the phono jacks as shown in Fig. 7.63. The circuit connections are given below. Note that the jacks are labeled with **com**, **R** and **L**. The first two pins on each side are "com" and the second two pins on each side are R and L, respectively. You will use one phono jack for each of the input and output signals.

FIGURE 7.63 (Step 31) Connect the phono jacks and wires as shown. Note in the lower jack, the right wire should just be in one of the holes.

Put back your oscilloscope connections at V_{in} and V_{out} of your transistor circuit.

On your breadboard, place a wire from "com" of the jacks to ground of your circuit.

Instead of using the Rigol for the input to your amplifier, use a wire from an "R" pin of your input jack as V_{in}.

Place a wire from V_{out} of your amplifier to an "R" pin of your output jack. The output should be an amplified and inverted version of the input music waveform. Using the CH2 menu, invert the output so that both channels have the same polarity. This confirms that your amplifier is working properly. Save a screenshot like that shown in Fig. 7.64.

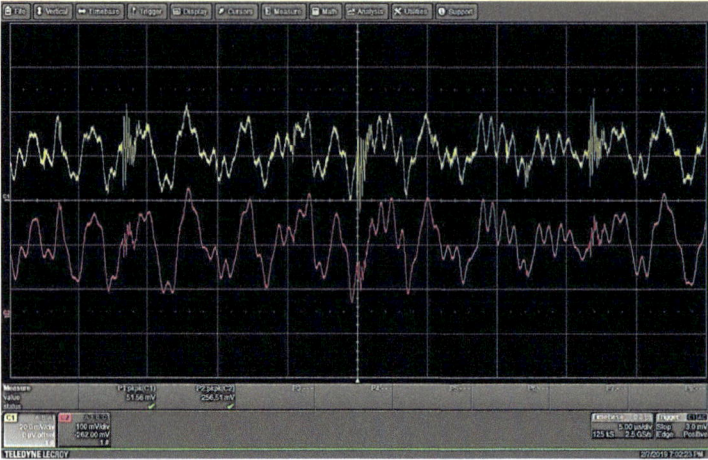

FIGURE 7.64 (Step 31) Waveform of input and output music from the BJT amplifier. The lower trace is the input and the top trace is the output. The input trace is inverted so that it will match the output of the inverting BJT amplifier. Note that the scale of the input on CH1 is 200 mV/div and that of the output on CH2 is 1.5 V/div, giving us a gain of about 7.5.

32. **Hearing the amplified music.** Listen to your source: The amplifier cannot produce enough current to drive earbuds. Use the speaker cable from your *music source* (phone, CD player, MP3 player, etc.) to the powered speakers. Turn the speaker volume all the way up and the source volume to a fairly low level to hear the music through the speaker at a fairly low volume. This makes sure that everything is working properly, and you get some idea of the speaker quality.

That was easy. Now you will insert the BJT or MOSFET amplifier between your music source and the powered speakers. Remove the speaker cable from your music source. Plug the extension cable from the music source to the amplifier *input* jack. Plug the speaker input cable into the amplifier output jack. Listen to the music on the speakers. Does it sound louder in the speakers? That shows that you have a voltage gain greater than 1.

33. It is left as an exercise for you to use the op-amp amplifier to listen to music.

Congratulations! You have built and tested your own curve tracer and amplifiers using semiconductor-based devices and integrated circuits. You also used them to amplify music. This is an important step in your understanding of basic electronic circuits.

If you want to delve further on your own, you might want to investigate variations of op-amp circuits called "integrators" and "differentiators."

This figure from Lab 3 shows how the generator was powered by a bipolar power supply. In this lab, you will power your op-amp the same way with −5 V and +5 V.

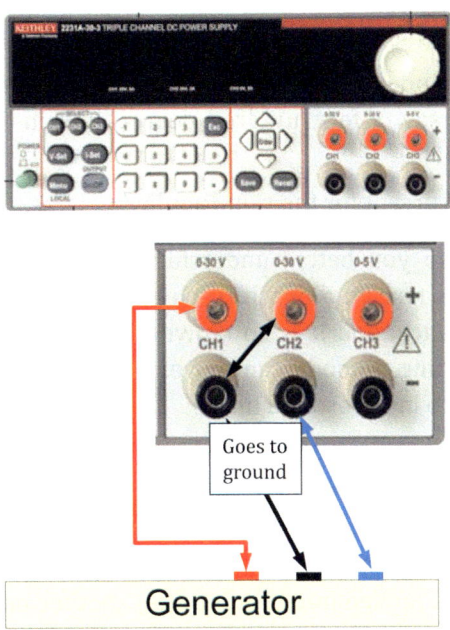

(Top) DC power supply and (bottom) power wiring diagram to the 3-phase generator.

Problems

1. This question ties in with material in Chapter 1. The following diagram shows a bar of silicon connected to a voltage source at room temperature. V_a = 2 V, L = 15 μm, h = 3 μm, w = 2 μm and resistivity ρ = 0.05 Ω·cm. (1 μm = 1 micrometer = 1 micron = 10^{-6} m = 10^{-3} mm.)

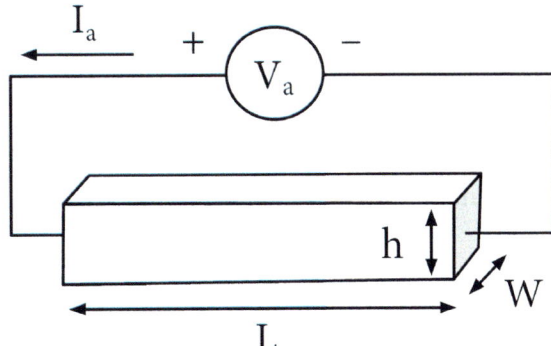

   (a) What is the resistance, R, of the slab?

   (b) What is I_a?

   (c) As given in Eq. 1.4, resistivity, ρ (rho), = $1/\sigma$, where σ (sigma) is the conductivity. σ depends on the number of available charge carriers and how easily they move in the material:

$$\sigma = 1/\rho = qn\mu_n + qp\mu_p,$$

where q is the charge of an electron or proton, and is positive; μ_n is the **electron mobility** and μ_p is the **hole mobility**. These parameters describe how easily electrons and holes, respectively, move through a material.

Qualitatively, what happens to the resistivity and current if the temperature of the Si slab is increased by, say, a few hundred degrees Celsius? This is the principle behind a device called a **thermistor**.

(d) What happens to the current for the same applied voltage?

2. Here is an analogy to help you better understand electrons and holes in a semiconductor. Imagine a steep cliff that is next to a beach (such a cliff and the land above it is called a bluff). There is a party on the beach, and it is so crowded that none of the partiers can even move from where they are standing. Nobody can move. Also, there is nobody at the top of the bluff where there is lots of area to move around in.

The cliff is high and steep, so only a few of the partiers think of climbing it. Someone decides that they can climb up the cliff to the top where they are free to move wherever they want. This leaves some open space on the beach, so now people can move around a bit. When the climber leaves, one person moves into her empty space, so the empty space moves in the opposite direction as the person who moved into it. After that, another person moves into his now-empty space. This continues because one empty space is available for partiers to move to, and the empty space migrates along the beach. As more people climb the cliff to the top, there is still lots of room at the top for all of the climbers to move around as they want, and there is also more room on the beach with empty spaces for other partiers to move into, so there is more motion of people there on the beach.

This situation is analogous to what happens in a semiconductor when an electron moves from the valence band to the conduction band, and both the hole and electron carry current.

Which part of this analogy are the electrons, holes, band gap, conduction band, and valence band? Explain.

3. Electrons in the valence band need more thermal energy to make it to the conduction band when the bandgap is larger. That is equivalent to a higher cliff in the beach analogy of the previous question. The larger the bandgap, the fewer the number of valence electrons that can jump to the conduction band at a given temperature. Also, you know that higher concentration of conduction electrons and holes leads to higher conductivity.

(a) Based on this information, order the following by *increasing* bandgap: **metal, insulator, and semiconductor**.

(b) Electrons need energy to break free of the bonds, i.e., to become an electron–hole pair, EHP (climb up the cliff). Thermal energy is kT, where k is Boltzmann's constant, and T is the temperature in Kelvin. As temperature changes, the value of kT changes, of course. As an analogy, it is more likely that water will shoot out of the surface of the ocean on a very windy day with more and higher waves (higher T) than on a calmer day (lower T) when the waves are lower. The bandgap energy, E_g, must be surmounted by the valence electrons to become conduction electrons. We talk about the bandgap energy in terms of the ratio of E_g to kT, E_g/kT, which can be thought of as the **number of kT**. The number of electrons in the conduction band decreases exponentially as the number of kT increases as temperature decreases, so temperature matters a lot.

Given this information, create a table with the following information: the number of kT of the bandgap of Si (E_g = 1.12 eV) and SiC (E_g = 3.23 eV) at the following temperatures: 77 K, 300 K, and 475 K. The last number is how hot it can get inside a car engine compartment or perhaps on a large electric motor.

Also, many equations used in semiconductor calculations include a term $e^{-Eg/kT}$. Include that for each temperature and bandgap as well. Can you see that the term $e^{-Eg/kT}$ gets ridiculously small for large bandgaps, meaning the thermally generated EHPs are, in some cases, practically nonexistent? The only reason that there are any thermally generated electrons and holes in any semiconductor is that there are so many atoms available from which an EHP can be created. Here are constants for you to use:

$$e = \text{Euler's number} = 2.718$$

$$k = 8.62 \times 10^{-5} \text{ eV/K}$$

4. Thermal energy kT was discussed in the previous problem. Consider the fact that more thermally-generated EHPs are created as the temperature increases. For intrinsic semiconductors (those without any chemical impurities), the concentrations of electrons and holes are equal. That concentration is called the **intrinsic carrier concentration** designated as n_i, and, therefore, $n = p = n_i$. Also, consider that there are more EHPs for narrower bandgaps.

Use data from the following table (N_C and N_V are explained in a full course on semiconductors):

Name of parameter	Si	GaAs
Energy gaps, E_g	1.12 eV	1.42 eV
Effective conduction band density of states, N_C	2.86×10^{19} cm^{-3}	4.7×10^{17} cm^{-3}
Effective valence band density of states, N_V	2.66×10^{19} cm^{-3}	7×10^{18} cm^{-3}

You can calculate the number of intrinsic electrons and holes as:

$$n_i = \sqrt{(N_C N_V)}e^{E_g/2kT}, \text{ where}$$

$$e = \text{Euler's number} = 2.718$$

$$k = 8.62 \times 10^{-5} \text{ eV/K}$$

As a check, at room temperature in Si, n_i = 1.5 × 10^{10} cm^{-3}. And in GaAs, n_i = 2.1 × 10^6 cm^{-3}. (Your calculations might be slightly different.) That may seem like a large number, but there are 5 × 10^{22} Si atoms cm^{-3} and 4.22 × 10^{22} GaAs atoms cm^{-3}, so you can see that very few of the valence electrons create an EHP.

Military specifications (MilSpecs) of semiconductor operation requires a device to operate from –55 °C to 125 °C. To check if a semiconductor device can operate over this range of temperatures, you need for n_i to be small relative to conventional doping concentrations of, say, 10^{14}/cm^3 or greater.

(a) Create a table for n_i at three temperatures for both Si and GaAs: $T = -55$ °C, 27 °C (room temperature) and 125 °C. (Don't forget to use Kelvin.) Do you think that these temperatures are excessive for these two semiconductors?

(b) Why does n_i depend so strongly on temperature (increases six to eight orders of magnitude over less than twice the thermal energy)?

5. Indicate whether each of the following statements is true or false.

(a) Semiconductors are either intrinsic, n-type or p-type.

(b) In a semiconductor material, raising a valence band electron to the conduction band creates one conduction electron and one hole.

(c) Doping is required to make a semiconductor either n-type or p-type.

(d) Donor atoms (i.e., donors) add holes and acceptors add electrons.

(e) Donors used as dopants for Si have 5 valence electrons.

(f) Acceptors used as dopants for Si have 4 valence electrons.

(g) When thermal energy increases, there are more electron–hole pairs.

(h) At higher temperatures, the bandgap is equivalent to **more kTs**.

(i) Thermal energy increases as temperature increases.

(i) The concentration of electron–hole pairs increases as temperature increases.

(k) For the same temperature, semiconductors having smaller bandgap have fewer electron–hole pairs.

(l) When the valance band is totally full of electrons and the conduction band is empty of electrons, the semiconductor material is very conductive.

6. For the following circuit, V_a increases from 0 V to 2 V. Assume the diode's turn-on voltage V_t is 0.7 V, and that the resistor has a high enough resistance to limit the current and prevent damage to the diode.

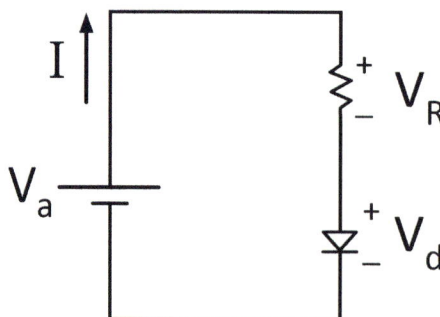

(a) Describe in words what happens to V_R and V_d as V_a increases from 0 to about 2 V.

(b) Sketch both V_d and V_R as a function of V_a on the same plot. Put in numbers where applicable.

7. In the following circuits, the diode is made of silicon and has a turn-on voltage of 0.7 V. Its breakdown voltage is $V_{br} = 20$ V. The resistor R_S is 100 Ω. Assume the voltmeter is ideal.

(a) If $V_a = 5$ V, what does the voltmeter read in the following circuit?

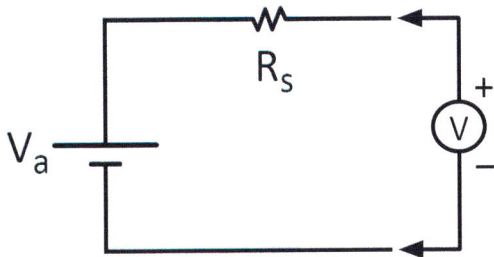

(b) Now, or the same circuit as part (a), $V_a = -5$ V. What does the voltmeter read?

(c) What does the voltmeter read in the circuit of the following diagram for $V_a = 5$ V?

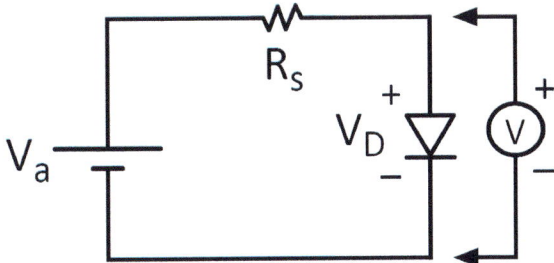

(d) For the circuit and conditions of (c) above, what is the voltage drop across R_S? What current flows?

(e) For Question (c), what does the voltmeter read if $V_a = -5$ V? What is the current through R_S?

8. For the circuit shown in the following diagram, $R = 500$ Ω, and the diode is made of GaAs; it has a turn-on voltage of $V_t = 1.3$ V and breakdown voltage of $V_{br} = -30$ V.

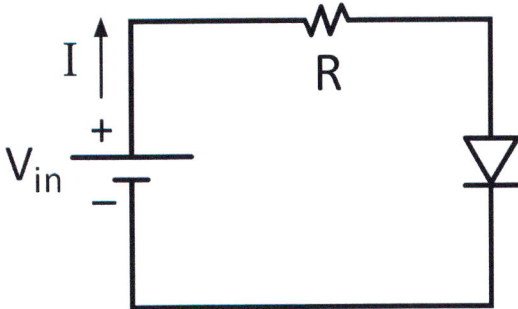

What is the current when V_{in} = (a) 0.7 V; (b) 4 V; (c) –4 V; (d) –40 V?

9. In Fig. 7.39 from Step 2, there are two overlapping curves, one yellowish-white and one light pink. They overlap everywhere but the top. The vertical scale is 1 V/div. Explain the following:

(a) What is the yellowish curve and why does it have the shape that it does?

(b) What is the pink curve and why does it partially overlap the yellowish curve on the positive side?

(c) What is the approximate voltage on CH1 in the flat region, and what does it signify? Why does the pink curve get flat at the top on the positive side?

(d) Why does the pink curve overlap the yellowish curve on the negative side?

10. Refer to Fig. 7.40. There are two overlapping curves, one yellow and one pink. The vertical scale is 2 V/div. Explain the following:

    (a) What is the cause of the flat part on the pink curve for positive voltages?

    (b) What is the cause of the rounded part on the pink curve for negative voltages?

11. Refer to Fig. 7.41 of the current versus voltage of two back-to-back diodes. Draw the related circuit diagram based on Fig. 7.38. Why can't both diodes be in breakdown at the same time?

12. Make a list of things that will be different when you replace the Si diode with a light emitting diode in Step 4.

13. In step 5 of the lab you are going to plot the I-V characteristics of a diode. In terms of V_{C1}, V_{C2} and R_S what will be the X and Y axes?

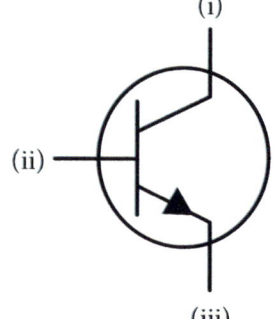

(i)

14. (a) What type of transistor is the symbol in the adjoining diagram?

    (b) Is this an npn or pnp transistor?

    (c) Identify all three terminals on the figure.

    (d) What direction does the current flow in each terminal under normal bias, i.e., in use as an amplifier? (This is called the **active region**.) Draw arrows to show the current direction and label each current.

    (e) For $\beta = 225$ and $i_b = 1.6$ mA, what is i_c?

(ii)

(iii)

15. Refer to the BJT amplifier in Fig. 7.47, top. Indicate whether each statement is true or false.

    When the input signal is positive and increases:

    (a) more current enters the base of the NPN BJT.

    (b) more current flows through the transistor from collector to emitter.

    (c) less current flows from source to drain.

    (d) the output voltage decreases, i.e., becomes more negative.

16. A sinusoid with an amplitude of 1 V and frequency of 1 Hz is the input of a voltage amplifier with a gain of $A_V = -2$. Use Desmos (or your favorite plotting program) or sketch by hand V_{in} and V_{out} vs. time on the same plot.

17. For the circuit shown in the following diagram, beta, β, is 150 and the current into the base, I_b, is 10 μA_{DC} (DC microamp). What is V_{out}? *Hint*: First find the collector current then find the voltage drop across R_L.

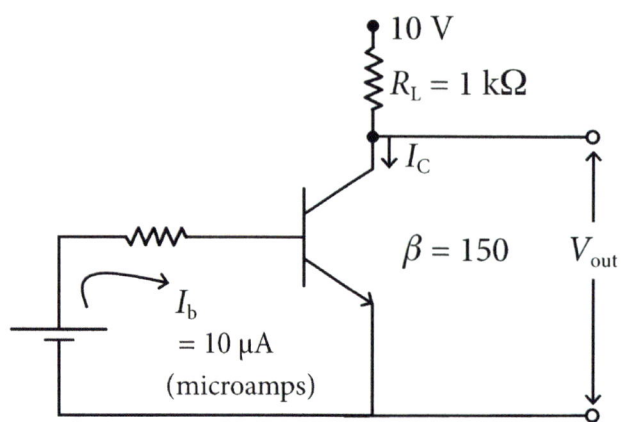

10 V

$R_L = 1$ kΩ

I_C

$\beta = 150$ V_{out}

I_b
$= 10$ μA
(microamps)

18. The following diagram shows a BJT amplifier circuit like that in Fig. 7.47, top. Here, V_{CC} = 12 V, R_1 = 4 kΩ and R_2 = 2 kΩ. V_{CC} is a DC voltage that provides all the power to run this amplifier. Resistors R_1 and R_2 set the **bias**, V_B, for the transistor input, which is often called the **Q-point**. (Q stands for quiescent, or not changing.) The DC voltage bias at the base of this transistor circuit, node B, sets the voltage into the transistor around which the AC signal changes. V_{in} may or may not have a DC component along with the AC signal.

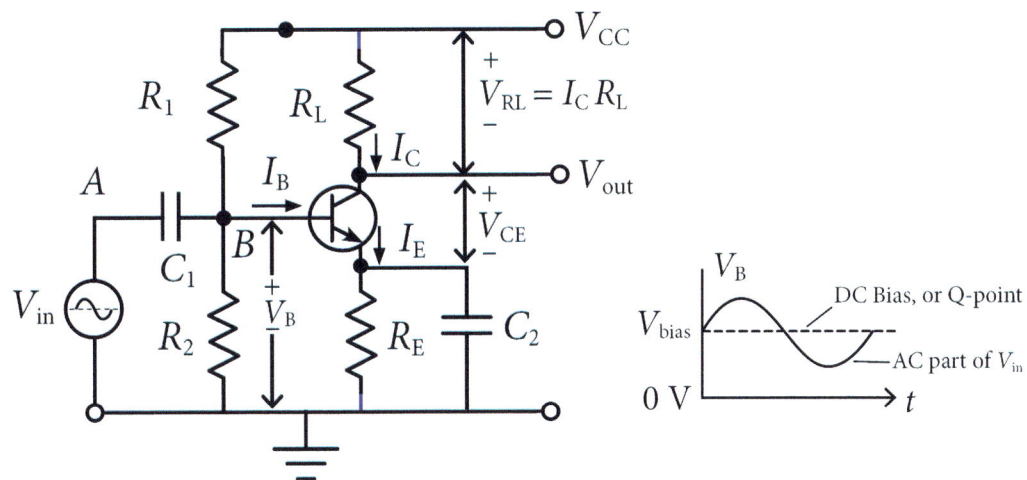

(a) For this problem, assume that almost no current flows into the base of the transistor. What DC bias V_B is present at the base, point B, i.e., the Q-point? Note that C_1 blocks any DC voltage from V_{CC} from appearing at point A. *Hint*: think **voltage divider**.

(b) Again, notice the capacitor C1 between points A and B. Review Chapter 4, Fig. 4.31. Is V_{in} AC- or DC-coupled to the base of the transistor?

(c) In words, explain your answer to part (b).

(d) Due to the **superposition theorem**, the signal at B is the sum of the DC biasing voltage and the AC voltage V_{in}, which passes through the DC-blocking capacitor C1. (If you are not familiar with this theorem, you can simply take this as a fact.) In your own words, what is that total signal at node B, the base of the transistor?

(e) Now let's put numbers to this. Assume V_{in} is an AC signal of 50 mV$_{PP}$ riding on a DC voltage of 0.5 V, as shown in the following diagram. Assume as well that the frequency is high enough that C_1 is essentially a short circuit, and also that the Q-point at the base is 1 V. Draw the waveform at node B.

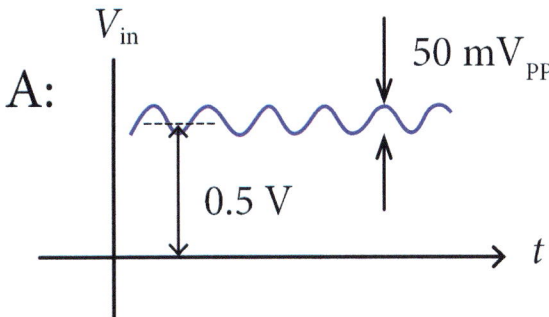

19. Refer to Fig. 7.24, which is a drawing of the cross-section of an **n-channel** MOSFET. The source and drain are n-type and they are embedded in a block of p-type material, referred to as the **body**. The source and drain are p-n junctions to the body, which are just like the diode you have learned about.

    The gate is a capacitor with a layer of insulating oxide between the top plate and the p-type channel. This MOSFET is called 'n-channel' because it is 'on' (i.e., conducts current from the drain to the source) when there is a positive voltage on the gate causing an excess of electrons to form underneath the gate. This layer of electrons is like a wire that connects the source to the drain so current can flow.

    (a) There is a positive voltage on the drain relative to the source, as shown. Because V_{DS} is positive, *when the device is on*, electrons will flow from the source through the channel to the drain. Because the sign of current is the opposite of the movement of electrons, positive current will flow from the drain, through the channel to the source. However, *suppose it is off*, meaning that the gate voltage does not have a positive voltage on it, and therefore there are no electrons under the gate. Why doesn't the MOSFET conduct current from the drain to source? Your answer should be in terms of forward and reverse biased diodes.

    (b) This question ties in with material in Chapter 1. Suppose the gate has a positive voltage of 1 V, the oxide thickness is 2 nm (2×10^{-3} μm), and the size of the gate is 0.01 × 0.1 micron. (One μm is 10^{-3} mm.) Although you might not appreciate this, assume that the gate oxide is the most modern material, namely $LaAlO_3$, which has a relative permeability $\varepsilon_r = 30$. (This material is used for the gate dielectrics in modern integrated circuits.) How many electrons collect under the gate? Your answer will be in coulombs, but then you have to divide by the number of coulombs per electron.

20. Refer to Figs. 7.27 and 7.28. Indicate whether each statement is true or false.

    (a) The op-amp integrated circuits like the one used in the amplifier generally have very low input impedance.

    (b) A 741 op-amp is fabricated from millions or sometimes billions of transistors.

    (c) The amplifier made in lab from the 741 op-amp, and shown in Fig. 7.28, is an inverting amplifier.

    (d) The internal, or 'open-loop', voltage gain, G, of an op-amp such as the 741 cannot be greater than the total voltage gain of the op-amp circuit, $A_v = -R_f/R_{in}$.

    (e) The voltage difference between V⁻ and V⁺ is multiplied by the gain factor $A_v = -R_f/R_{in}$ to yield V_{out}.

21. Figure 7.28 shows an op-amp amplifier circuit. For that circuit, $R_{in,c} = 14$ kΩ , $R_f = 35$ kΩ and $V_{in,c} = 3$ V. Find $I_{in,c}$, I_{out}, A_v, $V_{out,c}$, and I_{out},c.

22. Re-sketch the schematic in Fig. 7.61 replacing the op-amp symbol with the pinout diagram in Fig. 7.58. Mark the pins with their pin numbers. Include the power-supply voltages and ground.

23. This question illustrates the size scale of transistors in modern ICs.

    The 64-core AMD Epyc Rome central processing unit (CPU) chip has about 40 billion transistors on a single chip that has an area of 1088 mm². (The smallest feature, i.e., part of a transistor, on the chip is about 7 nm.) Let us assume that the surface of that chip is all transistors and they are uniformly dispersed across the chip.

(a) Under these assumptions, what is the area of a transistor in square microns, μm^2? (One μm is 10^{-3} mm.) Also, how many transistors is that per μm^2?

(b) The diameter of a human hair is often taken as about 100 μm, which is often used as a convenient reference for small things. How many of these transistors could fit on the cross-sectional surface of a human hair?

(c) Cigarette smoke particles exhaled from the lungs hours after smoking are between 0.1 and 1 micron in diameter. Suppose a circular particle having a diameter of 1 micron falls on the chip during processing one of the transistor layers. How many transistors are destroyed by the dust particle? This demonstrates the need for extreme cleanliness in cleanrooms, and why air filters for exhaled air is used in modern cleanrooms.

Chapter 8

Lighting and Renewable Energy

8.1   Introduction and Overview

In this chapter, you will learn:

1. How an incandescent bulb produces light
2. What blackbody radiation is
3. How a fluorescent light bulb produces light
4. How a light-emitting diode produces light
5. Some of the properties of the solar spectrum at the Earth
6. About properties of human vision and light sensitivity
7. How a solar cell converts light to electricity
8. About the rapid rise in solar and wind power

In the lab you will perform a variety of experiments on various kinds of light bulbs to compare and contrast their emission and efficiency properties and compare with manufacturers' reported values. Additionally, you will measure the output of solar cells under various conditions.

8.1.1   Rapid Changes in Energy Technologies

Energy is the foundation of civilization. We rely on a steady supply of energy for transportation, manufacturing, living comforts, safety, communications, etc. Without a large and steady supply of energy, modern society as we know it would not exist. It is that simple. A continuous supply of energy goes hand-in-hand with affordability. If energy were too expensive, then it would be available only to a few, and on a limited basis. Also, if it were too dirty and dangerous, then it would undermine the very purpose of having it, i.e., to benefit humankind.

Because energy is such an important topic, it tends to be heavily politicized. There is significant disagreement about what problems should be given priority and what solutions to pursue. Among these, we could include global warming, fossil fuels, most notably coal-fired power plants, nuclear energy and its perceived dangers, geopolitical ramifications of global sources of energy and global conflict, allocation of dollars for energy-related research, electric vehicles, alternative energy sources, and

An Introduction to Electrical Engineering with Lab Activities
Gary H. Bernstein
Copyright © 2026 Jenny Stanford Publishing Pte. Ltd.
ISBN 978-981-5129-30-4 (Hardcover), 978-1-003-71345-6 (eBook)
www.jennystanford.com
DOI: 10.1201/9781003713456-8

more. Energy-related issues are in the news daily. This laboratory activity addresses two of the myriad issues that affect global energy policy, namely the use of electricity for lighting, and renewable energy, specifically solar energy, and more specifically, solar cells.

In order to be better stewards of energy in our society, we must be vigilant to both generate and use energy wisely. By that, I mean that we must use the least amount of energy possible without *unnecessarily* degrading our standard of living, and we must generate energy as cheaply as possible, all the while being careful to have the least possible impact on the environment. Ideally, we would do no damage to our planet in serving the needs of society. In nearly every case there are important tradeoffs, but some choices are better than others. The specific decisions to be made are not only technical but also societal and political. Those decisions will play themselves out over the next decades in the centers of government, research laboratories, and courtrooms – and in the marketplace.

In the first two decades of the 21st century, the world has seen remarkable changes to energy patterns that include the acceptance of more efficient lighting, more solar energy and more wind energy. These trends will continue throughout the next decades, and changes come so quickly that any data provided in this chapter will be out of date within only a few years. Still, such data presented here will be the best available as of 2023 and it is left to you, the reader, to search for your most recent data. This chapter uses the technologies of incandescent, fluorescent and light-emitting-diode (LED) lighting as well as solar cells as the basis for delving into several basic science and engineering concepts as well as the area of alternative energy.

FIGURE 8.1 Our dependence on electric lighting is unquestionably great. *Source*: https://commons. wikimedia.org/wiki/File:Downtown_Los_Angeles-02.jpg.

Figure 8.1 suggests how widespread our dependence on electric lighting has become. Some lighting technologies discussed in this chapter are rapidly becoming outdated. As of 2023, LED lighting has largely overtaken outmoded lighting technologies including mostly incandescent and fluorescent. Figure 8.2 shows a projection to 2030 of the relative use of these three lighting technologies. The chart shows that LED lighting is overtaking the marketplace, incandescent lighting is rapidly disappearing, and fluorescent lighting is projected to stay relatively steady at a small level.

So, why base a chapter on outmoded technologies? First, lighting is a very practical subject, and worth knowing about. Second, it is interesting to see where lighting technology has led us in only a few years. Third, electrical engineers are employed in industries related to the development of new materials and

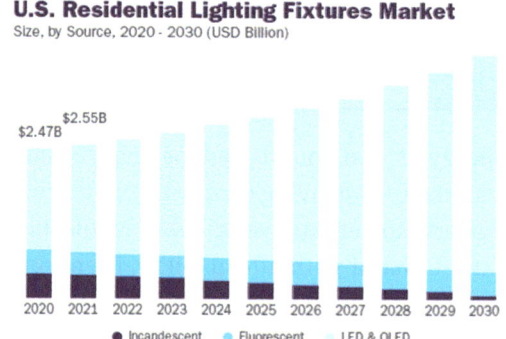

FIGURE 8.2 Relative residential market for LED, fluorescent and incandescent lighting. Data is 2020 and predictions are until 2030. Grand View Research, Inc.; U.S. Residential Lighting Fixtures Market Research Report; April 2023 (https://www.grandviewresearch.com/industry-analysis/us-residential-lighting-fixtures-market-report).

methods of creating lighting, and in the lighting industry as a whole. The development of new materials for electric lighting is an ongoing area of intense research in industry and at universities. Fourth, and not least, the discussions about lighting and other technologies in this chapter offer the opportunity to study some basic physics and engineering that is important not only in lighting technology but also photonics in general, which is a major part of the panoply of electronics materials and devices research and development. Optical devices and sensors play a huge role in communications and computing and are firmly planted within the scope of electrical engineering. Furthermore, plasmas are introduced in the context of fluorescent bulbs; plasmas are critical to the manufacturing of advanced integrated circuits, among other contexts in which they appear. Variations of LEDs form the solid-state laser, which is commonplace in commercial and consumer goods. In short, this chapter, although based on the ordinary light bulb, is foundational to advanced technology across the field of electrical engineering.

8.1.2   World Electrical Energy Usage

Total electricity usage worldwide in 2023 was about 2.5×10^{13} kilowatt-hours, kWh (*source*: https://www.statista.com/statistics/280704/world-power-consumption/). Of that, the fraction used for lighting is predicted to drop from about 13% today to 8% by 2030 (https://worldgbc.org/article/energy-efficiency-is-a-cant-do-without/), due to the newest lighting technologies presented in this chapter. This is still a staggering amount of energy, and deserves serious consideration and understanding in an effort to decrease the world's energy usage any way that we can.

This chapter deals also with the subject of solar cells. Although there are myriad types of **alternative**, or **renewable** energy sources (these terms are used interchangeably) it would be hard to argue that anything is as renewable as sunlight. At sea level, the maximum available solar power is about 1 kW per square meter if the Sun is directly overhead. Since a typical home typically uses a maximum of 20 kW, we can appreciate that, under perfect conditions, just 20 sq. meters of sunlight per home (about the size of a medium room in a house), if converted completely into electricity would be enough to satisfy our domestic needs. Well, too bad that the Sun is almost never directly overhead and doesn't shine much on cloudy days, or at night, and we can use only a limited fraction of solar energy for electricity, and there is not yet a universally accepted massive, scalable way to store the energy that is captured for use when the Sun isn't shining, and Northern latitudes generally don't get much sunlight, and solar farms use a considerable amount of space. Okay, enough of the bad points. The good thing is that solar energy is clean, and renewable, and we have cheap technologies that work pretty well to convert light to electricity, and those technologies are improving and getting cheaper all the time, so we should use it if we possibly can.

Solar cells are not only a promising renewable energy technology that is increasingly becoming economically important, and is an active area of research and development in industry and universities, but also the basic principles of solar cells are much like those of other semiconductor devices and are therefore important to know about. In this chapter, you will learn the underlying principles of the direct conversion of light to electricity. Near the end of this chapter, we will also discuss other alternative energy sources, including other ways to use sunlight for power.

Another theme of this chapter is that of human vision and color and light perception. Electrical engineers are engaged in a vast spectrum of activities including the development of optical and image sensors and systems. Therefore, light and human vision play a role in much of what many EEs do.

Of interest: The Sun does not ever shine directly overhead over much of the planet. In fact, where it does happen, it is only for a moment on only two days per year. That happens between the Tropic of Cancer and the Tropic of Capricorn, on either side of the Equator. Think about the 23.5 degree tilt of the Earth relative to the plane of Earth's orbit around the Sun and try to understand it clearly. This is central to understanding the economics of solar energy everywhere on Earth.

8.2   Light Bulbs: Introduction to Light Emission

As already mentioned, we will rely very much on fundamental physical properties to form the foundation of the engineered systems that will be discussed. One aspect of practicing electrical engineering is that it is often just one level removed from physics and/or math. If you enjoy understanding how the natural world works, you will enjoy learning the physical aspects of various electrical engineering topics such as these. If you look forward to having a positive impact on the lives of your fellow human beings, you will appreciate the relationship between using physics and engineering to effect important technological advancements.

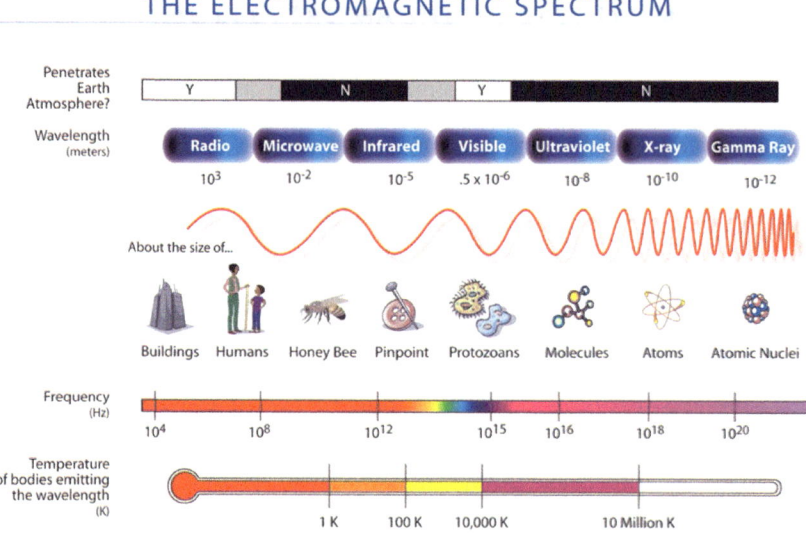

FIGURE 8.3 Electromagnetic spectrum focused on the visible light frequency range.

Chapter 6 was about electromagnetic (EM) radiation in the form of radio waves. This chapter is about EM radiation in the form of light – it is basically the same stuff. As shown in Chapter 6, and repeated here (Fig. 8.3), electromagnetic radiation can exist over a vast range of wavelengths. When we can see it, we often call it just "light," but sometimes distinguish those wavelengths from others by calling them "visible light." The frequency and wavelength of all electromagnetic radiation are related by:

$$\lambda f = c, \tag{8.1}$$

where f is the frequency of the light in units of cycles per second, or hertz, λ is wavelength, in length units to match your choice of the speed of light, and c is the speed of light in some medium such as, for example, the vacuum of space, water or glass. This equation tells us that within some medium, as frequency increases, wavelength decreases. The electric and magnetic fields of the light wave (see Chapter 6) oscillate at very high frequencies; very high frequencies have very short wavelengths, as you can see in Fig. 8.3. For this treatment we will assume the light travels through a vacuum or air, for which c is very close to 3×10^8 m/s. In Fig. 8.3, we see that the range of wavelengths of visible light is actually quite narrow compared to the vast range of wavelengths on the chart, which covers 16 orders of magnitude. The bottom of the plot shows the relationship between the temperature of an object (such as a glowing ember) and the wavelength of light that it emits. This will be discussed further below.

The visible spectrum of light ranges from red to violet, with wavelengths λ ranging from about 390 nm to 700 nm (f = 790 THz to 430 THz), respectively, which is less than a factor of two in wavelength or frequency. It seems rather surprising that the richness of our perceived visible world comes down to such a narrow range. As can be seen by comparing the visible spectrum in Fig. 8.3

to the solar spectrum in Fig. 8.4, our visible color perception closely matches the peak of the spectrum of sunlight. It makes sense that we evolved to use the light that was available to us, and we developed very sensitive light sensors because it could give us information about dangers approaching from much greater distances than touch, smell or sound can provide, thus helping humans to survive and pass that color vision on to further generations.

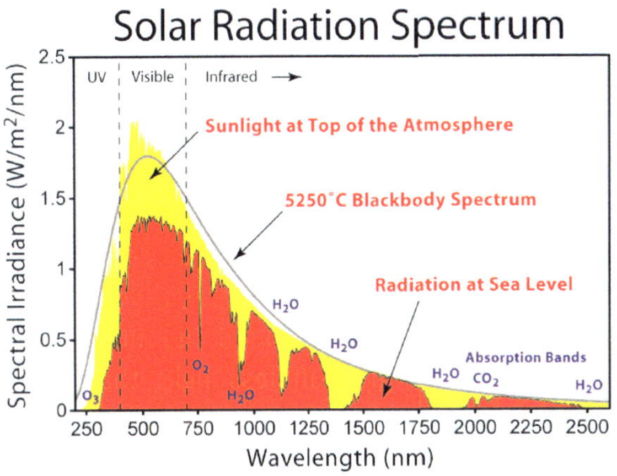

Longer wavelengths in the infrared (IR) correspond to radiation of heat. Imagine you are sitting ten feet from a campfire. Your eyes see the flame, whose high-temperature radiates its peak

FIGURE 8.4 Solar radiation spectrum. *Source*: https://commons.wikimedia.org/wiki/File:Solar_Spectrum.png.

wavelength in the visible, but your hands feel, or in some sense 'see,' the longer wavelengths that come from the flame. We know that snakes can see longer wavelengths, into the IR, to detect the heat from prey, and birds and bees can see shorter wavelengths, well into the ultraviolet, to spot flowers, all because it helped in their specific evolutionary paths.

In the natural world, the vast majority of objects have color because they *reflect* certain wavelengths that already exist in the solar spectrum. A few living objects *produce* light by the chemical process called **bioluminescence**, including some insects (e.g., fireflies), bacteria, deep-sea fish, and plants (including some mushrooms), among others. However, as alluded to above, all things that have a temperature above zero Kelvins emit electromagnetic radiation, which, again, we commonly call "light" when we can see it, but it is still there even when we can't. This

Of interest: It is not a coincidence that our vision sensitivity matches the peak of the solar spectrum. It is, however, a nearly miraculous coincidence that the peak of the solar spectrum coincides with a peak in the transmittance of light through air on Earth. If the peaks did not coincide, we would not be able to use the peak of the 'visible' light spectrum falling on Earth, but likely would still use some of the light. The world would be a dim place, and we might have evolved with bigger ears to better use available sound waves.

phenomenon, called **blackbody radiation**, is partly due to the quantum-mechanical fact that light energy comes in little "bundles," called **quanta** (singular: **quantum**).

When light of a certain wavelength arrives at an object, it can transfer energy only in units of its associated quantum of energy. At any wavelength, the packets of electromagnetic energy are **quantized**, or lumped, in units of:

$$E = hf = \frac{hc}{\lambda}, \tag{8.2}$$

where E in this equation is energy, f is the frequency of the electromagnetic oscillation, and h is Planck's constant ($h = 6.63 \times 10^{-34}$ J-s). (Note that energy, E, is a scalar, so it is not in bold font as is the electric field, $\mathbf{E}$, which is a vector). For example, a photon can't give up only half of hc/λ. It's "all or nothing at all."

Strictly speaking, a perfect **blackbody** radiator is one that absorbs all light incident on it. If so, then how can we talk about "blackbody *radiation*," which implies that light comes from something that absorbs all light, and seems to be an oxymoron? The answer is that although all incident light is, in fact, absorbed, it must also emit light in order to maintain a constant equilibrium temperature.

The emission spectrum has a peak at some wavelength and a shape that depends only on its temperature, as shown in Fig. 8.5. The spectrum shown is called the **blackbody radiation spectrum**. It's amazing that the shape of the curve depends ONLY on the temperature of the object, and not its shape, size, or, to a great extent, what it is made of. In reality, some materials give off less light than that of a perfect black body according to its **emissivity**, ε, a positive quantity that is less than 1. Some emissivities can be much lower than 1. The more absorptive a material is (like a black piece of cloth),

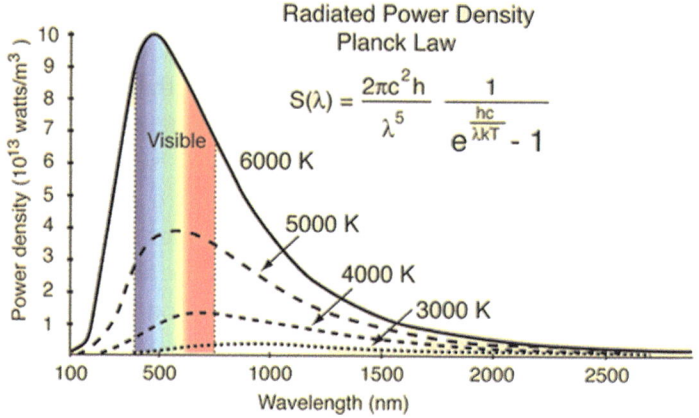

FIGURE 8.5 Blackbody radiation spectrum at various temperatures. Reproduced with permission from Dr Rod Nave, HyperPhysics Project, Department of Physics and Astronomy, Georgia State University.

the more it radiates, so $\varepsilon \sim 1$. The more reflective it is, the less it radiates, so ε is small, say, 0.1. Note the similarity of the spectral shape of a black body radiator at 6000 K of Fig. 8.5 to the solar spectrum of Fig. 8.4, shown at 5250 °C, or about 5500 K. So, the Sun behaves much like a blackbody radiator.

The peak of emission by objects that are around **room temperature**, defined specifically as 300 K (yes, "room temperature" is actually a technical term), is in the far infrared, having wavelength around 10 μm. This is well outside our visible range of wavelengths. Otherwise, we could see objects glowing in total darkness. The word "darkness" implies that there is no external source of *visible* light, but if we could see infrared, we wouldn't need it, because we would be able to see everything glowing at its blackbody temperature. Wouldn't that be cool (or hot, so to speak)? If you were in a totally dark kitchen and the stove-top burner started to warm up, assuming you have normal color vision, you would start to see it glow only when it got to about 800 K. Note that light can be emitted from an object for other reasons, such as electron transitions between energy levels within atoms, but there is always an underlying blackbody spectrum.

So, the Sun is glowing with an optical spectrum that corresponds to its "surface" temperature (see side bar for more about what we see as the surface of the sun) – it is actually glowing with blackbody radiation, but for us, it is just (largely) the light that we see. Figure 8.4 compares the solar spectrum with blackbody radiation at 5250°C (5523 K), which is approximately the temperature of the Sun's photosphere.

> *Of interest:* Some modern night vision goggles convert infrared light emitted by the blackbody radiation of objects into visible light. Older styles either required an external infrared source, or amplified tiny amounts of visible light.

(Solar temperatures differing by a few hundred deg. C or so appear in various sources since various regions of the Sun near the center or toward the edges can present different temperatures. 5800 K is frequently used.) We will need the concept of blackbody radiation when we discuss light bulbs as well as the solar spectrum when we discuss solar cells, so this is very relevant.

8.3 The Big Three Residential Lighting Technologies

Historically, there have been basically three choices in residential light bulbs – **incandescent**, **fluorescent**, and **light-emitting diode**, or **LED**. As of 2023, the sale of most incandescent bulbs in the

United States has been banned, as is the case with most of the world. There are other lighting sources used in industrial settings and highways, including those called **high-density discharge** (such as mercury vapor and high-pressure sodium lights) but they will not be discussed here. Each of the 'big three' (my expression) light sources operates on totally different underlying physical mechanisms that determine their cost, light emission properties, and energy efficiency. All of these characteristics are important for public acceptance, and therefore have huge implications on our national energy policy, for reasons discussed in the introduction.

8.3.1 Photometry

Before we begin a discussion of the big-three bulbs, we need to discuss various characteristics of light sources, including efficiency and the perceived amount of light they produce. **Photometry** is the measurement of visible light. The prefix "lum" is a clue that a word pertains to what light we can see, and not all of the light is actually *present* since usually there is more light present than what fits inside the visible spectrum. The measurement of all of the light, visible or not, is called **radiometry**. **Photometric** units of light emission include the **lumen** and the **lux**. The **luminous flux** ("flux" = "flow"), in units of lumens, of a light source tells us how much *total visible light* it emits, i.e., integrated over all of the directions that the light is emitted into. Photons that we cannot see also carry energy, but if we can't use that energy to see, then it is wasted for the purposes of illumination. The luminous power efficiency is called **luminous efficacy**, which is the amount of visible light, in lumens, produced per watt of input power (lm/W).

Of interest: The Sun is a ball of hot, electrically charged gas, called a plasma (discussed at length later in this chapter). There is no 'surface' to the Sun *per se*. Rather, the gas simply gets less dense further out from the core. The Sun's energy output is generated by nuclear fusion in the core of the Sun, and flows outwards by transport of radiation through the *radiative zone* around the central core. The next layer outwards is the *convection zone* in which the plasma behaves like boiling water in a pot, transferring heat towards the surface by the motion of hot gas. The layer that we see and call 'the surface of the Sun' is called the *photosphere*. Below the photosphere, the Sun is actually opaque to all of the light that is generated – no light escapes to get to the Earth. The thin photosphere, only 300 miles or so thick, is heated by the convection zone; at the photosphere, the density has finally fallen off enough that light generated by blackbody radiation can escape the Sun and travel to Earth. That is why we see the Sun as being at around 5800 K, but inside it is much hotter. Beyond the photosphere is a thin layer called the *chromosphere* where temperature actually starts to go up again. The last, outer layer of lowest density is the *corona*, which is oddly at much higher temperature than the photosphere and chromosphere – up to ten million K. These outer layers of the Sun's atmosphere are where the explosions called solar flares and huge mass eruptions occur. Solar physicists are not yet entirely sure why the corona is so hot, but they believe that there may be mini flares occurring all the time that release enough energy to heat it.

Figure 8.6 shows a fountain that sprays water in all directions equally. If we imagine it as a full globe, then the total of all the water flowing out of the fountain in all directions of a sphere is like the total visible light output from a common, kind-of-spherical light bulb. Light that is spread around a room by reflecting off of the ceiling and walls may be useful to many people in the room, say at a party, but that same amount of light directed onto the surface of a table is more useful when, say, you are reading a book. The most common unit of light output provided to the consumer of a light bulb is the lumen, which, again, is the total amount of visible light produced by the bulb. It may turn out that most of the light is emitted in directions that aren't useful, so a spherical

FIGURE 8.6 Water is "emitted" from this fountain the same way light is emitted from a round light bulb, i.e., approximately equally in all directions.

light bulb will not produce light as intensely as one having the same luminous flux that is collected and then directed into a small area.

The density of the water flow, in units of, say, gallons/square meter/second coming from the water fountain in Fig. 8.6 and landing on some surface is a flux density. For some water nozzles, we can imagine collecting, or focusing, all of that water to spray into a narrower cone of flux (see Fig. 8.7). Now, the density of the water flow is increased as compared with the same amount of total water flowing out of a sphere. Instead of water, the density of visible light landing on the surface is **illuminance** in units of lumens per square meter, lm/m², which is called the **lux**. Clearly, for light coming from a spherical light bulb, the light intensity in lux decreases with the distance from the source, since all common light sources diverge from their emission point. Lenses in optical systems can be used to cause light to converge. Also, laser beams are nearly collimated, and diverge only slightly with distance. Figure 8.8, top, shows light that is collected, reflected and collimated by a reflecting surface that is behind the bulb.

FIGURE 8.7 Water is "emitted" from this nozzle the same way light is emitted from a light-emitting diode, i.e., approximately in one preferred direction. Image by Ralph from Pixabay.

Mirrors are a type of lens, and are often inexpensively used to collect and collimate light into a smaller area with much higher illuminance.

These water examples serve to illustrate that light may come out of a lamp somewhere in between either uniformly from all angles, like a spherical incandescent bulb, or in a narrow beam, like a flashlight. Another way to put this is that if you buy a bulb that puts out a certain number of lumens, it is your job to decide how you want to distribute those lumens – either bundled into small, bright area (high illuminance), or spread over a larger, dimmer area (low illuminance).

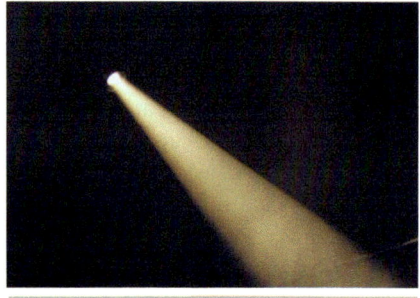

FIGURE 8.8 The spotlight (top, https://simple.wiktionary.org/wiki/spotlight#/media/File:Spotlight_(14871682502).jpg) reflects all of its light into the forward direction. Although light-emitting diodes, the small yellow squares in the bottom image, direct their light in only a narrow direction, some LED light bulbs use a diffuser shell to distribute their light in all directions, much like a common incandescent bulb.

Note that the light meter you will use in this chapter's lab activity can read out only the light intensity in lux, and cannot measure total lumens, simply because it has only a small detector area. If you could measure the light emitted at all angles by, say, moving the detector over a full sphere, it could tell you the total lumens, but that isn't practical. Real measurements are done using an **integrating sphere** (Fig. 8.9) that diffuses the light around the inside so that the detector measures the same illuminance that is present everywhere inside the surface of the sphere. Then all that is needed is to do a trivial integration by multiplying the uniform lux reading by the inside surface area of the sphere. The system described below is a 'poor man's' version of one of those.

In the lab activities, you will measure illuminance in lux directly from the big three bulbs. However, not all light bulbs emit light in the same way. This measurement will be done only over the small area of the detector. In order to make a comparison of the luminous flux of the various light bulbs, we will make a very

FIGURE 8.9 Integrating sphere. The inside surface diffuses the light so that the illuminance (lux) is uniform on the entire inner surface. Multiplying the measured illuminance by the surface area yields the luminous flux (lumens). *Source*: https://commons.wikimedia.org/wiki/File:Luminance_Chamber.jpg.

crude integrating sphere. We will do our best to collect all of the light from the bulbs and reflect it downward so that the flux density is reasonably uniform, like that shown in Fig. 8.7, and all the photons are approximately accounted for in the measurement. It will be relatively crude, but it's all that we have available in our lab. By collecting as much light as we can so that only a small amount is lost to emission outside our measurement area, or absorption by the reflector, we will be able to roughly compare the brightness ratings of the bulbs, and therefore their luminous efficacy, but we will not actually measure the total lumens.

The second important photometric rating is a bulb's **color temperature**, which is related to the color spectrum of its light. The color temperature is straightforward if you understand blackbody radiation. The color temperature of a light source is the temperature of a black body that emits with a similar *perceived* color as the bulb itself. Put another way, if you put a black body next to the bulb and change the black body's temperature until its color matches that of the bulb, then the color temperature of the bulb is the temperature of the black body. Color temperature applies to all lighting sources, even if they are relatively cool, like LEDs and fluorescent bulbs.

Incandescent bulbs, discussed more in the next section, give off light from a very hot metal wire, or **filament**, whose temperature is usually at around 2700 K. The hot filament (Figs. 8.10 and 8.11) is at the actual temperature that corresponds to its color temperature. That is to say, it actually is a black body. Its real temperature and perceived color temperature are, therefore, identical. And yes, incandescent bulbs are very hot; some, such as halogen bulbs, are even hotter, and will burn your skin if you accidentally touch them.

In the case of a bulb that mimics daylight, it would have a perceived color temperature similar to the Sun, i.e., around 5800 K. For fluorescent and LED bulbs, the actual temperature of the bulb is nothing like their perceived blackbody temperature. (There is no known material that isn't melted at 5800 K, or about 10,000 °F). They are not hot and do not generally emit a spectrum like a blackbody spectrum, but we can perceive them as having a color that is similar to a black body at a particular temperature. In order to give us a choice of the color of the light that we are buying, bulbs are sold not only with a lumens rating but also with a color temperature rating. Note reddish/yellowish light corresponds to a lower blackbody temperature, but is called "warm," and blue light corresponds to a higher blackbody temperature, but is called "cool." These descriptors presumably evoke thoughts of a warm fire and cold ice or a cold sky.

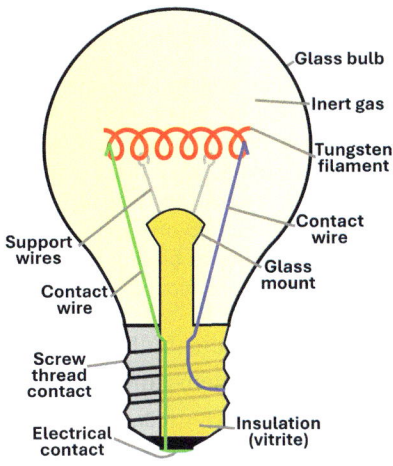

FIGURE 8.10 Incandescent light bulb construction.

FIGURE 8.11 Filament in an incandescent light bulb. The word filament means "slender thread," which in this case is made of tungsten. The filament spirals as a coiled-coil to increase the total length.

A little about color perception. Our eyes have three different color receptors, called **cones**, whose responses are peaked in the red, green and blue regions. All colors that we perceive are due to some combination of the firing of these receptors. For that reason, we might perceive a particular color from two objects to be the same that have overall different spectra. In that case, we might perceive a bulb to give off "white" light, but the spectrum could be very different from natural sunlight.

Medical personnel, artists and others often need to accurately evaluate colors without undue influence of a particular spectrum produced by whatever bulb happens to be in the room. The extent to which a bulb can accurately produce the same spectrum as the Sun is the **color rendering index**, **CRI**. CRI is an average over several colors and whose value approaches 100, meaning exactly that of the Sun (within the visible spectrum). CRI values as high as 99 are available commercially.

8.3.2 Incandescent Light Bulbs

Incandescent light bulbs are the grandmother of all electric lighting technology (not the actual first, but the first wide-scale commercially successful), and is generally attributed to Thomas Edison. (You might want to read more about Edison and/or the invention of the light bulb. They are interesting stories). **Incandescent** means that light is emitted by heating. In a common household incandescent bulb, electric current flows through a thin, tungsten filament (Figs. 8.10 and 8.11) that is heated to such a high temperature that it glows, just like the Sun in our previous example, but at a lower temperature. The light, again, is its blackbody radiation. The problem is that 90% of the energy used by the bulb is wasted as invisible infrared light, and only 10% of the energy is used as visible light, as shown in Fig. 8.12, so it has low luminous efficacy, about 10 lm/W. In the winter and indoors, this isn't such a problem, since they make nice little space heaters, but overall, this is a major waste of energy. It is so bad that they have been largely phased out in favor of the more efficient lights – fluorescent at first and now LEDs.

> *For fun:* There is an idiom about people who 'generate more heat than light.' It means that although they are doing a lot of talking, they don't actually say anything useful. Of course, everyone who uses that idiom knows they are really referring to blackbody radiator at a temperature of about 2700 K.

Incandescent bulbs come in various wattages, mostly 60, 75 and 100 W. A 60 W bulb produces about 800 lumens, and a 100 W bulb produces about 1600 lumens, for an improved luminous efficacy. In the lab activities, one bulb that you will use is a "halogen" bulb. A halogen bulb is incandescent, but the filament is surrounded by a small amount of halogen gas, such as iodine or bromine. The gas causes the evaporated tungsten from the filament to redeposit back to the filament, so the filament lasts longer. Now the bulb can be run hotter without burning out quickly, and emit light (again, blackbody radiation) that is more in the visible and less in the infrared, which we cannot see. By running hotter, it shifts the spectrum of the incandescent bulb in Fig. 8.12 to the left, and increases its magnitude (refer to Fig. 8.5) so that the peak wavelength of blackbody emission corresponds better with the spectrum of visible light. Therefore, the luminous efficacy is increased, to about 20 lm/W. The halogen lamp you will use has an outer glass bulb that is in the shape of a regular incandescent bulb. The actual small halogen bulb is inside of that.

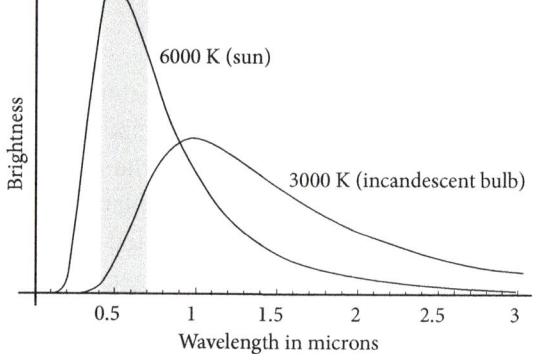

FIGURE 8.12 Emission spectrum of an incandescent bulb. As shown, a lot of energy is emitted into the long wavelengths that we can't see. (In this figure, the incandescent bulb operates on the hotter side, at 3000 K.) *Source*: https://www.laserfocusworld.com/test-measurement/research/article/16549510/photonic-frontiers-high-efficiency-photovoltaics-photovoltaics-takes-small-steps-on-journey-to-greater-efficiency. Reproduced with permission.

So, how much money does it cost to run one of these inefficient incandescent bulbs? More than you might think. Here is an important, simple, rule-of-thumb that is based on a typical cost of electricity of 12 cents per kWh: **one watt for 1 year costs one dollar** (we could call that amount of energy a "watt-year" if we wanted to). For example, a 60 W light bulb forgotten and left on in the attic for a full year would set you back about $60 in electricity costs. Ow, that hurts. In this case, you would be lucky that incandescent bulbs have a lifetime of only about 1000 h, so you wouldn't be stuck with the whole cost of the mistake.

An example of an "energy vampire" is the small "wall wart," i.e., those ubiquitous power supplies for small electronic appliances, such as phone chargers. Even unused, older ones can consume one or two watts of standby power, although newer ones waste about 0.25 W; each one, then, can cost, say, a buck or so per year if left unused in the wall outlet, and many of those scattered around the home can add up to a big waste of energy and money. Also, instant-on remote-controlled devices, like TVs, cable boxes, and more can often use a standby power of several watts. The average house wastes about $100/yr in standby power.

8.3.3 Fluorescent Light Bulbs

The word **fluorescent** implies the emission of light due to stimulation by other electro-magnetic radiation. Briefly, a high-energy, short-wavelength photon hits a material and it emits a lower-energy, longer-wavelength photon. A fluorescent glass tube, or bulb, (Fig. 8.13) contains argon gas with a small amount of mercury vapor mixed in. A high-voltage power supply, called the **ballast**, creates a strong electric field inside the bulb. There is also a pair of small heated filaments at the ends, like in an incandescent bulb, that emit electrons due to the high electric field. These electrons gain energy from the field and bombard the gas atoms, transferring energy from the electric field to the gas atoms, effectively causing them to get very hot, and causing electrons to be knocked out of the outer shells of many of them, including the mercury atoms. As more free electrons are created this way, they cause further ionization of more gas molecules (due to the high electric field) until a steady-state ionized gas is formed. The state of matter of a mixture of ions and electrons is called a **plasma** (Fig. 8.14). Once the plasma is "struck," meaning that the gas is

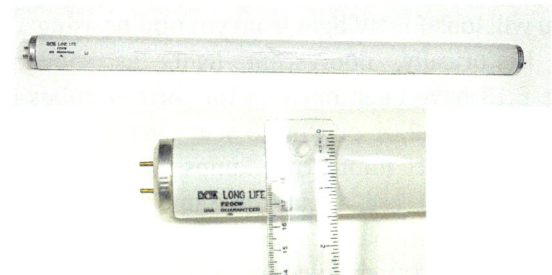

FIGURE 8.13 Traditional fluorescent lightbulb tube. This one is 2-feet long. Various diameters and lengths are available. T5 tubes are 5/8" diameter, T8 tubes are 1" diameter, and shown is the most-common variety, namely T12, which is 1.5" diameter. T12 tubes are available in lengths up to 8 feet.

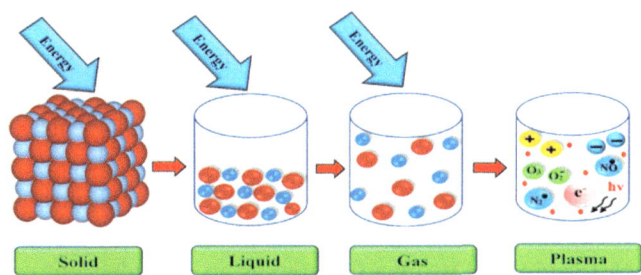

FIGURE 8.14 There are four states of matter arranged in order of energy. When 'hot' enough, a gas ionizes and becomes a plasma. *Source*: https://www.sciencedirect.com/science/article/pii/S27727-53X23000692?via%3Dihub.

well-ionized, it is self-sustaining and presents a very-low resistance, so the ballast is also used to limit the current to a level that is safe for the bulb.

When the high-energy electrons in the plasma randomly collide with the mercury atoms, they transfer some energy to the outer-shell electrons, which jump to higher energy levels. When those same outer-shell electrons return to their lower energy levels, they give up that energy in the form of ultraviolet (UV) light. Every kind of atom has its own characteristic energy levels. The emitted wavelengths are characteristic of the electronic transitions in mercury atoms, and are *not* related to blackbody radiation.

You may already know that UV light (including that from the Sun) is bad for our skin and eyes, and can cause cataracts if we are exposed to too much of it over our lifetimes. Besides being potentially harmful, we can't see it, so the UV light produced by the fluorescent bulb is not useful in its original form. That's where the fluorescence comes in. Fluorescent bulbs are white in color because the entire inner surface of the glass tube is coated with a **phosphor**, which is a material that **fluoresces**, i.e., gives off light in response to another wavelength of light that hits it. In this case, higher-energy UV light hits the phosphor, and, again through electron energy transitions, the energy is converted to lower-energy visible light, which is transmitted outwards from the phosphor through the glass tube. The phosphor coating screens out nearly all of the UV light from escaping. In some bulbs, cracked phosphor coatings can allow a small amount of UV to leak out, but it is deemed safe under normal use. (In this week's lab, you will look for UV light from your bulbs using a UV light meter.)

> *Safety:* The danger from ultraviolet light exposure for humans is usually referenced to the UVAB spectrum, which includes wavelengths from about 280 nm to 400 nm. The meter used in this lab measures UVAB intensity. Under full sun (AM1, discussed below), the meter reads about 5 mW/cm² of UVAB. You may use this number to compare with UV readings that you get in the lab activities, and decide how to interpret those intensities vis-à-vis danger from exposure. You will find that typical bulbs emit very little UVAB. Considering the distances from the bulb that they are typically used, that exposure is insignificant.

Historically, fluorescent lights as shown in Fig. 8.13 have been made in the form of tubes that are typically 2, 4 or even 8 feet long. Since about 1995, compact fluorescent lamps (CFL) have been available that are about the same size and form factor as traditional *incandescent* bulbs, but look like either short, straight tubes or swirly ice cream cones (Fig. 8.15). These bulbs usually screw into the standard, medium, household base, called an E26 base (E for Edison), or the smaller candelabra base called the E12 base (E for, well, ..., you know). CFLs have luminous efficacy of about 60 lm/W to 70

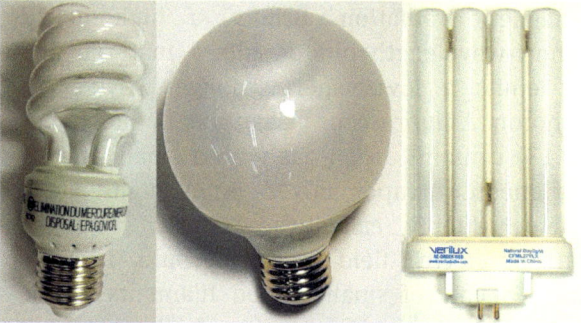

FIGURE 8.15 Compact fluorescent bulbs come in various shapes and with various bases.

lm/W, and are sold with their incandescent-equivalent wattage prominently displayed. For example, a CFL label might show "60 W" when it actually uses about 13 W because it produces an amount of light that is comparable to the same 800 lumens as a standard 60 W incandescent bulb. As we shall see, even this relatively recent advancement in lighting has largely given way to LED bulbs, as shown in Fig. 8.2.

Besides using about ¼ of the energy of incandescents, CFLs also have much longer lifetimes, about 10,000 hours compared to about 1000 hours for a common incandescent bulb. In the best case, the cost savings of using a CFL versus an incandescent can be many, many times the cost of the bulb itself. As another example of cost savings, consider just the outdoor lighting around your home. At my home, I once had six outdoor 60 W incandescent bulbs that were on for a yearly average of 12 h per day; the cost of the electricity for just those bulbs was about $180/yr. After switching to 13 W CFLS, the operating cost dropped to about $40/yr. After a few more years I replaced those with 8.5 W LED bulbs, for a total electricity cost of about $26/yr. Now consider that cost and energy savings repeated for every house in the neighborhood, city, state, country and world! Clearly, from an energy, global-warming, air-pollution standpoint, a widespread conversion from incandescent bulbs to efficient

> *Safety:* If a CFL bulb happens to break, it is recommended to open the windows or turn on a fan, clean it up carefully, sweeping it into a plastic bag, seal it and discard appropriately.

sources such as CFLs or, even better, LEDs, as described next, is having have a measurable impact on the world's energy burden due to lighting.

"But Professor Bernstein," you say, "what about the mercury in CFLs? Isn't that bad for the environment?" Well, yes, mercury is definitely a bad thing in the environment. However, environmentalists were okay with using CFLs because about 40% of all electricity was at that time generated by burning coal, and burning coal for the electricity to run a competing incandescent bulb over a CFL's lifetime emitted lots more mercury into the environment than was in a CFL bulb (about one milligram of mercury). Besides, CFLs can, and should, be recycled at many hardware stores, as long as they aren't broken. All of that said, it is a good thing for many reasons that CFLs are being phased out in favor of LEDs.

CFLs come in a variety of color temperatures, as set by the choice of phosphor. They can mimic (give the same appearance to our eyes as) the 'warm' glow of an incandescent bulb, at around 2700 K, or daylight at around 5000 K. This 'light appearance' is printed clearly on the label as the equivalent color temperature for the consumer to choose, as discussed above. You can see in Fig. 8.16 that the color temperature is translated as "warm" or "cool" for those not among the cognoscenti, as you now are. So, now you understand all of the "Lighting Facts" on the bulb labeling and can be a super-smart consumer of lighting technology. We will use that new-found knowledge with the last of the Big Three, the light-emitting diode, or LED, light bulb.

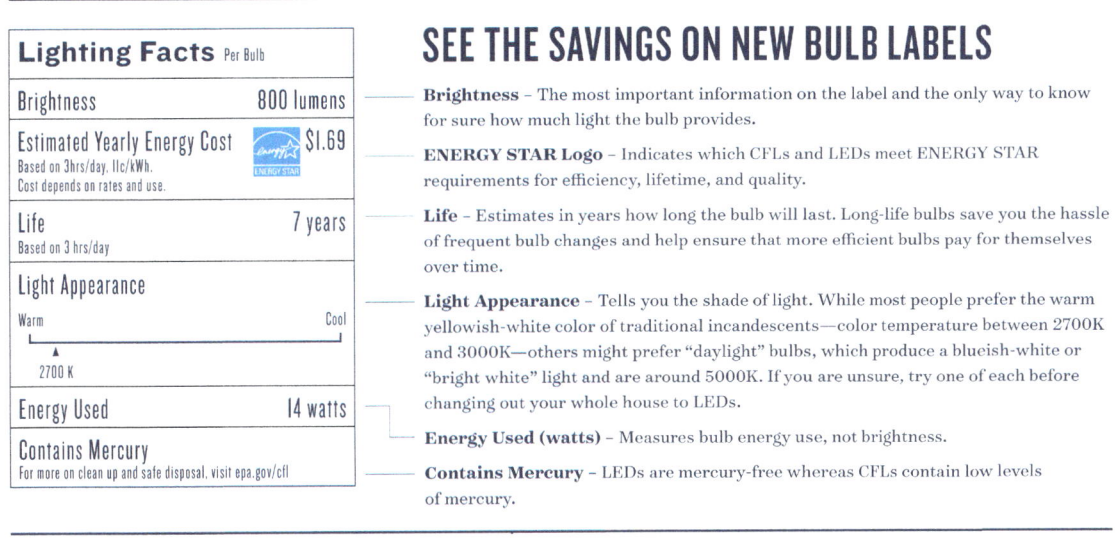

FIGURE 8.16 Light bulb parameters. This one is for a CFL bulb. *Source*: NRDC Lightbulb Buying Guide 2019.

8.3.4   LED Light Bulbs

LEDs are the newest practical, general-purpose lighting technology. The LED has been commercially available as a light emitter since about 1970. At first, they were dim, but over time the brightness and efficiency have improved so much that they are now used in flashlights, light bulbs, outdoor security lights, video lighting, and even car headlights. Until fairly recently, they were more expensive than CFLs while offering a modest improvement in energy savings. For example, a "60 W" (i.e., gives off 800 lumens) LED bulb that uses about 8 W had become generally available in about 2010 at a cost at about $10 each, but by 2017 the retail cost had dropped to under $1.00 – a very sharp drop indeed! Now it is routine to find LED bulbs in all price ranges with a wide variety of features.

The rated lifetime of LEDs, at around 50,000 h, is much longer than either incandescent bulbs or CFLs. This is the equivalent of continuous operation for almost 6 years. Because of their longevity, municipalities have replaced stop lights with LED lights, mostly due to the cost in maintenance compared to older incandescent bulbs. Since those lights operate around the clock (cycling, of course between red, yellow and green), there is also the significant energy cost savings.

LEDs are diodes in every way, except that they give off light. So, how do LEDs emit light? In Chapter 7, you learned about electrons and holes, and a little bit about diodes. You know that n-type material carries current mostly by electrons, which we therefore refer to as the **majority carriers**. There are very few holes, which in n-type material are the **minority carriers**. Conversely for p-type material, holes are the majority carriers and electrons are the minority carriers.

You know that n-type material carries current mostly by electrons, which we therefore refer to as the **majority carriers**. In forward bias, conduction electrons from the n-side, where the electrons are majority carriers, flow

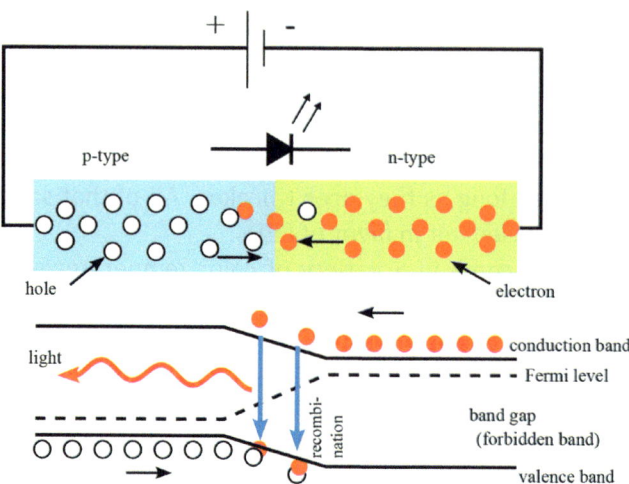

FIGURE 8.17 Top picture shows the electrons and holes before they move across the junction. The bottom picture shows the energy band diagram in which electrons fall across the bandgap near the junction. In the process of radiative recombination, electrons lose energy to fill the holes, and that energy is converted to light. *Source*: https://en.wikipedia.org/wiki/Band_diagram#/media/File:PnJunction-LED-E.svg.

across the **p-n junction** (the region where the p- and n- materials touch) toward the p-side that is full of holes, where they now find themselves as minority carriers; conversely, the majority holes in the p-type material flow across the junction toward the n-side that is full of majority conduction electrons where the holes are then the minority carriers. This is illustrated in Fig. 8.17.

Majority and minority carriers recombine when they are in the same place, meaning that the electrons lose energy and fall into, or fill up, the holes, resulting in both the electrons and holes disappearing. Well, conduction electrons and holes don't actually disappear, but what does happen is that when a mobile conduction electron combines with a mobile hole, the result is an immobile covalently-bonded electron that no longer participates in current conduction, so both the mobile electron and hole are effectively gone. Here is a rough analogy: If you have a pile of dirt and a hole in the ground, and then put the dirt in the hole, then both the pile and the hole are gone. The dirt hasn't disappeared but rather fills in where the hole was. If you are wondering how current can flow if charge carriers disappear: the holes and electrons are replenished at the contacts at the ends of the device and continue to flow towards, and then across, the junction to provide the forward bias current.

Electron–hole recombination gives up energy as the electron falls across the bandgap and fills the hole, as shown in Fig. 8.17. Energy is always conserved. Electrons that lose energy can do so in the form of either heat or light. In a conventional diode used in a circuit, for example in a power supply (Chapter 10), the conduction electron loses energy to heat. A light-emitting diode is a diode in every way we have so far discussed, but the materials are very special; in forward bias, when the electron combines with a hole, the falling electron gives up its energy to create a photon rather than heat, so light is emitted. This process is called **radiative recombination**.

Why do some materials recombine radiatively and others not? That is truly complicated. In short, electrons are waves with a certain wavelength in the material. So are the holes. The electron and hole waves have to match when they recombine. In some materials, called **direct-gap** semiconductors, the electron and hole waves match and the electron can directly fill the hole without any other energy process needed, so the energy goes straight to light. In **indirect** semiconductors, the electron does not 'fit' into the hole perfectly. In order to recombine, a phonon is necessary. Recall that a phonon is a 'jiggle' of the atomic lattice, and is the actual heat of a material. The electron must first create a phonon (heat) that jiggles the lattice, changing the wavelength of the hole, before it can fit into the hole (match the wavelengths), so its energy goes entirely to heat, rather than light. Only certain semiconductor materials are direct-gap, so a considerable amount of materials science and engineering goes into the development of new LEDs.

Radiative recombination is also the foundation of solid-state lasers, like those used in laser pointers, at the checkout counters of grocery stores, LiDAR (light detection and ranging) systems for self-driving cars, and for laser surgery, to name but a very few applications. Semiconductor diode lasers start off as LEDs and become lasers when they are shaped in a way that reflects and concentrates the light energy inside the diode until it leaks out as a laser beam.

LEDs are made from combinations of In, Ga, As, N, Al, and P, (and not Si), called **compound semiconductors**, e.g., InGaN (indium gallium nitride). Different combinations of materials result in direct-gap semiconductors with different bandgaps. The energy gained by the electron falling across the bandgap of energy E_g goes directly into the creation of the photon. Using Eq. 8.2, $E = hf = hc/\lambda$, where E is now E_g, we see that larger bandgaps result in shorter wavelengths, so it is possible to create LEDs that give off light of different colors. In the beginning, LEDs were available only in red. Then LEDs with larger bandgaps were developed, and yellow and green yellow LEDs came along, and finally blue LEDs were perfected in 1994. The 2014 Nobel Prize in physics was given for the invention of the blue LED. This was specifically for the fact that the blue light revolutionized the world of lighting technology. Why? The early LEDs gave off longer visible wavelengths, which had become progressively shorter with new technology development, eventually reaching the blue, and even UV, LED wavelengths of today. This led to the creation of the holy grail of LED lighting, namely the white-light source. It also made possible higher-density optical storage media, such as Blu-ray disks.

Now we have the same sort of situation as we did with the CFLs – we can bombard a phosphor with shorter-wavelength, higher-energy photons and emit longer-wavelength photons that are suitable for creating white light. (See Fig. 8.18 for a plot of the spectra of various light sources). Without white light, there would be little call for LEDs as a lighting technology. A blue or UV LED is converted by the phosphor to some of the longer wavelengths, and tuned to give a color temperature closer to daylight. Some blue is let through to contribute the shorter wavelengths, and the phosphor provides the green, yellow and red parts of the spectrum. All of this light comes at even higher efficiency than the CFLs, and are getting better all the time. LED technology has become prevalent, and as manufacturing costs have gone down and options have increased, they have vastly overtaken fluorescent and incandescent lighting, resulting in vast savings in energy, pollution, and overall operating costs.

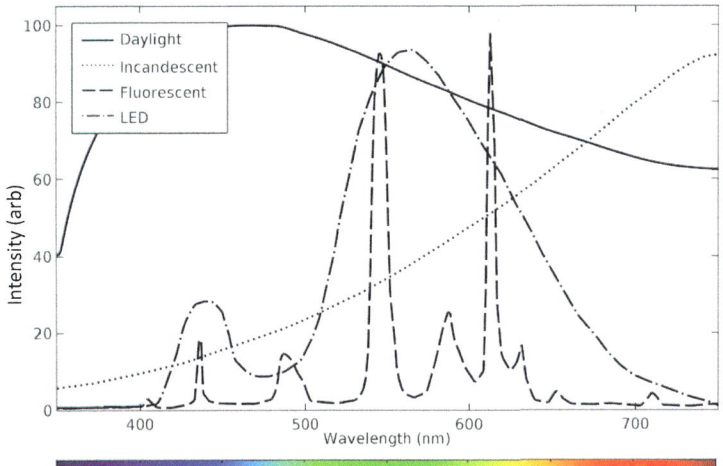

FIGURE 8.18 Spectra of various light sources and colors of the various wavelengths emitted by the bulbs. These spectra are simulated using computer modeling, and are representative of the various light sources. Red and infrared are toward the right and blue and UV are to the left. The spectrum from the Sun (daylight) overlaps well with the colors that humans can see. Incandescent light is weak in the visible and strong in the IR, where we cannot see. The spectrum of fluorescent bulbs has many peaks due to specific atomic energy transitions in the phosphor. The LED bulb has two broad peaks due to the bandgap energy and impurities in the crystal, as well as properties of the phosphor. The combinations of the visible wavelengths give us the color temperature, where shorter blue wavelengths are necessary to make white. Image credit: COMSOL. https://www.comsol.com/blogs/calculating-the-emission-spectra-from-common-light-sources.

8.4 Alternative Energy and Solar Cells – The Other Side of the Coin

So far, we have discussed mostly how to improve efficiency through better lighting in order to decrease the world's energy consumption. Now we discuss the other side of the coin – how to generate electricity in a way that is less impactful on the environment. Typically, this is through **renewable energy**, since anything that is renewable on a human timescale is less likely to have a permanent impact on the environment. (The term **alternative energy** implies that it is still in development and is not yet mainstream. The two terms are often used interchangeably. We can look forward to the day when 'alternative' energy isn't alternative at all, but rather is mainstream.)

Renewable energy sources include wind (often using massive wind turbine generators), **photovoltaics (PV)**, also called **solar** (which comes in many flavors), tidal (putting turbines in the ocean where tides are strong, or capturing the water at high tide and releasing it through turbines at low tide), and geothermal (from heat available deep in the Earth). The renewable hydroelectric power has been an important source of energy since the earliest days of electrical (and mechanical) power generation, and it is often counted among the renewable energies, but is not generally referred to as "alternative."

Every one of these options comes with its own disadvantages and advantages. A brief collection of cons include: Wind power is not steady or reliable and is considered by many to be unsightly in pristine natural environments, and can affect bird populations; solar power is useful only when the Sun shines and can be relatively expensive in Northern latitudes; tidal power faces challenges with ecological impacts, and the fact that it is useful only certain times of the day (but it is guaranteed at those times!); geothermal energy is derived from heating water driven deep into hot subterranean rocks, and is therefore limited to only certain geographic locations that are often far from population centers. It is expensive to drill and to operate such installations.

Sunlight is attractive as an energy source because it is rich in usable energy and is largely available. Solar energy can be generated by using mirrors to reflect and concentrate the light onto solar furnaces that can reach temperatures upwards of 3500°C. This heat energy can be used directly for manufacturing, or, at lower temperatures, to boil water or other chemicals to turn turbine generators. In this section, however, we are concerned mostly with large solar panels of many solar cells that convert sunlight directly into electricity.

In order to use solar power, we must consider how much power per unit area (called the **solar irradiance**) falls on the solar collector, and how much of that can be used for electricity. If the collector is exoatmospheric, i.e., in space around the Earth, for example on a satellite, the solar irradiance is referred to as being that of **air mass zero**, or **AM0**. At AM0, the energy is transmitted from the Sun to the collector without any absorption from Earth's atmosphere. At AM0, power density from solar radiation is 1366 W/m^2, which is a pretty significant amount of power. Unfortunately, only about 20–40% of that power can be converted to electricity, depending on the type of solar cell used.

The solar irradiance through an amount of air equal to one atmosphere, such as that taken at sea level with the Sun directly overhead, is referred to as **air mass one**, or **AM1**. At AM1, approximately 1 kW of power falls on a square meter. The atmosphere absorbs a significant amount of the power at various wavelengths due to the excitation of molecular vibrations in the various gasses comprising the atmosphere. Figure 8.4 shows the spectrum "at the top of the atmosphere" (AM0) where it is the unabsorbed blackbody radiation. The figure also shows the spectrum "at sea level" (AM1) where many wavelengths are attenuated or completely absorbed.

If the Sun is not directly overhead, the distance of atmosphere that the light passes through is increased according to the factor $1/\cos(\theta)$ where θ is the angle of the Sun relative to the perpendicular, and the available power is decreased from AM1 by the same factor. Other angles are given other air mass designations according to the same factor of $1/\cos(\theta)$. In fact, the Sun is directly overhead for AM1 only between the two tropics around the equator, and at that only two days per year. A satellite is exposed, therefore, to about 30% more power at AM0 than it would be at AM1.

Since a typical residential home consumes a maximum of about 20 kW, we can appreciate how much power the Sun provides to the Earth. If we could harness all of the solar power at the best case of AM1 as electricity, we could fully power our homes with less than 15 × 15 sq. ft. of sunlight. We will see, however, that there are considerable limitations to using solar energy, but we can still consider solar energy to be a vast, largely untapped resource.

The most common way to convert solar energy to electricity is to allow it to fall on large sheets of silicon solar cells that directly convert the photon energy to electrical energy. Figure 8.19 shows a single solar cell of a few inches on a side, and Fig. 8.20 shows collections of solar cells connected together in series and parallel combinations into panels. A solar cell is a relative of our friends the rectifying diode and light-emitting diode, but it is run 'in reverse,' meaning that instead of current driving electrons and holes that give up their energies to make light, the light gives energy to electrons and holes that drive current in an external circuit. As mentioned in Chapter 7, semiconductors get their properties from their crystalline properties. The best solar cells are **monocrystalline**, meaning

FIGURE 8.19 A single silicon solar cell.

the whole cell is a perfect crystal. Cheaper, less-efficient solar cells are **polycrystalline**, meaning that they are formed from a collection of tiny crystallites. It is much cheaper to make solar cells that way, although they are not as efficient. By far, polycrystalline cells are the most common, and responsible for the recent, huge upswing in solar installations.

All inexpensive, mass-produced solar cells are made from silicon. For reasons that we will not go into here, silicon is quite capable of generating electricity from light, but not light from electricity, as in LEDs. When light is incident on a semiconductor, it penetrates the surface into the bulk material and can provide energy to a valence electron (See Chapter 7), which normally does not contribute to current. The absorbed photon energy promotes a valence electron across the bandgap energy to the conduction band, meaning that it frees the electron from the bonding state in the lattice, thus allowing it to conduct electricity. The electron missing from the valence

FIGURE 8.20 A bank of large solar cell panels.

band leaves a hole behind, which also carries current. The process by which sunlight creates these electron–hole pairs, EHPs, is called **optical generation**. If an electron and a hole move together, it is equivalent to no current flowing because there is no *net* flow of charge; the electron and hole must be separated so that they go in opposite directions for there to be a net flow of current.

Refer to Fig. 8.17. The narrow region around the p-n junction where the two materials touch is called the **depletion region**. It is the slanted region in the middle of the band diagram at the bottom of Fig. 8.17. There is a strong electric field within the depletion region. As you know, an electron and a hole will go in opposite directions inside of an electric field, but both of them carry *current in the same direction*. When an electron or hole generated by sunlight from an EHP finds itself inside that narrow region of electric field, it is swept out because of the electric field. That is equivalent to a source of EMF that drives a current in an external circuit. The process of optical generation is shown in Fig. 8.21. When the electrons and holes are separated by the field within the depletion region at the p-n junction, they are separated and flow the opposite way than they do in a forward-biased diode. Hence, the solar cell acts like a battery, producing power to the circuit rather than consuming it as heat.

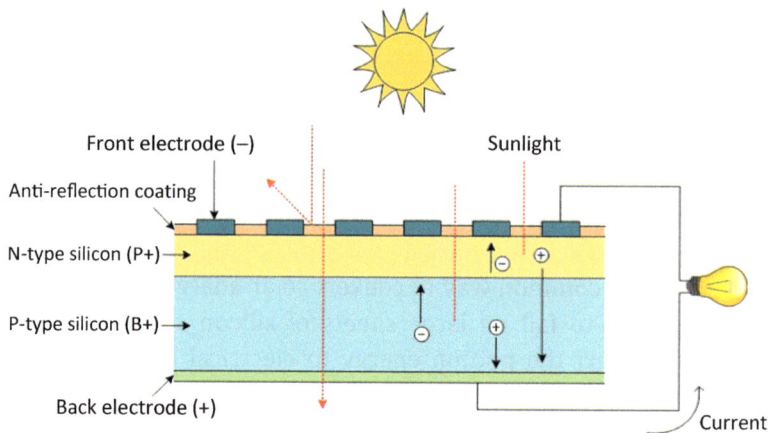

FIGURE 8.21 Solar cells convert photons from the Sun into electrical energy.

Since the voltage supplied by the solar cell is due to the reverse processes of that of forward bias, the open-circuit voltage of a silicon solar cell tends to be in the range of about 0.5 V to 0.7 V. In order to obtain higher voltages, many solar cells are wired in series. In summary, when sunlight creates EHPs, those new electrons and holes that find themselves in the depletion region are separated by the electric field, thus creating a voltage and current. This same phenomenon of generating EHPs with photons, but on a smaller scale, underlies the operation of optical sensors used in digital cameras, laser communication over the internet, Blu-Ray players, and much more, so they are an important part of many electronic systems.

Figure 8.22 is an aerial view of the Bhadla Solar Park in India, which is, as of 2023, the world's largest photovoltaic installation. Covering 14,000 acres, or 22 square miles, it came on line completely in 2020. All large commercial solar panels are made of silicon using processes that are far cheaper than those used in making other semiconductor devices,

FIGURE 8.22 As of 2023, the world's largest solar farm is the Bhadla Solar Park in Rajasthan, India. Peak power produced is 2245 MW. Photo courtesy of Azure Power.

such as integrated circuits. It is imperative that a large-scale power installation be inexpensive enough that it can supply enough power for long enough at a given cost to be profitable in the energy market. Such factors as the global cost of oil, new methods of manufacturing solar cells, government regulations, tax incentives, and much more can have a go/no-go effect on such decisions.

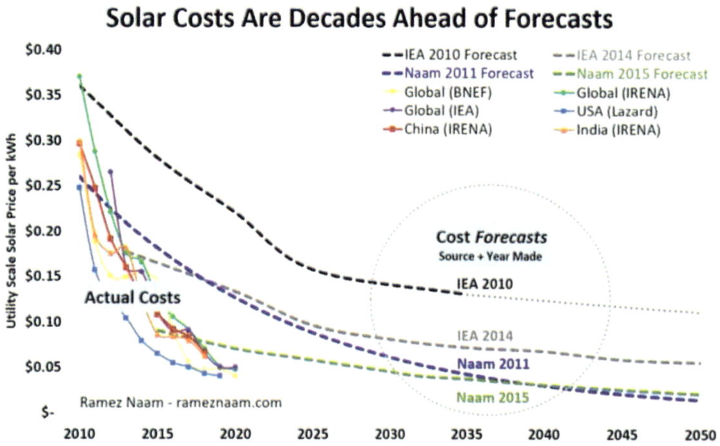

Figure 8.23 shows trends in the cost of solar energy, including predictions of future costs. In fact, prices have dropped much faster than had been anticipated. Note that the wholesale price of a kWh

FIGURE 8.23 The cost of solar energy has decreased significantly over time, and has finally overtaken the cost of fossil fuels. As a result, the number of solar cell installations has increased dramatically, and is far ahead of predictions made a decade ago. Reproduced with permission from Ramez Naam (https://rameznaam.com/).

of solar electricity is now well below the 12 cents per kWh assumed in the watt-year rule (which is fairly typical in the United States), but that does not take into account many other factors such as location and useable hours per year. Still, the overall cost trends ever downwards, and the very competitive cost of solar energy has caused the rate of new global solar farm installations to increasing dramatically, as shown in Fig. 8.24.

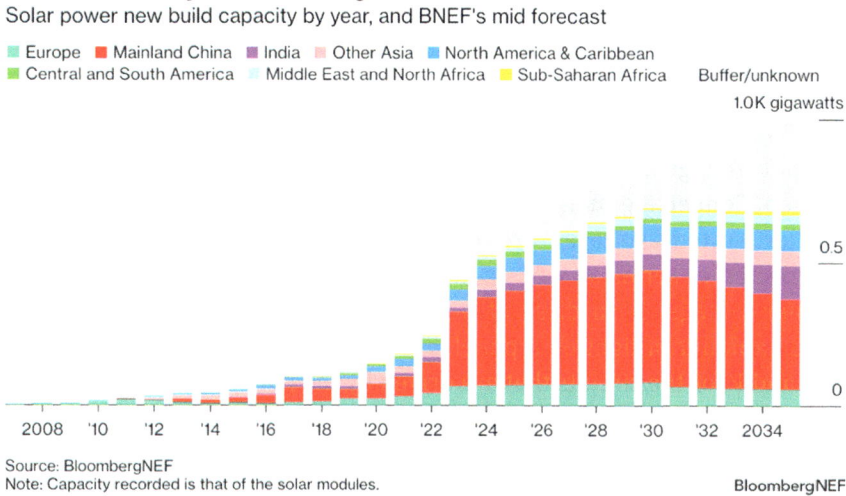

Global PV Industry to Build 592 Gigawatts This Year
Solar power new build capacity by year, and BNEF's mid forecast

Source: BloombergNEF
Note: Capacity recorded is that of the solar modules.   BloombergNEF

FIGURE 8.24 Global photovoltaic installed power as of 2024, and predictions. Reproduced with permission from BloombergNEF.

8.4.1 | Limits to Solar Cell Efficiency

Solar cell efficiency is the ratio of power output from the cell to power input from the Sun. It would be great if we could convert all of the Sun's optical power to electrical power, but it isn't even close. It would be a very big deal to increase the efficiency of solar cells used in the giant solar farms discussed above by even 1%. This would translate to millions of tons of CO_2 not being put into the atmosphere, many billions of dollars more profit, more solar installations coming online, and less land used. Electrical engineers, as well as other types of engineers and scientists, are working toward increasing practical solar cell efficiency.

So, why aren't solar cells more efficient? Unfortunately, not all of the solar spectrum (Fig. 8.25) has the right energy to create the electron–hole pair necessary for electrical power. It is important to keep in mind that the electric field at the junction is several microns below the surface of the solar cell. The long-wavelength, i.e., low-energy, photons do not have enough energy to lift valence electrons across the bandgap, thus creating EHPs. Since they aren't absorbed, they pass through the solar cell as if it were transparent. Short-wavelength, i.e., high-energy, photons create EHPs too near to the surface to be of use because the electrons and holes recombine at the surface before they can **diffuse**, i.e., randomly wander, to the junction and be separated by it. A simple silicon solar cell can convert only about 20% of the incident optical power to electrical power. There are many tricks to making solar cells more efficient than simple poly-crystalline ones: make them monocrystalline, use different

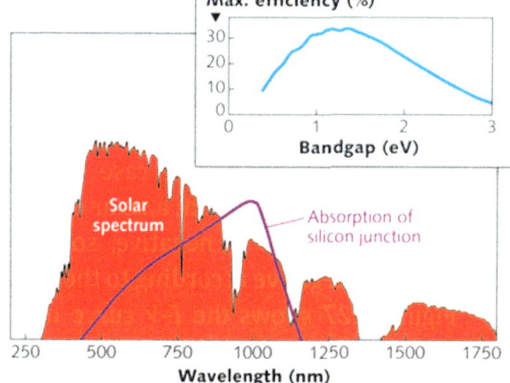

FIGURE 8.25 Not all of the solar spectrum is used by the solar cell. This figure shows the wavelengths that can be absorbed, and thus converted to electricity, by Si. *Source*: https://www.laserfocusworld.com/test-measurement/research/article/16549510/photonic-frontiers-high-efficiency-photovoltaics-photovoltaics-takes-small-steps-on-journey-to-greater-efficiency. Reproduced with permission.

materials in layers whose combined bandgaps better match the solar spectrum, move the positions of the junctions relative to the surface, and more. These more exotic cells are complicated and more expensive, but the best of these can reach higher efficiencies of 40% to 50%. Such exotic solar cells are used in satellites where cost is not the primary factor.

As discussed above, a silicon solar cell produces an open-circuit voltage of about 0.5 V to 0.7 V. By "open-circuit" we mean that it just sits there not being used to power a load, i.e., it drives no current. Solar cells made of other semiconductors have different bandgaps and have different open-circuit voltages.

"Under load," meaning when it is powering a load, i.e., current is flowing from it, a solar cell's output voltage and current depend on its efficiency and its size, as well as the load that is being driven. Connecting many of them in parallel is the equivalent of making solar cells with greater area. If higher system voltages are desired, as in large solar panel installations, many solar cells are connected in series as well as in parallel, so that a large panel can produce considerable current at a higher voltage. Regardless of how they are arranged, the output power, $P = IV$, is the same for a given amount of light on a panel, but the arrangement makes it easier to tailor a desired combination of I and V for the circuits that are driven by the panel.

8.4.2 Solar Cell Current–Voltage Characteristic

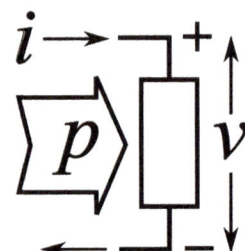

In the dark, a solar cell has a current–voltage curve that is the same as a regular diode. (You will see that in this week's lab activities). However, with light shining on it, the current is shifted negatively for a given voltage, which signifies that current comes out of the diode rather than goes into it (which you will also see). This means that it is producing power, rather than using it, as does a conventional diode. The direction of power flow, $P = IV$, having a positive or negative sign depends on the signs of I and V. In electrical engineering, the agreed-upon convention, called the **passive sign convention** (PSC), is that a passive device, i.e., one that only uses power and cannot produce power, is taken as having positive current flowing into the positive terminal, as shown in Fig. 8.26. For a passive device, such as a resistor or diode, the power is then positive, and therefore *positive power implies that it is consumed by the device.*

FIGURE 8.26 Passive sign convention (PSC) showing power flowing from a circuit into a passive device.

An *active* device, on the other hand, can produce power and deliver that to the circuit. In order for that to be the case, current must flow out of the positive voltage terminal, as in the case of a battery, power supply, or as discussed next, a solar cell, in which case I is negative and $P = IV$ is negative, so power flowing out of a device is negative according to the PSC.

Figure 8.27 shows the *I–V* curve of a solar cell both in the dark (dashed line) and under illumination (solid line). In the first quadrant, labeled "(1)," current flows into the positive terminal of the diode, so I and V are both positive, and, by the PSC, the diode only consumes power. When light is shining on it, the curve shifts downwards so that now current and voltage are in quadrant (4). In quadrant (4), I is negative (it flows out of the positive terminal of the diode) and V is positive, so $P = IV$ is negative, and by the PSC, the diode, now solar cell, is producing power that can be used by the load.

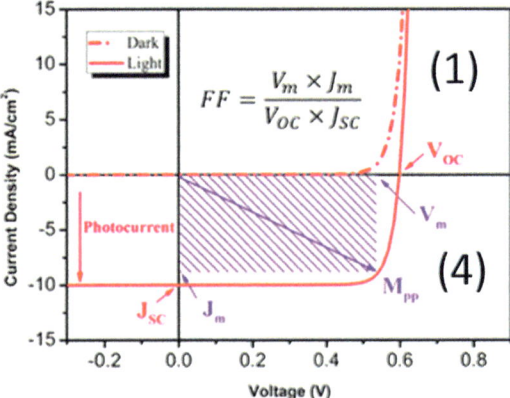

FIGURE 8.27 Solar cell *I–V* curves in the dark and illuminated. The area of the hatched region is the product of the operating voltage and current. The larger that is, the better the solar cell. Reproduced with permission from *Phys. Chem. Chem. Phys.,* 2013, 15, 8972–8982.

We can measure the output voltage of the solar cell with no load, i.e., open-circuited, using a simple voltmeter. That voltage, V_{oc}, is the largest voltage possible from the solar cell under those lighting conditions and can be thought of as the Thevenin voltage of the solar cell. Because it is an open circuit (the ideal voltmeter having infinite input impedance), there is no current out of the cell. Likewise, if we put a wire across the output, i.e., a short circuit, we will get the most current, J_{sc}, (where "J" signifies a current density in the solar cell rather than a current, since we don't want to always have to account for the size of the solar cell). This current is like the Thevenin current of the solar cell. Because wires have essentially zero resistance, the voltage at the output will be nearly zero when it is short-circuited. You can find both V_{oc} and J_{sc} on the I–V curve of Fig. 8.27, and you will also do so in this week's lab.

> *Deeper understanding:* How to think about I–V curves and fill factor: It takes a certain mindset to interpret an I–V curve. It is an absolute statement about the current that flows through it when a particular voltage is across that device. If you want, you can think of putting an ideal voltage source across the device and changing its value while you read off the current through it. Or, you can think of any one point on the I–V curve as occurring due to some external circuit condition. In the case of a solar cell, you get an entire I–V curve for some illumination. On that curve, you get a particular I and V point, or **operating point**, for a particular load resistance, as indicated by the point M_{pp} in Fig. 8.27. So, for a certain amount of light, the load determines the operating point and, therefore, the fill factor.

You may have already guessed that we do not operate a solar cell in either open- or short-circuit conditions. We need to drive some kind of load, or what would be the point? (We can think of the load as a simple resistor). For open and short-circuit conditions, $I = 0$ or $V = 0$, respectively, and $P = IV = 0$. In each of these two cases, one of the two terms would be zero, and each case represents no useable power from the solar cell. Therefore, we get the most efficiency out of the solar cell by making the load resistance somewhere between an open and short circuit that gets the most power from the solar cell, as represented by the blue-hatched region in Fig. 8.27. The values of I and V for which the blue rectangle is the largest is the largest $P = IV$ product, and is therefore the maximum power that can be extracted from the cell under that given lighting condition. The blue rectangle formed by $J_m \times V_m$ is some fraction of the ideal but unobtainable rectangle formed by $J_{sc} \times V_{oc}$. The ratio of the areas of those two rectangles $(J_m \times V_m)/(J_{sc} \times V_{oc})$ is called the **fill factor**. We choose a load that lets us operate the cell at the largest fill factor in order to get the most power out possible. Fill factors can be as high as in the 80% range but can suffer from poor-quality material.

8.5 Additional Considerations about Measurements, Spectra and Solar Cells

Here is important information about photons, visible light, and light meters. Photons carry energy according to Eq. 8.2, $E = hf = hc/\lambda$. The shorter the wavelength of a photon, the higher the energy. By definition, a "black" surface will absorb all the photons, regardless of wavelength, and convert all of the photon energy to heat. Sensors that collect and measure the heat from all wavelengths of the light, or radiation, (now we use the word "light" to mean all wavelengths, visible or not) are called **bolometers**. So, all the energy carried by all of the wavelengths of the radiation is measured by a bolometer. However, our eyes are sensitive to only certain wavelengths of light. As presented above, measurements that yield information about our *perceived* response to light, in lumens, are photometric, whereas measurements that yield information about the total

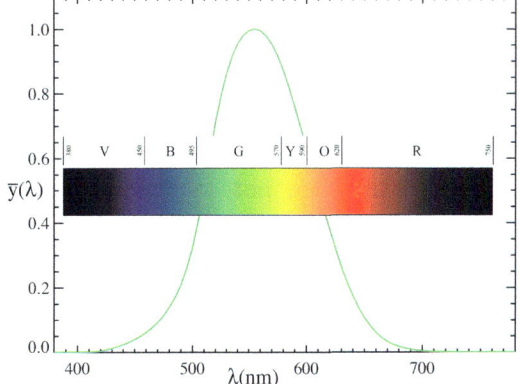

Figure 8.28 Sensitivity of human vision versus wavelength, i.e., spectral sensitivity, under well-lit conditions (photopic).

energy, in watts, in all wavelengths of light are radiometric. When it comes to light bulbs and lighting, we are mostly concerned with photometric units.

When the light is bright enough for us to have our best color vision, the condition is referred to as **photopic**. The sensitivity of human vision to light wavelengths is shown in Fig. 8.28. For comparison, colors and their wavelengths are shown over the curve. Comparing the two figures, you can see that human color vision is most sensitive to greenish light at 555 nm. The purpose of a light bulb is to allow us to see, so using a bolometer to measure the total (radiometric) radiation at all wavelengths from the bulb would be optimistic regarding its ability to provide visible light. In fact, if all the radiation were infrared, the bolometer might measure a very large value, but we would still be 'in the dark.'

Light meters like that which you will use in the lab are filtered to absorb, scale and measure only the wavelengths of light that match the typical human vision sensitivity curve (photometric). The sensitivity of the sensor is shown in Fig. 8.29. The manual for the sensors you will use says, "Spectral response: CIE Photopic. (CIE human eye response curve)." CIE refers to a standard for converting an amount of light energy into how visible it is. Note that the photopic spectral sensitivity curve of Fig. 8.29 corresponds

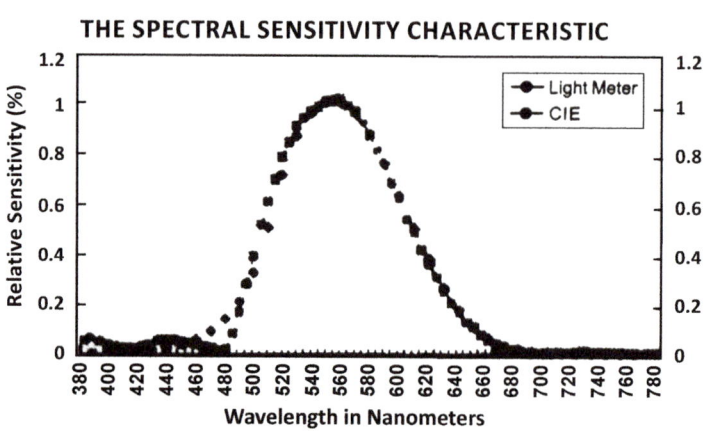

FIGURE 8.29 Spectral sensitivity curve of our light meter.

well with Fig. 8.28. The point of this discussion is that the meter reads out light output in lux, or *visible* (photometric) light intensity, so we can relate it back to the *usefulness* of the light bulb as advertised by the manufacturer in units of lumens, which is total *visible* light output.

The Si solar cell is, in some sense, just a form of a Si light detector, but optimized to produce such a large power output that we can actually power things with it. As such, we need to understand its sensitivity to various wavelengths of both visible and invisible photons. Figures 8.25 and 8.30 show the sensitivity of a typical *unfiltered* Si photodetector (red curve), which we will assume applies as well to our solar cell. Also, the photopic curve is inset for comparison to our light vision. As you can see, the sensitivity of the solar cell is not maximum in the visible light wavelengths, i.e., the very ones from the light bulbs that we measured in the previous steps. To put it another way, what 'looks bright' to us does not 'look bright' to the solar cell. Solar cells are more sensitive to the longer, invisible infrared wavelengths, which are more present coming from the hot incandescent bulbs than from the cooler CFL and LED bulbs. Therefore, we expect our solar cells to give stronger output when illuminated by the heated bulbs, even though we may not perceive them as being as bright as the other bulbs. Look for that in this week's lab activities.

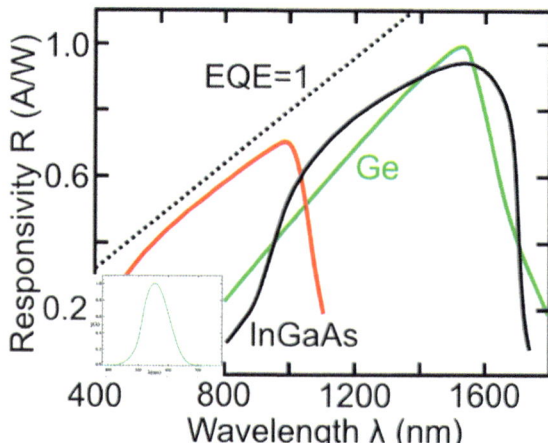

FIGURE 8.30 Responsivity of various types of detectors to wavelength compared to photopic vision. Note that the Si detector, of which the Si solar cell is one, is much more sensitive to longer wavelengths than is the human eye. The photopic curve of visible light is placed correctly with wavelength on this figure. Note also that the bandgaps of Ge and InGaAs are both narrower than that of Si, and can, therefore, detect light of longer wavelengths. Adapted from Kalt H., Klingshirn C. F. (2024). *Light-Absorbing Devices*, Springer Nature.

In Fig. 8.30, **responsivity** is electrical output, such as voltage or current, per total (radiometric) optical power input, typically in V/W or A/W. QE stands for **quantum efficiency**, which is the number of electron–hole pairs generated per photon incident on the detector. "Ideal photodiode QE = 100%"

implies that it is assumed that every photon creates an EHP. The reason that ideal responsivity is a straight line increasing to the right is that longer wavelengths (λ_g is the wavelength that corresponds to the bandgap) have less energy per photon ($E = h\nu = hc/\lambda$). If each photon at lower energy still creates an electron–hole pair (i.e., QE = 100%), then you will get more current per watt, and thus get increasing responsivity with longer wavelengths. However, the actual responsivity of the Si photodiode (red curve) eventually drops off with longer wavelength because the photons no longer have enough energy to create an electron–hole pair, i.e., their energy is less than that of the bandgap. At the longer wavelengths, the light is present but does not contribute to power output, so the responsivity drops abruptly. At short wavelengths, the electron–hole pairs recombine at the surface and therefore do not reach the p-n junction, are not separated, and do not contribute to output current, so again, responsivity goes down.

8.6 Other Alternative Energy Considerations: Wind and the Power Grid

Although we cannot discuss in our limited space every alternative energy source, we would be remiss in not at least mentioning wind power, the use of which has increased tremendously along with solar in the past several years. You read about 3-phase generators in Chapter 3, where it was assumed that the power source was either hydroelectric or nuclear, or provided by burning some kind of fuel, such as coal or natural gas. Wind power uses a similar generator, but no fuel is used – just add wind (which is another form of solar energy, of course). Figure 8.31 shows the various components of a wind turbine.

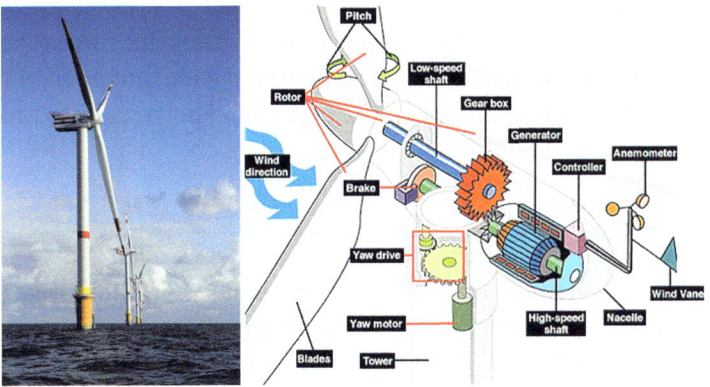

FIGURE 8.31 The various components of a wind turbine. The heart of the turbine is a generator. The resulting AC voltage is converted to DC and then synchronized to the power grid by a power inverter.

Figure 8.32 shows global solar and wind power addition forecasted through 2032. Solar and wind power additions combined far exceed every other type, and wind is being added 25% more quickly than solar power. Figure 8.33 shows the Fowler Ridge wind farm in the Northwestern part of Indiana, about 130 miles Southwest of Notre Dame. Each Fowler turbine produces between 1.5 MW and 2.5 MW, and the 500 or so turbines produce a total of 600 MW and are distributed over an area of roughly 100 sq. miles. In 2021, the Fowler Ridge wind farm was the 13th largest wind

GLOBAL NON-HYDROPOWER RENEWABLE ELECTRICITY CAPACITY (GW)
Sources: Global X ETFs with information derived from sources specified in the footnotes section titled "Forecast Analysis Derived from the Following Sources".

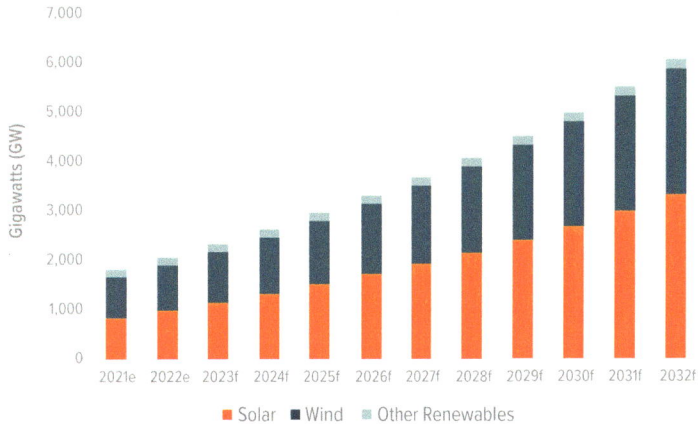

Note: e = estimate, f = forecast.

FIGURE 8.32 Total global renewable energy capacity as of 2022 and forecasts to 2032. Combined, solar and wind account for about 25% of US generating capacity. Both are increasing rapidly with that fraction increasing by about 15 percent per year globally. https://www.globalxetfs.com/renewable-energy-poised-to-drive-growth-in-the-power-sector/.

farm in the United States, but is now 23rd, which is an example of how quickly new wind power is being added.

You might ask, "How is the power from a wind turbine added to the grid?" For that matter, how is DC power from solar panels added to the grid? In order for any power plant to couple electrically to the grid, it must do so at the correct voltage, frequency and phase at which the grid is operating. Wind turbines do not turn at the same amplitude, frequency and phase all the time, and certainly not in synchrony with the grid. Solar panels don't even produce AC power – just DC. In order to condition the power, wind power is converted to DC (which is already the case for solar) and used to power a circuit called a **power inverter**, which electronically converts the DC power to AC power at the correct frequency, phase and amplitude

FIGURE 8.33 The Fowler Ridge wind farm in Benton County, Indiana, produces 600 MW. Its ranking has fallen rapidly, and as of 2023 is the 23rd largest wind farm in the world.

to match the grid. This conversion process wastes about 5% of the input power in the conversion process, which is part of the overall efficiency of the entire power-generation process. Other advanced electronics control the direction into the wind and pitch of the blades, much like a sophisticated aircraft, in order to make maximum use of the wind power, and to protect the blades in the presence of high winds.

8.6.1 Moving Away from Fossil Fuels

The urgency of increasing the availability of alternative energy sources such as solar and wind is undeniable. 2024 was, at the time, the hottest year in recorded history. It is likely that as you read this, even hotter years will have occurred. An important milestone was achieved at the COP28 Climate Summit in Dubai. On December 12, 2023, after much debate, 200 nations, including those that produce most of the world's oil, agreed to shift away from the use of fossil fuels and to significantly reduce methane emissions, which are far worse per molecule than CO_2 for global warming. They also agreed to triple the amount of renewable energy over the next 7 years. Such a change will make the graphs shown in Figs. 8.23 and 8.24 completely irrelevant. We can only hope to see these promises honored over the next decade, and that even these changes will not be too little, too late for massive societal disruptions.

8.7 Lab 8 Activities

This week's lab will have you doing a direct comparison of various types of light bulbs, and will compare their light output and efficiency. After that, you will experiment with solar cells to investigate their electrical characteristics, much as you did the diodes in Lab 7, and some optical properties.

What properties might you want to know about a light bulb? We have discussed their color, amount of light output and power consumption, but you can also measure their power factor and amount of ultraviolet (UV) light in the spectrum. You will measure all of those things, except for the color temperature, which will be given to you. However, you can download an app for your smartphone that uses the color sensor in the camera to provide a reading of color temperature. This may not be so accurate, but standalone color temperature meters are fairly costly, so we do not have them in this lab. You are encouraged to (safely) download an app to compare with the actual ratings of the bulb.

Your first task will be to fill out the following table using the tools provided, as discussed below. Feel free to make your own table in Excel rather than use this cramped one. Be sure that you have all the data before leaving the lab for the day.

Type of Bulb				
Model				
Labeled Lifetime				
Labeled Watts				
Line Voltage				
Current into Bulb				
VA (apparent power)				
Power Factor				
Measured Watts				
Bulb Surface temp. deg. F				
Labeled Lumens				
Measured Lux				
UV Output				
Stated Color Temp. Kelvin				
Measured Color Temp. Kelvin				
Measured Power Efficiency Lux/W				
Measured Effic. Lumens/W				

You will be provided four bulbs: (a) incandescent, (b) halogen (a brighter, hotter incandescent), (c) compact fluorescent lamp (CFL) and (d) light-emitting diode (LED) lamp. The color temperatures for the bulbs will be provided to you based on the commercial labeling, but you may choose to run your phone app to compare.

You will also be provided four reflecting lamp fixtures. These fixtures comprise the outer metal reflector and bulb socket as one unit. ***Set all lamp fixtures down on their sides; do NOT place any fixtures bulb-downwards as the bulbs can break or cause heat damage to the table tops.***

Figure 8.34 shows all of the equipment to be used in this week's lab. You should not have to remove any of the bulbs from the fixtures. You also need to do the experiments without causing light pollution to other stations, and want to avoid erroneous readings on your bench. Therefore, you will be provided a metal shield, or 'can,' that supports the lamp holder and also has a cutout at the bottom from which to take readings. Be careful of any possible sharp edges. The lamp fixture collects and reflects most of the light downwards, and the metal can both shields the light meter from external light and reflects most of the stray light radiating outwards down toward the meter. It also shields your meter from the rest of the room light.

FIGURE 8.34 Equipment used in this week's lab. Shown are the lamp fixtures, shield (can), power meter, UV meter, lux meter, solar cells and small DC motor.

You will be provided with the following pieces of measurement equipment, as shown in Fig. 8.34: 1. Visible-light meter that measures light intensity in lux (red with curly wire); 2. UV light meter (left of can); 3. "Kill-a-Watt" meter that measures power consumption of household appliances (right of can with buttons and plug in it); 4. Small solar cells for the second part of the lab (flat in front of can); 5. Small electric motor to power by the solar cells (yellow to left of solar cells).

Before we get started, it should be pointed out that **the bulbs can get very hot!!** Even the fixture will get too hot to touch after some time. Therefore, do not leave the incandescent or halogen bulbs on any longer than necessary. Do NOT touch any of the bulbs with your fingers while it is on – you don't know if it is hot, and you don't want to get burned by it. You might notice that the actual halogen bulb is small and is well inside the glass outer bulb, but nevertheless, the outer glass bulb gets extremely hot. (The actual temperature of the halogen bulb inside is well in excess of 600°F). Also note that the CFLs change properties as they warm up, so be aware that leaving them on for a few minutes may be necessary to get good data. Now let's get started with today's lab experiments.

1. **Observing color temperature.** Plug the power meter (Kill-a-Watt) into the short extension cord and to the power outlet. Select the incandescent bulb and fixture and plug it into the power meter. (Later you will repeat these steps for different bulbs). Turn on the light. Make a note of the color of the bulb, i.e., 'yellowish,' 'reddish,' 'bluish,' or 'white,' or something else depending on your own color perception. Also, characterize that color somewhere between "warm" and "cold" as the manufacturers do. How does your perception compare to the manufacturer's claim of color temperature and 'warmth' of the bulb? Turn off the light.

FIGURE 8.35 (Step 2) Lux meter head placed inside the light shield.

Optional: Use your smartphone app to read the color temperature. How well does it correspond to the manufacturer's color temperature claim? *Turn off the fixture until you need it again.*

2. **Using kill-a-Watt to investigate incandescent bulb properties.** Place the sensor part of the lux meter on the table and place the light shield over it so that the wire comes out through the cutout at the bottom (Fig. 8.35). Keeping the other fixtures safely at the back of the bench, place the fixture with this bulb into the top opening of the light shield. It should fit snugly and flat on the ends of the screws. Do not stick your hand inside the shield. Be careful of sharp edges.

Turn on the light. Using the buttons on the Kill-a-Watt power meter (Fig. 8.36), you can select the line voltage (make a note of it), line frequency (should be nearly or exactly 60 Hz – make a note), current through the bulb, apparent power in VA, power factor, and real power consumed. Use these settings to fill in the table for the incandescent bulb.

Discuss your power factor reading. What is the best power factor and what is the worst? Can you explain the power factor for the incandescent bulbs?

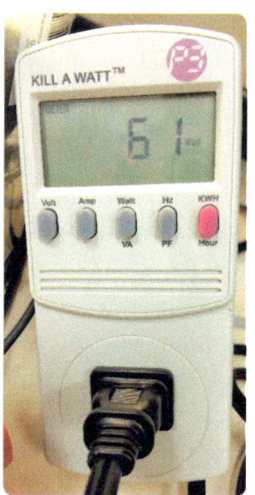

FIGURE 8.36 (Step 2) Kill-a-Watt power meter.

3. **Measuring light intensity using lux meter.** Make sure that the light meter is in the center of the shield. Center the light meter by sliding the shield on the table around the sensor without bumping the cable. Be reasonably careful that the sensor is in the center of the shield and stays there during the measurement. Slide the shield around gently on the bench to maximize the meter reading.

The bulbs all emit light in all directions, so we want to collect as much light as we can on the light sensor. The reflector and shield are an attempt to collect as much light as possible emitted by the bulb by reflecting it down to the sensor. We are making the very crude assumption that our lux reading is proportional to the total visible light emitted, i.e., visible brightness, in lumens, of the bulb. Again, this is a crude assumption, but we are only looking for trends among the bulbs, so it is good enough for our purposes.

Use the light meter to measure the maximum visible light intensity, in lux, that falls on the white disk on the light meter. Turn off the bulb as soon as you have the measurement so that the measurement setup does not get too hot.

4. **Measuring UV.** Use your UV meter to measure the ultraviolet output of the bulb. To use it properly, set the units to microwatts per square centimeter, i.e., $\mu W/cm^2$. Also, you will need to properly zero the meter in order to get accurate measurements. To do this, set the units and then place the cap over the sensor. Press the **Zero** button until it reads "0000." Now remove the cap, carefully slide the meter into the shield, and perform the measurement. You will have to look through the opening in the shield to see the meter reading. Given the optical spectrum of the bulb, do you expect to read much if any UV light from it? Refer to Fig. 8.18 for your answer.

5. **Measuring bulb temperature. Don't forget that the CFL, incandescent and halogen bulbs can get very hot.** In this part you will use the temperature probe to measure the surface temperature of the light bulbs. You will use the Extech MN35 (Fig. 8.37) DMM and its associated thermocouple probe. The tip of the probe contains a thermocouple, which is a welded junction between two different metals. That produces a small voltage due to the **Seebeck effect** when it is heated. The voltage is proportional to temperature and is measured by the meter. Note that the wires are very delicate so be very careful and do not bend them or pull them apart.

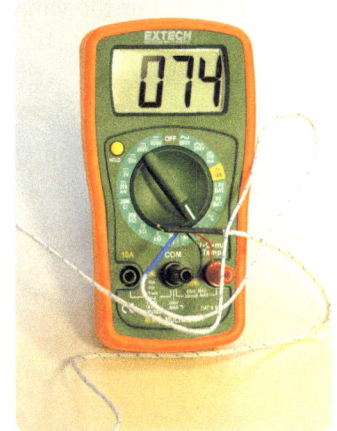

FIGURE 8.37 (Step 5) Extech MN35 DMM with temperature probe. You must be very careful of the delicate thermocouple at the end of the wire. It is easily broken.

Be sure the incandescent bulb is on for at least 2 min before taking your measurements. Holding the fixture from the top plastic part, very carefully, touch the tip of the temperature probe to the surface of the bulb. Probe around for the hottest spot on the bulb – different locations can be at very different temperatures. Make a note of your experimental technique. What maximum temperature do you read? Add that reading to the table. Note that bath water above 120°F can cause scalding. Discuss in your lab report.

6. **Investigate halogen bulb.** Repeat Steps 1 to 5 for the halogen bulb. Halogen bulbs get very, very hot! Again, BE CAREFUL TO NOT GET BURNED! For the halogen bulb, keep it turned on only when you are doing measurements, but at least for 2 min to get a good temperature reading off of it. Turn it off whenever you aren't using it otherwise.

Regarding the UV output, did you measure UV light from the halogen? Is it the same as the somewhat cooler incandescent bulb? If it is different, why do you suppose that is? Use the blackbody spectrum to answer this question.

7. **Investigate CFL bulb.** Now you are ready to measure the CFL bulb, but we can have a little bit more fun with it. Note that CFLs take a minute or so to warm up. Remember, there is a bit of mercury inside the bulb. The mercury gas emits UV light that causes the phosphor to fluoresce at the wavelength of the bulb's visible light output. While the CFL bulb warms up, the small mercury drop inside vaporizes, but that takes a minute or so. Therefore, if you are using your cell phone

app for color temperature, turn it on briefly to get a reading while it is cold. Then, put the fixture on the shield and take intensity readings as a function of time, say every 5 s. After it has warmed up and stabilized, take another color temperature reading. Is it different? How so? How did the brightness change with time?

Repeat all of the measurements in steps 1 to 5 with the CFL bulb warmed up. Discuss your UV light reading with respect to the discussion in the write-up about how light is produced from the plasma in the bulb, and the role of UV light inside as well as the quality of the phosphor coating inside the bulb.

8. **Investigate LED bulb.** Now we are left with the LED bulb. Repeat Steps 1–5. Note how little power it uses and how cool it remains.

9. **Analyzing your data.** Make sure that you haven't missed anything to be filled out on the table. Make a note of the lux value you obtained for each bulb. That reading is assumed to be proportional to the total lumens output of the bulb. Calculate both lux/watt from your experiment and compare to lumens/watt from the manufacturer's numbers. The manufacturer's ratings tell you how much light you get for a given amount of power usage (lumens/watt), and your measurements should give approximately the same trends with errors due to the poor light collection. Put the values of the four bulbs in order and discuss what you find in the lab report. You might notice that your lux readings for the CFL and LED bulbs do not correspond to the relative brightness (in lumens) of those bulbs. Upon inspection of the construction of those bulbs, comment on why this may have happened.

Say something about how your results pertain to power usage in the world for the widespread use of various light bulbs. If you were responsible for a large corporation with dozens or hundreds of large buildings, would you choose one particular type of bulb over another? Assume that the costs of the bulbs are approximately the same, which has become the case in the marketplace. Discuss what choices you would make.

You are now finished with the 'Consumer Reports' type investigation of the light bulbs. Hopefully your results mostly agreed with the manufacturer's claims. Also, perhaps you now have a deeper appreciation for the physics of light emission from various sources, and the importance of lighting in general.

The next section is about the properties of solar cells (SC) and how they produce electrical power from light. Since we don't have a 6000 K source available on your bench that can produce light with exactly the solar blackbody spectrum, we will rely on the bulbs and their optical properties to power your SC.

10. **Inspecting your solar cell (SC).** The ones provided to you are Si solar cells with two smaller ones in series inside the package, each having an output of about 0.5 V, resulting in a maximum output of about 1 V. The output voltage can be a little higher or a lot lower depending on the intensity of the light on it. There is a red and a black wire indicating positive and negative voltage output, much like a battery.

11. **Obtaining an I–V curve for your solar cell in the dark.** Turn the SC upside down so that the active surface is downwards toward the lab bench, i.e., it is in the dark. Perform the following steps to obtain the *I–V* curve of the SC in the dark. Figure 8.38 shows how your screen should look at the end of this step. Set the waveform generator to 20 Hz with an output voltage of 20 V_{pp}. Connect the output directly to the SC putting the positive to the red wire.

Use a voltage probe on the scope to measure the voltage across the SC to CH1 of the scope. Be sure to connect the positive probe lead to the red lead of the SC. Set CH1 coupling to **DC 1 MΩ**. *Note that the output of the waveform generator will be loaded down by the SC and will not reach 10 Vp, but that will not prevent you from obtaining a good SC I–V curve.*

Attach the current probe to CH4 and turn the arrow on the probe head to face toward the SC on the red wire. Use your X-Y skills from Lab 7 to plot the *I–V* curve of the SC. In case you have forgotten: choose X-Y mode in the **Display** menu, and set input X as CH1 and input Y as CH4. Set the CH1 and CH4 scales until you see a diode-like *I–V* curve (See Fig. 8.38).

In the dark, the SC should be producing zero current. You will notice that the curve is offset from zero current. It will help to choose CH4, select the **CP030** probe, **Degauss Probe** (get rid of residual magnetism) and follow instructions to set the **Auto Zero**. If you still need to make

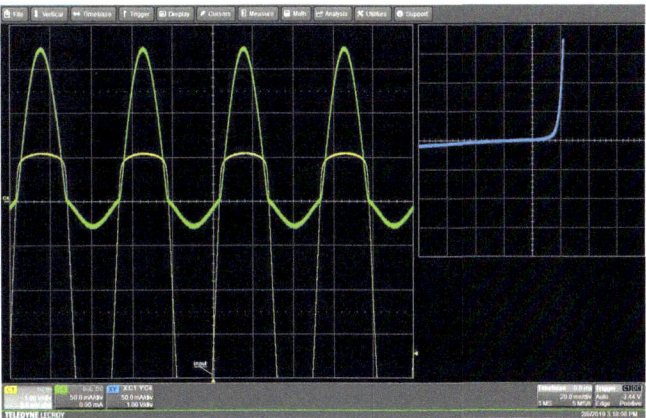

FIGURE 8.38 (Step 11) (Left) Current through the SC and voltage across the SC (Right) SC *I–V* curve in the dark. The scale at the bottom shows that CH1, *X*, is 1 V/div and CH2, *Y*, is 50 mA/div.

adjustments, use the CH4 offset to move the curve to pass through the origin. Capture the plot for your lab report. (If you do not have access to a current probe, you should use a small shunt resistor and the diode-measurement techniques given in the Chapter 7 lab activities.)

12. **Creating an I–V curve for your solar cell with incandescent bulb.** Turn the SC over to let light land on it. Put a single plain white piece of paper over it to act as a simple light attenuator. (The bulb is too bright to do this experiment). Put the shield can over the paper and the SC. Place the incandescent bulb on the shield and turn on the bulb. Keep the bulb on only as long as necessary, since the properties of solar cells change with temperature. (They become less efficient as they get hotter). It should look now like Fig. 8.39. Notice that the

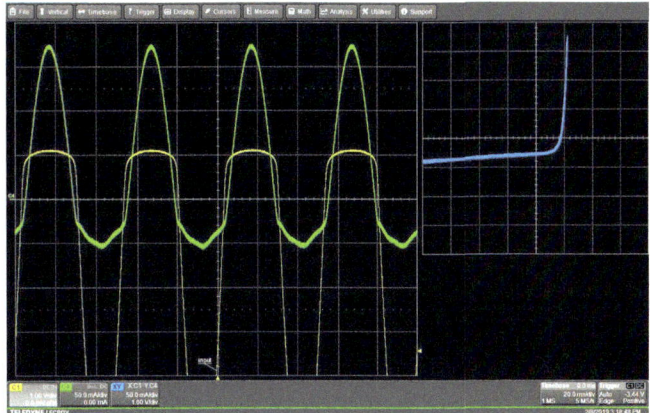

FIGURE 8.39 (Step 12) Solar cell under illumination. Use this screen to estimate fill factor and associated values.

I–V curve is identical to that found in Step 11, but is shifted down, indicating that it is producing power.

13. **Estimating fill factor.** Capture this plot for the lab report. For the lab report, read off the short-circuit current and open-circuit voltage. Also, estimate the fill factor of the SC under these lighting conditions. What is the maximum power that the SC can produce under these conditions? It is the area of the largest rectangle inscribed inside the *I–V* curve.

14. *Power the SC with the rest of the bulbs. In the following steps, turn off the halogen bulb as quickly possible after getting the measurement to avoid heating or damaging the SC.* Without changing any settings, replace the incandescent bulb with each of the other three bulbs. Capture each curve. Extract the same data as you did in Steps 12 and 13 for each bulb. Explain your results in terms of the Si responsivity curve in Fig. 8.30, and the color spectrum of the bulbs based on their true spectra as shown in Fig. 8.18. Explain the relative values that you obtain. Hint: The amount of power from the SC can be explained by a statement

made previously, that what looks bright to us does not necessarily look bright to the SC. More explicitly, the answer lies in the amount of invisible IR that each bulb puts out.

15. **Using SCs to do work.** Remove the piece of paper. Put two SCs in series with the DC motor, as shown in Fig. 8.40, but using clip leads as needed. Use the series solar cells to drive the motor with the two cells in series and with the incandescent bulb held in your hand without the shield. Move the bulb closer and further to get the most power out. What do you observe? Try this with the other bulbs for comparison. Turn out the light as soon as you are done to let the SCs cool.

FIGURE 8.40 (Step 15) DC motor with two solar cells in series. Follow the colors of the lead wires as shown.

16. **Measuring the effect of heat on the SC output voltage.** Now you will use the thermocouple again. Let the solar cells cool. For this step, you will use only one SC. Use a small piece of tape (available on the front desk) to press the temperature probe tip to the surface of one SC near the center. Try to give it good contact to the surface.

    Here you will measure the temperature of the SC while you simultaneously use the Fluke DMM to read the open-circuit SC voltage. Using the shield and the halogen bulb to light up and heat up the SC, you and your lab partner should each take readings – one of open-circuit voltage and one of surface temperature at the same times, and record them in your notebook. Here, make sure that the voltage and temperature correspond well, so read them off and write them down at the same instants in time. This can take several minutes while the SC heats up and the voltage drops to a low value. For your lab report, plot how V_{OC} of a SC degrades as its temperature increases.

    What do you conclude about the efficiency of a SC as its temperature increases? This illustrates one challenge of using SCs in the bright sunshine and obtaining the highest possible power output. Many strategies are used in practice to keep solar cells from excessive heating.

Congratulations, you have completed this lab and have now been exposed to a range of properties of light bulbs and solar cells. Now you know how solar cells work and why they are so important. I hope you can now appreciate many of the practical issues of choosing something as commonplace as a light bulb, but having extremely important implications to the environment and the economy. As the years go by, you will undoubtedly see ever more solar cell installations and witness the dominance of alternative energy sources.

Problems

1. X-rays have wavelengths ranging from approximately 0.01 nm to 10 nm. What frequencies does this correspond to? What is the range of photon energies?

2. Typically, some physical quantity like sound, gravity or light decreases as the inverse square of the distance from a small, or "point" source. This is called the "inverse square law." Assume that a light source is sufficiently small enough so that from far enough away we can approximate it as a point source that is shining light equally in all directions. Suppose that the light source

is providing only half the illuminance that you need on a surface. If the light source is currently 1 m away, how far away from the surface does it need to be to provide the correct illuminance?

3. (a) What does a lux meter measure?

   (b) In terms of the use of the lux meter, explain the purposes of the "can" and reflector in measuring the properties of the light bulbs.

4. A typical IMAX screen is 22 m wide and 16.1 m tall. An IMAX projector bulb produces about 600,000 lumens. Suppose we use the projector (without any film) to illuminate the screen. What is the illuminance (in lux) on the screen?

5. It is common for LED flashlights to incorporate a lens that focuses the light to tighter or wider areas in a cone of illumination. Suppose a flashlight is rated at 10,000 lumens and at a distance of D = 10 meters the beam is concentrated in a circle of uniform illumination with diameters between d = (i) 3 meters and (ii) 10 meters. 10,000 lumen flashlights are readily available on the internet, which can damage your retina if you put your eye too close to it, so such a flashlight should be handled with care.

   (a) What range of illuminance (denoted as E_V) can be achieved by the beam distances, (i) and (ii)? Your answers should be in units of lux.

   (b) The light in part (a) forms the shape of a cone with the apex of the cone at the flashlight. A half-angle is the angle between the line that is the axis of the cone and any edge of the cone. The circle of light is the base of the cone. What are the half-angles, in radians, of the cones of illumination for part (a)?

   (c) A luminous intensity of greater than 100,000 lux can cause permanent damage to the retina. For the half-angle of (i) in parts (a) and (b), what is the distance below which you risk permanent damage if you stare into the flashlight?

   Note: Such safety guidelines depend on many factors, including how long one is exposed to the light, so this example is meant only to illustrate the point that some commonly available flashlights are actually dangerous and should be handled with care. (Flashlights with as low as 1000 lumens can also be dangerous under similar conditions.)

6. Suppose we measure the power being emitted by some light source and find that the energy per second (power) at an X-Ray wavelength of 15 nm and an infrared wavelength of 5 μm are equal.

   (a) In words, describe qualitatively how the number of photons per second at each of the two wavelengths compare.

   (b) Quantitatively, what is the ratio of the number of photons at the two wavelengths?

   (c) For a radiometric power of 1 W, how many photons at each wavelength are incident on the detector per second?

7. Wien's displacement law gives the peak wavelength of blackbody radiation as a function of the object's temperature:

$$\lambda_{peak} = \frac{b}{T},$$

where b = 2.898 nm-K

(a) Use Desmos to plot the peak wavelength as a function of temperature from 2000 K to 5600 K. Use temperature in Kelvin and wavelength in nm, as already assumed in the equation.

(b) Approximately what are the temperatures for the peaks to be at (i) red, (ii) orange, (iii) yellow and (iv) green? Indicate these values on your plot. Note that answers may vary depending on which wavelengths you choose for the colors, but trends will be apparent.

8. Explain why the color temperature of an incandescent bulb is the same as the actual temperature of the filament?

9. The rover Sojourner, which operated on Mars in 1997, had a solar cell array made up of 2 cm × 4 cm cells arranged in 13 strings with 18 cells per string. The solar cells had 18% efficiency and produced a maximum total power of 16.5 W on Mars. What was the maximum power per square meter of sunlight reaching the rover?

10. Refer to Fig. 8.27, which is the *I–V* characteristic of a solar cell in terms of current density, *J*. Since total current is simply current density x area, the plot would show total current versus voltage if we are given the area of a particular solar cell. For some total area, a solar cell has the following parameters: V_{oc} = 0.5 V, $|I_{SC}|$ = 150 mA, V_m = 0.4 V, $|I_m|$ = 140 mA. Sketch the *I–V* curve of this cell for total current rather than current density. Label the significant points and their given values. What is the cell's fill factor?

11. You are the lead engineer on a new solar farm installation. Assume you have chosen the following (high quality) Vertex S solar panels (from https://static.trinasolar.com/us) for your installation. For this question, base your calculations on data from the spec sheet: https://static.trinasolar.com/sites/default/files/EN_Datasheet_Vertex_DE09.pdf.

(a) How many cells make up one of these panels?

(b) The data provided in the spec sheet are for the full panels and not individual cells. Read the spec sheet carefully including fine print. Under what conditions were the "Electrical Data (STC)" for the full panels obtained? STC stands for "standard test conditions" and is guidance for comparing the tested performance of solar cells and panels across the industry.

(c) What are the NOCT test conditions? NOCT stands for "nominal operating cell temperature." NOCT test conditions better reflect the real-world use of solar cells.

(d) For the solar panel producing 306 W under NOCT conditions, what are the open-circuit voltage and short-circuit current?

(e) Assume that the 120 cells are connected as follows: two strings of 60 cells each, connected in parallel; the 60 cells in each string are connected in series. Using the NOCT data at 306 W, convert the voltage values to single-cell values. The short-circuit current of one cell will take some careful thought.

(f) The Maximum Power Voltage, V_{MPP}, and Maximum Power Current, I_{MPP}, are the voltage and current, respectively, that provide the maximum fill factor. What are those values for the panels under NOCT conditions at 306 W?

(g) What are the per/cell values for the maximum-power condition?

(h) Figure 8.27 includes current densities, J, rather than total current, I. Using the values for voltage and current from parts (e) and (g), what is the fill factor for these cells?

(i) Draw an I–V curve of this solar cell like that of Fig. 8.27 with currents, rather than current densities. Include the data for each cell as calculated above and the fill factor. The drawing may be rough, but the fill factor and other data should be indicated on your drawing.

(k) What is the total area of the panel?

(l) What is the efficiency of this solar cell under real-world conditions? Remember, this is in actual use, which is not optimized for the best solar-cell performance. Also, you must use the NOCT test conditions to calculate the efficiency.

12. Notice that the I–V curve of a solar cell with light is the same as the curve without light but shifted down. Explain using the passive sign convention that when a solar cell is operating at the lower right quadrant that it is supplying power, but not when it is in the upper right or the lower left. Also, using the same reasoning, explain why a solar cell acting as a diode without light cannot supply power.

13. Review Fig. 8.39.

(a) On the solar cell I–V curve on the right side of the screen, what do the axes correspond to?

(b) Is the SC illuminated or not? How do you know?

(c) On the left of the screen, what is the green curve that is the highest in the positive direction and is triangular and short and triangular in the negative direction? How is it related to the I–V curve on the right?

(d) What is the yellow curve that is short and rounded at the top, positive values and triangular at the bottom? How is it related to the I–V curve on the right?

14. Consider the cost of electricity for outdoor lighting. Let's estimate how much money would be saved using an LED bulb over a CFL bulb if the bulb was left on for 12 hours every day for 11 years. Since the cost of bulbs is relatively low, we will ignore it. Let's assume that both bulbs are equivalent to a 60 W incandescent bulb. However, the CFL bulb consumes 13 W and the LED bulb consumes only 8 W. Finally, suppose you used incandescent bulbs for the entire 11 years. How much will the electricity cost over the 11 years for each bulb? Remember that we can ballpark the cost of 1 W for 1 year as $1 if that one watt were running 24 hours every day.

15. Choose the correct statement(s) or choice(s):

(a) A blackbody radiator at the temperature of the sun emits electromagnetic radiation at:
  (i) only certain discrete, or separate, wavelengths
  (ii) only visible wavelengths
  (iii) only infrared wavelengths
  (iv) only ultraviolet wavelengths
  (v) a broad range of wavelengths from ultraviolet to infrared

(b) The standard incandescent bulb temperature is 2700 K.
  (i) An incandescent bulb operating at 2700 K emits strongly in the infrared.

(ii)   An incandescent bulb operating at 2700 K emits strongly in the UV.

(iii)   A halogen bulb is more efficient because it operates at a lower temperature than 2700 K.

(iv)   An incandescent bulb operating at 2700 K makes an efficient light source because it mostly overlaps with the human visible spectrum.

(c)   A CFL light bulb produces light by which of the following processes:

(i)   blackbody radiation

(ii)   radiative electron/hole recombination

(iii)   ultraviolet light created by electron energy transitions in a plasma

(iv)   visible light emitted by a phosphor

(d)   The fill factor of a solar cell:

(i)   can only be based on rectangles and no other shape.

(ii)   is always less than 100%

(iii)   is sometimes greater than 100% and sometimes less

(iv)   is always greater than 100%.

16.   Indicate True or False for each statement:

(a)   A halogen bulb emits light by blackbody radiation.

(b)   A light-emitting diode emits light by radiative electron/hole recombination.

(c)   A fluorescent bulb internally produces ultraviolet light created by electron energy transitions in a plasma.

(d)   In a CFL, phosphors emit visible light when heated by the plasma.

(e)   The current–voltage characteristic of a silicon solar cell in the dark is the same as that of a silicon diode.

(f)   A solar cell generates power through the creation of electron–hole pairs that move together through the device to create current in an external circuit.

(g)   Solar cells can be combined in series and parallel to increase the overall solar-to-energy efficiency.

(h)   A fill factor that is greater than 1 is typical of a very efficient solar cell.

Chapter 9

Digital Signal Processing and MP3 Music Compression

9.1 Introduction and Overview

This chapter deals with how computers interact with our physical world. The example we will explore is the process of digitizing, or **encoding**, audio, specifically music, and playing it back on our digital music players. This technology is quite sophisticated, and affects many of us every day. By the end of this chapter you will have an appreciation for the technology that makes a plethora of technologies possible, including digital music, voice recognition, portable health monitoring, digital cameras, and on and on.

In this chapter, you will learn about:

1. Continuous versus discrete signals
2. Binary numbers
3. How music was recorded before and after digital technology
4. How signals are converted from analog to digital and back
5. Theoretical limits to how music and other signals can be digitized
6. Various audio file compression algorithms

Also, in this lab activity you will experiment with analog and digital representations of music, going back and forth between the two. You will use a custom system with which you can set the number of bits of resolution and sampling rate in order to visualize the resulting waveforms and frequency spectra.

9.2 Continuous and Discrete Signals

We live in an **analog** world, meaning one that changes continuously – quiet to loud, dark to light, slow to fast, soft to hard, here to there, etc. In this chapter, we mostly discuss analog *voltage* waveforms, like that shown in Fig. 9.1.

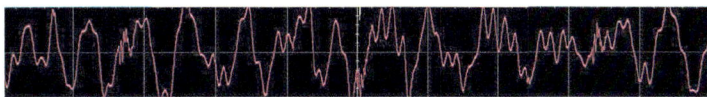

FIGURE 9.1 Continuously varying, i.e., analog, voltage waveform (this one is of music).

An Introduction to Electrical Engineering with Lab Activities
Gary H. Bernstein
Copyright © 2026 Jenny Stanford Publishing Pte. Ltd.
ISBN 978-981-5129-30-4 (Hardcover), 978-1-003-71345-6 (eBook)
www.jennystanford.com
DOI: 10.1201/9781003713456-9

The opposite of analog is **discrete** (not to be confused with the word "discreet"), in which only certain values of a variable, such as voltage and current, are allowed, with measurable gaps between them. A good analogy is a staircase compared to a ramp. Walking up a ramp can be done with the tiniest of steps (continuous or analog), whereas you are constrained to take each step of a fixed height when you go up a flight of stairs (discrete or digital).

> *Of interest:* Time is continuous; there is no 'smallest increment of time' of which all durations of time are composed. In fact, as technology increases, time intervals can be measured in smaller and smaller increments from picoseconds (10^{-12} s) to femtoseconds (10^{-15} s) to even attoseconds (10^{-18} s). The changes in these parameters are continuous, existing over a full range of values.

Digital computers cannot recognize the analog world that we live in. They are built of transistors that are either on or off, and nothing in between. Computers operate inter-nally using binary digits and binary arithmetic and there-fore can recognize only discrete numbers. This chapter is about how analog signals, such as music or photographs, are converted to digital values, so that computers can process them, and then back to our analog world.

In an analog world, we might think that we can add as many digits as we want in order to represent the **resolution** (see sidebar below) of any given measurement, but there are limits due to the presence of **noise**. As you turn up the gain of, say, an oscilloscope trace (i.e., magnify the vertical trace) with no input voltage, eventually you just see a random, scraggly looking waveform that is just noise, as suggested by Fig. 9.2(a). That same noise

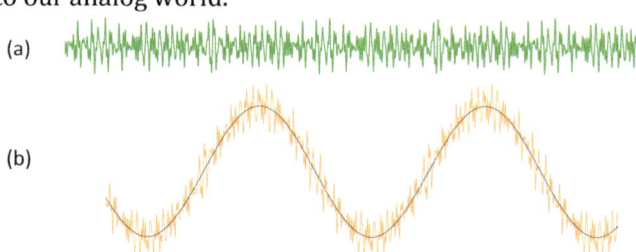

FIGURE 9.2 Noise. (a) Noise alone, as you would see on an oscilloscope with gain turned all the way up and no signal. (b) "Clean" sine wave superimposed with a "noisy" sine wave.

would be evident on a signal that has an amplitude not much larger than the noise itself, as shown in Fig. 9.2(b). The noise might be due to the scope amplifier itself. There are many sources of noise in the world, but one main source is the random fluctuations of electrons due to heat in the devices.

> *Required:* The following terms are often confused and used incorrectly. Pay attention to these definitions:
>
> **Resolution**: The smallest part that can be separated in a measurement. More digits implies higher resolution.
>
> **Precision**: How repeatable the measurement is. If you get nearly the same value each time you measure, it has high precision.
>
> **Sensitivity**: The smallest change that can be detected. If an instrument has poor sensitivity, it doesn't matter how many digits of resolution it has.
>
> **Accuracy**: How 'correct' the measurement is relative to an agreed-upon standard. A measurement can have high resolution, high precision and high sensitivity, but could be mis-calibrated and give 'wrong' readings, in which case its accuracy would be low.

This is called **Johnson–Nyquist noise**. There is also the phenomenon of **interference** in which an unwanted signal is mixed with the signal you are trying to measure. That signal might be caused by magnetic fields from a nearby elevator, or perhaps someone running a drill press in another part of the building, where electrical fluctuations or interference from the motor is carried on the electrical wires.

Noise and interference are not the same things, but often they seem the same, and the terms are sometimes used interchangeably. In this entire discussion, you must keep in mind that noise plus interference is always present, but we will mostly be discussing the desired signal characteristics. Also, one can spend an entire career studying noise and interference, and ways to improve circuitry to be less susceptible. So, when we discuss the minimum voltage that can be measured by an electronic system, keep in mind that it is always a competition with the noise that is inevitably present.

Figure 9.3 shows a meter having "4½ digits of resolution," where the first digit is only a 0 or 1, and is the "half digit" (this has nothing to do with binary arithmetic). If, for example, the first digit shown represents 1 V, then the fourth full digit to the right tells us that the smallest voltage that can be measured at that setting is 0.1 mV (100 µV). Figure 9.4 shows an 8½ digit DMM in which the 8th and last digit can resolve no lower than 0.01 µV on the 1 V setting.

The resolution of computers is limited by the number of transistors that form the digital bits used in its internal architecture. You know that computers can do a lot of things that seem to be analog – play music, display beautiful images, measure acceleration, temperature and light intensity, and much more. However, the discreteness of computer-generated signals is very real, and often easily measured using analog methods. For example, when you turn up the volume on a digital radio or MP3 player control, you can often discern two different adjacent volumes. You are offered either the quieter one or the louder one, but if you want one in between, then you are out of luck. The manufacturer has decided for you how finely (what resolution) you are able to control your volume. Dealing with the discreteness of digital systems is a major theme in electrical and computer engineering, and is the subject of this chapter.

FIGURE 9.3 4½-digit digital multimeter. *Source*: https://en.wikipedia.org/wiki/Multimeter#/media/File:Fluke87-V_Multimeter.jpg.

FIGURE 9.4 High-resolution DMM.

9.3　The Binary Number System

Although many of you are probably already familiar with this, let's do a short review of binary numbers and get it out of the way since much of this discussion will be about the digital representation of signals. A number such as decimal '134' is really a shorthand notation for adding powers of the **radix**, or **base**, of the numbering system. In decimal, the radix or base is 10, using the digits 0 through 9. The number 134_{10} is actually shorthand for $1 \times 10^2 + 3 \times 10^1 + 4 \times 10^0 = 100 + 30 + 4 = 134$.

Binary numbering uses a radix of 2, consisting of the **binary digits**, or **bits**, 0 and 1. For example, the decimal number 45_{10} is represented in binary as 100010_2. The binary number 101101_2 is really shorthand notation for $1 \times 2^5 + 0 \times 2^4 + 1 \times 2^3 + 1 \times 2^2 + 0 \times 2^1 + 1 \times 2^0 = 32_{10} + 8_{10} + 4_{10} + 1_{10} = 45_{10}$.

Put another way, if we have n binary digits to represent a group of things, then we can represent 2^n unique items with those n digits. Let's say $n = 8$ bits, where a group of 8 bits is called a **byte**. Those 2^8 things would be represented by 00000000, 00000001, 00000010, 00000011, 00000100 ... to 11111111. The left-most bit is the **most-significant bit, MSB**, having the largest value of all the bits, and the right-most bit is the **least-significant bit, LSB**, have the smallest value. For $n = 8$ this would be 256 unique combinations. (In this case, we get $45_{10} = 00101101_2$ by adding two leading zeros). We could say those represent the numbers 0 to 255, or we could say they are just 256 different things.

In this course we are most often discussing voltages and currents. Suppose you have a continuously varying voltage signal that looks like the continuous line in Fig. 9.1. This waveform might have been created by the analog output of a microphone. That analog signal could be approximated by

256 voltage levels, each one represented by one of the binary numbers that is 8 bits long, as suggested in Fig. 9.5. The number of bits used is called the **bit depth**. Telephone-quality speech, in fact, uses $n = 8$ bits, or 256 levels, to represent speech. However, in music reproduction, more bits result in finer spacing between voltage levels, and higher audio fidelity, but the idea is the same – a binary number represents a voltage, *as if a voltmeter provided a number for the instantaneous voltage, and that number were converted to one of the many possible binary numbers*. For example, digital compact disks, or CDs, have excellent sound reproduction and break the possible voltages into 65,536, or 2^{16}, levels, i.e., 16 bits. Other audio formats use even more bits.

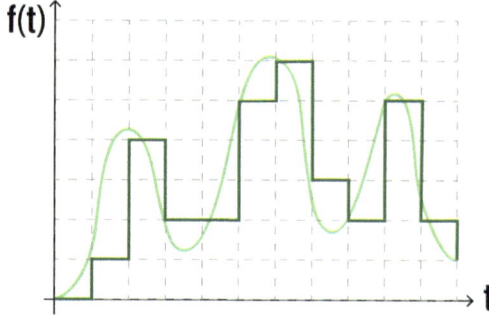

FIGURE 9.5 Discretized analog waveform. The discrete voltage at each measurement time step is that which is closest to the original analog signal.

9.4   Analog Sound Recording

In order to better understand digital signal processing through digital recording and modern music reproduction, it is instructive to start with a simpler explanation of older analog recording and then later discuss the differences. Figure 9.6 shows a turntable with its tonearm, at the end of which is the cartridge that holds a needle, or stylus. The platter spins the record, and the stylus rides in microscopic grooves on the surface of the record. Figure 9.7 is a high-magnification image of an analog phonograph record on which the needle, or stylus, rides in the grooves. Most commonly, the stylus holds a magnet embedded in a tiny coil. As the magnet wiggles back and forth, and up and down, the magnetic field changes, and through Faraday's Law (EMF = $-d\Phi/dt$) it generates a time-varying voltage signal that reproduces the music. You can see that the grooves look much like the trace on an oscilloscope screen. In spite of all of the steps necessary to create and then play back an LP, as discussed next, it is remarkably effective at reproducing extremely accurate musical performances.

Let's now trace the process by which an analog phonograph record is created, and sound is reproduced in a stereo record player. (There are no digital phonograph records, although there can be individual recording steps that are digital.) Note that even though this is 'old' technology, there is a ton of electrical engineering to be discussed, nearly all of which is still highly relevant.

1. A singer sings, causing sound to travel in the air. Sound is carried by changes in air pressure caused by the singer's vocal cords. (Figure 9.8). The changes in air pressure propagate in longitudinal

FIGURE 9.6 Old-school vinyl "long-play" (LP) record on a Phase Linear 8000 linear-tracking turntable, the best in my collection. (If you are curious, the record is the amazing Beethoven 7th Symphony on Deutsche Grammophon.)

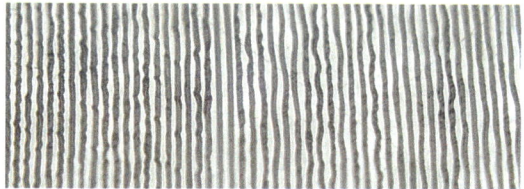

FIGURE 9.7 Photomicrograph of grooves (dark lines) on a phonograph record. The shape of the grooves carries the analog music signal, much like what would be seen on an oscilloscope. The frequency and amplitude can be easily discerned at various moments of the music. Looking closely near the center of the image, you can see that there is silence recorded as smooth lines. Also looking closely, you can see that the grooves have a triangular depth to their cross-sectional shapes. Stereo (left and right channels) is encoded by the motion of the stylus in both the horizontal and vertical directions of the V-shaped groove, exciting voltage on separate coils. These record grooves presumably gave us the terms "groovy" and "in the groove."

(i.e., forward and backward) pressure waves to a microphone. The air pressure at the microphone as a function of time is the analog waveform to be eventually reproduced by the speakers after the complete sound reproduction process.

2. The air pressure waves carry the sound of the singer's voice to a microphone; a thin diaphragm moves in response to the pressure waves with nearly the exact same time dependence (Fig. 9.9).

3. The movement of the diaphragm pushes on a small coil surrounding a magnet, just like a speaker, but operating in reverse. Again, due to Faraday's Law, the changing magnetic field on the moving coil creates a changing voltage at the output of the microphone.

4. That microphone voltage is amplified and recorded either digitally (the subject of this week's lab) or, as in the old days, as an analog signal onto magnetic tape. (Fig. 9.10). For an analog tape recording, the pattern of magnetic density on the tape as a function of position is now a duplicate of the analog voltage from the microphone, and hence the air pressure, and hence the original sound.

5. The tape is played back on a tape player by sliding it past a play-head coil (Faraday's Law yet again), creating a voltage that is again a reproduction of the original voice.

6. That voltage is amplified and filtered (see sidebar below on RIAA equalization) and used to drive a motor that is connected to a machine (Fig. 9.11) that cuts grooves in a soft plastic "master" phonograph record. (Hence the term to "cut a record." Really.)

7. Those grooves have a pattern that represents what was first the voltage variations from the microphone, and then the magnetic density on the magnetic tape that represents the music. It is almost exactly an oscilloscope trace cut into the surface of the record! The original master is plated up with metal, the metal is removed, and that master is used to make submasters, or "stampers."

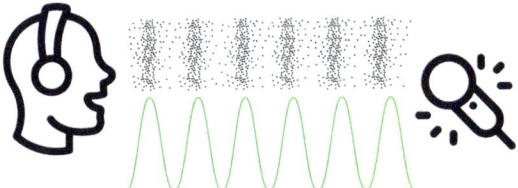

FIGURE 9.8 Artist in a recording studio. Sound is vibrations of air that travel to the microphone, causing small voltage changes.

Cross-Section of Dynamic Microphone

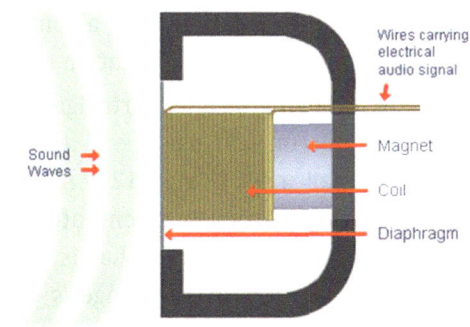

FIGURE 9.9 Basic internal structure of a common microphone. It is very much the same structure as a speaker (Chapter 1), except the "mic" (pronounced "mike") converts sound to a voltage rather than vice versa.

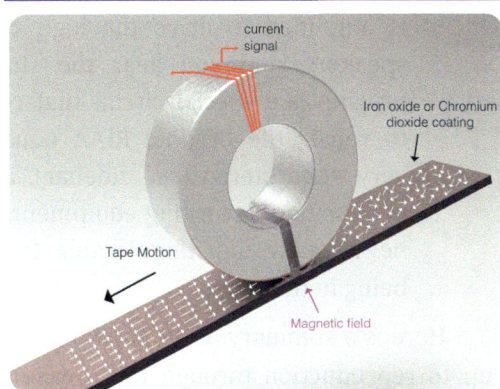

FIGURE 9.10 Rendering of a magnetic recording head whose magnetic field varies with the analog music. During recording, the tape head imprints an analog pattern onto the magnetic particles on the moving tape, as shown. During playback, the particles induce a voltage in the windings that correspond to the original music.

Of interest: Because the amplitude of low frequencies must be higher than that of the high frequencies to be heard by us humans, the unattenuated low frequency component of the music would cause very large groove amplitudes that would take up more space on the surface, and less music would fit on a recording. To address this, since about 1950, phonograph recordings are filtered according to the RIAA (Recording Industry Association of America) equalization standard. RIAA **equalization** attenuates the low frequencies (as in Chapter 4) on the physical vinyl disk and then selectively amplifies them in the phono pre-amplifier to achieve the original larger amplitude upon playback. The RIAA standard is not ideal, but is a good compromise for the recording industry.

8. The submaster stamper records are used on a record press to stamp out millions of copies of the record (that is, if a record goes "platinum") in vinyl material. (Sometimes the entire field of phonograph records is referred to simply as "vinyl." See, for example, www.vinylengine.com)

9. A record player uses a "tonearm" that holds a phono cartridge at the far end. Protruding from the cartridge is a fine stylus or "needle." The record is played on a home turntable by placing the stylus in the grooves, and the record is spun at a constant speed. (See the historical note about rotation speeds.)

FIGURE 9.11 Master record cutting machine. *Source*: https://en.m.wikipedia.org/wiki/File:Neumann_VMS-70_Cutting_Lathe.png.

Moving Magnet Principle

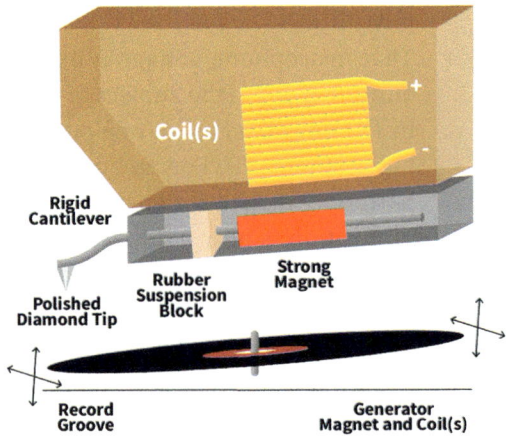

FIGURE 9.12 Components of a stereo phonograph cartridge.

10. In a moving magnet cartridge, the stylus has a tiny magnet on its inside end that is located within a small coil (Fig. 9.12). As the stylus wiggles inside the grooves, it generates (again by Faraday's Law) a small voltage that is still a nearly perfect reproduction of the singer's voice. It works by the same basic physics as the microphone, except instead of the diaphragm oscillating, now the needle and magnet oscillate.

11. The small voltage is first amplified by a phono pre-amp since it is such a weak signal, and then is amplified to drive the high currents used by speakers, and you hear the singer. The phono preamp is a special circuit that contains the filter to repair the original RIAA equalization. With a very good stereo (see sidebar), and presuming professional recording equipment, the result can be virtually indistinguishable from the person being in the room.

Of interest: You might ask how stereo, *i.e.*, two different channels of music, is encoded on a vinyl record. The arrows in Fig. 9.12 offer a clue. The grooves have two different sides, as seen in Fig. 9.7, that are at 45 degrees to each other. One surface is the left channel and the other surface is the right. As the needle rides in the groove it moves in a complex manner, both up and down and left and right, due to both surfaces, and two separate coils in the cartridge create two different vector components of motion. The two channels are sent out of two pairs of wires from the cartridge to the phono preamp where they are interpreted for the left and right channels.

Here is a summary of the total path of the waveform up to reproduction through the speakers. Just for fun, let's throw in one more step for someone "taping" their record for playback on a car cassette tape player of the 1970s to 1990s, after which the sound was still quite good: Air pressure (sound); microphone; electronics; magnetic density versus distance on a tape (recorded on tape); electronics (play back); record cutter; pressing; phono stylus motion; electronics; magnetic tape density (record to cassette tape); electronics (play back); speakers. Whew.

It is impressive that at each stage of the process the quality was so good (meaning very little noise or distortion was introduced into the analog signal) that the music could still be an excellent reproduction of the original. At each analog reproduction step, there was the opportunity to introduce noise or distortion into the next copy, and only high-end audio gear could be counted on to do the job well enough. It would be rare if a regular consumer could copy music without introducing some noise, even if it is small. This fact virtually eliminated the possibility of music piracy on a large scale. However, today, now that the music is stored, or encoded, as digital bits, random noise in reproduction is completely eliminated – every copy is as perfect as the original. This has been the cause of a major shift in copyright laws in the recording industry (https://psmag.com/news/music-copyright-law-is-finally-out-of-the-analog-age).

Historical: The use of phonograph records had gone way down due to the move to digital music, but some audio purists are now helping that medium make a small comeback. It is still just a tiny fraction of all music sold, but it is hanging in there. More and more music shops are adding record sections.

Why are records sometimes referred to as 'LPs'? The rotational speed of records today is either 45 revolutions per minute, rpm, or 33.3 ('thirty-three and a third') rpm. The '45s,' or 'singles,' are the small records with one song per side typically found in old-style juke boxes. Older records, prior to about the mid-1950s, spun at 78 rpm. Those also held one song per side. In fact, since about 3 minutes of music could be held on one side of a '78,' that pretty much defined the length of today's standard song. A group of songs on 78s came in a book like a photo album with each disk in a sleeve. The new 33s spun slower and had much narrower grooves, making it possible to put about six songs on each side. These new records were called 'long play,' 'LP,' or 'albums.'

9.4.1 Modern Sound Recording

Well, that was a fun trip down memory lane, but more importantly, it is instructive to know the long list of steps in the process from recording to playback, and all of the opportunities for noise to be introduced into the system.

How about today? Starting in the 1970s, key steps in the music recording and reproduction process have been replaced by digital, rather than analog, steps. Analog magnetic tapes are notorious for carrying, along with the music, a slight amount of "tape hiss." This sound, due to the tiny magnetic regions on the tape itself, is like the gentle blowing of air. In the (then) newly developed digital recording, the analog signal was passed from the microphone straight into circuit that changed it to digital bits, i.e., it was **digitized** into a sequence of 0s and 1s onto a tape instead of the older analog replica of the sound that was accompanied by the hiss baggage. A noisy '1' and noisy '0' are still a '1' and a '0,' unless the noise is extreme, which is rare in practice. This eliminated the problem of tape hiss since the sound was, in the last step, created by digital-to-analog electronics, and not directly from the tape itself. Also, as mentioned above, copying the digital form of the music from tape to tape resulted in a perfect replica of the digital bits, with no addition of any noise or distortion.

Of interest: 'Dolby' is the trademark name for an electronic filter that reduces noise on magnetic tapes. Noise, called "hiss," results from the random orientations of tiny magnetic particles in all magnetic tape. The Dolby system boosts the music at the frequencies of the hiss during recording. When the tape is played back, those same frequencies are **attenuated**, or made quieter. That way, the music is returned to the original volume across the spectrum, but the tape hiss is decreased.

Of interest: In the 1970s, some records advertised that they were recorded digitally and reconverted back to analog at the stage of pressing the record. Those were, and still are, somewhat prized vinyl recordings.

Of interest: Although the method of stamping records seems quaint by today's technology standards, there is an important technology called "nanoimprint lithography" for mass production of nanostructures that is the direct descendant of the record-making process. http://www.set-sas.fr/en/cat424981--NPS300.html?Cookie=set

What has been described so far is a hybrid process that used both analog and digital methods. In the recording there was one step of **analog-to-digital (A/D** or **ADC**, pronounced "ay-to-dee" or "ay-dee-cee", respectively) **conversion**, i.e. it was **digitized**, before storing on tape. The mastering of the music, meaning the changes made by the recording engineer, was done using digital algorithms operating on the bits. Finally, a **digital-to-analog** (**D/A** or **DAC**, pronounced "dee-to-ay" or "dak", respectively) **converter** was used to change the bits into an analog signal that moved the record-cutting arm in an analog fashion. These records were played back using the same analog steps numbered from Step 6 above. The key difference is that somewhere in the process, the music 'lived' only as 0s and 1s, and not as analog voltages.

A very high-end home stereo system. *Source*: https://commons.wikimedia. org/wiki/File:Avantgarde_Acoustic_ Duo_speakers_at_HighEnd- 2009_%283556459639%29.jpg.

The analog chain was completely broken as modern digital electronics became available to consumers. The early 1980s brought the music compact disk, or CD, and CD players. That development marked the true end of the analog recording age as such, since now the music was converted immediately after the microphone into digital signals (Step 4), stored on either tape or in computer memory (the digital values don't care at all how they are stored so long as the 0s and 1s remain distinctly different), printed as digital bits onto the CD, played back as digital numbers, and converted back to analog only at the very last step before being amplified into the speakers at Step 11. So, the only analog processing was the microphone and the final conversion to analog into the amplifier to the speakers.

This, my friends, was the start of a whole new world of signal processing that you now take for granted in your cell phones with digital radio communications, digital conversion of voice signals, digital music players, and voice-to-text and voice command. Until a few years ago, voice recognition in particular was crude at best. Only fairly recently has it come into its own making the use of hands-off smart-phone and other control applications possible, and amazing products such as Apple's Siri, Amazon's Alexa and Google's Assistant, and many others. It is even possible to search music libraries from a digital fingerprint, such as with the program called Soundhound. These applications might sound

Of interest: The basic technology of a CD described here is common to CDROM disks, music CDs, DVD video, and Blu-ray (with some variations). Regardless of what the signal eventually represents, it is still just a collection of digital bits. With the availability of digital content over the internet, it has become far less common for consumers to purchase physical copies of media such as music and movies, and are now overwhelmingly streaming their entertainment from the internet. As with all technologies, these optical technologies will disappear as new ones are introduced.

quaint to you now, but in terms of technology development, they are extremely sophisticated.

9.4.2 How a CD Music Disk Works

It is worth discussing how a music or video disk works because it utilizes many modern electronic components. In conventional CD music, a bit depth of 16 is used to store 2^{16} = 65,536 possible digital voltage values for each channel (left and right for stereo). These bits are represented as depressions, or **pits**, on a layer within a plastic CD (Fig. 9.13). The surface is metallized to cause a laser beam to reflect. As the disk spins, the laser shines on the surface and specularly reflects (like a mirror) from the flat regions, called **lands**, but is scattered away by the pits. A light sensor detects the reflected flashes of light as a sequence of 0s and 1s, represented and caused by the pits and lands. The sensor's electronics converts the light to digital bits, and a computer

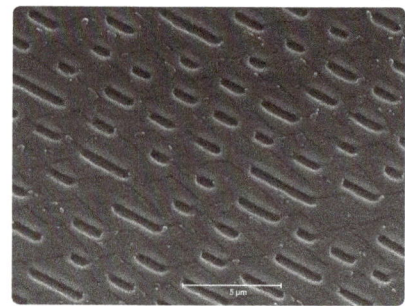

FIGURE 9.13 Scanning electron microscope picture of pits in a plastic CD or DVD disk. The pits, shown, and lands, regions between the pits, store digital data files of music, photos, movies, etc.

converts that by DAC into analog music that is then amplified by the stereo amplifier.

What about MP3 music on an (now outdated) MP3 player or modern smartphone? In this age of ultraminiaturized electronics (see Chapter 7), the data representing the music is stored in semiconductor memory, converted by a DAC chip, and amplified by the same or additional chip to a power level that can drive earbuds. So, the entire system from what used to be the CD, the disk player, amplifier, and speakers (taking up considerable space and many watts of power!) now all fit, with earbuds, neatly in a small part of the palm of your hand (Fig. 9.14) and can be clipped to your sleeve during a jog, or as a chip and app on your smartphone. And, the sound can be amazingly good (as you may well know)!

FIGURE 9.14 Common music player. Hopefully you will begin to appreciate the incredible level of sophistication that this technology represents.

Of course, music is only one application of analog to digital conversion to provide data to a computer. There is virtually no end to the applications. To name a few, digital cameras convert light to bits, sensors in cars feed a huge amount of data to onboard computers, cell phones incorporate a large variety of analog sensors, Wi-Fi and Bluetooth convert and transmit radio waves as digital bits that are received and converted to music, voice and video, personal health monitors track your activity through sensing acceleration and then calculating position and speed, etc. etc. Now that you can appreciate the importance to your daily life of converting analog signals to digital, we are ready to get down to the nuts and bolts of ADC and DAC techniques!

Of interest: CDs are stamped out in mass quantities just as records used to be. However, after stamping the bit pattern into soft plastic, a reflective aluminum layer is added, and then a protective, transparent surface is placed over it.

9.5 Analog-to-Digital Conversion

You have learned (Chapter 4) that the Fourier transform of a signal reveals the frequency components that make it up. The frequency components are sine waves, and the frequency spectrum tells us how much of each sine-wave frequency is contained in the signal. Music is a time-varying signal. It is generally not periodic, in that it does not repeat, but it still has frequency content over some bandwidth, often from about 20 Hz to 20 kHz. Going forward, we will discuss the process of converting analog music signals to digital ones that can be stored and processed by a computer. It is necessary to understand how this works only for simple sine waves, since every part of the spectrum is just sine waves, but at different frequencies. Once we understand that process, we will discuss how the process affects sine waves of various frequencies.

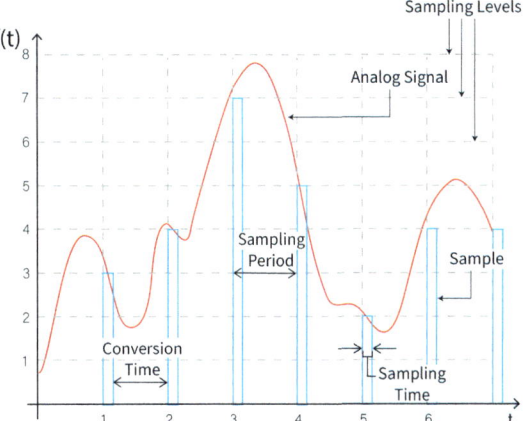

FIGURE 9.15 The process of sampling shown here for three bits. At some regular interval, a voltage measurement is taken, converted to digital bits and then stored in memory.

An ADC is a circuit that takes in an analog signal and outputs a sequence, or "string," of bits that *attempts* to represent that signal. The opposite function uses a DAC to reconstruct the analog signal from the digital bits, which will be discussed later. There are many types of circuits that perform ADC and DAC, but they will not be discussed here. Also, there are many ways to do A/D conversion, but here we will concentrate on the most basic method, which is called **pulse-code modulation (PCM)**.

Think of the PCM ADC process this way. Take the analog signal; pick a point in time; use a voltmeter to determine the voltage. This process is called **sampling**, as shown in Fig. 9.15. Hold the voltage value while the sampling process can work, which is called **sample-and-hold**. Represent that sampled voltage by a digital string of n (say 8 or 16) bits (a digital number); output that string of bits (Fig. 9.16); save that data; go to the next time step and repeat. The output will be a string of digital numbers, each one representing the signal at a discrete time. The bit string of bit-depth n can come out of the ADC chip either as a *serial* (one bit at a time) string on a *single pin* of the chip, or in parallel (all the bits at once) on n *pins*. The ADC you will use this week is a parallel bus with a bit depth, n, of 8 bits.

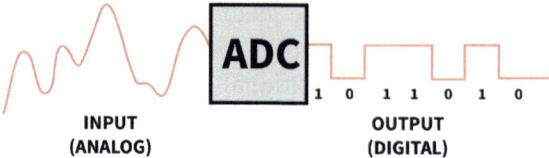

FIGURE 9.16 ADC process showing an analog signal in and a string of bits out. The bits could also come out in parallel.

Note that only 2^n possible digital numbers can be generated from a bit-depth of n. When converting back to analog by the DAC, there can be only 2^n discrete voltage levels, which in general are not exactly equal to the sampled voltages. There are no possible voltage values in between these 2^n possibilities, and we say that the resulting analog signal is **quantized**. So, the digitized samples do not represent the precise analog voltage values. In the example of Fig. 9.15, there are only 3 bits, or only $2^3 = 8$ possible levels, so the discretized voltages from the DAC are likely to not match the original voltages very well. Music quantized into only 8 discrete voltage levels would not sound good enough for high-fidelity applications.

The process of converting a continuous-valued function to one of only a finite number of discrete values is called **quantization**, and the finite number of bits results in **quantization error**, or **quantization distortion**. Higher **bit resolution** (i.e., higher bit depth or more bits) corresponds to more possible digital values, more discretized voltages from the DAC, and smaller quantization error, i.e., smaller differences between the actual voltages and the quantized values.

Keep in mind that there are two voltages to keep track of – the voltages at the input to the ADC and voltages produced by the DAC. In general, they have absolutely nothing to do with each other. The input and output voltages can be totally unrelated in time or space. For example, a digital recording could be made at a particular time in a music studio and played back years later around the world. So, the ADC voltages and the DAC voltages depend only on the local circuitry. For simplicity, we could assume that the voltages input to the ADC are in the same range as the voltages output by the DAC. Then, every statement about voltage steps for n bits would apply equally to both inputs and outputs.

Let's use an 8-bit ADC and DAC as an example, similar to the chips used in this week's lab. Eight bits can represent $2^8 = 256$ distinct numbers. We can assign any range of input voltages (to the ADC) or output voltages (from the DAC) to the range of digital numbers; let's say that both the input and output voltages range from 0 V to 5 V. For both, that maximum voltage is called the **reference voltage**, or V_{ref}, or "**full-scale**" voltage, V_{fs}. Note that there are $2^n - 1$ *steps or intervals* for 2^n *levels*. It might help to notice that going from the lowest level to the next level up uses two levels, but is only one step or interval. Therefore, no matter how many bits there are, there is always one less step than there is levels. V_{ref} along with n define the voltage step size represented by the rightmost digit, the LSB, in both the ADC and DAC. The smallest change in input voltage that can be interpreted by the ADC is the **resolution of the ADC**. The smallest voltage difference between output voltages from the DAC is the **resolution of the DAC**. Assuming the lowest level is 0 V, for both the ADC and DAC,

$$\text{Res}_{ADC} = \frac{V_{ref,ADC}}{2^n - 1} \text{ and}$$

$$\text{Res}_{DAC} = \frac{V_{ref,DAC}}{2^n - 1}. \tag{9.1}$$

In our example, the smallest step size allowed by 8 bits is 5/255 V = 0.0196 V, or 19.6 mV per step. (There are other ways to do the arithmetic, but we'll use this scheme in this chapter). In our example, a change of 19.6 mV causes the LSB of the ADC to change. For example, this might occur from 10110100 to 10110101. For the same V_{ref}, the quantized voltage output from the DAC changes in steps of 19.6 mV.

The MSB, the leftmost bit, represents either the *bottom half* of the voltage range, for MSB = 0, or the *top half* of the voltage range, for MSB = 1. This common case of modulation in which the steps are of *equal size* is referred to as **linear pulse-code modulation (LPCM)**. The ratio of the maximum voltage to the resolution is called the **dynamic range**, and is specified in dB. Recall from Chapter 4 that dB units always express a ratio of two things having the same units. In this case, take the ratio of the V_{ref} to the minimum step size. The dynamic range is, therefore

$$DR_{ADC} = 20\log_{10}\left(\frac{(2^n - 1)\cdot Res_{ADC}}{Res_{ADC}}\right)\left(\frac{(2^n - 1)\cdot Res_{ADC}}{Res_{ADC}}\right) \approx 6.02 \times n \text{ dB, and}$$

$$DR_{DAC} = 20\log_{10}\left(\frac{(2^n - 1)\cdot Res_{DAC}}{Res_{DAC}}\right)\left(\frac{(2^n - 1)\cdot Res_{DAC}}{Res_{DAC}}\right) \approx 6.02 \times n \text{ dB.} \qquad (9.2)$$

There are many ways to characterize an ADC or DAC. Using 16-bit audio as an example, we can think of the bit resolution in terms of: the *number of bits* (16-bit resolution), the *number of quantization levels* (2^{16} = 65,536), the *voltage resolution*, i.e., the LSB or smallest representable voltage, given by $V_{ref}/(2^n - 1)$, (e.g., 1 V/65,535), the *percentage of full scale per step*, 100%/($2^n - 1$) (e.g., 100%/65,535), and the *dynamic range* ($20\log_{10}(2^{16})$ = 6.02 × 16 = 96 dB).

For an actual ADC chip, we have to specify what range of voltages the ADC expects to see. Again, the maximum value that can be digitized is the reference voltage, V_{ref}. If the input does not actually use all of the available voltage values that V_{ref} provides, i.e., does not use all of the dynamic range, then the digitized signal will not use all of the bits of quantization available to it. For example, let's say you want to digitize the output of a microphone. It outputs a voltage with a maximum value of, say, 50 mV$_{pp}$ due to a very loud input sound. If $n = 8$ and $V_{ref} = 5$ V, then the LSB = 19.6 mV and you would barely use even two bits of dynamic range. Thus, the signal could not be captured with any accuracy. Now, you take the output of the microphone and put it through an amplifier with a gain of, say, 100 (a voltage gain of 20log(100) = 40 dB) that boosts the signal to a maximum of 5 V. Now, all of the bits of the ADC with $V_{ref} = 5$ V will be used to capture the music, meaning that you are using all of the dynamic range of the ADC.

9.6 Digital-to-Analog Conversion

So, you have digitized the music from the microphone and saved all the bits in a digital file on your laptop computer. Now you want to convert it back to the music that was the original source. The DAC function is the inverse of the ADC function in that the DAC takes a string of bits and creates a discretized or quantized voltage waveform. In this process, think of a power supply voltage set by a string of digital bits. As each string is read, the voltage changes at the output. Figure 9.17 shows the process

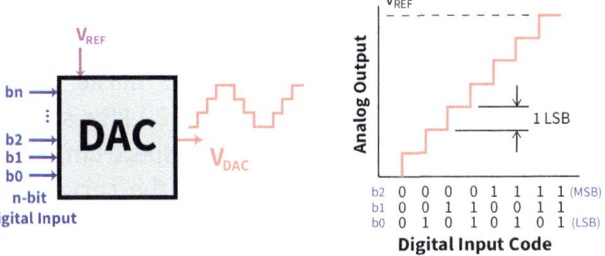

FIGURE 9.17 The DAC chip takes in a string of bits and outputs a voltage. Only 2^n voltages can be produced. For the 3 bits shown, the smallest change in voltage is 1/8 of the full-scale.

of digital-to-analog conversion. A string of bits goes into the DAC, either in series (one bit of the string at a time), or in parallel (all bits in the string at once), as shown, and as will be the case in this week's activities. The DAC chip converts that binary number to a voltage and puts that voltage at the output pin of the chip. The DAC holds that voltage until the next time step. This process is called **zero-order hold**.

Figure 9.17 shows that given a limited number of bits, here it is 3 bits, only a limited number of possible discrete voltages can be produced, which here is $2^3 = 8$. If the horizontal axis were time, the output voltage would represent the result of a linear ramp converted to a staircase function (just like the actual ramp and staircase analogy at the beginning of this discussion). In the figure, the progression of the bit strings is: 000, 001, 010, etc. The rightmost bit, the LSB, controls the smallest change in voltage, and the leftmost bit, the MSB, controls the largest voltage change. The largest voltage available, or full-scale voltage, at the output of the DAC is its V_{ref}.

Figure 9.18 shows a discretized sine wave after quantization by an ADC and then output by a DAC. As the number of bits increases, the digitized voltages approach the original sine wave. 16 bits yields a wave that in this image is essentially indistinguishable from an analog sine wave. Using some very rough numbers, one voltage step for 16 bits on the figure would take up in the vertical direction about 1/400 of the breadth of a human hair. In contrast, the discretized steps for 3 bits are easily resolved.

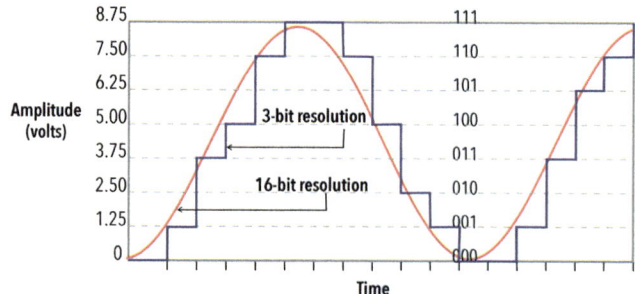

FIGURE 9.18 After ADC and DAC: Discretized sine wave showing the output of a DAC for 3 bits and 16 bits. The 16-bit version is much closer to the original sine wave because the voltage steps are so much smaller.

9.7 The Nyquist–Shannon Sampling Theorem

Here is an important question, and perhaps the central question regarding the entire ADC/DAC process: *If all of the voltages, and hence information, between the discrete sampling times are lost during the sampling process, then can we somehow reproduce that signal exactly using only the sampled values?* Amazingly, the answer is YES! The reason that this is true lies in the **Nyquist–Shannon sampling theorem**.

The Nyquist–Shannon sampling theorem applies only if the frequency components of the sampled signal are restricted to be within a certain bandwidth, i.e., the signal is **band-limited.** It also assumes that there is no quantization error, i.e., the samples are at the precise values of the voltage waveform. Still, it's kind of hard to believe that in spite of all those flat spots between the samples we can recover actual, great-sounding music! For the case of music specifically, there are no frequency components greater than 20 kHz, so the signal is band-limited. Figure 9.19 represents a frequency spectrum (expressed in hertz) whose maximum frequency is the end of the curve on the right, at B, which is indicated as "f_B." You may think of that as all of the frequencies in the music, from the bass guitar to the hi-hat cymbals.

The Nyquist–Shannon sampling theorem says that the original waveform can be exactly reproduced if the sampling occurs at a frequency

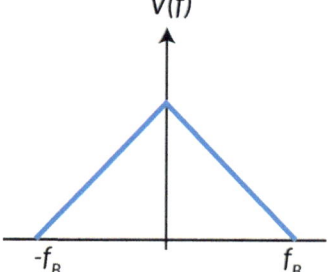

FIGURE 9.19 Representation of music out to a frequency of f_B. This signal is 'band-limited' with a bandwidth of f_B. Don't forget that negative frequencies are a mathematical construct needed for theoretical treatments of signals. Graphically we will keep those negative frequencies when we dive deeper into sampling.

of at least 2f_B, i.e., a rate that is at least twice as high as the highest frequency component. If we sample at a frequency less than twice the highest frequency component, then the output waveform will be distorted. Sampling at higher than twice the sampling rate only improves the process.

We call the minimum sampling rate that avoids distortion the **Nyquist sampling rate (NSR)**, which is exactly twice the highest frequency component. (Don't be confused by the word 'rate'; it just means 'frequency.') Alternatively, whatever the sampling rate actually is (since it is optional), we call half that actual sampling rate

Know this: Here is how to keep the terms 'Nyquist sampling rate' and 'Nyquist frequency' straight. If you have a signal with a particular bandwidth, that sets the **Nyquist sampling rate** at twice that bandwidth. If you are sampling a signal with some sampling rate, then half of that sampling rate is the **Nyquist frequency**. It all depends on your starting point. Put another way, give me a frequency, I tell you the Nyquist sampling rate. Give me a sampling rate, I tell you the Nyquist frequency.

the **Nyquist frequency (NF)**, **Nyquist limit**, or **folding frequency**. For excellent music reproduction, the bandwidth should be at least 20 kHz for which the NSR would be 40 kHz. Now is a good time to point out that WAV files on music CDs have a sampling rate of 44.1 kHz for a Nyquist frequency of 22.05 kHz.

Figure 9.20 shows a sine-wave component of the signal that is **oversampled**, meaning it is sampled more frequently than the NSR. This might represent, say, a 1 kHz signal sampled 20,000 times per second, so it is sampled 10 times faster than Nyquist tells us is necessary. With a little thought, you will realize that if a signal is sampled at the NSR, then every frequency component below the Nyquist frequency is actually oversampled, so oversampling could not be a bad thing, or none of this would work. Again, in general, higher sampling rates are better.

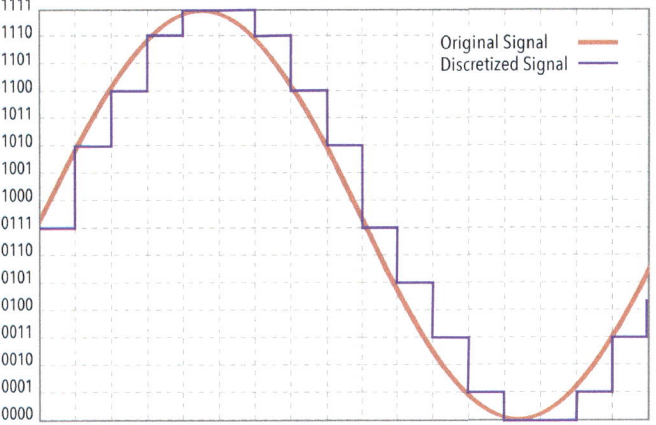

FIGURE 9.20 A sine wave that is oversampled, meaning sampled above the Nyquist sampling rate for this signal frequency.

The Nyquist theorem says that there must be something we can do to recover the exact signal from the discrete waveform. That would convert the discretized waveform in Fig. 9.20 into a smooth, continuous sine wave, like the original signal shown in the figure. In fact, that thing is to pass it through a low-pass filter (LPF) to smoothen out all of the steps. You may recall that sharp discontinuities require high-frequency components, which we saw in Chapter 4. From a time-domain perspective, passing a square wave through an LPF makes the sharp edges rise more slowly, essentially smoothening the square wave, as you saw in the time-domain step-response analysis of Chapter 4. So, if we pass this quantized sine wave through a LPF, then the steps will have some rise and fall times, and the waveform will become smoother, and (hopefully) approach a perfect sine wave. If you look ahead to Fig. 9.37, you will see the result of doing exactly that. The top of the figure shows the discretized sine wave, and the bottom shows that the discretized waveform becomes a very clean sine wave after filtering.

Now let's take a frequency-domain perspective. Recall that we studied the spectrum of a square wave in Chapter 4. We saw that there is a fundamental frequency component, and (odd) harmonics, or multiples, of that frequency. If we pass a square wave with frequency f_B through an LPF that blocks all of the higher harmonics but allows only the fundamental frequency at f_B to get through, then what will we have? A sine wave at the frequency f_B!

As pointed out in Chapter 4, things can change rapidly in the time domain only if there are high-frequency components available to make it happen. The vertical steps in the quantized waveform are there due to the existence of high-frequency harmonics. It is those high-frequency components that we want to filter out. Getting rid of those high-frequency components in the frequency domain is the same thing as smoothening out the sharp edges. So you see, all we need to do in order to recover all of the frequency components of the original signal is to pass the quantized waveform through a LPF that blocks all of the frequency components above the fundamental frequency of the highest frequency component! (Well, it isn't really *quite* that simple, but almost). This LPF is called a **reconstruction filter** because it literally reconstructs the original waveform from the discretized waveform. (In practical electronic systems, it can be done in a variety of ways, with the same result).

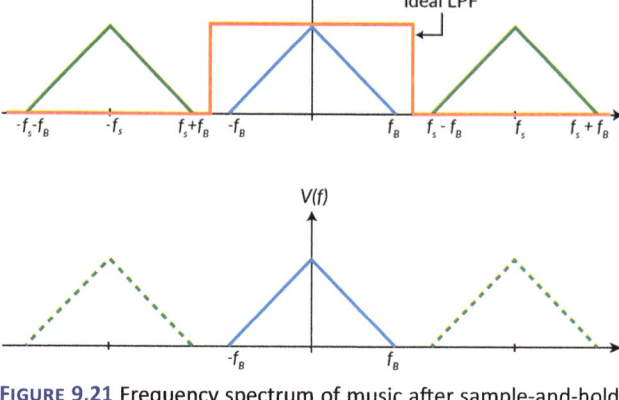

FIGURE 9.21 Frequency spectrum of music after sample-and-hold. Sampled waveforms come with extra baggage in the form of extra frequency components. Using a low-pass filter can recover the original frequency spectrum.

For somewhat more complicated reasons, the sampling process creates a spectrum of frequencies that is the original bandwidth plus other perfect copies at multiples of the sampling frequency. This is (approximately) shown in Fig. 9.21, where both the positive and negative frequency components are included. Let's call the sampling rate, or frequency, f_s. *Each copy is centered at a multiple of f_s.* If $f_s > 2f_B$ (i.e., the sampling rate is greater than the NSR), then great – all of the copies are separate and distinct from each other in the frequency domain. (Lower sampling rates are discussed below.)

Let's put some numbers to it. Suppose that you are sampling a single sine wave having a frequency of 1 kHz. Recall that in the frequency domain, a single frequency is a delta function at that frequency. If you sample the 1 kHz sine wave at 10 kHz, then, accounting for the frequencies at both –1 kHz and +1 kHz that are mathematically part of the spectrum, the first copied frequencies will appear at 10 kHz –1 kHz = 9 kHz, and 10 kHz + 1 kHz = 11 kHz

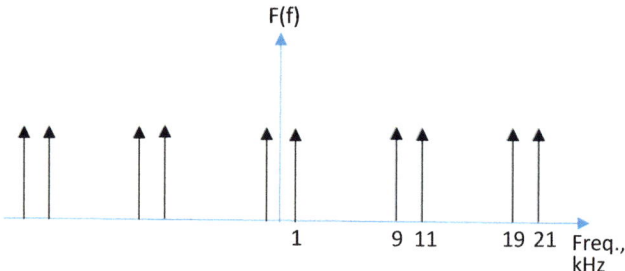

FIGURE 9.22 Frequency spectrum of a 1 kHz sine wave sampled at 10 kHz.

in the frequency spectrum. (This should remind you of the Chapter 6 AM radio spectrum, but without the carrier, and in fact this is a form of amplitude modulation). The second copy of the frequencies will be at 19 kHz and 21 kHz, and so on. This is shown in Fig. 9.22.

Remember, negative frequencies are pre-sent in a mathematical sense, but we see only positive frequencies in the frequency spectrum in actual practice. Your spectrum analyzer will show peaks at 1 kHz (because the original signal is present), 9 kHz and 11 kHz (the first copy of the spectrum), 19 kHz and 21 kHz (the second copy of the spectrum), etc., as shown in Fig. 9.22. You will get to see this happening in this week's lab activity. You might wish to jump ahead and look at Fig. 9.36 to see the spectrum in practice.

As Fig. 9.21 (bottom) illustrates, after we LPF the full spectrum with the reconstruction filter, we get rid of all of the higher frequencies so that they don't interfere with our final waveform. Hence, we can recover the original signal exactly, just as mentioned above with the square wave example.

Now you ask, "What if we sample at a rate that is *less* than the Nyquist rate? What happens then?" Well, now you are asking for trouble. Then you would get an effect called **aliasing**. If $f_s < 2f_B$, i.e., the SR is less than the NSR, then the frequency spectra will overlap, as shown in Fig. 9.23, middle, which is a demonstration of aliasing.

The word 'alias' is a false identity, which is a clue to what happens in the case of aliasing, as demonstrated here. In general, aliasing is the appearance of low-frequency waveforms that do not really exist due to sampling below the NSR. Aliasing, just like everything else regarding waveforms that we have discussed, is not just an electrical phenomenon, but a *generalized wave phenomenon*. Aliasing occurs in many things in daily life. In fact, old movies of a stagecoach where the wheel looks like it is turning slowly forward or backward, even though the coach is going quickly, is an example of aliasing. (This is officially called "the wagon wheel effect.") During the filming, each frame of the film is one sample of the rotation of the wheel. If the camera samples the position of the wheel only slightly more frequently than it turns (SR < NSR), then due to aliasing, the wheel will look as if it is moving slowly. More precisely, at each frame, or sample, some part on the wheel has turned once but is

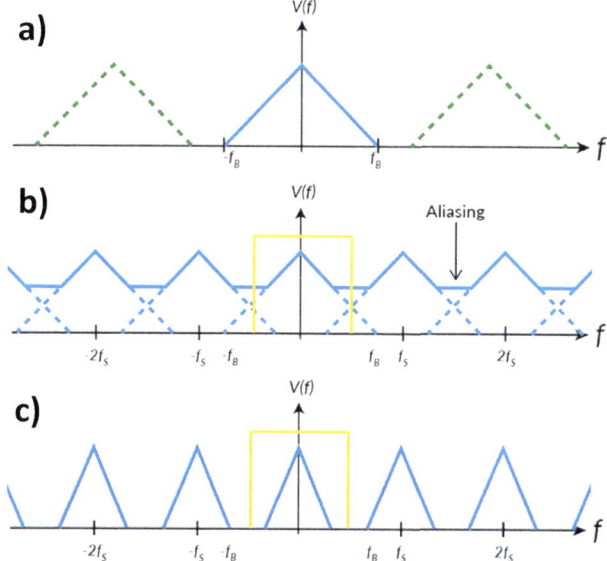

FIGURE 9.23 (a) Signal bandwidth; (b) aliasing due to sub-Nyquist sampling shown in the frequency domain. The rectangle is the reconstruction filter, and it cannot filter out the baseband signal due to aliasing; (c) no aliasing when the bandwidth at the input is reduced below the NF.

captured, or sampled as that part of the wheel is just slightly ahead or behind where it was in the previous frame. Therefore, it *appears* to be turning slowly, but it is actually turning quickly, as most of the movement is lost between camera frames. So, the actual rotation is taking on an alias – it is acting as if it is something that it is not!

Another, related, example of aliasing: A strobe light that makes a rotating fan look like it is nearly standing still is an extreme case of aliasing in which the fan standing still is equivalent to a DC component in the frequency spectrum. Aliasing can also occur on a still image. You may have noticed that many complicated patterns emerged in the waveforms on your oscilloscope because the screen pixels were not small enough and spaced closely enough together to show each data point. Those patterns show up across the width of the screen, which is a low 'frequency,' and are also aliasing.

Let's be a little more quantitative about aliasing. Figure 9.23 shows the *aliased* frequency spectrum of a signal that is sampled below the Nyquist sampling rate. Part a) shows the bandwidth of the original signal. Part b) shows that when sampled below the NSR, the resulting copies of the signal

frequencies overlap, including the baseband, so that there is interference between the upper frequencies of the baseband, which is our original signal, and the lower frequencies of the first copy. Part c) of Figure 9.23 shows the full spectrum in which the signal bandwidth is reduced so that the sampling rate is now above the new NSR. (Can you see that?) As seen here, when the signal is sampled below the NSR, the frequency bands overlap, so the frequency range that was the baseband now includes parts of the first copied spectrum. Therefore, it is not possible to place a low-pass filter, i.e., reconstruction filter, to separate out the original signal from all of the copies. In this case, the reconstructed

Deeper understanding: The chariot race in the film Ben Hur has several clear instances of aliasing. Go to https://www.youtube.com/watch?v=frE9rXnaHpE and look at the following times (among others): 0 s, 27 s, 96 s. It helps to slow it down to 25% of the normal speed. At 108 s you can see the aliasing change slightly as the wheel's rotation frequency changes

signal would be a distorted version of the original baseband signal due to those extra frequency components caused by aliasing.

The bottom part of the figure shows the full spectrum in which the *signal bandwidth is reduced* so that the sampling rate is now above the new NSR. (Can you see that?) In practice, the bandwidth of the original signal can be reduced by the use of an **anti-aliasing filter**. By decreasing the bandwidth of the baseband relative to the sampling rate, the copies can be made to be distinct, and can be separated by a reconstruction filter because the overlap is gone. Conversely, we could have left the baseband alone and increased the sampling rate in order to spread the copies further apart and then allow filtering to isolate the original baseband signal. If this were music, it would be less objectionable to miss out on some of the hi-hat cymbals rather than hear them distorted.

Let's use the 1 kHz sine wave again, now to illustrate aliasing in the time domain, which is actually harder to understand than in the frequency domain. First the frequency domain review: If you sample it at 1.5 kHz, which is below the NSR, then, accounting for the frequencies at both −1 kHz and +1 kHz, the first copied frequencies will appear at 1.5 kHz −1 kHz = 0.5 kHz, and 1.5 kHz + 1 kHz = 2.5 kHz in the frequency spectrum. Note now that the lowest resulting frequency is at 0.5 kHz (an alias!), which is *below* our original 1 kHz sine wave, and *an LPF cannot separate it* from the original baseband 1 kHz sine wave, so the resulting signal is not the same as the original signal. You will see this in this week's lab activity as well.

Figure 9.24 sheds some light on the aliasing phenomenon in the time domain. It shows a (red) sine wave of frequency '1' sampled by less than twice every period. In fact, it is sampled about once every $T = 9/10$ of a period in this picture, which is a frequency of $f = 10/9$. The black dots show the sampled values. Notice that when the dots are connected, it forms another (blue) sine wave with a lower frequency of about $0.11 = 10/9 - 10/10$. The figure demonstrates that if the sampling rate is below the Nyquist rate for

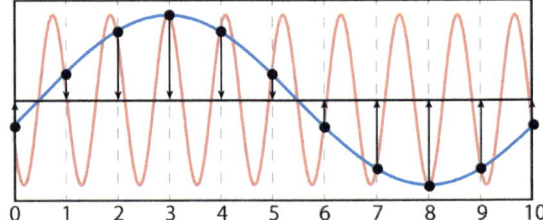

FIGURE 9.24 Sampling at a frequency less than twice the frequency of the sine wave. The resulting values trace out a sine wave of lower frequency, which results in aliasing.

the waveform, which is the case here, those sampled values fit both the lower and the higher frequencies shown, since the dots fit into both sine waves.

Figure 9.25 shows the same thing from the perspective of the sample-and-hold steps like that shown in Fig. 9.20. When building the quantized sine wave by the DAC, the lower frequency component will show up in the frequency spectrum, even though the actual sampled sine wave was the higher frequency. In this way, the input frequency spectrum is copied to a lower frequency, as suggested by Fig. 9.23a. Again, you will see this happening in this week's lab.

You may also understand that sampling as shown in Fig. 9.24 is multiplication of a sine wave by a series of delta functions. Multiplication is the same as amplitude modulation, and we already saw in Chapter 6 that this results in both higher and lower frequencies in the resulting wave. The same effect is going on here, but don't worry too much about understanding it. It is complicated and mentioned only to tie it in with previous course material and motivate future material in a digital signal processing, DSP, course.

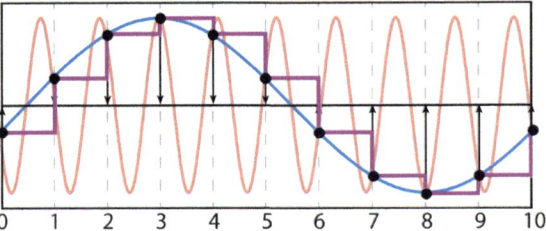

FIGURE 9.25 This figure shows the sampling with the sample-and-hold output, and clearly demonstrates that filtering would produce a lower frequency.

As discussed, when a signal is sampled at a rate less than the NSR, the copies of the spectrum do not extend completely past the base-band, but instead overlap in some range of frequencies, as shown previously in Fig. 9.23b. Now, the reconstruction filter would either include extra, unintended frequency components, or could be selected to have a cutoff frequency at a lower frequency than the upper range of the original spectrum. Either way, the result is a distorted version of the original signal. In summary, in order to achieve a faithful reproduction of a signal after sampling, you must choose a sampling frequency that is at least twice the highest frequency component of that signal's baseband, and then use a low-pass filter to reject all of the higher frequency components that are above the

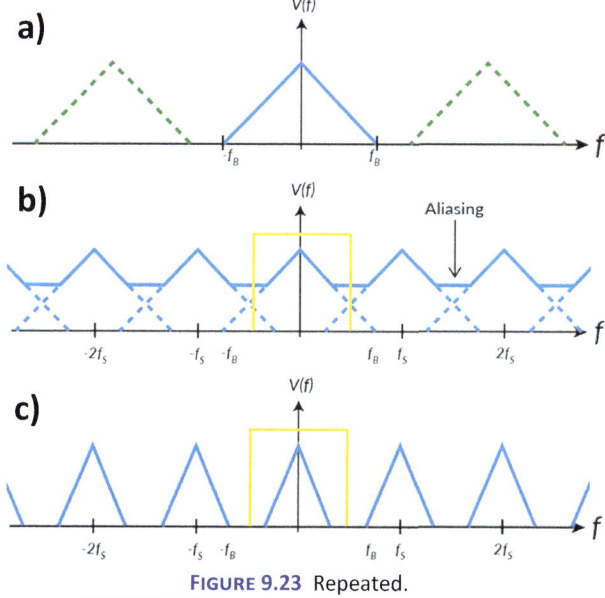

FIGURE 9.23 Repeated.

baseband. The filters shown in Fig. 9.23 are ideal in that they fall off perfectly vertically. This is never the case in reality. Since LPFs are not perfect, and have some slope as they fall off at higher frequencies, which you know from Chapter 4, it is best to oversample to separate the copies and to allow 'room' between the baseband and higher frequencies for the filter to fall off within.

This week's lab will allow you to observe many of the properties of sampled signals as discussed above. It will be fun!

9.8   Audio File Formats and Compression

One goal of this discussion is to give you a working understanding of the music format that you might use, and of the music files that you might encounter. Although the music CD and MP3 players are now largely the stuff of yard sales, the principles are the same, and we will use them as prototypical music storage formats. As a rule, music that is sampled at very high frequencies sounds best, but can result in very large digital files, so they must be converted to smaller files in order to be useful on portable electronic devices and to be transmitted efficiently over the internet for streaming. **Compression algorithms** convert large files into smaller files by discarding information that is either actually useless or may hardly be noticed when removed. You are likely most familiar with compressed music in the form of **MP3** files, but there are many other types of audio file formats (actually close to 50 of them!) that we will not get into here. "MP3" is an abbreviation for **MPEG-1 or MPEG-2 Audio Layer III**, where MPEG is short for "Moving Picture Experts Group." In other words, MP3 was originally developed for movie soundtracks.

Compression algorithms differ mostly in how they attempt to store the music as efficiently as possible, i.e., take up the least amount of storage space on your phone's internal memory, or your flash memory in your MP3 player. You may have noticed that typically (but not necessarily) MP3 songs take up about one MB of storage per minute of music. So, you might get about 20,000 regular (say, 3 min) songs into 64 GB of memory on your phone. That's a lot for just about anyone, so compressing the data is a very good thing in that regard.

Let's compare that to the original, uncompressed CD audio storage file format. Those LPCM files are called "WAV" files, which is an abbreviation for "waveform audio file format" and whose file extension is ".wav." You may be aware that a music CD (Fig. 9.26) can hold up to about 70 min or 80 min of music. Note that most commercial music CDs have much less than that, but that is strictly a marketing decision made by the manufacturer.

FIGURE. 9.26 CDs and players like this have gone out of style in favor of music downloaded from the internet. Ironically, as sales of CDs fall, those of vinyl records are increasing, although they remain very small. Soon, you may feel nostalgic to listen to music the "old-fashioned" way of putting an actual disk into a CD player.

Let's calculate for ourselves how many minutes of uncompressed music can fit on a typical 700 MB CD. The 0s and 1s stored on the disk don't know if they are a spreadsheet, photographs, music, or whatever. If they represent music, then we can use Nyquist as our guide to how many minutes of music can be stored. We calculate that as follows: 700 MB × 1,048,576 bytes per MB (this is the conventional notation for data storage, because $2^{20} = 1,048,576$ is the definition of a "megabyte"), ÷ 2 channels ÷ 44,100 samples per second ÷ 60 s per minute ÷ 2 bytes (16 bits) per sample. This comes to 69 min, which is about typical of CD music storage advertised by common CD manufacturers as the capacity of their blank CDs. So, you see that WAV files use about 10 MB per minute of music while MP3 files use roughly 1/10 of that (1/11 is a better estimate, as you will see below).

So, to save space, why don't we just always use MP3 instead of WAV files, and come to think of it, how do we get around the Nyquist criterion if the larger WAV files are sampled just over the Nyquist rate? The answer is that it is possible to shrink the amount of memory used to store the bits without losing too much of the underlying information about the frequency spectrum.

So, let's talk about data compression. If the original file can be perfectly recovered from the compressed file with no loss of information, then it is called **lossless compression. Lossy compression** provides a much greater amount of compression than lossless, but you cannot get back to exactly the original file. How far from the original file you do get is a tradeoff between how much space you want to save, and how much of the music (or photograph or whatever) quality you are willing to give up.

Here is a trivial example of compression: "The quick brown fox jumps over the lazy dog" might become "Th qck brwn fx jmps vr th lzy dg." The original sentence was **encoded** into a compressed sentence based on the assumption that we could fill in the missing vowels (among other languages, Hebrew is often written that way). In the case of language, the assumption of lossless compression might not hold. If we have a smart **decoder** that knows exactly how to insert the vowels to reproduce the original sentence, then it is lossless compression. If we lose the precise meaning of some of the words, then it is lossy compression. (Bible scholars often argue about the intended vowels in the Old Testament). An example of lossless compression closer to data files might be that of compressing "aaaaabbb" into "5a3b." Also, "Zipped" files are an example of lossless compression.

Of course, music is much more subtle than that. MP3 is a compression algorithm that is based on what is known about human cognition of sound, a field called **psychoacoustics**. For example, some data could be thrown out if we wish to tolerate that certain features of a hi-hat cymbal would be lost, which might be hard to detect by the average listener, but perhaps not by a professional jazz drummer. It is lossy in the sense that the real hi-hat sound can never be recovered. How is the compression accomplished while retaining what we perceive as excellent-sounding music? Once music is digitized at the highest possible resolution, sophisticated **digital signal processing** algorithms can be used to perform complex functions that go well beyond analog functions, which would be limited to filtering or amplifying, plus a few other things. Some features of the MP3 compression algorithm based on psychoacoustics are the following:

Frequency Limits: Frequencies outside the limits of human hearing are discarded.

Absolute Threshold of Hearing: Some sounds are not loud enough to be heard, but still take up data in the original digitized file, and so are discarded.

Simultaneous Masking: If two frequency components are close together, but one is much louder, then the quieter one is masked, i.e., is inaudible. Discarding the masked signal saves data.

Temporal Masking: The human mind rejects some sounds either immediately before or after an abrupt, loud sound. Discarding these smaller components saves data.

It is safe to say that MP3 is adequate for some, or most people, but is inferior to WAV files for some others. So, it is a good thing that MP3 listeners have a choice of the quality of the MP3 files that they download from a streaming service. A WAV file consumes 1,411,200 bits (not bytes) per second, while a conventional MP3 file ultimately uses 128,000 bits/s; that is, MP3 files take up about 1/11 of the data of a WAV file, as mentioned above. More discerning listeners may choose higher data rates that provide less lossy compression, and therefore better approach the lossless music quality. In fact, bit rates from 32 kb/s to 320 kb/s are available at sampling frequencies from 8 kHz to 48 kHz, depending on whether the MPEG-1 or MPEG-2 Audio Layer III format is chosen. When you download music or convert a disk from CD to MP3, you may choose something other than the standard level of compression based on your personal level of listening sophistication. Now, when you have a choice of music quality to pay for on your streaming service, you will be able to make better-informed choices.

9.9   Sigma-Delta Audio Modulation

So far, we have discussed pulse-code modulation in which each sample of the music is translated into a 16-bit string. Music is commonly sampled at a rate of 44.1 kHz for WAV files sold as typical music CDs. In order to prevent aliasing of frequencies above the Nyquist frequency, an antialiasing filter is used to block audio frequencies above 20 kHz. 96 kHz is used for special applications, such as archiving of sound files, so larger bandwidths are possible.

An alternate scheme for achieving better performance than PCM at 44.1 kHz is that of **sigma-delta** ($\Sigma\Delta$) (or **delta-sigma**, $\Delta\Sigma$) audio modulation, and is used for special audiophile-quality music disks. A super audio compact disc (**SACD**) is a less common format for audiophile CDs. SACDs perform delta-sigma conversion to store data in **Direct Stream Digital©** (**DSD©**) format. (DSD is a trademark of the Sony and Philips Corporations). In delta-sigma, each voltage is converted

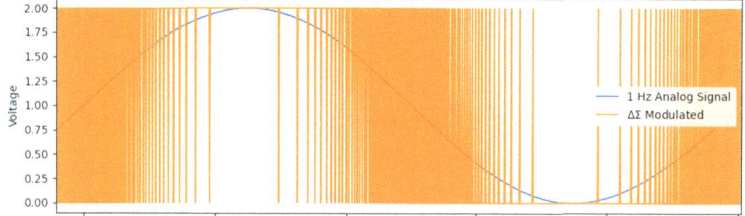

FIGURE 9.27 Sigma-delta modulation. The bits are mostly ones where the waveform is a maximum and mostly zeros where the waveform is a minimum. They are of equal amount as the waveform passes through the intermediate values, i.e., 1 volt on this diagram. The original analog waveform output of the DAC is reconstructed from the digital waveform by passing it through a LPF. *Source*: https://en.wikipedia.org/wiki/Delta-sigma_modulation#/media/File:Delta_sigma.png.

to a stream of very short (in time) 0's and 1's, like a square wave, as shown in Fig. 9.27. The 1's are always the same voltage and always the same width in time. Higher input voltages (say, the peak of a sine wave) cause the '1' bits to be closer together in time and lower voltages (the lowest parts of the sine wave) cause them to be further apart. This is sort of a digital version of frequency modulation, described in Chapter 6. As the audio waveform progresses in time, a stream of bits is produced for which every instant of time produces some density of '1' bits relating to the instantaneous amplitude. The waveform is continuously sampled and produces the bit stream.

You can loosely say that the duty cycle of the square wave is higher when the 1's are closer together and lower when the 1's are farther apart. Another way to look at it is if it were a square wave whose duty cycle changed with time. Lots of 1's in a certain amount of time is a high duty cycle, and very few 1's is a very low duty cycle. You can see why this is also called **pulse-density modulation**.

Each of the output bits is called a "sample," whether it is a 0 or a 1, although that term does not correspond well to sampling in LPCM WAV files. The music on an SACD is sampled (again, just the 1's and 0's) at 2.8224 MHz, 1 bit per sample as described above. It turns out that the SACD can reproduce a bandwidth of 100 kHz with a dynamic range of 120 dB.

The waveform represented as a string of digital bits is converted back to analog, i.e., reconstructed, by passing the data through a LPF that outputs the average value of the voltages at any given moment. A simple capacitor could do this. More 1's in a row, meaning originally a higher voltage in the audio waveform, results in a larger average voltage (and so on for low voltage), so the string of 1's and 0's is converted back to the original relative analog voltage. Because of the very high sampling frequency, quantization error is reduced compared to conventional PCM. Also, circuitry for this scheme is much simpler than that of conventional ADC and DAC. In fact, most common CD players internally convert the LPCM WAV files using delta-sigma prior to converting back to analog.

9.10   Audacity Open-Source Software and Our ADC/DAC Board

Now that you know something of music encoding, you might want to explore further on your own. A great sandbox to play in is a free, open-source program called Audacity (Fig. 9.28), which has been around since 2000. Audacity is user-friendly, but is more useful if you know the terminology and concepts. This week's lab is a great introduction that should teach you about a few of Audacity's advanced features.

The lab activities are based around a custom, 8-bit demonstrator board that will allow you to experiment with various analog inputs, such as sine waves, triangle waves, etc., and even music, and then change the sampling rate and bit resolution. It has an output jack for listening to music that you will digitize.

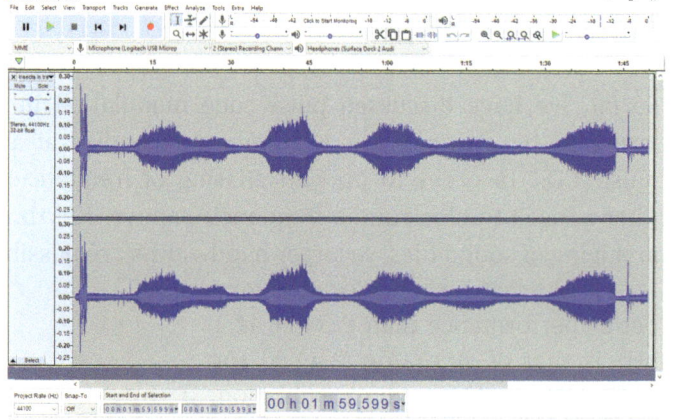

FIGURE 9.28 Screenshot from Audacity showing the analog display of the sound of insects in trees in both left and right channels. It is essentially an oscillograph of the sound file. Notice how the insects get very loud and then very quiet over a time of about 30 seconds.

A graphic equalizer (see Chapter 4) is used as a quick-and-dirty reconstruction filter after the output of the demonstrator, although it would be better to use a high-quality low-pass filter with a steeper cutoff profile. One more feature not discussed until here deals with the **reference voltage**, V_{ref}. The reference voltage sets the range over which all 256 bits are spread out. V_{ref} is set at 5 V, but the board has the capability to input a different value of V_{ref} if desired. That is not done in this lab activity.

You have gained a good introductory knowledge of what happens when music is recorded and played back on a digital device. With that background, hopefully you will never experience your recorded music the same way again, much as you will never look at power lines the same way after studying Chapter 3.

9.11 Lab 9 Activities

You have read carefully the background materials for ADC and DAC in this document. You know, therefore, that you can affect the appearance of the digitized waveform by changing the bit resolution, i.e., number of bits, and the frequency at which you sample the waveform. The magic number is the Nyquist sampling rate (or frequency) – the higher the sampling rate the better the reproduction of the original signal. Also, the more bits used, the better the reproduction due to reduced quantization error. You will change these parameters, from 8 bits to 1 bit, and from well above to below the Nyquist sampling rate and see how these changes affect the quality of the output in the time domain. Also, you will use an EQ as a low-pass filter, i.e., the reconstruction filter, to remove the higher-frequency copies and reproduce your original waveform more accurately. This will be part of exploring the frequency-domain representation of sampling.

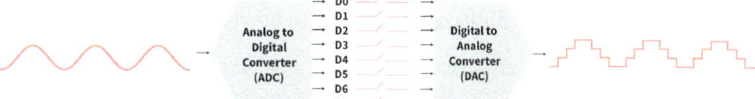

FIGURE 9.29 Functional diagram of the ADC/DAC demonstrator board.

By now, you are becoming somewhat expert with the digital oscilloscopes; you understand the inputs and displays, functions, noise averaging, triggering, frequency spectrum analyzer, and more. In this lab you will not get precise instructions except where necessary. This lab is simply too complicated to give precise details, so instead, you will sometimes be given general instructions and you will decide how to obtain and display the data. A new feature added this week is the use of the digital data probes, which should be of interest to both electrical and computer engineering students. Those instructions are provided within the lab steps.

This course provides for you a custom-made platform for experimenting with ADC and DAC concepts. Figures 9.29 and 9.30 show the functional and physical layouts of the board, respectively. We call the board the 'demonstrator.' The demonstrator has the following features:

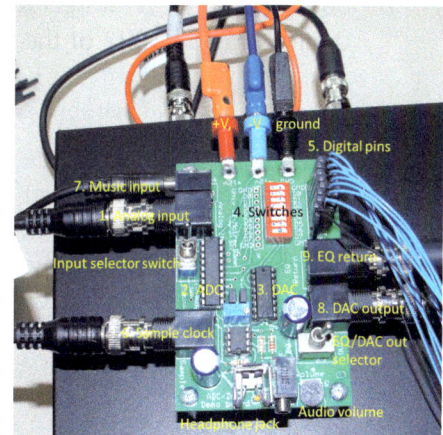

FIGURE 9.30 The demonstrator board.

1. An input for an analog signal.
2. An 8-bit ADC having parallel output bits.
3. An 8-bit DAC that allows conversion of the sampled signal back to analog form. Note that just as drawn in Figure 9.29, the output of the ADC is fed directly into the DAC. This is for demonstration purposes only. In reality, there would be some kind of data storage and digital processing that would happen before the digital file was converted back to analog. Here, it is convenient to simply feed it directly back to the DAC.
4. A set of 8 switches between the output of the ADC to the input of the DAC that allows you to turn off some of the bits, essentially converting the ADC from 8 bits to fewer bits of resolution.
5. A set of 8 pins to allow the monitoring of the digital output from the ADC.
6. In input for a clock signal that allows you to experiment with different sampling rates.
7. An additional input port for music from an MP3 or CD player.
8. An output port for taking the output from the DAC to a reconstruction filter, which in this case is the graphic equalizer (EQ).
9. An input port from the EQ back to the box to allow earphones to be connected for purposes of hearing differences in sound quality. A volume control is included for the sound output.

1. **Syncing your waveforms.** Set up your sync signal from CH1 of the waveform generator, WG, to the external trigger input of the oscilloscope (**Ext**). (In case you forgot: select **Trigger Setup**, **Source**, **Other** and **EXT**.)

2. Locate the digital probe set from the probe box. It has 11 wires per set, and two sets. Plug it into the flat socket at the lower right of the scope. The words on the plug should face up.

3. Carefully sort in order **D0** to **D7**. Plug them onto the vertical pins in the correct positions corresponding to the labels on the demonstrator board and the probes. Also, plug both ground wires (the narrow black probes) into the **GND** pins. The **CLK** probe is not used.

4. **Setting up the demonstrator – signal input to ADC.** Send the output of CH1 from the WG to CH1 of the scope and then through a tee to **Analog In** of the demonstrator. This is the waveform that you will digitize for now. Later you will use your MP3 source.

5. **Setting up the demonstrator – sampling clock input.** Send the output of C2 of the WG, which will be your clock signal, to CH4 of the scope and tee it off from there to **Sample** at the demonstrator. This clock frequency is the sampling rate.

6. **Setting up the demonstrator – observing the DAC output.** Take the output (**Out**) from the demonstrator to CH2 of the scope. The output of the demonstrator will be a quantized version of the input, which is on CH1. Note that the steps will be either small or large depending on the conditions that you choose.

7. **Setting up the demonstrator – bipolar power supply.** Use the same bipolar power supply method that you have used in previous labs, which is as follows. Set CH1 and CH2 outputs to 12 V. Using color-coded wires, apply +/−12 V_{DC} to the red and black power pins on the demonstrator, respectively. Now connect a short, black banana-to-banana jumper between the positive terminal of CH2 to the negative of CH1. Earth ground will be provided by the scope and WG inputs. This is just like back-to-back batteries with ground placed in the center, creating a single **bipolar power supply** to run the demonstrator board.

8. **Setting up the demonstrator – setting input voltage range.** Set C1 of the WG to a sine wave at 1 kHz. The sine wave will be shifted by a DC offset: Choose **Ampl**. Set **HiLevel** to 5 V and **LoLevel** to 0 V. Now your input to the ADC is a sine wave that ranges from 0 V to 5 V at 1 kHz. Observe it on CH1 of the scope.

   You may not have noticed it before: The small letters and markers on the left of the screen show the ground, or zero voltage, level of the waveform. Watching these will help you know how you are offsetting your positive sine wave on the screen as you move it up and down. This will be helpful later.

9. **Setting up the demonstrator – setting clock waveform and voltage range.** Set C2 of the WG to a square wave from 0 V to 5 V (using the same method as above) at a frequency of 100 kHz with a duty cycle of 50%. This is your sampling clock. Observe it on CH4 of the scope. Take a screenshot of all waveforms.

10. **Setting up the demonstrator – setting all bits to 'on.'** Note that there are numbers printed on the 8-pin DIP (dual inline package) switch on the demonstrator. When the switches are depressed toward the side printed with the numbers, the switches are closed, and the digital output of the ADC goes straight into the DAC. When a switch is depressed toward the side that reads "open," then that switch is open and the signal from the ADC does not get to the DAC. This decreases the number of bits that are used by the DAC. The number of bits, n, used to digitize a signal is called the **bit depth**. So, using the scope stylus as a tool, carefully switch on, i.e., close, all of the switches 1 to 8, thus ensuring that you are using all 8-bits of the DAC. Switch 1 is the MSB and switch 8 is the LSB.

11. **Setting up the demonstrator – turning off the input bias voltage.** Find the switch labeled **SW9 Sine-->**. For now, set the switch toward **Sine-->**. That switch allows the input to go to the ADC without an additional offset. Later, when you input music, you will need an offset to make all of the music signal positive, and this switch provides that when it is set in the opposite position.

    Set the coupling on each scope channel to **DC1MΩ.** This will give cleaner steps than you will get with AC coupling.

12. **Setting up the demonstrator – setting input voltage range for DAC to cover full output voltage range.** Confirm that you see the following:

    CH1: Sine wave at 1 kHz from 0 V to 5 V;

    CH2: Another sine wave from the output of the DAC, but this one is discretized.

    CH4: Square wave at 100 kHz from 0 V to 5 V (a total blur of pixels).

    The discretized output on CH2 should look fairly clean when using a sampling rate of 100 kHz, and also range from 0 to somewhere between 3 V to 4 V.

    Turn off all channels except CH2 to keep the screen uncluttered. Normally you would define the maximum values of the input and output voltages as V_{ref}, assuming that the lowest value is zero volts. Here we will be just a little bit more accurate in determining the upper and lower limits, and the voltage steps at the output of the DAC.

    Slowly increase the **HiLevel** of the input sine wave (CH1) until the output of the DAC (CH2) just clips, then back away by small amounts until the clipping on CH2 just goes away. You might find that the **HiLevel** is not exactly 5 V.

    Now do the same with the **LoLevel** of the input. Decrease it until CH2 is clipped on the bottom, and then back away. You may find that the LoLevel needs to be slightly higher than 0 V. Use cursors to measure the peak-to-peak amplitude of the discretized waveforms on CH1 and CH2. What are they?

    The input full range divided by 2^n-1 steps is the smallest voltage step of your ADC and the output full range divided by 2^n-1 steps is the smallest voltage step of your DAC. Calculate the ADC and DAC resolutions using $n = 8$ and the modified Eqs. 9.1 and 9.3: Res = $(V_{max} - V_{min})/(2^n-1)$. What are they? Using Eq. 9.2, what is the dynamic range, in dB, of the ADC? Make a record of your measurements with screenshots.

13. **Observing Res$_{DAC}$, the LSB voltage step.** Make sure you are triggering on EXT from C1 of the WG. Expand CH2 until you can see the smallest steps, as suggested by Fig. 9.20. This takes some fidgeting with the vertical and horizontal scales and vertical and horizontal offsets. The smallest step is the least significant bit, LSB, change.

    You can do two things to help you see Res$_{DAC}$, i.e., the LSB voltage step: you can decrease the frequency of the input sine wave so that it changes more gradually, thus changing

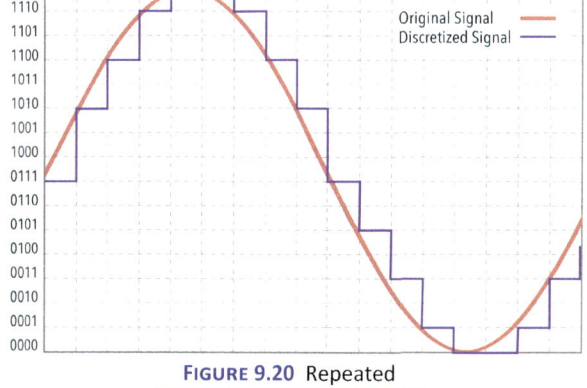

FIGURE 9.20 Repeated

less between samples. And, you can increase the clock frequency as needed, thus sampling so quickly that the voltage has less chance to change by much.

    Do both of these until you can see the smallest step size in the output (as shown in Fig. 9.31). First, decrease the frequency to 30 Hz. You will need to slowly expand the curve, constantly adjusting to bring it back on the screen with every change. You should be able to see the minimum step anywhere on the curve with a sampling rate of 100 kHz – particularly in the middle where the slope is greatest. Press **Stop** before you make a measurement to get a jitter-free curve. Take the average of 10 to 15 steps and see how closely that matches your estimate of Res$_{DAC}$, i.e., the minimum step size due to changes in the LSB. The easiest way to do that is to view about 10 or so steps on the screen and set the top and bottom cursors at the extremes. Ignore the many steps that were not of uniform size, with large jumps. Many inexpensive DACs are not very linear in their output. Read off ΔY over a good range and divide by the number of steps, as shown in

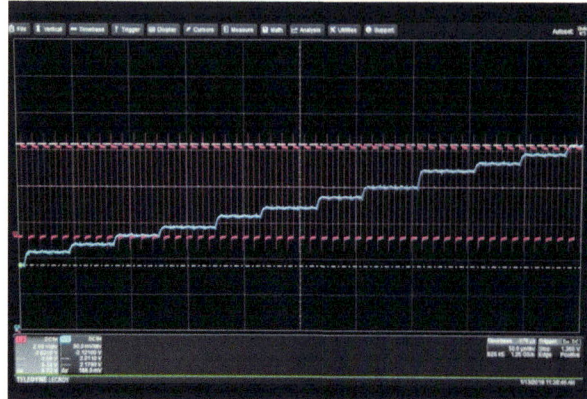

FIGURE 9.31 (Step 13) Digitized signal expanded enough to show the voltage steps due to the LSB.

Fig. 9.31. You should get a step size of ($V_{max} - V_{min}$)/255 V that is roughly in the range of 12 mV. What value do you measure? Include all the details. Compare this measurement to the value of Res$_{DAC}$ you predicted in Step 12. How close are they?

14. **Viewing the effect of lower bit depth.** Change bit depth in the DAC from 8 to 7 by switching off the least significant bit, LSB, which is number 8 on the switch. When you remove the LSB, you cut the number of possible voltage values in half and the minimum step size doubles. Do you see a change? Describe what happens. Measure the new minimum change the same way you did in the previous step. What is it?

15. **Exploring bit depth.** Keep switching off the bits in order. You are changing your 8-bit DAC into a 7-bit, 6-bit, etc. DAC. Explore what happens, including as you change the frequency. Use your discretion about how best to show the results. By the time you get to 4 bits, the output should become noticeably discrete. Describe what happens. See Fig. 9.32.

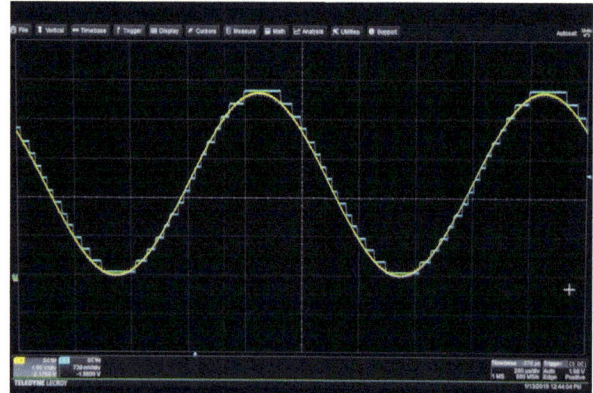

FIGURE 9.32 (Step 15) Demonstration of bit depth of 4 bits.

16. **Observing oversampling.** Set all 8 bits back on. Use an input frequency of 1 kHz and sampling rate of 100 kHz. View CH1, CH2 and CH4. We have no expectation or need for the output voltage of your DAC (CH2) to match the input voltage to the ADC (CH1). In this case, V$_{ref}$ for the ADC and DAC are not the same (approx. 5 V and 3–4 V, respectively), so the two waveforms do not have the same amplitude. To help you interpret your results more easily, press the **V/mV** button and adjust the vertical scale on CH2 until the output of the DAC is the same height as the input to the ADC. This corresponds to Fig. 9.20, repeated in Step 13.

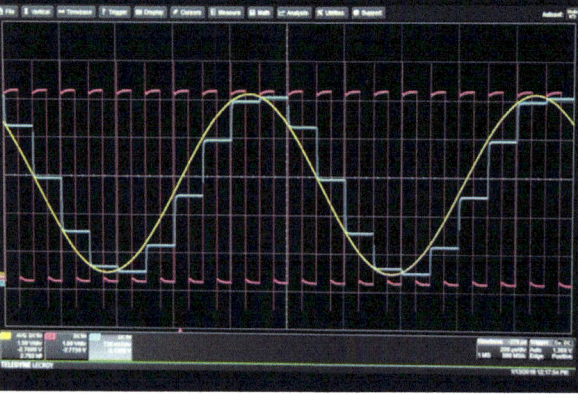

FIGURE 9.33 (Step 16) Sine wave digitized at low sampling rate showing sample-and-hold at the rising edges of the clock.

Decrease the sampling rate slowly from 100 kHz toward the Nyquist frequency of 2 kHz while adjusting the timebase and observe what happens.

Notice that the steps occur on the leading, or rising, edge of the clock. This is the correct operation of this DAC. You will see something like Fig. 9.33. Explain what you see as the sampling rate starts high and then decreases.

17. **Observing undersampling/aliasing.** You don't need to observe the clock on CH4 anymore. Turn off CH4 to avoid clutter. Continue to decrease the sampling rate below the Nyquist frequency, i.e., below 2 kHz. As the sampling frequency approaches 1 kHz, continue to adjust the timebase to get a few sampled waveforms on the screen. You should see that the steps start to form a sine wave that is at a lower frequency than the original 1 kHz sine wave, as shown in Fig. 9.34. That is aliasing, which happens when you make Nyquist angry.

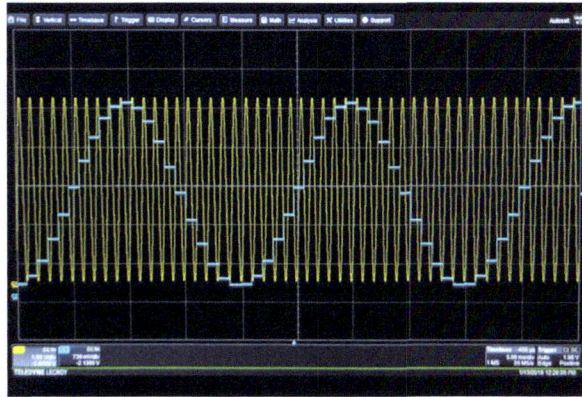

FIGURE 9.34 (Step 17) Clear display of aliasing in the time domain due to sampling at a frequency that is less than the signal frequency.

You decide which waveforms and what data you want to display. The important thing is to illustrate the behavior of the sampling process above and below the NSR. Describe what you did and what you saw.

18. **Setting up the digital probe.** In order to see the digital bits from the ADC you must use some new features of the scope.

    a. Go back to a 1 kHz sine wave. Reset all dip switches to on. Press **Dig** on the panel under CH4. You will see all 16 digital channels but want only the first 8.

    b. Turn off D8-D15 on lower left corner of display button that reads **Display D8-D15**.

    c. Press **Group Height** and select '4' so that only half of the screen shows the digital data. (Fig. 9.35).

    d. Selecting **Position** lowers the display by the chosen amount (e.g., 2).

    e. Choose the **Display Mode** at lower left.

    f. Choose **Lines and Bus** in order to see the hex values that represent the value of the 8 bits from the ADC at any time. You may have to either decrease the sampling frequency or expand the horizontal trace to see the hex numbers.

A quick tutorial on hexadecimal (hex) numbers: We can define numbers (here I mean base 10) using any symbols we want, but you are already familiar with 0–9. We could use the letter "A" to represent "10_{10}," B for "11_{10}," C for "12_{10}," etc. Now we could represent "0_{10}–15_{10}" as "0-F." "Who cares?" you might wonder. Well, it is very convenient to fold up four bits, e.g., 1001_2, into a single letter. 4 bits can represent 16 numbers, 0_{10}–15_{10}. Rather than use 4-binary digits, i.e., bits, we can use one hex digit. So, an 8-bit string becomes only two digits long. For example, hex 5_h is 0101_2 and hex C_h is 1100_2, so $5C_h$ represents 01011100_2, which represents the decimal

FIGURE 9.35 (Step 18) Appearance of scope display with digital data of ADC displayed.

number 92_{10}. On your data bus on the screen, the bottom row shows two hex numbers that tell you the 8 bits that digital display is showing. You will be able to see higher and lower numbers as representations of the activities of the MSB and downwards toward the LSB.

19. **Playing around with timing.** Using the digital data display, explore what happens as you change the sampling frequency and time base.

20. **Playing around with bits.** Using the digital data display, explore what happens as you change the number of bits, turning off the LSB and then more bits. What happens on the display to the channels that are turned off?

21. **Exploring the frequency spectrum of a sampled waveform.** You are now ready to explore sampling in the frequency domain. Activate the spectrum analyzer. Note: In all of these demonstrations you will have large peaks and a background of small peaks. The larger peaks are the frequency components of the discretized sine wave. The smaller peaks are artifacts due to quantization noise, and are often more than 20 dB smaller than the large peaks.

Next you will demonstrate the pheno-menon shown in Fig. 9.22. Go back to a 1 kHz sine wave with all 8 bits on. Set the scale to dBV or dBmV with a spectrum start frequency of 0 and stop frequency of 50 kHz (Fig. 9.36). Start by sampling at 10 kHz.

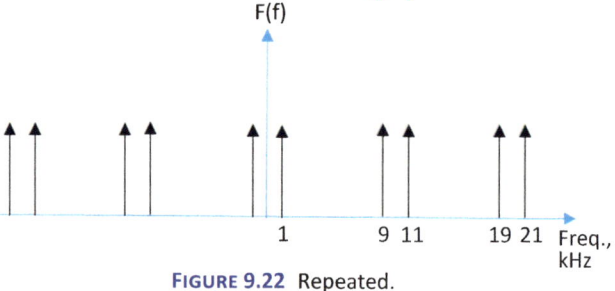

FIGURE 9.22 Repeated.

First, confirm that the original waveform (CH1) provides a single frequency peak, and then change the spectrum input to CH2 to see what the sampled output looks like. Change the sampling frequency slowly up and down by 1 kHz at a time. Notice that the copied pairs stay separated by 2 kHz, but the pairs move closer and further apart from each other, as the theory predicts. Did you see that? Why are there two peaks copied together? How do the pairs of peaks move apart as you change the sampling frequency?

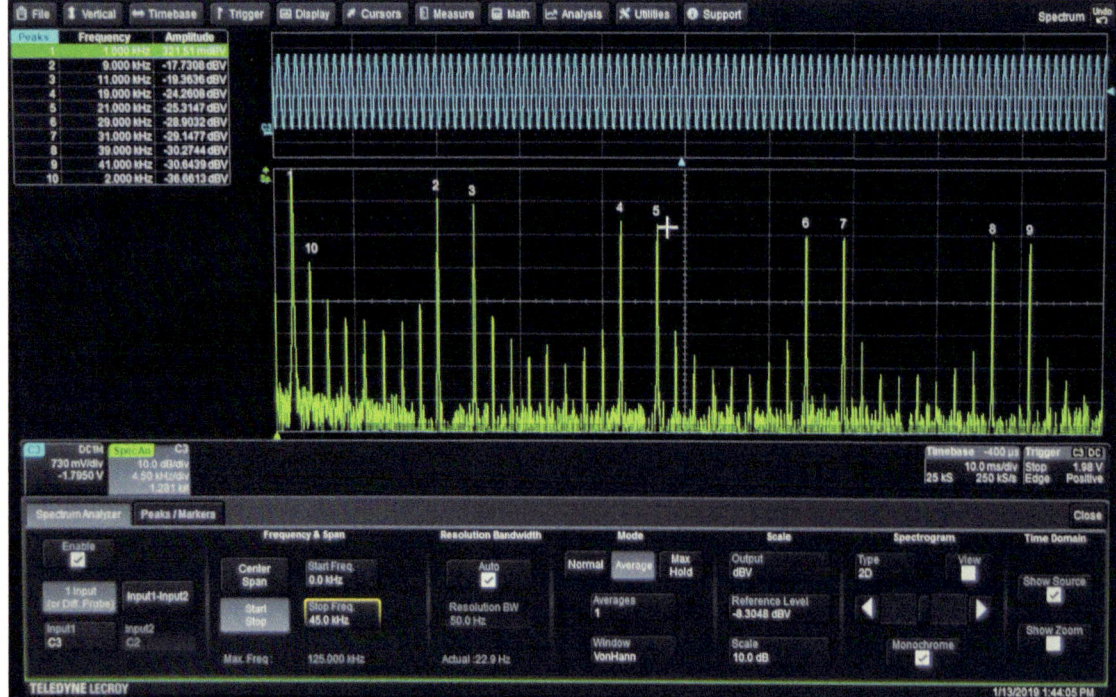

FIGURE 9.36 (Step 21) Frequency spectrum of digitized sine wave with copies. This particular spectrum is for sampling a 1 kHz sine wave at 10 kilosamples/second.

Experiment with different (a) input frequencies, (b) bit depth, and (c) sampling frequencies. Pause for a while to play with this. You should basically understand everything that happens. Show several screenshots of what you experimented with. Provide a general explanation that ties together everything that you observed.

22. **Using the reconstruction filter.** In these next steps, you will experiment with bit depth, sampling frequency and the reconstruction filter. Before you proceed to this step, there is something you should know about the EQ: This particular EQ loads down the output from the demonstrator, thus changing its characteristics at the input to the EQ. It would smoothen the edges of your discrete waveforms. Therefore, you will insert the PA1011 power amplifier in series with the output of the demonstrator before it goes into the EQ.

Set up the amplifier as you did in Lab 5: The PA does not have an on/off switch; instead you just plug in the power supply to turn it on. First, *unplug* the power supply from the back of the PA; power off the WG; connect the USB cable from the PA to the WG; insert the power cord into the electrical outlet and the plug the power supply into the PA (display lights should come on); power on the WG. Press **Utility** on the WG; scroll to PA **Setup**; set **Gain** to 1× and **Switch** to ON. Press **Utility** again to exit. The **Output** and **Link** lights should be green. The menus on the WG can be used to turn the PA output on and off and to toggle the PA gain between 1× and 10× output gains. Make sure it is set to 1×. If you have trouble getting the PA to function correctly, tell your TA.

Reset all of the voltages on C1 and C2 of the WG as you had them at the first few steps of today's activities. Now, plug the output of the demonstrator into the input of the PA. Tee off the output of the PA sending one cable to CH2 of the scope and the other to the input of the EQ using the BNC to ¼ inch phono plug adapter. Send the output of the EQ into CH3 of the scope. For reference, the table summarizes your scope channels.

Scope Channel	Input
CH1	Output of WG or Music Player
CH2	Output of PA
CH3	Output of EQ
CH4	Clock Signal

Generate a quantized waveform as you did in the previous step. Turn on the EQ using the power switch on the back. (The output waveform shifts up or down because the EQ is AC coupled, so the DC level is blocked.)

This is a little different: Set all the sliders on the EQ to be flat at the middle line above 0 dB. Do it carefully. That position is +7.5 dB. Now when decreasing the sliders to –15 dB, it is a total attenuation for each slider of –22.5 dB, which is more attenuation than if you had all the sliders set to 0 dB, and makes a better filter. The output of the filter will be a little bit distorted due to inaccuracies in the sliders, but don't worry about that. Calculate the voltage attenuation of the peaks that is represented by both –15 dB and –22.5 dB. You might need to review this from Chapter 4.

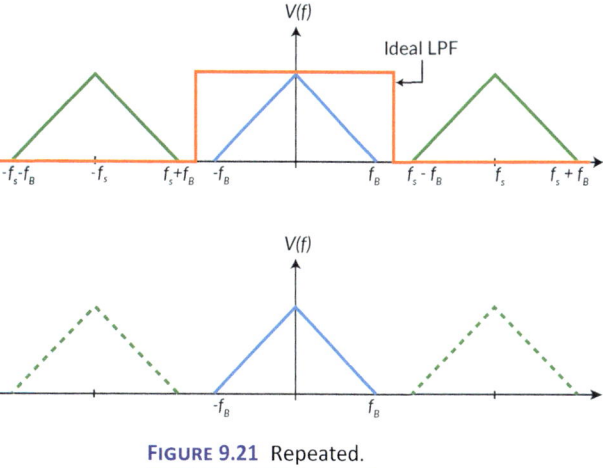

FIGURE 9.21 Repeated.

Use the same parameters as in Step 21: 1 kHz input, 10 kHz sampling rate, and spectrum bandwidth of 50 kHz. Watch the spectrum as you decrease the sliders to −15 dB. Start with the highest frequencies and decrease them one band at a time from the right, thus making a low-pass filter. Notice how the steps smoothen out and the wave becomes more like a sine wave while at the same time the high-frequency components reduce as expected. You have just made a low pass filter that functions like the reconstruction filter shown in Fig. 9.21. Describe what happened as you removed more and more of the high-frequency components. Figure 9.37 shows a 'before' and 'after' filtering comparison. Don't forget that although the peaks are still visible, the scale is logarithmic, and they are down 22.5 dB from their original magnitude.

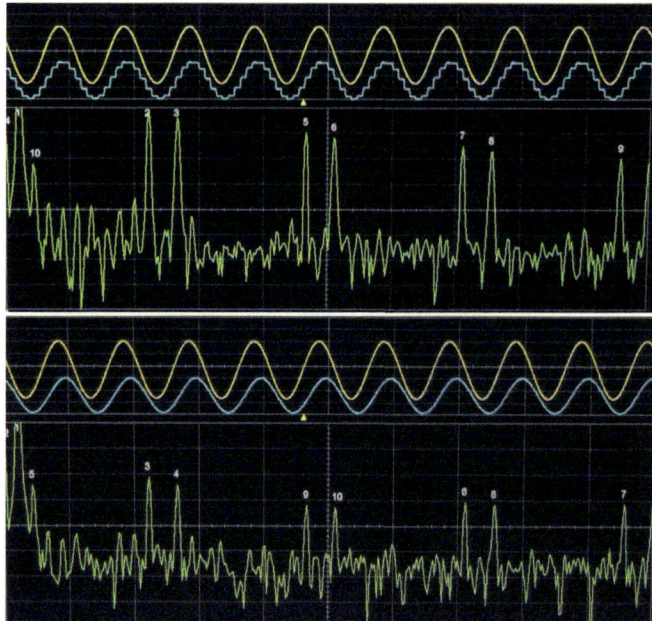

FIGURE 9.37 (Step 22) (Top) Waveform and spectrum of digitized sine wave before reconstruction filter and (bottom) after reconstruction filter. The high-frequency components are decreased and the sine wave is made smoother.

What is the last frequency that you removed before CH3 looks acceptably like a sine wave? Notice that if you keep removing more frequencies, you will just attenuate the sine wave. That would be more distortion in more complicated signals, however.

23. **Comparing original, discretized and reconstructed waveforms.** Time for music. As always, DO NOT HAVE YOUR EARBUDS IN YOUR EARS BEFORE YOU CONFIRM THE LOUDNESS OF THE SIGNAL THAT YOU ARE LISTENING TO!!

Set CH1, CH2, and CH3 of the scope to **AC**1MΩ coupling. Remove the cables and tee from C1 of the WG and CH1 of the scope. We don't need to observe the sampling clock, so remove those cables and tee and connect C2 of the WG, i.e., the sampling clock, directly to the **Sample Input** of the demonstrator.

Connect the output of your MP3 player, phone, or provided music player into the music input jack of the demonstrator. Because the **Analog In BNC** jack is directly in parallel with, i.e., connected to, the audio input jack, you can use the **Analog In BNC** connector as a music *output* by connecting **Analog In BNC** to CH1 of the scope.

Set switch **SW9** the input selector switch, on the demonstrator in the

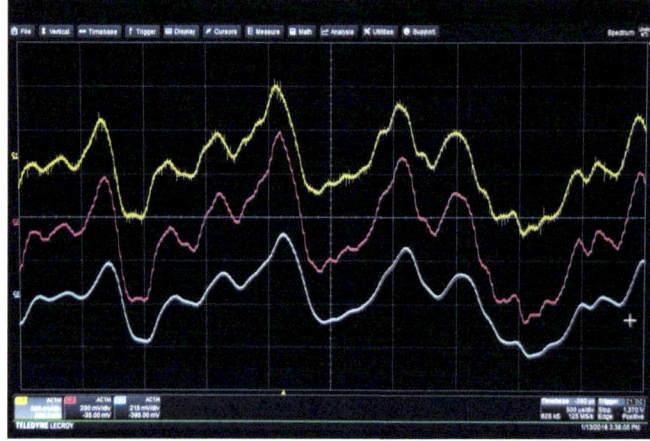

FIGURE 9.38 (Step 23) Scope traces of the three main music waveforms. Bottom: Original analog signal (with some noise due to the demonstrator board on it). Middle: Digitized analog music showing quantization steps. Top: Reconstructed signal after low-pass filtering. You might notice that the reconstruction filter removes the high-frequency noise from the original music signal – an added benefit!

opposite direction from where it has been, toward the input connectors. That adds a DC offset to your AC music so that your music is centered at the midrange of the ADC, and does not go to negative voltages, so the ADC can use all of its dynamic range. Put a tee at the **Input** of the EQ. From there, connect to **Analog Out** of the demonstrator and CH2 of the scope. Connect a tee to the EQ **Output**. From there go to EQ **Return** of the demonstrator and CH3 of the scope.

Summary: Music source is on CH1, quantized music is on CH2, and re-constructed music is on CH3. Start low and increase your music volume while watching the waveforms. Play with the filters to smoothen the music waveform on CH3 to best match the original music on CH1.

Compare with Fig. 9.38. Were you able to reconstruct the original music waveform by adjusting the EQ as an LPF? You can try changing other frequencies across the bandwidth in an effort to improve the result. Comment on what you learned from this step.

24. **Experiencing discretized music.** To listen to your reconstructed music: Turn the **Input Gain** control on the EQ all the way counterclockwise and the volume control at the demonstrator all the way clockwise. Plug your earbuds or headset into the jack next to the **VOLUME** control. Use the splitter if two people will be listening at the same time.

Set the **EQ/OUT** switch on the demonstrator to **EQ**. Set the sampling rate to 44 kHz to start. That should provide a full audio spectrum of frequencies. Bring the earbuds slowly up to your ears in order to determine that the volume is not too loud. Increase the volume at the demonstrator and then at the EQ if you want it louder. It is best to keep the music source volume from your player as low as possible to avoid saturating the input to the ADC and causing distortion. Try different output levels to get the best sound. You can compare the quantized and reconstructed sounds by pressing the **EQ Bypass**.

25. **Exploring the discretization process through music.** Now ... play with it. Change the sampling frequency above and below Nyquist. At what frequency does it start to really sound distorted? Change the bit depth. How low can you go? Compare how it sounds with all of the variations, including with the reconstruction filter and without. Change the filter cutoff frequency with the sliders. You could spend hours exploring at this point. Have fun. Take notes. Tell us ALL about it in your lab report.

Congratulations! Think about all you have accomplished so far in the course. You have gone from basic soldering all the way to understanding the fundamentals of the signal processing theory behind an MP3 player. This is not easy stuff, and the experiments you did would be the envy of many more advanced engineering students!

Problems

1. (a) How many unique combinations can be represented by ten binary digits (bits)?

   (b) What are the positive numbers in base-ten that can be represented by those ten bits? (There are ways to represent negative numbers that we will not discuss here.)

2. A rhythm-and-blues song recorded with CD audio-quality PCM (WAV files) is 5 minutes and 15 seconds long. How much disk space does the song occupy? Assume 44,100 samples per second and 16 bits per sample, and the song is in stereo, i.e., has two channels of music. Express your answer three ways: bits, bytes and MB (where 1 MB = 2^{20} bytes).

3. Consider a 3-bit digital-to-analog converter, DAC, for which the digital input of 000 corresponds to zero volts at the output, and 111 corresponds to 10 V at the output.

   (a) Draw all the levels and label them with their digital values before you start to answer the rest of the questions.

   (b) How many voltage levels can it produce at the output?

   (c) How many steps or voltage intervals are in the output?

   (d) What is the minimum step, or interval, size?

   (e) What is the voltage value of the least significant bit?

   (f) What is the voltage value of the most significant bit?

   (g) What is the output due to a digital input of 011?

   (h) What is the output due to a digital input of 100?

4. Indicate all correct answers: Adding more bits to an ADC will result in:

   (a) higher sampling rate

   (b) higher resolution

   (c) less quantization distortion

   (d) more quantization error

5. Indicate all correct answers: For an analog signal represented by a string of digital bits, the least significant bit represents:

   (a) the smallest difference in voltage

   (b) the largest difference in voltage

   (c) the same difference in voltage as all the other bits

6. Fill in the missing values in the table below regarding linear pulse-code modulation.

	Formula	8-bit DAC	16-bit DAC
Number of bits			
Number of discrete voltages			
Number of steps in the output			
Percentage resolution			
Step size (0–5 V range)			

7. Consider a 16-bit DAC having $V_{ref} = 5$ V. How many volts will the output of the DAC change when the following bits are changed from 0 to 1? Show your work and your reasoning.

   (a) The LSB?

   (b) The MSB?

   (c) The third bit (up from the LSB), as in 0000000000000100.

   (d) In many digitization schemes, not discussed in this text, the LSB is defined as $V_{ref}/(2^n)$ rather than $V_{ref}/(2^n - 1)$. What is the value of the LSB now?

(e) What percentage difference is there in the LSBs between these two definitions? Define percentage difference as 100% x (MAX – MIN)/MAX values.

(f) For large numbers of bits, such as 16 bits here, or higher, do you think it matters which definition and discretization scheme is used?

8. The **method of successive approximations** is one way to perform quantization of a voltage, i.e., convert the voltage to a digital number of some desired number of bits. The method of successive approximations assigns a voltage value to each bit in the digital string and finds the value of the full string that comes closest to the input voltage. It uses a circuit called a **comparator** to test whether the result is greater or less than the input voltage every time it tests a value for a particular bit.

The overall method is to:

- divide the dynamic range into two pieces,
- decide into which part the input voltage fits,
- and successively continue to divide the remaining regions in half and compare, until you reach the number of available bits.

Even more succinctly, "divide and then check" repeatedly until you run out of bits.

The following is an example: Assume a dynamic range of possible voltages of V = 0 to 5 V and an input of V_{in} = 2 V. We will assume 4 bits, numbered 3, 2, 1 and 0 from left to right. The leftmost bit, number 3, is the **most significant bit**, **MSB**, meaning that it carries the most voltage weight, and the rightmost bit, number 0, is the **least significant bit**, **LSB**, meaning that it carries the least voltage weight.

Bit 3: The first step is to compare our input voltage to the middle voltage of the whole range, i.e., $A = V_{min} + \frac{1}{2}(V_{max} - V_{min})$. If the input is less than the middle value A, then bit 3 is 0; otherwise it's 1. In other words, we determine if V_{in} is in the top half or the lower half of the full range.

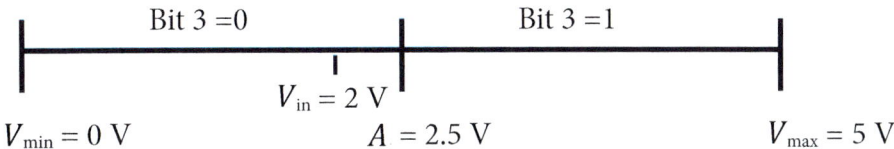

For Bit 3, $A = 0V + \frac{1}{2}(5V - 0V) = 2.5$, and our input (2 V) is less than 2.5 V, so we conclude that bit 3 is 0.

Bit 2: Since bit 3 is 0, we divide the bottom half in two. The bottom half has a range from 0 V to A, so we compare V_{in} with $B = V_{min} + \frac{1}{2}(A - V_{min}) = 0V + \frac{1}{2}(2.5V - 0V) = 1.25V$.

(*If bit 3 had been 1*, then we would have divided the top half in two and compare. The top half has a range from A to V_{max} so we would compare V_{in} to $B = A + \frac{1}{2}(V_{max} - A) = 2.5V + \frac{1}{2}$ (5 V – 2.5 V) = 3.75 V.) This is where B falls relative to V_{in}:

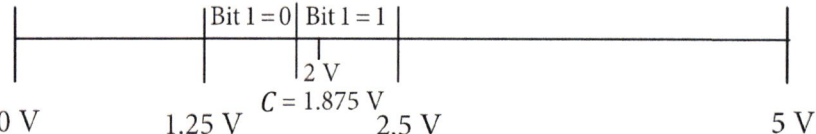

Continuing with our example, the input V_{in} (2 V) is greater than 1.25 V, so we conclude that bit 2 is 1.

Bit 1: Next, because we've narrowed down where our input is to the range of 1.25–2.5 V, we now divide that range in two and get the value $C = 1.25 \text{ V} + \frac{1}{2} (2.5 \text{ V} - 1.25 \text{ V}) = 1.875 \text{ V}$ to compare the input with.

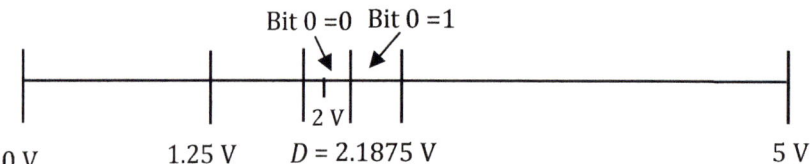

The input (2 V) is greater than 1.875 V so we conclude that bit 1 is 1.

Bit 0: We now have a range of 1.875–2.5 V to look at, so we divide that range in two and get the midpoint $D = 1.875 \text{ V} + \frac{1}{2} (2.5 \text{ V} - 1.875 \text{ V}) = 2.1875 \text{ V}$.

Finally, our input (2 V) is less than 2.1875 V, so we conclude that bit 0 is 0.

For V_{in} = 2 V, we summarize the result as:

Bit #		Bit value (1 or 0)
3 (MSB)	A = 2.5 V	0
2	B = 1.25 V	1
1	C = 1.875 V	1
0 (LSB)	D = 2.1875 V	0

(a) Repeat the process and find the 4-bit string for V_{in} = 3.7 V.

(b) Perform the digital-to-analog conversion of the bit string found in part (a) to find the voltage at the output of a 4-bit digital-to-analog converter having the same range of values, i.e., 0 to 5 V. You will compare that to the input of the ADC that was 3.7 V.

Take your answer from part (a) and fill in the table. To do this, write the bit string in the bit value column. Then, calculate the weight of each bit (the bit weight of bit 3 has been done for you). Then, calculate the product of the bit weight and the bit value. Finally, add up the total products to get the output voltage of the 4-bit DAC.

Bit #	Bit Value	Bit Weight (V)	BW×BV (V)
3 (MSB)		$\dfrac{5\,V}{2^{(4-\text{bit}\#)}} = \dfrac{5\,V}{2^{(4-3)}} = 2.5\,V$	
2			
1			
0 (LSB)			
Total			

(c) What is the quantization error of this ADC/DAC conversion process for this voltage?

9. Explain in words, what is the Nyquist-Shannon sampling theorem? What is the Nyquist sampling rate?

10. Indicate whether the statements about the reconstruction filter are true or false:

(a) It is used at the input of the ADC to limit the bandwidth of the sampled signal.

(b) It is a low-pass filter.

(c) It has a bandwidth that is larger than that of the baseband signal.

(d) It must have a sufficiently narrow bandwidth and be steep enough that it does not include any part of the baseband copies.

(e) It causes the discretized signal at the output of the DAC to be smoothened out.

11. Indicate whether the statements are true or false:

(a) The minimum sampling rate to avoid aliasing and therefore distortion in a reconstructed waveform is the Nyquist sampling rate (NSR).

(b) The Nyquist sampling rate is half the highest frequency component of the waveform.

(c) Oversampling causes aliasing.

(d) Half the sampling rate is the Nyquist frequency (NF).

(e) If the Nyquist frequency is less than the highest frequency component of the signal, then the resulting waveform after reconstruction will be distorted due to aliasing.

12. Consider a signal having a bandwidth from 100 Hz to 15 kHz. It is sampled at 50 kHz. Indicate whether the statements are true or false:

(a) The Nyquist sampling rate is 50 kHz.

(b) The Nyquist sampling rate is 30 kHz.

(c) The resulting reconstructed waveform will be aliased.

(d) The Nyquist frequency is 50 kHz.

(e) The Nyquist frequency is 25 kHz.

(f) If a frequency component at 40 kHz were added to the waveform, it would be aliased at the given sampling rate.

13. If a signal has a bandwidth of 300 Hz, with the lowest frequency at 100 Hz, what should be the minimum sampling rate for this signal according to the Nyquist theorem? (Be careful to pay attention to the word "bandwidth.")

14. Recall that the Nyquist frequency of a signal is the highest frequency that can be sampled without aliasing at a given sampling rate.

   (a) A signal $x(t) = 4 \sin(2\pi * 300t)$ mV is sampled at 1 kHz. What is the Nyquist frequency?

   (b) Will the resulting waveform be aliased?

   (c) Now the signal is $x(t) = 2 \sin(2\pi \times 700t)$ mV for the same sampling rate. Would that output be aliased? Explain your answer.

15. For each of the following figures, select the correct answers, Yes or No:

   (a) Oversampled: Yes or No? Undersampled: Yes or No? Aliased: Yes or No?

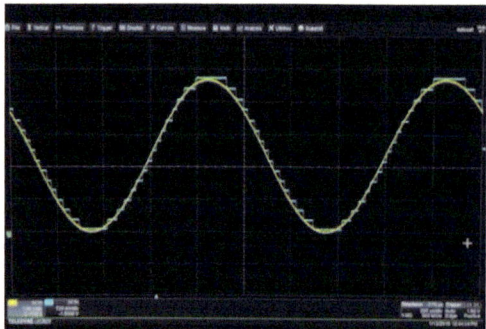

   (b) Oversampled: Yes or No? Undersampled: Yes or No? Aliased: Yes or No?

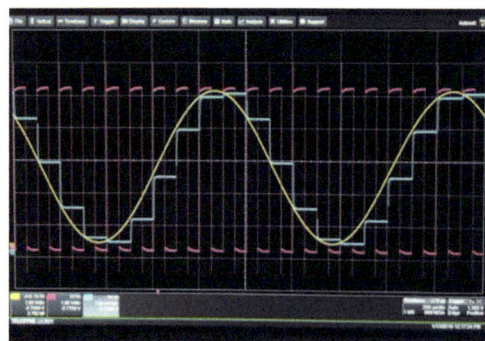

   (c) Oversampled: Yes or No? Undersampled: Yes or No? Aliased: Yes or No?

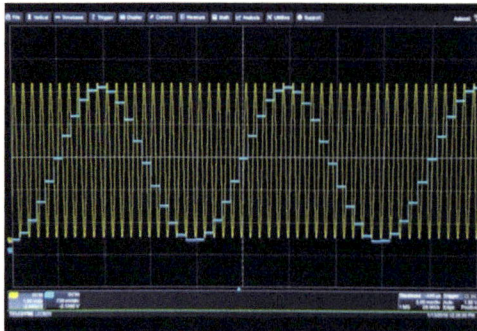

16. This problem illustrates linear pulse-code modulation, LPCM, with aliasing. Below are two sine waves with frequency of 1 Hz and amplitude of 1 V. Each lighter horizontal grid line is a quantization level (at 0.1 V per level). Draw over each of the two graphs the discretized waveform after quantization using sample-and-hold. Your answers should look something like Figs. 9.32, 9.33 and 9.34.

(a) Sampling rate of 10 Hz (every 0.1 s). Don't forget to include $t = 0$ as a sampling time. At each sample time, you will need to carefully assess which quantization voltage level is closest to the waveform voltage. Show the resulting waveform and period, on the graph in seconds, and state the fundamental frequency of the resulting waveform. You may stop making your plot after two full periods.

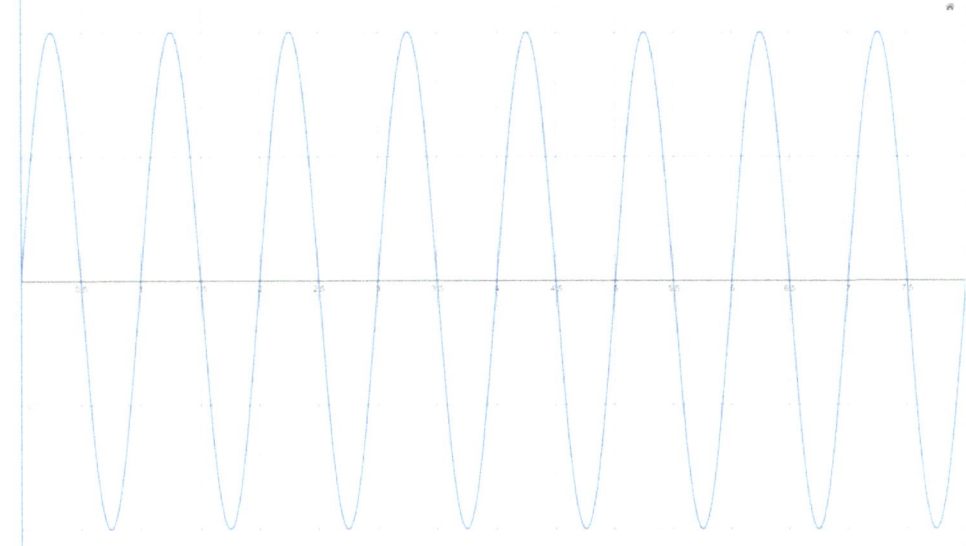

(b) Repeat part (a) for sampling at 1.11 Hz (every 0.9 s). Show the resulting waveform and its period on the graph in seconds, and state the fundamental frequency of the resulting waveform. Can you see the effect of undersampling and the resulting aliasing?

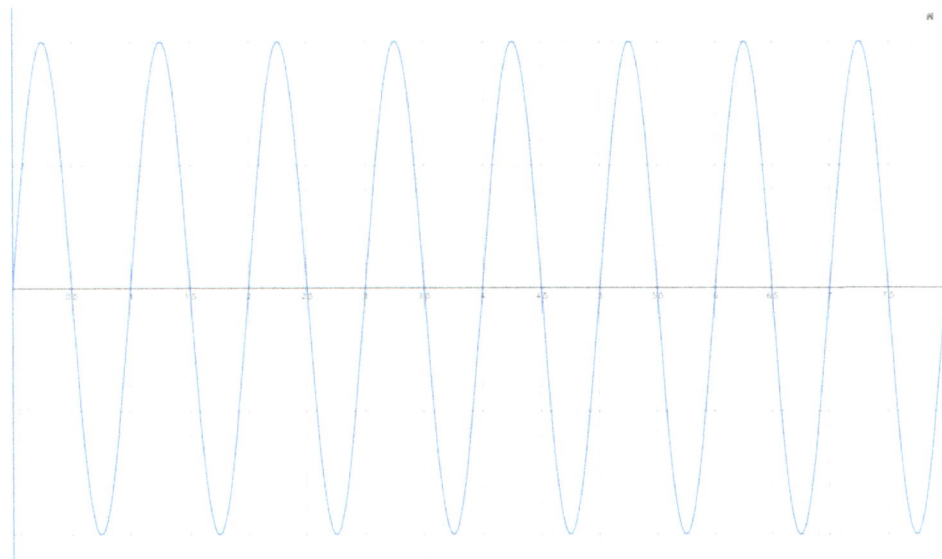

17. Refer to Figs. 9.22 and 9.36.

(a) A sine wave having a frequency of 1.5 kHz is sampled at a rate of 6 kHz.

    (i)   What is the baseband frequency, f_B?

    (ii)   What is the sampling frequency, f_S?

    (iii) What is the Nyquist sampling rate, NSR?

    (iv) What is the Nyquist frequency, NF?

    (v)   Will the spectrum be aliased? Why or why not?

    (vi) Draw the frequency spectrum of the resulting quantized waveform. **Label all the frequencies in the spectrum including the sampling frequency.** Recall from Chapter 4 that the spectrum of the 1.5 kHz sine waveform is just a delta function at 1.5 kHz. Include at least two complete copies of the quantized baseband signal in the spectrum (show the graph out to frequencies of at least 13.5 kHz). The spectrum you draw will include only positive frequencies, but recall the original baseband signal has negative frequency components, as shown in Fig. 9.22.

(b) Repeat part (a) for a sampling rate of 1.7 kHz. This takes more thought and care. In part (vi), show the frequency spectrum out to frequencies of at least 3.4 kHz.

18. Consider the following baseband frequency spectrum. It is sampled at 150 kHz.

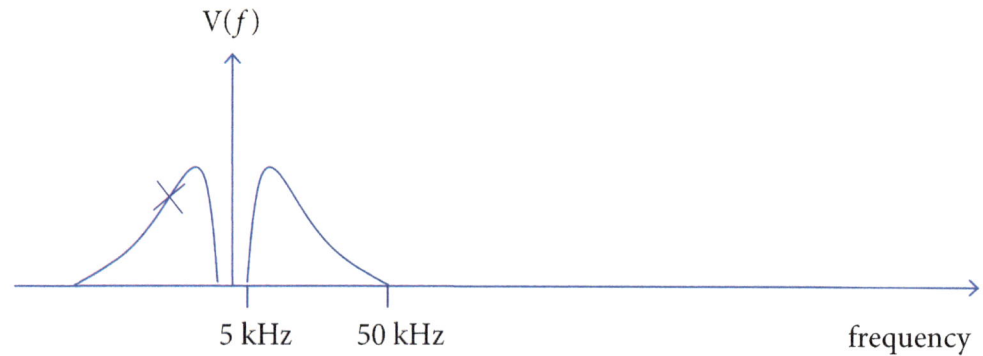

(a) Draw the frequency spectrum of the discretized signal at the output of the DAC with two copies in addition to the baseband spectrum. Label all relevant frequencies.

(b) Show one possible reconstruction filter in your drawing of part (a).

(c) What is the Nyquist frequency (a number in kHz)?

(d) What is the Nyquist sampling rate (a number in kHz)?

(e) Assume that you use an ideal reconstruction filter. What range of cutoff frequencies, f_c, are acceptable in order to recover the baseband? Your answer should be in the form "............ kHz to .... kHz".

(f) As you lower the sampling frequency from 150 kHz, at what sampling frequency does aliasing start to occur?

19. Your music has a bandwidth of 20 kHz. You are sampling the music at the following frequencies:

(a) 30 kHz, (b) 44.1 kHz, (c) 90 kHz.

Draw the resulting frequency spectrum for each sampling frequency on a plot like that provided below. The baseband is a triangle, as shown. For each plot of (a) and (b) include at least two copies. Part (c) needs only one copy. Also, on each plot draw two reconstruction filters that attempt to recover the baseband signal – one with a shallow slope rolls off at, say, 20 dB/decade, with f_c = 20 kHz and an ideal filter (also called a "brick-wall" filter) with f_c = 20 kHz. The same two filters should be shown on each plot. The filters don't need to be accurate, just show one as very steep and one as less steep, but the ideal filter should be placed accurately. Can you see why higher sampling rates are better with regards to the reconstruction filter? Say something about aliasing of the reconstructed waveform, including the effect of the steepness of the reconstruction filter. (Note: the answer to (b) will suggest why "brick-wall" reconstruction filters are not used in commercial digital music players to filter the music with high fidelity. Other methods are used instead, but the basic concept is the same.)

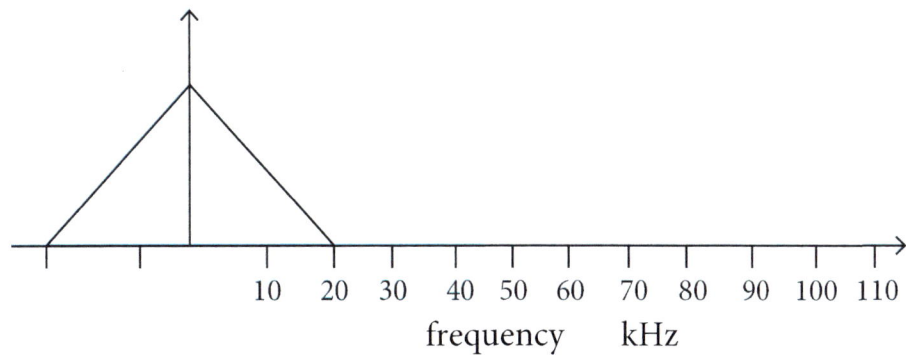

20. Indicate all correct choices:

MP3 files achieve compression by:

(a) lossy compression

(b) lossless compression

(c) eliminating very high frequencies that humans cannot hear

(d) eliminating very quiet sounds that cannot be heard

(e) eliminating some of the instruments or voice that people are not interested in

21. Review Section 9.9 about **delta-sigma or sigma-delta**, modulation at 2.8224 MHz.

(a) What is the shortest duration of a 1-bit that this scheme can support? Hint: the bits occur at 2.8224 million bits/second.

(b) The sampling rate for conventional LPCM WAV files is 44.1 kHz. How many times larger is the SACD sampling rate compared to that?

(c) Draw a picture that represents the delta-sigma ADC scheme method. You should show an input sine wave and the form of the modulated output, including how the 1's and 0's change as a function of the value of the sine wave voltage. The drawing should just capture the main idea of the method and does not need to be to scale.

(d) The SACD disk contains 4.7 Gbytes, GB, (2^{30} bytes per GB) of storage (which corresponds to DVD video disks) and is stereo (2 channel). How many minutes of uncompressed music can fit onto the SACD?

22. In this chapter's lab, you will be using the demonstrator board to convert an analog signal to digital, and subsequently convert the digital signal to analog. These questions will help you prepare for the lab.

(a) What sets the sampling rate of the demonstrator board in the lab activities?

(b) Can you change it?

(c) If so, how? Be specific in your answer.

(d) In the lab, how should the 8 pin DIP switch be set to allow the signals from the ADC go to DAC?

(e) What does the "EQ bypass" button do?

(f) What is the hexadecimal number BD_h in binary? What is it in decimal?

Chapter 10

Batteries, Energy Storage, and Power Supplies

10.1 Introduction and Overview

In this chapter, you will learn:

1. The difference between energy and power
2. The importance of energy storage and, in particular, batteries
3. The principles of battery operation
4. The various types of battery chemistry and form factors
5. Internal resistance of voltage sources
6. How to test a battery – why current must be drawn during testing
7. The design of a simple power supply
8. The role of batteries in electric cars and the grid

10.2 Introduction to Electrochemistry in Batteries

Batteries have historically been the hidden, under-appreciated workhorses of the electrical circuits and systems world. Not anymore. Due to the rise of electric vehicles, EVs, over internal combustion engines, as well as the rise of various alternative energy technologies, batteries have come front and center on the world technology stage. The future of our modern age of transportation and our success at reducing greenhouse gasses will depend on the development of new batteries and other energy storage technologies.

As of 2024, battery research and development has experienced a massive surge of resources and attention. The number of viable types of electrical battery storage has skyrocketed, having various chemistries and physical attributes. All of them are vying for a piece of the renewable-energy pie that is ramping up and expected to dominate energy production as fossil fuel use shrinks over the next couple of decades. This chapter does not attempt to expound on the variety of such battery technologies, but rather to present a motivation for the extreme importance of batteries in the coming new world of energy, and to present chemistry principles that drive battery operation. The remaining part of the chapter is more electrical-engineering oriented in that it presents practical designs of power supplies, which are basic to nearly every electrical system you will encounter.

An Introduction to Electrical Engineering with Lab Activities
Gary H. Bernstein
Copyright © 2026 Jenny Stanford Publishing Pte. Ltd.
ISBN 978-981-5129-30-4 (Hardcover), 978-1-003-71345-6 (eBook)
www.jennystanford.com
DOI: 10.1201/9781003713456-10

Nowhere in the news is a single technology that depends on batteries more present than that of electric cars, whose acceptance in the marketplace rises and falls on their batteries' capacity to travel long distances. The Tesla company is currently the preeminent EV manufacturer, being the first to bring modern electric cars to the market, and (as of 2024) continues to dominate worldwide electric car sales. To meet this demand, Tesla, in collaboration with Panasonic, had from the start produced its own batteries, rather than rely on external suppliers. Tesla's largest battery manufacturing plant is in Nevada and called "Gigafactory 1." It is the largest of its kind in the world, producing 37 GWh of energy capacity per year. (The GWh will be discussed below.) That is enough to supply batteries for something like 500,000 eVs per year.

A close second to the need for batteries in cars is that of battery backup of the electrical grid. Consider the biggest problem with solar and wind power – it is not reliable. For sure the Sun does not shine at night (or very much on cloudy days), and the wind does not blow all the time. If your home were totally dependent on these sources of renewable energy, you might not be able to catch the news on TV many evenings, let alone take a hot shower, cook your meals or dry your clothes. The energy-storage space is changing rapidly as battery costs decline, and batteries become commonly available *at a lower cost*. The great thing is that there is a 'virtuous cycle' with battery development and sales – the more they are accepted into the marketplace, the more they

The bigger picture: Energy storage for demand-response applications takes many forms. In each one, the energy is stored in some large system during times when energy production exceeds demand, which is becoming commonplace. After sundown, when solar is no longer possible, or winds are light, that energy is released through generators that put power onto the grid. The list of possible energy-storage methods is substantial, including large banks of chemical/electrical batteries (discussed in this chapter); pumping water uphill to a reservoir (called pumped hydro, and in use for more than 100 years); raising and lowering massive weights using a crane ('gravity storage'); moving massive weights on railroad cars up inclines over miles of track; spinning large rotors called fly-wheels; heating molten salt ('thermal energy storage'); heating sand or concrete; compressing or liquifying air; and hydrogen (made from water) that drives fuel cells (which convert chemical to electrical energy).

Gravity storage is coming on strong in early 2024 as China's Energy Vault is building up to seven gravity storage facilities. The advantages are that they can be scaled to any reasonable size without excessive land use. Pumped hydro has been very effective, but works only in areas where two entire reservoirs of different altitudes can be built. Gravity storage can be built anywhere. Another major effort ramping up as of early 2024 is by Rondo Energy in the U.S. and Siam Cement Group in Thailand. Their 'heat battery' production is on track to reach 90 GWh of per year. One year's worth of heat battery production will provide four hours of grid energy storage for about 18 million homes.

In every storage system there are inefficiencies in storing and then using the energy, compared with using the energy directly. This is called the "round-trip efficiency.' Round-trip efficiencies range from about 70% to as high as 95%. Obviously, this is an important consideration that affects the economies of various options.

are sold and the lower is their manufacturing cost, resulting in even more sales and even lower prices. The holy grails of battery technology had for years been to offer an EV driving range of greater than 300 miles on a single charge and to provide backup power for utilities (when the Sun and wind are nowhere to be found), either regionally, locally, or at the household level. Meeting these goals will cause energy users to be less affected by global politics, emissions will be cut drastically (assuming, reasonably, that electric power is derived from renewable sources), and energy costs will decrease. As of early 2024, the maximum driving range (for Tesla and Mercedes-Benz) is about 400 miles. However, the Chinese company Nio claims that they will soon produce battery packs capable of providing over 600 miles of range on a single charge. Since the average internal-combustion-engine, ICE, vehicle has a range less than 400 miles, EVs will not only be competitive, but *advantageous* to drive. As range increases, the worry of not finding a charging station and potentially being stranded, a phenomenon called 'range anxiety,' will be significantly decreased, and EVs will be even more quickly accepted by consumers.

As wind and solar power become dominant compared with the generation of electricity by burning fossil fuels, the power industry will be transformed. This is becoming a game changer for the world's economy. You might ask "What will become of the giant utility companies if new, smaller companies can produce, use and sell their own power?" The presence of increasing amounts of solar and wind power is already presenting issues of power control and pricing. This is an ongoing issue for which no clear solution is in sight.

Regarding power utilities and grid energy storage, there are increasingly more locations in which extremely large rechargeable battery energy storage stations are used as backup for the power grid to smooth over low spots in energy generation for wind and solar or in general for peaks in energy consumption. This process is, in general, referred to as "demand response," "grid storage," or "peak shaving." As of 2024, the largest chemical/electrical battery installation is the Moss Landing Energy Storage Facility, in Monterey County, California. That single installation can store 3000 MWh of energy (more on those units later) and produce up to 750 MW of power. This is enough energy to power, on average, about 600,000 homes for up to 4 h.

On a more-personal scale, smaller and lighter batteries that hold more energy would allow longer cell phone and laptop/tablet operating times, and could further shrink the size and weight of portable devices. In short, battery technology is a seriously limiting factor in the advancement of technology. So, that tiny rechargeable battery in your smartwatch demands a little more respect, doesn't it?

Figure 10.1 shows a variety of the most common household batteries. From left to right: D cell (less common these days, was often used in incandescent-type flashlights and other power-hungry appliances and toys); C cell (smaller version of D cell, even less common); AA, or "double-A," cell (common in small appliances like wall clocks, remote controls, LED flashlights,

FIGURE 10.1 Various battery cell sizes. *Source*: https://en.wiki2.org/wiki/Battery_sizes.

etc.); AAA cell, or "triple-A," (overtaking all of the previous ones in popularity as electronics become smaller and less power-hungry); AAAA, or "quad-A" (very small appliances, not all that common); N cell (not at all common, but sometimes found in small clocks); "transistor battery" (more about this below); and two "coin" or "button" cells (used in watches, small toys, small remote controls, etc.).

It is important to know that the voltage rating of a battery is set by the particular kind of chemistry inside, e.g., nickel-cadmium, NiCd: 1.2 V; nickel-metal hydride, NiMH: 1.2 V; alkaline: 1.5 V, and lithium-ion: 3.6 V. This is why it is common to see these numbers, or multiples of them, in the specifications of batteries. The term "battery" can refer to either a single chemical "cell" or to several cells in series to achieve higher voltages. Often, a battery label does not indicate what's inside the outer shell. The word "battery" has an interesting history. Initially, a group of individual chemical cells connected in series was called a "pile" and then later referred to as a "battery" by Benjamin Franklin (who was a great EE) after a grouping of cannon, which was commonly called a "battery." Now we typically refer to all electrochemical cells as "batteries."

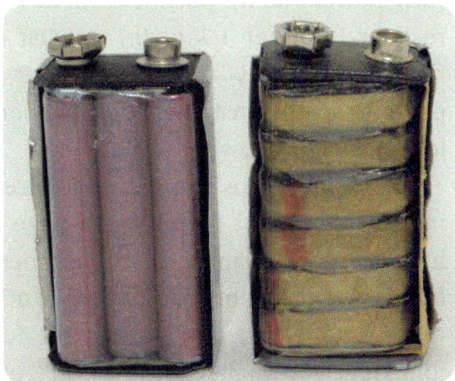

FIGURE 10.2 9 V batteries used in early transistor radios and still in common use today. The outer shell is removed to reveal two types of batteries.

The **fuel cell**, which is an important emerging alternative-energy technology, and produces electricity directly from a slow chemical reaction much like burning, is not discussed here.

One example of "what's inside is not obvious" is the "9 V battery" (Figs. 10.1 and 10.2) found in many consumer items, including your AM radio kit. The technical term for this battery is the PP3 battery. Historical note: in the past, these were called "transistor batteries" because of their use in the original miniature "transistor radios" of the 1960s that required a small battery form factor and higher voltage. (See Fig. 10.3.)

Inside the shell of a 9 V battery is either a stack (i.e., series) of six layers of cells (Fig. 10.2 right), each of 1.5 V, or in some cases a series combination of six separate 1.5 V LR61 cells (Fig. 10.2, left), which are a little smaller even than the quad-A battery shown in Fig. 10.1. Needless to say, there is a seemingly endless variety of batteries out there, with various sizes, voltages, form factors, energy density, storage capacity, recharging capability and more.

FIGURE 10.3 An example of a vintage (about 1965) "transistor radio," so-called because they were the first to use transistors instead of vacuum tubes. They were the driving force behind the development of 9V "transistor batteries." Note the similarities of the components to those of the radio you built in Chapter 1.

10.3 Energy and Power

The Moss Landing battery storage station will store a whopping 3000 MWh (**megawatt-hours**) of energy. So, what kind of unit is a MWh? In order to understand, you must be clear about the difference between energy and power. Let's get back in touch with our high-school physics selves. The joule, j, is the SI unit of energy. It is the energy expended pushing with the force of 1 newton, N, for 1 meter, m, so, a joule is a N-m. It is also the energy used to pass one ampere of current through one ohm for 1 s.

Recall that electrical power, in units of watts, as dissipated by a resistance is $I^2R = V^2/R = VI$, and is the *rate* at which energy is transferred in j/s. If we were to dissipate 1 W for 1 s or 2 W for ½ second, it would still be a joule, even though the latter case was higher power for less time. Power matters a lot since most physical events are time-dependent. It matters if we drip the energy through that resistor (low power) or we push it all in at once (high power). If we transfer that same joule of energy to the resistor in either 1 s or 0.1 s, what difference would it make? Putting that much energy in a short time doesn't allow time for the heat to dissipate out of the resistor (or any physical thing that uses power), so the instantaneous temperature of the resistor will rise more than if it is heated for a longer time at a lower rate.

Let's say there is a power failure and you don't know how long it will last, and say you run an incandescent light bulb using a large battery until the battery runs all the way down. As you already have learned (Chapter 8), an incandescent bulb is just a tungsten filament resistor that gives off light by making it so hot that it glows. You could make the filament run brightly for less time or dimly for more time. Either way, it uses the full energy stored in the battery, but if the power failure lasts a long time, you might choose to run it at lower power for longer.

A more extreme example of this is the bank of lasers at the National Ignition Facility (NIF) at Lawrence Livermore National Laboratory in Lawrence, California. (See sidebar.) There, a collection of laser beams is pointed at a single pea-sized pellet of deuterium and tritium and used to initiate nuclear fusion. The light crushes the pellet with a pressure of 3×10^{11} atmospheres, increasing the density to 100 times that of lead. This causes so much heat with the atoms pushed so closely together, that a fusion reaction results. To accomplish this feat, powerful laser beams are split into 192 paths that deliver 2 MJ (megajoules) in about 4 ns (nanoseconds). During that instant of time, the NIF is producing about 500 TW (terrawatts), which is about 30× more power than is being used by the entire planet! That enormous amount of power, even for a very short time, heats the fuel so hot that the nuclei fuse, giving off, hopefully, more energy than it takes to heat them. It was announced in December of 2022 that the NIF had achieved "breakeven" in which more power came out of the pellet than went in as laser power. Note, though, that the laser power is produced at the expense of tremendous electrical power, so 'breakeven' power is much less than the 'wall-plug' power need to run the tests. A practical facility must produce more power than is used by the entire facility, and not just the confinement power, which, as in this case, is the laser power incident on the pellet.

Although laser fusion is not economically practical, it is extremely valuable as a research tool to demonstrate that nuclear fusion is possible. Currently there are dozens of public and private research programs aimed at developing practical nuclear fusion using a variety of methods. When fusion technology is finally perfected, it will change the world vastly more than even improved battery technology would, or for that matter arguably more than any technology in history. Some estimates are that it will happen by 2050. That is a familiar trope, since the estimate has been 'in the next 30 years' for the past 50 years. However, never in history has so much money, both public and private, been invested in its development.

The big picture: NIF target chamber external view (top) and internal view (middle). (Bottom) Doc Ock in Spiderman 2 was trying to accomplish essentially the same thing, which is to confine the plasma for the atoms to be so tightly packed that the nuclei fuse, releasing enormous amounts of energy. The lasers in the NIF create a pressure of 300 billion Earth atmospheres, causing the atoms to undergo fusion.

Notice that if we multiply power × time, we get energy. The 3000 MWh Moss Landing battery installation stores 3000×10^6 watts (i.e., j/s) × 1 h × 3600 s/h = 1.08×10^{13} J of energy. The facility is rated at up to 750 MW, so theoretically could supply that power for 4 h, which is the standard backup rating specification for these large installations.

Another rating for batteries is in terms of current that can be passed at the battery voltage for a given amount of time. For example, the capacity of a triple-A alkaline battery is about 1000 mAh (milliamp-hour), or 1 Ah. How is a mAh related to energy? First of all, you need to recall that the rated voltage of an alkaline battery, which is set by the chemistry and not the size of the battery, is 1.5 V. Since power is $I \times V$, then we know that the normal operating current flows at a potential of about 1.5 V. Under ideal conditions, the stored energy is the ideal power × time, or 1 A × 1.5 V ×

3600 s (1 h), which gives us 5400 J. That giant Moss Landing storage station, then, is equivalent to roughly 2 billion AAA alkaline batteries. How big would that be? The diameter of a AAA battery about 1 cm, so a single layer of AAA batteries (standing vertically) would cover an area of about 100,000 sq. meters (if densely packed), or the equivalent of about 20 football fields. BTW, it wouldn't matter if we chose AAA, AA, C or D-sized batteries for this silly example since they are all about the same length, and the energy density of the alkaline battery material would be the same regardless of the form factor. (Just for fun, if we estimate the cost of the AAA batteries to be about $.20 each, then you could use the battery station made of alkaline batteries *once* for about 400 million dollars.) The NiMH battery has comparable energy density to alkaline, so the volume of batteries needed by the Moss Landing facility is about the same. However, they pack those 20 football fields of batteries into modules that take up a much smaller footprint.

Now let's look at energy density based on the amount of charge that flows as a battery supplies power. If 1000 mA can be passed for 1 h (or at a lower current for a longer time) by the AAA battery example above, how many electrons have flowed, and therefore how many ions must have exchanged charge inside that AAA battery to become fully discharged? 1000 mA, or 1 Amp, is 1 C/s which is 6.24×10^{18} electrons/s. So, in 1 h, a single AAA battery can move 6.24×10^{18} electrons/s $\times$ 3600 s, or 2.25×10^{22} electrons. Since each ion in the alkaline battery is doubly ionized, the number of ions that exchange charge is half the number of electrons. Converting to moles, we find that the number of ions is about 0.04 moles. Zn and MnO_2 electrodes have molar masses of 65 g/mol and 40 g/mol, respectively. The combined mass of the electrodes that participate in carrying current is, therefore, 105 g/mol $\times$ 0.04 moles = 4.2 grams. A triple-A alkaline battery weighs about 11 g. These rough numbers make reasonable sense – about 40% of its mass participates in the electrochemical process. The rest of the mass is the case and the electrolyte, which passes charge, but does not contribute charge any energy to the process. These are all just rough numbers, but they demonstrate the basic chemistry of the transfer of chemical energy and bonding into electrical energy. The details of the process are presented in Section 10.4.

10.4 Battery Types and Form Factors

There exist commercially a large number of kinds of battery chemistries because there are so many aspects that can be optimized for the customer, including nominal voltage, energy density, peak power capacity, reliability, recharge capability, storage time, cost, and much more (see http://web.mit.edu/battery_specifications.pdf and https://en.wiki2.org/wiki/Battery_size). Disposable batteries are referred to in the trade as **primary batteries**, and rechargeable ones as **secondary batteries**. Two common types of primary batteries include zinc-carbon (cheap, "heavy-duty" battery) and alkaline; secondary batteries include NiMH, NiCd, and lithium-ion. Your computer laptop battery is likely to be of the Li-ion type. Beware of the

For fun: Jules Verne took many technical liberties in his novel *Journey to the Center of the Earth*, in which the three characters travel underground for months, mostly in total darkness were it not for their crude Ruhmkorff arc-lamp lights. I estimate that in 1864, the explorers would have used something like 150,000 pounds of batteries to make the full trip underground, among other things like food and water. Okay, never mind the fact that the air pressure would have crushed them to death at the depths of their trek. Hey, it was 1864, and Ruhmkorff lamps were state-of-the-art at the time. It's still a good read, and actually quite frightening in parts.

"heavy-duty" Zn-C battery type because the term is certainly a misnomer compared to alkaline batteries. Zn-C batteries have about 1/3 of the **energy density** (joules/liter) of alkaline batteries, so unless you get them very cheap and don't mind changing them often, you might want to avoid buying them. Figure 10.4

is a comparison of the energy density of various technologies at low and high drain rates. Energy density refers to how much energy fits into a certain volume. This matters, of course, for applications such as cell phones and flashlights, where only a limited volume is available to store all of the energy that you will take with you when you head out on a walk. More importantly, as discussed above, more energy packed into an electric car for the same weight and size translates directly to higher range, which further translates to

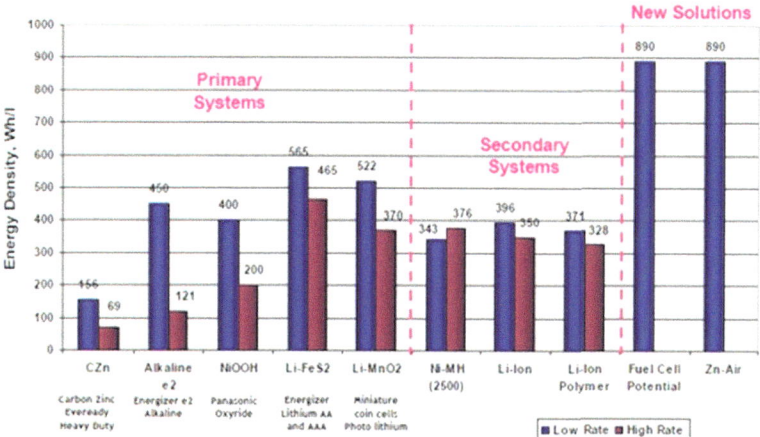

FIGURE 10.4 Energy density of various battery technologies at low and high drain rates. *Source*: http://robotics.stackexchange.com/questions/554/quadcopter-lipo-battery-weight-capacity-trade-off.

faster public acceptance and a faster path to all-electric vehicles that use renewable energy rather than the burning of fossil fuels. This *really* matters.

10.5 How an Electrochemical Cell (Battery) Works

Batteries are becoming increasingly important to supporting the new energy economy. Electrical engineers will more frequently be engaged on projects that involve batteries, both large and small, so it is worth the effort to gain at least a basic understanding of how the batteries work internally. To this end, you will learn some electrochemistry fundamentals in the following sections.

10.5.1 Charge Transport

Current is the motion of electrical charges, with units of coulombs/s passing a plane. In a metal wire, those moving charges are electrons, which is the most common notion of current. When an electron enters a copper wire at one end, and current therefore is flowing, Kirchoff's current law tells us that an electron must leave the wire at the other end, since there are only two nodes to the wire as a circuit branch. The electron that enters, in fact, moves very slowly; the current detected at the ends of the wire is due to Coulomb repulsion between all the electrons on the wire. The electron entering at one end pushes all of the electrons so that one pops off the wire at the far end. By this mechanism, called **dielectric relaxation**, electric current is propagated down the wire very quickly. (See sidebar about "Newton's cradle.")

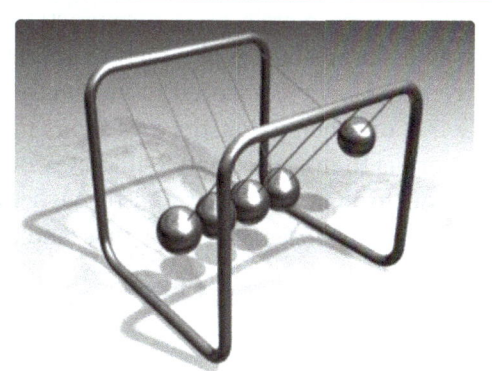

For future reference: This contraption, called "Newton's cradle," is often regarded as a desktop diversion for the habitually bored. It demonstrates conservation of mass and energy in an elastic collision. When a ball on the left strikes the row of balls, only the rightmost ball leaves the group, and all the others stay in place. I like to use this as a way to visualize how current passes through a metal wire where an electron enters at one end and another electron leaves the other end nearly instantaneously. If you take a semiconductor devices course, you will learn that this is how majority carriers (but *not* minority carriers) move through semiconductor devices. *Image source*: https://commons.wikimedia.org/wiki/Category:Newton%27s_cradle#/media/File:Newtons_cradle_animation_new.gif.

Electrons move through materials under the influence of an electric field in a very indirect way; they collide with atoms and bounce around much more than they manage to make forward progress. However, they do eventually make their way slowly forward. The **drift velocity** of the electron is the average speed that the electron moves through the wire due to the electric field after ignoring all of the random collisions. The velocity of the *signal propagation*, which is the current flow or voltage changes, along a wire might be a significant fraction of the speed of light (depending very strongly on what sort of wire we are talking about); however, the drift velocity of the electron that enters the wire is only on the order of a millimeter per minute, which is far slower than a snail. The reason that wires can carry a lot of current is because there is a *huge density* of free electrons, and not because they move quickly! Each atom of a copper wire contributes a mobile electron to the conduction process, which, when considering the density of the atoms (about 6×10^{22} cm^{-1} in Cu), is an awful lot of available electrons that can carry current.

Likewise, charged particles, i.e., positive or negatively charged ions, can carry current if they move. You can think of an ion as a tiny train car whose cargo is one or more electrical charges. The ease with which a particle, be that electron or ion, moves through a material due to some electric field is called its **mobility**. As slowly as the electron moves through the wire due to its low mobility, the mobility of electrons in a metal is still very high compared to that of an ion moving in a solid. Ions move through most solids such as metals or semiconductors only under

> *Safety:* Be very careful working around car batteries if you are trying to "jump" a dead battery from car to car. There are scenarios in which you can be badly injured, and they do happen occasionally. This is usually from explosions resulting from ignition of hydrogen generated in the battery. Suggestion: go to this web site, and see if you can understand from a circuits perspective why each step of the process is as it is:
>
> https://www.dummies.com/home-garden/car-repair/how-to-jump-start-a-car/

extreme circumstances like high temperature plus very high electric fields or currents, and only then extremely slowly compared to an electron. In short, you cannot (generally) use ions to carry current in wires or other solids. However, ions move relatively easily through fluids. For this reason, it is very dangerous to mix electricity with baths or showers, since the ions in the water can easily carry enough current through the water and through wet skin to kill a person. (In fact, you should avoid touching electrical appliances with wet hands, and certainly not while standing in water.)

10.5.2 Electrochemical Cells

As in a fluid, ions can move and carry current rather easily in wet, viscous, permeable media. The term "**electrolyte solution**" is used to describe a solution of dissolved, ionized solids in a solvent. Current is carried in electrolyte solutions inside of a battery, whereas the current outside the battery is the flow of electrons in the wires that feed the electrochemical reactions inside the battery. Here we discuss the basic operation of a generalized electrochemical cell and extrapolate that to simple batteries. You should note, however, that one of the newest classes of batteries is the "solid-state" variety. In solid-state batteries the solid electrolyte is physically different from most solids in that it is engineered to have open spaces through which ions can easily move. Electrons do not move through the solid electrolyte. Electrolyte materials in solid-state batteries are either ceramics (like that of a common coffee cup) or polymers (which are plastic materials commonly made into such items as contact lenses, fishing lines and cell-phone cases). These new solid-state batteries have numerous advantages over traditional 'wet'-electrolyte batteries and are nearing insertion into electric vehicles, which, as claimed by Toyota, is anticipated about the year 2028. For simplicity, the following discussion refers to traditional wet cells.

Figure 10.5 shows a **Daniell cell** (also called an **electrochemical cell**), which is basically a simple battery. The following explains how this apparatus functions, and is the foundation for our understanding of the operation of batteries in general. All batteries have two **electrodes**, namely the anode and cathode. The **cathode** is the terminal at which electrons enter the battery, and at which electrons are contributed to the charge carriers inside the battery. The **anode** is the terminal from which electrons are removed from neutral electrode atoms and leave the battery.

In this example there are solid zinc, Zn, and copper, Cu, electrodes. The Zn electrode, the anode, is immersed in a bath of zinc sulfate, (sulfate: SO_4^{2-}) electrolyte. The Cu electrode, the cathode, is immersed in a bath of copper sulfate electrolyte. So, the baths are at the beginning both already rich in **anions** (negatively charged ions) from

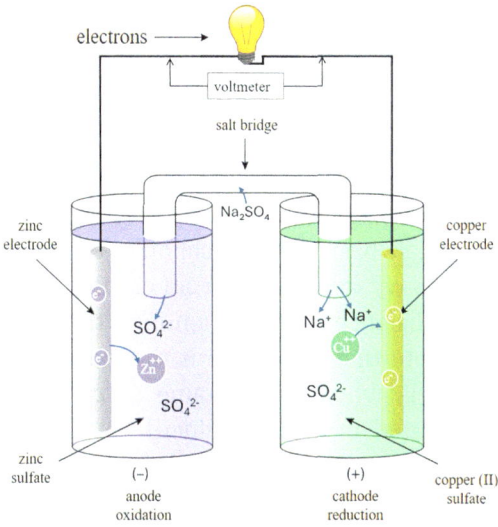

FIGURE 10.5 Daniell cell showing basic battery operation.

sulfuric acid, SO_4^{2-}, and either of the **cations** (positively charged ions) Zn^{+2} or Cu^{+2} from the Zn and Cu electrodes, respectively. This would be the state of a fresh, "charged," battery.

There are many possible variations of Daniell cells having combinations of electrode metal and solutions that can make this operate as a battery. What drives this battery is the difference in energy that it takes to strip two electrons off of Zn compared to that of Cu. It so happens that it is easier for a Zn atom to be **oxidized**, i.e., lose electrons, in this case to go to a Zn^{2+} state, than it is for a copper atom to go to a Cu^{2+} state. The technical way of saying this is that the **reduction potential** of Zn, which is −0.76 V, is less than that of Cu, which is +0.34 V (assuming 1 molar concentration). Put more simply, electrons want to leave Zn more than they want to leave Cu. As a result, more electrons will naturally build up on the Zn electrode and, in an effort to get away from each other (Coulomb repulsion), will be forced to travel as current from the Zn electrode to the Cu electrode. That is the same thing as saying that a **potential** (i.e., a voltage) develops across the two electrodes, and it is a battery. This electrical potential can drive a load, the lightbulb in Fig. 10.5, at some voltage set by the battery chemistry. The potential is the sum of the two reduction potentials, which in the case of the Daniell cell is 0.34 − (−0.76) = 1.1 V. (Other factors can affect the battery voltage, so this is not always exactly the measured voltage.) Other combinations of electrode materials have different relationships that result in different open-circuit voltages of batteries. This is why different battery chemistries result in different terminal voltages, such as 1.2 V, 1.5 V, etc. (Factoid for filing away: this is the same phenomenon that causes the built-in voltage of a semiconductor diode, discussed in Chapter 7.)

When a Zn atom loses its two electrons at the anode, it separates from the Zn rod and goes into solution as an ion. If the cell is open-circuited, by, say removing the lightbulb, the electrons would quickly build up and prevent more Zn from being oxidized and entering the solution. When no current flows, the battery chemistry stops. This phenomenon is called **charge neutrality**. Charge neutrality is an extremely important concept. Nature does not like charge to build up in a confined region. If that happens, Coulomb forces try to bring opposite charges together or push like charges apart, which tends to return the region to charge neutrality. As charges build up, potentials build up, making it harder for more charges to accumulate. Examples of high potentials due to loss of neutrality are shocks caused by static electricity in the winter, hair standing when dried after a shower, and a balloon that sticks to a wall after being rubbed on clothes. These are tiny versions of giant Van de

Graaff generators that can create potentials of up to 5 MV. In a battery, as soon as the electrodes charge up to the differences in the reduction potentials, then the charges will no longer be able to overcome that energy barrier, and charges stop accumulating, and ions stop going into the electrolyte solution. That difference in electrode potentials is the open-circuit voltage of the battery. Charge neutrality is the underlying physical basis for Kirchhoff's current law: like charges will always try to get away from each other until the resulting electric fields balance the forces amongst the charges.

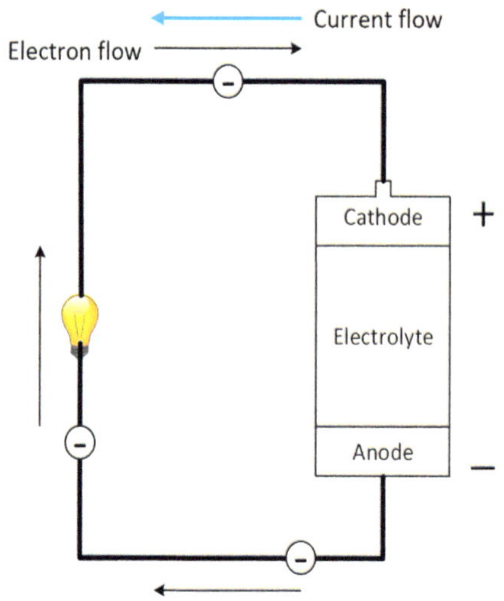

Note that the anode is the electrode at which *electrons leave* the cell, which is the same as *current* flowing *into* the cell as illustrated in (the simplified) Fig. 10.6. Current flows *out* of the device at its cathode, which is the same as saying that electrons enter at the cathode. The Zn electrode is the anode, i.e., the negative terminal, of the battery, where current flows in. As the electrons flow through the load, i.e., the lightbulbs in Figs. 10.5 and 10.6, they flow into the Cu electrode of the Daniell cell, the cathode, where current flows out.

Electrons leave the cell having come from the neutral Zn atoms in the solid piece of Zn that is the anode, causing some Zn to go into solution as Zn^{2+} ions, i.e., **oxidizing** them. At the cathode,

FIGURE 10.6 Graphic showing direction of electron flow relative to cathode and anode of a battery. Don't forget that current is defined to be positive in the opposite direction to electron flow. If you work only with positive ion flow, then that is a fine convention, but EEs work mostly with electron flow, so we have to keep that convention straight in our minds.

the electrons continue to flow into the solution by attaching onto, or **reducing**, the Cu ions that are already in solution, having been put there when the freshly charged cell was originally built. In this case, two electrons transfer to a Cu^{2+} ion, changing it from 2+ to zero charge. A sequence of charge exchange like that described here is called a **redox** reaction.

Note that a Cu atom will stick to the solid Cu electrode. (Just in case you are really thinking carefully about this, since electrons cannot move by themselves through a liquid, reduction will happen only for ions that are physically in contact with the Cu electrode, but not yet chemically bonded to it.) Hence, as the battery is used up, the Zn electrode erodes away by going into solution, and the Cu electrode gets thicker as the Cu^{2+} ions in the solution **plate** onto the Cu electrode. This is the same chemistry used for electroplating metals, which is common in the jewelry industry. Good, but less expensive, jewelry can have a thin layer of electroplated silver or gold on the surface. It looks good, since that thin layer is what we see, but it does not last a long time before it is rubbed off in constant use.

So far, we have ignored the electrolyte **salt bridge**. The salt bridge in this example is a thick goo (a paste) of sodium sulfate, Na_2SO_4, in the form of Na^+ and SO_4^{2-} ions that is necessary for the battery to operate. (Other electrolytes can be used, such as KNO_3 or $NaNO_3$.) The bridge is to the ions what a wire is to electrons. Every component in the battery maintains charge neutrality, meaning that it does not tolerate having an imbalance of positive and negative charges. To maintain charge neutrality, charge must flow around the loop, which also satisfies Kirchoff's current law. In a wire, as one electron enters one end, one electron leaves the other. For every Cu^{2+} that receives two electrons through the cathode, two electron's worth of charge have entered the Cu beaker and must be neutralized. This happens through the salt bridge by having two (positive) Na^+ ions enter the beaker to the right from the salt bridge, thus countering the extra two (negative) electrons. The same sort of thing must be happening simultaneously in the left beaker; as two electrons leave the Zn electrode through the wire to the lightbulb, two electrons worth of charge must enter the beaker, which happens by one SO_4^{2-} ion entering from the salt bridge. In short, charge neutrality is maintained in both beakers.

The battery-plus-lightbulb circuit is open-circuited without the salt bridge, i.e., as soon as the loop is broken, the battery will develop its open circuit voltage, currents both in the form of mobile ionic charges and electrons in the outer circuit won't flow any longer, and the chemical reaction will come to a stop at the voltage set by the differences in the reduction potentials, which in this case is 1.1 V.

One important aspect of the salt bridge is that it allows current (in the form of mobile ion charges) to pass through it, but *does not allow* the Zn and Cu electrolyte fluids to mix. Were the electrolyte fluids to mix, that would, in a sense, short-circuit the battery in such a way that all the energy would be used up inside the fluid, causing it to heat up, but not generating any electrical current outside the cell. The salt bridge is made of a gummy paste that does not allow the fluids to be exchanged between the cells, but is conductive enough for ions to move in and out at the ends.

What causes the end of a battery's life? Three things can be used up: the copper ions in the original copper nitrate bath, Zn on the Zn electrode, and Na_2SO_4 in the salt bridge. As any of these components decreases, there are fewer ions available to flow, so the ability to carry current decreases. This is the same thing as saying that the **internal resistance** (see Chapter 2) of the battery increases. At the same time, the EMF of the battery decreases (as captured in the **Nernst equation**, which relates battery potential to concentrations, and is not discussed here.) So, when either all of the Cu^{2+} reactants that started out in the cathode solution, or all of the material in the Zn anode electrode, or all of the ions in the salt bridge are used up, then the battery is depleted, or "dead," as we EEs like to call it.

Many batteries work on the basic principle of the Daniell cell. Because it would not be very convenient to walk around with tiny beakers of acid in your cell phones, this wet battery isn't very practical. However, when it comes to very high currents, the liquid electrolyte solution conducts charges very well, making it a good technology for the lead-acid batteries used to start cars with internal-combustion engines. These are large and heavy sealed containers of lead electrodes and sulfuric acid that can produce the hundreds of amps needed to turn a cold engine on winter

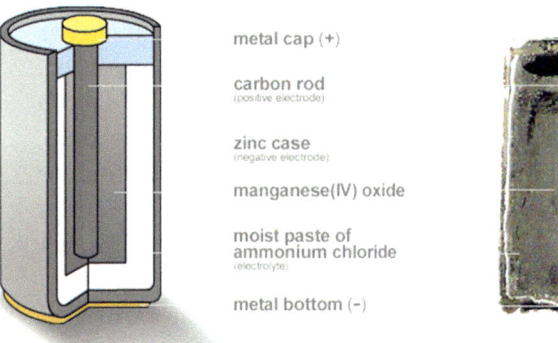

FIGURE 10.7 Internal construction of a "heavy duty" Zn-C battery. By Mcy jerry, CC BY 2.5, https://commons.wikimedia.org/w/index.php?curid=22406581.

days in order to get it started. It might occur to you to ask how a battery can have just one type of electrode metal, as in a "lead-acid" battery. That answer is that the lead at the electrodes is in two different forms, namely lead dioxide at the cathode and lead metal at the anode. The lead is porous with lots of cavities that increase the surface area, and is called 'sponge lead.' These two forms of lead have different reduction potentials, and six cells are placed in series to obtain an open-circuit voltage of about 12.6 V.

Typical batteries allow current to flow through pasty, damp layers, and don't contain actual fluid electrolytes, as in the Daniell cell and the lead-acid battery. Figure 10.7 shows a "heavy duty" Zn-C battery. Try to identify the various parts of the battery and relate it back to the Daniell cell.

The rechargeable lithium-ion batteries used in your cell phone, laptop and most electric cars operate by a similar principle, but with a few modifications. One distinguishing feature of Li-ion batteries is the concept of ion **intercalation** into the anode and cathode. Think of the electrodes as parking garages with energetically favorable parking spaces available for the Li atoms to 'park,' or intercalate, as shown in Fig. 10.8. The anode is carbon-based, and the lithium ions intercalate between sheets of graphite, which is carbon in the form of atomically thin layers. The cathode is typically some lithium oxide material, such as $LiCoO_2$. In a fully charged battery, as much Li as possible is intercalated in the anode, and upon discharge, each lithium atom gives up one electron to the circuit, becoming

Li⁺, and **drifts**, i.e., is pushed by an electric field, through the electrolyte to the cathode. Here it finds an energetically favorable place to intercalate inside the cathode and is then neutralized by one electron in the cathode. The situation is reversed during charging. In a Li-ion battery, the anode and cathode are separated by a porous separator that is saturated with liquid electrolyte. The porous separator permits ions to pass through, while providing a mechanical buffer between the anode and cathode to prevent a short circuit. The electrolyte is typically made of lithium salts such as $LiPF_6$ or $LiClO_4$ dissolved in solvents like ethylene carbonate and propylene carbonate.

Li-ion batteries are susceptible to catching fire. As such, the U.S. postal service has strict rules about shipping batteries. In 2013, several Boeing 787 Dreamliner jets caught fire in their first year of operation due to faulty battery operation. There have also been numerous recalls of laptop computers due to Li-ion battery fires. In late 2016, Samsung was forced to recall their very popular Galaxy Note 7 phones, which would sometimes catch fire due to battery failure. This failed product launch cost Samsung about $10B.

High-energy-density batteries that are less susceptible to burning is an intense area of research. Why do Li-ion batteries sometimes catch on fire? For example, a small filament, or dendrite, of metallic Li can form during many charge and discharge cycles – similar to the plating of copper onto the cathode of the wet battery discussed above. The short circuit can create enough heat for the battery to catch fire, including the flammable solvent. Research is ongoing to remove the flammable solvent from the electrolyte.

There are many potential replacements for Li-ion batteries. For a new battery technology to succeed, it should have some advantage in volume density, energy mass density, safety, charging time, reliability, number of possible recharging cycles, cost, rate of self-discharge (how long they stay charged), toxicity of materials, temperature range of operation, availability of materials, or ease of manufacturing. Clearly there is a long list of considerations in selecting a battery for a particular application, so any battery that has advantages in a few of these areas might be a candidate for insertion into a certain market. For example, high energy-volume density is important for miniaturized devices, where a smaller size makes for a better product, but not so important for grid storage, where a larger installation to house larger batteries may not be an issue.

Some (to name only a few) new battery technologies currently under development include: lithium-sulfur (high energy density, available materials, easy to manufacture); solid-state (high energy density

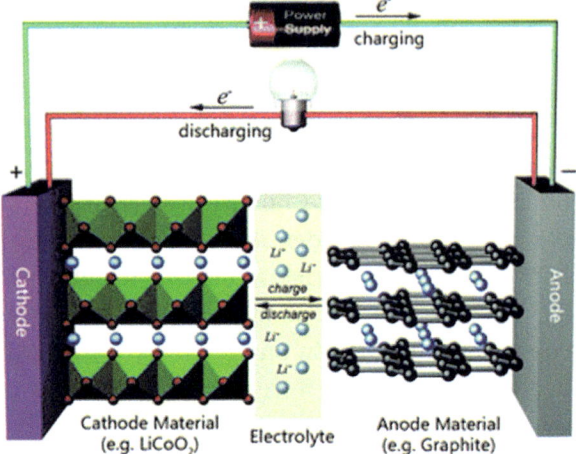

FIGURE 10.8 Graphic showing how a Li ion moves between cathode and anode during charge and discharge cycles. Reproduced with permission from *Chem. Commun.*, 2011, 47, 1384–1404.

Something to ponder: You've heard of 'conservation of energy,' which says that you don't get energy from just nothing. (You might have also heard of 'zero-point energy,' which we are ignoring here.) Yet, you take a bunch of stuff, throw it together, and you get energy from a battery. Where did that energy come from? Arguably, the big bang. Of course, all energy must be traceable back to the big bang, but this is a more-detailed look at the progression of that energy to the music that is made possible by a fresh alkaline battery: big bang – subatomic particles form – hydrogen and helium form – gravity condenses gasses – stars form – fusion within stars form some elements, supernovae form others – those elements condense into planets, most importantly (to us) Earth – we take those elements and make batteries from them. It is inherent in the quantum-mechanical atomic structure of these atoms formed in those stars that electrons are bound with a certain amount of energy to the nucleus, resulting in different reduction potentials for different electrode materials. When we let the electrons out of the battery, they flow energetically 'downhill,' powering your electronic devices.

and safe); sodium-ion (low temperatures and cheap); cobalt-free lithium-ion (available materials); zinc-based (cheap, available materials, low self-discharge rate); and iron-air (cheap to manufacture, available materials, many recharge cycles). Not stated explicitly is that every battery has some disadvantages, or that technology would displace Li-ion batteries and take over the entire battery space, which has not yet happened. Research is ongoing at a rapid pace, and changes are on the near horizon. As discussed above, solid-state batteries are a good candidate for electric cars because of their safety and high energy density.

One unique new battery technology in a class by itself is the 'flow battery.' They are unique in that the anode and cathode are fluids rather than solids. As the fluids flow through the battery they are discharged. "Recharging" the battery is as simple as filling the tank with fresh, charged fluids. The old fluid is taken away to be recharged. Or, the fluid can be recharged inside the battery, like any other rechargeable battery. Flow batteries are subject to the same list of possible advantages and disadvantages as other batteries. They are already in use in some grid-scale storage applications but have some way to go in reducing cost before they are widely adopted.

The bottom of this Tesla electric car is lined with batteries. This is called the 'skateboard' configuration. Prediction: internal combustion engines will be largely just a bad dream by about 2040 after existing car battery technology has been sufficiently improved or replaced by some better source of electrical energy. Photo: https://insideevs.com/news/450534/tesla-leaves-skateboard-design-new-structural-battery-packs/

10.6   Ideal Voltage Sources

What do we want from our batteries and, for that matter, our benchtop power supplies? We want them to give us a constant voltage across our load no matter how much current we want to pass through the load. For example, if you have a 1.5 V AA ("double A") battery and want to drive a load of 1.5 kohm, you would expect to drive 1 mA through the load without the battery voltage "sagging." That amount of current isn't very high, and any reasonable battery (D, C, AA, AAA) can do that. However, what if you want to drive a load having very *low* resistance, say the electric starter motor of a car, that requires a *lot* of current? Can you do this using 8 × 1.5 V alkaline AA batteries in series (12 V)? The starter motor uses the car's lead-acid battery to "turn over" the engine so it can begin

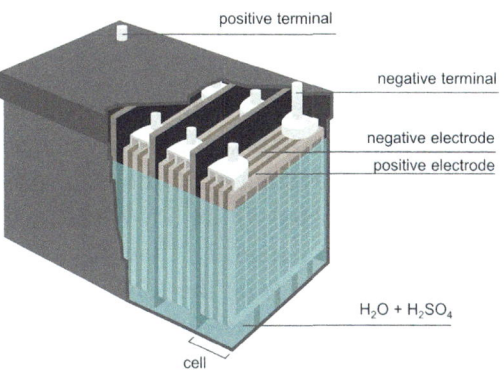

FIGURE 10.9 A '12 V' lead-acid car battery needs to generate very large currents. The fully-charged open-circuit voltage of a car battery is actually about 12.6 V. *Source*: https://commons.wikimedia.org/wiki/File:Scheme_of_a_lead-acid_battery.jpg.

the compression and firing sequence that starts the internal combustion engine, after which it runs on gasoline and uses an alternator to recharge the car's battery and run all of the electrical systems in the car. Car starter motors can temporarily draw well over 100 amps (and perhaps ten times that for a big diesel truck engine). For a "12 V" lead-acid battery like that shown in Fig. 10.9, there must physically be enough reactants in the battery to generate all of the current, and large metal conductors to carry that current without excessive resistance. This helps to explain why the weight of a lead-acid car battery is about 15 to 20 kg, compared to that of 8 AA alkaline batteries at only about 160 g. As a consequence of its small size, the AA battery simply doesn't have enough reactant in it to generate that much current, or the mechanical structure to carry it. An AA battery would either melt or explode if you somehow forced 100 A through it. (My money is on 'explode.')

We can express all of that with a very simple circuit concept called **internal resistance**. Here is a basic treatment of it, but you might want to review this in Chapter 2. We model the battery as an **ideal voltage source,** V_{in}, in series with an internal resistance, R_{in}. An *ideal* voltage source is one that can produce any current asked of it regardless of how low the load resistance is. (Note that a very low load resistance is called a "large load" because low resistances draw more current from the source, which is physically more challenging for the voltage source.) The circuit schematic is shown in Fig. 10.10.

Let's start the analysis by asking a very simple question: "What voltage will we read with an open circuit (no load, i.e., R_{load} is infinity) using a voltmeter?" (Don't forget that voltmeters have very high input resistances, which is shown here as infinite resistance.) Sometimes in circuit analysis a question is really as simple as it sounds, which is the case here. If the battery is not loaded, i.e., is open circuit, the current through the internal resistor is zero, so the voltmeter will read just the voltage of the ideal voltage source. That is the "open circuit voltage" or "no-load voltage" of the battery.

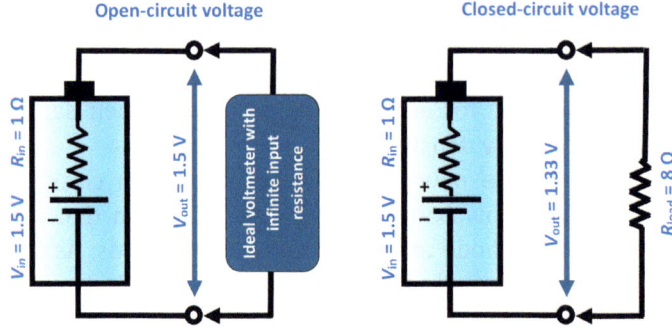

FIGURE 10.10 Circuit models of a battery as an ideal voltage source with internal resistance.

Now let's add a load, R_{load}, starting with a high resistance and working our way down, as shown in Fig. 10.10, right side. In this example, the internal resistance shown in Fig. 10.10 is 1 ohm. We keep the voltmeter at the terminals of the battery, which will now read the ideal source voltage minus the voltage drop on R_{in}. The more current that flows, the larger will be the voltage drop on the internal resistance, and the lower will be the voltage measured at the terminals.

You can use the very simple voltage-divider equation to analyze this circuit, where the voltage is divided between the internal resistance and the load resistance. The voltage read by the meter will be 1.5 V × $(R_{load}/(R_{load} + R_{in}))$. If R_{load} is much greater than R_{in}, then V_{out} will be nearly equal to V_{in}. Such is the case for the 1.5 kohm load with 1 mA flowing through it, as mentioned above. In that case, V_{out} would be 1.5 V × (1000/1001), which is nearly equal to 1.5 V. However, as shown in Fig. 10.10, suppose R_{load} is 8 ohms and R_{in} is 1 ohm. Then V_{out} would sag to 1.33 V. Now the current would be $I = V/R_{total} = V_{in}/9 = 167$ mA. So, you can see that for this battery (which is pretty well representative of a fully charged AAA alkaline battery), the voltage sags with a low-resistance (high) load and relatively high currents. As batteries wear out for the reasons discussed above, their internal resistance increases, and their voltage under load decreases. For this reason, using a voltmeter to test a battery is a blunt instrument; it is best to use a dedicated battery tester that applies an appropriate load for the type of battery, and is calibrated to tell what the charge state of the battery is, as discussed in the sidebar.

Practical tip about battery Testers: You have learned so far that the voltage of a battery is set largely by its internal chemistry. The voltage can drop as the battery is depleted, but this is not the largest effect on a spent battery. Instead, battery testers monitor the internal resistance by measuring current flow under load. As the internal resistance increases, the load current will drop. Voltmeters can only measure the open-circuit voltage, which changes noticeably only when the battery is close to being depleted. So, a voltmeter is a rather blunt instrument for monitoring battery life.

If you short-circuit the battery terminals (not advised), then you could get a maximum current of about 1.5 A (for this hypothetical battery), with a terminal voltage equal to that across a resistance of just the wire of the short circuit (a low voltage, but not exactly zero); the voltmeter would read approximately 0 V. Short-circuiting the battery would lower its lifetime dramatically, which is why in this week's lab we use a fast circuit to do it for us. We want to minimize the time spent in short-circuit mode.

Think about this: what is the most dangerous thing about a car battery like that shown in Fig. 10.9? Is it that (a) you could be electrocuted by it, (b) there could be an explosion under short circuit or during charging, or (c) it could drop on your foot? Let's think about (a) getting electrocuted by it. The reason we are generally not electrocuted by household batteries is that our skin and bodies have too much resistance to allow much current to flow through our hearts, so even a 9 V battery, which can produce more than 6 A of short-circuit current, doesn't present much of a danger to us. (It is generally regarded that 100 mA to 200 mA directly through the heart is fatal. I have never heard

Safety: Shorting the terminals of a car battery with a wrench is a really bad thing to do due to the electrolyte rapidly boiling from the heat, and the battery rupturing. (http://en.wikipedia.org/wiki/Automotive_battery#Exploding_batteries) This picture shows what happens if you mix up the polarity of jumper cables. It becomes a short circuit to the 24 V of two car batteries in series. That is why it is useful to repeat over and over again the mantra, "plus-to-plus and minus-to-minus, plus-to-plus and minus-to-minus..."

http://tywkiwdbi.blogspot.com/2011/10/next-time-you-jumpstart-your-car.html

of anyone being injured by a single, 9 V battery.) A car battery is nominally 12 V, which I think we can agree isn't much more than 9 V. So, if we were to touch both terminals of a car battery with our hands, would we be at much risk of electrocution? Hint: don't be fooled simply because the battery is very large and can produce hundreds of amps. In fact, no, you would not be electrocuted since the current from a battery is set not only by its internal resistance but also by the load resistance, i.e., our skin. We people are a small load (high resistance), so we would not experience much more current from a car battery than from a 9 V battery. If we were short circuits, i.e., the current went easily through our body and our heart, thus posing a real danger to us, both the 9 V battery and the car battery have low enough internal resistance and would be capable of producing enough current to hurt us.

So, the answer to the question above is (b) or (c), but not necessarily (a). Therefore, be careful when you jumpstart a car from another car (see sidebar), and don't drop a car battery on your foot. Accidentally touching the terminals of a car battery with your hands is not the biggest concern when working under the hood. Now, shorting the terminals with a wrench is another matter – that *has* resulted in serious incidents.

10.7 Electrical Power Supplies

Now that we have covered the topic of batteries, we turn to more practical sources of power for your lab bench, as well as those used in any number of electrical systems that are plugged into the wall. Many electrical appliances work off of AC voltage and current, e.g., the washing machine motor, air conditioner compressor, electric heater, toaster, food processor, furnace blower, etc. Others, such as the digital controller on your washing machine, clock radio, TV set, stereo amplifier, computer,

USB anything, etc., work off of DC voltage and current. Of course, we don't use batteries to operate all of those. If we did, the world would be a giant landfill of spent batteries.

Most electronic devices use some sort of DC power supply, either embedded somewhere inside, or connected from the outside. The challenge of a power supply is to convert alternating current, AC, to direct current, DC, in such a way that it can supply the required current without the rated voltage sagging, and without adding any unwanted noise.

To be clear, we can think loosely of AC as what comes out of a wall outlet (you might wish to review Chapter 3) – a sinusoid at 60 Hz, and 120 V_{RMS}, but technically it is any waveform for which current flows in both directions at various times. What we mean by "DC" is a little less clear. Technically, it means that the current flows in only one direction, no matter what that waveform happens to look like. It could be a square, triangle, or even sine wave, or any arbitrary waveform, as long as current only goes one way, and never flows in the opposite direction. So, if it charged a battery in one direction, it might spend some time *not charging* the battery, but would never spend any time *discharging* the battery. The other sense of DC is a *constant* value of voltage or current, with no time variation associated with it, much like the output voltage of a battery. We will use both meanings throughout this discussion, and leave it to you to understand which one is intended in each case.

We will focus on creating a "clean," i.e. constant with no variation, DC voltage source from what starts out as the purely-AC wall voltage. Many electrical systems use low DC voltages, such as 5 V, 12 V, or 15 V, so it is most common to first drive a step-down transformer from the line voltage to decrease the AC voltage, and then put that through a diode in series with the load, as shown in Fig. 10.11. Without the diode, AC current would normally flow into a load equally in both directions. However, the series diode allows the current to flow

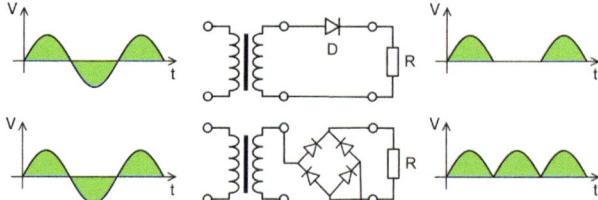

FIGURE 10.11 Unfiltered half-wave and full-wave rectifiers allow current to flow in only one direction. (Top) Half-wave rectifier using a simple diode. *Source*: https://en.wikipedia.org/wiki/Rectifier#/media/File:Halfwave.rectifier.en.svg. (Bottom) Full-wave rectifier using a diode bridge.

nearly unchanged one direction and blocks the current from flowing in the other direction.

The output waveform shown in Fig. 10.11 (top) is approximately a sine wave with the negative parts missing, i.e., set to zero. A silicon diode drops about 0.6 volts to 0.7 volts when it conducts current in the forward direction, as discussed in Chapter 7. Therefore, the half-wave voltage after rectification is about 0.7 V lower in voltage. This circuit is called a **half-wave rectifier**. If our goal were to merely charge a battery, the half-wave rectifier would work great – simple and cheap. However, if we were using this to power a radio or stereo, it would add horrible noise to the sound, as if in some sense the radio sound were turning on and off 60 times per second, so we need a better design for a DC power supply.

The next step in developing a clean DC voltage source would be to try to fill in those dead times in the waveform that were stopped by the diode. The simplest circuit that does this is called a **full-wave rectifier**, or **bridge rectifier**, and is shown at the bottom of Fig. 10.11. This is a bit more complicated, but with a little effort you can trace the flow of current through one diode, the load, and then another diode for each half wave of the bidirectional sine-wave source. Turn your attention now to Fig. 10.12. Ignore the capacitor for now. For each half wave, the current flows through a different pair of diodes, but always flows the same way through the load, as follows: When the source (again, the output of the secondary of a transformer) is positive (the top of the bridge is positive w.r.t. the bottom of the bridge), the current flows through D1, load, and D2 back to the source, as indicated by the blue arrows. When the source is negative (the top of the bridge is negative w.r.t. the bottom of the bridge), the current flows through D3, load, and D4 back to the source, as indicated by the gold arrows. For each half wave, two diodes are forward-biased and conduct current and the other two are reverse biased and block current. Importantly, the current goes the same way through the load on each half cycle.

Note that now the voltage drops are due to two diodes in series, so the total voltage drop due to the diodes is something like 1.2 V to 1.4 V (the voltage drop depends slightly on the current, which depends on the load resistance). The waveforms shown in both Figs. 10.11 and 10.12 are idealized and do not reflect this voltage drop. Or, you can consider them to be very large output voltages for which 0.7 V and 1.4 V, respectively, are negligible and are insignificant on the plots. You will see those voltage drops clearly during this week's lab activities.

Now we have a DC waveform with less dead time, but a radio would still sound really bad - there would still be a lot of hum on the audio output. We really need to smoothen it so it more-closely approximates a constant DC voltage, again like a battery. In both full- and half-wave rectified cases, we can think of it *in the frequency domain* as a DC component plus a bunch of AC components (as discussed in Chapter 4). The rectified AC waveform is always positive, so it has a nonzero

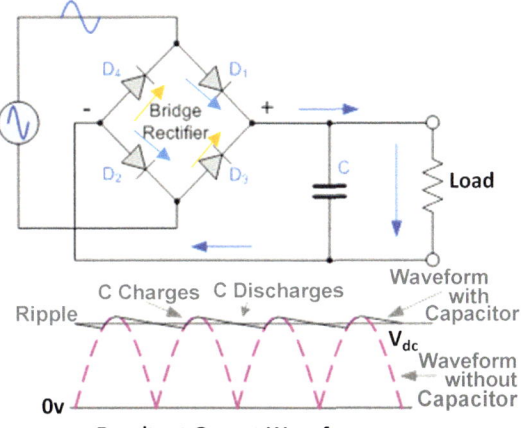

Resultant Output Waveform

FIGURE 10.12 A bridge rectifier and a capacitor can be used to convert an AC power supply into a DC voltage. (Top) Circuit diagram of full-wave bridge rectifier and (bottom) output voltage of both unfiltered and filtered circuits. Note that the unfiltered output voltage shown does not include the diode voltage drops. *Source*: http://electronics.stackexchange.com/questions/73863/cap-value-for-full-wave-rectifier-circuit.

average value. Recall from Chapter 2 that a sinusoid can be described by its peak, peak-to-peak, average or RMS values. In this case, we care about the average value, which for an unrectified sine or cosine is zero. So, if a DC voltage is added to a sine wave, the sine wave is raised up by the DC voltage, and the average voltage is that of the DC component alone since the average of the AC component alone is zero. To recover only the DC value, i.e., to make the waveform smooth and flat, we can filter out (most of) the AC components of the voltage waveform with a capacitor, as shown in Fig. 10.12, leaving (mostly) the DC voltage behind. This will look like a constant voltage with a little bit of waviness.

Thinking back to Chapter 4, the capacitor in parallel with the load is a low-pass filter, LPF. The capacitor is much like an open circuit to low frequencies, so all of the DC voltage input to the LPF appears across, or is transferred to, the load. High-frequency AC components are largely short-circuited through the capacitor to ground, and therefore do not appear much across the load. Also, larger values of C cause lower frequencies to be shorted to ground. So, the larger C is, the fewer AC frequency components appear across the load. Later you will see that the quality of the filter does, in fact, depend on both frequency and capacitance. For these reasons, the capacitor is called the "filter capacitor" in the power supply circuit. With the filter capacitor in the circuit, the radio will sound good with the hum removed.

Conversely, and without loss of correctness, we can think of the filter capacitor in the following way in the time domain. Refer to the solid line in the waveform that is the bottom of Fig. 10.12. The filter capacitor stores charge as the voltage is increasing, and releases it, like a short-lived battery, when the voltage from the AC source (transformer) decreases. Just as you learned about RC time constants in Chapter 4, depending on the load resistance, R_L, and the value of the capacitance, C, once the capacitor is charged it can fill in the voltage drops during each half wave. Recall that the larger the time constant, R_LC, the longer it takes to discharge the capacitor. A long RC time constant results from either a very large R_L (small load), or a very large C. If the load is a high resistance, it doesn't draw much current, so the capacitor can have a small value. However, if the load is a low resistance, meaning that it draws a lot of current (for example from a very powerful stereo amplifier like the Pioneer SX-1980 in the sidebar), the filter capacitor has to be very large.

You might ask, "How can two voltages compete to drive the load at the same time, namely the output of the transformer and voltage on the capacitor? The answer is that the diode bridge cannot pass current from the capacitor back to the transformer, as both diodes would be reverse biased. By design, the transformer has a much lower internal resistance, R_{trans}, than the load, so $R_{trans}C \ll R_L C$. As the transformer voltage increases, it relatively quickly charges the capacitor while simultaneously driving the load. This implies that during charging the voltage across the capacitor/load exactly follows the output of the transformer (minus the diode drops, of course). Then, as the transformer voltage drops below that on the capacitor, the capacitor takes over and drives the load. All of the current from the capacitor goes through the load because none of it can go backwards through the diodes.

The ripple on the DC voltage shown in Fig. 10.12 is called ....wait for it.... **ripple**. If you don't want to hear any hum on your music from your stereo, the power supply must have very low ripple, which is the job of the filter capacitor. This is one reason that vintage stereo amplifiers are often "recapped" (the old capacitors are replaced) to bring the sound back to its original, noise-free condition. In general, the **ripple factor**, γ (gamma), of a waveform is a measure of the quality of the output

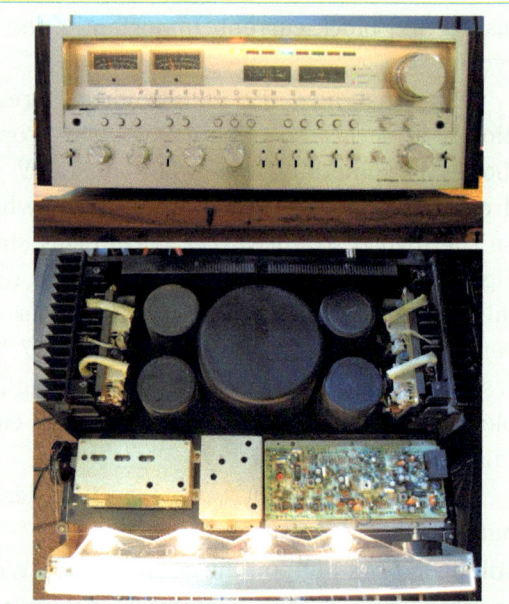

Concept reinforcement: **Top-front panel and bottom-inside views of the much-coveted Pioneer SX-1980 stereo receiver from the 1980s. At 270 W per channel, this was among the most-powerful stereo amplifier/receivers ever sold. The bottom picture shows the large round step-down transformer in the center. That is surrounded by four large filter capacitors required to keep the ripple to a minimum when driving low-resistance speakers at high current. Zener diodes (discussed below) are used as voltage regulators at 35 V in this amplifier.**

http://mattsvintageaudiorepair.blogspot.com/2012/05/pioneer-sx-1980-score-and-restoration.html

of a power supply, and is the ratio of the RMS value of the combined AC components of the power supply voltage divided by the DC component of the waveform:

$$\gamma = V_{ACrms}/V_{DC}. \tag{10.1}$$

Again, you can think of ripple as the AC component "riding" on the average, or DC, voltage of the waveform. The Activities section explains in detail how to obtain the ripple factor using the DMM and the oscilloscope.

The ripple factor of a full-wave rectified power supply is, as a percentage, approximated as:

$$\gamma = \frac{1}{fCR_L 4\sqrt{3}} \times 100\%. \tag{10.2}$$

Lower ripple factor is better. Figure 10.12 shows that the shape of the ripple is complicated. Again, γ is the ratio of the RMS value of the ripple (which isn't straightforward to calculate since it isn't a pure sinusoid) to the average value of the voltage, i.e., the DC component of the full waveform (which is also not simple to calculate). In the frequency domain you would need to find the RMS value of all of the frequency components of the ripple and divide by the DC component. However, a rough estimate of the ripple factor could be made by simply assuming that the AC component is the RMS value of a sine wave having the amplitude of the ripple, and that the DC value is approximately in the middle of the ripple.

All of that said, it is relatively easy to actually measure the ripple factor of a rectified and filtered AC waveform. Here is how to use a DMM to measure ripple factor. DMMs come in two flavors: average responding and true-RMS. True-RMS digital multimeters are discussed in Chapter 2. The true-RMS type will read the actual RMS value of the waveform regardless of its shape. Therefore, using a true-RMS DMM in AC mode on the ripply DC voltage would tell you the RMS value of the ripple. (In AC mode, the DMM puts a capacitor in series with the measurement, and that capacitor blocks the DC component and lets only the AC component through, as long as the frequency is high enough based on the meter's specifications.) Then, the DC reading would tell you the average, or DC, component alone. Using these two measurements, a value for the ripple factor can be obtained by simply taking the ratio of the two numbers.

Thinking in the time domain, it makes sense that as the frequency goes up, the capacitor needs to supply current to the load for less time between charging cycles, so there is less time that the capacitor needs to discharge, leading to a lower ripple factor; also, higher R_L or higher C creates a longer time constant, so again, the voltage drops less during each power cycle, again giving less ripple. As discussed above, in frequency-domain terms, we can say that the impedance of the filter capacitor goes down as the frequency goes up, so the ripple frequency component is more efficiently shorted to ground, but looks like an open circuit to the DC lowest-frequency components.

A further improvement to obtaining pure DC from AC is to add a **voltage regulator**, which actively maintains a constant DC voltage in spite of ripple on its input or changing load conditions. This gets us closer to an ideal voltage source. Voltage regulators are generally integrated circuits that internally measure the output voltage and change their internal parameters to keep the same output voltage. This process of maintaining a controlled output in the face of a changing input is called **feedback**, which is itself an important component of the branch of engineering known as **control systems**.

A simple voltage regulator can be made from a kind of diode called a **Zener diode**. Figure 10.13 shows the *I–V* curve of a Zener diode. As discussed in Chapter 7, a Zener is a conventional diode that has a very steep *I–V* curve in its reverse direction where it "breaks down." (This is a semiconductor phenomenon that is beyond this course.) In the figure, the intersection of the extrapolated *I–V* curve on the voltage axis is labelled "V_Z." In order not to make things too complicated, we will refer to the actual voltage drop across the Zener diode for any given current in breakdown as V_Z as well, where the distinction can be inferred by the context.

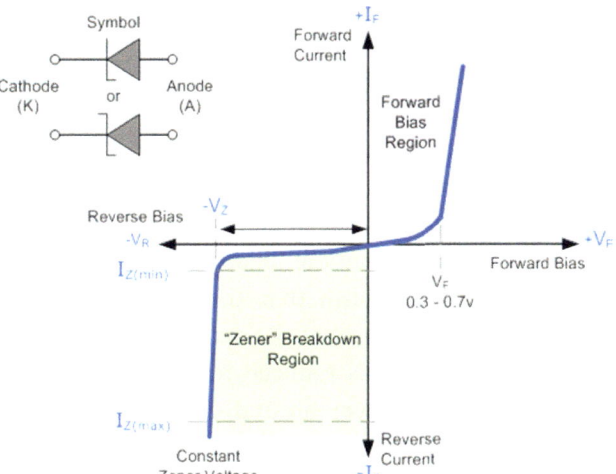

FIGURE 10.13 *I–V* curve of a Zener diode. The curve shows that Zener diodes will "break down" if a large enough reverse voltage is applied. Note that as the reverse current changes over a large range, the reverse voltage hardly changes, and can be considered to be constant at V_Z. The voltage on a load in parallel with the Zener diode will be fixed at V_Z. That is the source of the voltage regulation. *Source:* http://www.electronics-tutorials.ws/diode/diode_7.html.

Recall that a diode is not damaged by reverse-bias breakdown, unless the current gets so large that it is destroyed by the heat generated inside of it. At any moment, the power dissipated by the Zener diode is

$$P = IV = IV_Z, \tag{10.3}$$

since in breakdown the voltage is nearly constant at V_Z in breakdown. As long as the power (heat) dissipated by the diode is removed or minimized, it can operate along the very steep vertical part of the *I–V* curve. So, do not confuse "breakdown" with "destruction."

Figure 10.14 shows a simple full-wave rectified power supply with a Zener diode as a voltage regulator. Notice that the Zener diode is in the circuit "backwards," meaning that it is intended for current to flow through it in the reverse direction. Refer back to Fig. 10.13 showing the steep reverse breakdown region of a Zener diode. Since V_Z is in parallel with the load, the load voltage is V_Z, which depends only slightly on the current through the Zener, so the output voltage of the power supply is regulated by the Zener. As a circuit designer, you are limited in your choice of power supply voltages to the available values of V_Z.

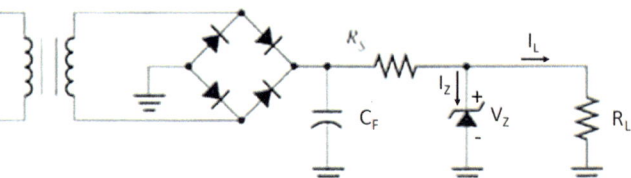

FIGURE 10.14 Schematic of simple Zener-regulated power supply. The current and voltage through the Zener diode must be exactly that shown in the I-V curve of the Zener diode, shown in Figure 10.13. The Zener can take a lot of current through it without changing V_Z by much, so the output voltage of the power supply is regulated by V_Z.

The operating principle of the circuit is as follows: The output voltage from the diode bridge goes to a filter capacitor before going on to the rest of the circuit. The filter capacitor does the first stage of regulation exactly as discussed above. There will be ripple at the capacitor, but that ripple is further smoothened by the Zener regulation that comes afterwards.

If the voltage at the output of the capacitor is greater than the breakdown voltage of the Zener (as it must be for all of this to work correctly), then the Zener diode will **clamp** its voltage to V_Z, and the excess voltage will be dropped by the series resistor, R_S. Without R_S, the Zener diode and load would have to be the same as the capacitor voltage, and the Zener diode could not provide regulation, as described next.

There must be enough current for the Zener diode to remain well in its breakdown region so that it shunts current from the load through itself in order to maintain a voltage drop of approximately V_Z. In order to accomplish this, the bridge voltage, range of load resistances and R_S must be chosen properly, which is part of the overall circuit design. By doing this, the ripple of the power supply after the filter capacitor is reduced to smaller variations along the voltage axis of the diode curve as the operating point moves up and down along the Zener breakdown *I–V* curve. If R_L changes, as loads have a habit of doing, V_Z stays more or less constant at the nominal value of V_Z as some of the current that goes through R_S is shunted through the Zener. Pretty neat, huh? Circuits can be fun.

10.8   Lab 10 Activities

1. **Measuring battery V_{OC}.** Measure (and record, of course) the open-circuit voltage, V_{OC}, for each of the batteries provided to you. It is easiest to do this with the DMM. Note that V_{OC} is a poor substitute for knowing the health of a battery as it drops very little even for a moderately discharged battery. You can tell if a battery is charged or discharged only under load, which is what you are doing today.

2. **Getting to know the short-circuit timer.** In the next several steps, you will create a short-circuit as a load to your batteries. That is not a normal thing to do, and is almost always advised against. In these steps you will do them to get a value for the internal resistance of a fresh battery. Small batteries do not present a danger of exploding, especially for being only briefly

shorted. The battery will dissipate at most a few watts, say, $P \approx 1.5 \text{ V} \times 1 \text{ A} = 1.5 \text{ W}$, while shorted and deplete very quickly, so it will just get warm if left shorted. A 9-volt battery can get fairly warm to hot if left shorted for any moderate length of time, since $P \approx 9 \text{ V} \times 6 \text{ A} = 54 \text{ W}$. That is a lot for a small package.

Before you begin the next set of short-circuit current experiments, be aware that you are using a load having on the order of around or less than a mere 1 ohm, so you must be careful to account for all of the resistances in your circuit. Start by using clip leads from the DMM. Short the leads and measure the resistance. Now, press **REL** to zero out the display so that the resistance that you now read will be what is attached to the leads, and the resistance of the leads themselves is subtracted out. Note: if you leave the resistance mode, you will have to repeat the REL function.

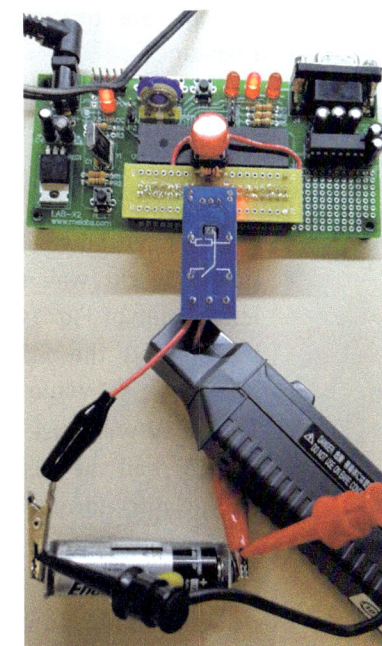

FIGURE 10.15 (Step 2) Short-circuit tester with leads and battery attached.

Figure 10.15 shows the test circuit that closes the contacts of a relay when the red button is pushed. When the power is off, the relay is in the open position. A relay in which unpowered contacts are open-circuited is called "normally open." Were the relay closed, or shorted, when unpowered, it would be a "normally closed" relay. In Fig. 10.16, the single-pole double-throw (SPDT) relay has three terminals. When not powered, the CA connection is normally closed and the CB connection is normally open. When it is energized, i.e., current flows through the coil, the switch moves from CA to CB so that CB goes from normally open to shorted, and CA goes from being normally closed when not energized to an open circuit when energized. Discuss why this is called a "single-pole double-throw" relay.

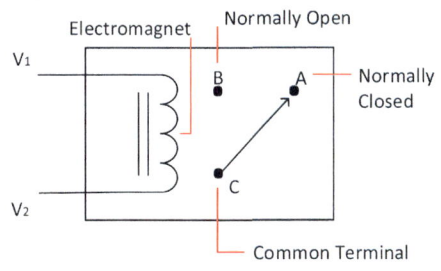

FIGURE 10.16 (Step 2) Schematic of relay.

3. **Measuring series resistance of the setup.** You will need to take into account the resistance of the relay and its external leads in determining the internal resistance of the battery. Also, whenever metal touches metal, as in the contacts, there is an opportunity to introduce resistance due to oxides and other contaminants. These external resistances are in series with the battery's internal resistance, as shown in Fig. 10.17, so they must be subtracted from the total resistance that you get in the next steps. To do this, you will in the next steps connect the leads of the DMM so that you measure the resistance of the wires, including the alligator clips and the relay contacts.

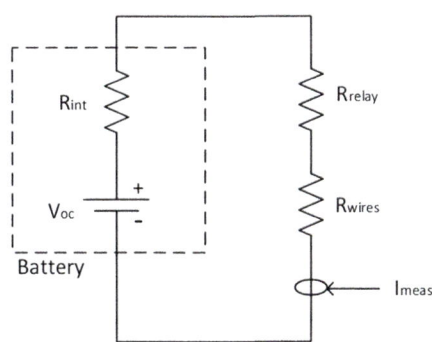

FIGURE 10.17 (Step 3) Equivalent circuit of the battery and relay showing the internal resistance of the battery and total resistance of the relay and wires, lumped in with the relay.

Do the following procedure in order to close the relay for 10 s, which will allow you to take a resistance measurement of the closed relay plus leads.

a. Attach leads to the Fluke DMM and set it to read resistance.

b. Short the leads and press **REL** as you did above to subtract out the resistance of the leads from the DMM.

c. Hook the ohmmeter leads to the tester clips.

d. Connect the power to the current tester using the power brick (wall wart), as shown in the upper left corner of Fig. 10.18. Press and hold the **Reset** button and the **SW1** button at the same time. Release **Reset** while continuing to hold down **SW1**.

FIGURE 10.18 (Step 3) Short-circuit tester showing all the switches. SW1 is at the top center. Reset is at lower left.

e. The relay will close the contacts for 10 s allowing you to read the resistance. After 10 s, the device will go into normal operation, just as if you had not initiated the contact test. You will use this resistance in Step 5.

Make a note of the resistance that you measured. Do you consider that to be a small or large resistance? Based on what?

Be aware that the resistance of the relay and its wires is not exactly zero – it is in fact about 0.05 ohms. (is that close to what you measured?) When you calculate the internal resistance of the batteries, it will include the relay resistance. You will need to subtract the relay resistance to get a better estimate of the internal resistance.

4. **Using a conventional battery tester.** During the actual measurement, you will short the battery for only 100 ms at a time, which is okay, but repetition will eventually deplete the battery. In this experiment you will have the battery connected to the relay. The relay is normally open, so you will not drain the battery when you are not doing the measurement. Nevertheless, to be safe, *disconnect the battery from the relay unless you are actually measuring the short-circuit current.*

Use the battery tester to test the health of each of your batteries. Battery testers apply a load and measure current, so use it sparingly. Use your intuition to decide how to operate it. If any batteries test less than "Good," ask your TA for a fresh battery. Note that battery testers are not precision instruments, so batteries that read about the same amount of 'good' may give significantly different short-circuit currents.

In the next step, you will directly measure the short-circuit current of all the batteries. Before you begin, discuss with your lab partner how much current you guess that each battery will actually be able to provide over the test duration of 100 ms. As you do the measurements, compare your guesses to the measured short-circuit current.

5. **Measuring the short-circuit current.** To measure the internal resistance of the battery, R_{in}, (see Figs. 10.17 and 10.19) you will use the timer provided to perform a pulsed measurement of the short-circuit current, I_{SC}. By pressing the large red button, the circuit controls a relay that short circuits the two small alligator clips for 100 ms. (In this circuit, a microprocessor is programmed to perform the timing and provide the voltage to the relay.) During that time you can measure the short-circuit current, I_{SC}, directly using the current probe and the oscilloscope.

Find and attach the 9 V battery clip to the 9 V battery. To use the circuit, clip the alligator clips of the timer circuit to the battery holder's leads; pressing the red button closes the relay contacts, and the battery will be briefly short-circuited through the relay contacts, allowing the most possible current to flow. **LED1** will briefly light during the measurement, and then **LED2** will light for a second or so telling you that you cannot press the button immediately or hold it down. We'll call that time the 'lock-out time.'

Attach the current probe to CH4. Degauss your current probe as you have done in previous labs. This is explained in Chapter 3.

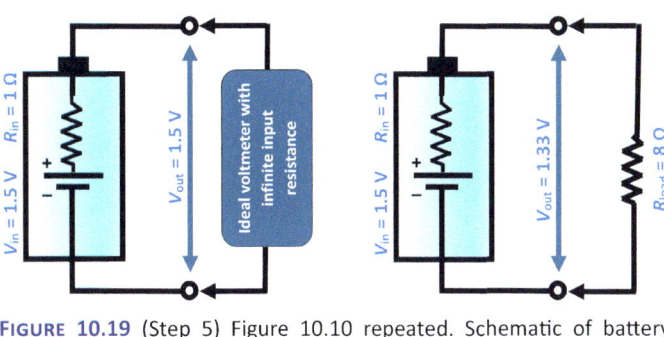

FIGURE 10.19 (Step 5) Figure 10.10 repeated. Schematic of battery showing internal ideal power supply and internal resistance. On the left, the load is the nearly open circuit of a voltmeter or scope. On the right shows a practical load on the battery. Your load during the short-circuit test will be some small fraction of 1 ohm that you measured in Step 3.

Set up the current probe so that the arrow points away from the positive terminal of the battery. Set the scale to 5 A/div to get started (1 A/div for coin cell). It is necessary to have the full waveform on the screen for the trigger function to work. You can change it later to help you do the current measurement.

In order to measure the 100 ms current pulse, you need to set up the trigger to capture that event. Press **Trigger Setup**. Select CH4. Set **Level** to 300 mA by either typing it in or using the level knob. Choose the trigger event to occur on the positive slope. Choose **Single**, and set the time base to 20 ms/div. Move the horizontal position knob to set the trace to start on the left half of the screen. Press the red button. You will get a pulse to show up on the screen, but then you need to adjust the

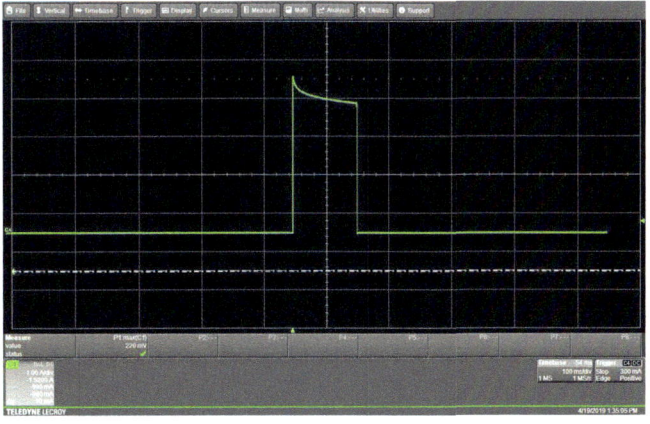

FIGURE 10.20 (Step 5) Current waveform from battery.

vertical gain to get the entire pulse amplitude onto the screen in order to do your measurement. Pressing **Single** again resets the trigger. If you do not get a pulse, you might suspect that you have a polarity wrong. If the pulse goes negative, it will not trigger the scope. Compare your waveform to that of Fig. 10.20.

Repeat for all of the batteries. You will have to make adjustments for the coin cell, and you may get the polarities reversed at first. Also, for the coin cell, try pressing **Noise Filter (ERes)** and choosing 2 bits. This averages point-by-point rather than averaging scans and will give a cleaner waveform. *When finished with the 9V battery measurement, remove the clip so that the leads do not inadvertently short out the battery.*

How did the data compare to your guesses? Based on Fig. 10.17 and your measurement of I_{SC} and V_{OC}, what is the internal resistance, R_{in}, of all of the batteries? Don't forget to subtract the relay resistance. For your lab report, make up a table that has each battery and its V_{OC}, I_{SC} and R_{in}.

Note: One of the biggest time sinks in the lab is to make mistakes while breadboarding your circuits. For the next experiments, turn on the light over your work area. Go slowly. Confirm each step with your lab partner. Concentrate on not making any mistakes. Use extra wires clipped to the scope probes rather than clipping the probes directly to your circuit components. That will be less cluttered and prevent the probe from pulling out your components. Turn out the light if there is interference on the scope.

6. **Setting up the AC input to your DC power supply.** In this part of today's lab you will build simple power supplies using half- and full-wave rectifiers. As you read in this chapter, the filtered rectifier circuits use a capacitor in parallel with the load. The capacitor charges during the part of the cycle when the input sine wave is near its peak, and it discharges during the rest of the cycle, giving rise to the ripple. As you will see, there is considerable current flow into the capacitor during its charging – so much so that providing that current is challenging to the equipment we have on hand. In order to get the most current that we can, we will use the three-phase generator (3PG) to source the current into a step-up transformer. We can say that the internal, or output, resistance of the 3PG is much smaller than that of any of our other available sources.

FIGURE 10.21 (Step 6) Transformer used to step up the voltage by 2X.

Power up the 3PG with +/– 16 V making sure to get the polarities and ground correct as you did in Chapter 3, the op-amp supply voltages in Chapter 7, and the demonstrator board of Chapter 9. SET THE CURRENT LIMIT ON EACH CHANNEL OF THE DC POWER SUPPLY TO 100 mA. Connect the output of the waveform generator to the input of the 3PG. Use a BNC-to-banana-plug to measure the output of CH1 of the 3PG making sure to put the grounded plug into the white jack, which is also earth ground. Start with a waveform generator output (input to 3PG) of 2.5 V_{pp} and using your scope, increase that value until the output of CH1 of the 3PG is 9 V_p, or 18 V_{pp}. You may have to make adjustments to the amplitude as you did in Lab 3. REMOVE THE PROBE FROM THE GENERATOR since that could later cause a grounding problem.

In order to boost the voltage, use the transformer in reverse, switching the conventional secondary with the primary as a step-up transformer. To do this, put the output from the 3PG across the (normally) "secondary" winding of the transformer, i.e., jacks X6 and X10, and take the output from (normally) "primary" jacks H2 and H4 (Fig. 10.21). You may remember that this provides step-up voltage of approximately two times the input voltage. Using the scope probe that you just removed, make a final adjustment of the 3PG output to get 36 V_{pp} at the output of the transformer (H2 to H4). You get the peak voltage of 18 V_p from the transformer because you lose half due to the conversion from 18 V_{pp} to 9 V_p, but get a factor of 2 back due to the step-up transformer to 18 V_p. We want that relatively high peak voltage to minimize the relative effects of the large LED diode drop that you will see later. We don't want it to be any higher than what is specified here or later the diode will go into reverse breakdown. REMOVE THE SCOPE PROBE FROM THE SECONDARY OF THE TRANSFORMER AND THE SCOPE again to avoid ground issues after you build your circuit.

7. **Measuring the turn-on voltage of a diode.** Here you will learn how to test a diode using the diode-test function of a DMM. See Figs. 10.22 and 10.23. Figure 10.23 shows the anode and the cathode of a diode. *Usually* the anode (positive) leg of the LED is longer than the cathode (negative) leg, but not always. Be sure you check this for *your* diode in this step.

Clip the diode in either direction to the leads plugged into the two upper right ports labeled with the diode symbol and the LO jack beneath. Press Shift and then the button with the diode symbol. Testing the diode in both directions, you will read two things with a good diode. In forward bias, with the anode (top of diode symbol in Fig. 10.22) connected to the red jack, it will display its turn-on voltage at 1 mA, but keep in mind that the diode will drop slightly more forward voltage with higher forward current. In reverse-bias, with the cathode (terminal with the line in Fig. 10.22, bottom) connected to the red jack, the meter will display **OL** for "overload." From this you learn three things: (a) the diode is in good operating condition, (b) which leads are the anode (+ terminal) and cathode (– terminal), and (c) the turn-on voltage. When in doubt today, you can quickly and easily test your diodes. One leg of the diode is longer. Is it the anode or cathode? What is the turn-on voltage of the LEDs that you are using today? Make a note of that since that voltage will show up in all of the rectifier circuits.

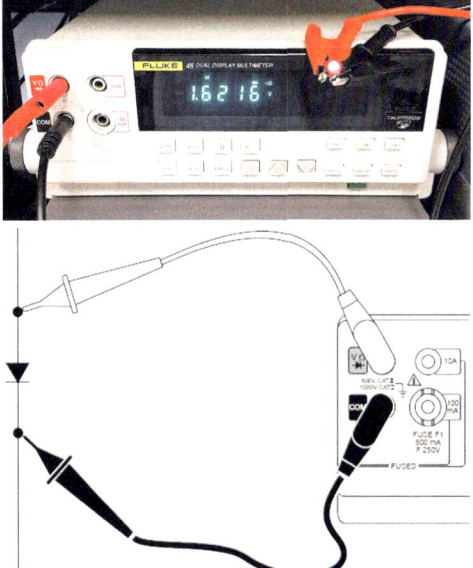

FIGURE 10.22 (Step 7) (Top) DMM showing measurement of LED diode turn-on voltage. Notice that the LED is lit in forward bias. (Bottom) Leads with diode showing forward bias arrangement. The anode is at the top and the cathode is at the bottom in this figure.

8. **Build the half-wave rectifier circuit shown in Fig. 10.24,** using a sine wave from the transformer (H2 and H4) and a single LED diode.

    Note that this circuit will work regardless of the direction in which you insert the diode, but we want the polarity to be correct, so build the circuit as shown in the schematic (anode toward the transformer) because later you will add an electrolytic capacitor, which is DC-polarity dependent.

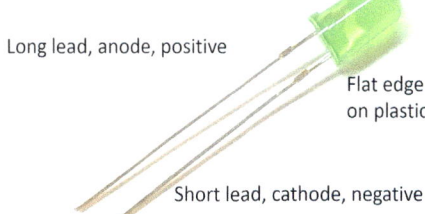

FIGURE 10.23 (Step 7) LED with anode and cathode wires shown. Recall that the anode is the positive side of the diode for forward bias: current flows through the diode from anode to cathode. Be careful to put the long and short wires in the right places in your circuits.

    Use a 4.7 kohm resistor as the load. In this lab you will always use the LED as the diode so that you can see it turning on and off at low frequencies, or just see that it is energized at higher frequencies.

    Don't forget that although Si diodes drop about 0.7 V in forward bias, LEDs drop about 1.8 V or more depending on the type and color of the diode. *Therefore, the flat spots on the output voltage curve that you saw in Chapter 7 may be wider than they would be using the Si diode.* Looking ahead, Fig. 10.25 shows that as the output sine wave increases, it must get to at least 1.8 V before the LED is forward biased and current begins to flow in the forward direction. That detail is left out of Figs. 10.11 and 10.12, but is shown clearly in Figs. 10.25 and 10.29, the second of which you will see suffers from two diode voltage drops, or about 3.6 V.

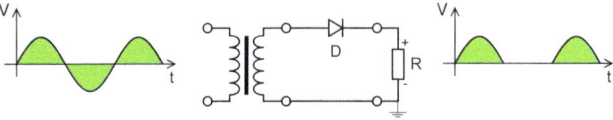

FIGURE 10.24 (Step 8) Half-wave rectifier circuit. Pay attention to the polarity of the supply based on the direction of the diode.

9. **Viewing the output of the unfiltered half-wave rectifier.** Change the frequency to 5 Hz to observe the change in blinking frequency. Do NOT go below 5 Hz with the source attached to the circuit and the output turned on. Somewhat surprisingly, the transformer operates at this low frequency, although not as well as at 50 Hz to 60 Hz. The LED should be blinking on and off one time per period, i.e., five times per second.

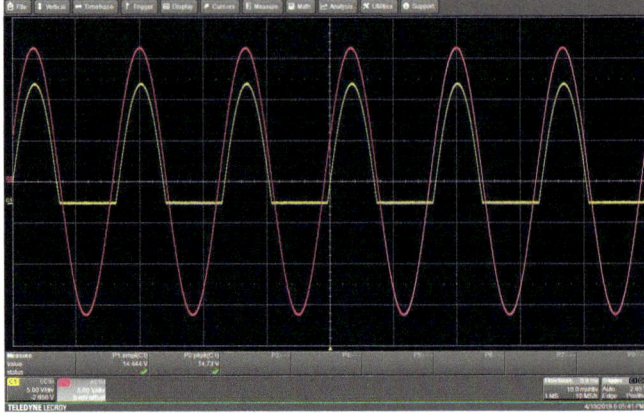

FIGURE 10.25 (Step 9) Scope traces of output of the transformer and the half-wave rectified sine wave measured using DC coupling. The rectified waveform is shown here shifted down by 2.65 V in order to not obscure the transformer waveform. Carefully comparing the peak voltages will show the difference due to one diode voltage drop.

Now increase the frequency to 60 Hz, which is the frequency of line voltage. Display the input (i.e., the transformer output) on CH1 and the output waveform (i.e., across the load) on CH2 of the oscilloscope, like that shown in Fig. 10.25. Make sure that the inputs are set to **Coupling** and **DC1MΩ** in order to measure DC voltages. Measure accurately the peak voltage for the input voltage and the output voltage. Do they differ by one LED forward voltage drop as expected? Do you think this makes a particularly good power supply?

Suppose you were to substitute a 1.5 V battery for the load resistor in the circuit of Fig. 10.23 with the positive terminal up. Could you charge a battery with this? Why would the battery NOT discharge through the transformer during the times when the output of the transformer goes negative? Discuss with your lab partner and put some notes in your lab notebook.

10. **Measuring the ripple factor of the unfiltered half-wave rectifier.** Next you will measure the ripple factor of your power supply. Recall that the ripple factor is $\gamma = 100\% \times$ (RMS value of the AC components) ÷ (the DC component). Because the output current all goes in the same direction, even though it changes considerably, it is technically a DC signal, which you can analyze in terms of the ripple factor. It is a DC voltage with some (well, a lot) ripple on it. This is in theory difficult to calculate, but in practice straightforward using either your oscilloscope or the true-RMS-reading DMM.

View the load voltage on CH1 and the transformer voltage on CH2. If you didn't already know it, now is a good time to learn about the DC- and AC-coupled modes (with 1 Mohm impedance). Ignore the "50 ohm" mode. When AC-coupled, a capacitor is placed in series with the input. The cap blocks the DC component and allows the AC to get through, which you learned in Chapter 4 is exactly what a high-pass filter does. Set the CH1 mode to **AC1MΩ**. In this case, the ripple is shifted to be centered at 0 V, i.e., will appear at a DC offset of zero volts. This offset is due to the removal of the DC component of the waveform, and not due to an offset on the screen set by a shift in the vertical position on the oscilloscope screen. Note that the shift observed in Fig. 10.25 is due only to adjusting the vertical position control.

Use the **Measure** function for V_{RMS} and get the RMS value of the ripple. When it is DC-coupled, you can use the **Measure** function set to **Mean** voltage and get the DC component. In both the DC-coupled or AC-coupled mode, you can press the **Find Scale** button at the bottom of the screen and it will put the trace fully on the screen. Now you have what you need to obtain the ripple factor (after you multiply by 100%).

Here is how you do the same thing with the DMM. This is easier to do, but will not, of course, allow you to see and appreciate the waveform. In the DC voltage measurement setting, the DMM gives the average DC value. In AC mode it does the same thing as the scope – it blocks the DC, and because it is "true-RMS-reading," it will give you the actual RMS value of the ripple voltage. (Less sophisticated DMMs that are not true-RMS-reading meters approximate the RMS value by assuming the waveform is a sine wave. Here it is clearly not.) Both of these measurements are the same as what you did using the scope to be used in the ripple factor calculation.

Since both the DMM and the scope are high-impedance devices, you can use both of them to measure the ripple factor at the same time and compare their values by placing the DMM on the load along with the scope probes. Changing between AC and DC voltage measurement will easily yield the RMS and DC components of your waveform. Obtain the ripple factor for your unfiltered half-wave rectifier circuit using *both* methods just discussed. Note: The ripple factor for an unfiltered half-wave rectifier is about as bad as it gets, and is in fact greater than 1. Compare the two readings from the scope and the DMM and confirm that they give you very nearly the same values of ripple factor.

11. **Filtering the half-wave rectifier.** Turn off the output from the function generator AC source. Now you will add a filter capacitor to the output of the half-wave-rectified power supply to reduce the ripple factor. Choose a 100 µF electrolytic capacitor. You must be mindful of the polarity of the electrolytic filter capacitor, or it could burst. Add the capacitor in parallel with the load, as shown in Fig. 10.26. The negative lead of the cap is

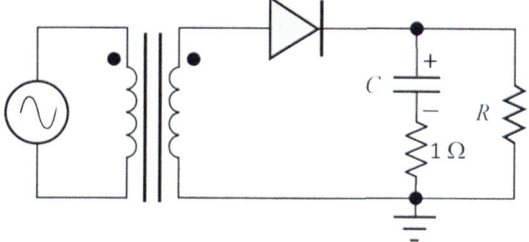

FIGURE 10.26 (Step 11) Half-wave rectifier circuit with a filter capacitor. The positive lead of the electrolytic capacitor is closest to the diode (top).

printed on the side in the form of light-colored bar, which is actually a minus sign. Put it in the circuit to match the polarity of the voltage shown across the capacitor in Fig. 10.26. Check it twice for correct polarity. If you and your lab partner agree that the diode and the capacitor are connected correctly then you may turn on the waveform generator. In order to measure the current through the capacitor as it charges near the peak of the cycle, add a 1 Ω sense resistor between the capacitor and ground and view that voltage drop on CH3.

12. **Viewing the output of the filtered half-wave rectifier.** Your current waveform on CH3 (whose value is $V_{CH3}/1\ \Omega$) should look like that in the light blue trace of Fig. 10.27. View the transformer voltage and load voltage as you did in Step 9, along with the current. Note that that current is so

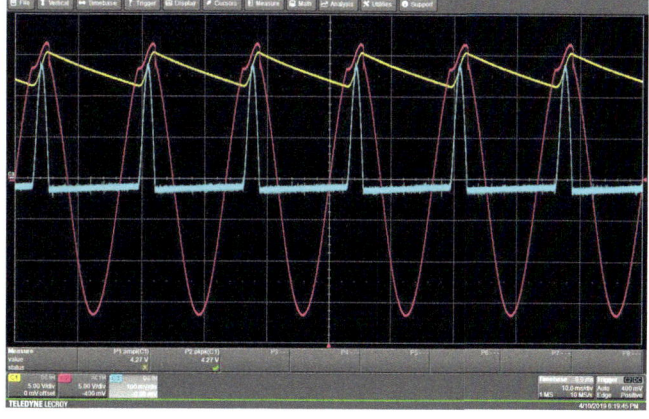

FIGURE 10.27 (Step 12) Scope traces of voltage output of the transformer (magenta), the filtered half-wave rectified sine wave (yellow), and the current flowing into the capacitor (light blue). Notice that the current flows only when the voltage from the transformer is greater than the voltage at the load, during which the capacitor is charging. Current flows into the capacitor (positive) during the rising part of the ripple and flows out (negative) during the falling part of the ripple. The transformer voltage is distorted because it is not an ideal voltage source, and its output drops while providing the large charging current.

large that it loads down the secondary of the transformer. The voltage drop on the internal resistance of the transformer distorts the waveform at the transformer. What is the peak current? This current is actually limited by the transformer and would be somewhat higher if the transformer could provide more current. Do you see that the capacitor charges only during the rising part of the ripple, when the transformer voltage is greater than the load current?

13. **Measuring the ripple factor of the filtered half-wave rectifier.** Now the ripple is relatively small compared to the DC offset. Using AC coupling you can turn up the sensitivity on the scope input and see the features of the ripple. If you tried to do this with DC coupling, it would be far off of the screen and you wouldn't be able to view it. This is described in Fig. 4.31, which you might need to review.

All of the following affect the ripple factor: R_L, C of the filter capacitor, and the frequency of the input. Change each of them within reason and compare ripple factors. By now it should be fast and easy to measure the ripple factor with the scope, and maybe even faster with the DMM. You may use your choice of instruments to take the data. Make a table showing the values that you chose and the resulting half-wave ripple factors, including that with no filter capacitor. Refer to Eq. 10.2. Do the changes give you trends that you might expect, for example, that higher frequency leads to less ripple? How does the current behave as you decrease the ripple factor using any of the methods in the previous step? Why?

14. **Building the unfiltered full-wave rectifier.** So far you have experimented only with the half-wave rectifier using a single diode. Now you will build a bridge using four diodes as part

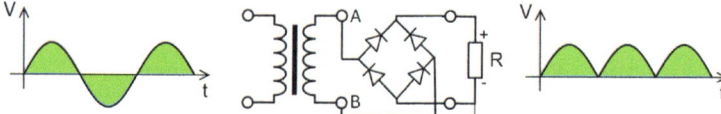

FIGURE 10.28 (Step 14) Full-wave rectifier circuit without filter capacitor. Be sure to put the diodes in correctly keeping track of the anode and cathode leads.

of a full-wave rectified power supply. First remove all probes so that later you can get all of the grounding correct. Now, build the full-wave bridge rectifier circuit without the filter capacitor, as shown in Fig. 10.28. As before, use the LEDs as the diodes so that you can see them light up as they pass current. Do not turn on the power until both you and your lab partner have confirmed that the diodes are inserted exactly correctly with the anodes and cathodes facing the correct directions.

15. **Viewing the output of the unfiltered full-wave rectifier.** Power your circuit using the same amplitude as in Step 9 and a frequency of 5 Hz from the function generator. The LED pairs should blink on and off one time per period, i.e., five times per second. Looking carefully you will be able to tell that they are flashing in pairs and not all at the same time. Can you understand from that how the current flows in two diodes at a time at each half cycle? In your notebook, draw the circuit and the directions of current flow for each half cycle.

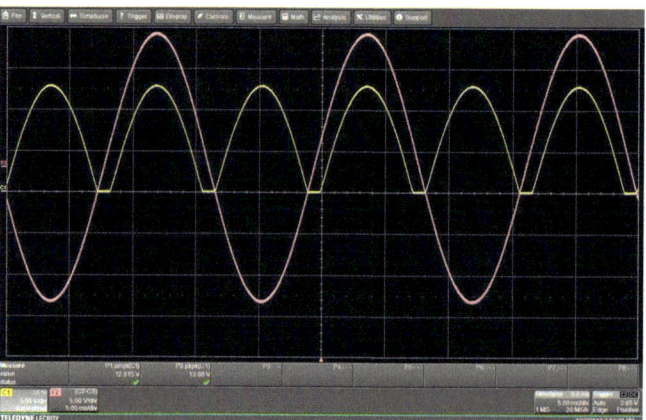

FIGURE 10.29 (Step 15) Oscillograph of input sine wave from transformer secondary as a math function and the voltage on the resistor load. Note that the difference in voltage is *two diode voltage drops*.

Now increase the frequency to 60 Hz. *Using the following instructions*, display both the input (i.e., the stepped-up voltage from the transformer) and output waveforms on the oscilloscope. *Note that the input (secondary of transformer) is not tied anywhere to ground, so you cannot directly display the voltage across the transformer output using the scope probes at the same time that you display the load voltage because you would have to place ground in two different locations, and that would cause short circuits and blow out your LEDS.*

Add a scope probe (CH1) across the load, and include the ground lead as shown in Fig. 10.28, which will establish earth ground for the load circuit. To display the transformer secondary voltage, place one probe, CH3, from A in Fig. 10.28 to ground and a second probe, CH2, from B to ground (as in Labs 2 and 5), and then use a function to display $V_A - V_B$. This function now shows the sinusoidal transformer voltage input to the bridge rectifier, as shown in Fig. 10.29.

Compare the transformer waveform to the waveform across the load, as shown in Fig. 10.29. Use the cursors to confirm that the rectified power supply waveform at the load is that of the input (i.e., at the transformer) reduced by two forward diode voltage drops (approx. 2 × 1.8 V for the LEDs shown here).

What is the main difference in the output between the half-wave rectifier without filtering and the full-wave bridge rectifier without filtering? What accounts for that difference?

16. **Measuring the ripple factor of the unfiltered full-wave rectifier.** Measure the ripple of your unfiltered full-wave rectified power supply. Is it better than the ripple factor in the unfiltered half-wave rectifier? Why?

17. **Filtering the full-wave rectifier.** Turn off the power to your circuit. Add the 100 µF capacitor with the 1 Ω sense resistor (closest to ground) in parallel with the load, as you did in Step 11. Again, do not proceed to turn on the power until you and your lab partner have both confirmed that the electrolytic capacitor is inserted correctly.

18. **Measuring the ripple factor of the filtered full-wave bridge rectifier.** Repeat Step 13 for the F.W. rectifier and observe various ripple factors for the *same* various conditions as you used in Step 13. (See Fig. 10.30 where the capacitor current is on CH4.) Are the ripple factors smaller or larger than the values you obtained for the half-wave rectifier? Compare them for the same conditions between each circuit, filtered and unfiltered. Show your data in the form of a table containing data from both Steps 13 and 18. Now you can use Eq. 10.2 $(\gamma = 1/(fCR_L4\sqrt{3}))$ to calculate the ripple factor and compare to your measured data. Add that data to the table. What effect of the transformer current loading do you see in the oscillographs?

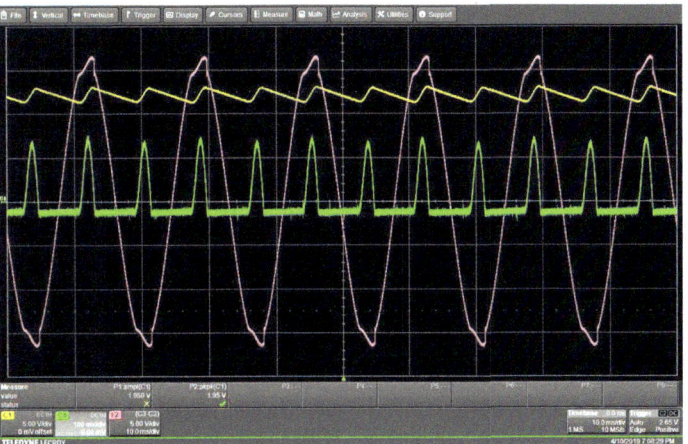

FIGURE 10.30 (Steps 18 and 19) Scope traces of voltage output of the transformer (pink), the filtered half-wave rectified sine wave (yellow), and the current flowing into the capacitor (green). The green trace is the current flowing into the capacitor that recharges it on every half cycle. Current flows into the capacitor (positive) during the rising part of the ripple and flows out (negative) during the falling part of the ripple. The peak current is less here than in the case of the half-wave rectifier because the capacitor recharges every half cycle, rather than full cycle, and the amount of discharge at each half-cycle is lower.

19. **Viewing current flow in the filtered full-wave rectifier.** Repeat Step 12 to observe the current through the capacitor using the current probe. Does it charge where you expected it to, i.e., on each half cycle? How does the current behave as you decrease the ripple factor using any of the methods in the previous step? Why?

20. **Building the Zener-regulated full-wave rectifier.** Figure 10.13, repeated here, shows the steep reverse breakdown region of a Zener diode. The use of a Zener diode as a voltage regulator was explained in the chapter. The Zener diode in your kit has a breakdown voltage of about 4.6 V or so. The breakdown *I–V* curve is shown in Fig. 10.31 using the curve tracer from Chapter 7.

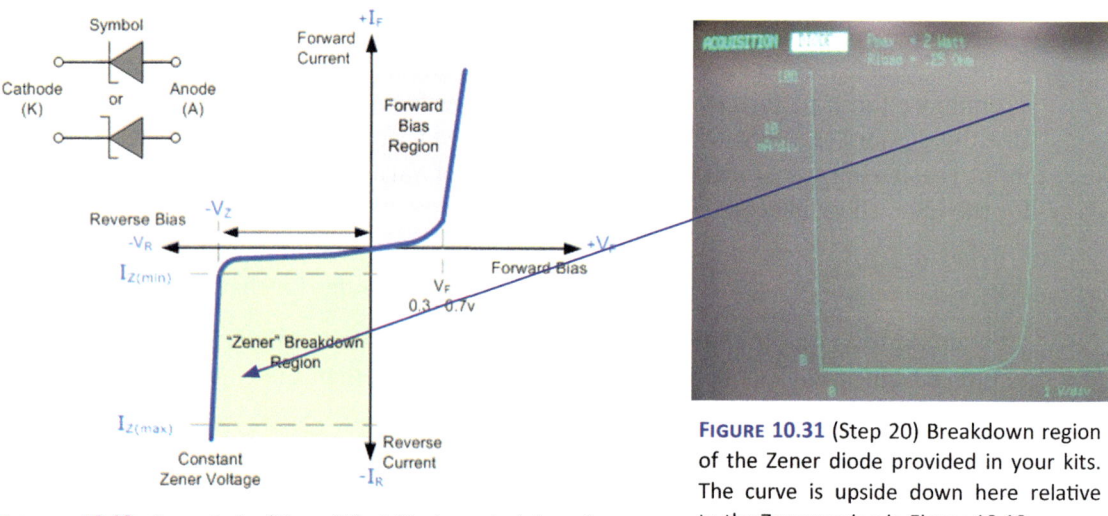

FIGURE 10.31 (Step 20) Breakdown region of the Zener diode provided in your kits. The curve is upside down here relative to the Zener region in Figure 10.13.

FIGURE 10.13 Repeated: (Step 20) *I–V* characteristic of a diode showing the Zener breakdown voltage. A Zener diode is characterized by a very steep breakdown region, which allows voltage regulation.

Build the circuit shown in Fig. 10.32. Use the same input voltage, load resistance, R_L, of 4.7 kohm and 10 μF electrolytic capacitor with the 1 Ω sense resistor, as usual. Use a series resistance, R_S, of 470 ohms. Use the voltage difference method in Step 15 to view the input and output voltages simultaneously. Measure the ripple factor of this circuit with the Zener diode in place. Carefully state your AC and

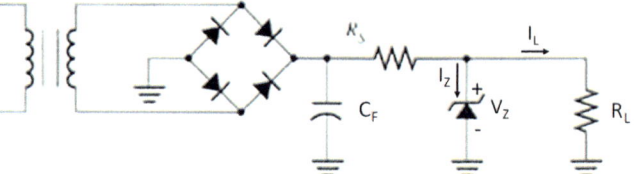

FIGURE 10.32 (Step 20) Figure 10.14 repeated. Schematic of simple Zener-regulated power supply. Build this circuit with the component values given in Step 20. As you might guess, all places with separate ground symbols are the same electrical potential, since they are all ground, and can be connected by a wire or tie point of a breadboard.

DC components of the ripple factor. Compare the two ripple factors for the Zener diode and the F.W. rectified power supplies. Discuss.

21. **Testing the Zener-regulated full-wave rectifier.** Use the sense resistor to measure the current in the Zener diode (shown here on CH4). Can you see that it conducts considerable current in an effort to maintain a constant output voltage? Compare your results to Fig. 10.33.

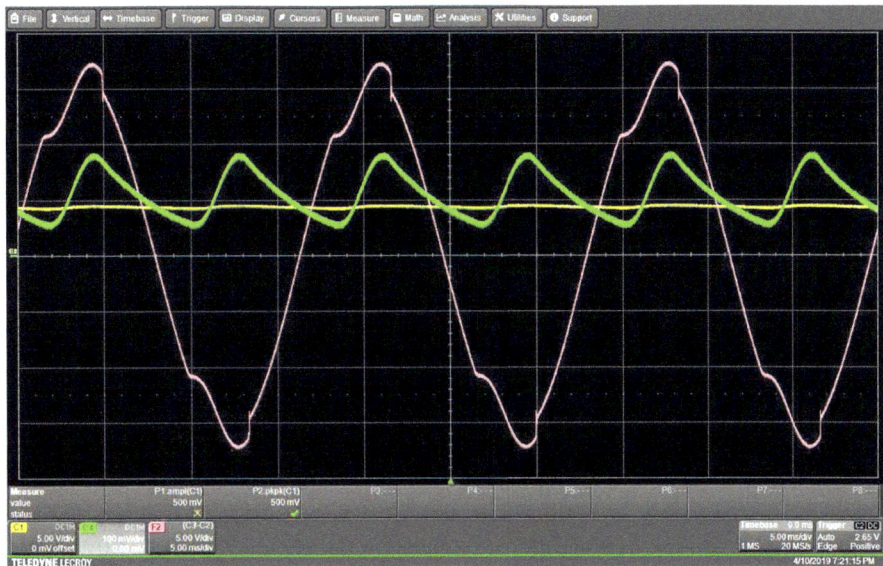

FIGURE 10.33 (Step 21) Zener diode voltage regulator circuit waveforms. The math function F2 (pink) is the transformer output, CH1 (yellow) is the voltage across the load and has very little ripple. CH4 (green) is the current through the Zener diode. Note that the Zener current changes according to its *I–V* curve, passing whatever current does not go through the load. The voltage across the Zener diode is the same as the load voltage, and is fairly constant at its rated value of V_Z.

Congratulations! You have completed this lab and now better understand some basic properties of batteries and the concept of internal resistance. You also built several different types of power supplies. This is very practical information and will be useful to you as you build future electronics projects.

Problems

1. Pumped hydro storage, PHS, is an important technique used to support renewable energy. A PHS comprises two bodies of water: a "head" water pond and a "tail" water pond. During periods of low power demand, energy from some external source, such as a thermal plant, nuclear plant, wind or solar power, uses excess power to pump water from the lower "tail" water pond up to the "head" water pond. In periods of high power demand, water from the head pond flows down through turbines into the tail pond, producing electricity. Even though some energy is wasted to store energy in the head water reservoir, this allows power-generating stations to operate at peak efficiency and saves money over the long term. PHS accounts for 67% of utility-scale energy storage in the U.S. Note: the capacitor in the full-wave rectifier works sort-of the same way as the PHS, on a much smaller scale. When the voltage from the transformer is high enough, current flows into the capacitor and the capacitor charges. When the voltage across the transformer drops, current flows out of the capacitor into the load, and the stored energy in the capacitor is used up.

   The Bath County Pumped Storage Station in Virginia (https://en.wikipedia.org/wiki/Bath_County_Pumped_Storage_Station), originally completed in 1985, is now the largest of its kind in the world, and is often called the "largest battery in the world." The head and tail reservoirs are separated by about 1,260 ft. of elevation. It has a maximum output rating of 3,000 MW and total storage capacity of 24,000 MWh.

(a) How long can the Bath County storage facility provide power at maximum output?

(b) The average American household uses about 10,400 kWh of energy per year. What is the average power consumption of a household?

(c) Assuming the average power used per home that you just calculated, how many average houses could the stored water in the Bath County facility power at maximum output?

(d) How much energy, in joules, is stored by the Bath County facility?

2. (a) Summarize in one paragraph how the Daniell cell of Figure 10.5 works. Be sure to point out charge neutrality whenever your description requires that it be maintained.

(b) What components would cause the battery to be completely depleted, i.e., be dead, when used up?

3. A typical 1.5 V AA alkaline battery has a capacity of about 2500 mA-h. (In fact, the voltage drops as the battery depletes, but for simplicity we ignore that here). The battery cannot actually provide 2500 mA, and the actual highest possible current could not flow for one full hour. Rather, some much lower current times the long time that it flows at 1.5 V provided by the battery would add up (integrate) to the theoretical maximum of 2500 mA-h. The same can be said for the power that is provided if the battery capacity is expressed in Watt-hours, W-h.

(a) How many W-h does 2500 mA-h correspond to for *this* battery? (Hint: battery voltage matters.)

(b) How many joules does the 2500 mA-h correspond to for this battery?

4. Rechargeable batteries operate differently from the Daniel cell discussed in this chapter. Say you are watching a movie on your cell phone at full volume. The large data rates, illumination of the screen, and audio require about 2 A. Assume that the voltage of a typical $LiPF_6$ ("LiPo", lithium hexafluorophosphate) battery is about 3.7 V averaged over a full discharge cycle. Also, assume that it has a battery capacity of about 15 W-h. One lithium ion must be moved from the intercalated anode to the cathode for each electron that flows through the external circuit.

A typical cell phone battery of this type has a mass of about 75 g. In fact, only a small fraction of the battery mass is the Li. The rest of the mass comes from:

- Cathode (e.g., $LiCoO_2$, NMC)
- Anode (graphite with intercalated Li, LiC_6)
- Electrolyte and solvents, $LiPF_6$
- Separator film
- Casing and current collectors

Answer these questions to analyze the mass of the Li that moves through the battery.

(a) How much total charge (in coulombs) flows through the external circuit during full discharge?

(b) How many lithium ions are needed to supply this charge, assuming each carries one elementary charge?

(c) Given that the molar mass of lithium is 6.94 g/mol, and that 1 mole = 6.022×10^{23} ions, calculate the total mass of lithium that must be transported to carry this charge.

5. You have a wall clock that isn't working and suspect that it is because the AA alkaline battery is dead. Using an ideal voltmeter (not a battery tester), you measure the open-circuit voltage, V_{OC}, which is due to the chemistry of the battery, to be 1.5 V. Rather than concluding that the battery is good, you then use an inexpensive battery tester that measures voltage under the condition that current flows from the battery. The battery tester is simply an ideal voltmeter in parallel with a resistance, R_L, that serves as a load to draw current from the battery.

   Batteries are *non-ideal* voltage sources – they have some internal resistance, R_{in}, and ideal voltage source, V_{in}. R_{in} increases as the battery is depleted. When the internal resistance increases, the measured, external voltage decreases. A battery tester measures the battery voltage, V_{BT}, at its terminals under the battery-tester load and compares that voltage to that of a fully charged battery. Let's see how that works out.

   You attach the battery tester that includes a load resistance of R_L = 15 Ω to your depleted battery and now measure the voltage across the 15 Ω load as 1.2 V.

   (a) Draw the equivalent circuit model of the depleted battery including R_{in}, its ideal internal voltage source, V_{in}, and the battery tester, with its load resistance (remember – the battery tester is the load in parallel with the ideal voltmeter).

   (b) What is the voltage V_{in} of the depleted battery as determined by measuring the voltage with the original ideal voltmeter (not battery tester)? How does it compare to the voltage measured by the battery tester?

   (c) Based on the voltage reading of the battery tester, what is the internal resistance of the depleted battery? (Hint: voltage divider.)

   (d) How much current would flow through the battery tester if R_{in} were 0 Ω, which would be ideal for a fresh battery? (It is never actually 0 Ω.)

   (e) How much current flows through the load including the actual R_{in} that you calculated in (c)?

   (f) Alkaline AA batteries typically have a capacity of about 2500 mA-h. (That is ideally the total charge that can flow through the battery before being depleted.) Based on your answer to part (d), what fraction of the stored battery energy is used in testing it for one second? Does testing a fresh battery for one second deplete it by much?

6. Compare a half-wave rectifier and a full-wave rectifier with filter capacitors having the same capacitances and driving loads with the same resistance. Which rectifier will have the higher ripple factor? Qualitatively, explain why. *Hints*: Which of the two will give the filter capacitor a longer time to discharge and how does this affect the voltage drop?

7. A full-wave rectifier is to be used with a 60 Hz input, a 50 Ω load, and has a ripple factor (not expressed here as a percent) of 0.003.

   (a) What capacitance provides this ripple factor?

   (b) Would you increase or decrease the capacitance it to get a better ripple factor?

   (c) Would increasing the load resistance increase or decrease the ripple factor? Support your answer qualitatively, and not just by using the equation.

   (d) Would increasing the frequency for the same capacitance and load increase or decrease the ripple factor? Draw a picture of ripple at two frequencies that supports your answer, rather than just relying on the equation.

8. The legendary Pioneer SX-1980 stereo receiver/amplifier provides up to 270 W of audio power into an 8 Ω speaker for each of the two stereo channels. The DC voltage from the power supply is 80 V. Each channel uses a 22,000 microfarad filter capacitor to decrease the ripple on the output signal.

   (a) What is the ripple factor for each channel of the amplifier? Worst case, assume it is being used in a country for which line frequency is 50 Hz. (Note: this ripple does not show up on the music for well-designed amplifiers. That is left to one of your future courses.)

   (b) What is the RMS voltage of the ripple?

9. Assume that the diodes in a full-wave bridge rectifier are essentially short circuits when conducting, as they are during half of the input's cycle, and open circuits during the other half cycle. Redraw the full-wave bridge rectifier shown here, substituting the diodes with short and open circuits. Do this for the positive half cycle of V_{in} and then redraw it for the negative half cycle of V_{in}.

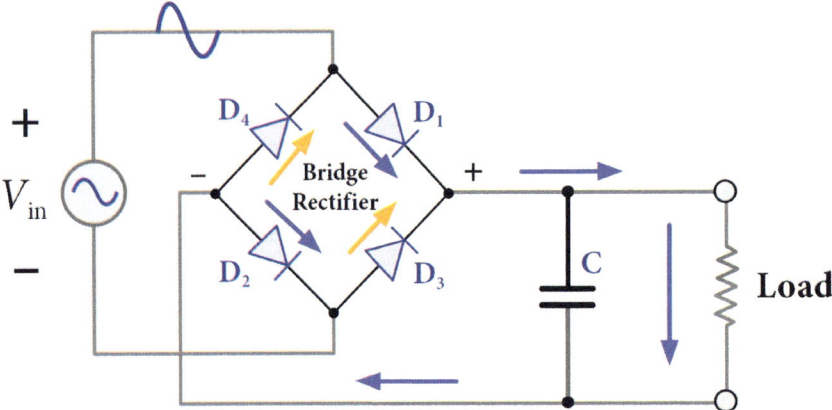

10. Shown here is a version of the Zener diode regulated full-wave rectifier in which the filter capacitor is removed. The voltage at the output of the transformer is V_a and the voltage across the Zener diode is V_d. The voltage at which the Zener diode breaks down is V_Z. Also shown are the waveforms V_a and V_d. The solid line at V_d is the load voltage, and the dashed line is the envelope of V_a.

   Draw on the lower right waveform of V_d where the Zener diode is in breakdown, where it is in reverse bias but not yet breakdown, and where it is in forward bias.

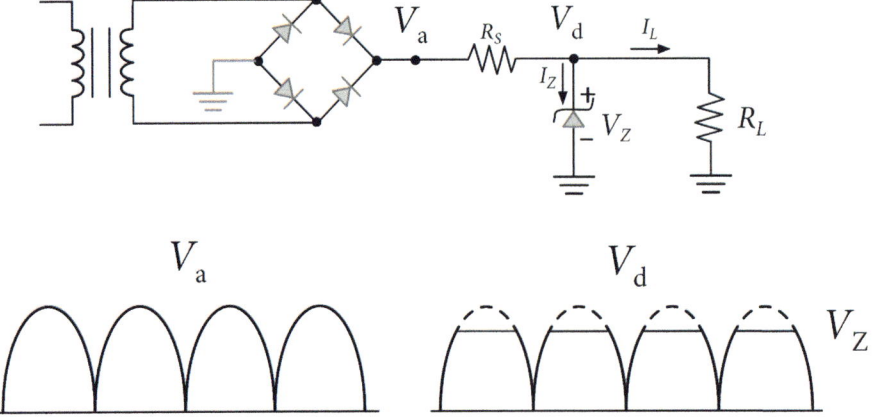

11. In the following figure, the voltage at the output of the bridge is V_a and the voltage across the Zener diode is V_d. Let's analyze what happens to the load voltage as we increase the input voltage to the transformer. Let's make the crude approximation that the current at the reverse breakdown voltage V_Z is infinitely steep, meaning that no matter how much current flows in the diode in reverse breakdown, the voltage is exactly V_Z. Let's also assume that R_S is much less than R_L.

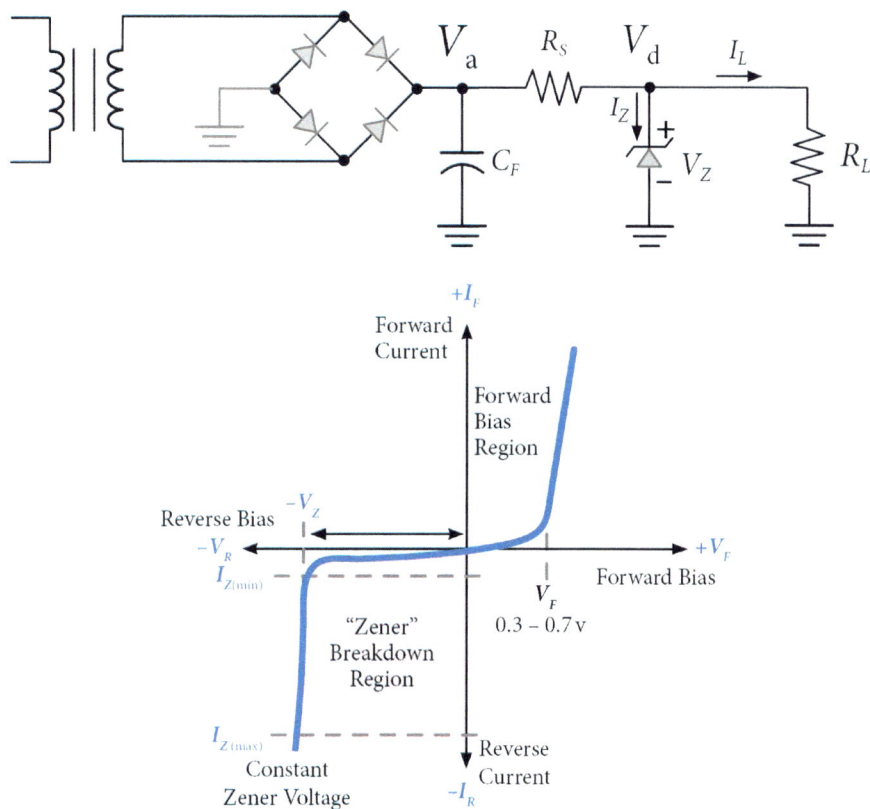

(a) Assume that the amplitude of the sine wave at the output of the transformer, V_a, is of low amplitude, such that V_a is less than V_Z. What will the Zener diode behave like? Hint: the reverse leakage current of a diode is extremely small.

(b) In that case, how is the behavior different from that of a simple full-wave bridge rectifier discussed in Fig. 10.12?

(c) For the case discussed so far, what will the ripple factor for R_L be as compared to the full-wave bridge rectifier circuit? (Describe it, don't calculate it.)

(d) Now assume that the voltage V_a is large enough that at no time does V_a, including ripple, drop below V_Z. Draw a picture of the waveform at point "A" and include V_Z in the picture.

(e) Now $V_a > V_Z$ at all times. Because $R_S \ll R_L$, essentially $V_{RL} = V_a$ and the Zener diode is conducting. In this case, what is the voltage drop across the Zener diode, V_d?

(f) Write a formula for the current through R_S.

(g) Write a formula for the current through the load, R_L. Hint: use ohms law, and note that the Zener diode is in parallel with R_L.

(h) For this ideal case, in which the Zener current is infinitely steep when $V_d = V_Z$, what is the ripple factor of this circuit?

12. Indicate all correct answers: The open-circuit voltage of a fully-charged, commercially available battery is set by:

    (a)   The difference between the reduction potentials of the electrodes

    (b)   The manufacturers' specifications

    (c)   The current being drawn by the battery

    (d)   The size of the batteries (e.g. D cell versus AA cells).

13. Indicate all correct answers: Consider a full-wave bridge rectifier circuit with the resistive load filtered by a capacitor. All parameters are unchanged except where noted:

    (a)   As the capacitance is increased, ripple factor decreases.

    (b)   As the load resistance is increased, ripple factor decreases.

    (c)   As the frequency increases, ripple factor does not change.

14. Indicate all correct answers: A Zener diode is different from a conventional diode and is useful as a voltage regulator because:

    (a)   It has a very low turn-on voltage relative to other silicon diodes.

    (b)   It has a very high reverse-breakdown voltage relative to other silicon diodes.

    (c)   Its current–voltage curve is steep in the breakdown region relative to other silicon diodes.

15. Indicate all correct statements about breakdown in a Zener diode:

    (a)   It can destroy the diode if the current is too high at the breakdown voltage.

    (b)   The breakdown voltage equals the current times the load resistance.

    (c)   The breakdown voltage is approximately constant with diode current.

Index